Percentage Points of the t Distributions

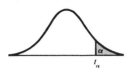

$t_{.100}$	$t_{.050}$	$t_{.025}$	$t_{.010}$	$t_{.005}$	d.f.
3.078	6.314	12.706	31.821	63.657	1
1.886	2.920	4.303	6.965	9.925	2
1.638	2.353	3.182	4.541	5.841	3
1.533	2.132	2.776	3.747	4.604	4
1.476	2.015	2.571	3.365	4.032	5
1.440	1.943	2.447	3.143	3.707	6
1.415	1.895	2.365	2.998	3.499	7
1.397	1.860	2.306	2.896	3.355	8
1.383	1.833	2.262	2.821	3.250	9
1.372	1.812	2.228	2.764	3.169	10
1.363	1.796	2.201	2.718	3.106	11
1.356	1.782	2.179	2.681	3.055	12
1.350	1.771	2.160	2.650	3.012	13
1.345	1.761	2.145	2.624	2.977	14
1.341	1.753	2.131	2.602	2.947	15
1.337	1.746	2.120	2.583	2.921	16
1.333	1.740	2.110	2.567	2.898	17
1.330	1.734	2.101	2.552	2.878	18
1.328	1.729	2.093	2.539	2.861	19
1.325	1.725	2.086	2.528	2.845	20
1.323	1.721	2.080	2.518	2.831	21
1.321	1.717	2.074	2.508	2.819	22
1.319	1.714	2.069	2.500	2.807	23
1.318	1.711	2.064	2.492	2.797	24
1.316	1.708	2.060	2.485	2.787	25
1.315	1.706	2.056	2.479	2.779	26
1.314	1.703	2.052	2.473	2.771	27
1.313	1.701	2.048	2.467	2.763	28
1.311	1.699	2.045	2.462	2.756	29
1.282	1.645	1.960	2.326	2.576	inf.

From "Table of Percentage Points of the *t*-Distribution." Computed by Maxine Merrington, *Biometrika*, Vol. 32 (1941), p. 300. Reproduced by permission of Professor E. S. Pearson.

Mathematical
Statistics
with Applications

Mathematical Statistics with Applications

Fourth Edition

William Mendenhall
Professor Emeritus, University of Florida

Dennis D. Wackerly
University of Florida

Richard L. Scheaffer
University of Florida

PWS–KENT Publishing Company
Boston

PWS–KENT
Publishing Company

20 Park Plaza
Boston, Massachusetts 02116

PWS–KENT Publishing Company is a division of Wadsworth, Inc.

Library of Congress Cataloging-in-Publication Data

Mendenhall, William.
  Mathematical statistics with applications.

  Includes index.
  1. Mathematical statistics.   I. Wackerly, Dennis D.,
        II. Scheaffer, Richard L.   III. Title.
QA276.M426   1989   519.5   89-8631
ISBN 0-534-92026-8

Printed in the United States of America.
90 91 92 93 94—10 9 8 7 6 5 4 3 2 1

Sponsoring Editor   *Michael Payne*
Production Editor   *Susan M. C. Caffey*
Manufacturing Coordinator   *Margaret Sullivan Higgins*
Interior Designer   *Susan M. C. Caffey*
Cover Designer   *Hannus Design Associates*
Typesetter   *A.P.S. Salisbury*
Cover Printer   *New England Book Components, Inc.*
Printer and Binder   *Arcata Graphics/Halliday Lithography*

PREFACE

This text was written for use with a one-year sequence of courses (9 quarter or 6 semester hours) on mathematical statistics for undergraduates. The intent of the text is to present a solid undergraduate foundation in statistical theory and, at the same time, to provide an indication of the relevance and importance of the theory in solving practical problems in the real world. We think a course of this type is suitable for most undergraduate disciplines, including mathematics, where contact with these applications may provide a refreshing and motivating experience. The only mathematical prerequisite is a thorough knowledge of first-year college calculus.

Talking with students taking or having completed a beginning course in mathematical statistics reveals a major flaw that exists in many courses. The student can take the course and leave it without a clear understanding of the nature of statistics. Many see the theory as a collection of topics, weakly or strongly related, but fail to see that statistics is a theory of information with inference as its goal. Further, they may leave the course without an understanding of the important role played by statistics in scientific investigations. Why this is true (assuming that you agree with us) is a matter for conjecture, but several reasons suggest themselves.

First, all mathematical statistics courses require, of necessity, a large amount of time devoted to the theory of probability. Since at least 50% of the total course time is devoted to this subject, and it occurs at the beginning of the course, it is not surprising that a student may think of probability and statistics as being synonymous.

A second reason is that the objective of statistics often is not defined clearly at the beginning of the course, and no attempt is made to relate the probabilistic half of the course to the ultimate objective, inference.

Third, the utility of statistics cannot be revealed until late in the course because of the large amount of material that must preface it.

A fourth and final possibility is that the interest of some instructors leads them to present the material as a sequence of topics in applied mathematics rather than as a cohesive course in statistics. In any case, regardless of the reasons why students fail to form a clear picture of their subject, our text is an attempt to cope with the problem.

We think this text differs from others in three ways. First, we have preceded the presentation of probability with a clear statement of the objective of statistics and its role in scientific research, and we hold this objective before the student throughout the text. As the student proceeds through the theory of probability (Chapters 2 through 7), he or she is reminded frequently of the role that major topics play in the objective of the course, statistical inference. We attempt to strongly emphasize statistical inference as the sole and dominating theme of the course. The second feature of the text is connectivity. We try to explain not only how major topics play a role in statistical inference but also how the topics are related one to another. These connective discussions most frequently appear in chapter introductions and conclusions. Finally, we think the text is unique in its practical emphasis throughout the text, by the exercises and by the useful statistical methodology presented at the end of the text. We hope to reinforce an elementary but sound theoretical foundation with some very useful methodological topics contained in the last five chapters.

The book can be used in a variety of ways and adjusted to the tastes of the students and instructor. The difficulty of the material can be increased or decreased by controlling the assignment of exercises, by eliminating some topics, and by varying the amount of time to be devoted to each. A stronger applied flavor can be added by the elimination of some topics, for example, some sections of Chapter 7, and by devoting more time to the applied chapters at the end.

While retaining the features that various users of the text found attractive in the previous editions, we have introduced some revisions into the fourth edition of *Mathematical Statistics with Applications*. For the most part, these revisions were motivated by suggestions from instructors who use the text. The order of presentation of the material in Chapter 3 has been changed. Immediately after defining the probability distribution for a discrete random variable, we define the mean and variance of discrete random variables and present some of the elementary results that are useful in finding expected values of functions of random variables. After these preliminary discussions, we present the common discrete distributions, simultaneously developing means and variances in each case. This permits a more unified presentation of the key results for the various distributions. Because the means and variances typically are functions of the parameters used to define the various distributions, the parameters are better motivated and made more meaningful. Chapter 3 has been expanded to include a section (optional) on the negative binomial distribution. Similar changes in the order of presentation were made in Chapter 4, which deals with continuous random variables. Chapter 13 has been expanded to include sections on explicit models for the analysis of variance of completely randomized and randomized block designs. As in most introductory texts, we focus primarily on the randomized block design with fixed (nonrandom) block effects. However, we have added a series of supplemental exercises that deal with the randomized block design with random block effects. Thus, if the instructor is so inclined, it would be relatively easy to discuss the similarities and differences of these two versions of the randomized block design. Other, less obvious, changes have been made throughout the text.

We have maintained the use of the upper-tail normal tables because users of the text find them to be more convenient. We also have maintained the format of the table of the F distributions that we introduced into the previous edition of the text. This table of the F distributions provides critical values corresponding to upper-tail areas of .100, .050, .025, .010, and .005 in a single table. Because tests based on statistics possessing the F distribution occur quite often, this table facilitates the computation of attained significance levels, or p-values, associated with observed values of these statistics.

This edition contains over 200 new exercises. Some of the newly added exercises are of a theoretical nature, whereas others contain data from documented sources that deal with actual research in a variety of fields. The exercises based on documented data sets and other realistic experimental situations allow the student to see the practical uses of the various statistical and probabilistic methods presented in the text. In attempting these exercises, students gain insight into the real-life applications of the mathematically derived techniques studied in the body of the text. We maintain our belief that this insight makes learning the necessary theory more enjoyable and produces a deeper understanding of the required mathematical methods. As in previous editions, the more challenging exercises are marked clearly with an asterisk (*).

We also have maintained our practice of providing easy access to often-used information. Because the normal and t tables are the most frequently used statistical tables in the text, copies of these tables are given in Appendix III, and inside the front cover of the text. Users of previous editions have often remarked favorably about the utility of tables of the common probability distributions, means, variances, and moment-generating functions provided in Appendix II and inside the back cover of the text. In addition, we have included some frequently used mathematical results in a supplement to Appendix I. These results include the binomial expansion of $(x + y)^n$, the series expansion of e^x, sums of geometric series, definitions of the gamma and beta functions, and so on. As before, each chapter begins with a "local" table of contents, containing the titles of the major sections in that chapter.

The authors wish to thank the many colleagues, friends, and students who have made helpful suggestions concerning the revisions of this text. In particular, we are indebted to P. V. Rao, J. G. Saw, J. Witmer, Malay Ghosh, and Andrew Rosalsky for their technical comments. We wish to thank E. S. Pearson, W. H. Beyer, I. Olkin, R. A. Wilcox, C. W. Dunnett, A. Hald, and John Wiley and Sons for their kind permission to use the tables reprinted in Appendix III. Thanks are also due to the reviewers of the text: John K. Scheidt, *University of Wisconsin—La Crosse*; Martin Stuart Levy, *University of Cincinnati*; James R. Bock, *California State University—Northridge*; Kevin M. Beam, *United States Military Academy*; Walter Freiberger, *Brown University*; George Sturm, *Grand Valley State University*; Melvin D. Lax, *California State University—Long Beach*; and Jean-Francois Richard, *Duke University*. Finally, we wish to thank

our wives and families for their patience and understanding throughout the time
that this work was in progress.

William Mendenhall
Dennis D. Wackerly
Richard L. Scheaffer

Note to the Student ∽

As the title *Mathematical Statistics with Applications* implies, this text is
concerned with statistics, both in theory and application, and only deals with
mathematics as a necessary tool to give you a firm understanding of statistical
techniques. The following suggestions for using the text will increase your
learning and save your time.

The connectivity of the textbook is provided by the introductions and
summaries in each chapter. These sections explain how each chapter fits into the
overall picture of statistical inference and how each chapter relates to the
preceding ones.

Within the chapters, important concepts are set off as definitions. These
should be read and reread until they are clearly understood, because they form
the framework on which everything else is built. The main theoretical results are
set off as theorems. Although it is not necessary to understand the proof of each
theorem, a clear understanding of the meaning and implications of the theorems
is essential.

It also is essential that you work many of the exercises—for at least three
reasons. First, you can be certain that you understand what you have read only
by putting your knowledge to the test of working problems. Second, many of the
exercises are of a practical nature and shed light on the applicability of
probability and statistics. Third, some of the exercises present new concepts and
thus extend the material covered in the chapter.

W. M.
D. D. W.
R. L. S.

CONTENTS

WHAT IS STATISTICS?

1.1 Introduction

Statistical techniques are employed in almost every phase of life. Surveys are designed to collect early returns on election day to forecast the outcome of an election, and consumers are sampled to provide information for predicting product preference. The research physician conducts experiments to determine the effect of various drugs and controlled environmental conditions on humans in order to infer the appropriate treatment of a particular disease. The engineer samples a product quality characteristic along with various controllable process variables to assist in locating important variables related to product quality. Newly manufactured fuses are sampled before shipping to decide whether to ship or hold individual lots. The economist observes various indices of economic health over a period of time and uses the information to forecast the condition of the economy next fall. Statistical techniques play an important role in achieving the objective of each of these practical problems, and it is to the theory underlying this methodology that this textbook is devoted.

A prerequisite to a discussion of theory is a definition of *statistics* and a statement of its objectives. *Webster's New Collegiate Dictionary* defines statistics as "a branch of mathematics dealing with the collection, analysis, interpretation, and presentation of masses of numerical data." Kendall and Stuart [3] state: "Statistics is the branch of scientific method which deals with the data obtained by counting or measuring the properties of populations." Fraser [1], commenting on experimentation and statistical applications, states that "statistics is concerned with methods for drawing conclusions from results of the experiments or processes." Freund [2], among others, views statistics as encompassing "the entire science of decision making in the face of uncertainty," and Mood, Graybill, and Boes [5] define statistics as "the technology of the scientific method" and add that statistics is concerned with "(1) the design of experiments and investigations, (2) statistical inference." *A superficial examination of these definitions suggests a bewildering lack of agreement, but all possess common elements. Each implies collection of data with inference as the objective. Each requires the selection of a subset of a large collection of data, either existent or conceptual, in order to infer the characteristics of the complete set. Thus statistics is a theory of information with inference making as its objective.*

The large body of data that is the target of our interest is called a *population*, and the subset selected from it is a *sample*. The preferences of voters for a gubernatorial candidate, Jones, expressed in quantitative form (1 for "prefer" and 0 for "do not prefer") provide a real, finite, and existing population of great interest to Jones. Indeed, he may wish to sample the complete set of eligible voters in order to determine the fraction favoring his election. The voltage at a particular point in the guidance system for a spacecraft may be tested in the only three systems that have been built in order to estimate the voltage characteristics for other systems that might be manufactured some time in the future. In this case the population is *conceptual*. We think of the sample of three as being representative of a large population of guidance systems that could be built and would possess characteristics similar to the three in the sample. As another example, measurements on patients in a medical experiment represent a sample from a conceptual population consisting of all patients similarly afflicted today as well as those who will be afflicted in the near future. You will find it useful to clearly define the populations of interest for each of the statistical problems described earlier in this section and to clarify the inferential objective for each.

It is interesting to note that billions of dollars are spent each year by American industry and government for data from experimentation, sample surveys, or other collection procedures. Consequently, we see that this money is expended solely for information about a phenomenon susceptible to measurement in an area of business, science, or the arts. The implication of this statement provides a key to the nature of the very valuable contribution that statistics makes to research and development in all areas of society. Information useful in inferring some characteristic of a population (either existing or conceptual) can be purchased in a specified quantity and will result in an inference (estimation or decision) with an associated degree of goodness. For example, if Jones arranges for a sample of voters to be interviewed, the information in the sample can be used to estimate the true fraction of all voters favoring Jones's election. In addition to the estimate itself, Jones should also be concerned with the likelihood (chances) that the estimate provided is close to the true fraction of eligible voters favoring his election. Intuitively, the larger the number of eligible voters in the sample, the higher will be the likelihood of an accurate estimate. If a decision is made regarding the relative merits of two manufacturing processes based upon examination of samples of products from each process, we should be interested in the decision and the likelihood that the decision is correct. In general, we can say that *statistics is concerned with the design of experiments or sample surveys to obtain a specified quantity of information at minimum cost and the optimal utilization of this information in making an inference about a population. The objective of statistics is to make an inference about a population based on information contained in a sample and to provide an associated measure of goodness for the inference.*

Exercise

1.1   For each of the following situations discuss the nature of the population of interest, the inferential objective, and how you might go about collecting a sample.

(a)   A city engineer wants to estimate the average weekly water consumption for single-family dwelling units in the city.

(b)   The National Highway Safety Council wants to estimate the proportion of automobile tires with unsafe tread among all tires manufactured by a certain company in a specified year.

(c)   A political scientist wants to determine if a majority of adult residents of a state favor a unicameral legislature.

(d)   A medical scientist wants to estimate the average length of time until the recurrence of a certain disease.

(e)   An electrical engineer wants to determine if the average length of life of transistors of a certain type is greater than 500 hours.

1.2 Characterizing a Set of Measurements: Graphical Methods

In the broadest sense, making an inference implies the partial or complete description of a phenomenon or physical object. Little difficulty is encountered when appropriate and meaningful descriptive measures are available, but this is not always the case. For example, it is easy to characterize a person by using height, weight, color of hair and eyes, and other descriptive measures of one's physiognomy. Locating a set of descriptive measures to characterize an oil painting would be a comparatively more difficult task; characterizing a population, which consists of a set of measurements, is equally challenging. Consequently, a necessary prelude to a discussion of inference making is the acquisition of a method for characterizing a set of numbers. The characterization must be meaningful so that knowledge of the descriptive measures will enable us to clearly visualize the set of numbers. In addition, we would hope that the characterization possesses practical significance, so that knowledge of the descriptive measures for a population might solve a practical nonstatistical problem. We will develop our ideas on this subject by examination of a process that generates a population.

Consider a study to determine important variables affecting profit in a business that manufactures custom-made machined devices. Some of these variables might be the dollar size of the contract, the type of industry with which the contract is negotiated, the degree of competition in acquiring contracts, the salesperson who estimates the contract, fixed dollar cost, and the foreman who is assigned the task of organizing and conducting the manufacturing operation. The statistician will wish to measure the response or dependent variable, profit per contract, for several jobs (the sample) along with measurements on the variables that might be related to profit, the independent variables. His or her objective will be to utilize information in the sample to infer the approximate relationship of the independent variables just described to the dependent variable, profit, and to measure the strength of this relationship. The manufacturer's objective will be to determine optimum conditions for maximizing the profit in the business.

The population of interest in the manufacturing problem is conceptual and consists of all measurements of profit (per unit of capital and labor invested) that might be made on contracts, now and in the future, for fixed values of the independent variables (size of the contract, measure of competition, etc.). The profit measurements would vary from contract to contract in an apparent random manner as a result of variations in materials, time to complete individual segments of the work, and other uncontrollable variables affecting the job. Consequently, we view the population as a *distribution* of profit measurements, with the form of the distribution depending upon specific values of the independent variables. To say that we seek to determine the relationship between the dependent variable "profit" and a set of independent variables therefore is translated to mean that we seek to determine the effect of the independent variables on the conceptual distribution of population measurements.

An individual population (or any set of measurements) can be characterized by a *relative frequency distribution*, which is also called a *relative frequency histogram*. A graph is constructed by subdividing the axis of measurement into intervals of equal width. A rectangle is constructed over each interval, with the height of the rectangle proportional to the fraction of the total number of measurements falling in each cell. For example, to characterize the ten measurements 2.1, 2.4, 2.2, 2.3, 2.7, 2.5, 2.4, 2.6, 2.6, and 2.9, we could divide the axis of measurement into intervals of equal width (.2 unit), commencing with 2.05. The relative frequencies (fraction of total number of measurements), calculated for each interval, are shown in Figure 1.1. Note that the figure gives a clear pictorial description of the set of ten measurements.

Observe that we have not given precise rules for selecting the number, widths, or locations of the intervals used in constructing a histogram. This is because the selection of these items is somewhat at the discretion of the person who is involved in the construction.

Although arbitrary, adherence to a few guidelines can be very helpful in selecting the intervals. *Points of subdivision of the axis of measurement should be*

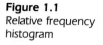

Figure 1.1
Relative frequency
histogram

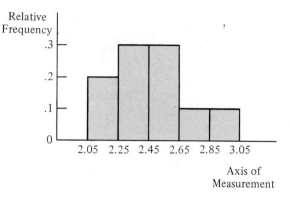

chosen so that it is impossible for a measurement to fall on a point of division. This eliminates a source of confusion and is easily accomplished, as indicated in Figure 1.1. The second guideline concerns the width and consequently the minimum number of intervals needed to describe the data. Generally speaking, we wish to obtain information on the form of the distribution of the data. Many times the form will be mound-shaped, as illustrated in Figure 1.2. (Others prefer to refer to distributions like these as bell-shaped or normal.) Many intervals for a small amount of data would result in little summarization and would present a picture essentially very similar to the data in its original form. The larger the amount of data, the greater will be the number of intervals that can be included and still present a satisfactory picture of the data. *We would suggest spanning the range of the data with from 5 to 20 intervals and using the larger number of intervals for larger quantities of data.*

Figure 1.2
Relative frequency
distribution

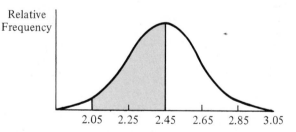

Many feel that the description of data is primarily an end in itself. No one would debate that people are concerned with history, but a major reason for this interest is to secure a glimpse of the future. We therefore could suggest that the description of a sample, say the national corporate production figures for iron ore, is to be used to forecast ore production next year and further in the future. Knowledge of today's production may have immediate uses, but a longer-range objective is forecasting (inference).

Finally, we point to a probabilistic interpretation that can be derived from the frequency histogram, Figure 1.1. We have already stated that the area of a rectangle over a given interval is proportional to the fraction of the total number of measurements falling in that interval. We may now wish to extend this idea one step further. If a measurement were selected at random from the original set, the probability that it would fall in a given interval is proportional to the area under the histogram lying over that interval. (At this point we rely on the layperson's concept of probability. This term will be discussed in greater detail in Chapter 2.) For example, for the data used to construct Figure 1.1, the probability of selecting a measurement in the interval 2.05 to 2.45 is .5, because half the measurements fall in the specified interval. Correspondingly, the area under the histogram of Figure 1.1 over the interval from 2.05 to 2.45 is .5. It is clear that this interpretation would apply to the distribution of any set of measurements, even populations. If Figure 1.2 gave the frequency distribution of profit (in millions of dollars) for a conceptual population of profit responses for contracts at a given setting of the independent variables (size of contract, measure of competition, etc.), the probability that the next contract (at the same setting of the independent variables) yields a profit that will lie in the interval 2.05 to 2.45 million is proportional to the shaded area under the distribution curve.

Exercises

1.2   The price to earnings (P/E) ratio of a stock is the ratio of a stock's most recent price-per-share to the stock's earnings-per-share (averaged over a twelve-month period), multiplied by 100. The P/E ratios for 44 securities selling at earnings multiples much higher than the average market price follow. Construct a relative frequency histogram for the data.

25.8	14.8	76.0	17.2	19.8	17.1
14.3	30.5	20.0	19.5	19.2	23.4
16.9	39.7	18.3	15.8	50.6	17.8
18.1	20.0	45.5	44.3	15.1	16.8
40.2	16.2	14.9	18.5	23.6	
21.3	15.2	17.7	14.7	17.4	
19.7	14.5	15.5	20.0	15.7	
15.6	20.8	19.4	18.7	16.4	

Source Forbes, September 28, 1981.

1.3   Of great importance to residents of central Florida is the amount of radioactive material present in the soil of reclaimed phosphate mining areas. Measurements of the amount of ^{238}U in 25 soil samples were as follows (measurements in picocuries per gram):

.74	6.47	1.90	2.69	.75
.32	9.99	1.77	2.41	1.96
1.66	.70	2.42	.54	3.36
3.59	.37	1.09	8.32	4.06
4.55	.76	2.03	5.70	12.48

Construct a relative frequency histogram for these data.

1.4   The percentages of change in crude oil production from 1976 to 1977 for 30 countries in North America, South America, Europe, and the Middle East are as follows:

-1.4	-13.6	8.1	-2.2	$-.2$
9.4	$-.4$	-2.7	-10.6	-4.5
$.3$	-4.1	-7.5	-6.5	-7.0
$.0$	-5.0	-4.1	-40.3	-20.7
8.0	-2.1	-2.0	205.2	-7.6
12.7	19.1	4.5	1.8	6.3

Source   Data adapted from *The World Almanac and Book of Facts* (New York: Newspaper Enterprise Association, 1979), p. 113.

Construct a relative frequency histogram for these data.

1.3 Characterizing a Set of Measurements: Numerical Methods

The relative frequency histogram presented in Section 1.2 provides useful information regarding a set of measurements, but it is not adequate for purposes of inference, mainly because it is not well defined. That is, many similar histograms could be formed from the same set of measurements. To make inferences about a population based on information contained in a sample and measure the goodness of the inferences, we need rigorously defined quantities with which to measure the sample information. We can then mathematically derive certain properties of these sample quantities and make probability statements regarding the goodness of our inferences.

The quantities we define are *numerical descriptive measures* of a set of data. We seek some numbers that describe the frequency distribution for any set of measurements. We will confine our attention to two types of descriptive numbers, *measures of central tendency* and *measures of dispersion or variation*.

The most common measure of central tendency used in statistics is the arithmetic mean. (Because this is the only type of mean discussed in this text, we shall omit the word *arithmetic*.)

Definition 1.1

> The *mean* of a set of n measured responses $y_1, y_2, \ldots, y_n$ is given by
>
> $$\bar{y} = \frac{1}{n} \sum_{i=1}^{n} y_i$$

The symbol $\bar{y}$ refers to a sample mean. The mean of all responses in a population will be denoted by the symbol μ. Note that we usually cannot measure μ; rather, μ is an unknown constant that we may want to estimate from sample information.

The mean of a set of measurements only locates the center of the distribution of data; by itself it does not provide an adequate description of a set of measurements. Two sets of measurements could have widely different frequency distributions with the same mean, as pictured in Figure 1.3. The difference between distributions I and II in the figure is in the variation or dispersion of measurements to either side of the mean. An adequate description of data requires that we define measures of data variability.

Figure 1.3
Frequency distributions with the same mean but different amounts of variation

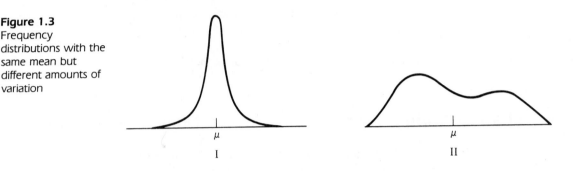

The most common measure of variability used in statistics is the variance, which is a function of the deviations (or distances) of the sample measurements from their mean.

Definition 1.2

> The *variance* of a set of measurements $y_1, y_2, \ldots, y_n$ is the average of the square of the deviations of the measurements about their mean. Symbolically the sample variance is
>
> $$s'^2 = \frac{1}{n} \sum_{i=1}^{n} (y_i - \bar{y})^2$$

The corresponding population variance is denoted by the symbol σ^2. The larger the variance of a set of measurements, the greater will be the variation within the set. The variance is of value in comparing the relative variation of two sets of measurements, but it gives information concerning only the variation in a single set when interpreted in terms of the *standard deviation*.

Definition 1.3

> The *standard deviation* of a set of measurement is the positive square root of the variance; that is,
>
> $$s' = \sqrt{s'^2}$$

The corresponding population standard deviation is denoted by σ.

Although closely related to the variance, the standard deviation can be used to give a fairly accurate picture of data variation for a single set of measurements. It can be interpreted using Tchebysheff's theorem, which will be presented in a later chapter, and by the empirical rule, which we will now explain.

Many distributions of data in real life are mound-shaped. That is, they can be approximated by a bell-shaped frequency distribution known as a normal curve. Data possessing mound-shaped distributions will possess very definite characteristics of variation, which are expressed in the following statement:

Empirical Rule:

For a distribution of measurements that is approximately normal (bell shaped), it follows that the interval with endpoints

$\mu \pm \sigma$ contains approximately 68 % of the measurements.

$\mu \pm 2\sigma$ contains approximately 95 % of the measurements.

$\mu \pm 3\sigma$ contains almost all of the measurements.

As was mentioned in Section 1.2, once the frequency distribution of a set of measurements is known, probability statements regarding the measurements can be made. These probabilities were shown as areas under a frequency histogram. Thus the probabilities contained in the empirical rule are areas under the normal curve shown in Figure 1.4.

Figure 1.4
Normal curve

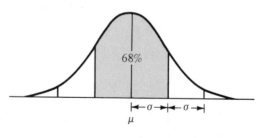

The use of the empirical rule will be illustrated by the following example. Suppose the scores on an achievement test given to all high school seniors in a certain state are known to have, approximately, a normal distribution with a mean μ of 64 and a standard deviation σ of 10. It can then be deduced that approximately 68 % of the scores are between 54 and 74, 95 % of the scores are between 44 and 84, and almost all of the scores are between 34 and 94. Thus knowledge of the mean and the standard deviation gives us a fairly good picture of the frequency distribution of scores.

Suppose that a single high school student is randomly selected from those who took the test. What is the probability that his score will be between 54 and 74? On looking at the relative frequency of observations in this interval, we find that 0.68 is a reasonable answer to the probability question.

The utility and value of the empirical rule is great because of the very common occurrence of approximately normal distributions of data in nature—more so because the rule applies to distributions that are not exactly normal but just mound shaped. You will find that approximately 95% of a set of measurements will be within 2σ of μ for a variety of distributions.

Exercises

1.5   The following results on summations will help us in calculating the sample variance s'^2. For any constant c,

(a) $\displaystyle\sum_{i=1}^{n} c = nc$

(b) $\displaystyle\sum_{i=1}^{n} cy_i = c\sum_{i=1}^{n} y_i$

(c) $\displaystyle\sum_{i=1}^{n} (x_i + y_i) = \sum_{i=1}^{n} x_i + \sum_{i=1}^{n} y_i$

Use (a), (b), and (c) to show that

$$s'^2 = \frac{1}{n}\left[\sum_{i=1}^{n} y_i^2 - \frac{1}{n}\left(\sum_{i=1}^{n} y_i\right)^2\right]$$

1.6   Use the result of Exercise 1.5 to calculate s' for the $n = 6$ sample measurements 1, 4, 2, 1, 3, and 3.

1.7   Refer to Exercise 1.2.

(a)   Calculate $\bar{y}$ and s' for the data given there.

(b)   Calculate the interval $\bar{y} + ks'$ for $k = 1$, 2, and 3. Count the number of measurements falling within each interval and compare this result with the number that you would expect according to the empirical rule.

1.8   Refer to Exercise 1.3 and repeat parts (a) and (b) of Exercise 1.7.

1.9   Refer to Exercise 1.4 and repeat parts (a) and (b) of Exercise 1.7.

1.10   In Exercise 1.4, the extremely large value of 205.2 is for the United Kindgom and was due to a major oil strike in the North Sea. Eliminate this value and calculate $\bar{y}$ and s' for the remaining 29 observations. Also, calculate the intervals $\bar{y} + ks'$ for $k = 1$, 2, and 3, count the number of measurements in each, and compare this result with the results predicted by the empirical rule. Compare the answers here to those found in Exercise 1.9. Note the effect of a single large observation on $\bar{y}$ and s'.

1.11   The *range* of a set of measurements is the difference between the largest and the smallest values. The empirical rule suggests that the standard deviation of a set of measurements may be roughly approximated by one-fourth of the range (i.e., range/4). Calculate this approximation to s' for the data sets in Exercises 1.2, 1.3, and 1.4. Compare the result in each case to the actual calculated value of s'.

1.12   Weekly maintenance costs for a certain factory, recorded over a long period of time and adjusted for inflation, tend to have an approximately normal distribution with an average of \$420 and a standard deviation of \$30. If \$450 is budgeted for next week, what is an approximate probability that this budgeted figure will be exceeded?

1.13   A manufacturer of a new food additive for beef cattle claims that 80% of the animals fed on a diet including this additive should have monthly weight gains in excess of 20 pounds. A large sample of measurements on weight gains from this diet exhibits an approximately normal distribution with a mean of 22 pounds and a standard deviation of 2 pounds. Do you think the sample information contradicts the manufacturer's claim? (Calculate the probability of a weight gain exceeding 20 pounds.)

1.4 How Inferences Are Made

The mechanism instrumental in making inferences can best be seen by analyzing our own intuitive inference-making procedures.

Suppose that two candidates are running for a public office in our community and that we wish to determine whether our candidate, Jones, is favored to win. Thus the population of interest is the set of responses from all eligible voters who will vote on election day, and we wish to determine whether the fraction favoring Jones exceeds .5. For the sake of simplicity suppose that all eligible voters will go to the polls and that we randomly select a sample of twenty from the courthouse rosters. All twenty are contacted and all favor Jones. What do you conclude with regard to Jones's prospects for winning the election?

There is little doubt that most of us would immediately infer that Jones will win the election, and we find that subsequent study will substantiate our good judgment. This was indeed an easy inference to make, but the inference itself is not our immediate goal. Rather, we wish to examine the mental processes that were employed in reaching a conclusion about the prospective behavior of a large voting population on the basis of a sample of only twenty people.

Winning means acquiring more than 50% of the votes. Did we conclude that Jones would win because we expect the fraction favoring Jones in the sample to correspond identically to the fraction in the population? We know that this is not true and a simple experiment will verify the fact that the sample fraction favoring Jones need not be the same as the fraction of the population that favors Jones. If a balanced coin is tossed, it is intuitively obvious that the true proportion of the times it will turn up heads is .5. Yet if we sample the outcomes for our coin by tossing it twenty times, the proportion of heads will vary from sample to sample. That is, on one occasion we might observe twelve heads out of twenty flips for a sample proportion of $12/20 = .6$. On another occasion we might observe eight heads out of twenty flips for a sample proportion of $8/20 = .4$. In fact, the sample proportion of heads could be 0, .05, .10, ..., 1.0.

Did we conclude that Jones must win because it would be impossible for twenty out of twenty to favor Jones if in fact less than 50% of the electorate intended to vote for him? The answer to this question is certainly "no," but it provides the key to our hidden line of logic. *It is not impossible to draw twenty out of twenty favoring Jones (even though less than 50% of the electorate favor him), but it is highly improbable.* Thus our intuitive feel for probability suggested that it was highly improbable that twenty out of twenty in the sample would select Jones if, in fact, he is to be a loser. We therefore concluded that he would win.

This example illustrates the potent role played by probability in making inferences. Probabilists assume that they know the structure of the phenomenon in question, the population, and they use the theory of probability to make an inference about a sample. Thus they assume that they know the structure of a population generated by random drawings of five cards from a standard deck, and they use probability to make an inference concerning a draw that will yield three aces and two kings. Statisticians use probability to make the trip in reverse, from the sample to the population. Observing a draw of five aces in a sample, they immediately infer that the deck (which generates the population) is loaded and not standard, because the probability of drawing five aces from a standard deck is zero. This is an exaggerated case but it makes the point. Probability is the mechanism used in making inferences. Basic to inference is the problem of calculating the probability of the observed sample.

One final comment is in order. Any of you who did not think that the sample justified an inference that Jones would win should not feel too chagrined. One can be easily misled when making intuitive evaluations of the probabilities of events. Any of you who decided that the probability that twenty voters out of twenty would favor Jones was very low, assuming Jones to be a loser, was correct. It would not be difficult to concoct an example in which an intuitive assessment of probability would be in error. Thus intuitive assessments of probabilities are unsatisfactory, and we must have a rigorous theory of probability at hand in order to develop the mechanism for inference.

1.5 Theory and Reality

It is essential that you grasp the difference between theory and reality. Theories are ideas proposed to explain phenomena in the real world and, as such, are approximations or models for reality. These models, or explanations of reality, are presented in verbal forms in some less quantitative fields and as mathematical relationships in others. Whereas a theory of social change might be expressed verbally in sociology, describing the motion of a vibrating string is presented in a precise mathematical manner in physics. When we choose a mathematical model for a physical process, we hope that the model reflects faithfully, in mathematical terms, the attributes of the physical process so that mathematical methods can be used to arrive at conclusions about the process itself. If we could develop an equation to predict the position of a vibrating string, the quality of the prediction would depend upon how well the equation fit the motion of the string. The process of finding a good equation is not necessarily simple and usually requires several simplifying assumptions (uniform string mass, no air resistance, etc.). The final criterion for deciding whether a model is good is whether it yields good and useful information. The motivation for using mathematical models lies primarily in their utility.

This text is concerned with the theory of statistics and hence a model of reality. We will postulate theoretical frequency distributions for populations and will develop a theory of probability and inference in a precise mathematical manner. The net result will be a theoretical or mathematical model for the acquisition and utilization of information in real life. It will not be an exact representation of nature, but this should not disturb us. Like the utility of other theories, its utility will be measured by its ability to assist us in understanding nature and in solving problems in the real world.

1.6 Summary

The objective of statistics is to make an inference about a population based on information contained in a sample. The theory of statistics is a theory of information concerned with its quantification, with the design of experiments or procedures for data collection that will minimize the cost of a specified quantity of information, and with the use of this information in making inferences. Most important, we have viewed the making of an inference about the unknown population as a two-step procedure. First, we seek the best inferential procedure for the given situation and, second, we desire a measure of its goodness. For example, every estimate of a population characteristic based on information contained in the sample might have associated with it a probabilistic bound on the error of estimation.

A necessary prelude to making inferences about a population is the ability to describe a set of numbers. Frequency distributions provide a very graphic and useful method for characterizing a conceptual or real population of numbers. More useful to inference making are numerical descriptive measures.

The mechanism for making inferences is the theory of probability. The probabilist reasons from a known population to the outcome of a single experiment, the sample. In contrast, the statistician utilizes the theory of probability to calculate the probability of an observed sample and to infer from this the characteristics of an unknown population. Thus probability is the foundation of the theory of statistics.

Finally, we have noted the difference between theory and reality. In this text we study the mathematical theory of statistics, which is an idealization of nature. It is rigorous, mathematical, and subject to study in a vacuum completely isolated from the real world. Or it can be tied very closely to reality and can be useful in making inferences from data in all fields of science. In this text we will be utilitarian. We will not regard statistics as a branch of mathematics but as an area of science concerned with the development of a practical theory of information. We will consider statistics as a separate field analogous to physics, not as a branch of mathematics but as a theory of information that utilizes mathematics heavily.

Subsequent chapters will expand upon the topics that we have just discussed. We will begin with a study of the mechanism employed in making inferences, the theory of probability. This theory will provide a theoretical model for the generation of experimental data and will provide the basis for our study of statistical inference.

References and Further Readings

1.   Fraser, D. A. S. *Statistics, an Introduction.* New York: Wiley, 1958.

2.   Freund, J. E., and Walpole, R. E. *Mathematical Statistics.* 4th ed. Englewood Cliffs, N.J.: Prentice-Hall, 1987.

3.   Kendall, M. G., and Stuart, A. *The Advanced Theory of Statistics.* 4th ed. Vol. 1. New York: Hafner Press, 1977.

4.   Mendenhall, W. *Introduction to Probability and Statistics.* 7th ed. N. Scituate, Mass.: Duxbury Press, 1987.

5.   Mood, A. M.; Graybill, F. A.; and Boes, D. *Introduction to the Theory of Statistics.* 3d ed. New York: McGraw-Hill, 1974.

Supplementary Exercises

1.14   Prove that the sum of the deviations of a set of measurements about their mean is equal to zero; that is,

$$\sum_{i=1}^{n} (y_i - \bar{y}) = 0$$

1.15   An important consideration in planning urban mass transit systems in metropolitan areas is the cost of construction per mile. Studies at the University of Pennsylvania's Department of City and Regional Planning estimate the mean construction cost per mile of subways in 1980 at $45 million (*Journal of Advanced Transportation*, Spring 1981). If the distribution of subway construction costs is bell shaped with a standard deviation of $10 million per mile, what percentage of subway systems constructed in 1980 had a cost per mile between $35 million and $65 million?

1.16   Use this sample of $n = 15$ measurements:

8	10	4	3	8
5	6	8	7	6
7	7	5	11	7

(a)   Use the range of the measurements to obtain an estimate of the standard deviation.

(b)   Construct a frequency histogram for the data. Use the histogram to obtain a visual approximation to $\bar{y}$ and s'.

(c)   Calculate $\bar{y}$ and s'. Compare these results with the calculation checks provided by (a) and (b).

(d)   Construct the intervals $\bar{y} \pm ks'$, $k = 1, 2, 3$, and count the number of measurements falling in each interval. Compare the fractions falling in the intervals with the fractions that you would expect according to the empirical rule.

1.17   The following data give the times to failure for $n = 88$ radio transmitter-receivers:

16	224	16	80	96	536	400	80
392	576	128	56	656	224	40	32
358	384	256	246	328	464	448	716
304	16	72	8	80	72	56	608
108	194	136	224	80	16	424	264
156	216	168	184	552	72	184	240
438	120	308	32	272	152	328	480
60	208	340	104	72	168	40	152
360	232	40	112	112	288	168	352
56	72	64	40	184	264	96	224
168	168	114	280	152	208	160	176

(a)   Use the range to approximate s' for the $n = 88$ lengths of time to failure.

(b)   Construct a frequency histogram for the data. [Note the tendency for the distribution to tail outward (skew) to the right.]

(c)   Use an electric calculator (or computer) to calculate $\bar{y}$ and s'. (Hand calculation is much too tedious for this exercise.)

(d)   Calculate the intervals $\bar{y} \pm ks'$, $k = 1, 2, 3$, and count the number of measurements falling in each interval. Compare your results with the empirical rule results. Note that the empirical rule provides a rather good description of these data, even though the distribution is highly skewed.

1.18   Compare the ratio of the range to s' for the three sample sizes ($n = 6, 15,$ and 88) for Exercises 1.6, 1.16, and 1.17. Note that the ratio tends to increase as the amount of data increases. The greater the amount of data, the greater will be the tendency for it to contain a few extreme values that will inflate the range and have little effect on s'. We ignored this phenomenon and suggested that you use 4 as the ratio for finding a guessed value of s' in checking calculations.

1.19   A set of 340 examination scores exhibiting a bell-shaped relative frequency distribution has a mean of $\bar{y} = 72$ and a standard deviation of $s' = 8$. Approximately how many of the scores would you expect to fall in the interval from 64 to 80? The interval from 56 to 88?

1.20   A machine produces bearings with a mean diameter of 3.00 inches and a standard deviation of 0.01 inch. Bearings with diameters in excess of 3.02 inches or less than 2.98 inches will fail to meet quality specifications. Approximately what fraction of this machine's production will fail to meet specifications? What assumptions did you make concerning the distribution of bearing diameters in order to answer this question?

1.21   A recent study by the Highway Loss Data Institute reported that the average loss-payment per insurance claim by automobile owners during the first half of 1974 was $495 with a standard deviation of $75 (*Money*, September 1974)[†]. Assume that the distribution of loss-payment per claim is mound-shaped.

(a)   Describe the distribution of loss payments during this period.

(b)   Approximately what fraction of the loss payments exceeded $570 during this period?

1.22   A fertilizer production process, over the past year, has shown an average daily yield of 60 tons with a variance in daily yields of 100. If the yield should fall below 40 tons tomorrow, should this result cause you to suspect an abnormality in the process? (Calculate the probability of obtaining below 40 tons.) What assumptions did you make concerning the distribution of yields?

*1.23   Let $k \geq 1$. Show that for any set of n measurements the fraction included in the interval $\bar{y} - ks'$ to $\bar{y} + ks'$ is at least $(1 - 1/k^2)$. [Hint:

$$s'^2 = \frac{\sum\limits_{i=1}^{n} (y_i - \bar{y})^2}{n}$$

In this expression replace all deviations for which $|y_i - \bar{y}| \geq ks'$ with ks'. Simplify.] This result is known as *Tchebysheff's theorem*.

1.24   A personnel manager for a certain industry has records of the number of employees absent per day. The average number absent is 5.5 and the standard deviation is 2.5. Since there are many days with zero, one, or two absent and only a few with more than ten absent, the frequency distribution is highly skewed. The manager wants to publish an interval in which at least 75% of these values lie. Use the result in Exercise 1.23 to find such an interval.

1.25   For the data discussed in Exercise 1.24 give an upper bound to the fraction of days when there are more than thirteen absentees.

*1.26   The Economic Recovery Act, a federal bill passed in 1981, included several provisions aimed at stimulating increased savings. Part of the law included increased contribution limits with regard to Individual Retirement Accounts (IRAs). The maximum amount that an individual can contribute to an IRA was raised from $1500 to the lesser of $2000 or 100% of the individual's compensation for the year. If the mean and standard deviation of the amounts contributed to IRAs are $1070 and $460 respectively, describe the distribution for the population of all IRAs.

[†] *Source* Reprinted from the September 1974 issue of *Money* by special permission; © 1974; Time-Life, Inc. All rights reserved.

* Exercises preceded by an asterisk are optional.

PROBABILITY

2.1 Introduction

To most people, *probability* is a loosely defined term employed in everyday conversation to indicate the measure of one's belief in the occurrence of a future event. We accept this as a meaningful and practical interpretation of the term but seek a clearer understanding of the context in which it is used, how it is measured, and how probability assists in making inferences.

The concept of probability is necessary when dealing with physical, biological, or social mechanisms generating observations that cannot be predicted with certainty. For example, the blood pressure of a human at a given point in time cannot be predicted with certainty, and we never know the exact load that a bridge will endure before collapsing into a river. Such random events cannot be predicted with certainty, but the relative frequency with which they occur in a long series of trials is often remarkably stable. Events possessing this property are called *random* or *stochastic* events. This stable relative frequency provides an intuitively meaningful measure of one's belief in the occurrence of a random event for a future observation. It is impossible, for example, to predict with certainty the occurrence of heads for the single toss of a balanced coin, but we would be willing to state with a fair measure of confidence that the fraction of heads in a long series of trials would be very near .5. That this relative frequency is used as a measure of belief in the outcome for a single toss is evident when we consider a gambler's objective. He risks his money on the single toss of a coin, not a long series of tosses. The relative frequency of a head in a long series of tosses, which he calls "the probability of a head," gives him a measure of his chances of winning on a single toss. If the coin were unbalanced and gave 90% heads in a long series of tosses, the gambler would say that the "probability of a head" is .9, and his belief in the occurrence of a head on a single toss of the coin would be fairly strong.

The preceding example possesses some very realistic and practical analogies. In many respects all humans are gamblers. The research physician gambles time and money on a research project and she is concerned with her success on a single flip of this symbolic coin. Similarly, the investment of capital in a new manufacturing plant is a gamble that represents a single flip of a coin on which the entrepreneur has high hopes for success. The fraction of similar investments that are successful in a long series of trials is of interest to the entrepreneur only insofar as it provides a measure of belief in the successful outcome of his single individual investment.

The relative frequency concept of probability is intuitively meaningful, but it does not provide a rigorous definition of probability. Many other concepts of probability have been proposed, including that of subjective probability, which allows the probability of an event to vary depending upon the person performing the evaluation.

A clarification of the practical meaning of probability is not essential for the development of a theory, but it is absolutely necessary if it is to be used in achieving the very practical inference-making goal that we have specified as the objective of statistics. We will devote no more time to this point, because the practical meaning of probability is a matter of philosophy that can be argued endlessly. For our purposes we accept an interpretation based on relative frequency as a meaningful measure of one's belief in the occurrence of an event, and we will now examine the link that probability provides between observation and inference.

2.2 Probability and Inference

The role that probability plays in making inferences will be discussed in detail after an adequate foundation has been laid in the theory of probability. At this point we will present a preview to motivate an elementary treatment of this theory using an example and an appeal to your intuition.

The example that we have selected is similar to that presented in Section 1.4 but is simpler and less practical. It was chosen because of the ease with which we can visualize the population and sample and because it provides an observation-producing mechanism for which a probabilistic model will be constructed in Section 2.3.

Consider a gambler who wishes to make an inference concerning the balance of a die. The conceptual population of interest is the set of numbers that would be generated if the die were rolled over and over again, ad infinitum. If the die were perfectly balanced, one-sixth of the measurements in the population would be 1s, one-sixth 2s, one-sixth 3s, and so on. The corresponding frequency distribution is shown in Figure 2.1.

Figure 2.1
Frequency distribution for the population generated by a balanced die

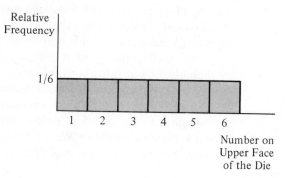

Using the scientific method, the gambler proposes a hypothesis *that the die is balanced*, and he seeks observations from nature to contradict the theory, if false. A sample of ten tosses is selected from the population by rolling the die ten times. All ten tosses result in 1s. The gambler looks upon this output of nature with a jaundiced eye and concludes that his hypothesis is not in agreement with nature and hence that the die is unbalanced.

The reasoning employed by the gambler identifies the role that probability plays in making inferences. The gambler rejected his hypothesis (concluding that the die is unbalanced) not because it is *impossible* to throw ten 1s with a balanced die but because it is highly *improbable*. His evaluation of the probability was most likely subjective. That is, the gambler may not have known how to calculate the probability of ten 1s, but he had an intuitive feeling that this event would occur with very low frequency if the die were balanced and the ten tosses were repeated over and over again for a long series of trials. The point to note is that his decision was based on the probability of the observed sample.

The need for a theory of probability that will provide a rigorous method for finding a number (a probability) that will agree with the actual relative frequency of occurrence of an event in a long series of trials is apparent if we imagine a different result for the gambler's sample. Suppose, for example, that instead of ten 1s, he observed five 1s along with a 3, 4, 6, 3, and a 2. If the die is balanced, is this result so improbable as to cause us to reject our hypothesis and conclude that the die is loaded in favor of 1s? Unlike the case in which we had ten 1s in ten tosses, in this case it is not so easy to decide whether the probability of five 1s in ten tosses is large or small if we must simply rely on experience and intuition to make the evaluation. The probability of throwing four 1s in ten tosses would be even more difficult to guess.

We will not deny that many experimental results are obviously inconsistent with and lead to a rejection of the hypothesis in question, but many samples fall in a gray area that requires a rigorous assessment of probability. Indeed, it is not difficult to show that the intuitive evaluation of probability often leads to an answer that is substantially in error and results in incorrect inferences about the target population. For example, if there are twenty people in a room, most people would guess that it is very unlikely that there would be two or more persons with the same birthday. Yet, under certain reasonable assumptions, it can be shown that the probability of such an occurrence is larger than .4. (A calculation of this probability is completed in Example 2.18.)

We need a theory of probability that will permit us to calculate the probability (or a quantity proportional to the probability) of observing specified outcomes, assuming that our hypothesized model is correct. This topic will be developed in detail in subsequent chapters. Our immediate goal is to present an introduction to the theory of probability, which provides the foundation for modern statistical inference. We will begin by reviewing set notation, which will be used in constructing a probabilistic model for an experiment.

2.3 A Review of Set Notation

To proceed with an orderly development of probability theory, we need some basic concepts of set theory. We will use capital letters, A, B, C, ..., to denote sets of points. If the elements in the set A are a_1, a_2, and a_3, we will write

$$A = \{a_1, a_2, a_3\}$$

Let S denote the set of all elements under consideration; that is, S is the *universal set*. For any two sets A and B we will say that A is a *subset* of B, or A is contained in B (denoted $A \subset B$), if every point in A is also in B. The *null* or *empty set*, denoted by ϕ, is the set consisting of no points. Thus ϕ is a subset of every set.

Sets and relationships between sets can be conveniently portrayed by using *Venn diagrams*. The Venn diagram in Figure 2.2 shows two sets, A and B,

Figure 2.2
Venn diagram for
$A \subset B$

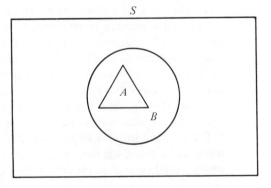

in the universal set S. Set A is the set of all points inside the triangle; set B is the set of all points inside the circle. Note that in Figure 2.2 $A \subset B$.

Consider now two arbitrary sets of points. The *union* of A and B, denoted by $A \cup B$, is the set of all points in A or B or both. That is, the union of A and B contains all points that are in at least one of the sets. The Venn diagram in Figure 2.3 shows two sets A and B, where A is the set of points in the left-hand circle and B is the set of points in the right-hand circle. The set $A \cup B$ is the shaded region consisting of all points inside either circle (or both). The key word for expressing the union of two sets is *or* (meaning A or B or both).

Figure 2.3
Venn diagram for
A ∪ B

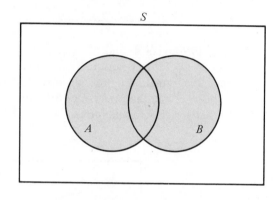

The *intersection* of A and B, denoted by $A \cap B$ or by AB, is the set of all points in both A and B. The Venn diagram of Figure 2.4 shows two sets A and B, with AB consisting of the points in the shaded region where the two sets overlap. The key word for expressing the intersections is *and* (meaning A and B simultaneously).

Figure 2.4
Venn diagram for AB

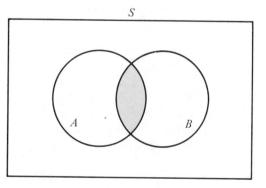

If A is a subset of S, then the *complement* of A, denoted by $\bar{A}$, is the set of points that are in S but not in A. Figure 2.5 is a Venn diagram illustrating that the shaded area in S but not in A is $\bar{A}$. Note that $A \cup \bar{A} = S$.

Figure 2.5
Venn diagram for Ā

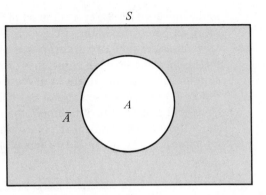

Two sets, *A* and *B*, are said to be *disjoint* or *mutually exclusive* if $A \cap B = \phi$. That is, mutually exclusive sets have no points in common. The Venn diagram in Figure 2.6 illustrates two sets *A* and *B* that are mutually exclusive. Note that, referring to Figure 2.5, it is easy to see that *A* and *Ā* are mutually exclusive sets for any set *A*.

Figure 2.6
Venn diagram for
mutually exclusive
sets A and B

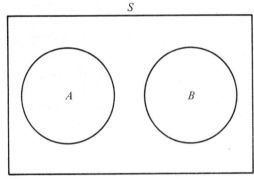

Consider the die-tossing problem of Section 2.2 and let *S* denote the set of all possible numerical observations for a single toss of a die. That is, $S = \{1, 2, 3, 4, 5, 6\}$. Let $A = \{1, 2\}$, $B = \{1, 3\}$, and $C = \{2, 4, 6\}$. Then $A \cup B = \{1, 2, 3\}$, $A \cap B = \{1\}$, and $\bar{A} = \{3, 4, 5, 6\}$. Also, note that *B* and *C* are mutually exclusive, whereas *A* and *C* are not.

We will not attempt a thorough review of set algebra, but we mention two equalities of considerable importance. These are the *distributive laws*, given by

$$A \cap (B \cup C) = (A \cap B) \cup (A \cap C)$$
$$A \cup (B \cap C) = (A \cup B) \cap (A \cup C)$$

In the next section we will proceed with an elementary discussion of probability theory.

Exercises

2.1   Suppose a family contains two children of different ages, and we are interested in the sex of these children. Let F denote that a child is female and M that the child is male, and let a pair such as FM denote that the older child is female and the younger male. There are four points in the set S of possible observations:

$$S = \{FF, FM, MF, MM\}$$

Let A denote the subset of possibilities containing no males, B the subset containing two males, and C the subset containing at least one male. List the elements of A, B, C, $A \cap B$, $A \cup B$, $A \cap C$, $A \cup C$, $B \cap C$, $B \cup C$, and $C \cap \bar{B}$.

2.2   Draw Venn diagrams to verify for yourself that, for any two sets A and B, $\overline{(A \cup B)} = \bar{A} \cap \bar{B}$ and $\overline{(A \cap B)} = \bar{A} \cup \bar{B}$.

2.3   Suppose two dice are tossed and the numbers on the upper faces are observed. Let S denote the set of all possible pairs that can be observed. [These pairs can be listed, for example, by letting (2, 3) denote that a 2 was observed on the first die and a 3 on the second.] Define the following subsets of S:

  A: The number on the second die is even.

  B: The sum of the two numbers is even.

  C: At least one number in the pair is odd.

List the points in A, $\bar{C}$, $A \cap B$, $A \cap \bar{B}$, $\bar{A} \cup B$, and $\bar{A} \cap C$.

2.4   A group of five applicants for a pair of identical jobs consists of three men and two women. The employer is to select two of the five applicants for the jobs. Let S denote the set of all possible outcomes for the employer's selection. Let A denote the subset of outcomes corresponding to the selection of two men and B the subset corresponding to the selection of at least one woman. List the outcomes in A, $\bar{B}$, $A \cup B$, $A \cap B$, and $A \cap \bar{B}$. (Denote the different men and women by M_1, M_2, M_3 and W_1, W_2, respectively.)

2.5   From a survey of sixty students attending a university, it was found that nine were living off campus, thirty-six were undergraduates, and three were undergraduates living off campus.

  (a)   Find the number of these students who were undergraduates, were living off campus, or both.

  (b)   Find the number of these students who were undergraduates living on campus.

  (c)   Find the number of these students who were graduate students living on campus.

2.4 A Probabilistic Model for an Experiment: The Discrete Case

In Section 2.2 we referred to the die-tossing *experiment* when we observed the number appearing on the upper face. We will use the term *experiment* to include observations obtained from completely uncontrollable situations (such as observations on the daily price of a particular stock) as well as those made under controlled laboratory conditions. We have the following definition:

Definition 2.1

> An *experiment* is the process by which an observation is made.

Examples of experiments include coin and die tossing, measuring the IQ score of an individual, or determining the number of bacteria per cubic centimeter in a portion of processed food.

When an experiment is performed, it can result in one or more outcomes, which are called *events*. In our discussions events will be denoted by capital letters. If the experiment consists of counting the number of bacteria in a portion of food, some events of interest could be these:

A: There are 110 bacteria present.

B: There are more than 200 bacteria present.

C: The number of bacteria present is between 100 and 300.

Some events associated with a single toss of a balanced die are these:

A: Observe an odd number.

B: Observe a number less than 5.

C: Observe a 2 or a 3.

E_1: Observe a 1.

E_2: Observe a 2.

E_3: Observe a 3.

E_4: Observe a 4.

E_5: Observe a 5.

E_6: Observe a 6.

You can see that there is a distinct difference among some of the events associated with the die-tossing experiment. For example, if you observe event A (an odd number), at the same time you will have observed E_1, E_3, or E_5. Thus event A, which can be decomposed into three other events, is called a *compound event*. In contrast, the events E_1, E_2, E_3, E_4, E_5, and E_6 cannot be decomposed and are called *simple events*. A simple event can happen only in one way, whereas a compound event can happen in more than one distinct way.

The results of set theory are very useful for expressing the relationships between various events associated with an experiment. Because sets are collections of points, let us think of associating a distinct point, called a *sample point*, with each and every simple event associated with an experiment.

Definition 2.2

> A *simple event* is an event that cannot be decomposed. Each simple event corresponds to one and only one *sample point*. The letter E with a subscript will be used to denote a simple event or the corresponding sample point.

Thus we can think of a simple event as a set consisting of a single point; namely, the single sample point with which the event is associated.

Definition 2.3

> The *sample space* associated with an experiment is the set consisting of all possible sample points. A sample space will be denoted by S.

We can easily see that the sample space S associated with the die-tossing experiment consists of six sample points corresponding to the six simple events E_1, E_2, E_3, E_4, E_5, and E_6. That is, $S = \{E_1, E_2, E_3, E_4, E_5, E_6\}$. A Venn diagram exhibiting the sample space for the die-tossing experiment is given in Figure 2.7.

Figure 2.7
Venn diagram for the sample space associated with the die-tossing experiment

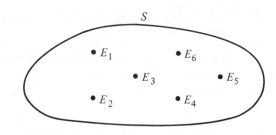

For the microbiology example of counting bacteria in a food specimen, let E_0 correspond to observing 0 bacteria, E_1 correspond to observing 1 bacterium, and so on. Then the sample space is

$$S = \{E_0, E_1, E_2, \ldots\}$$

because no integer number of bacteria can be ruled out as a possible outcome.

Both of the sample spaces we examined have the property that they consist of either a finite or a countable number of sample points. In the die-tossing example there are six (a finite number) sample points. The number of sample points associated with the bacteria-counting experiment is infinite, but the number of distinct sample points can be put into a one-to-one correspondence with the integers (that is, it is countable). Such sample spaces are said to be discrete.

Definition 2.4

> A *discrete sample space* is one that contains a finite or countable number of sample points.

When an experiment is conducted a single time, you will observe one and only one simple event. For example, if you toss a die and observe a 1, you cannot at the same time observe a 2. Thus the single sample point E_1 associated with observing a 1 and the single sample point E_2 associated with observing a 2 are distinct, and the sets $\{E_1\}$ and $\{E_2\}$ are mutually exclusive sets. Thus events E_1 and E_2 are mutually exclusive events. Similarly, all distinct simple events correspond to mutually exclusive sets of simple events and are thus mutually exclusive events.

For experiments with discrete sample spaces, *compound events* can be viewed as collections (sets) of sample points or, equivalently, as unions of the sets of single sample points corresponding to the appropriate simple events. For example, the die-tossing event A (observe an odd number) will occur if and only if one of the simple events E_1, E_3, or E_5 occurs. Thus

$$A = \{E_1, E_3, E_5\} \qquad \text{or} \qquad A = E_1 \cup E_3 \cup E_5$$

Similarly, B (observe a number less than 5) can be written as

$$B = \{E_1, E_2, E_3, E_4\} \qquad \text{or} \qquad B = E_1 \cup E_2 \cup E_3 \cup E_4$$

The rule for determining which simple events to include in a compound event is very precise. *A simple event E_i is included in event A if, and only if, A occurs whenever E_i occurs.*

Definition 2.5

An *event* in a discrete sample space S is a collection of sample points; that is, any subset of S.

Figure 2.8 gives a Venn diagram illustrating a representation of the sample space and events A (observe an odd number) and B (observe a number less than 5) for the die-tossing experiment. Notice that it is very easy to visualize the relationship between events if a Venn diagram like that in Figure 2.8 is developed.

Figure 2.8
Venn diagram for the die-tossing experiment

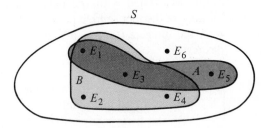

By Definition 2.5, any event in a discrete sample space S is a subset of S. In the example concerning counting bacteria in a portion of food, the event B (the number of bacteria is more than 200) can be expressed as

$$B = \{E_{201}, E_{202}, E_{203}, \ldots\}$$

where E_i denotes the simple event that there are i bacteria present in the food sample and $i = 0, 1, 2, \ldots$.

A probabilistic model for an experiment with a discrete sample space can be constructed by assigning a probability to each simple event in the sample space S. In doing so, we will select this number, a measure in our belief in its occurrence on a single repetition of the experiment, so that it will be consistent with the relative frequency concept of probability. Although relative frequency does not provide a rigorous definition of probability, any definition that is to be applicable to the real world should agree with our intuitive notion of the behavior of relative frequencies of events.

On analyzing the frequency concept of probability, we see that three conditions must hold.

1. The relative frequency of occurrence of any event must be greater than or equal to zero. A negative frequency does not make sense.

2. The relative frequency of the whole sample space S must be unity. Because every possible outcome of the experiment is a point in S, it follows that S must occur every time the experiment is performed.

3. If two events are mutually exclusive, the relative frequency of their union is the sum of their respective relative frequencies.

As an example of condition 3, if the experiment of tossing a balanced die yields a 1 on 1/6 of the tosses, it should yield a 1 or a 2 on $1/6 + 1/6$ or 1/3 of the tosses. These three concepts form the basis of our definition of probability.

Definition 2.6

Suppose that an experiment has associated with it a sample space S. To every event A in S (A is a subset of S) we assign a number, $P(A)$, called the *probability* of A, so that the following axioms hold:

Axiom 1: $P(A) \geqslant 0$.

Axiom 2: $P(S) = 1$.

Axiom 3:   If $A_1, A_2, A_3, \ldots$ form a sequence of pairwise mutually exclusive events in S, then

$$P(A_1 \cup A_2 \cup A_3 \cup \cdots) = \sum_{i=1}^{\infty} P(A_i)$$

We can easily show that Axiom 3, which is stated in terms of an infinite sequence of events, implies a similar property for a finite sequence. Specifically, if A_1, $A_2, \ldots, A_n$ are pairwise mutually exclusive events, then

$$P(A_1 \cup A_2 \cup A_3 \cup \cdots \cup A_n) = \sum_{i=1}^{n} P(A_i)$$

Note that the definition states only the properties a probability must satisfy; it does not tell us how to assign specific probabilities to events. For example, suppose that a coin has yielded eight heads in ten previous tosses. Consider the experiment of one more toss of the same coin. There are two possible outcomes, head or tail, and hence two simple events. The definition of probability allows us to assign to these simple events any two nonnegative numbers that add to one. For example, each simple event could have probability 1/2. In light of the past history of this coin, a more reasonable solution would be to give probability of .8 to the outcome involving a head. Specific assignments of probabilities must be done in a manner that is consistent with reality if the probabilistic model is to serve a useful purpose.

For discrete sample spaces it suffices to assign probabilities to each simple event. If a balanced die is used for the die-tossing example, it seems reasonable to assume that all simple events would have the same relative frequency in the long run. We will assign a probability of 1/6 to each simple event; $P(E_i) = 1/6$, $i = 1$, $2, \ldots, 6$. This assignment of probabilities agrees with Axiom 1. To see that Axiom 2 is satisfied, write

$$P(S) = P(E_1 \cup E_2 \cup \cdots \cup E_6) = P(E_1) + P(E_2) + \cdots + P(E_6) = 1$$

The second equality follows because Axiom 3 must hold. Axiom 3 really tells us that now we can calculate the probability of any event by summing the probabilities of the simple events contained in that event. Event A was defined to be "observe an odd number." Hence

$$P(A) = P(E_1 \cup E_3 \cup E_5) = P(E_1) + P(E_3) + P(E_5) = 1/2$$

EXAMPLE 2.1   A manufacturer has five seemingly identical computer terminals available for shipping. Unknown to her, two of the five are defective. A particular order calls for two of the terminals and is filled by randomly selecting two of the five that are available.

(a) List the sample space for this experiment.

(b) Let A denote the event that the order is filled with two nondefective terminals. List the sample points in A.

(c) Construct a Venn diagram for the experiment, illustrating event A.

(d) Assign probabilities to the simple events in such a way that the information in the problem is used and the axioms in Definition 2.6 are met.

(e) Find the probability of event A.

Solution

(a) Let the two defective terminals be labeled D_1 and D_2 and the three good terminals be labeled G_1, G_2, and G_3. Any single sample point will consist of a listing of the two terminals selected for shipment. The simple events may be denoted by

$$
\begin{aligned}
E_1 &= \{D_1, D_2\} & E_6 &= \{D_2, G_2\} \\
E_2 &= \{D_1, G_1\} & E_7 &= \{D_2, G_3\} \\
E_3 &= \{D_1, G_2\} & E_8 &= \{G_1, G_2\} \\
E_4 &= \{D_1, G_3\} & E_9 &= \{G_1, G_3\} \\
E_5 &= \{D_2, G_1\} & E_{10} &= \{G_2, G_3\}
\end{aligned}
$$

Thus with our notation there are ten sample points in S, and $S = \{E_1, E_2, \ldots, E_{10}\}$.

(b) Event $A = \{E_8, E_9, E_{10}\}$.

(c)

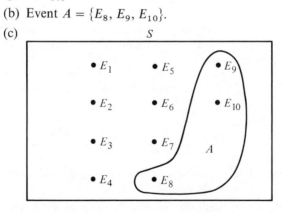

(d) Because the terminals are selected at random, any pair of terminals is as likely to be selected as any others. Thus $P(E_i) = 1/10$, $i = 1, 2, \ldots, 10$, is a reasonable assignment of probabilities.

(e) Because $A = E_8 \cup E_9 \cup E_{10}$, then

$$
P(A) = P(E_8) + P(E_9) + P(E_{10}) = 3/10
$$

The next section contains an axiomatic description of the method for calculating $P(A)$ that we just used.

Before we proceed, let us note that there are experiments for which the sample space is not countable and hence is not discrete. Suppose, for example, that the experiment consists of measuring the blood glucose level of a diabetic patient. The sample space for this experiment would contain an interval of real numbers, and any such interval contains an uncountable number of values. Thus the sample space is not discrete. Situations like the latter will be discussed in Chapter 4. The remainder of the present chapter will be devoted to developing methods for calculating the probabilities of events defined on discrete sample spaces.

Exercises

2.6 A vehicle arriving at an intersection can turn right, turn left, or continue straight ahead. The experiment consists of observing the movement of a single vehicle through the intersection.

(a) List the sample space for this experiment.

(b) Assuming that all sample points are equally likely, find the probability that the vehicle turns.

2.7 An oil prospecting firm hits oil or gas on 10% of its drillings. If the firm drills two wells, the four possible simple events and three of their associated probabilities are

Simple event	Outcome of 1st drilling	Outcome of 2nd drilling	Probability
E_1	Hit (oil or gas)	Hit (oil or gas)	.01
E_2	Hit	Miss	?
E_3	Miss	Hit	.09
E_4	Miss	Miss	.81

(a) Find the probability that the company will hit oil or gas on the first drilling and miss on the second.

(b) Find the probability that the company will hit oil or gas on at least one of the two drillings.

2.8 Volunteers coming into a blood center are such that 1 in 3 have O^+ blood, 1 in 15 have O^-, 1 in 3 have A^+ and 1 in 16 have A^-. The name of one person who previously has donated blood is selected from the records of the center. What is the probability that the person selected has

(a) O^+ blood?

(b) type O blood?

(c)   type A blood?

(d)   neither type A nor type O blood?

2.9   Hydraulic assemblies for landing coming from an aircraft rework facility are each inspected for defects. Historical records indicate that 8% have defects in shafts only, 6% have defects in bushings only, and 2% have defects in both shafts and bushings. One of the hydraulic assemblies is selected randomly. What is the probability that the assembly has

(a)   a bushing defect?

(b)   a shaft or bushing defect?

(c)   exactly one of the two types of defects?

(d)   neither type of defect?

2.10   Suppose two balanced coins are tossed and the upper faces observed.

(a)   List the sample points for this experiment.

(b)   Assign a reasonable probability to each point. (Are the points equally likely?)

(c)   Let A denote the event that *exactly* one head is observed and B the event that *at least* one head is observed. List the sample points in A and B.

(d)   From your answer to (c) find $P(A)$, $P(B)$, $P(A \cap B)$, $P(A \cup B)$, and $P(\bar{A} \cup B)$.

2.11   A business office orders paper supplies from one of three vendors, V_1, V_2, and V_3. Orders are to be placed on two successive days, one order per day. Thus (V_2, V_3) might denote that vendor V_2 gets the order on the first day and vendor V_3 gets it on the second day.

(a)   List the sample points in this experiment of ordering paper on two successive days.

(b)   Assume the vendors are selected at random each day and assign a probability to each sample point.

(c)   Let A denote the event that the same vendor gets both orders and B the event that V_2 gets at least one order. Find $P(A)$, $P(B)$, $P(A \cup B)$, and $P(A \cap B)$ by summing the probabilities of the sample points in these events.

2.5 Calculating the Probability of an Event: The Sample-Point Method

Finding the probability of an event defined on a sample space containing a finite or denumerable (countably infinite) set of sample points can be approached in two ways, which we shall call the *sample-point* and the *event-composition* methods. Both methods utilize the sample space model but they differ in the *sequence* of steps necessary to obtain a solution and in the tools that are of assistance in each. Separation of the two procedures may not be palatable to the unity-seeking theorist, but it is extremely useful in finding the probability of an event, which may not be an easy task for a beginner. In this section we consider the sample-point method. The event-composition method requires additional results and will be presented in Section 2.9.

The sample-point method is outlined in Section 2.4. The steps used to find the probability of an event are the following:

1. Define the experiment.
2. List the simple events associated with the experiment and test each to make certain that they cannot be decomposed. This defines the sample space S.
3. Assign reasonable probabilities to the sample points in S, making certain that $P(E_i) \geq 0$ and $\Sigma P(E_i) = 1$.
4. Define the event of interest, A, as a specific collection of sample points. (A sample point is in A if A occurs when the sample point occurs. Test *all* sample points in S to locate those in A.)
5. Find $P(A)$ by summing the probabilities of the sample points in A.

We will illustrate these steps with three examples.

EXAMPLE 2.2   Consider the problem of selecting two applicants for a job out of a group of five and imagine that the applicants vary in competence, 1 being the best, 2 second best, and so on for 3, 4, and 5. These ratings are, of course, unknown to the employer. Define two events A and B as

> A: The employer selects the best and one of the two poorest applicants (applicants 1 and 4 or 1 and 5).
>
> B: The employer selects at least one of the two best.

Find the probabilities of these events.

Solution   The steps are as follows:

1. The experiment involves randomly selecting two applicants out of five.
2. The ten simple events, with $\{i, j\}$ denoting the selection of applicants i and j, are

$$
\begin{array}{llll}
E_1\colon \{1, 2\} & E_5\colon \{2, 3\} & E_8\colon \{3, 4\} & E_{10}\colon \{4, 5\} \\
E_2\colon \{1, 3\} & E_6\colon \{2, 4\} & E_9\colon \{3, 5\} & \\
E_3\colon \{1, 4\} & E_7\colon \{2, 5\} & & \\
E_4\colon \{1, 5\} & & &
\end{array}
$$

3. A random selection of two out of five should give each pair an equal chance for selection. Hence we will assign each sample point a probability equal to 1/10. That is,

$$P(E_i) = 1/10, \qquad i = 1, 2, \ldots, 10$$

4. Checking the sample points, we see that B occurs whenever $E_1, E_2, E_3, E_4, E_5, E_6,$ or E_7 occurs. Hence these sample points are included in B.

5. Finally, $P(B)$ is equal to the sum of the probabilities of the sample points in B, or

$$P(B) = \sum_{i=1}^{7} P(E_i) = \sum_{i=1}^{7} \frac{1}{10} = \frac{7}{10} \qquad \text{↶}$$

Similarly, we see that event A is the union of E_3 and E_4. Hence $P(A) = 2/10$. Note that the solution of this and similar problems would be of importance to a company personnel director.

EXAMPLE 2.3

A balanced coin is tossed three times. Calculate the probability that exactly two of the three tosses turn out to be heads.

Solution

The five steps of the sample-point method are as follows:

1. The experiment consists of observing the outcomes (heads or tails) for each of three tosses of a coin.

2. A simple event for this experiment can be symbolized by a three-letter sequence of H's and T's, representing heads and tails, respectively. The first letter in the sequence represents the observation on the first coin. The second letter represents the observation on the second coin, and so on. The eight simple events in S are

$$
\begin{array}{llll}
E_1: \ HHH & E_3: \ HTH & E_5: \ HTT & E_7: \ TTH \\
E_2: \ HHT & E_4: \ THH & E_6: \ THT & E_8: \ TTT
\end{array}
$$

3. Since the coin is balanced, we would expect the simple events to be equally likely; that is,

$$P(E_i) = 1/8, \qquad i = 1, 2, \ldots, 8$$

4. The event of interest, A, is the event that exactly two of the tosses result in heads. An examination of the sample points will verify that

$$A = \{E_2, E_3, E_4\}$$

5. Finally,

$$P(A) = P(E_2) + P(E_3) + P(E_4) = 1/8 + 1/8 + 1/8 = 3/8 \qquad \text{↶}$$

Both Examples 2.2 and 2.3 involve working with sample spaces S wherein the simple events are equally likely. It is important to realize that the sample points in the sample space for an experiment are not always equally likely. An example to illustrate this point follows.

EXAMPLE 2.4   The odds are two to one that when A and B play tennis, A wins. Suppose that A and B play two matches. What is the probability that A wins at least one match?

Solution
1. The experiment consists of observing the winner (A or B) for each of two matches.
2. The sample space for the experiment consists of four sample points:

$$E_1:\ \ AA \qquad E_2:\ \ AB \qquad E_3:\ \ BA \qquad E_4:\ \ BB$$

where, for example, AB denotes the event that player A wins the first game and player B wins the second.

3. Because A has a better chance of winning any match, it does not seem appropriate to assign equal probabilities to the sample points. As you will see in Section 2.9, under certain conditions it is reasonable to make the following assignment of probabilities:

$$P(E_1) = 4/9 \qquad P(E_2) = 2/9 \qquad P(E_3) = 2/9 \qquad P(E_4) = 1/9$$

Note that even though the probabilities assigned to the simple events are not all equal, $P(E_i) \geqslant 0$ for $i = 1, 2, 3, 4$ and $\sum_S P(E_i) = 1$.

4. The event of interest is that A wins at least one game. Thus if we denote the event of interest as C, it is easily seen that

$$C = E_1 \cup E_2 \cup E_3$$

5. Finally,

$$P(C) = P(E_1) + P(E_2) + P(E_3) = 4/9 + 2/9 + 2/9 = 8/9$$

The sample-point method for solving a probability problem is direct and powerful and in some respects is a bulldozer approach. It can be applied to find the probability of any event defined over a sample space containing a finite or denumerable set of sample points, but it is not resistant to human error. Errors are frequently committed by incorrectly diagnosing the nature of a simple event and by failing to list all the sample points in S. A second complication occurs because many sample spaces contain a very large number of sample points, and a complete itemization of each is both tedious and time consuming. Fortunately, many sample spaces generated by experimental data contain subsets of sample points that are equiprobable. (The sample spaces for Examples 2.2 and 2.3 possess this property.) When this occurs, we need not list the points but only count the number in each subset. If counting methods are unavailable, an orderly method should be concocted for listing the sample points (notice the listing schemes for Examples 2.2 and 2.3). The listing of large numbers of sample points can be accomplished by using an electronic computer.

These comments point to the tools that reduce the effort and error associated with the sample-point approach for finding the probability of an event. The tools are orderliness, the electronic computer, and the mathematical theory of counting, called *combinatorial analysis.* Computer programming and applications are a subject in themselves and form a topic for separate study. The mathematical theory of combinatorial analysis also is a broad subject, but some quite useful results can be detailed in a very short space. Hence our next topic concerns some elementary results in combinatorial analysis and their application to the sample-point approach for the solution of probability problems.

Exercises

2.12 In Exercise 2.6 we considered a situation where cars entering an intersection each could turn right, turn left, or go straight. An experiment consists in observing two vehicles moving through the intersection.

(a) How many sample points are there in the sample space? List them.

(b) Assuming that all sample points are equally likely, what is the probability that at least one car turns left?

(c) Again assuming equally likely sample points, what is the probability that at most one vehicle turns?

2.13 Four equally qualified people apply for two identical positions in a company. One and only one applicant is a member of a minority group. The positions are filled by choosing two of the applicants at random.

(a) List the possible outcomes for this experiment.

(b) Assign reasonable probabilities to the sample points.

(c) Find the probability that the applicant from the minority group gets selected for a position.

2.14 A boxcar contains six complex electronic systems. Two of the six are to be randomly selected for thorough testing and then classified as defective or not defective.

(a) If two of the six systems are actually defective, find the probability that at least one of the two systems tested will be defective. Find the probability that both are defective.

(b) Find the probabilities indicated in (a) for the case in which four of the six systems are actually defective.

2.15   A retailer sells only two styles of stereo consoles that experience shows are in equal demand. Four customers in succession come into the store to order stereos. The retailer is interested in their preferences.

    (a)   List the possibilities for preference arrangements among the four customers (that is, list the sample space).

    (b)   Assign probabilities to the sample points.

    (c)   Let A denote the event that all four customers prefer the same style. Find $P(A)$.

2.16   Two baseball teams, I and II, are of equal abilities and play each other in a four-game series. The outcome of each game is recorded.

    (a)   List the points in the sample space.

    (b)   Let A be the event that team I wins exactly three times. List the sample points in A.

    (c)   Make a reasonable assignment of probabilities to the sample points and find $P(A)$.

2.17   Patients arriving at a hospital outpatient clinic can select one of three stations for service. Suppose that physicians are assigned randomly to the stations and that the patients therefore have no station preference. Three patients arrive at the clinic and their selection of stations is observed.

    (a)   List the sample points for the experiment.

    (b)   Let A be the event that each station receives a patient. List the sample points in A.

    (c)   Make a reasonable assignment of probabilities to the sample points and find $P(A)$.

2.6 Tools for Use when Counting Sample Points

The following represent a few results from the theory of combinatorial analysis that are particularly useful when applying the sample-point method for finding the probability of an event. In particular, these results will often enable you to count the total number of sample points in the sample space S and in an event of interest, thereby providing a check on your listing of simple events. In addition, the number of simple events in a sample space is often very large and an enumeration of every sample point may be extremely time consuming or even practically impossible. Thus counting the number of points in the sample space and in the event of interest may be the only efficient way to proceed in calculating the probability of an event. Indeed, if the sample points in a sample space are equiprobable, and it can be determined that there are a total of N, the probability of each sample point is $P(E_i) = 1/N$, and the probability of an event A containing exactly n_a sample points easily is seen to be n_a/N.

The first result from combinatorial analysis that we present is often called the *mn rule* and is stated as follows:

Theorem 2.1

> With m elements $a_1, a_2, a_3, \ldots, a_m$ and n elements $b_1, b_2, \ldots, b_n$ it is possible to form mn pairs containing one element from each group.

Proof   Verification of the theorem can be seen by observing the rectangular table in Figure 2.9. There will be one square in the table for each $a_i b_j$ pair and hence a total of *mn squares.*

Figure 2.9
Table indicating
the number of pairs
(a_i, b_j)

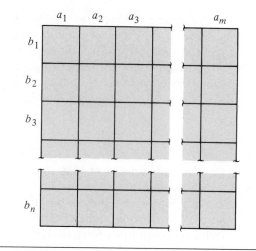

The *mn* rule can be extended to any number of sets. Given three sets of elements, $a_1, a_2, \ldots, a_m; b_1, b_2, \ldots, b_n;$ and $c_1, c_2, \ldots, c_p,$ the number of distinct triplets containing one element from each set is equal to *mnp.* The proof of the theorem for three sets involves a reapplication of Theorem 2.1. Thus we think of the first set as an (a_i, b_j) pair. We unite these pairs with elements of the third set, $c_1, c_2, \ldots, c_p.$ There are *mn* pairs (a_i, b_j) and p elements $c_1, c_2, \ldots, c_p.$ Then through Theorem 2.1 it can be seen that the number of ways that they can be paired to form triplets $a_i b_j c_k$ is $(mn)(p) = mnp.$

EXAMPLE 2.5   An experiment involves tossing a pair of dice and observing the numbers on the upper faces. Find the number of sample points in S, the sample space for the experiment.

Solution   A sample point for this experiment can be represented symbolically as an ordered pair of numbers representing the outcomes on the first and second die, respectively. Thus (4, 5) would denote the event that the uppermost face on the first die was a 4 and the second die resulted in a 5. The sample space S would consist of the set of all possible pairs (x, y), where x and y are both integers between 1 and 6.

The first die can result in one of six numbers. These represent $a_1, a_2, \ldots,$ a_6. Similarly, the second die can fall in one of six ways, and these correspond to $b_1, b_2, \ldots, b_6$. Then $m = n = 6$ and the total number of sample points in S is $mn = (6)(6) = 36.$

EXAMPLE 2.6   Refer to the coin-tossing experiment in Example 2.3. We found for this example that the total number of sample points was eight. Use the *mn* rule (or an extension of it) to confirm this result.

Solution   Each sample point in S was identified by a sequence of three letters, where each position in the sequence could contain one of two letters, an H or a T. The problem therefore involves the formation of triples, selecting an element (an H or a T) from each of three sets. For this example the sets are identical and all contain two elements (H and T). Thus the number of elements in each set is $m = n = p = 2$, and the total number of triples that can be formed is $mnp = (2)^3 = 8$.

EXAMPLE 2.7   Consider an experiment that consists of recording the birthday for each of twenty randomly selected persons. If we ignore leap years and assume that there are only 365 possible distinct birthdays, find the number of points in the sample space S for this experiment. If we assume that each of the possible sets of birthdays is equiprobable, what is the probability that each person in the twenty has a different birthday?

Solution   Let us number the days of the year 1, 2, ..., 365. A sample point for this experiment can be represented by an ordered sequence of twenty numbers, where the first number denotes the number of the day that is the first person's birthday, the second number denotes the number of the day that is the second person's birthday, and so on. We are concerned with the number of twenty-tuples that can be formed, selecting a number representing one of the 365 days in the year from each of twenty sets. The sets are all identical and each contains 365 elements. Repeated applications of the *mn* rule tell us there are $(365)^{20}$ such twenty-tuples. Thus the sample space S contains $N = (365)^{20}$ sample points. Although we could not feasibly list all the sample points, if we assume them to be equiprobable, $P(E_i) = 1/(365)^{20}$ for each simple event.

     If we denote the event that each person has a different birthday by A, the probability of A can be calculated if we can determine n_a, the number of sample points in A. A sample point will be in A if the corresponding twenty-tuple is such that no two numbers are the same. Thus the set of numbers from which the first element in a twenty-tuple in A can be selected contains 365 numbers, the set from which the second element can be selected contains 364 numbers (all but the one selected for the first element), the set from which the third can be selected contains 363 (all but the two selected for the first two elements), ..., the set from which the twentieth element can be selected contains 346 elements (all but those selected for the first nineteen positions). An extension of the *mn* rule yields

$$n_a = (365) \times (364) \times \cdots \times (346)$$

Finally, we may determine that

$$P(A) = \frac{n_a}{N} = \frac{365 \times 364 \times \cdots \times 346}{(365)^{20}} = 0.5886$$

Notice that for Examples 2.5 and 2.6 the numbers of sample points in the appropriate sample spaces are both relatively small and that representations for these sample spaces easily could be written down. For instances like these the *mn* rule provides an easy method to verify that the sample spaces contain the correct number of points. In contrast, it is not feasible to list the sample space in Example 2.7, but the *mn* rule can be used to calculate the probability of the event of interest.

We have seen that the sample points associated with an experiment often can be represented symbolically as a sequence of numbers or symbols. In some instances it will be clear that the totality of sample points is the number of distinct ways that the respective symbols can be arranged in sequence. The following theorem can be used to determine the number of ordered arrangements that can be formed.

Definition 2.7

> An ordered arrangement of *r* distinct objects is called a *permutation*. The number of ways of ordering *n* distinct objects taken *r* at a time will be designated by the symbol P_r^n.

Theorem 2.2

$$P_r^n = n(n-1)(n-2)\cdots(n-r+1) = \frac{n!}{(n-r)!}$$

Proof   We are concerned with the number of ways of filling *r* positions with *n* distinct objects. Applying the extension of the *mn* rule, we see that the first object can be chosen in one of *n* ways. After the first is chosen, the second can be chosen in $(n-1)$ ways, the third in $(n-2)$, and the *r*th in $(n-r+1)$ ways. Hence the total number of distinct arrangements is

$$P_r^n = n(n-1)(n-2)\cdots(n-r+1)$$

Expressed in terms of factorials,

$$P_r^n = n(n-1)(n-2)\cdots(n-r+1)\frac{(n-r)!}{(n-r)!} = \frac{n!}{(n-r)!}$$

where $n! = n(n-1)\cdots(2)(1)$ and $0! = 1$.

EXAMPLE 2.8   Opening a combination lock requires the selection of the correct set of four different digits in sequence. The digits are set by rotating the tumbler in alternating clockwise and counterclockwise directions. Assume that no digit is used twice. Give the total number of possible combinations.

Solution   The total number of lock combinations would equal the number of ways of arranging $r = 4$ out of the possible 10 digits. Thus

$$P_4^{10} = \frac{10!}{6!} = (10)(9)(8)(7) = 5040$$

EXAMPLE 2.9   Suppose that an assembly operation in a manufacturing plant involves four steps, which can be performed in any sequence. If the manufacturer wishes to experimentally compare time to assembly for each of the sequences, how many different sequences will be involved in the experiment?

Solution   The total number of sequences will be the number of ways of arranging the $n = 4$ steps taken $r = 4$ at a time. This will equal

$$P_4^4 = \frac{4!}{(4-4)!} = \frac{4!}{0!} = 24$$

The next result from combinatorial analysis can be used to determine the number of subsets of various sizes that can be formed by partitioning a set of n distinct objects.

Theorem 2.3

> The number of ways of partitioning n distinct objects into k distinct groups containing $n_1, n_2, \ldots, n_k$ objects, respectively, is
>
> $$N = \frac{n!}{n_1! n_2! \cdots n_k!} \qquad \text{where} \qquad \sum_{i=1}^{k} n_i = n$$

Proof N is the number of distinct arrangements of n objects in a row for a case in which rearrangement of the objects within a group does not count. For example, the letters a to l are arranged in three groups, where $n_1 = 3$, $n_2 = 4$, and $n_3 = 5$:

$$abc|defg|hijkl$$

The number of distinct arrangements of the n objects, *assuming all distinct*, is $P_n^n = n!$ (from Theorem 2.2). Then P_n^n equals the number of ways of partitioning the n objects into k groups (ignoring order within groups) multiplied by the number of ways of ordering the $n_1, n_2, \ldots, n_k$ elements within each group. This application of the extended mn rule gives

$$P_n^n = (N) \cdot (n_1! n_2! n_3! \cdots n_k!)$$

where $n_i!$ is the number of distinct arrangements of the n_i objects in group i. Solving for N, we have

$$N = \frac{n!}{n_1! n_2! \cdots n_k!}$$

We will sometimes use the notation

$$\binom{n}{n_1 n_2 \cdots n_k} = \frac{n!}{n_1! n_2! \cdots n_k!}$$

EXAMPLE 2.10   A labor dispute has arisen concerning the alleged unequal distribution of twenty laborers to four different construction jobs. The first job (considered to be abominable employment) required six laborers; the second, third, and fourth utilized four, five, and five, respectively. The dispute arose over an alleged random distribution of the laborers to the jobs which placed all four members of a particular ethnic group on job 1. In considering whether the assignment represented injustice, a mediation panel desired the probability of the observed event. Determine the number of sample points in the sample space S for this experiment. That is, determine the number of ways the twenty laborers can be divided into groups of the appropriate size to fill all the jobs. Find the probability of the observed event if it is assumed that the laborers are randomly assigned to jobs.

Solution   The number of ways of assigning the twenty laborers to the four jobs is equal to the number of ways of partitioning the twenty into four groups of $n_1 = 6, n_2 = 4, n_3 = n_4 = 5$. Then

$$N = \frac{20!}{6! 4! 5! 5!}$$

By a *random assignment* of laborers to the jobs we mean that the N sample points are equiprobable with probability equal to $1/N$. If A denotes the event of interest and n_a the number of sample points in A, the sum of the probabilities of the sample points in A will be $P(A) = n_a(1/N) = n_a/N$. The number of sample points in A, n_a, is the number of ways of assigning laborers to the four jobs if it is known that the four of the specified ethnic group all go to job 1. Then there are sixteen laborers remaining to assign randomly and

$$n_a = \frac{16!}{2!4!5!5!}$$

(there are two openings remaining for job 1). It follows that

$$P(A) = \frac{n_a}{N} = 0.0031$$

Thus if laborers are randomly assigned to jobs, the probability that the four members of the ethnic group all go to the undesirable job is very small. We would have reason to doubt that the jobs were randomly assigned.

In many situations the sample points are identified by an array of symbols in which the arrangement of symbols is *unimportant*. The sample points for the selection of applicants, Example 2.2, imply a selection of two applicants out of five. Each sample point is identified as a pair of symbols and the order of the symbols identifying the sample points is irrelevant.

Definition 2.8

> The number of *combinations* of n objects taken r at a time is the number of subsets, each of size r, that can be formed from the n objects. This number will be denoted by C_r^n or $\binom{n}{r}$.

Theorem 2.4

> $$\binom{n}{r} = C_r^n = \frac{P_r^n}{r!} = \frac{n!}{r!(n-r)!}$$

Proof   The selection of r objects from a total of n is equivalent to partitioning the n objects into $k = 2$ groups, the r selected and the $(n - r)$ remaining. It thus is a special case of the general partitioning problem of Theorem 2.3, where $k = 2$, $n_1 = r$, and $n_2 = (n - r)$. Therefore,

$$\binom{n}{r} = C_r^n = \frac{n!}{r!(n-r)!}$$

EXAMPLE 2.11   Find the number of ways of selecting two applicants out of five and hence the total number of sample points in S for Example 2.2.

Solution

$$\binom{5}{2} = \frac{5!}{2!3!} = 10$$

(Note that this agrees with the number of sample points listed in Example 2.2.)

EXAMPLE 2.12 Find the number of ways of selecting exactly one of the two best applicants in a selection of two out of five. Then find the probability of that event.

Solution

Let A be the event that exactly one of the two best is selected and let n_a denote the number of sample points in A. Then n_a equals the number of ways of selecting one of the two best out of a possible two (call this number m) times the number of ways of selecting one of the three low-ranking applicants out of a possible three (call this number n). Then $m = \binom{2}{1}$, $n = \binom{3}{1}$, and, applying the mn rule,

$$n_a = \binom{2}{1} \cdot \binom{3}{1} = \frac{2!}{1!1!} \cdot \frac{3!}{1!2!} = 6$$

(This number can be verified by counting the sample points in A from the listing in Example 2.2.)

In Example 2.11 we found the total number of sample points in S to equal $N = 10$. If each selection is equiprobable, $P(E_i) = 1/10$, $i = 1, 2, \ldots, 10$, and

$$P(A) = \sum_{E_i \subset A} P(E_i) = n_a \left(\frac{1}{10} \right) = \frac{6}{10} = \frac{3}{5}$$

EXAMPLE 2.13 A company orders supplies from M distributors and wishes to place n orders $(n < M)$. Assume that the company places the orders in a manner that allows every distributor an equal chance of obtaining any one order and there is no restriction on the number of orders that can be placed with any vendor. Find the probability that a particular distributor, say distributor I, gets exactly k orders $(k \le n)$.

Solution

Because any of the M distributors can be selected to receive any one of the orders, there are M ways that each order can be placed. Then the number of different ways that the n orders can be placed is $M \cdot M \cdot M \cdots M = (M)^n$. Consequently, there are $(M)^n$ sample points in S. All these points are equally likely; hence $P(E_i) = 1/(M)^n$.

Let A denote the event that distributor I receives exactly k orders from among the n. The k orders assigned to distributor I can be chosen from the n in $\binom{n}{k}$ ways. It remains to determine the number of ways the remaining $(n - k)$ orders can be assigned to the other $M - 1$ distributors. Because each of these

$(n - k)$ orders can go to any of the $(M - 1)$ distributors, this assignment can be made in $(M - 1)^{n-k}$ ways. Thus A contains

$$n_a = \binom{n}{k}(M - 1)^{n-k}$$

sample points, and because the sample points are equally likely, then

$$P(A) = \frac{\binom{n}{k}(M - 1)^{n-k}}{(M)^n}$$

Theorems 2.1 through 2.4 provide a few of the many useful counting rules found in the theory of combinatorial analysis. A few additional theorems appear in the exercises at the end of the chapter. If you are interested in extending your knowledge of combinatorial analysis, refer to one of the numerous texts on this subject.

We now will direct our attention to the concept of conditional probability. Conditional probability will play an important role in the event-composition approach for finding the probability of an event and sometimes will be useful in finding the probabilities of sample points (for sample spaces with unequal probabilities attached to the sample points).

Exercises

2.18 An airline has six flights from New York to California and seven flights from California to Hawaii per day. If the flights are to be made on separate days, how many different flight arrangements can the airline offer from New York to Hawaii?

2.19 An assembly operation in a manufacturing plant requires three steps that can be performed in any sequence. How many different ways can the assembly be performed?

2.20 An experiment consists of tossing a pair of dice.

(a) Use the combinational theorems to determine the number of sample points in the sample space S.

(b) Find the probability that the sum of the numbers appearing on the dice is equal to 7.

2.21 A brand of automobile comes in five different styles, with four types of engines, with two types of transmissions, and in eight colors.

(a) How many autos would a dealer have to stock if he included one for each style-engine-transmission combination?

(b) How many would a distributing center have to carry if all colors of cars were stocked for each combination in part (a)?

2.22 How many different seven-digit telephone numbers can be formed if the first digit cannot be zero?

2.23   A personnel director for a corporation has hired ten new engineers. If three (distinctly different) positions are open at a Cleveland plant, in how many ways can she fill the positions?

2.24   A fleet of eight taxis is to be dispatched to three airports in such a way that two go to airport A, five go to airport B, and one goes to airport C. In how many distinct ways can this be accomplished?

2.25   Refer to Exercise 2.24. Exactly one of the taxis is in need of repair. Assuming that taxis are allocated to airports at random, what is the probability that the taxi requiring repair is dispatched to airport C?

2.26   Probability played a role in the rigging of the April 24, 1980, Pennsylvania state lottery (*The Los Angeles Times*, September 8, 1980). To determine each digit of the three-digit winning number, each of the numbers $0, 1, 2, \ldots, 9$ is placed on a ping pong ball, the ten balls are blown into a compartment and the number selected for the digit is the one on the ball that floats to the top of the machine. To alter the odds, the conspirators injected a liquid into all balls used in the game except those numbered 4 and 6, making it almost certain that the lighter balls would be selected and determine the digits in the winning number. Then they proceeded to buy lottery tickets bearing the potential winning numbers. How many potential winning numbers were there (666 was the eventual winner)?

2.27   An experimenter wishes to investigate the effect of three variables, pressure, temperature, and the type of catalyst, on the yield in a refining process. If the experimenter intends to use three settings each for temperature and pressure and two types of catalysts, how many experimental runs will have to be conducted if he wishes to run all possible combinations of pressure, temperature, and types of catalysts?

2.28   Five firms, $F_1, F_2, \ldots, F_5$, each offer bids on three separate contracts, C_1, C_2, and C_3. Any one firm will be awarded at most one contract. The contracts are quite different, so an assignment of C_1 to F_1, say, is to be distinguished from an assignment of C_2 to F_1.

    (a)   How many sample points are there altogether in this experiment involving assignment of contracts to the firms? (No need to list them all.)

    (b)   Under the assumption of equally likely sample points, find the probability that F_3 is awarded a contract.

2.29   Four students are to be randomly selected to fill certain student government posts, from a group of three undergraduate and five graduate students. Find the probability that exactly two undergraduates will be among the four chosen.

2.30   Two cards are drawn from a standard fifty-two-card playing deck. What is the probability that the draw will yield an ace and a face card?

2.31   A manufacturer has nine distinct motors in stock, two of which came from a particular supplier. The motors must be divided among three production lines, with three motors going to each line. If the assignment of motors to lines is random, find the probability that both motors from the particular supplier are assigned to the first line.

2.32   In 1976 the eight-member Human Relations Advisory Board of Gainesville, Florida, considered the complaint of a woman who claimed discrimination, based on sex, on the part of a local company. The board, composed of five women and three men, voted 5–3 in favor of the plaintiff, the five women voting in favor of the plaintiff, the three men against. The attorney representing the company appealed the board's decision by claiming sex bias on the part of the board members. If there was no sex bias among the board members, it might be reasonable to conjecture that any group of five board members would be as likely to vote for the complainant as any other group of five. If this were the case, what is the probability that the vote would split along sex lines (five women for, three men against)?

2.33   A balanced die is tossed six times and the number on the uppermost face is recorded each time. What is the probability that the numbers recorded are 1, 2, 3, 4, 5, and 6 in any order?

2.34   Refer to Exercise 2.33. Suppose that the die has been altered so that the faces are 1, 2, 3, 4, 5, and 5. If the die is tossed five times, what is the probability that the numbers recorded are 1, 2, 3, 4, and 5 in any order?

2.35   Prove that $\binom{n+1}{k} = \binom{n}{k} + \binom{n}{k-1}$.

2.7 Conditional Probability and the Independence of Events

The probability of an event will vary depending upon the occurrence or nonoccurrence of one or more related events. For example, Florida sport fishermen are vitally interested in the probability of rain. The probability of rain on a given day, ignoring the daily atmospheric conditions or any other events, is the fraction of days in which rain occurs over a long period of time. This would be called the *unconditional probability* of the event "rain on a given day." Now suppose that we wish to consider the probability of rain tomorrow. It has rained almost continuously for two days in succession and a tropical storm is heading up the coast. What is the probability of rain? This probability is conditional on the occurrence of several events, and a Floridian would tell you that it is much larger than the unconditional probability of rain.

The unconditional probability of a 1 in the toss of a single, balanced die is 1/6. The conditional probability of a 1, given that an odd number has fallen, is 1/3. That is, 1, 3, and 5 occur with equal frequency. Knowing that an odd number has occurred, the relative frequency of occurrence of a 1 is 1/3. Thus the conditional probability of an event is the probability (relative frequency of occurrence) of the event given the fact that one or more events have already occurred. A careful perusal of this example will indicate agreement of the following definition with the relative frequency concept of probability.

Definition 2.9

The *conditional probability of an event* A, given that an event B has occurred, is equal to

$$P(A|B) = \frac{P(A \cap B)}{P(B)}$$

provided $P(B) > 0$. [The symbol $P(A|B)$ is read "probability of A given the occurrence of B."]

Further confirmation of the consistency of Definition 2.9 with the relative frequency concept of probability can be obtained from the following construction. Suppose that an experiment is repeated a large number, N, of times, resulting in both A and B, $A \cap B$, n_{11} times; A and not B, $A \cap \bar{B}$, n_{21} times; B and not A, $\bar{A} \cap B$, n_{12} times; and neither A nor B, $\bar{A} \cap \bar{B}$, n_{22} times. We present these results in Table 2.1.

Table 2.1

Table for events A and B

	A	$\bar{A}$
B	n_{11}	n_{12}
$\bar{B}$	n_{21}	n_{22}

Note that $n_{11} + n_{12} + n_{21} + n_{22} = N$. Then it follows that

$$P(A) \approx \frac{n_{11} + n_{21}}{N} \qquad P(B) \approx \frac{n_{11} + n_{12}}{N} \qquad P(A|B) \approx \frac{n_{11}}{n_{11} + n_{12}}$$

$$P(B|A) \approx \frac{n_{11}}{n_{11} + n_{21}} \qquad P(A \cap B) \approx \frac{n_{11}}{N}$$

where $\approx$ is read *approximately equal to*.

With these probabilities it is easy to see that

$$P(B|A) \approx \frac{P(A \cap B)}{P(A)} \qquad \text{and} \qquad P(A|B) \approx \frac{P(A \cap B)}{P(B)}$$

Hence Definition 2.9 is consistent with the relative frequency concept of probability.

EXAMPLE 2.14   Use Definition 2.9 to find the probability of a 1, given the occurrence of an odd number, in a single toss of a balanced die.

Solution   Define these events:

A: Observe a 1.

B: Observe an odd number.

We seek the probability of A given that the event B has occurred. The probability that both A and B occur implies the observance of both a 1 and an odd number, and hence $P(A \cap B) = 1/6$. Also, $P(B) = 1/2$. Then

$$P(A|B) = \frac{P(A \cap B)}{P(B)} = \frac{1/6}{1/2} = \frac{1}{3}$$

Note that this result is in complete agreement with our earlier intuitive evaluation of this probability.

Suppose that the occurrence of an event A is unaffected by the occurrence or nonoccurrence of event B. When this occurs, we would be inclined to say that the event A is independent of B. This event relationship is expressed by the following definition.

Definition 2.10

Two events A and B are said to be *independent* if

$$P(A \cap B) = P(A)P(B)$$

Otherwise, the events are said to be *dependent*. Note that this definition is equivalent to stating that two events A and B are independent if $P(A|B) = P(A)$ or $P(B|A) = P(B)$.

The notion of independence as a probabilistic concept is in agreement with our everyday usage of the word if we carefully consider the events in question. Most would agree that "smoking" and "lung cancer" are not independent events and would intuitively feel that the probability of lung cancer, given that a person smokes, is greater than the (unconditional) probability of lung cancer. In contrast, the events "rain today" and "rain a month from today" may well be independent.

EXAMPLE 2.15   Consider the following events in the toss of a single die:

A: Observe an odd number.

B: Observe an even number.

C: Observe a 1 or 2.

(*a*)   Are A and B independent events?

(*b*)   Are A and C independent events?

Solution

(a) To decide whether A and B are independent, we must see whether they satisfy the conditions of Definition 2.10. The probability of an odd number given that an even number has occurred is zero. That is, $P(A|B) = 0$. Also, $P(A) = 1/2$. Then it is clear that $P(A|B) \neq P(A)$, and hence A and B are not independent events.

(b) Are A and C independent? Note that $P(A|C) = 1/2$ and $P(A) = 1/2$. Therefore, $P(A|C) = P(A)$, and A and C are independent.

EXAMPLE 2.16   Three brands of coffee, X, Y, and Z, are to be ranked according to taste by a judge. Define the following events:

> A: Brand X is preferred to Y.
>
> B: Brand X is ranked best.
>
> C: Brand X is ranked second best.
>
> D: Brand X is ranked third best.

If the judge actually has no taste preference and thus randomly assigns ranks to the brands, is event A independent of events B, C, and D?

Solution   The six equally likely sample points for this experiment are given by

$$E_1:\ XYZ \qquad E_3:\ YXZ \qquad E_5:\ ZXY$$
$$E_2:\ XZY \qquad E_4:\ YZX \qquad E_6:\ ZYX$$

where XYZ denotes that X is ranked best and Y second best. Then

$$P(A) = P(E_1) + P(E_2) + P(E_5) = 1/2$$

$$P(A|B) = \frac{P(A \cap B)}{P(B)} = \frac{P(E_1) + P(E_2)}{P(E_1) + P(E_2)} = 1$$

$$P(A|C) = 1/2$$

$$P(A|D) = 0$$

Thus A is independent of C but is not independent of B or D.   ∽

Exercises

2.36   If two events, A and B, are such that $P(A) = .5$, $P(B) = .3$, and $P(A \cap B) = .1$, find the following:

(a) $P(A|B)$

(b) $P(B|A)$

(c) $P(A|A \cup B)$

(d) $P(A|A \cap B)$

(e) $P(A \cap B|A \cup B)$

2.37   For a certain population of employees the percentages passing and failing a job competency exam, listed according to sex, were as shown in the accompanying table. That is, of all the people taking the exam, 24% were in the male-pass category, 16% were in the male-fail category, and so forth. An employee is to be selected randomly from this population. Let A be the event that the employee scores a passing grade on the exam and

M the event that a male is selected. Are the events *A* and *M* independent? Are the events $\bar{A}$ and *F* independent?

	Sex		
Outcome	Male (M)	Female (F)	Total
Pass (A)	24	36	60
Fail ($\bar{A}$)	16	24	40
Total	40	60	100

2.38 A policy requiring all hospital employees to take lie detector tests may reduce losses due to theft, but some employees regard such actions as a violation of their rights. Reporting on a particular hospital that uses this procedure, the *Orlando Sentinel Star* (August 3, 1981)* notes that lie detectors have accuracy rates that vary from 92 to 99%. To gain some insight into the risks that employees face when taking a lie detector test, suppose that the probability is .05 that a particular lie detector concludes that a person is lying who, in fact, is telling the truth and suppose that any pair of tests are independent.

(a) What is the probability that a machine will conclude that each of three employees is lying when all are telling the truth?

(b) What is the probability that the machine will conclude that at least one of the three employees is lying when all are telling the truth?

2.39 A survey of consumers in a particular community showed that 10% were dissatisfied with plumbing jobs done in their homes. Half the complaints dealt with plumber *A*. If plumber *A* does 40% of the plumbing jobs in the town, find the following probabilities:

(a) that a consumer will obtain an unsatisfactory plumbing job, given that the plumber was *A*.

(b) that a consumer will obtain an satisfactory plumbing job, given that the plumber was *A*.

2.40 A study of the posttreatment behavior of a large number of drug abusers suggests that the likelihood of conviction within a two-year period after treatment may depend upon the offender's education. The proportions of the total number of cases falling in four education-conviction categories follow:

	Status within two years after treatment		
Education	Convicted	Not convicted	Totals
10 years or more	.10	.30	.40
9 years or less	.27	.33	.60
Totals	.37	.63	1.00

Suppose that a single offender is selected from the treatment program. Define the events:

   A: the offender has 10 or more years of education,
   B: the offender is convicted within two years after completion of treatment.

Find the probabilities for the events

   (a) A
   (b) B
   (c) AB
   (d) $A \cup B$
   (e) $\bar{A}$
   (f) $\overline{A \cup B}$
   (g) $\overline{AB}$
   (h)   event A given that event B has occurred
   (i)   event B given that event A has occurred.

2.41   Suppose A and B are mutually exclusive events, with $P(A) > 0$ and $P(B) > 0$. Are A and B independent? Give a proof for your answer.

2.42   If $P(A) > 0$, $P(B) > 0$, and $P(A) < P(A|B)$, show that $P(B) < P(B|A)$.

2.8 Two Laws of Probability

The following two laws give the probabilities of unions and intersections. As such, they will play an important role in the event-composition approach to the solution of probability problems.

Theorem 2.5

(The Multiplicative Law of Probability) The probability of the intersection of two events A and B is

$$P(A \cap B) = P(A)P(B|A)$$
$$= P(B)P(A|B)$$

If A and B are independent, then

$$P(A \cap B) = P(A)P(B)$$

Proof   The multiplicative law follows directly from Definition 2.9, the definition of conditional probability.

Note that the multiplicative law can be extended to find the probability of the intersection of any number of events. Thus, twice applying Theorem 2.5, we obtain

$$P(A \cap B \cap C) = P[(A \cap B) \cap C] = P(A \cap B)P(C|A \cap B)$$
$$= P(A)P(B|A)P(C|A \cap B)$$

The probability of the intersection of any number of, say, k events, can be obtained in the same manner:

$$P(A_1 \cap A_2 \cap A_3 \cap \cdots \cap A_k)$$
$$= P(A_1)P(A_2|A_1)P(A_3|A_1 \cap A_2) \cdots P(A_k|A_1 \cap A_2 \cap \cdots \cap A_{k-1})$$

The additive law of probability gives the probability of the union of two events.

Theorem 2.6

(The Additive Law of Probability) The probability of the union of two events A and B is

$$P(A \cup B) = P(A) + P(B) - P(A \cap B)$$

If A and B are mutually exclusive events, $P(A \cap B) = 0$ and

$$P(A \cup B) = P(A) + P(B)$$

Proof   The motivation for the proof of the additive law can be seen by inspecting the Venn diagram in Figure 2.10.

Figure 2.10
Venn diagram for the union of A and B

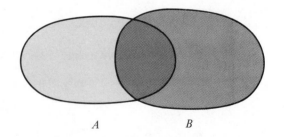

$$A \qquad\qquad B$$

Consider the events $A = (A \cap \bar{B}) \cup (A \cap B)$ and $B = (\bar{A} \cap B) \cup (A \cap B)$, where $[(A \cap \bar{B})$ and $(A \cap B)]$ and $[(\bar{A} \cap B)$ and $(A \cap B)]$ are mutually exclusive pairs of events. Consider also $A \cup B = (A \cap \bar{B}) \cup (A \cap B) \cup (\bar{A} \cap B)$, where all three of these events are mutually exclusive. Then, by Axiom 3,

$$P(A \cup B) = P(A \cap \bar{B}) + P(A \cap B) + P(\bar{A} \cap B)$$
$$P(A) = P(A \cap \bar{B}) + P(A \cap B)$$

[or $P(A \cap \bar{B}) = P(A) - P(A \cap B)$], and

$$P(B) = P(\bar{A} \cap B) + P(A \cap B)$$

[or $P(\bar{A} \cap B) = P(B) - P(A \cap B)$]. Substituting $P(A \cap \bar{B})$ and $P(\bar{A} \cap B)$ into the expression for $P(A \cup B)$, we obtain the desired result:

$$P(A \cup B) = P(A) + P(B) - P(A \cap B)$$

The probability of the union of three events can be obtained by making use of Theorem 2.6. Observe that

$$
\begin{aligned}
P(A \cup B \cup C) &= P[A \cup (B \cup C)] \\
&= P(A) + P(B \cup C) - P[A \cap (B \cup C)] \\
&= P(A) + P(B) + P(C) - P(B \cap C) - P[(A \cap B) \cup (A \cap C)] \\
&= P(A) + P(B) + P(C) - P(B \cap C) - P(A \cap B) - P(A \cap C) \\
&\quad + P(A \cap B \cap C)
\end{aligned}
$$

because $(A \cap B) \cap (A \cap C) = A \cap B \cap C$.

Another useful result expressing the relationship between the probability of an event and its complement is immediately available from the axioms of probability.

Theorem 2.7

> If A is an event, then
>
> $$P(A) = 1 - P(\bar{A}).$$

Proof Observe that $S = A \cup \bar{A}$. Because A and $\bar{A}$ are mutually exclusive events, it follows that $P(S) = P(A) + P(\bar{A})$. Therefore, $P(A) + P(\bar{A}) = 1$ and the result follows.

As we will see in Section 2.9, it is sometimes easier to calculate $P(\bar{A})$ than to calculate $P(A)$. In such cases it is easier to find $P(A)$ by the relationship $P(A) = 1 - P(\bar{A})$ than to find $P(A)$ directly.

Exercises

2.43 If A_1, A_2, and A_3 are three events and $P(A_1 A_2) = P(A_1 A_3) \neq 0$ but $P(A_2 A_3) = 0$, show that

$$P(\text{at least one } A_i) = P(A_1) + P(A_2) + P(A_3) - 2P(A_1 A_2).$$

2.44 If A and B are independent events, show that A and $\bar{B}$ are also independent. Are $\bar{A}$ and $\bar{B}$ independent?

2.45 A smoke detector system uses two devices, A and B. If smoke is present, the probability that it will be detected by device A is .95; by device B, .90; and by both devices .88.

(a) If smoke is present, find the probability that the smoke will be detected by either device A or B or both devices.

(b) Find the probability that the smoke will be undetected.

2.46 Two events A and B are such that $P(A) = .2$, $P(B) = .3$, and $P(A \cup B) = .4$.

(a) Find $P(A \cap B)$.

(b) Find $P(\bar{A} \cup \bar{B})$.

(c) Find $P(\bar{A}\bar{B})$.

(d) Find $P(\bar{A}|B)$.

2.47 If A and B are independent events with $P(A) = .5$ and $P(B) = .2$, find the following:

(a) $P(A \cup B)$

(b) $P(\bar{A} \cap \bar{B})$

(c) $P(\bar{A} \cup \bar{B})$

2.48 Consider the following portion of an electric circuit with three relays. Current will flow from point a to point b if there is at least one closed path when the relays are activated. The relays may malfunction and not close when activated. Suppose that the relays act independently of one another and close properly when activated with probability .9.

(a) What is the probability that current will flow when the relays are activated?

(b) Given that current flowed when the relays were activated, what is the probability that relay 1 functioned?

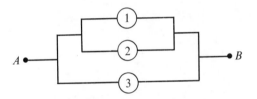

2.49 With relays operating as in Exercise 2.48, compare the probability of current flowing from a to b in the series system shown

with the probability of flow in the parallel system shown.

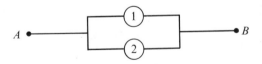

2.50 Suppose that A and B are independent events such that the probability that neither occurs is a and the probability of B is b. Show that $P(A) = \dfrac{1 - b - a}{1 - b}$.

2.51   Diseases I and II are prevalent among people in a certain population. It is assumed that 10% of the population will contract disease I sometime during their lifetime, 15% will contract disease II eventually, and 3% will contract both diseases.

(a)   Find the probability that a randomly chosen person from this population will contract at least one disease.

(b)   Find the conditional probability that a randomly chosen person from this population will contract both diseases, given that he or she has contracted at least one disease.

2.52   Refer to Exercise 2.26. Hours after the rigging of the Pennsylvania state lottery was announced, Connecticut state lottery officials were stunned to learn that their winning number for the day was 666 (*Los Angeles Times*, September 21, 1980).

(a)   All evidence indicates that the Connecticut selection of 666 was due to pure chance. What is the probability that a 666 would be drawn in Connecticut, given that a 666 had been selected in the April 24, 1980, Pennsylvania lottery?

(b)   What is the probability of drawing a 666 in the April 24, 1980, Pennsylvania lottery (remember, this drawing was rigged) and a 666 in the September 19, 1980, Connecticut lottery?

2.53   If A and B are two events, prove that $P(A \cap B) \geq 1 - P(\bar{A}) - P(\bar{B})$. Note: This is a simplified version of the *Bonferroni inequality*.

2.54   If the probability of injury on each individual parachute jump is .05, use the result in Exercise 2.53 to provide a lower bound for the probability of landing safely on both of two jumps.

2.9 Calculating the Probability of an Event: The Event-Composition Method

We learned in Section 2.4 that sets (events) can often be expressed as unions, intersections, or complements of other sets. The event-composition method for calculating the probability of an event makes use of these ideas by expressing the event of interest, say, event A, as a composition (unions and/or intersections). The laws of probability are then applied to find $P(A)$. We will illustrate this method with an example.

EXAMPLE 2.17   Of the voters in a certain city, 40% are Republicans and 60% are Democrats. Among the Republicans 70% are in favor of a particular bond issue, whereas 80% of the Democrats favor the issue. If a voter is selected at random in the city, what is the probability that he or she will favor the bond issue?

Solution   Let F denote the event "favor the bond issue," R the event that a Republican is selected, and D the event that a Democrat is selected. Then $P(R) = .4$, $P(D) = .6$, $P(F|R) = .7$, and $P(F|D) = .8$. Now

$$P(F) = P[(F \cap R) \cup (F \cap D)] = P(F \cap R) + P(F \cap D)$$

because $(F \cap R)$ and $(F \cap D)$ are mutually exclusive events. Figure 2.11 will help you visualize the result that $F = (F \cap R) \cup (F \cap D)$. Now

$$P(F \cap R) = P(F|R)P(R) = (.7)(.4) = .28$$

Similarly, $$P(F \cap D) = P(F|D)P(D) = (.8)(.6) = .48$$

Thus $$P(F) = .28 + .48 = .76$$ ∽

Figure 2.11
Venn diagram for events of Example 2.17

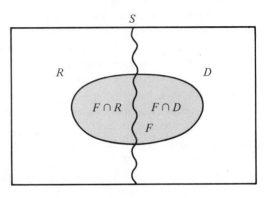

EXAMPLE 2.18   In Example 2.7 we considered an experiment wherein the birthdays of twenty randomly selected persons were recorded. Under certain conditions we found that $P(A) = .5886$, where A denotes the event that each person has a different birthday. Let B denote the event that at least one pair of individuals share a birthday. Find $P(B)$.

Solution   The event B is the set of all sample points in S that are not in A, that is, $B = \bar{A}$. Therefore,

$$P(B) = 1 - P(A) = 1 - .5886 = .4114$$

(Most would agree that this probability is surprisingly high!)   ∽

Let us refer to Example 2.4, which involves the two tennis players, and let D_1 and D_2 denote the events that player A wins the first and second games, respectively. The information given in the example implies that $P(D_1) = P(D_2) = 2/3$. Further, if we make the assumption that D_1 and D_2 are independent, it follows that $P(D_1 \cap D_2) = 2/3 \times 2/3 = 4/9$. In that example we identified the simple event E_1, which we denoted AA, as meaning that player A won both games. With the present notation,

$$E_1 = D_1 \cap D_2$$

and thus $P(E_1) = 4/9$. The probabilities assigned to the other simple events in Example 2.4 can be verified in a similar manner.

The event-composition approach will not be successful unless the probabilities of the events that appear in $P(A)$ (after the additive and multiplicative laws have been applied) are known. If one or more of these probabilities is unknown, the method fails. Also note that it is frequently desirable to form compositions of mutually exclusive or independent events. Mutually exclusive events simplify the additive law and independence simplifies the multiplicative law of probability.

A summarization of the steps to be followed in the event-composition approach is as follows:

1. Define the experiment.

2. Clearly visualize the nature of the sample points. Identify a few to clarify your thinking.

3. Write an equation expressing the event of interest, say, A, as a composition of two or more events, using unions, intersections, and/or complements. Note that this equates point sets. Make certain that the event implied by the composition and event A represent the same set of sample points.

4. Apply the additive and multiplicative laws of probability to step 3 and find $P(A)$.

Step 3 is the most difficult because we can form many compositions that will be equivalent to event A. The trick is to form a composition in which all the probabilities appearing in step 4 are known.

The event-composition approach does not require listing the sample points in S, but it does require a clear understanding of the nature of a typical sample point. The major error students tend to make in applying the event-composition approach occurs in writing the composition. That is, the point-set equation that expresses A as union and/or intersection of other events is frequently incorrect. Always test your equality to make certain that the composition implies an event that contains the same set of sample points as those in A.

Some comparison of the sample-point and event-composition methods for calculating the probability of an event can be obtained by applying both methods to the same problem. We will apply the event-composition approach to the problem of selecting applicants that was solved by the sample-point method in Examples 2.11 and 2.12.

EXAMPLE 2.19   The experiment involves the selection of two applicants out of five. Find the probability of drawing exactly one of the two best applicants, event A.

Solution   Define the following two events:

B: Draw the best and one of the three poorer applicants.

C: Draw the second best and one of the three poorer applicants.

Then B and C are mutually exclusive events and $A = B \cup C$. Also, let

$$D_1 = B_1 \cap B_2$$

where

$B_1 =$ draw the best on the first draw

$B_2 =$ draw one of the three poorer applicants on the second draw

and

$$D_2 = B_3 \cap B_4$$

where

$B_3 =$ draw one of the three poorer applicants on the first draw

$B_4 =$ draw the best on the second draw

Note that $B = D_1 \cup D_2$.

Similarly, we could let $G_1 = C_1 \cap C_2$ and $G_2 = C_3 \cap C_4$, where C_1, C_2, C_3, and C_4 are defined like B_1, B_2, B_3, and B_4, with the words *second best* replacing *best*. Note that D_1 and D_2 and G_1 and G_2 are pairs of mutually exclusive events and

$$A = B \cup C$$
$$A = (B_1 \cap B_2) \cup (B_3 \cap B_4) \cup (C_1 \cap C_2) \cup (C_3 \cap C_4)$$

Applying the additive law of probability to these four mutually exclusive events, we have

$$P(A) = P(B_1 \cap B_2) + P(B_3 \cap B_4) + P(C_1 \cap C_2) + P(C_3 \cap C_4)$$

Applying the multiplicative law, we have

$$P(B_1 \cap B_2) = P(B_1)P(B_2|B_1)$$

The probability of drawing the best on the first draw is

$$P(B_1) = 1/5$$

Similarly, the probability of drawing one of the three poorest on the second draw, given that the best was drawn on the first selection, is

$$P(B_2 | B_1) = 3/4$$

Then $$P(B_1 \cap B_2) = P(B_1)P(B_2 | B_1) = (1/5)(3/4) = 3/20$$

The probabilities of all the other intersections in $P(A)$, $P(B_3 \cap B_4)$, $P(C_1 \cap C_2)$, and $P(C_3 \cap C_4)$, are obtained in exactly the same manner and all equal 3/20. Then

$$P(A) = P(B_1 \cap B_2) + P(B_3 \cap B_4) + P(C_1 \cap C_2) + P(C_3 \cap C_4)$$
$$= (3/20) + (3/20) + (3/20) + (3/20) = 3/5$$

This answer is identical to that obtained in Example 2.12, where $P(A)$ was acquired by using the sample-point approach.   ∽

EXAMPLE 2.20   It is known that a patient will respond to treatment of a particular disease with probability equal to .9. If three patients are treated in an independent manner, find the probability that at least one will respond.

Solution   Define the following events:

A: At least one of the three patients will respond.

B_1:   The first patient will not respond.

B_2:   The second patient will not respond.

B_3:   The third patient will not respond.

Then observe that $\bar{A} = B_1 \cap B_2 \cap B_3$ and that $S = A \cup \bar{A}$, where A and $\bar{A}$ are complementary events and hence mutually exclusive. Then

$$P(S) = P(A) + P(\bar{A})$$
$$1 = P(A) + P(B_1 \cap B_2 \cap B_3)$$

or $$P(A) = 1 - P(B_1 \cap B_2 \cap B_3)$$

Applying the multiplicative law, we have

$$P(B_1 \cap B_2 \cap B_3) = P(B_1)P(B_2 | B_1)P(B_3 | B_1 \cap B_2)$$

where $P(B_2 | B_1) = P(B_2)$ and $P(B_3 | B_1 \cap B_2) = P(B_3)$

because the events are independent.

Substituting $P(B_i) = .1$, $i = 1, 2, 3$, we obtain

$$P(A) = 1 - (.1)^3 = .999$$

Note that we have demonstrated the utility of complementary events. This result is important because it is frequently easier to find the probability of the complement $P(\bar{A})$ than to find $P(A)$ directly. ∽

EXAMPLE 2.21   Observation of a waiting line at a medical clinic indicates that the probability that a new arrival will be an emergency case is $p = 1/6$. Find the probability that the rth patient is the first emergency case. (Assume that conditions of arriving patients represent independent events.)

Solution   The experiment consists of watching patient arrivals until the first emergency case appears. Then the sample points for the experiment are

E_i: The ith patient is the first emergency case, $i = 1, 2, \ldots$.

Since only one sample point falls in the event of interest,

$$P(r\text{th patient is the first emergency case}) = P(E_r)$$

Now define A_i to denote the event that the ith arrival is not an emergency case. Then we can represent E_r as the intersection

$$E_r = A_1 \cap A_2 \cap A_3 \cap \cdots \cap A_{r-1} \cap \bar{A}_r$$

Applying the multiplicative law, we have

$$P(E_r) = P(A_1)P(A_2|A_1)P(A_3|A_1 \cap A_2)\cdots P(\bar{A}_r|A_1 \cap \cdots \cap A_{r-1})$$

and since the events $A_1, A_2, \ldots, A_{r-1}$, and $\bar{A}_r$ are independent, it follows that

$$P(E_r) = P(A_1)P(A_2)\cdots P(A_{r-1})P(\bar{A}_r) = (1-p)^{r-1}p$$
$$= (5/6)^{r-1}(1/6) \qquad r = 1, 2, 3, \ldots$$

Note that

$$P(S) = P(E_1) + P(E_2) + P(E_3) + \cdots + P(E_i) + \cdots$$
$$= (1/6) + (5/6)(1/6) + (5/6)^2(1/6) + \cdots + (5/6)^{i-1}(1/6) + \cdots$$
$$= \frac{1/6}{1-(5/6)} = 1$$

This result follows from the formula for the sum of an infinite number of terms of a decreasing geometrical progression, $a/(1 - R)$, where a denotes the first term and R the common ratio. This formula is useful in many simple probability problems.

EXAMPLE 2.22   A monkey is to be taught to recognize colors by tossing one red, one black, and one white ball into boxes of the same respective colors, one ball to a box. If the monkey has not learned the colors, and merely tosses one ball into each box at random, find the following probabilities:

(*a*)   that there are no color matches,

(*b*)   that there is exactly one color match.

Solution   This problem can be solved by listing sample points because only three balls are involved, but a more general method of solution will be illustrated. Define the following events:

A_1:   A color match occurs in the red box.

A_2:   A color match occurs in the black box.

A_3:   A color match occurs in the white box.

There are $3! = 6$ equally likely ways of randomly tossing the balls into the boxes with one ball in each box. Also, there are only $2! = 2$ ways of tossing the balls into the boxes if one particular box is required to have a color match. Hence

$$P(A_1) = P(A_2) = P(A_3) = 2/6 = 1/3$$

Similarly, it follows that

$$P(A_1 \cap A_2) = P(A_1 \cap A_3) = P(A_2 \cap A_3) = P(A_1 \cap A_2 \cap A_3) = 1/6$$

We can now answer parts (a) and (b) by the event-composition method.
    (a) Note that

$$
\begin{aligned}
P(\text{no color matches}) &= 1 - P(\text{at least one color match}) \\
&= 1 - P(A_1 \cup A_2 \cup A_3) \\
&= 1 - [P(A_1) + P(A_2) + P(A_3) - P(A_1 \cap A_2) \\
&\quad - P(A_1 \cap A_3) - P(A_2 \cap A_3) + P(A_1 \cap A_2 \cap A_3)] \\
&= 1 - [3(1/3) - 3(1/6) + (1/6)] = 2/6 = 1/3
\end{aligned}
$$

(b) We leave it to you to show that

$$P(\text{exactly one match}) = P(A_1) + P(A_2) + P(A_3)$$
$$- 2[P(A_1 \cap A_2) + P(A_1 \cap A_3) + P(A_2 \cap A_3)]$$
$$+ 3[P(A_1 \cap A_2 \cap A_3)]$$
$$= (3)(1/3) - (2)(3)(1/6) + (3)(1/6) = 1/2$$

The best way to learn how to solve probability problems is to learn by doing. Many exercises are provided at the end of the chapter and in the references to assist you in developing your ability to diagnose and solve probability problems.

Exercises

2.55 Of the items produced daily by a certain factory, 40 % come from line I and 60 % from line II. Line I has a defect rate of 8 % whereas line II has a defect rate of 10 %. If an item is chosen at random from the day's production, find the probability that it will not be defective.

2.56 An advertising agency notes that approximately 1 in 50 potential buyers of a product sees a given magazine ad and 1 in 5 sees a corresponding ad on television. One in 100 sees both. One in 3 actually purchases the product after seeing the ad, 1 in 10 without seeing it. What is the probability that a randomly selected potential customer will purchase the product?

2.57 Three radar sets, operating independently, are set to detect any aircraft flying through a certain area. Each set has a probability of .02 of failing to detect a plane in its area.

(a) If an aircraft happens into the area, what is the probability that it goes undetected?

(b) If an aircraft happens into the area, what is the probability that it is detected by all three radar sets?

2.58 Consider one of the radar sets of Exercise 2.57. What is the probability that it will correctly detect exactly three aircraft before it fails to detect one if aircraft arrivals are independent single events occurring at different times?

2.59 A lie detector will show a positive reading (indicate a lie) 10 % of the time when a person is telling the truth and 95 % of the time when the person is lying. Suppose two people are suspects in a one-person crime and (for certain) one is guilty.

(a) What is the probability that the detector shows a positive reading for both suspects?

(b) What is the probability that the detector shows a positive reading for the guilty suspect and a negative reading for the innocent?

(c) What is the probability that the detector is completely wrong, that is, that it gives a positive reading for the innocent suspect and a negative reading for the guilty?

(d) What is the probability that it gives a positive reading for either or both of the two suspects?

2.60   A certain football team has a probability of .75 of winning when playing any of the other four teams in its conference. If the games are independent, what is the probability the team wins each of its conference games?

2.61   A communications network has a built-in safeguard system against failures. In this system if line I fails, it is bypassed and line II is used. If line II also fails, it is bypassed and line III is used. The probability of failure of any one of these three lines is .01, and the failures of these lines are independent events. What is the probability that this system of three lines does not completely fail?

2.62   A state auto inspection station utilizes two inspection teams. Team 1 is lenient and passes all automobiles of a recent vintage; team 2 rejects all autos on a first inspection because their "headlights are not properly adjusted." Four unsuspecting drivers take their autos to the station for inspection on four different days and randomly select one of the two teams.

   (a)   If all four cars are new and in excellent condition, what is the probability that three of the four will be rejected?

   (b)   What is the probability that all four will pass?

2.63   An accident victim will die unless in the next 10 minutes he receives an amount of type A, Rh positive blood that can be supplied by a single donor. It requires 2 minutes to type a prospective donor's blood and 2 minutes to complete the transfer of blood. Many untyped donors are available and 40% of them have type A, Rh positive blood. What is the probability that the accident victim will be saved if only one blood-typing kit is available? Assume that the typing kit is reusable but can only process one donor at a time.

2.64   Suppose that two defective refrigerators have been included in a shipment of six refrigerators. The buyer begins to test the six refrigerators one at a time.

   (a)   What is the probability that the last defective refrigerator is found on the fourth test?

   (b)   What is the probability that no more than four refrigerators need to be tested to locate both of the defective refrigerators?

   (c)   When given that exactly one of the two defective refrigerators has been located in the first two tests, what is the probability that the remaining defective refrigerator is found in the third or fourth test?

2.65   A new secretary has been given n computer passwords, only one of which will permit access to a computer file. Because the secretary has no idea which password is correct, he chooses one of the passwords at random and tries it. If the password is incorrect, he discards it and randomly selects another password from among those remaining, proceeding in this manner until he finds the correct password.

   (a)   What is the probability that he obtains the correct password on the first try?

   (b)   What is the probability that he obtains the correct password on the second try? the third try?

   (c)   A security system has been set up so that if three incorrect passwords are tried before the correct one, the computer file is locked and access to it denied. If $n = 7$, what is the probability that the secretary will gain access to the file?

2.10 Bayes's Rule

The event-composition approach to solving probability problems sometimes is facilitated by viewing the sample space, S, as a union of mutually exclusive subsets and using the following *law of total probability*.

Theorem 2.8

Assume that $S = B_1 \cup B_2 \cup \cdots \cup B_k$ where $P(B_i) > 0$, $i = 1, 2, \ldots, k$ and $B_i \cap B_j = \emptyset$ for $i \neq j$. Then for any event A

$$P(A) = \sum_{i=1}^{k} P(B_i)P(A|B_i).$$

Proof   Note that any subset A of S can be written as

$$A = A \cap S = A \cap (B_1 \cup B_2 \cup \cdots \cup B_k)$$
$$= (A \cap B_1) \cup (A \cap B_2) \cup \cdots \cup (A \cap B_k)$$

We then observe that

$$P(A) = P(A \cap B_1) + P(A \cap B_2) + \cdots + P(A \cap B_k)$$
$$= P(B_1)P(A|B_1) + P(B_2)P(A|B_2) + \cdots + P(B_k)P(A|B_k)$$
$$= \sum_{i=1}^{k} P(B_i)P(A|B_i)$$

Using the result of Theorem 2.8, it is a simple matter to derive the result known as *Bayes's Rule*.

Theorem 2.9

(Bayes's Rule) Assume that $S = B_1 \cup B_2 \cup \cdots \cup B_k$ where $P(B_i) > 0$, $i = 1, 2, \ldots, k$ and $B_i \cap B_j = \emptyset$ for $i \neq j$. Then

$$P(B_j|A) = \frac{P(B_j)P(A|B_j)}{\displaystyle\sum_{i=1}^{k} P(B_i)P(A|B_i)}.$$

Proof   The proof follows directly from the definition of conditional probability and the law of total probability. Note that

$$P(B_j|A) = \frac{P(A \cap B_j)}{P(A)} = \frac{P(B_j)P(A|B_j)}{\displaystyle\sum_{i=1}^{k} P(B_i)P(A|B_i)}$$

EXAMPLE 2.23   An electronic fuse is produced by five production lines in a manufacturing operation. The fuses are costly, are quite reliable, and are shipped to suppliers in 100-unit lots. Because testing is destructive, most buyers of the fuses test only a small number of fuses before deciding to accept or reject lots of incoming fuses.

All five production lines produce fuses at the same rate and normally produce only 2% defective fuses, which are dispersed randomly in the output. Unfortunately, production line 1 suffered mechanical difficulty and produced 5% defectives during the month of March. This situation became known to the manufacturer after the fuses had been shipped. A customer received a lot produced in March and tested three fuses. One failed. What is the probability that the lot was produced on line 1? What is the probability that the lot came from one of the four other lines?

Solution   Let $P(L_i)$ denote the probability that a fuse was drawn from line i and let D denote the event that a fuse was defective. Then

$$P(L_1) = .2$$
$$P(L_1 \cap D) = P(L_1)P(D|L_1) = (.2)(3)(.05)(.95)^2 = .0270750$$
$$P(\bar{L}_1 \cap D) = P(\bar{L}_1)P(D|\bar{L}_1) = (.8)(3)(.02)(.98)^2 = .0460992$$
$$P(D) = P(L_1 \cap D) + P(\bar{L}_1 \cap D) = .0270750 + .0460992 = .0731742$$
$$P(L_1|D) = \frac{P(L_1 \cap D)}{P(D)} = \frac{.0270750}{.0731742} = .37$$

and

$$P(\bar{L}_1|D) = 1 - P(L|D) = 1 - .37 = .63$$

Exercises

2.66   In a certain population of voters 40% are Republicans and 60% are Democrats. It is reported that 30% of the Republicans and 70% of the Democrats favor a certain election issue. A person is chosen at random from this population and is found to favor the issue in question. Find the conditional probability that this person is a Democrat.

2.67   A diagnostic test for a certain disease is said to be 90% accurate in that, if a person has the disease, the test will detect it with probability .9. Also, if a person does not have the disease, the test will report that he or she does not have it with probability .9. Only 1% of the population has the disease in question. If a person is chosen at random from the population and the diagnostic test indicates that she has the disease, what is the conditional probability that she does, in fact, have the disease? Are you surprised by the answer? Would you call this diagnostic test reliable?

2.68   Males and females are observed to react differently to a given set of circumstances. It has been observed that 70% of the females react positively to these circumstances, whereas only 40% of males react positively. A group of twenty people, fifteen female and five male, was subjected to these circumstances, and the subjects were asked to describe their reactions on a written questionnaire. A response picked at random from the twenty was negative. What is the probability that it was that of a male?

2.69 A study of Georgia residents suggests that those who worked in shipyards during World War II were subjected to a significantly higher risk of lung cancer (*The Wall Street Journal*, September 21, 1978).[†] It was found that approximately 22% of those persons who had lung cancer worked at some prior time in a shipyard. In contrast, only 14% of those who had no lung cancer worked at some prior time in a shipyard. Suppose that the proportion of all Georgians living during World War II who have or will have contracted lung cancer is .04%. Find the percentage of Georgians living during the same period who will contract (or have contracted) lung cancer, given that they have some prior time worked in a shipyard.

2.70 As items come to the end of a production line, an inspector chooses which items are to go through a complete inspection. Ten percent of all items produced are defective. Sixty percent of all defective items go through a complete inspection, and 20% of all good items go through a complete inspection. Given that an item is completely inspected, what is the probability it is defective?

2.71 A student answers a multiple-choice examination question that offers four possible answers. Suppose that the probability that the student knows the answer to the question is .8 and the probability that the student will guess is .2. Assume that if the student guesses, the probability of selecting the correct answer is .25. If the student correctly answers a question, what is the probability that the student really knew the correct answer?

2.72 Two methods, *A* and *B*, are available for teaching a certain industrial skill. The failure rate is 20% for *A* and 10% for *B*. However, *B* is more expensive and hence is used only 30% of the time. (*A* is used the other 70%.) A worker was taught the skill by one of the methods but failed to learn it correctly. What is the probability that she was taught by method *A*?

2.73 Of the travelers arriving at a small airport, 60% fly on major airlines, 30% fly on privately owned planes, and the remainder fly on commercially owned planes not belonging to a major airline. Of those traveling on major airlines, 50% are traveling for business reasons, whereas 60% of those arriving on private planes and 90% of those arriving on other commercially owned planes are traveling for business reasons. Suppose that we randomly select one person arriving at this airport. What is the probability that the person

(a) is traveling on business?

(b) is traveling for business on a privately owned plane?

(c) arrived on a privately owned plane given that the person is traveling for business reasons?

(d) is traveling on business given that the person is flying on a commercially owned plane?

2.74   A personnel director has two lists of applicants for jobs. List 1 contains the names of five women and two men whereas list 2 contains the names of two women and six men. A name is randomly selected from list 1 and added to list 2. A name is then randomly selected from the augmented list 2. Given that the name selected is that of a man, what is the probability that a woman's name was originally selected from list 1?

2.75   Five identical bowls are labeled 1, 2, 3, 4, and 5. Bowl i contains i white and $5 - i$ black balls, $i = 1, 2, \ldots, 5$. A bowl is randomly selected and two balls are selected (without replacement) from the contents of the bowl.

(a)   What is the probability that both balls selected are white?

(b)   Given that both balls selected are white, what is the probability that bowl 3 was selected?

2.11 Numerical Events and Random Variables

Events of major interest to the scientist, engineer, or businessperson are those identified by numbers, called *numerical events*. The research physician is interested in the event that ten of ten treated patients survive an illness; the businessperson is interested in the event that sales next year will reach $5 million. Because the value of a numerical event will vary in repeated samplings, it is called a *random variable*.

To each point in the sample space we will assign a real number denoting the value of a numerical event. The numbers will vary from one sample point to another, but some points may be assigned the same number. Thus we have defined a variable that is a function of the sample points in S. If we let Y denote this variable, then $Y = a$ is the numerical event that contains all sample points assigned the number a. Indeed, the sample space S can be partitioned into subsets so that points within a subset are assigned the same value of Y. These subsets are mutually exclusive. The partitioning of S is symbolically indicated in Figure 2.12 for a random variable that can assume values 0, 1, 2, 3, 4.

Figure 2.12
Partitioning S into subsets that define the events Y = 0, 1, 2, 3, 4

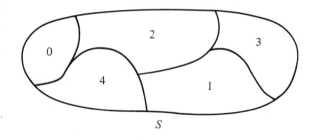

Definition 2.11

A *random variable* is a real-valued function for which the domain is a sample space.

EXAMPLE 2.24   Define an experiment as tossing two coins and observing the results. Let Y equal the number of heads observed. Identify the sample points in S, assign a value of Y to each sample point, and identify the sample points associated with each value of the random variable.

Solution   Let H and T represent head and tail, respectively; and let an ordered pair of symbols identify the outcome for the first and second coins, respectively. (Thus HT implies a head on the first coin and a tail on the second.) Then the four sample points in S are E_1: HH, E_2: HT, E_3: TH, and E_4: TT. The values of Y assigned to the sample points depend on the number of heads implied by each point. For E_1: HH, two heads were observed, and E_1 is assigned the value $Y = 2$. Similarly, we assign the values $Y = 1$ to E_2 and E_3 and $Y = 0$ to E_4. Summarizing, the random variable Y can take three values, $Y = 0, 1, 2$, which are events defined by specific collections of sample points:

$$Y = 0: E_4 \qquad Y = 1: E_2, E_3 \qquad Y = 2: E_1$$

Let y denote an observed value of the random variable Y. We then set the probability that $Y = y$ equal to the sum of the probabilities of the sample points that are assigned the value y.

EXAMPLE 2.25   Compute the probabilities for each value of Y in Example 2.24.

Solution $Y = 0$ results only from sample point E_4. If the coins are balanced, the sample points are equally likely and hence

$$P(Y = 0) = P(E_4) = 1/4$$

Similarly,

$$P(Y = 1) = P(E_2) + P(E_3) = 1/2$$
$$P(Y = 2) = P(E_1) = 1/4$$

A more detailed examination of random variables will be undertaken in the next two chapters.

Exercises

2.76   Refer to Exercise 2.57. Let the random variable Y represent the number of radar sets that detect a particular aircraft. Compute the probabilities associated with each value of Y.

2.77   Refer to Exercise 2.64. Let the random variable Y represent the number of defective refrigerators found after three refrigerators have been tested. Compute the probabilities for each value of Y.

2.78   Refer again to Exercise 2.64. Let the random variable Y represent the number of the test on which the last defective refrigerator is located. Compute the probabilities for each value of Y.

2.79   A spinner can land in any of four positions, A, B, C, and D, with equal probability. The spinner is used twice and the position noted each time. Let the random variable Y denote the number of positions that the spinner did *not* land on. Compute the probabilities for each value of Y.

2.12 Random Sampling

As our final topic in this chapter, we move from theory to application and examine the type of experiments conducted in statistics. A statistical experiment involves the observation of a sample selected from a larger body of data, existing or conceptual, called a *population*. The measurements in the sample, viewed as observations on one or more random variables, are then employed to make an inference about the characteristics of the target population.

How are these inferences made? An exact answer to this question is deferred until later but a general observation follows from our discussion in Section 2.2. There we learned that the probability of the observed sample plays a major role in making an inference and evaluating its credibility.

Without belaboring the point, it is clear that the method of sampling will affect the probability of a particular sample outcome. For example, suppose that a fictitious population contains only $N = 5$ elements, from which we plan to sample $n = 2$. You could mix the elements thoroughly and select two in such a way that all pairs of elements possess an equal probability of selection. A second sampling procedure might require selecting a single element, replacing it in the population, and then drawing a single element again. The two methods of sample selection are called *sampling without* and *with replacement*, respectively.

If all the $N = 5$ population elements are distinctly different, the probability of drawing a specific pair, sampling without replacement, is 1/10. The probability of drawing the same pair, sampling with replacement, is 2/25. You can easily verify these results.

The point that we make is that the method of sampling, known as the *design of an experiment*, affects both the quantity of information in a sample as well as the probability of observing a specific sample result. Hence every sampling procedure must be clearly described if we wish to make valid inferences from sample to population.

The study of the design of experiments, the various types of designs along with their properties, is a course in itself. Hence at this early stage of study we only introduce the simplest sampling procedure, *simple random sampling*. The notation of simple random sampling will be needed in subsequent discussions of the probabilities associated with random variables, and it will inject some realism into our discussion of statistics. This is because simple random sampling is often employed in practice. Now let us define the term *random sample*.

Definition 2.12

> Let N and n represent the numbers of elements in the population and sample, respectively. If the sampling is conducted in such a way that each of the $\binom{N}{n}$ samples has an equal probability of being selected, the sampling is said to be random and the result is said to be a *random sample*.

Perfect random sampling is difficult to achieve in practice. If the population is not too large, we might write each of the N numbers on a poker chip, mix the total, and select a sample of n chips. The numbers on the poker chips would specify the measurements to appear in the sample.

Tables of random numbers have been formed by computer to expedite the selecting of random samples. An example of such a table is Table 12, Appendix III. A random number table is a set of integers $(0, 1, \ldots, 9)$ generated so that, in the long run, the table will contain all ten integers in approximately equal proportions, with no trends in the patterns in which the digits were generated. Thus if one digit is selected from a random point on the table, it is equally likely to be any of the digits 0 through 9.

Choosing numbers from the table is analogous to drawing numbered poker chips from the mixed pile, as mentioned earlier. Suppose we want a random sample of three persons to be selected from seven persons. We could number the people from 1 to 7, put the numbers on chips, thoroughly mix the chips, and then draw three out. Analogously, we could drop a pencil point on a random starting point in Table 12, Appendix III. Suppose the point falls on the fifteenth line of column 9 and we decide to use the right-most digit of the group of five, which is a 5 in this case. This process is like drawing the chip numbered 5. We may now proceed in any direction to obtain the remaining numbers in the sample. If we decide to proceed down the page, the next number (immediately below the 5) is a 2. So our second sampled person would be number 2. Proceeding, we next come to an 8, but there are only seven elements in the population. Thus the 8 is ignored and we continue down the column. Two more 5s then appear, but they must both be ignored since person 5 has already been selected. (The chip numbered 5 has been removed from the pile.) Finally, we come to a 1, and our sample of three is completed with persons numbered 5, 2, and 1.

Note that any starting point can be used in a random number table, and we may proceed in any direction from the starting point. However, if more than one sample is to be used in any problem, each should have a unique starting point.

In many situations the population is conceptual, as in an observation made during a laboratory experiment. Here the population is envisioned to be the infinitely many measurements that would be obtained if the experiment were to be repeated over and over again. If we wish a sample of $n = 10$ measurements from this population, we repeat the experiment ten times and hope that the results represent, to a reasonable degree of approximation, a random sample.

Although the primary purpose of this discussion was to clarify the meaning of a random sample, we would like to mention that some sampling techniques are only partly random. For instance, if we wish to determine the voting preference of the nation in a presidential election, we would not likely choose a random sample from the population of voters. By pure chance, all the voters appearing in the sample might be drawn from a single city, say, San Francisco, which might not be at all representative of the population. We would prefer a random selection of voters from smaller political districts, perhaps states, allotting a specified number to each state. The information from the randomly selected subsamples drawn from the respective states would be combined to form a prediction concerning the entire population of voters in the country. In general, we want to select a sample so as to obtain a specified quantity of information at minimum cost.

2.13 Summary

This chapter has been concerned with providing a model for the repetition of an experiment and, consequently, a model for the population frequency distributions of Chapter 1. The acquisition of a probability distribution is the first step in forming a theory to model reality and to develop the machinery for making inferences.

An experiment was defined as the process of making an observation. The concepts of an event, a simple event, the sample space, and the probability axioms have provided a probabilistic model for calculating the probability of an event. Numerical events and the definition of a random variable were introduced in Section 2.11.

Embedded among our theoretical abstractions of reality are some tools that assist in finding the probability of an event. Inherent in the model is the sample-point approach for calculating the probability of an event (Section 2.5). Counting rules useful in applying the sample-point method were discussed in Section 2.6. The concept of conditional probability, the operations of set algebra, and the laws of probability set the stage for the event-composition method for calculating the probability of an event (Section 2.9).

Of what value is the theory of probability? It provides the theory and the tools for calculating the probabilities of numerical events and hence the probability distributions for the random variables that will be discussed in

Chapter 3. The numerical events of interest to us appear in a sample, and we will wish to calculate the probability of an observed sample to make an inference about the target population. Probability provides both the foundation and the tools for modern statistical inference, and this is the objective of statistics.

References and Further Readings

1. Cramer, H. *The Elements of Probability Theory and Some of Its Applications.* 2d ed. Huntington, N.Y.: Krieger, 1973.

2. Feller, W. *An Introduction to Probability Theory and Its Applications.* Vol. 1, 3d ed. New York: Wiley, 1968.

3. Feller, W. *An Introduction to Probability Theory and Its Applications.* Vol. 2, 2d ed. New York: Wiley, 1971.

4. Meyer, P. L. *Introductory Probability and Statistical Applications.* 2d ed. Reading, Mass.: Addison-Wesley, 1970.

5. Parzen, E. *Modern Probability Theory and Its Applications.* New York: Wiley, 1960.

6. Riordan, J. *An Introduction to Combinatorial Analysis.* Princeton, N.J.: Princeton University Press, 1980.

Supplementary Exercises

2.80   Show that

$$\binom{3}{0} + \binom{3}{1} + \binom{3}{2} + \binom{3}{3} = 2^3$$

Note that in general

$$\sum_{i=0}^{n} \binom{n}{i} = 2^n$$

2.81   Let S contain four sample points, E_1, E_2, E_3, and E_4.

  (a)   List all possible events in S (include the null event).

  (b)   Use the results of Exercise 2.80 to give the total number of events in S.

  (c)   Let A and B be the events $\{E_1, E_2, E_3\}$ and $\{E_2, E_4\}$, respectively. Give the sample points in the following events: $A \cup B$, $A \cap B$, $\bar{A} \cap \bar{B}$, and $\bar{A} \cup B$.

2.82   A patient receiving a yearly physical examination must have eighteen checks or tests performed. The sequence in which the tests are conducted is important because the time lost between tests will vary and depend on the sequence. If an efficiency expert were to study the sequences to find the one that required the minimum length of time, how many sequences would be included in her study if all possible sequences were admissible?

2.83 Five cards are drawn from a standard fifty-two–card playing deck. What is the probability that all five cards will be of the same suit?

2.84 Refer to Exercise 2.83. What is the probability of a "full house"? (Recall that a full house is a hand of five cards containing one pair and three of another kind, e.g., two kings and three tens.)

2.85 A bin contains three components from supplier A, four from supplier B, and five from supplier C. If four of the components are randomly selected for testing, what is the probability that each supplier would have at least one component tested?

2.86 A large group of people is to be checked for two common symptoms of a certain disease. It is thought that 20% of the people possess symptom A alone, 30% possess symptom B alone, 10% possess both symptoms, and the remainder have neither symptom. For one person chosen at random from this group, find these probabilities:

(a) that the person has neither symptom,

(b) that the person has at least one symptom,

(c) that the person has both symptoms, given that he has symptom B.

2.87 Refer to Exercise 2.86. Let the random variable Y represent the number of symptoms possessed by a person chosen at random from the group. Compute the probabilities associated with each value of Y.

2.88 A certain population of men possesses the three characteristics of being married (A), having a college degree (B), and being a citizen of a specified state (C) according to the fractions given in the accompanying Venn diagram. That is, 5% of the men possess all three characteristics, whereas 20% have a college education but are not married and not citizens of the specified state. One man is chosen at random from this population.

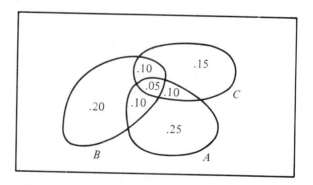

(a) Find the probability that he is married.

(b) Find the probability that he has a college degree and is married.

(c) Find the probability that he is not from the specified state but is married and has a college degree.

(d) Find the probability that he is not married or does not have a college degree, given that he is from the specified state.

2.89 The accompanying table lists accidental deaths by age and certain specific types for the United States in 1976.

Age	Type of Accident			
	All Types	Motor Vehicle	Falls	Drowning
Under 5	4,692	1,532	201	720
5 to 14	6,308	3,175	121	1,050
15 to 24	24,316	16,650	463	2,090
25 to 34	13,868	7,888	426	1,060
35 to 44	8,531	4,224	534	520
45 to 54	9,434	4,118	931	500
55 to 64	9,566	3,652	1,340	420
65 to 74	8,823	3,082	1,997	270
75 and over	15,223	2,717	8,123	197
Total	100,761	47,038	14,136	6,827

Source The World Almanac and Book of Facts, 1979 Edition, copyright © Newspaper Enterprise Association Inc., 1978, New York, NY 10166.

(a) A randomly selected person from the United States was known to have an accidental death in 1976.

(*i*)   Find the probability that he was over the age of 15.

(*ii*) Find the probability that the cause of death was a motor vehicle accident.

(*iii*) Find the probability that the cause of death was a motor vehicle accident, given that the person was between 15 and 24 years old.

(*iv*) Find the probability that the cause of death was a drowning accident, given that it was not a motor vehicle accident and the person was 34 or under.

(b) From these figures can you determine the probability that a person selected at random from the U.S. population had a motor vehicle accident in 1976?

2.90 The accompanying table is derived from 1970 U.S. census population figures. (The figures

Area	1970 U.S. Population		
	Total	White	Negro and Other
Urban	149,325	128,773	20,552
inside urbanized areas	118,447	100,952	17,495
central cities	63,922	49,547	14,375
urban fringe	54,525	51,405	3,120
outside urbanized areas	30,878	27,821	3,057
Rural	53,887	48,976	4,911
Total	203,212	177,749	25,463

Source The World Almanac and Book of Facts, 1979 Edition, copyright © Newspaper Enterprise Association Inc., 1978, New York, NY 10166.

are in thousands.) If a person is selected at random from the U.S. population, find the probabilities, based on the 1970 figures, for the following situations:

(a)   that the person is white,

(b)   that the person lives in a central city area,

(c)   that the person lives in an urban fringe, given that he is white,

(d)   that the person is white, given that he lives in an urban fringe,

(e)   that the person lives outside an urbanized area, given that he is not white,

(f)   that the person either is not white living in a central city or is white and lives outside an urbanized area.

2.91   A machine for producing a new experimental electronic tube generates defectives from time to time in a random manner. The supervising engineer for a particular machine has noticed that defectives seems to be grouping (hence appearing in a nonrandom manner) and thereby suggesting a malfunction in some part of the machine. One test for nonrandomness is based on the number of *runs* of defectives and nondefectives (a run is an unbroken sequence of either defectives or nondefectives). The smaller the number of runs, the greater will be the amount of evidence indicating nonrandomness. Of twelve tubes drawn from the machine, the first ten were not defective and the last two defective ($NNNNNNNNNNDD$). Assume randomness.

(a)   What is the probability of observing this arrangement (resulting in two runs) given that ten of the twelve tubes are not defective?

(b)   What is the probability of observing two runs?

2.92   Refer to Exercise 2.91. What is the probability that the number of runs R is $R \le 3$?

2.93   Assume that there are nine parking spaces next to one another in a parking lot. Nine cars need to be parked by an attendant. Three of the cars are expensive sports cars, three are large domestic cars, and three are imported compacts. Assuming that the attendant parks the cars at random, what is the probability that the three expensive sports cars are parked adjacent to one another?

2.94   Relays used in the construction of electric circuits function properly with probability 0.9. Assuming that the circuits operate independently, which of the following designs yields the higher probability that current will flow when the relays are activated?

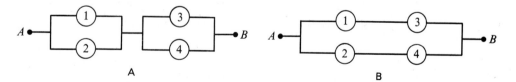

2.95   Eight tires of different brands are ranked from 1 to 8 (best to worst) according to mileage performance. If four of these tires are chosen at random by a customer, find the probability that the best tire among those selected by the customer is actually ranked third among the original eight.

2.96   Three names are to be selected from a list of seven names for use in a particular public opinion survey. Find the probability that the first name on the list is selected for the survey.

2.97   An AP news service story, printed in the *Gainesville Sun* of May 20, 1979, states the following with regard to debris from Skylab striking someone on the ground: "The odds are 1 in 150 that a piece of Skylab will hit someone. But 4 billion people ... live in the zone in which pieces could fall. So any one person's chances of being struck are one in 150 times 4 billion—or one in 600 billion." Do you see any inaccuracies in this reasoning?

2.98   Let A and B be any two events. Which of the following statements, in general, are false?

$$P(A|B) + P(\bar{A}|\bar{B}) = 1 \qquad P(A|B) + P(A|\bar{B}) = 1 \qquad P(A|B) + P(\bar{A}|B) = 1$$

2.99   Three events, A, B, and C, are said to be independent if

$$P(AB) = P(A)P(B) \qquad P(BC) = P(B)P(C)$$
$$P(AC) = P(A)P(C) \qquad P(ABC) = P(A)P(B)P(C)$$

Suppose that a balanced coin is independently tossed two times. Define the following events:

   A: Head appears on the first toss.

   B: Head appears on the second toss.

   C: Both tosses yield the same outcome.

Are A, B, and C independent?

*2.100   Suppose that the probability of exposure to the flu during an epidemic is .6. Experience has shown that a serum is 80% successful in preventing an inoculated person from acquiring the flu, if exposed. A person not inoculated faces a probability of .90 of acquiring the flu if exposed. Two persons, one inoculated and one not, are capable of performing a highly specialized task in a business. Assume that they are not at the same location, are not in contact with the same people, and cannot expose each other. What is the probability that at least one will get the flu?

*2.101   Two gamblers bet $1 each on the successive tosses of a coin. Each has a bank of $6.

   (a)   What is the probability that they break even after six tosses of the coin?

   (b)   What is the probability that one player, say, Jones, wins all the money on the tenth toss of the coin?

*2.102   Suppose that the streets of a city are laid out in a grid with streets running north-south and east-west. Consider the following scheme for patrolling an area of 16 blocks by 16 blocks. A patrolman commences walking at the intersection in the center of the area. At the corner of each block he randomly elects to go north, south, east, or west.

   (a)   What is the probability that he will reach the boundary of his patrol area by the time he walks the first 8 blocks?

   (b)   What is the probability that he will return to the starting point after walking exactly 4 blocks?

*Exercises preceded by an asterisk are optional.

*2.103 Suppose that n indistinguishable balls are to be arranged in N distinguishable boxes so that each distinguishable arrangement is equally likely. If $n \geq N$, show that the probability that no box will be empty is given by

$$\frac{\binom{n-1}{N-1}}{\binom{N+n-1}{N-1}}$$

*2.104 It can be shown that the following set equations, called the *distributive laws*, hold:

$$A \cap (B \cup C) = (A \cap B) \cup (A \cap C)$$
$$A \cup (B \cap C) = (A \cup B) \cap (A \cup C)$$

Use the first equation to prove that, for events A, B, and C,

$$P[A \cup B)|C] = P(A|C) + P(B|C) - P(AB|C)$$

DISCRETE RANDOM VARIABLES AND THEIR PROBABILITY DISTRIBUTIONS

3.1 Basic Definition

As stated in Section 2.11, a random variable is a real-valued function defined over a sample space. You will recall that a random variable has the effect of changing events in a sample space into numerical events. For example, the event of interest in an opinion poll regarding voter preferences is not the particular people sampled or the order in which preferences were obtained but the *number* of voters favoring a certain candidate or issue. This event gives rise to a random variable, the number of voters in the sample who favor a certain candidate or issue, which can take on only a finite number of values with nonzero probability. That is, the observed value of the random variable of interest must be an integer between zero and the sample size. A random variable of this type is said to be discrete.

Definition 3.1

> A random variable Y is said to be *discrete* if it can assume only a finite or countably infinite* number of distinct values.

* Recall that a set of elements is countably infinite if the elements of the set can be put into one-to-one correspondence with the positive integers.

A less formidable characterization of discrete random variables can be obtained by considering some practical examples. The number of bacteria per unit area in the study of drug control on bacterial growth is a discrete random variable, as is the number of defective television sets in a shipment of 100. Indeed, discrete random variables most often represent counts associated with real phenomena.

Let us now consider the relation of Chapter 2 to Chapter 3. Why study the theory of probability? The answer is that the probability of an observed sample is needed to make inferences about a population. The sample observations will frequently be numerical counts; that is, values of discrete random variables, and hence it is imperative that we know the probabilities of these numerical events. Since certain types of random variables occur so frequently in practice, it is useful to have at hand the probability for each value of a random variable. This collection of probabilities is called its *probability distribution*. We will find that many types of experiments exhibit similar identifying characteristics and generate random variables with the same probability distribution. Consequently, knowledge of the probability distributions for certain common types of experiments will eliminate the need for solving the same probability problem over and over again.

3.2 The Probability Distribution for a Discrete Random Variable

Notationally, we will use uppercase letters, such as Y, to denote random variables and lowercase letters, such as y, to denote particular values a random variable may assume. For example, let Y denote any one of the six possible values that could be observed on the upper face when a die is tossed. After the die is tossed, the number actually observed will be denoted by the symbol y. Note that Y is a random variable but the specific observed value, y, is *not* random.

The expression $(Y = y)$ can be read, *the set of all points in S assigned the value y by the random variable Y.*

It is now meaningful to talk about the probability that Y takes on the value y, denoted by $P(Y = y)$. As in Section 2.11 this probability is defined as a sum of probabilities of certain sample points.

Definition 3.2

> The probability that Y takes on the value of y, $P(Y = y)$, is defined to be the *sum of the probabilities of all sample points in* S *that are* assigned the value y. We will sometimes denote $P(Y = y)$ by $p(y)$.

Note that $p(y)$ is nothing more than a function that assigns probabilities to each value y, and hence it is sometimes called the *probability function* for Y.

Definition 3.3

> The *probability distribution* for a discrete variable Y can be represented by a formula, a table, or a graph, which provides the probabilities $p(y)$ corresponding to each and every value of y.

Note that $p(y) \geq 0$ for all y, but the probability distribution for a discrete random variable assigns nonzero probabilities to only a countable number of distinct y values. Any value y not explicitly assigned a positive probability is understood to be such that $p(y) = 0$. We illustrate these ideas with an example.

EXAMPLE 3.1 A foreman in a manufacturing plant has three men and three women working for him. He wants to choose two workers for a special job. Not wishing to show any biases in his selection, he decides to select the two workers at random. Let Y denote the number of women in his selection. Find the probability distribution for Y.

Solution The foreman can select two workers from six in $\binom{6}{2} = 15$ ways. Hence S contains fifteen sample points, which we assume to be equally likely because random sampling was employed. Thus $P(E_i) = 1/15$, $i = 1, 2, \ldots, 15$. The values for Y that have nonzero probability are 0, 1, and 2. The number of ways of selecting $Y = 0$ women is $\binom{3}{0}\binom{3}{2}$ because the foreman must select zero workers from the three women and two from the three men. Thus there are $\binom{3}{0}\binom{3}{2} = 1 \cdot 3 = 3$ sample points in the event $Y = 0$ and

$$p(0) = P(Y = 0) = \frac{\binom{3}{0}\binom{3}{2}}{15} = \frac{3}{15} = \frac{1}{5}$$

Similarly,

$$p(1) = P(Y = 1) = \frac{\binom{3}{1}\binom{3}{1}}{15} = \frac{9}{15} = \frac{3}{5}$$

$$p(2) = P(Y = 2) = \frac{\binom{3}{2}\binom{3}{0}}{15} = \frac{3}{15} = \frac{1}{5}$$

Note that $Y = 1$ is by far the most likely outcome, which should seem reasonable since the number of women equals the number of men in the original group.

The table for the probability distribution of Example 3.1 is given as Table 3.1, which can be graphed as shown in Figure 3.1. If we regard the width at each bar in Figure 3.1 as one unit, then the area in a bar is equal to the probability that Y takes on the value over which the bar is centered. This concept of areas representing probabilities was introduced in Section 1.2.

Table 3.1
Probability distribution for Example 3.1

y	$p(y)$
0	1/5
1	3/5
2	1/5

Figure 3.1
Probability histogram for Table 3.1

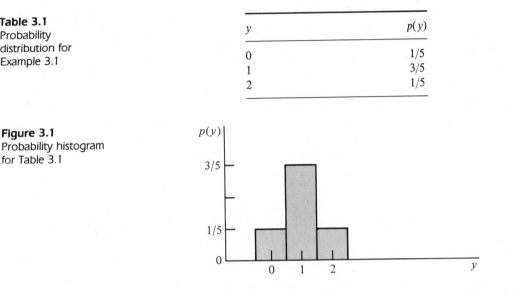

The most useful method of representing discrete probability distributions is by means of a formula. For Example 3.1 we see that the formula for $p(y)$ can be written as

$$p(y) = \frac{\binom{3}{y}\binom{3}{2-y}}{\binom{6}{2}}, \qquad y = 0, 1, 2$$

Note that the probabilities associated with all distinct values of a discrete random variable must sum to 1. In summary, the following properties must hold for any discrete probability distribution:

Theorem 3.1

For any discrete probability distribution, the following must be true:

1. $0 \le p(y) \le 1$ for all y.

2. $\sum_y p(y) = 1$, where the summation is over all values of y with nonzero probability.

As mentioned in Section 1.5, the probability distributions we derive are *models*, not exact representations, for the frequency distributions of populations of real data that occur (or would be generated) in nature. Thus they are models for real distributions of data similar to those discussed in Chapter 1. For example, if we were to randomly select two workers from among the six described in Example 3.1, we would observe a single y value. In this instance the observed y value would be 0, 1, or 2. If the experiment were repeated many times, many y values would be generated. A relative frequency histogram for the resulting data, constructed in the manner described in Chapter 1, would be very similar to the probability histogram of Figure 3.1. Such simulation studies are very useful. By repeating some experiments over and over again, we can generate measurements on discrete random variables that possess frequency distributions very similar to the probability distributions derived in this chapter and hence reinforce the conviction that our models are quite good characterizations of the physical situations they are supposed to depict.

Exercises

3.1   When testing private wells in a county for two impurities commonly found in drinking water, it was found that 20% of the wells had neither impurity, 40% had impurity A, and 50% had impurity B. (Obviously, some had both impurities.) If a well is randomly chosen from those in the county, find the probability distribution for Y, the number of impurities found in the well.

3.2   A group of four components is known to contain two defectives. An inspector tests the components one at a time until the two defectives are located. Once she locates the two defectives, she stops testing, but the second defective is tested to ensure accuracy. Let Y denote the number of the test on which the second defective is found. Find the probability distribution for Y.

3.3   Consider a system of water flowing through valves from A to B. (See the accompanying diagram.) Valves 1, 2, and 3 operate independently, and each correctly opens on signal with a probability of .8. Find the probability distribution for Y, the number of open paths from A to B after the signal is given. (Note that Y can take on the values 0, 1, and 2.)

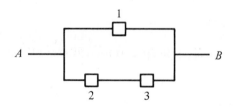

3.4   A problem in a test given to small children asks them to match each of three pictures of animals to the word identifying that animal. If a child assigns the three words at random to the three pictures, find the probability distribution for Y, the number of correct matches.

3.5   Five balls, numbered 1, 2, 3, 4, and 5, are placed in an urn. Two balls are randomly selected from the five, and their numbers noted. Find the probability distribution for the following:

(a)   the *largest* of the two sampled numbers,

(b)   the *sum* of the two sampled numbers.

3.6   In order to verify the accuracy of their financial accounts, companies use auditors on a regular basis to verify accounting entries. Suppose that the company's employees make erroneous entries 5% of the time. If an auditor randomly checks three entries:

(a)   Find the probability distribution for Y, the number of errors detected by the auditor.

(b)   Construct a probability histogram for $p(y)$.

(c)   Find the probability that the auditor will detect more than one error.

3.7   A rental agency, which leases heavy equipment by the day, has found that one expensive piece of equipment is leased, on the average, only one day in five. If rental on one day is independent of rental on any other day, find the probability distribution of Y, the number of days between a pair of rentals.

3.8   Persons entering a blood bank are such that 1 in 3 have type O^+ blood and 1 in 15 have type O^- blood. Consider three randomly selected donors for the blood bank. Let X denote the number of donors with type O^+ blood and Y denote the number with type O^- blood. Find the probability distributions for X and Y. Also find the probability distribution for $X + Y$, the number of donors who have type O blood.

3.3 The Expected Value of a Random Variable or a Function of a Random Variable

We have observed that the probability distribution for a random variable is a theoretical model for the empirical distribution of data associated with a real population. If the model is an accurate representation of nature, the theoretical and empirical distributions are equivalent. Consequently, as in Chapter 1, it is natural that we attempt to find the mean and the variance for a random variable and thereby to acquire descriptive measures for the probability distribution $p(y)$.

Definition 3.4

Let Y be a discrete random variable with probability function $p(y)$. Then the *expected value* of Y, $E(Y)$, is defined to be[†]

$$E(Y) = \sum_y yp(y)$$

[†] To be precise, the expected value of a discrete random variable is said to exist if the sum, as given earlier, is absolutely convergent; that is, if

$$\sum_y |y|p(y) < \infty$$

This absolute convergence will hold for all examples in this text and will not be mentioned each time an expected value is defined.

If $p(y)$ is an accurate characterization of the population frequency distribution, then $E(Y) = \mu$, the population mean.

Definition 3.4 is completely consistent with the definition of the mean of a set of measurements, Definition 1.1. For example, consider a discrete random variable Y that can assume values 0, 1, 2 with probability distribution $p(y)$ as shown in Table 3.2 and probability histogram as shown in Figure 3.2. A visual inspection will reveal the mean of the distribution to be located at $y = 1$.

Table 3.2
Probability
distribution for Y

y	$p(y)$
0	1/4
1	1/2
2	1/4

Figure 3.2
Probability
distribution for Y

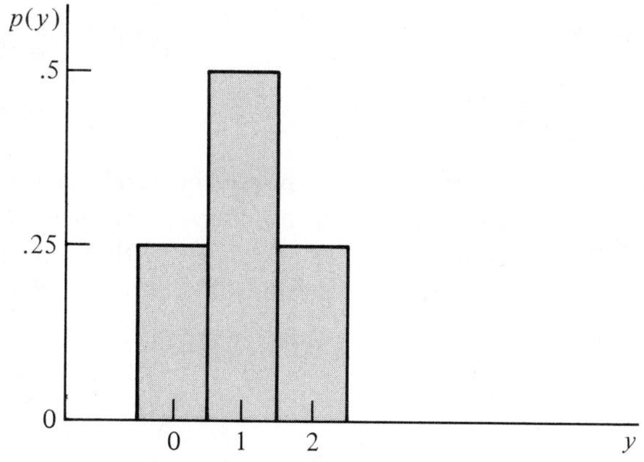

To show that $E(Y) = \sum_y yp(y)$ is the mean of the probability distribution $p(y)$, suppose that the experiment were conducted 4 million times and a corresponding number of measurements on Y observed. We wish to find the mean value of Y. Noting $p(y)$, in Figure 3.2, we would expect *approximately* 1 million of the 4 million repetitions to result in the outcome $Y = 0$, 2 million in $Y = 1$, 1, and 1 million in $Y = 2$. Averaging the 4 million measurements, we obtain

$$\mu \approx \frac{\sum_{i=1}^{n} y_i}{n} = \frac{(1{,}000{,}000)(0) + (2{,}000{,}000)(1) + (1{,}000{,}000)(2)}{4{,}000{,}000}$$

$$= (0)(1/4) + (1)(1/2) + (2)(1/4) = \sum_{y=0}^{2} yp(y) = 1$$

Thus $E(Y)$ is an average and Definition 3.4 is consistent with the definition of a mean given in Definition 1.1.

Similarly, we frequently are interested in the mean or expected value of a function of a random variable Y. For example, molecules in space move at varying velocities, where Y, the velocity of a given molecule, is a random variable. The energy imparted upon impact by a moving body is proportional to the square of the velocity. Consequently, to find the mean amount of energy transmitted by a molecule upon impact, we must find the mean value of Y^2. More important, we note in Definition 1.2 that the variance of a set of measurements is the mean of the square of the deviation of a set of measurements about their mean, or the mean value of $(Y - \mu)^2$.

Theorem 3.2

Let Y be a discrete random variable with probability function $p(y)$ and $g(Y)$ be a real-valued function of Y. Then the expected value of $g(Y)$ is given by

$$E[g(Y)] = \sum_y g(y)p(y).$$

Proof   We prove the result in the case where the random variable Y takes on the finite number of values $y_1, y_2, \ldots, y_n$. Because the function $g(y)$ may not be one to one, suppose that $g(Y)$ takes on values $g_1, g_2, \ldots, g_m$ (where $m \le n$). It follows that $g(Y)$ is a random variable such that for $i = 1, 2, \ldots, m$,

$$P[g(Y) = g_i] = \sum_{\substack{\text{all } y_j \text{ such that} \\ g(y_j) = g_i}} p(y_j) = p^*(g_i).$$

Thus, by Definition 3.4,

$$E[g(Y)] = \sum_{i=1}^{m} g_i p^*(g_i)$$

$$= \sum_{i=1}^{m} g_i \sum_{\substack{\text{all } y_j \text{ such that} \\ g(y_j) = g_i}} p(y_j)$$

$$= \sum_{i=1}^{m} \sum_{\substack{\text{all } y_j \text{ such that} \\ g(y_j) = g_i}} g_i p(y_j)$$

$$= \sum_{j=1}^{n} g(y_j) p(y_j)$$

Now let us return to our immediate objective, finding numerical descriptive measures (or *parameters*) to characterize $p(y)$. Because $E(Y)$ provides its mean, we next seek its variance and standard deviation. You will recall from Chapter 1 that the variance of a set of measurements is the average of the square of the deviations of a set of measurements about their mean. Thus we wish to find the mean value of the function $g(Y) = (Y - \mu)^2$.

Definition 3.5

> The variance of a random variable Y is defined to be the expected value of $(Y - \mu)^2$. That is,
>
> $$V(Y) = E[(Y - \mu)^2]$$
>
> The *standard deviation* of Y is the positive square root of $V(Y)$.

If $p(y)$ is an accurate characterization of the population frequency distribution (and to simplify notation, we will assume this to be true), then $E(Y) = \mu$, $V(Y) = \sigma^2$, the population variance, and σ is the population standard deviation.

EXAMPLE 3.2   For the probability distribution in Table 3.3 for a random variable Y, find its mean, variance, and standard deviation.

Table 3.3
Probability
distribution for Y

y	$p(y)$
0	1/8
1	1/4
2	3/8
3	1/4

Solution   By Definitions 3.4 and 3.5,

$$\mu = E(Y) = \sum_{y=0}^{3} yp(y) = (0)(1/8) + (1)(1/4) + (2)(3/8) + (3)(1/4) = 1.75$$

$$\sigma^2 = E[(Y - \mu)^2] = \sum_{y=0}^{3} (y - \mu)^2 \, p(y)$$

$$= (0 - 1.75)^2(1/8) + (1 - 1.75)^2(1/4) + (2 - 1.75)^2(3/8) + (3 - 1.75)^2(1/4)$$

$$= .9375$$

$$\sigma = \sqrt{\sigma^2} = \sqrt{.9375} = .97$$

The probability histogram is shown in Figure 3.3. Locate μ on the axis of measurement and observe that it does locate the "center" of the nonsymmetrical

Figure 3.3
Probability histogram
for Example 3.2

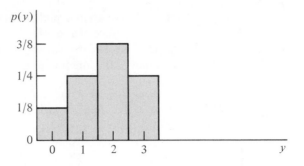

probability distribution $p(y)$. Also note that the interval $(\mu \pm \sigma)$ contains the discrete points $Y = 1$ and $Y = 2$, which comprise 5/8 of the probability. Thus the empirical rule provides a reasonable approximation to the probability of a measurement falling in this interval. (Keep in mind that the probabilities are concentrated at the points $Y = 0, 1, 2, 3$ because Y cannot take intermediate values.)

It will be helpful to acquire a few additional tools and definitions before attempting to find the expected values and variances of a more complicated discrete random variable, such as the binomial or Poisson. Hence we present three useful expectation theorems that follow directly from the theory of summation. (Other useful techniques are presented in Sections 3.4 and 3.9.) For each theorem we assume that Y is a discrete random variable with probability distribution $p(y)$.

The first theorem states the rather obvious result that the mean or expected value of a nonrandom quantity c is equal to c.

Theorem 3.3

Let c be a constant. Then $E(c) = c$.

Proof   By Theorem 3.2,

$$E(c) = \sum_y cp(y) = c\sum_y p(y)$$

But $\sum\limits_y p(y) = 1$ (Theorem 3.1) and hence $E(c) = c(1) = c$.

The second theorem states that the expected value of the product of a constant c times a function of a random variable is equal to the constant times the expected value of the function.

Theorem 3.4

> Let $g(Y)$ be a function of the random variable Y and let c be a constant. Then
>
> $$E[cg(Y)] = cE[g(Y)]$$

Proof   By Theorem 3.2,

$$E[cg(Y)] = \sum_y cg(y)p(y) = c \sum_y g(y)p(y) = cE[g(Y)]$$

The third theorem states that the mean or expected value of a sum of functions of a random variable Y is equal to the sum of their respective expected values.

Theorem 3.5

> Let $g_1(Y), g_2(Y), \ldots, g_k(Y)$ be k functions of the random variable Y. Then
>
> $$E[g_1(Y) + g_2(Y) + \cdots + g_k(Y)]$$
> $$= E[g_1(Y)] + E[g_2(Y)] + \cdots + E[g_k(Y)]$$

Proof   We will demonstrate the proof only for the case $k = 2$, but analogous steps will hold for any finite k. By Theorem 3.2,

$$E[g_1(Y) + g_2(Y)] = \sum_y [g_1(y) + g_2(y)]p(y)$$
$$= \sum_y g_1(y)p(y) + \sum_y g_2(y)p(y)$$
$$= E[g_1(Y)] + E[g_2(Y)]$$

Theorems 3.3, 3.4, and 3.5 can be used immediately to develop a theorem useful in finding the variance of a discrete random variable.

Theorem 3.6

> $$V(Y) = \sigma^2 = E[(Y - \mu)^2] = E(Y^2) - \mu^2$$

Proof
$$\sigma^2 = E[(Y - \mu)^2] = E(Y^2 - 2\mu Y + \mu^2)$$
$$= E(Y^2) - E(2\mu Y) + E(\mu^2) \quad \text{(by Theorem 3.5)}$$

Noting that μ is a constant and applying Theorems 3.4 and 3.3 to the second and third terms, respectively, we have

$$\sigma^2 = E(Y^2) - 2\mu E(Y) + \mu^2$$

But $\mu = E(Y)$ and therefore

$$\sigma^2 = E(Y^2) - 2\mu^2 + \mu^2 = E(Y^2) - \mu^2$$

Theorem 3.6 greatly reduces the labor in finding the variance of a discrete random variable. We will demonstrate the usefulness of this result in an example.

EXAMPLE 3.3   Use Theorem 3.6 to find the variance of the random variable Y in Example 3.2.

Solution   The mean, $\mu = 1.75$, was found in Example 3.3. Then

$$E(Y^2) = \sum_y y^2 p(y) = (0)^2(1/8) + (1)^2(1/4) + (2)^2(3/8) + (3)^2(1/4) = 4$$

Also, by Theorem 3.6,

$$\sigma^2 = E(Y^2) - \mu^2 = 4 - (1.75)^2 = .9375$$

EXAMPLE 3.4   The manager of an industrial plant is planning to buy a new machine of either type A or type B. The number of daily repairs Y_1 required to maintain a machine of type A is a random variable with mean and variance both equal to $.10t$, where t denotes the number of hours of daily operation. The number of daily repairs Y_2 for a machine of type B is a random variable with mean and variance both equal to $.12t$. The daily cost of operating A is $C_A(t) = 10t + 30Y_1^2$; for B it is $C_B(t) = 8t + 30Y_2^2$. Assume that the repairs take negligible time and that each night the machines are tuned so that they operate essentially like new machines at the start of the next day. Which machine minimizes the expected daily cost if a workday consists of (a) 10 hours and (b) 20 hours?

Solution   The expected daily cost for A is

$$\begin{aligned} E[C_A(t)] &= E[10t + 30Y_1^2] = 10t + 30E(Y_1^2) \\ &= 10t + 30\{V(Y_1) + [E(Y_1)]^2\} = 10t + 30[.10t + (.10t)^2] \\ &= 13t + .3t^2 \end{aligned}$$

Similarly,

$$E[C_B(t)] = E[8t + 30Y_2^2] = 8t + 30E(Y_2^2)$$
$$= 8t + 30\{V(Y_2) + [E(Y_2)]^2\} = 8t + 30[.12t + (.12t)^2]$$
$$= 11.6t + .432t^2$$

Thus for part (a),

$$E[C_A(10)] = 160 \qquad \text{and} \qquad E[C_B(10)] = 159.2$$

which results in the choice of machine B.
    For part (b)

$$E[C_A(20)] = 380 \qquad \text{and} \qquad E[C_B(20)] = 404.8$$

In conclusion, B is more economical for short time periods because of its smaller hourly operating cost. However, for long time periods A is more economical because it tends to be repaired less frequently.   ∽

The purpose of this section was to introduce the concept of an expected value and to develop some useful theorems for finding means and variances of random variables or functions of random variables. In the following sections, we present some specific types of discrete random variables along with formulas for their probability distributions and means and variances. As you will see, actually deriving some of these expected values requires skill in the summation of algebraic series and knowledge of a few tricks. We will illustrate some of these tricks in some of the derivations in the next sections.

Exercises

3.9   Let Y be a random variable distributed as shown in the accompanying table. Find $E(Y)$, $E(1/Y)$, $E(Y^2 - 1)$, and $V(Y)$.

y	$p(y)$
1	.4
2	.3
3	.2
4	.1

3.10   A manufacturing company ships its product in two different sizes of truck trailers, an $8 \times 10 \times 30$ and an $8 \times 10 \times 40$. If 30% of its shipments are made by using the 30-foot trailer and 70% by using the 40-foot trailer, find the mean volume shipped per trailer load (assume that the trailers are always full).

3.11   The number N of residential homes that a fire company can serve depends on the distance r (in city blocks) that a fire engine can cover in a specified (fixed) period of time. If we assume that N is proportional to the area of a circle r blocks from the firehouse, then

$$N = C\pi r^2,$$

where C is a constant, $\pi = 3.1416\ldots$, and r, a random variable, is the number of blocks that a fire engine can move in the specified time interval. For a particular fire company, $C = 8$, the probability distribution for r is as shown in the accompanying table, and $p(r) = 0$ for $r \le 20$ and $r \ge 27$.

r	21	22	23	24	25	26
$p(r)$	.05	.20	.30	.25	.15	.05

Find the expected value of N, the number of homes that the fire department can serve.

3.12   A single fair die is tossed once. Let Y be the number facing up. Find the expected value and variance of Y.

3.13   In a gambling game a person who draws a jack or a queen is paid $15 and $5 for drawing a king or an ace from an ordinary deck of fifty-two playing cards. A person who draws any other card pays $4. If a person plays the game, what is the expected gain?

3.14   Approximately 10% of the glass bottles coming off a production line have serious flaws in the glass. If two bottles are randomly selected, find the mean and variance of the number of bottles that have serious flaws.

3.15   Two construction contracts are to be randomly assigned to one or more of three firms: I, II, and III. Any firm may receive more than one contract. If each contract will yield a profit of $90,000 for the firm, find the expected profit for firm I. If firms I and II are actually owned by the same individual, what is the owner's expected total profit?

*3.16   A heavy-equipment salesperson can contact either one or two customers per day with probability 1/3 and 2/3, respectively. Each contact will result in either no sale or a $50,000 sale, with the probabilities .9 and .1, respectively. Give the probability distribution for daily sales. Find the mean and standard deviation of the daily sales.

3.17   A potential customer for a $20,000 fire insurance policy possesses a home in an area that, according to experience, may sustain a total loss in a given year with probability of .001 and a 50% loss with probability .01. Ignoring all other partial losses, what premium should the insurance company charge for a yearly policy in order to break even on all $20,000 policies of this type?

3.18   Refer to Exercise 3.2. If the cost of testing a component is $2 and the cost of repairing a defective is $4, find the expected total cost for testing and repairing the lot.

3.19   Let Y be a discrete random variable with mean μ and variance σ^2. If a and b are constants, use Theorems 3.3 through 3.6 to prove that

(a) $E(aY + b) = aE(Y) + b = a\mu + b$

(b) $V(aY + b) = a^2 V(Y) = a^2 \sigma^2$

---

* Exercises preceded by an asterisk are optional.

3.20 The manager of a stockroom in a factory has constructed the following probability distribution for the daily demand (number of times used) for a particular tool.

y	0	1	2
$p(y)$	.1	.5	.4

It costs the factory $10 each time the tool is used. Find the mean and variance of the daily costs for use of the tool.

3.4 The Binomial Probability Distribution

Some experiments consist of the observation of a sequence of identical and independent trials, each of which can result in one of two outcomes. Each item leaving a manufacturing production line is either defective or nondefective. Each shot in a sequence of firings at a target can result in a hit or a miss, and each of n persons questioned prior to a local election will either favor candidate Jones or not. In this section we are concerned with experiments, known as *binomial experiments*, that exhibit the following characteristics.

Definition 3.6

A *binomial experiment* is one that possesses the following properties:

1. The experiment consists of n identical trials.
2. Each trial results in one of two outcomes. We will call one outcome a success S and the other a failure F.
3. The probability of success on a single trial is equal to p and remains the same from trial to trial. The probability of a failure is equal to $(1 - p) = q$.
4. The trials are independent.
5. The random variable of interest is Y, the number of successes observed during the n trials.

Determination of whether a particular experiment is a binomial experiment can only be accomplished by examining the experiment for each of the characteristics just listed. Note that the random variable of interest is the number of successes observed in the n trials. It is important to note at an early stage that a success is not necessarily good in the everyday sense of the word. In our discussions success is merely a name for one of the two possible outcomes on a single trial of an experiment.

EXAMPLE 3.5   An early warning detection system for aircraft consists of four identical radar units operating independently of one another. Suppose that each has a probability of .95 of detecting an intruding aircraft. When an intruding aircraft enters the scene, the random variable of interest is Y, the number of radar units that do not detect the plane. Is this a binomial experiment?

Solution   To decide whether this is a binomial experiment, we must determine whether each of the five requirements in Definition 3.6 is met. Note that the random variable of interest is Y, the number of radar units that do not detect an aircraft. Because the random variable of interest in a binomial experiment is always the number of successes, the present experiment can be binomial only if we call the event *do not detect* a success. We now examine the experiment for the five characteristics of the binomial experiment.

1. The experiment consists of four trials. One trial represents determining whether a particular radar unit detects the aircraft.

2. Each trial results in one of two outcomes. For previously established reasons, S denotes that the aircraft was not detected, F that it was detected.

3. Because all the radar units detect aircraft with equal probability, the probability of an S on each trial is the same, and $p = P(S) = P(\text{do not detect}) = .05$.

4. The trials are independent because the units operate independently.

5. The random variable of interest is Y, the number of successes in four trials.

Thus the experiment is a binomial experiment, with $n = 4$, $p = .05$, and $q = 1 - .05 = .95$.   ∽

EXAMPLE 3.6   Suppose that 40% of a large population of registered voters favor candidate Jones. A random sample of $n = 10$ voters will be selected, and Y, the number favoring Jones, is to be observed. Does this experiment meet the requirements of a binomial experiment?

Solution   If each of the ten people is selected at random from the population, then we have ten nearly identical trials, with each trial resulting in a person either favoring Jones (S) or not favoring Jones (F). The random variable of interest is then the number of successes in the ten trials. For the first person selected the probability of favoring Jones (S) is .4. But what can be said about the probability that the second person will favor Jones? If the population of voters is large, removal of one person will not substantially change the fraction of voters favoring Jones and the probability that the second person favors Jones will be very close to .4. In general, if the population is large and the sample size is small relative to the population size, the probability of success will stay approximately the same from trial to trial regardless of the outcomes on previous trials, and the trials will be approximately independent. Thus sampling problems of this type are approximately binomial.

                                                                                 $\backsim$

    If the sample size is Example 3.6 was large relative to the population size (say 20% of the population), the probability of selecting a supporter of Jones on a later selection would be significantly altered by the preferences of persons selected earlier in the experiment, and the experiment would not be binomial. The hypergeometric probability distribution, the topic of Section 3.7, is the appropriate probability model to be used when the sample size is large relative to the population size.

    You may wish to refine your ability to identify binomial experiments by reexamining the exercises at the end of Chapter 2. Several of the experiments in those exercises are binomial or approximately binomial experiments.

    The binomial probability distribution $p(y)$ can be derived by applying the sample-point approach in finding the probability that the experiment yields y successes. Each sample point in the sample space can be characterized by an n-tuple involving the letters S and F, corresponding to success and failure. A typical sample point would thus appear as

$$SSFSFFFSFS \ldots FS$$

where the letter in the ith position (proceeding from left to right) indicates the outcome of the ith trial.

    Now let us consider a typical sample point implying y successes and hence contained in the numerical event $Y = y$. This sample point,

$$\underbrace{SSSSS \ldots SSS}_{y}\underbrace{FFF \ldots FF}_{n-y}$$

is the intersection of n *independent* events, y successes, and $(n - y)$ failures, and hence its probability is

$$ppppp \cdots pppqqq \cdots qq = p^{y}q^{n-y}$$

Every other sample point in the event $Y = y$ will appear as a rearrangement of the S's and F's in the sample point described earlier and therefore will contain y S's and $n - y$ F's and be assigned the same probability. Since the number of distinct arrangements of the y S's and $(n - y)$ F's is (from Theorem 2.3)

$$\frac{n!}{y!(n - y)!} = \binom{n}{y}$$

it follows that $p(y) = \binom{n}{y}p^{y}q^{n-y}$, $y = 0, 1, 2, \ldots, n$. The result that we have just derived is the formula for the *binomial probability distribution*.

Definition 3.7

A random variable Y is said to have a *binomial distribution* based on n trials with success probability p if and only if

$$p(y) = \binom{n}{y} p^y q^{n-y}, \quad y = 0, 1, 2, \ldots, n \text{ and } 0 \leq p \leq 1.$$

Figure 3.4 portrays $p(y)$ graphically as probability histograms, the first for $n = 10, p = .1$; the second for $n = 10, p = .5$; and the third for $n = 20, p = .5$.

Figure 3.4
Binomial probability
histograms

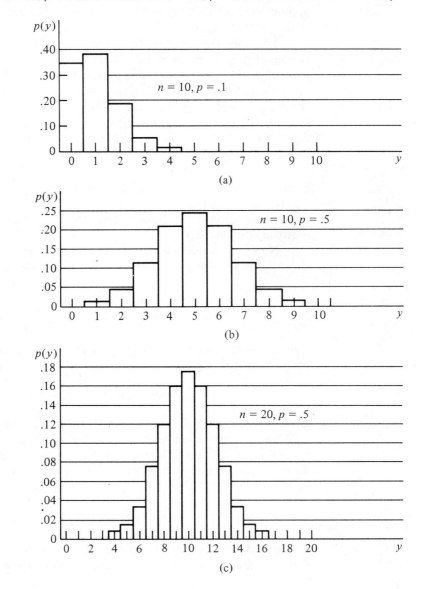

(a)

(b)

(c)

Before we proceed, let us reconsider the representation for the sample points in this experiment. We have seen that a sample point can be represented by a sequence of n letters, each of which is either S or F. If the sample point contains exactly one S, the probability associated with that sample point is pq^{n-1}. If another sample point contains 2 S's (and $n-2$ F's), the probability of this sample point is p^2q^{n-2}. The point to notice is that the sample points for a binomial experiment are not equiprobable unless $p = .5$.

The term *binomial experiment* derives from the fact that the probabilities $p(y)$, $y = 0, 1, 2, \ldots, n$, are terms of the binomial expansion

$$(q + p)^n = \binom{n}{0}q^n + \binom{n}{1}p^1q^{n-1} + \binom{n}{2}p^2q^{n-2} + \cdots + \binom{n}{n}p^n$$

You will observe that $\binom{n}{0}q^n = p(0)$, $\binom{n}{1}pq^{n-1} = p(1)$, and in general, $p(y) = \binom{n}{y}p^yq^{n-y}$. It also follows that $p(y)$ satisfies the necessary properties for a probability function, because $p(y)$ is positive for $y = 0, 1, \ldots, n$ and

$$\sum_y p(y) = \sum_{y=0}^n \binom{n}{y}p^yq^{n-y} = (q + p)^n = 1$$

[because $(q + p) = 1$].

The binomial probability distribution has many applications, because the binomial experiment occurs in sampling for defectives in industrial quality control, in the sampling of consumer preference or voting populations, and in many other physical situations. We will illustrate with a few examples. Other practical examples will appear in the exercises at the end of the chapter.

EXAMPLE 3.7   Suppose that a lot of 300 electrical fuses contains 5% defectives. If a sample of five fuses is tested, find the probability of observing at least one defective.

Solution   It is reasonable to assume that Y, the number of defectives observed, has an approximate binomial distribution because the lot is large. Thus

$$P(\text{at least one defective}) = 1 - p(0) = 1 - \binom{5}{0}q^5$$

$$= 1 - (.95)^5 = 1 - .774 = .226$$

Note that there is a fairly large chance of seeing at least one defective, even though the sample is quite small.   ∽

EXAMPLE 3.8   Experience has shown that 30% of all persons afflicted by a certain illness recover. A drug company has developed a new vaccine. Ten people with the illness were selected at random and injected with the vaccine; nine recovered shortly thereafter. Suppose that the vaccine was absolutely worthless. What is the probability that at least nine of ten injected by the vaccine will recover?

Solution

Let Y denote the number of people who recover. If the vaccine is worthless, the probability that a single ill person will recover is $p = .3$. Then the number of trials is $n = 10$ and the probability of *exactly* nine recoveries is

$$P(Y = 9) = p(9) = \binom{10}{9}(.3)^9(.7) = .000138$$

Similarly, the probability of exactly ten recoveries is

$$P(Y = 10) = p(10) = \binom{10}{10}(.3)^{10}(.7)^0 = .000006$$

Then

$$P(Y \geq 9) = p(9) + p(10) = .000138 + .000006 = .000144$$

If the vaccine is ineffective, the probability of observing at least nine recoveries is extremely small. Either we have observed a very rare event, or the vaccine is indeed very useful in curing the illness. We would adhere to the latter point of view.

A tabulation of binomial probabilities in the form $\sum\limits_{y=0}^{a} p(y)$, presented in Table 1, Appendix III, will greatly reduce the computations associated with some of the exercises. The references at the end of the chapter list several, more extensive tabulations of binomial probabilities. We illustrate the use of Table 1 in the following example.

EXAMPLE 3.9

The large lot of electrical fuses of Example 3.7 is supposed to contain only 5% defectives. If $n = 20$ fuses are randomly sampled from this lot, find the probability that at least three defectives will be observed.

Solution

Letting Y denote the number of defectives in the sample, we assume the binomial model for Y, with $p = .05$. Thus

$$P(Y \geq 3) = 1 - P(Y \leq 2)$$

and using Table 1, Appendix III, we have

$$P(Y \leq 2) = \sum_{y=0}^{2} p(y) = .925$$

It follows that

$$P(Y \geq 3) = 1 - .925 = .075$$

This probability is quite small, which leads us to conclude that if we did indeed observe more than three defectives out of twenty fuses, we might suspect the 5% reported defect rate to be in error.

We have seen that there are many distinct contexts in which we encounter random variables with binomial distributions. The mean and variance associated with a binomial random variable are derived in the following theorem. As you will see in the proof of the theorem, it is necessary to evaluate the sum of some arithmetic series. In the course of the proof we illustrate some of the many techniques that are available for summing such series. In particular, we make use of the fact that $\sum_{y} p(y) = 1$ for any discrete random variable.

Theorem 3.7

> Let Y be a binomial random variable based on n trials and success probability p. Then
>
> $$\mu = E(Y) = np \quad \text{and} \quad \sigma^2 = V(Y) = npq.$$

Proof By Definitions 3.4 and 3.7,

$$E(Y) = \sum_{y} yp(y) = \sum_{y=0}^{n} y \binom{n}{y} p^y q^{n-y}.$$

Note that the first term in the sum is 0 and hence that

$$E(Y) = \sum_{y=1}^{n} y \frac{n!}{(n-y)!\, y!} p^y q^{n-y}$$

$$= \sum_{y=1}^{n} \frac{n!}{(n-y)!(y-1)!} p^y q^{n-y}.$$

The summands in this last expression bear a striking resemblance to binomial probabilities. In fact, if we factor np out of each term in the sum and let $z = y - 1$,

$$E(Y) = np \sum_{y=1}^{n} \frac{(n-1)!}{(n-y)!(y-1)!} p^{y-1} q^{n-y}.$$

$$= np \sum_{z=0}^{n-1} \frac{(n-1)!}{(n-1-z)!\, z!} p^z q^{n-1-z}.$$

$$= np \sum_{z=0}^{n-1} \binom{n-1}{z} p^z q^{n-1-z}.$$

Note that $p(z) = \binom{n-1}{z}p^z q^{n-1-z}$ is the binomial probability function based on $(n-1)$ trials. Thus $\sum_z p(z) = 1$, and it follows that

$$\mu = E(Y) = np.$$

From Theorem 3.6 we know that $\sigma^2 = V(Y) = E(Y^2) - \mu^2$. Thus σ^2 can be calculated if we find $E(Y^2)$. Finding $E(Y^2)$ directly is difficult because

$$E(Y^2) = \sum_{y=0}^{n} y^2 p(y) = \sum_{y=0}^{n} y^2 \binom{n}{y} p^y q^{n-y} = \sum_{y=0}^{n} y^2 \frac{n!}{y!(n-y)!} p^y q^{n-y}$$

and the quantity y^2 does not appear as a factor of $y!$. Where do we go from here? Note that

$$E[Y(Y-1)] = E(Y^2 - Y) = E(Y^2) - E(Y)$$

and therefore

$$E(Y^2) = E[Y(Y-1)] + E(Y) = E[Y(Y-1)] + \mu$$

In this case

$$E[Y(Y-1)] = \sum_{y=0}^{n} y(y-1) \frac{n!}{y!(n-y)!} p^y q^{n-y}$$

The first and second terms of this sum equal zero (when $y = 0$ and $y = 1$). Then

$$E[Y(Y-1)] = \sum_{y=2}^{n} \frac{n!}{(y-2)!(n-y)!} p^y q^{n-y}$$

(Notice the cancellation that led to this last result. The anticipation of this cancellation is what actually motivated the consideration of $E[Y(Y-1)]$.) Again, the summands in the last expression look very much like binomial probabilities. Factor $n(n-1)p^2$ out of each term in the sum and let $z = y - 2$ to obtain

$$E[Y(Y-1] = n(n-1)p^2 \sum_{y=2}^{n} \frac{(n-2)!}{(y-2)!(n-y)!} p^{y-2} q^{n-y}$$

$$= n(n-1)p^2 \sum_{z=0}^{n-2} \frac{(n-2)!}{z!(n-2-z)!} p^z q^{n-2-z}$$

$$= n(n-1)p^2 \sum_{z=0}^{n-2} \binom{n-2}{z} p^z q^{n-2-z}$$

Again note that $p(z) = \binom{n-2}{z}p^z q^{n-2-z}$ is the binomial probability function based on $(n-2)$ trials. Then $\sum_{z=0}^{n-2} p(z) = 1$ (again using the device illustrated in the derivation of the mean) and

$$E[Y(Y-1)] = n(n-1)p^2$$

Thus

$$E(Y^2) = E[Y(Y-1)] + \mu = n(n-1)p^2 + np$$

and

$$\sigma^2 = E(Y^2) - \mu^2 = n(n-1)p^2 + np - n^2 p^2$$
$$= np[(n-1)p + 1 - np] = np(1-p)$$

Thus $\sigma^2 = npq$, where $q = 1 - p$.

In addition to providing formulas for the mean and variance of a binomial random variable, the derivation of Theorem 3.8 illustrates the use of two fairly common tricks. Namely, to use the fact that $\sum p(y) = 1$ if $p(y)$ is a valid probability function, and to find $E(Y^2)$ by finding $E[Y(Y-1)]$. These techniques also will be useful in the next sections, where we consider other discrete probability distributions and the associated means and variances.

A frequent source of error in applying the binomial probability distribution to practical problems is the failure to define which of the two possible results of a trial is the success. As a consequence, q may be used erroneously in place of p. Carefully define a success and make certain that p equals the probability of a success for each application.

Exercises

3.21 The manufacturer of a low-calorie dairy drink wishes to compare the taste appeal of a new formula (formula B) with that of the standard formula (formula A). Each of four judges is given three glasses in random order, two containing formula A and the other containing formula B. Each judge is asked to state which glass she most enjoyed. Suppose that the two formulas are equally attractive. Let Y be the number of judges stating a preference for the new formula.

(a) Find the probability function for Y.

(b) What is the probability that at least three of the four judges state a preference for the new formula?

(c) Find the expected value of Y.

(d) Find the variance of Y.

3.22   A complex electronic system is built with a certain number of backup components in its subsystems. One subsystem has four identical components, each with a probability of .2 of failing in less than 1000 hours. The subsystem will operate if any two of the four components are operating. Assume that the components operate independently.

(a)   Find the probability that exactly two of the four components last longer than 1000 hours.

(b)   Find the probability that the subsystem operates longer than 1000 hours.

3.23   The probability that a patient recovers from a stomach disease is .8. Suppose twenty people are known to have contracted this disease.

(a)   What is the probability that exactly fourteen survive?

(b)   What is the probability that at least ten survive?

(c)   What is the probability that at least fourteen but not more than eighteen survive?

(d)   What is the probability that at most sixteen survive?

3.24   A multiple-choice examination has fifteen questions, each having five possible answers, only one of which is correct. Suppose one of the students who takes the examination answers the questions by guessing. What is the probability that he answers at least ten questions correctly?

3.25   Many employers are finding that some of the people they hire are not who and what they claim to be. Detecting job applicants who falsify their application information has spawned some new businesses: credential checking services. *U.S. News and World Report* (July 13, 1981) reported on this problem, and noted that one service in a 2-month period found that 35% of all credentials examined were falsified. Suppose that you hired five new employees last week and that the probability that a single employee would falsify the information on his or her application form is .35. What is the probability that at least one of the five application forms has been falsified? Two or more?

3.26   Many utility companies have begun to promote energy conservation by offering discount rates to consumers who keep their energy usage below certain established subsidy standards. A recent EPA report notes that 70% of the island residents of Puerto Rico have reduced their electricity usage sufficiently to qualify for discounted rates. If five residential subscribers are randomly selected from San Juan, Puerto Rico, find the probability that:

(a)   all five qualify for the favorable rates,

(b)   at least four qualify for the favorable rates.

3.27   A new surgical procedure is succesful with a probability of p. Assume the operation is performed five times and that the results are independent of one another.

(a)   What is the probability that all five operations are successful if $p = .8$?

(b)   What is the probability that exactly four are successful if $p = .6$?

(c)   What is the probability that less than two are successful if $p = .3$?

3.28   A fire-detection device utilizes three temperature-sensitive cells acting independently of each other in such a manner that any one or more may actuate the alarm. Each cell possesses a probability of $p = .8$ of actuating the alarm when the temperature reaches 100 degrees Celsius or more. Let Y equal the number of cells actuating the alarm when the temperature reaches 100 degrees. Find the probability distribution for Y. Find the probability that the alarm will function when the temperature reaches 100 degrees.

3.29   Construct probability histograms for the binomial probability distributions for $n = 5$, $p = .1, .5,$ and $.9$. (Table 1, Appendix III, will reduce the amount of calculation.) Note the symmetry for $p = .5$ and the direction of skewness for $p = .1$ and $.9$.

3.30   Use Table 1, Appendix III, to construct a probability histogram for the binomial probability distribution for $n = 20$ and $p = .5$. Note that almost all the probability falls in the interval $5 \le y \le 15$.

3.31   A missile protection system consists of n radar sets operating independently, each with a probability of .9 of detecting a missile entering a zone that is covered by all of the units.

   (a)   If $n = 5$ and a missile enters the zone, what is the probability that exactly four sets detect the missile? At least one set?

   (b)   How large must n be if we require that the probability of detecting a missile that enters the zone be .999?

3.32   A manufacturer of floor wax has developed two new brands, A and B, which she wishes to subject to a housewife evaluation to determine which of the two is superior. Both waxes, A and B, are applied to floor surfaces in each of fifteen homes. Assume that there is actually no difference in the quality of the brands.

   (a)   What is the probability that ten or more housewives would state a preference for brand A?

   (b)   What is the probability that ten or more housewives would state a preference for either brand A or brand B?

3.33   An oil exploration firm is formed with enough capital to finance ten explorations. The probability of a particular exploration being successful is .1. Assume the explorations are independent. Find the mean and variance of the number of successful explorations.

3.34   Refer to Exercise 3.33. Suppose the firm has a fixed cost of $20,000 in preparing equipment prior to doing its first exploration. If each successful exploration costs $30,000 and each unsuccessful exploration costs $15,000, find the expected total cost to the firm of its ten explorations.

3.35   A particular sale involves four items randomly selected from a large lot that is known to contain 10% defectives. Let Y denote the number of defectives among the four sold. The purchaser of the items will return the defectives for repair, and the repair cost is given by $C = 3Y^2 + Y + 2$. Find the expected repair cost. [Hint: The result of Theorem 3.6 implies that for any random variable Y, $E(Y^2) = \sigma^2 + \mu^2$.]

3.36   Ten motors are packaged for sale in a certain warehouse. The motors sell for $100 each, but a double-your-money-back guarantee is in effect for any defectives the purchaser may receive. Find the expected net gain for the seller if the probability of any one motor being defective is 0.08. (Assume the quality of any one motor is independent of that of the others.)

3.37   A particular concentration of a chemical found in polluted water has been found to be lethal to 20% of the fish that are exposed to the concentration for 24 hours. If twenty fish are placed in a tank containing this concentration of chemical in water.

   (a)   Find the probability that exactly fourteen survive.

   (b)   Find the probability that at least ten survive.

   (c)   Find the probability that at most sixteen survive.

   (d)   Find the mean and variance of the number that survive.

3.38   Of the volunteers donating blood in a clinic, 80% have the Rhesus factor present in their blood.

(a)   If the five of the volunteers are randomly selected, what is the probability that at least one does not have the Rhesus factor?

(b)   If five are randomly selected, what is the probability that at most four have the Rhesus factor?

(c)   What is the smallest number of volunteers who must be selected if we want to be at least 90% certain that we obtain at least five donors with the Rhesus factor?

3.39   Goranson and Hall [2] explain that the probability of detecting a crack in an airplane wing is the product of p_1, the probability of inspecting a plane with a wing crack; p_2, the probability of inspecting the detail in which the crack is located; and p_3, the probability of detecting the damage.

(a)   What assumptions justify the multiplication of these probabilities?

(b)   Suppose $p_1 = 0.9$, $p_2 = 0.8$, and $p_3 = 0.5$ for a certain fleet of planes. If three planes are inspected from this fleet, find the probability that a wing crack will be detected in at least one of them.

3.5 The Geometric Probability Distribution

The random variable having the geometric distribution is defined for an experiment that is very similar to the binomial experiment. It also is concerned with identical and independent trials, each of which can result in one of two outcomes, success or failure. The probability of success is equal to p and is constant from trial to trial. However, instead of the number of successes that occur in n trials, the geometric random variable Y is the number of the trial on which the first success occurs. Thus the experiment consists of a series of trials that concludes with the first success. Consequently, the experiment could end with the first trial if a success is observed or it could go on indefinitely.

The sample space S for the experiment contains the countably infinite set of sample points.

E_1: S (success on first trial)

E_2: FS (failure on first, success on second)

E_3: FFS (failure on first and second, success on third)

E_4: $FFFS$

$\vdots$

E_k: $\underbrace{FFFF\ldots F}_{k-1}S$

Because the random variable Y is the number of trials up to and including the first success, $Y = 1$, $Y = 2$, and $Y = 3$ will contain E_1, E_2, and E_3, respectively, and, in general, the numerical event $Y = y$ will contain only E_y. Therefore,

$$p(y) = P(E_y) = P(\underbrace{FFFF \ldots F}_{y-1}S)$$

The probability of this intersection of y independent events gives the geometric probability distribution.

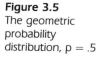

Definition 3.8

A random variable Y is said to have a *geometric probability distribution* if and only if

$$p(y) = q^{y-1}p, \qquad y = 1, 2, 3, \ldots, 0 \le p \le 1.$$

A probability histogram for $p(y)$, $p = .5$, is shown in Figure 3.5. Areas over intervals correspond to probabilities, as they did for the frequency distributions of data in Chapter 1, except that you must keep in mind that Y can assume only discrete values, $y = 1, 2, \ldots, \infty$.

Figure 3.5
The geometric
probability
distribution, p = .5

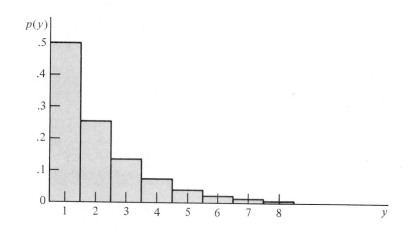

The geometric probability distribution is frequently used to model distributions of lengths of waiting times. For example, suppose that a commercial aircraft engine is serviced periodically so that its various parts are replaced at different points in time and hence are of varying ages. Then it might be reasonable to assume that the probability of engine malfunction, p, during any 1-hour interval of operation is the same as for any other 1-hour interval. The length of time to engine malfunction is the number of 1-hour intervals, Y, until the first malfunction. (For this application engine malfunction, or failure, in a given 1-hour period is defined to be a success. Note that, as in the case of the binomial experiment, either of the two outcomes of a trial can be defined as a success and that this event need not imply success as used in everyday conversation.)

EXAMPLE 3.10   Suppose that the probability of engine malfunction during any 1-hour period is $p = .02$. Find the probability that a given engine will survive 2 hours.

Solution   Letting Y denote the number of 1-hour intervals until the first malfunction, we have

$$P(\text{survive 2 hours}) = P(Y \geq 3) = \sum_{y=3}^{\infty} p(y)$$

Because $\sum_{y=1}^{\infty} p(y) = 1$,

$$P(\text{survive 2 hours}) = 1 - \sum_{y=1}^{2} p(y)$$

$$= 1 - p - qp = 1 - .02 - (.98)(.02) = .9604 \qquad \text{\reflectbox{$\wp$}}$$

If you examine the formula for the geometric distribution given in Definition 3.8, you will see that larger values of p (and hence the smaller the value of q) lead to higher probabilities for the smaller values of Y. Thus it seems that the mean value of Y will be inversely proportional to p. As we show in the next theorem, the mean of a random variable with a geometric distribution is actually equal to $1/p$.

Theorem 3.8

> If Y is a random variable with a geometric distribution
>
> $$\mu = E(Y) = \frac{1}{p} \quad \text{and} \quad \sigma^2 = V(Y) = \frac{1-p}{p^2}.$$

Proof $$E(Y) = \sum_{y=1}^{\infty} yq^{y-1}p = p \sum_{y=1}^{\infty} yq^{y-1}$$

This series is difficult to sum directly but can be easily summed by noting that

$$\frac{d}{dq}(q^y) = yq^{y-1}$$

and hence

$$\frac{d}{dq}\left(\sum_{y=1}^{\infty} q^y\right) = \sum_{y=1}^{\infty} yq^{y-1}$$

(The interchanging of derivative and sum here can be justified.) Substituting, we obtain

$$E(Y) = p \sum_{y=1}^{\infty} yq^{y-1} = p \frac{d}{dq}\left(\sum_{y=1}^{\infty} q^y\right)$$

This latter sum is the summation of the terms of an infinite geometric progression, $q + q^2 + q^3 + \cdots$, which is equal to $q/(1-q)$ (given in most mathematical handbooks). Therefore,

$$E(Y) = p \frac{d}{dq}\left(\frac{q}{1-q}\right) = p \frac{1}{(1-q)^2} = \frac{p}{p^2} = \frac{1}{p}$$

To summarize, the device is to express a series that cannot be summed directly as the derivative (or integral) of a series for which the sum can be readily obtained. Once summed, differentiate (or integrate) to complete the process.

The derivation of the variance is left as an exercise. See Exercise 3.53.

EXAMPLE 3.11   If the probability of engine malfunction during any 1-hour period is $p = .02$ and Y denotes the number of 1-hour intervals until the first malfunction, find the mean and standard deviation of Y.

Solution   As in Example 3.10, it follows that Y has a geometric distribution with $p = .02$. Thus $E(Y) = 1/p = 1/(.02) = 50$ and we expect to wait quite a few hours before encountering a malfunction. Because $V(Y) = \sigma^2 = (1-p)/p^2 = .98/.0004 = 2450$, it follows that the standard deviation of Y is $\sigma = \sqrt{2450} = 49.497$.

Exercises

3.40   Suppose that 30% of the applicants for a certain industrial job have advanced training in computer programming. Applicants are interviewed sequentially and are selected at random from the pool. Find the probability that the first applicant having advanced training in programming is found on the fifth interview.

3.41   Refer to Exercise 3.40. What is the expected number of applicants who need to be interviewed in order to find the first one with advanced training?

3.42   An oil prospector will drill a succession of holes in a given area to find a productive well. The probability that he is successful on a given trial is .2.

(a)   What is the probability that the third hole drilled is the first that yields a productive well?

(b)   If the prospector can only afford to drill at most ten wells, what is the probability that he fails to find a productive well?

3.43   Let Y denote a geometric random variable with probability of success p.

(a)   Show that for a positive integer a,

$$P(Y > a) = q^a$$

(b)   Show that for positive integers a and b,

$$P(Y > a + b \mid Y > a) = q^b = P(Y > b)$$

3.44   A certified public accountant (CPA) has found that nine of ten company audits contain substantial errors. If the CPA audits a series of company accounts, what is the probability that

(a)   the first account containing substantial errors is the third one to be audited?

(b)   the first account containing substantial errors will occur on or after the third audited account?

3.45   Refer to Exercise 3.44. What are the mean and standard deviation of the number of accounts that must be examined to find the first one with substantial errors?

3.46   Given that we have already tossed a balanced coin ten times and obtained zero heads, what is the probability that we must toss it at least two more times to obtain the first head?

3.47   The probability of a customer arrival at a grocery service counter in any 1 second is equal to .1. Assume that customers arrive in a random stream and hence that the arrival any 1 second is independent of any other.

(a)   Find the probability that the first arrival will occur during the third 1-second interval.

(b)   Find the probability that the first arrival will not occur until at least the third 1-second interval.

3.48   Of a population of consumers, 60% is reputed to prefer a particular brand, A, of toothpaste. If a group of consumers is interviewed, what is the probability that exactly five people have to be interviewed to encounter the first consumer who prefers brand A? At least five people?

3.49   In responding to a survey question that concerns a sensitive topic (such as, "Have you ever tried marijuana?"), many times people will not want to respond in the affirmative. Derive the probability distribution of Y, the number of people you would need to question in order to obtain a single affirmative response, if 80% of the population will truthfully answer "no" to your question and if, of those 20% who should truthfully answer "yes," 70% will lie.

3.50 Two people take turns tossing a fair die until one of them tosses a 6. That is, person A tosses first, B second, A third, and so on. Given that person B throws the first 6, what is the probability that B obtained the first 6 on her second toss (that is, the 4th toss overall)?

3.51 How many times would you expect to toss a balanced coin to obtain the first head?

3.52 Refer to Exercise 3.42. How many holes would the prospector expect to drill before he finds a productive well? Give your answer in intuitive interpretation.

*3.53 Find $E[Y(Y-1)]$ for a geometric random variable Y by finding $d^2/dq^2 \left(\sum_{y=1}^{\infty} q^y \right)$. Use this result to find the variance of Y.

3.6 The Negative Binomial Probability Distribution (Optional)

A random variable with a negative binomial distribution originates from a context that is very similar to the one that leads to the geometric distribution. Again, we focus on independent and identical trials, each of which results on one of two outcomes, success or failure. The probability of success is p and stays the same from trial to trial. The geometric distribution handles the case where we are interested in the number of the trial on which the first success occurs. What if we are interested in the number of the trial on which the second, third, or fourth success occurs? The distribution that applies to the random variable, Y, equal to the number of the trial on which the rth success occurs ($r = 2, 3, 4$, etc.) is the negative binomial distribution.

The following steps are very similar to those in the previous section. Let us select fixed values for r and y and consider events A and B where

$$A = \{\text{the first } (y-1) \text{ trials contain } (r-1) \text{ successes}\}$$

and

$$B = \{\text{trial } y \text{ results in a success}\}.$$

Because we assume that the trials are independent, it follows that A and B are independent events and previous assumptions imply that $P(B) = p$. Therefore,

$$p(y) = p(Y = y) = P(A \cap B) = P(A) \times P(B).$$

Note that $P(A)$ is 0 if $(y-1) < (r-1)$, or equivalently if $y < r$. If $y \geq r$, our previous work with the binomial distribution implies that

$$P(A) = \binom{y-1}{r-1} p^{r-1} q^{y-r}$$

Finally,

$$p(y) = \binom{y-1}{r-1} p^r q^{y-r}, \qquad y = r, \qquad r+1, \qquad r+2, \dots$$

Definition 3.9

> A random variable Y is said to have a *negative binomial probability distribution* if and only if
>
> $$p(y) = \binom{y-1}{r-1} p^r q^{y-r}, \quad y = r, \quad r+1, \quad r+2, \ldots, \quad 0 \le p \le 1.$$

EXAMPLE 3.12   A geological study indicates that an exploratory oil well drilled in a particular region should strike oil with probability 0.2. Find the probability that the third oil strike comes on the fifth well drilled.

Solution   Assuming independent drillings with probability 0.2 of striking oil with any one well, let Y denote the number of the trial on which the third oil strike occurs. Then it is reasonable to assume that Y has a negative binomial distribution with $p = 0.2$. Thus

$$P(Y = 5) = p(5) = \binom{4}{2}(0.2)^3(0.8)^2$$

$$= 6(0.008)(0.64) = .0307$$

The mean and variance of a random variable with a negative binomial distribution can be derived directly from Definitions 3.4 and 3.5, using techniques like those previously illustrated. However, summing the resulting infinite series is very tedious. The derivation of the mean and variance of a negative binomial random variable will be much easier after we have developed some of the techniques of Chapter 5. For now, we state the following theorem without proof.

Theorem 3.9

> If Y is a random variable with a negative binomial distribution
>
> $$\mu = E(Y) = \frac{r}{p} \quad \text{and} \quad \sigma^2 = V(Y) = \frac{r(1-p)}{p^2}.$$

EXAMPLE 3.13   A large stockpile of used pumps contains 20% that are in need of repair. A maintenance worker is sent to the stockpile with three repair kits. She selects pumps at random and tests them one at a time. If the pump works, she sets it aside for future use. However, if the pump does not work, she uses one of her repair kits on it. Suppose that it takes 10 minutes to test a pump if it is in working condition and 30 minutes to test and repair a pump that does not work. Find the mean and variance of the total time it takes the maintenance worker to use her 3 repair kits.

Solution   Let Y denote the number of the trial on which the third nonfunctioning pump is found. It follows that Y has a negative binomial distribution with $p = 0.2$. Thus $E(Y) = 3/(0.2) = 15$ and $V(Y) = 3(0.8)/(0.2)^2 = 60$. Since it takes an additional 20 minutes to repair each defective pump, the total time necessary to use the three kits is

$$T = 10Y + 3(20).$$

Using the result derived in Exercise 3.19, it follows that

$$E(T) = 10E(Y) + 60$$
$$= 10(15) + 60 = 210$$

and

$$V(T) = 10^2 V(Y)$$
$$= 100(60) = 6000.$$

Thus, it follows that the total time necessary to use all three kits has mean 210 and standard deviation $\sqrt{6000} = 77.46$.

Exercises

3.54   The employees of a firm that manufactures insulation are being tested for indications of asbestos in their lungs. The firm is requested to send three employees who have positive indications of asbestos on to a medical center for further testing. If 40% of the employees have positive indications of asbestos in their lungs, find the probability that ten employees must be tested in order to find three positives.

3.55   Refer to Exercise 3.54. If each test costs $20, find the expected value and variance of the total cost of conducting the tests necessary to locate the three positives.

3.56   Ten percent of the engines manufactured on an assembly line are defective. If engines are randomly selected one at a time and tested, what is the probability that the first nondefective engine is found on the second trial?

3.57   Refer to Exercise 3.56. What is the probability that the third nondefective engine is found

(a)   on the fifth trial?

(b)   on or before the fifth trial?

3.58   Refer to Exercise 3.56. Find the mean and variance of the number of the trial on which

(a)   the first nondefective engine is found,

(b)   the third nondefective engine is found.

3.59   Refer to Exercise 3.56. Given that the first two engines tested are defective, what is the probability that at least two more engines must be tested before the first nondefective is found?

3.60 The telephone lines serving an airline reservation office all are busy about 60% of the time.

    (a) If you are calling this office, what is the probability that you complete your call on the first try? the second try? the third try?

    (b) If you and a friend must both complete calls to this office, what is the probability that it takes a total of four tries for both of you to get through?

3.61 A geological study indicates that an exploratory oil well should strike oil with probability 0.2. What is the probability that

    (a) the first strike comes on the third well drilled?

    (b) the third strike comes on the fifth well drilled?

    (c) What assumptions did you make to obtain the answers to parts (a) and (b)?

    (d) Find the mean and variance of the number of wells that must be drilled if the company wants to set up three producing wells.

3.7 The Hypergeometric Probability Distribution

In Example 3.6 we considered a population of voters, 40% of whom favored candidate Jones. A sample of voters was selected and Y, the number favoring Jones, was to be observed. If the sample size n is small relative to the population size N, we concluded that Y would possess, approximately, a binomial distribution. When n is large relative to N, we determined that the probability of selecting a supporter of Jones on a later draw will be significantly affected by the observed preferences of persons selected on earlier draws and that the probability distribution for Y could not be adequately approximated by a binomial probability distribution. Consequently, we need to develop the probability distribution for Y where n is large relative to N.

Suppose that a population contains a finite number N of elements that possess one of two characteristics. Thus r of the elements might be red and $b = N - r$ black. A sample of n elements is randomly selected from the population and the random variable of interest is Y, the number of red elements in the sample. This random variable has what is known as the *hypergeometric probability distribution*. For example, the number of women workers Y in Example 3.1 has the hypergeometric distribution.

The hypergeometric probability distribution can be derived by using the combinatorial theorems, Section 2.6, and the sample-point approach. A sample point in S will correspond to a unique selection of n elements, some red and the remainder black. As in the binomial experiment, each sample point can be characterized by an n-tuple whose elements correspond to a selection of n elements from the total of N. If each element in the population were numbered from 1 to N, the sample point indicating the selection of items 5, 7, 8, 64, 17, ..., 87 would appear as the n-tuple

$$(5, 7, 8, 64, 17, \ldots, 87)$$

The total number of sample points in S will therefore equal the number of ways of selecting a subset of n elements from a population of N, or $\binom{N}{n}$. Because random selection implies that all sample points are equiprobable, the probability of a sample point in S will equal

$$P(E_i) = \frac{1}{\binom{N}{n}}, \quad \text{all } E_i \subset S$$

The total number of sample points in the numerical event $Y = y$ would be the number of sample points in S that contain y red and $(n - y)$ black elements. This number can be obtained by applying the mn rule (Section 2.6). The number of ways of selecting y red elements to fill y positions in the n-tuple representing a sample point is the number of ways of selecting y from a total of r, or $\binom{r}{y}$. [We use the convention $\binom{a}{b} = 0$ if $b > a$.] The total number of ways of selecting $(n - y)$ black elements to fill the $(n - y)$ positions in the n-tuple is the number of ways of selecting $(n - y)$ black elements from a possible $N - r$, or $\binom{N-r}{n-y}$. Then the number of sample points in the numerical event $Y = y$ is the number of ways of combining a set of y red and $(n - y)$ black elements. By the mn rule, this would be the product $\binom{r}{y} \cdot \binom{N-r}{n-y}$. Summing the probabilities of the sample points in the numerical event $Y = y$ (multiplying the number of sample points by the common probability per sample point), we obtain the hypergeometric probability function.

Definition 3.10

> A random variable Y is said to have a *hypergeometric probability distribution* if and only if
>
> $$p(y) = \frac{\binom{r}{y}\binom{N-r}{n-y}}{\binom{N}{n}} \quad \begin{array}{l} y \text{ integer } 0, 1, 2, \ldots, n \\ \text{subject to restrictions} \\ \\ y \leq r \\ n - y \leq N - r \end{array}$$

EXAMPLE 3.14   An important problem encountered by personnel directors and others faced with the selection of the best in a finite set of elements is indicated by the following situation. From a group of twenty PhD engineers, ten are selected for employment. What is the probability that the ten selected include all the five best engineers in the group of twenty?

Solution   For this example $N = 20$, $n = 10$, and $r = 5$. That is, there are only 5 in the set of 5 best engineers, and we seek the probability that $Y = 5$, where Y denotes the number of best engineers among the ten selected. Then

$$p(5) = \frac{\binom{5}{5}\binom{15}{5}}{\binom{20}{10}} = \left(\frac{15!}{5!10!}\right)\left(\frac{10!10!}{20!}\right) = \frac{21}{1292} = .0162$$

The mean and variance of a random variable with a hypergeometric distribution can be derived directly from Definitions 3.4 and 3.5. However, deriving closed form expressions for the resulting summations is very tedious. In Chapter 5 we will develop methods that permit a much simpler derivation of the results presented in the following theorem.

Theorem 3.10

> If Y is a random variable with a hypergeometric distribution (see Definition 3.10),
>
> $$\mu = E(Y) = \frac{nr}{N} \quad \text{and} \quad \sigma^2 = V(Y) = n\left(\frac{r}{N}\right)\left(\frac{N-r}{N}\right)\left(\frac{N-n}{N-1}\right)$$

EXAMPLE 3.15   A particular industrial product is shipped in lots of twenty. Testing to determine whether an item is defective is costly, and hence the manufacturer samples his production rather than using a 100% inspection plan. A sampling plan constructed to minimize the number of defectives shipped to customers calls for sampling five items from each lot and rejecting the lot if more than one defective is observed. (If rejected, each item in the lot is tested.) If a lot contains four defectives, what is the probability that it will be rejected? What is the expected number of defectives in the sample of size 5? What is the variance of the number of defectives in the sample of size 5?

Solution   Let Y equal the number of defectives in the sample. Then $N = 20$, $r = 4$, and $n = 5$. The lot will be rejected if $Y = 2, 3,$ or 4. Then

$$P(\text{rejecting the lot}) = P(Y \geq 2) = p(2) + p(3) + p(4)$$

$$= 1 - p(0) - p(1)$$

$$= 1 - \frac{\binom{4}{0}\binom{16}{5}}{\binom{20}{5}} - \frac{\binom{4}{1}\binom{16}{4}}{\binom{20}{5}}$$

$$= 1 - .2817 - .4696 = .2487$$

The mean and variance of the number of defectives in the sample of size 5 are

$$\mu = \frac{(5)(4)}{20} = 1 \quad \text{and} \quad \sigma^2 = 5\left(\frac{4}{20}\right)\left(\frac{20-4}{20}\right)\left(\frac{20-5}{20-1}\right) = .632.$$

Example 3.15 involves sampling a lot of N industrial products of which r are defective. The random variable of interest is Y, the number of defectives in a sample of size n. As noted in the beginning of this section, Y possesses approximately a binomial distribution when N is large and n is relatively small. Consequently, we would expect the probabilities assigned to values of Y by the hypergeometric distribution to approach those assigned by the binomial distribution as N becomes large and r/N, the fraction defective in the population, is held constant and equal to p. You can verify this expectation by using limit theorems encountered in your calculus courses to show that (proof omitted)

$$\lim_{N \to \infty} \frac{\binom{r}{y}\binom{N-r}{n-y}}{\binom{N}{n}} = \binom{n}{y}p^y(1-p)^{n-y}$$

where

$$\frac{r}{N} = p$$

Hence, for a fixed fraction defective $p = r/N$, the hypergeometric probability function converges to the binomial probability function as N becomes large.

Exercises

3.62 An urn contains ten marbles of which five are green, two are blue, and three are red. Three marbles are to be drawn from the urn, one at a time without replacement. What is the probability that all three marbles drawn will be green?

3.63 A warehouse contains ten printing machines, four of which are defective. A company selects five of the machines at random, thinking all are in working condition. What is the probability that all five of the machines are nondefective?

3.64 Refer to Exercise 3.63. The company repairs the defective ones at a cost of $50 each. Find the mean and variance of the total repair cost.

3.65 A corporation is sampling without replacement for $n = 3$ firms from which to purchase certain supplies. The sample is to be selected from a pool of six firms, of which four are local and two not local. Let Y denote the number of nonlocal firms among the three selected.

(a) Find $P(Y = 1)$.

(b) Find $P(Y \geq 1)$.

(c) Find $P(Y \leq 1)$.

3.66   A jury of six persons was selected from a group of twenty potential jurors, of whom eight were black and twelve were white. The jury was supposedly randomly selected, but it contained only one black member. Do you have any reason to doubt the randomness of the selection?

3.67   Refer to Exercise 3.66. If the selection process were really random, what would be the mean and variance of the number of black members selected for the jury?

3.68   Suppose that a radio contains six transistors, two of which are defective. Three transistors are selected at random, removed from the radio, and inspected. Let Y equal the number of defectives observed, where $Y = 0$, 1, or 2. Find the probability distribution for Y. Express your results graphically as a probability histogram.

3.69   Simulate the experiment described in Exercise 3.68 by marking six marbles, or coins, so that two represent defectives and four represent nondefectives. Place the marbles in a hat, mix, draw three, and record Y, the number of defectives observed. Replace the marbles and repeat the process until a total of $n = 100$ observations on Y has been recorded. Construct a relative frequency histogram for this sample and compare it with the population probability distribution, Exercise 3.68.

*3.70   The sizes of animal populations are often estimated by using a capture-tag-recapture method. In this method k animals are captured, tagged, and then released into the population. Some time later n animals are captured, and Y, the number of tagged animals among the n, is noted. The probabilities associated with Y are a function of N, the number of animals in the population, and the observed value of Y contains information on this unknown N. Suppose $k = 4$ animals are tagged and then released. A sample of $n = 3$ animals is then selected at random from the same population. Find $P(Y = 1)$ as a function of N. What value of N will *maximize* $P(Y = 1)$?

3.8 The Poisson Probability Distribution

Suppose that we want to find the probability distribution of the number of automobile accidents at a particular intersection during a time period of 1 week. At first glance this random variable, the number of accidents, may not seem even remotely related to a binomial random variable, but we will see that there is an interesting relationship.

Think of the time period, 1 week in this example, as split up into n subintervals, *each of which is so small that at most one accident could occur in it with probability different from zero.* Denoting the probability of an accident in any subinterval by p, we have, for all practical purposes,

$$P(\text{no accidents in a subinterval}) = 1 - p$$
$$P(\text{one accident in a subinterval}) = p$$

and

$$P(\text{more than one accident in a subinterval}) = 0$$

Then the total number of accidents in the week is just the total number of subintervals that contain one accident. If the occurrence of accidents can be regarded as independent from interval to interval, the total number of accidents has a binomial distribution.

Although there is no unique way to choose the subintervals, and we therefore know neither n nor p, it seems reasonable that as we divide the week into a greater number n of subintervals, the probability p of one accident in one of these shorter subintervals will decrease. Letting $\lambda = np$ and taking the limit of the binomial probability $p(y) = \binom{n}{y}p^y(1-p)^{n-y}$ as $n \to \infty$, we have

$$\lim_{n \to \infty} \binom{n}{y}p^y(1-p)^{n-y} = \lim_{n \to \infty} \frac{n(n-1)\cdots(n-y+1)}{y!} \left(\frac{\lambda}{n}\right)^y \left(1-\frac{\lambda}{n}\right)^{n-y}$$

$$= \lim_{n \to \infty} \frac{\lambda^y}{y!} \left(1-\frac{\lambda}{n}\right)^n \frac{n(n-1)\cdots(n-y+1)}{n^y} \left(1-\frac{\lambda}{n}\right)^{-y}$$

$$= \frac{\lambda^y}{y!} \lim_{n \to \infty} \left(1-\frac{\lambda}{n}\right)^n \left(1-\frac{\lambda}{n}\right)^{-y} \left(1-\frac{1}{n}\right)$$

$$\times \left(1-\frac{2}{n}\right) \cdots \left(1-\frac{y-1}{n}\right)$$

Noting that

$$\lim_{n \to \infty} \left(1-\frac{\lambda}{n}\right)^n = e^{-\lambda}$$

and all other terms to the right of the limit have a limit of 1, we obtain

$$p(y) = \frac{\lambda^y}{y!} e^{-\lambda}$$

(Note: $e = 2.718\ldots$.) Random variables possessing this distribution are said to be Poisson random variables. Hence Y, the number of accidents per week, should possess the Poisson distribution given earlier.

The convergence of the binomial probability function to the Poisson is of practical value, because the Poisson probabilities can be used to approximate their binomial counterparts for large n, small p, and $\lambda = np$ less than, roughly, 7. Exercise 3.78 will require the calculation of corresponding binomial and Poisson probabilities and will demonstrate the adequacy of the approximation.

The Poisson probability distribution often provides a good model for the probability distribution of the number Y of rare events that occur infrequently in space, time, volume, or any other dimension, where λ is the average value of y. As we have noted, it provides a good model for the probability distribution of the number Y of automobile accidents, industrial accidents, or other types of accidents in a given unit of time. Other examples of random variables with approximate Poisson distributions are the number of telephone calls handled by a switchboard in a time interval, the number of radioactive particles that decay in a particular time period, and the number of errors a typist makes in typing a page.

Definition 3.11

A random variable Y is said to have a *Poisson probability distribution* if and only if

$$p(y) = \frac{\lambda^y}{y!} e^{-\lambda}, \qquad y = 0, 1, 2, \dots \qquad \lambda > 0.$$

As we shall see in Theorem 3.11, the parameter λ that appears in the formula for the Poisson distribution is actually the mean of the distribution.

EXAMPLE 3.16   Show that the probabilities assigned by the Poisson probability distribution satisfy the requirements that $0 \leq p(y) \leq 1$ for all y and $\sum_y p(y) = 1$.

Solution   Because $\lambda > 0$, it is obvious that $p(y) > 0$ for $y = 0, 1, 2, \dots$ and that $p(y) = 0$ otherwise. Further,

$$\sum_{y=0}^{\infty} p(y) = \sum_{y=0}^{\infty} \frac{\lambda^y}{y!} e^{-\lambda} = e^{-\lambda} \sum_{y=0}^{\infty} \frac{\lambda^y}{y!} = e^{-\lambda} e^{\lambda} = 1$$

because the infinite sum $\sum_{y=0}^{\infty} \lambda^y / y!$ is a series expansion of e^{λ}. Sums of special series are given in the Chemical Rubber Company's handbook [7].

EXAMPLE 3.17   Suppose that a random system of police patrol is devised so that a patrol officer may visit a given beat location $Y = 0, 1, 2, 3, \dots$ times per half-hour period and that the system is arranged so that each location is visited on an average of once per time period. Assume that Y possesses, approximately, a Poisson probability distribution. Calculate the probability that the patrol officer will miss a given location during a half-hour period. What is the probability that it will be visited once? Twice? At least once?

Solution   For this example the time period is a half hour and the mean visits per half-hour interval is $\lambda = 1$. Then

$$p(y) = \frac{(1)^y e^{-1}}{y!} = \frac{e^{-1}}{y!}$$

The event that a given location is missed in one half-hour period corresponds to $y = 0$ and

$$p(0) = \frac{e^{-1}}{0!} = e^{-1} = 0.368$$

Similarly,

$$p(1) = \frac{e^{-1}}{1!} = e^{-1} = 0.368$$

and

$$p(2) = \frac{e^{-1}}{2!} = \frac{e^{-1}}{2} = .184$$

The probability that the location is visited *at least* once is the event that $Y \geq 1$. Then

$$P(Y \geq 1) = \sum_{y=1}^{\infty} p(y) = 1 - p(0) = 1 - e^{-1} = .632$$

EXAMPLE 3.18   A certain type of tree has seedlings randomly dispersed in a large area, with the mean density of seedlings being approximately five per square yard. If a forester randomly locates ten 1-square-yard sampling regions in the area, find the probability that none of the regions will contain any seedlings.

Solution   If the seedlings really are randomly dispersed, the number of seedlings per region, Y, can be modeled as a Poisson random variable with $\lambda = 5$. (The average density is five per square yard.) Thus

$$P(Y = 0) = p(0) = \frac{\lambda^0 e^{-\lambda}}{0!} = e^{-5}$$

The probability that $Y = 0$ on ten independently selected regions is $(e^{-5})^{10}$, because the probability of the intersection of independent events is equal to the product of the respective probabilities. The resulting probability is very, very small. Thus if this event actually occurred, we would seriously question the assumption of randomness, or the stated average density of seedlings, or both.

For your convenience, we provide in Table 3, Appendix III, the partial sums $\sum_{y=0}^{a} p(y)$ for the Poisson probability distribution for many values of λ between 0.02 and 15. This table is constructed in a manner similar to that used for the table of partial sums for the binomial distribution, Table 1, Appendix III. The following example illustrates the use of Table 3 and demonstrates the use of the Poisson probability distribution to approximate the binomial probability distribution.

EXAMPLE 3.19   Suppose that Y possesses a binomial distribution with $n = 20$ and $p = .1$. Find the exact value of $P(Y \leq 3)$ using the table of binomial probabilities, Table 1, Appendix III. Use Table 3, Appendix III, to approximate this probability, using a corresponding probability given by the Poisson distribution, and compare the exact and approximate values for $P(Y \leq 3)$.

Solution   From Table 1, Appendix III, the exact value of $P(Y \leq 3) = .867$. If W is a Poisson-distributed random variable with $\lambda = np = 20(.1) = 2$, previous discussions indicate that $P(Y \leq 3)$ is approximately equal to $P(W \leq 3)$. Table 3, Appendix III, gives $P(W \leq 3) = .857$. Thus you can see that the Poisson approximation is quite good, yielding a value that differs from the exact value by only .01.   ∽

In our derivation of the mean and variance of a random variable with the Poisson distribution, we again use the fundamental property that $\sum_y p(y) = 1$ for any discrete probability distribution.

Theorem 3.11

> If Y is a random variable with a Poisson distribution with parameter λ, then
>
> $$\mu = E(Y) = \lambda \quad \text{and} \quad \sigma^2 = V(Y) = \lambda.$$

Proof   By definition,

$$E(Y) = \sum_y yp(y) = \sum_{y=0}^{\infty} y \frac{\lambda^y e^{-\lambda}}{y!}$$

Note that the first term in this sum is equal to 0 (when $y = 0$) and hence

$$E(Y) = \sum_{y=1}^{\infty} y \frac{\lambda^y e^{-\lambda}}{y!} = \sum_{y=1}^{\infty} \frac{\lambda^y e^{-\lambda}}{(y-1)!}$$

As it stands, this quantity is not equal to the sum of a probability function $p(y)$ over all values of y, but we can change it to the proper form by factoring λ out of the expression and letting $z = y - 1$. Then the limits of summation become $z = 0$ (when $y = 1$) and $z = \infty$ (when $y = \infty$), and

$$E(Y) = \lambda \sum_{y=1}^{\infty} \frac{\lambda^{y-1} e^{-\lambda}}{(y-1)!} = \lambda \sum_{z=0}^{\infty} \frac{\lambda^z e^{-\lambda}}{z!}$$

Note that $p(z) = \lambda^z e^{-\lambda}/z!$ is the probability function for a Poisson random variable and $\sum_{z=0}^{\infty} p(z) = 1$. Therefore, $E(Y) = \lambda$. Thus the mean of a Poisson random variable is the single parameter λ that appears in the expression for the Poisson probability function.

We leave the derivation of the variance as an exercise (Exercise 3.82).

EXAMPLE 3.20   The number of industrial accidents at a particular manufacturing plant is found to average three per month. During the last month six accidents occurred. Would you regard this number as unusually large (highly improbable if μ were still equal to 3) and indicative of an increase in the mean μ?

Solution   The number of accidents Y would likely follow a Poisson probability distribution with $\lambda = 3$. The probability that Y is as large as 6 is

$$P(Y \geq 6) = \sum_{y=6}^{\infty} \frac{3^y e^{-3}}{y!}$$

The tedious calculations required to find $P(Y \geq 6)$ can be avoided by using Table 3, Appendix III, or the empirical rule. From Theorem 3.11,

$$\mu = \lambda = 3 \qquad \sigma^2 = \lambda = 3 \qquad \sigma = \sqrt{3} = 1.73$$

The empirical rule tells us that we should expect Y to take values in the interval $\mu \pm 2\sigma$ with a high probability.

Note that $\mu + 2\sigma = 3 + (2)(1.73) = 6.5$. The observed number of accidents, $Y = 6$, does not lie more than 2σ from μ, but it is close to the boundary. Thus the observed result is not highly improbable, but it may be improbable enough to warrant an investigation. See Exercise 3.129 for the exact probability $P(|Y - \lambda| \leq 2\sigma)$.

Exercises

3.71   Let Y denote a random variable having a Poisson distribution with mean $\lambda = 2$. Find

(a) $P(Y = 4)$.   (b) $P(Y \geq 4)$.

(c) $P(Y < 4)$.   (d) $P(Y \geq 4 | Y \geq 2)$.

3.72   In a particular department store customers arrive at a checkout counter according to a Poisson distribution at an average of seven per hour. During a given hour, what are the probabilities

    (a)   that no more than three customers arrive?

    (b)   that at least two customers arrive?

    (c)   that exactly five customers arrive?

3.73   The number of typing errors made by a particular typist has a Poisson distribution with an average of four errors per page. If more than four errors show on a given page, the typist must retype the whole page. What is the probability a certain page does not have to be retyped?

3.74   There are two entrances to a parking lot. Cars arrive at entrance I according to a Poisson distribution at an average of three per hour, and at entrance II according to a Poisson distribution at an average of four per hour. What is the probability that three cars arrive at the parking lot in a given hour? (Assume that the numbers of cars arriving at the two entrances are independent.)

3.75   The number of knots in a particular type of wood has a Poisson distribution with an average of 1.5 knots in 10 cubic feet of the wood. Find the probability that a 10-cubic-foot block of the wood has at most one knot.

3.76   The mean number of automobiles entering a mountain tunnel per 2-minute period is one. An excessive number of cars entering the tunnel during a brief period of time produces a hazardous situation. Find the probability that the number of autos entering the tunnel during a 2-minute period exceeds three. Does the Poisson model seem reasonable for this problem?

3.77   Assume that the tunnel is observed during ten 2-minute intervals, thus giving ten independent observations, $Y_1, Y_2, \ldots, Y_{10}$, on the Poisson random variable of Exercise 3.76. Find the probability that $Y > 3$ during at least one of the ten 2-minute intervals.

3.78   Consider a binomial experiment for $n = 20$, $p = .05$. Use Table 1, Appendix III, to calculate the binomial probabilities for $Y = 0, 1, 2, 3, 4$. Calculate the same probabilities by using the Poisson approximation with $\lambda = np$. Compare.

3.79   A salesperson has found the probability of a sale on a single contact is approximately .03. If the salesperson contacts 100 prospects, what is the approximate probability of making at least one sale?

3.80   According to the U.S. Fire Administration, 185 people died in 12,438 hotel and motel fires in 1979, roughly 1.5 deaths per 100 fires (*U.S. News and World Report*, August 3, 1981).

    (a)   If 200 hotel and motel fires occurred in a given region, what is the probability that the number of deaths would exceed 8? [Hint: The mean number of deaths per 200 fires would equal $2(1.5) = 3$.]

    (b)   If 200 hotel and motel fires occurred in a given region and the number of deaths exceeded 8, would you suspect that the region's mean death rate would exceed the national average? Explain.

3.81   The probability that a mouse inoculated with a serum contracts a certain disease is .2. Using the Poisson approximation, find the probability that at most three of thirty inoculated mice contract the disease.

3.82 Let Y have a Poisson distribution with mean λ. Find $E[Y(Y-1)]$ and then use this to show that $V(Y) = \lambda$.

3.83 In the daily production of a certain kind of rope, the number of defects per foot Y is assumed to have a Poisson distribution with mean $\lambda = 2$. The profit per foot when the rope is sold is given by X, where $X = 50 - 2Y - Y^2$. Find the expected profit per foot.

*3.84 A store owner is overstocked in a certain item and decides to use the following promotion to decrease the supply. The item has a marked price of \$100. For each customer purchasing the item during a particular day, the owner will reduce the price by a factor of one-half. Thus the first customer will pay \$50 for the item, the second will pay \$25, and so on. Suppose the number of customers purchasing the item during the day has a Poisson distribution with mean 2. Find the expected cost of the item at the end of the day. [Hint: The cost at the end of the day is $100(1/2)^Y$, where Y is the number of customers who have purchased the item.]

3.85 A food manufacturer uses an extruder (a machine that produces bite-size food like cookies and snack food) that yields revenue for the firm at a rate of \$200 per hour when in operation. However, the extruder breaks down an average of two times every day of operation. If Y denotes the number of breakdowns per day, the daily revenue generated by the machine is $R = 1600 - 50Y^2$. Find the expected daily revenue for the extruder.

3.9 Moments and Moment-Generating Functions

The parameters μ and σ are meaningful numerical descriptive measures that locate the center and describe the spread of $p(y)$, but they do not provide a unique characterization of the distribution. Many different distributions possess the same means and standard deviations. Consequently, we might consider a set of numerical descriptive measures that, under rather general conditions, uniquely determine $p(y)$.

Definition 3.12

> The ith *moment of a random variable* Y *taken about the origin* is defined to be $E(Y^i)$ and denoted by μ_i'.

Particularly, note that the first moment about the origin is $E(Y) = \mu_1' = \mu$ and that $\mu_2' = E(Y^2)$ is employed in Theorem 3.6 for finding σ^2.

A second useful moment of a random variable is one taken about its mean.

Definition 3.13

> The ith *moment of a random variable* Y *taken about its mean*, or the ith *central moment of* Y, is defined to be $E[(Y - \mu)^i]$ and denoted by μ_i.

In particular, $\sigma^2 = \mu_2$.

Let us concentrate on moments about the origin, μ'_i, $i = 1, 2, 3, \ldots$. Suppose that two random variables Y and Z possess finite moments with $\mu'_{1y} = \mu'_{1z}, \mu'_{2y} = \mu'_{2z}, \ldots, \mu'_{ky} = \mu'_{kz}$, where k can assume any integral value. That is, the two random variables possess identical corresponding moments about the origin. Under some fairly general conditions it then can be shown that Y and Z have identical probability distributions. Thus a major use of moments is to approximate the probability distribution of a random variable (usually an estimator or a decision maker). Consequently, the moments, μ'_i, $i = 1, 2, 3, \ldots$, are primarily of theoretical value for $i > 2$.

A third interesting expectation is the moment-generating function for a random variable, which, figuratively speaking, packages all the moments for a random variable in one simple expression. First we will define the moment-generating function and then explain how it works.

Definition 3.14

> The *moment-generating function* m(t) *for a random variable* Y is defined to be $E(e^{tY})$. We say that a moment-generating function for Y exists if there exists a positive constant b such that $m(t)$ is finite for $|t| \leq b$.

Why is $E(e^{tY})$ called the *moment-generating function for* Y? From a series expansion for e^{ty} we have

$$e^{ty} = 1 + ty + \frac{(ty)^2}{2!} + \frac{(ty)^3}{3!} + \frac{(ty)^4}{4!} + \cdots$$

Then assuming that μ'_i is finite, $i = 1, 2, 3, \ldots$, we have

$$E(e^{tY}) = \sum_y e^{ty} p(y) = \sum_y \left[1 + ty + \frac{(ty)^2}{2!} + \frac{(ty)^3}{3!} + \cdots \right] p(y)$$

$$= \sum_y p(y) + t \sum_y y p(y) + \frac{t^2}{2!} \sum_y y^2 p(y) + \frac{t^3}{3!} \sum_y y^3 p(y) + \cdots$$

$$= 1 + t\mu'_1 + \frac{t^2}{2!} \mu'_2 + \frac{t^3}{3!} \mu'_3 + \cdots$$

[This argument involves an interchange of summations, which is justifiable if $m(t)$ exists.] Thus $E(e^{tY})$ is a function of all the moments about the origin, μ'_i, $i = 1, 2, 3, \ldots$, where μ'_i is the coefficient of $t^i/i!$.

The moment-generating function possess two important applications. First, if we can find $E(e^{tY})$, we can find any of the moments for Y.

Theorem 3.12

If $m(t)$ exists, then for any positive integer k,

$$\frac{d^k m(t)}{dt^k}\bigg]_{t=0} = m^{(k)}(0) = \mu_k'$$

The essence of Theorem 3.12 is that if you find the kth derivative of $m(t)$ with respect to t then set $t = 0$, the result will be μ_k'. The proof of Theorem 3.12 follows.

Proof $d^k m(t)/dt^k$, or $m^{(k)}(t)$, is the kth derivative of $m(t)$ with respect to t. Now

$$m(t) = E(e^{tY}) = 1 + t\mu_1' + \frac{t^2}{2!}\mu_2' + \frac{t^3}{3!}\mu_3' + \cdots$$

Then

$$m^{(1)}(t) = \mu_1' + \frac{2t}{2!}\mu_2' + \frac{3t^2}{3!}\mu_3' + \cdots$$

Setting $t = 0$, we obtain

$$m^{(1)}(0) = \mu_1'$$

Similarly,

$$m^{(2)}(t) = \mu_2' + \frac{2t}{2!}\mu_3' + \cdots$$

Setting $t = 0$, we have

$$m^{(2)}(0) = \mu_2'$$

Continuing the differentiation of $m(t)$, we can see that

$$m^{(k)}(0) = \mu_k', \qquad k = 1, 2, \ldots$$

[These operations involve interchanging a derivative and an infinite sum, which can be justified if $m(t)$ exists.]

EXAMPLE 3.21   Find the moment-generating function $m(t)$ for the Poisson random variable.

Solution

$$m(t) = E(e^{tY}) = \sum_{y=0}^{\infty} e^{ty} p(y) = \sum_{y=0}^{\infty} e^{ty} \frac{\lambda^y e^{-\lambda}}{y!}$$

$$= \sum_{y=0}^{\infty} \frac{(\lambda e^t)^y e^{-\lambda}}{y!} = e^{-\lambda} \sum_{y=0}^{\infty} \frac{(\lambda e^t)^y}{y!}$$

To complete the summation, consult a mathematical handbook to find the Taylor series expansion

$$\sum_{y=0}^{\infty} \frac{(\lambda e^t)^y}{y!} = e^{\lambda e^t}$$

or employ the method of Theorem 3.11. Thus multiply and divide by $e^{\lambda e^t}$. Then

$$m(t) = e^{-\lambda} e^{\lambda e^t} \sum_{y=0}^{\infty} \frac{(\lambda e^t)^y e^{-\lambda e^t}}{y!}$$

The quantity inside the summation is the probability function for a Poisson random variable with mean λe^t. Hence

$$\sum_{y} p(y) = 1 \qquad \text{and} \qquad m(t) = e^{-\lambda} e^{\lambda e^t}(1) = e^{\lambda(e^t-1)} \qquad \qquad \backsim$$

Observe that the calculations in Example 3.21 are no more difficult than those in Theorem 3.11, where only the expected value for a Poisson random variable Y was calculated. Direct evaluation of the variance of Y through the use of Theorem 3.6 would require that $E(Y^2)$ be found by summing another series. Example 3.22, which follows, illustrates the use of the moment-generating function of the Poisson random variable to calculate the mean and the variance.

EXAMPLE 3.22   Use the moment-generating function of Example 3.21 and Theorem 3.12 to find $\mu'_1 = \mu$ and σ^2 for the Poisson random variable.

Solution   According to Theorem 3.12,

$$\mu = m^{(1)}(0) = \frac{d}{dt}(e^{\lambda(e^t-1)})]_{t=0} = e^{\lambda(e^t-1)} \cdot \lambda e^t]_{t=0} = \lambda$$

and

$$\mu'_2 = m^{(2)}(0) = \frac{d^2}{dt^2}(e^{\lambda(e^t-1)})]_{t=0} = \frac{d}{dt}(e^{\lambda(e^t-1)} \cdot \lambda e^t)]_{t=0}$$

$$= e^{\lambda(e^t-1)} \cdot (\lambda e^t)^2 + e^{\lambda(e^t-1)} \cdot \lambda e^t]_{t=0} = \lambda^2 + \lambda$$

Theorem 3.6 tells us that $\sigma^2 = E(Y^2) - \mu^2 = \mu'_2 - \mu^2$, and therefore

$$\sigma^2 = \lambda^2 + \lambda - (\lambda)^2 = \lambda$$

Notice how easily we obtained μ'_2 from $m(t)$. $\qquad \backsim$

The second and primary application of a moment-generating function is in proving that a random variable possesses a particular probability distribution $p(y)$. If $m(t)$ exists for a probability distribution $p(y)$, it is unique. That is, it is impossible for random variables with different probability distributions to have the same moment-generating functions. Also, it can be shown that if the moment-generating functions for two random variables Y and Z are equal, then Y and Z must have the same probability distribution. It follows that if we can recognize the moment-generating function of a random variable Y to be that of a known distribution, Y must have that distribution.

In summary, a moment-generating function is simply a mathematical device that sometimes (but not always) provides an easy way to find μ'_k and to prove the equivalence of two probability distributions.

Exercises

3.86 If Y has a binomial distribution with n trials and probability of success p, show that the moment-generating function for Y is

$$m(t) = [pe^t + q]^n$$

where $q = 1 - p$.

3.87 Differentiate the moment-generating function in Exercise 3.86 to find $E(Y)$ and $E(Y^2)$. Then find $V(Y)$.

3.88 If Y has a geometric distribution with probability of success p, show that the moment-generating function for Y is

$$m(t) = \frac{pe^t}{1 - qe^t}$$

where $q = 1 - p$.

3.89 Differentiate the moment-generating function in Exercise 3.88 to find $E(Y)$ and $E(Y^2)$. Then find $V(Y)$.

3.90 Find the distribution of the random variable Y for each of the following moment-generating functions:

(a) $m(t) = [(1/3)e^t + (2/3)]^5$

(b) $m(t) = \dfrac{e^t}{2 - e^t}$

(c) $m(t) = e^{2(e^t - 1)}$

3.91 Let $m(t) = (1/6)e^t + (2/6)e^{2t} + (3/6)e^{3t}$. Find the following:

(a) $E(Y)$,

(b) $V(Y)$,

(c) the distribution of Y.

3.92 If Y is a random variable with moment-generating function $m(t)$ and W is given by $W = aY + b$, show that the moment-generating function of W is $e^{tb}m(at)$.

3.93 Use the result in Exercise 3.92 to prove that, if $W = aY + b$, then $E(W) = aE(Y) + b$ and $V(W) = a^2V(Y)$.

*3.94 Let $r(t) = \ln m(t)$ and $r^{(k)}(0)$ denote the kth derivative of $r(t)$ evaluated for $t = 0$. Show that $r^{(1)}(0) = \mu_1' = \mu$ and $r^{(2)}(0) = \mu_2' - (\mu_1')^2 = \sigma^2$. [Hint: $m(0) = 1$.]

*3.95 Use the results of Exercise 3.94 to find the mean and variance of a Poisson random variable with $m(t) = e^{5(e^t - 1)}$. Note that $r(t)$ is easier to differentiate than $m(t)$ in this case.

3.10 Probability-Generating Functions (Optional)

An important class of discrete random variables is one in which Y represents a count and consequently takes integral values, $Y = 0, 1, 2, 3, \ldots$. The binomial, geometric, hypergeometric, and Poisson random variables all fall in this class. The following examples give practical situations that result in integral-valued random variables. One, involving the theory of queues (waiting lines), is concerned with the number of persons (or objects) awaiting service at a particular point in time. Knowledge of the behavior of this random variable is important in designing manufacturing plants where production consists of a sequence of operations, each of which requires a different length of time to complete. An insufficient number of service stations for a particular production operation can result in a bottleneck, the formation of a queue of products ready to be serviced, and a resulting slowdown in the manufacturing operation. Queuing theory also is important in determining the number of checkout counters needed for a supermarket and in the design of hospitals and clinics.

Integral-valued random variables also are very important in studies of population growth. For example, epidemiologists are interested in the growth of bacterial populations and the growth of the number of persons afflicted by a particular disease. The number of elements in each of these populations will be an integral-valued random variable.

A mathematical device very useful in finding the probability distributions and other properties of integral-valued random variables is the probability-generating function.

Definition 3.15

> Let Y be an integral-valued random variable for which $P(Y = i) = p_i$, $i = 0, 1, 2, \ldots$. The *probability-generating function* $P(t)$ for Y is defined to be
>
> $$P(t) = E(t^Y) = p_0 + p_1 t + p_2 t^2 + \cdots$$
>
> for all values of t such that $P(t)$ is finite.

The reason for calling $P(t)$ a probability-generating function is clear when we compare $P(t)$ with the moment-generating function $m(t)$. In particular, the coefficient of t^i in $P(t)$ is the probability p_i. Correspondingly, the coefficient of t^i for $m(t)$ is a constant times the ith moment, μ_i'. If we know $P(t)$ and can expand it into a series, we can determine $p(y)$ as the coefficient of t^y.

Repeated differentiation of $P(t)$ yields *factorial moments* for the random variable y.

Definition 3.16

> The Kth *factorial moment* for a random variable Y is defined to be
>
> $$\mu_{[K]} = E[Y(Y-1)(Y-2)\cdots(Y-K+1)]$$
>
> where K is a positive integer.

Note that $\mu_{[1]} = E(Y) = \mu$ and that $\mu_{[2]} = E[Y(Y-1)]$ was useful in finding the variance for the binomial random variable in Theorem 3.7.

Theorem 3.13

> $$\frac{d^K P(t)}{dt^K}\bigg]_{t=1} = P^{(K)}(1) = \mu_{[K]}$$

Proof

$$P(t) = p_0 + p_1 t + p_2 t^2 + p_3 t^3 + p_4 t^4 + \cdots$$

and

$$P^{(1)}(t) = p_1 + 2p_2 t + 3p_3 t^2 + 4p_4 t^3 + \cdots$$

Setting $t = 1$, we have

$$P^{(1)}(1) = p_1 + 2p_2 + 3p_3 + \cdots = \sum_{y=0}^{\infty} yp(y) = E(Y)$$

Then

$$P^{(2)}(t) = (2)(1)p_2 + (3)(2)p_3 t + (4)(3)p_4 t^2 + \cdots$$

and, in general,

$$P^{(K)}(t) = \sum_{y=K}^{\infty} y(y-1)(y-2)\cdots(y-K+1)p(y)t^{y-K}$$

Setting $t = 1$, we obtain

$$P^{(K)}(1) = \sum_{y=K}^{\infty} y(y-1)(y-2)\cdots(y-K+1)p(y)$$

$$= E[Y(Y-1)(Y-2)\cdots(Y-K+1)] = \mu_{[K]}$$

EXAMPLE 3.23   Find the probability-generating function for the geometric random variable.

Solution   Note that $p_0 = 0$, because Y cannot assume this value. Then

$$P(t) = E(t^Y) = \sum_{y=1}^{\infty} t^y q^{y-1} p = \sum_{y=1}^{\infty} \frac{p}{q}(qt)^y$$

$$= \frac{p}{q}[qt + (qt)^2 + (qt)^3 + \cdots]$$

The terms of the series are those of an infinite geometric progression. We can let $t \leq 1$, so that $qt \leq 1$. Then

$$P(t) = \frac{p}{q}\left(\frac{qt}{1-qt}\right) = \frac{pt}{1-qt}$$

(For summation of the series, consult a mathematical handbook.)   ∽

EXAMPLE 3.24   Use $P(t)$, Example 3.23, to find the mean of the geometric random variable.

Solution   From Theorem 3.13, $\mu_{[1]} = \mu = P^{(1)}(1)$. Then

$$P^{(1)}(t) = \frac{d}{dt}\left(\frac{pt}{1-qt}\right) = \frac{(1-qt)p - (pt)(-q)}{(1-qt)^2}$$

Setting $t = 1$, we obtain

$$P^{(1)}(1) = \frac{p^2 + pq}{p^2} = \frac{p(p+q)}{p^2} = \frac{1}{p}$$ ∽

Because we already have the moment-generating function to assist in finding the moments of a random variable, we might ask, of what value is $P(t)$? The answer is that it may be exceedingly difficult to find $m(t)$ but easy to find $P(t)$. Thus $P(t)$ simply provides an additional tool for finding the moments of a random variable. It may or may not be useful in a given situation.

Finding the moments of a random variable is not the major use of the probability-generating function. Its primary application is in deriving the probability function (and hence the probability distribution) for other related integral-valued random variables. For these applications, see Feller [1] and Parzen [6].

Exercises

*3.96   Let Y denote a binomial random variable with n trials and probability of success p. Find the probability-generating function for Y and use it to find $E(Y)$.

*3.97   Let Y denote a Poisson random variable with mean λ. Find the probability-generating function for Y and use it to find $E(Y)$ and $V(Y)$.

*3.98   Refer to Exercise 3.97. Use the probability-generating function found there to find $E(Y^3)$.

3.11 Tchebysheff's Theorem

We have seen in Section 1.3 and Example 3.20 that the empirical rule is of great help in approximating probabilities over certain intervals if the probability or population histogram is roughly bell-shaped and the mean and variance are known. In many instances the shapes of probability histograms differ markedly from a mound shape and the empirical rule may not yield useful approximations to probabilities of interest. The following result, known as Tchebysheff's theorem, can be utilized to give a lower bound to the probability that the random variable Y of interest falls in an interval of the form $\mu \pm k\sigma$.

Theorem 3.14
Tchebysheff's
Theorem

> Let Y be a random variable with finite mean μ and variance σ^2. Then for any positive constant k,
>
> $$P(|Y - \mu| < k\sigma) \geq 1 - \frac{1}{k^2}$$
>
> or
>
> $$P(|Y - \mu| \geq k\sigma) \leq \frac{1}{k^2}$$

Two important aspects of this result should be pointed out. First, the result applies for any probability distribution, whether the probability histogram is bell-shaped or not. Second, the results of the theorem are very conservative in the sense that the actual probability that Y is in the interval $\mu \pm k\sigma$ usually exceeds the lower bound for the probability $(1 - 1/k^2)$ by a considerable amount. You should verify that the results of the empirical rule do not contradict those given by Theorem 3.14. The proof of this theorem will be deferred to Section 4.10. The usefulness of this theorem is illustrated in the following example.

EXAMPLE 3.25   The number of customers per day at a certain sales counter denoted by Y, has been observed for a long period of time and found to have a mean of twenty customers with a standard deviation of two customers. The probability distribution of Y is not known. What can be said about the probability that Y will be between sixteen and twenty-four tomorrow?

Solution   We want to find $P(16 \leq Y \leq 24)$. From Theorem 3.14 we know that for any $k \geq 0$, $P(|Y - \mu| \leq k\sigma) \geq 1 - 1/k^2$, or

$$P[(\mu - k\sigma) \leq Y \leq (\mu + k\sigma)] \geq 1 - \frac{1}{k^2}$$

Because $\mu = 20$ and $\sigma = 2$, $\mu - k\sigma = 16$ and $\mu + k\sigma = 24$ if $k = 2$. Thus

$$P(16 \leq Y \leq 24) \geq 1 - \frac{1}{(2)^2} = \frac{3}{4}$$

In other words, tomorrow's customer total will be between sixteen and twenty-four with high probability (at least 3/4).

Note that if σ was 1, k would be 4 and

$$P(16 \leq Y \leq 24) \geq 1 - \frac{1}{(4)^2} = \frac{15}{16}$$

Thus we see that the value of σ has considerable effect on probabilities associated with intervals.   ∽

Exercises

3.99   Let Y be a random variable with a mean of 11 and a variance of 9. Using Tchebysheff's theorem, find the following:

(a)   a lower bound for $P(6 < Y < 16)$

(b)   the value of C such that $P(|Y - 11| \geq C) \leq .09$.

3.100   Suppose, as noted in Exercise 3.25, that approximately 35% of all applicants for jobs falsify the information on their application forms. If a company has 2300 employees,

(a)   what is the expected value of the number Y of application forms that have been falsified?

(b)   find the standard deviation of Y.

(c)   calculate the interval, $\mu \pm 2\sigma$.

(d)   suppose that the company had a credentials-checking firm verify the information on the 2300 application forms and that 249 application forms contained falsified information. Do you think that the company's application falsification rate is consistent with the contention that 35% of all job applicants falsify information on their applications? Explain.

3.101   The U.S. mint produces dimes with an average diameter of .5 inch and a standard deviation of .01. Using Tchebysheff's theorem, find a lower bound for the number of coins in a lot of 400 coins having diameter between .48 and .52.

3.102   For a certain soil the number of wireworms per cubic foot has a mean of 100. Assuming a Poisson distribution of wireworms, give an interval that will include at least 5/9 of the sample values on wireworm counts obtained from a large number of 1-cubic-foot samples.

3.103   Refer to Exercise 3.68. Using the probability histogram, find the fraction of the total population lying within two standard deviations of the mean. Compare your result with that of Tchebysheff's theorem.

3.104   A balanced coin is tossed three times. Let Y equal the number of heads observed.

   (a)   Use the formula for the binomial probability distribution to calculate the probabilities associated with $Y = 0, 1, 2,$ and 3.

   (b)   Construct a probability distribution similar to Figure 3.1.

   (c)   Find the expected value and standard deviation of Y, using the formulas $E(Y) = np$ and $V(Y) = npq$.

   (d)   Using the probability distribution of part (b), find the fraction of the population measurements lying within one standard deviation of the mean. Repeat for two standard deviations. How do your results agree with the results of Tchebysheff's theorem and the empirical rule?

3.105   Suppose that a coin was definitely unbalanced and that the probability of a head was equal to $p = .1$. Follow instructions (a), (b), (c), and (d) as stated in Exercise 3.104. Note that the probability distribution loses its symmetry and becomes skewed when p is not equal to $1/2$.

3.106   The *Wall Street Journal* (July 29, 1981),[†] in an article on hang gliding, notes that hang gliding can be dangerous. Of the estimated 36,000 hang-gliding enthusiasts in the United States, 29 were killed last year in hang-gliding accidents and, it is estimated, thousands more were injured. Suppose that a hang-glider pilot, randomly selected from among all hang-glider pilots, has a probability equal to .0006 of being killed in any one year. If there are 40,000 pilots next year,

   (a)   what is the expected number of fatalities?

   (b)   what is the standard deviation of the number Y of fatalities?

   (c)   is it likely that the number of fatalities would exceed 40?

3.107   A national poll of 1502 adults by pollster Louis Harris (*Environmental News*, EPA, September 1977) indicated that 71% "would rather live in an environment that is clean rather than in an area with a lot of jobs." If people really had no preference for environment over jobs, or vice versa, comment on the probability of observing this survey result (that is, observing 71% or more of a sample of 1502 in favor of environment over jobs). What assumption must be made about the sampling procedure in order to calculate this probability? [Hint: Recall Tchebysheff's theorem and the empirical rule.]

3.108   For a certain section of a pine forest, the number of diseased trees per acre, Y, has a Poisson distribution with mean $\lambda = 10$. The diseased trees are sprayed with an insecticide

---

† *Source* Reprinted by permission of *The Wall Street Journal*, © Dow Jones & Company, Inc. 1981. All rights reserved.

at a cost of $3.00 per tree, plus a fixed overhead cost for equipment rental of $50.00. Letting C denote the total spraying cost for a randomly selected acre, find the expected value and standard deviation for C. Within what interval would you expect C to lie with probability at least 0.75?

3.109 It is known that 10% of a brand of television tubes will burn out before their guarantee has expired. If 1000 tubes are sold, find the expected value and variance of Y, the number of original tubes that must be replaced. Within what limits would Y be expected to fall?

3.110 Refer to Exercise 3.55. In this exercise, we determined that the mean and variance of the costs necessary to find three employees with positive indications of asbestos poisoning were 150 and 4500, respectively. Do you think that it is highly unlikely that the cost of completing the tests will exceed $350?

3.12 Summary

This chapter has been concerned with discrete random variables and their probability distributions and expectations. Calculating the probability distribution for a discrete random variable implies evaluating the probability of a numerical event through the use of the probabilistic methods of Chapter 2. Probability distribution function were derived for the binomial, geometric, negative binomial, hypergeometric, and Poisson random variables.

The expected values of random variables and functions of random variables provided a method for finding the mean and variance of Y and consequently measures of centrality and variation for $p(y)$. Much of the remaining material in the chapter was devoted to the techniques for acquiring expectations, which sometimes involved summing apparently intractable series. Some of the techniques used to obtain closed form expressions for some of the resulting expected values included (1) Use of the fact that $\sum_y p(y) = 1$ for any discrete random variable and (2) $E(Y^2) = E[Y(Y-1)] + E(Y)$. Results obtained for the means and variances of several of the more common discrete distributions are summarized in Table 3.4. These results and more also are found in Appendix A2.1 and inside the rear cover of the book.

We then presented a discussion of the moment-generating function for a random variable. Although sometimes useful in finding μ and σ, the moment-generating function is of primary value to the theoretical statistician in deriving the probability distribution for a random variable. Appendix II gives the means, variances, and moment-generating functions for most of the common random variables.

The probability-generating function is another useful device for deriving moments and probability distributions of integer-valued random variables.

Finally, we gave Tchebysheff's theorem, which is useful for approximating probabilities when only the mean and variance are known.

To conclude this summary, we suggest relating this chapter to the objective of statistics, making an inference about a population based on

Table 3.4
Means and variances
for some common
random variables

Distribution	$E(Y)$	$V(Y)$
Binomial	np	npq
Geometric	$\dfrac{1}{p}$	$\dfrac{q}{p^2}$
Hypergeometric	$\dfrac{nr}{N}$	$\dfrac{nr(N-r)(N-n)}{N^2(N-1)}$
Poisson	λ	λ
Negative binomial	$\dfrac{r}{p}$	$\dfrac{r(1-p)}{p^2}$

information contained in a sample. Drawing the sample from the population is the experiment. The sample is a set of measurements on one or more random variables, and it is the observed event resulting from a single repetition of the experiment. Finally, making the inference about the population requires knowledge of the probability of the observed sample, which, as a basis, requires knowledge of the probability distributions of the random variables that generated the sample.

References and Further Readings

1. Feller, W. *An Introduction to Probability Theory and Its Applications*. Vol. 1, 3d ed. New York: Wiley, 1968.

2. Goranson, U. G., and Hall, J. "Airworthiness of Long-Life Jet Transport Structures," *Aeronautical Journal* 84, no. 838 (1980): 279–280.

3. Mendenhall, W. *An Introduction to Probability and Statistics*. 7th ed. N. Scituate, Mass.: Duxbury Press, 1987.

4. Mosteller, F.; Rourke, R. E. K.; and Thomas, G. B. *Probability with Statistical Applications*. 2d ed. Reading, Mass.: Addison-Wesley, 1970.

5. Parzen, E. *Modern Probability Theory and Its Applications*. New York: Wiley, 1960.

6. Parzen, E. *Stochastic Processes*. San Francisco: Holden-Day, 1962.

7. *Standard Mathematical Tables*. 17th ed. Cleveland: Chemical Rubber Company, 1969.

Supplementary Exercises

3.111   Sampling for defectives from large lots of manufactured product yields a number of defectives, Y, that follows a binomial probability distribution. A sampling plan consists in specifying the number of items to be included in a sample, n, and an acceptance number, a. The lot is accepted if $Y \leq a$ and rejected if $Y > a$. Let p denote the proportion of defectives in the lot. For $n = 5$ and $a = 0$ calculate the probability of lot acceptance if (a) $p = 0$, (b) $p = .1$, (c) $p = .3$, (d) $p = .5$, (e) $p = 1.0$. A graph showing the probability of lot acceptance as a function of lot fraction defective is called the *operating characteristic curve* for the sample plan. Construct the operating characteristic curve for the plan $n = 5$, $a = 0$. Note that a sampling plan is an example of statistical inference. Accepting or rejecting a lot based on information contained in the sample is equivalent to concluding that the lot is either good or bad, respectively. "Good" implies that a low fraction is defective and that the lot therefore is suitable for shipment.

3.112   Refer to Exercise 3.111. Use Table 1, Appendix III, to construct the operating characteristic curve for the following sampling plans:

(a) $n = 10$, $a = 0$,

(b) $n = 10$, $a = 1$,

(c) $n = 10$, $a = 2$.

For each, calculate P(lot acceptance) for $p = 0$, .05, .1, .3, .5, and 1.0. Our intuition suggests that sampling plan (a) would be much less likely to accept bad lots than plans (b) and (c). A visual comparison for the operating characteristic curves will confirm this intuitive conjecture.

3.113   A quality control engineer wishes to study the alternative sampling plans $n = 5$, $a = 1$ and $n = 25$, $a = 5$. On a sheet of graph paper construct the operating characteristic curves for both plans, making use of acceptance probabilities at $p = .05$, $p = .10$, $p = .20$, $p = .30$, and $p = .40$ in each case.

(a)   If you were a seller producing lots with fraction defective ranging from $p = 0$ to $p = .10$, which of the two sampling plans would you prefer?

(b)   If you were a buyer wishing to be protected against accepting lots with fraction defective exceeding $p = .30$, which of the two sampling plans would you prefer?

3.114   A starter motor utilized in a space vehicle possesses a high reliability and was reputed to start on any given occasion with probability .99999. What is the probability of at least one failure in the next 10,000 starts?

3.115   Refer to Exercise 3.68. Find μ, the expected value of Y, for the theoretical population by using the probability distribution obtained in Exercise 3.68. Find the sample mean $\bar{y}$ for the $n = 100$ measurements generated in Exercise 3.69. Does $\bar{y}$ provide a good estimate of μ?

3.116   Find the population variance σ^2 for Exercise 3.68 and the sample variance s'^2 for Exercise 3.69. Compare.

3.117   Toss a balanced die and let Y be the number of dots observed on the upper face. Find the mean and variance of Y. Construct a probability histogram and locate the interval $\mu \pm 2\sigma$. Verify that Tchebysheff's theorem holds.

3.118   One concern of a gambler is that he will go broke before he acquires his first win. Suppose that he plays a game in which the probability of winning is .1 (and is unknown to him). It costs him \$10 to play and he receives \$80 for a win. If he commences with \$30, what is the probability that he wins exactly once before he loses his initial capital?

3.119   The number of imperfections in the weave of a certain textile has a Poisson distribution with a mean of four per square yard.

(a)   Find the probability that a 1-square-yard sample will contain at least one imperfection.

(b)   Find the probability that a 3-square-yard sample will contain at least one imperfection.

3.120 The number of bacteria colonies of a certain type in samples of polluted water has a Poisson distribution with a mean of 2 per cubic centimeter.

   (a) If four 1-cubic-centimeter samples are independently selected from this water, find the probability that at least one sample will contain one or more bacteria colonies.

   (b) How many 1-cubic-centimeter samples should be selected in order to have a probability of approximately 0.95 of seeing at least one bacteria colony?

3.121 Use the fact that

$$e^z = 1 + z + \frac{z^2}{2!} + \frac{z^3}{3!} + \frac{z^4}{4!} + \cdots$$

to expand the moment-generating function for the binomial distribution

$$m(t) = (q + pe^t)^n$$

into a power series in t. (Acquire only the low-order terms in t.) Identify μ_i' as the coefficient of $t^i/i!$ appearing in the series. Specifically, find μ_1' and μ_2' and compare them with the results of Exercise 3.87.

3.122 Refer to Exercises 3.63 and 3.64. In what interval would you expect the repair costs on these five machines to lie? (Use Tchebysheff's Theorem.)

*3.123 The number of cars driving past a parking area in a 1-minute time interval has a Poisson distribution with mean λ. The probability that any individual driver actually wants to park his or her car is p. Assume that individuals decide whether to park independently of one another.

   (a) If one parking place is available and it will take you 1 minute to reach the parking area, what is the probability that a space will still be available when you reach the lot? (Assume that no one leaves the lot during the 1-minute interval.)

   (b) Let W denote the number of drivers who wish to park during a 1-minute interval. Derive the probability distribution of W.

3.124 An experiment consists of tossing a fair die until a 6 occurs four times. What is the probability that the process ends after exactly ten tosses with a 6 occurring on the ninth and tenth tosses?

3.125 Accident records collected by an automobile insurance company give the following information. The probability that an insured driver has an automobile accident is .15. If an accident has occurred, the damage to the vehicle amounts to 20% of its market value with a probability of .80, to 60% of its market value with a probability of .12, and to a total loss with a probability of .08. What premium should the company charge on a $4000 car so that the expected gain by the company is zero?

3.126 The number of people entering the intensive care unit at a particular hospital on any single day possesses a Poisson distribution with a mean equal to five persons per day.

   (a) What is the probability that the number of people entering the intensive care unit on a particular day is equal to two? Less than or equal to two?

   (b) Is it likely that Y will exceed ten? Explain.

3.127 A recent survey suggests that Americans anticipate a reduction in living standards and that a steadily increasing consumption no longer may be as important as it was in the past. Suppose that a poll of 2000 people indicated 1373 in favor of forcing a reduction in the size of American automobiles by legislative means. Would you expect to observe as many as 1373 in favor of this proposition if, in fact, the general public was split 50–50 on the issue? Why?

3.128   A supplier of heavy construction equipment has found that new customers are normally obtained by customer request for a sales call and that the probability of a sale of a particular piece of equipment is .3. If the supplier has three pieces of the equipment available for sale, what is the probability that it will take fewer than five customer contacts to clear the inventory?

3.129   Calculate $P(|Y - \lambda| \leq 2\sigma)$ for the Poisson probability distribution of Example 3.20. Does this agree with the empirical rule?

*3.130   A merchant stocks a certain perishable item. She knows that on any given day she will have a demand for either two, three, or four of these items with probabilities .1, .4, and .5, respectively. She buys the items for \$1.00 each and sells them for \$1.20 each. If any are left at the end of the day, they represent a total loss. How many items should the merchant stock so as to maximize her expected daily profit?

*3.131   Show that the hypergeometric probability function approaches the binomial in the limit as $N \to \infty$ and $p = r/N$ remains constant. That is, show that

$$\lim_{N \to \infty} \frac{\binom{r}{y}\binom{N-r}{n-y}}{\binom{N}{n}} = \binom{n}{y}p^y q^{n-y}$$

for $p = r/N$ constant.

3.132   A lot of $N = 100$ industrial products contains 40 defectives. Let Y be the number of defectives in a random sample of size 20. Find $p(10)$ by using (a) the hypergeometric probability distribution and (b) the binomial probability distribution. Is N large enough so that the binomial probability function is a good approximation to the hypergeometric probability function?

*3.133   For simplicity, let us assume that there are two kinds of drivers. The safe drivers, who are 70% of the population, have a probability of .1 of causing an accident in a year. The rest of the population are accident makers, who have a probability of .5 of causing an accident in a year. The insurance premium is \$400 times one's probability of causing an accident in the following year. A new subscriber has an accident during the first year. What should be his insurance premium for the next year?

*3.134   It is known that 5% of a population have disease A, which can be discovered by blood test. Suppose that N (a large number) people are to be tested. This can be done in two ways: (1) Each person is tested separately; (2) the blood samples of k people are pooled together and analyzed. (Assume that $N = nk$, with n an integer.) If the test is negative, all of them are healthy (that is, just this one test is needed). If the test is positive, each of the k persons must be tested separately (that is, totally, $k + 1$ tests are needed).

(a)   For fixed k what is the expected number of tests needed in (2)?

(b)   Find the k that will minimize the expected number of tests in (2).

(c)   How many tests does (b) save in comparison with (1)?

*3.135 Let Y have a hypergeometric distribution; that is,

$$p(y) = \frac{\binom{r}{y}\binom{N-r}{n-y}}{\binom{N}{n}}, \qquad y = 0, 1, 2, \ldots, n$$

(a) Show that

$$P(Y=n) = p(n) = \left(\frac{r}{N}\right)\left(\frac{r-1}{N-1}\right)\left(\frac{r-2}{N-2}\right)\cdots\left(\frac{r-n+1}{N-n+1}\right)$$

(b) Write $p(y)$ as $p(y|r)$. Show that if $r_1 < r_2$, then

$$\frac{p(y|r_1)}{p(y|r_2)} > \frac{p(y+1|r_1)}{p(y+1|r_2)}$$

(c) Apply the binomial expansion to each factor in the following equation:

$$(1 + a)^{N_1}(1 + a)^{N_2} = (1 + a)^{N_1 + N_2}$$

Now compare the coefficients of a^n in both sides to prove that

$$\binom{N_1}{0}\binom{N_2}{n} + \binom{N_1}{1}\binom{N_2}{n-1} + \cdots + \binom{N_1}{n}\binom{N_2}{0} = \binom{N_1 + N_2}{n}$$

(d) Using the result of part (c), conclude that

$$\sum_{y=0}^{n} p(y) = 1$$

3.136 Use the result derived in Exercise 3.135(c) and Definition 3.4 to directly derive the mean of a hypergeometric random variable.

3.137 Use the result of Exercises 3.135(c) and 3.136 to show that for a hypergeometric random variable

$$E[Y(Y - 1)] = \frac{r(r - 1)n(n - 1)}{N(N - 1)}.$$

chapter four

CONTINUOUS RANDOM VARIABLES AND THEIR PROBABILITY DISTRIBUTIONS

4.1 Introduction

A moment of reflection on statistical problems encountered in the real world should convince you that not all random variables fit the definition for discrete random variables. For example, observing the number of defectives in a lot of n manufactured items leads to a discrete random variable because the number of defectives is finite and takes one of the values $0, 1, 2, \ldots$, or n. On the other hand, consider the daily rainfall at a certain geographical point. Theoretically, with measuring equipment of perfect accuracy, we could associate each possible amount of rainfall with a unique point on a line interval. Thus each of the uncountable infinity of points on the line interval would represent a distinct possible value of the amount of rainfall. The type of random variable that takes on any value in an interval is called *continuous*, and the purpose of this chapter is to study probability distributions for continuous random variables. The yield of an antibiotic in a fermentation process is a continuous random variable, as is the length of life of a washing machine. The line segment over which these two random variables are defined is the positive half of the real line. This does not mean that, if we observed enough washing machines, we would eventually observe every positive real number as at least one outcome, rather that no positive real number can be ruled out as a possible outcome of an observation on length of life.

The probability distribution for a discrete random variable can always be given by assigning a positive probability to each of the possible values the variable may assume. Of course, we must ensure that the sum of the probabilities we assign is always equal to one. Unfortunately, the probability distribution for a continuous random variable cannot be specified in the same way. It is mathematically impossible to assign nonzero probabilities to all the points on a line interval and at the same time satisfy the requirement that the probabilities of the distinct possible values sum to one. So we must develop a different method for describing the probability distribution for a continuous random variable.

4.2 The Probability Distribution for a Continuous Random Variable

We will lead to a formal definition for a continuous random variable by first defining a distribution function (or *cumulative* distribution function).

Definition 4.1

> Let Y denote any random variable. The *distribution function* of Y, denoted by $F(y)$, is given by $F(y) = P(Y \leq y)$, $-\infty < y < \infty$.

As you will subsequently learn, the nature of the distribution function associated with a random variable will be used to determine whether the variable is continuous or discrete. Consequently, we will commence our discussion by examining the distribution function for a discrete random variable, noting its characteristics.

EXAMPLE 4.1   Suppose that Y has a binomial distribution with $n = 2$ and $p = 1/2$. Find $F(y)$.

Solution   The probability distribution for Y is given by

$$p(y) = \binom{2}{y}\left(\frac{1}{2}\right)^{y}\left(\frac{1}{2}\right)^{2-y}, \qquad y = 0, 1, 2$$

which yields

$$p(0) = 1/4 \qquad p(1) = 1/2 \qquad p(2) = 1/4$$

Then

$$F(y) = P(Y \leq y)$$
$$= 0 \qquad \text{for} \qquad y < 0$$
$$= 1/4 \qquad \text{for} \qquad 0 \leq y < 1$$
$$= 3/4 \qquad \text{for} \qquad 1 \leq y < 2$$
$$= 1 \qquad \text{for} \qquad y \geq 2$$

Note, for example, that

$$P(Y \leq 1.5) = P(Y \leq 1) = P(Y = 0) + P(Y = 1)$$
$$= (1/4) + (1/2) = 3/4$$

$F(y)$ is depicted graphically in Figure 4.1.

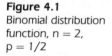

Figure 4.1
Binomial distribution
function, $n = 2$,
$p = 1/2$

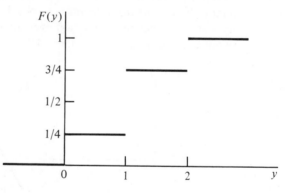

In Example 4.1 the points between 0 and 1 or between 1 and 2 contributed nothing to the cumulative probability depicted by the distribution function. As a result, the cumulative distribution function increased in jumps or steps at each of the possible values of Y. Functions that behave in such a manner are called *step functions. Distribution functions for discrete random variables are always step functions because the cumulative distribution function increases only at a countable number of points.*

Because the distribution function associated with any random variable is such that $F(y) = P(Y \leq y)$, from a practical point of view it is clear that $P(Y \leq -\infty) = F(-\infty)$ must equal zero. If we consider two values $y_1 < y_2$, then $P(Y \leq y_1) \leq P(Y \leq y_2)$—that is, $F(y_1) \leq F(y_2)$—and the function $F(y)$ (drop the subscript) is a monotonic, nondecreasing function. Further, it is clear that $P(Y \leq \infty) = F(\infty) = 1$. These three characteristics define the properties of any distribution function and are summarized in the following theorem.

Theorem 4.1

(Properties of a Distribution Function*) If $F(y)$ is a distribution function, then

1. $\lim\limits_{y \to -\infty} F(y) = F(-\infty) = 0.$

2. $\lim\limits_{y \to \infty} F(y) = F(\infty) = 1.$

3. $F(y_b) \geq F(y_a)$ if $y_b > y_a.$

You should check that the distribution function developed in Example 4.1 has these properties.

Let us now examine the distribution function for a continuous random variable. A random variable like the amount of daily rainfall is such that every interval of positive real numbers has a positive probability of including an observed value. Thus if Y, the random variable denoting daily rainfall, has distribution function $F(y)$, then $F(y_1) < F(y_2)$ when $0 < y_1 < y_2$. There is a positive probability of observing a rainfall measurement between y_1 and y_2, no matter how short this interval becomes. It follows that $F(y)$ in this case should be a smooth, increasing function over some interval of real numbers. Thus $F(y)$ would appear as shown in Figure 4.2.

Figure 4.2
Distribution function
for a continuous
random variable

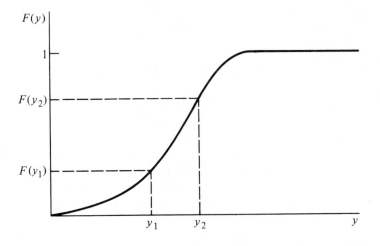

We are thus led to the definition of a continuous random variable.

Definition 4.2

Let Y denote a random variable with distribution function $F(y)$. Y is said to be *continuous* if $F(y)$ is continuous, for $-\infty < y < \infty.$†

* To be mathematically rigorous, if $F(y)$ is to have the properties that we ascribe, $F(y)$ also must be right continuous.

† To be mathematically precise, we also need the first derivative of $F(y)$ to exist and be continuous except for, at most, a finite number of points in any finite interval. The distribution functions for the continuous random variables discussed in this text satisfy this requirement.

For a continuous random variable Y we must have, for any real number y,

$$P(Y = y) = 0$$

If this were not true, and $P(Y = y_0) = p_0 > 0$, then $F(y)$ would have a discontinuity (jump) at the point y_0. Practically speaking, the fact that continuous random variables have zero probability at discrete points should not bother us. Consider the example of measuring daily rainfall. What is the probability that we will see a daily rainfall measurement of exactly 2.193 inches? It is quite likely that we would never observe that exact value even if we took rainfall measurements for a lifetime, although we may see many days with measurements between 2 and 3 inches.

The derivative of $F(y)$ is another function of prime importance in probability theory and statistics.

Definition 4.3

> Let $F(y)$ be the distribution function for a continuous random variable Y. Then $f(y)$, given by
>
> $$f(y) = \frac{dF(y)}{dy} = F'(y)$$
>
> wherever the derivative exists, is called the *probability density function* for the random variable Y.

It follows from Definitions 4.2 and 4.3 that $F(y)$ can be written as

$$F(y) = \int_{-\infty}^{y} f(t)\, dt$$

where $f(y)$ is the probability density function and t is used as the variable of integration. A graphical depiction of this relationship between the distribution and density functions is given in Figure 4.3.

The probability density function is a *theoretical model* for the frequency distribution of a population of measurements. For example, observations of the

Figure 4.3
The distribution
function

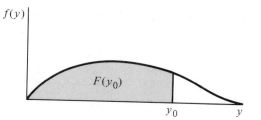

lengths of life of a particular brand of washing machines will generate measurements that can be characterized by a frequency histogram, as discussed in Chapter 1. Conceptually, the experiment could be repeated ad infinitum, thereby generating a frequency distribution (a smooth curve) that would characterize the population of interest to the manufacturer. This theoretical frequency distribution corresponds to the probability density function for the length of life of a single machine, Y.

Because the distribution function $F(y)$ for any random variable always has certain properties, it follows that density functions also will have some corresponding properties. Because $F(y)$ is a nondecreasing function, it follows that the derivative $f(y)$ is never negative. Further, we know that $F(\infty) = 1$ and therefore that $\int_{-\infty}^{\infty} f(t)\, dt = 1$. In summary, the properties of a probability density function are as given in the following theorem.

Theorem 4.2

(Properties of a Density Function) If $f(y)$ is a density function, then

1. $f(y) \geq 0$ for any value of y.

2. $\displaystyle\int_{-\infty}^{\infty} f(y)\, dy = 1$.

An example of a distribution function and a density function for a continuous random variable follows.

EXAMPLE 4.2   Suppose that $F(y) = y$, $0 \leq y \leq 1$, $F(y) = 0$, $y < 0$, and $F(y) = 1$, $y > 1$. Find the probability density function for y and graph it.

Solution   Because $F(y) = \int_{-\infty}^{y} f(t)\, dt = y$, $0 \leq y \leq 1$, then

$$\frac{dF(y)}{dy} = f(y) = \frac{d(y)}{dy} = \begin{cases} 1, & \text{for } 0 < y < 1 \\ 0, & \text{for } y < 0 \text{ or } y > 1 \end{cases}$$

and $f(y)$ is undefined at $y = 0$ and $y = 1$. A graph of $F(y)$ is shown in Figure 4.4.

Figure 4.4
Distribution function
F(y) for Example 4.2

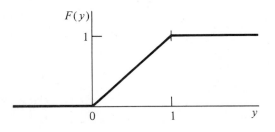

The graph of $f(y)$ for Example 4.2 is shown in Figure 4.5. Notice that the distribution and density functions given in Example 4.2 have all of the properties required to be bona fide distribution and density functions, respectively. Also note that $F(y)$ is a continuous function of y but that $f(y)$ is discontinuous at the points $y = 0$, 1. In general, the distribution function for a continuous random variable must be continuous, but the density function need not be everywhere continuous.

Figure 4.5
Density function f(y)
for Example 4.2

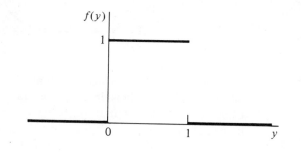

EXAMPLE 4.3   Let Y be a continuous random variable with probability density function given by

$$f(y) = \begin{cases} 3y^2, & 0 \le y \le 1 \\ 0, & \text{elsewhere} \end{cases}$$

Find $F(y)$. Graph both $f(y)$ and $F(y)$.

Solution   The graph of $f(y)$ appears in Figure 4.6. Because

$$F(y) = \int_{-\infty}^{y} f(t) \, dt$$

Figure 4.6
Density function for
Example 4.3

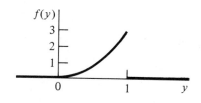

we have, for this example,

$$F(y) = \int_{0}^{y} 3t^2 \, dt = t^3 \big]_0^y = y^3 \qquad \text{for} \qquad 0 < y < 1$$

Note that $$F(y) = \begin{cases} 0, & \text{for } y \leq 0 \\ 1, & \text{for } y \geq 1 \end{cases}$$

The graph of $F(y)$ is given in Figure 4.7.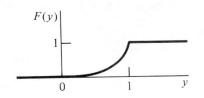

Figure 4.7
Distribution function
for Example 4.3

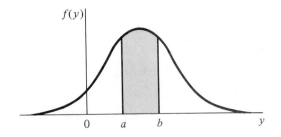

$F(y_0)$ gives the probability that $Y \leq y_0$. The next step is to find the probability that Y falls in a specific interval; that is, $P(a \leq Y \leq b)$. From Chapter 1 we know that probability corresponds to the area under the frequency distribution over the interval $a \leq y \leq b$. Because $f(y)$ is the theoretical counterpart of the frequency distribution, we would expect $P(a \leq Y \leq b)$ to equal a corresponding area under the density function $f(y)$. This indeed is true, because

$$P(a < Y \leq b) = P(Y \leq b) - P(Y \leq a) = F(b) - F(a) = \int_a^b f(y)\, dy$$

Because $P(Y = a) = 0$, we have the following result.

Theorem 4.3

> If the random variable Y has density function $f(y)$ and $a \leq b$, then the probability that Y falls in the interval $[a, b]$ is
>
> $$P(a \leq Y \leq b) = \int_a^b f(y)\, dy.$$
>
> where $f(y)$ is the probability density function for Y.

This probability is the shaded area in Figure 4.8.

Figure 4.8
$P(a \leq Y \leq b)$

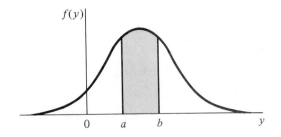

EXAMPLE 4.4   Given $f(y) = cy^2$, $0 \le y \le 2$, and $f(y) = 0$ elsewhere, find the value of c for which $f(y)$ will be a valid density function.

Solution   We require a value for c such that

$$F(\infty) = \int_{-\infty}^{\infty} f(y)\, dy = 1$$

$$F(\infty) = \int_{-\infty}^{\infty} f(y)\, dy = \int_{0}^{2} cy^2\, dy = \frac{cy^3}{3} \Big]_{0}^{2} = \left(\frac{8}{3}\right)c$$

Setting $F(\infty)$ equal to 1, $F(\infty) = (8/3)c = 1$, we find that $c = 3/8$.   ∽

EXAMPLE 4.5   Find the probability that $1 \le Y \le 2$ for Example 4.4.

Solution
$$P(1 \le Y \le 2) = \int_{1}^{2} f(y)\, dy = \frac{3}{8} \int_{1}^{2} y^2\, dy = \left(\frac{3}{8}\right) \frac{y^3}{3} \Big]_{1}^{2} = \frac{7}{8}.$$ ∽

Probability statements with regard to a continuous random variable Y will be meaningful only if, first, the integral defining the probability exists and, second, the resulting probabilities agree with the axioms of Chapter 2. These two conditions will always be satisfied if we consider only probabilities associated with a finite or countable collection of intervals. Because we almost always will be interested in probabilities of the form $P(a \le Y \le b)$, this consideration will cause us no practical difficulty. Some density functions that provide good models for population frequency distributions encountered in nature are presented in subsequent sections.

Exercises

4.1   Suppose Y possesses the density function

$$f(y) = \begin{cases} cy, & 0 \le y \le 2 \\ 0, & \text{elsewhere} \end{cases}$$

(a)   Find the value of c that makes $f(y)$ a probability density function.
(b)   Find $F(y)$.
(c)   Graph $f(y)$ and $F(y)$.
(d)   Use $F(y)$ to find $P(1 \le Y \le 2)$.
(e)   Use $f(y)$ and geometry to find $P(1 \le Y \le 2)$.

4.2   The length of time to failure (in hundreds of hours) for a certain transistor is a random variable Y with distribution function given by

$$F(y) = \begin{cases} 0, & y < 0 \\ 1 - e^{-y^2}, & y \geq 0 \end{cases}$$

   (a)   Show that $F(y)$ has the properties of a distribution function.

   (b)   Find $f(y)$.

   (c)   Find the probability that the transistor operates for at least 200 hours.

4.3   A supplier of kerosene has a 150-gallon tank that is filled at the beginning of each week. His weekly demand shows a relative frequency behavior that increases steadily up to 100 gallons and then levels off between 100 and 150 gallons. If Y denotes weekly demand in hundreds of gallons, the relative frequency of demand can be modeled by

$$f(y) = \begin{cases} y, & 0 \leq y \leq 1 \\ 1, & 1 < y \leq 1.5 \\ 0, & \text{elsewhere} \end{cases}$$

   (a)   Find $F(y)$.

   (b)   Find $P(0 \leq Y \leq 0.5)$

   (c)   Find $P(.5 \leq Y \leq 1.2)$.

4.4   A gas station operates two pumps, each of which can pump up to 10,000 gallons of gas in a month. The total amount of gas pumped at the station in a month is a random variable, Y (measured in 10,000 gallons), which has a probability density function given by

$$f(y) = \begin{cases} y, & 0 < y < 1 \\ 2 - y, & 1 \leq y < 2 \\ 0, & \text{elsewhere} \end{cases}$$

   (a)   Graph $f(y)$.

   (b)   Find $F(y)$ and graph it.

   (c)   Find the probability that the station pumps between 8,000 and 12,000 gallons in a month.

   (d)   Given that the station has pumped over 10,000 gallons for a particular month, find the probability that the station pumped over 15,000 gallons during the month.

4.5   As a measure of intelligence, mice are timed when going through a maze to reach a reward of food. The time (in seconds) required for any mouse is a random variable, Y, with a density function given by

$$f(y) = \begin{cases} \dfrac{b}{y^2}, & y \geq b \\ 0, & \text{elsewhere} \end{cases}$$

where b is the minimum possible time needed to traverse the maze.

(a)   Show that $f(y)$ has the properties of a density function.

(b)   Find $F(y)$.

(c)   Find $P(Y > b + c)$ for a positive constant c.

4.6   Let Y possess a density function

$$f(y) = \begin{cases} c(2 - y), & 0 \leq y \leq 2 \\ 0, & \text{elsewhere} \end{cases}$$

(a)   Find c.

(b)   Find $F(y)$.

(c)   Graph $f(y)$ and $F(y)$.

(d)   Use $F(y)$ in (b) to find $P(1 \leq Y \leq 2)$.

(e)   Use the geometric figure for $f(y)$ to calculate $P(1 \leq Y \leq 2)$.

4.7   The length of time required by students to complete a 1-hour exam is a random variable with a density function given by

$$f(y) = \begin{cases} cy^2 + y, & 0 \leq y \leq 1 \\ 0, & \text{elsewhere} \end{cases}$$

(a)   Find c.

(b)   Find $F(y)$.

(c)   Graph $f(y)$ and $F(y)$.

(d)   Use $F(y)$ in (b) to find $F(-1)$, $F(0)$, and $F(1)$.

(e)   Find the probability a student finishes in less than a half hour.

(f)   Given that a student needs at least 15 minutes to complete the exam, find the probability that she will require at least 30 minutes to finish.

4.8   Let Y have the density function given by

$$f(y) = \begin{cases} .2 & -1 < y \leq 0 \\ .2 + cy, & 0 < y \leq 1 \\ 0, & \text{elsewhere} \end{cases}$$

(a)   Find c.

(b)   Find $F(y)$.

(c)   Graph $f(y)$ and $F(y)$.

(d)   Use $F(y)$ in (b) to find $F(-1)$, $F(0)$, and $F(1)$.

(e)   Find $P(0 \leq Y \leq .5)$.

(f)   Find $P(Y > .5 | Y > .1)$.

4.9   Let the distribution function of a random variable Y be

$$F(y) = \begin{cases} 0, & y \leq 0 \\ \dfrac{y}{8}, & 0 < y < 2 \\ \dfrac{y^2}{16}, & 2 \leq y < 4 \\ 1, & y \geq 4 \end{cases}$$

(a)   Find the density function of Y.

(b)   Find $P(1 \leq Y \leq 3)$.

(c)   Find $P(Y \geq 1.5)$.

(d)   Find $P(Y \geq 1 \mid Y \leq 3)$.

4.3 The Expected Value for a Continuous Random Variable

The next step in the study of continuous random variables is to find their means, variances, and standard deviations and thereby to acquire numerical descriptive measures of their density functions. Particularly, we will learn subsequently that it is sometimes difficult to find the probability distribution for a random variable, Y, or a function $g(Y)$. We will soon see that integration over intervals for many density functions (the normal and gamma-type, for example) is very difficult. When this occurs, we can approximately describe the behavior of the random variable by using its moments along with the empirical rule and Tchebysheff's theorem (Chapters 1 and 3).

Definition 4.4

> The expected value of a continuous random variable Y is
>
> $$E(Y) = \int_{-\infty}^{\infty} y f(y) \, dy$$
>
> provided the integral exists.[†]

[†] Technically, $E(Y)$ is said to exist if

$$\int_{-\infty}^{\infty} |y| f(y) \, dy < \infty$$

This will be the case in all expectations we discuss, and we will not mention this additional condition each time we define an expected value.

If the definition of the expected value for a discrete random variable Y, $E(Y) = \sum_y yp(y)$, is meaningful, then Definition 4.4 also should agree with our intuitive notion of a mean. The quantity $f(y)\,dy$ corresponds to $p(y)$ for the discrete case, and integration evolves from and is analogous to summation. Hence $E(Y)$ in Definition 4.4 agrees completely with our notion of an average or mean.

As in the discrete case, there are times when we are interested in the expected value of a function of a random variable. A result that permits us to evaluate such an expected value is given in the following theorem.

Theorem 4.4

> Let $g(Y)$ be a function of Y, then the expected value of $g(Y)$ is given by
>
> $$E[g(Y)] = \int_{-\infty}^{\infty} g(y)f(y)\,dy$$
>
> provided the integral exits.

The proof of Theorem 4.4 is similar to that of Theorem 3.1 and is omitted. The expected values of three important functions of a continuous random variable, Y, evolve as a consequence of well-known theorems on integration. As expected, these results lead to conclusions analogous to those contained in Theorems 3.3, 3.4, and 3.5. As a consequence, the proof of Theorem 4.5 will be left as an exercise.

Theorem 4.5

> Let c be a constant and let $g(Y), g_1(Y), g_2(Y), \ldots, g_k(Y)$ be functions of a continuous random variable Y. Then the following results hold:
>
> **1.** $E(c) = c$.
> **2.** $E[cg(Y)] = cE[g(Y)]$.
> **3.** $E[g_1(Y) + g_2(Y) + \cdots + g_k(Y)] = E[g_1(Y)] +$
> $\qquad\qquad\qquad\qquad\qquad E[g_2(Y)] + \cdots + E[g_k(Y)]$.

As in the case of discrete random variables, we often seek the expected value of the function $g(Y) = (Y - \mu)^2$, which intuitively is the variance of the random variable Y. That is, as in Definition 3.5, $V(Y) = E(Y - \mu)^2$. It is a simple exercise to show that Theorem 4.5 implies that $V(Y) = E(Y^2) - \mu^2$.

EXAMPLE 4.6

In Example 4.4 we determined that $f(y) = (3/8)y^2$, $0 \le y \le 2$, $f(y) = 0$ elsewhere, is a valid density function. If the random variable Y has this density function, find $\mu = E(Y)$ and $\sigma^2 = V(Y)$.

Solution   According to Definition 4.4,

$$E(Y) = \int_{-\infty}^{\infty} y f(y)\, dy$$

$$= \int_{0}^{2} y \left(\frac{3}{8}\right) y^2\, dy$$

$$= \left(\frac{3}{8}\right)\left(\frac{1}{4}\right) y^4 \Big|_{0}^{2} = 1.5$$

The variance of Y can be found once we determine $E(Y^2)$. In this case,

$$E(Y^2) = \int_{-\infty}^{\infty} y^2 f(y)\, dy$$

$$= \int_{0}^{2} y^2 \left(\frac{3}{8}\right) y^2\, dy$$

$$= \left(\frac{3}{8}\right)\left(\frac{1}{5}\right) y^5 \Big|_{0}^{2} = 2.4$$

Thus, $\sigma^2 = V(Y) = E(Y^2) - [E(Y)]^2 = 2.4 - (1.5)^2 = 0.15.$ ✎

Exercises

4.10   Find the mean and variance of the random variable of Exercise 4.6.

4.11   Find the mean and variance of the random variable of Exercise 4.7.

4.12   Find the mean and variance of the random variable of Exercise 4.8.

4.13   If Y is a continuous random variable with mean μ and variance σ^2 and a and b are constants, use Theorem 4.5 to prove the following:

(a) $E(aY + b) = aE(Y) + b = a\mu + b.$

(b) $V(aY + b) = a^2 V(Y) = a^2 \sigma^2.$

4.14   For certain ore samples the proportion of impurities per sample, Y, is a random variable with a density function given by

$$f(y) = \begin{cases} (3/2)y^2 + y, & 0 \le y \le 1 \\ 0, & \text{elsewhere} \end{cases}$$

The dollar value of each sample is $V = 5 - 0.5Y$. Find the mean and variance of V.

4.15   The proportion of time per day that all checkout counters in a supermarket are busy is a random variable, Y, with a density function

$$f(y) = \begin{cases} cy^2(1 - y)^4, & 0 \le y \le 1 \\ 0, & \text{elsewhere} \end{cases}$$

(a)   Find the value of c that makes $f(y)$ a probability density function.

(b)   Find $E(Y)$.

4.16   The temperature, Y, at which a thermostatically controlled switch turns on has a probability density function given by

$$f(y) = \begin{cases} 1/2, & 59 \le y \le 61 \\ 0, & \text{elsewhere} \end{cases}$$

Find $E(Y)$ and $V(Y)$.

4.17   The proportion of time, Y, that an industrial robot is in operation during a 40-hour week is a random variable with probability density function

$$f(y) = \begin{cases} 2y, & 0 \le y \le 1 \\ 0, & \text{elsewhere} \end{cases}$$

(a)   Find $E(Y)$ and $V(Y)$.

(b)   For the robot under study the profit Y for a week is given by $X = 200Y - 60$, find $E(X)$ and $V(X)$.

(c)   Find an interval in which the profit should lie for at least 75% of the weeks that the robot is in use.

4.18   Daily total solar radiation for a certain location in Florida in October has a probability density function given by

$$f(y) = \begin{cases} 3/32(y - 2)(6 - y), & 2 \le y \le 6 \\ 0, & \text{elsewhere} \end{cases}$$

with measurements in hundreds of calories. Find the expected daily solar radiation for October.

4.19   Weekly CPU time used by an accounting firm has a probability density function (measured in hours) given by

$$f(y) = \begin{cases} 3/64y^2(4 - y), & 0 \le y \le 4 \\ 0, & \text{elsewhere} \end{cases}$$

(a)   Find the expected value and variance of weekly CPU time.

(b)   The CPU time costs the firm $200 per hour. Find the expected value and variance of the weekly cost for CPU time.

(c)   Would you expect the weekly cost to exceed $600 very often? Why?

4.20   The pH of water samples from a specific lake is a random variable, Y, with a probability density function given by

$$f(y) = \begin{cases} 3/8(7 - y)^2, & 5 \le y \le 7 \\ 0, & \text{elsewhere} \end{cases}$$

(a)   Find $E(Y)$ and $V(Y)$.

(b)   Find an interval shorter than (5, 7) in which at least 3/4 of the pH measurements must lie.

(c)   Would you expect to see a pH measurement below 5.5 very often? Why?

4.4 The Uniform Probability Distribution

Suppose that a bus always arrives at a particular stop between 8:00 and 8:10 A.M. and that the probability that the bus will arrive in any given subinterval of time is proportional only to the length of the subinterval. That is, the bus is as likely to arrive between 8:00 and 8:02 as it is to arrive between 8:06 and 8:08. Let Y denote the length of time a person must wait for the bus if that person arrived at the bus stop at exactly 8:00. If we carefully measured the number of minutes past 8:00 that the bus arrived for several mornings, a relative frequency histogram for the data could be developed. From the description just given, it should be clear that the relative frequency with which we observed Y between 0 and 2 would be approximately the same as the relative frequency with which we observed Y between 6 and 8. A reasonable model for the density function of Y is given in Figure 4.9. Because areas under curves represent probabilities for continuous random variables and $A_1 = A_2$ (by inspection), $P(0 \le Y \le 2) = P(6 \le Y \le 8)$, as desired.

Figure 4.9
Density function for Y

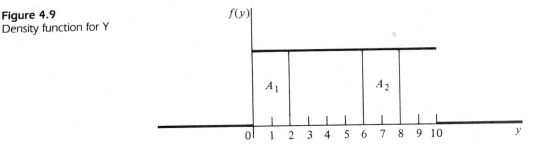

The random variable Y just discussed is an example of a random variable that has a uniform distribution. The general form for the density function of a random variable with a uniform distribution is as follows.

Definition 4.5

A random variable Y is said to have a *uniform probability distribution* if and only if the density function of Y is

$$f(y) = \begin{cases} \dfrac{1}{\theta_2 - \theta_1}, & \theta_1 \le y \le \theta_2 \\ 0, & \text{elsewhere} \end{cases}$$

In the bus problem we can take $\theta_1 = 0$ and $\theta_2 = 10$, because we are interested only in a particular 10-minute interval. The density function discussed in Example 4.2 is a uniform distribution with $\theta_1 = 0$ and $\theta_2 = 1$. The distribution function and density function for the random variable in Example 4.2 are graphically depicted in Figures 4.4 and 4.5, respectively.

Definition 4.6

> The constants that determine the specific form of a density function are called *parameters* of the density function.

The quantities θ_1 and θ_2 are parameters of the uniform density function and are clearly numerical descriptive measures of this theoretical density function. Both the range and the probability of Y falling in any given interval are dependent on θ_1 and θ_2.

Some continuous random variables in the physical, management, and biological sciences have approximate uniform probability distributions. For example, suppose we are counting events that have a Poisson distribution, such as telephone calls coming into a switchboard. If it is known that exactly one such event has occurred in a given interval, say, $(0, t)$, then the actual time of occurrence would be distributed uniformly over this interval.

EXAMPLE 4.7

Arrivals of customers at a certain checkout counter follow a Poisson distribution. It is known that during a given 30-minute period one customer arrived at the counter. Find the probability that the customer arrived during the last 5 minutes of the 30-minute period.

Solution

As just mentioned, the actual time of arrival would follow a uniform distribution over the interval of $(0, 30)$. If Y denotes the arrival time, then

$$P(25 \leq Y \leq 30) = \int_{25}^{30} \frac{1}{30}\, dy = \frac{30 - 25}{30} = \frac{5}{30} = \frac{1}{6}$$

Note that the probability of the arrival occurring in any other 5-minute interval would also be 1/6.   ∽

As we will subsequently see, the uniform distribution is also very important for theoretical reasons. Simulation studies are very valuable techniques for validating models in statistics. If we desire a set of observations on a random variable Y with distribution function $F(y)$, it often is possible to obtain the desired results by transforming a set of observations on a uniform random variable. For this reason most computer systems contain a random number generator, which will generate observed values for a uniform random variable.

Theorem 4.6

If Y is a random variable that possesses a uniform distribution with parameters θ_1 and θ_2, then

$$\mu = E(Y) = \frac{\theta_1 + \theta_2}{2} \quad \text{and} \quad \sigma^2 = V(Y) = \frac{(\theta_2 - \theta_1)^2}{12}$$

Proof   By Definition 4.4,

$$E(Y) = \int_{-\infty}^{\infty} yf(y)\, dy$$

$$= \int_{\theta_1}^{\theta_2} y \frac{1}{\theta_2 - \theta_1}\, dy$$

$$= \frac{1}{\theta_2 - \theta_1} \left(\frac{1}{2}\right) y^2 \Big|_{\theta_1}^{\theta_2} = \frac{\theta_2^2 - \theta_1^2}{2(\theta_2 - \theta_1)}$$

$$= \frac{\theta_2 + \theta_1}{2}.$$

Note that the mean of a uniform random variable is simply the value midway between the two parameter values, θ_1 and θ_2. The derivation of the variance is left as an exercise.

Exercises

4.21   Suppose Y has the uniform distribution over the interval $(0, 1)$.

(a)   Find $F(y)$.

(b)   Show that $P(a \le Y \le a + b)$, for $a \ge 0$, $b \ge 0$, $a + b \le 1$, depends only upon the value of b.

4.22   If a parachutist lands at a random point on a line between markers A and B, find the probability that she is closer to A than to B. Find the probability that her distance to A is more than three times her distance to B.

4.23   Suppose that three parachutists operate independently as described in Exercise 4.22. What is the probability that exactly one of the three lands past the midpoint between A and B.

4.24   A random variable Y has a uniform distribution over the interval (θ_1, θ_2). Derive the variance of Y.

4.25   The change in depth of a river from one day to the next, measured (in feet) at a specific location, is a random variable Y with the following density function:

$$f(y) = \begin{cases} k, & -2 \le y \le 2 \\ 0, & \text{elsewhere} \end{cases}$$

(a)   Determine the value of k.

(b)   Obtain the distribution function for Y.

4.26   The cycle time for trucks hauling concrete to a highway construction site is uniformly distributed over the interval 50 to 70 minutes. What is the probability that the cycle time exceeds 65 minutes if it is known that the cycle time exceeds 55 minutes?

4.27   Refer to Exercise 4.26. Find the mean and variance of the cycle times for the trucks.

4.28   The number of defective circuit boards coming off a soldering machine follows a Poisson distribution. During a specific 8-hour day, one defective circuit board was found.

(a)   Find the probability that it was produced during the first hour of operation on that day.

(b)   Find the probability that it was produced during the last hour of operation on that day.

(c)   Given that no defective circuit boards were produced during the first 4 hours of operation, find the probability that the board was manufactured during the fifth hour.

4.29   In determining the range of an acoustic source using the triangulation method, the time at which the spherical wave front arrives at a receiving sensor must be measured accurately. According to Perruzzi and Hilliard [6], measurement errors in these times can be modeled as possessing a uniform distribution from -0.05 to $+0.05$ microseconds.

(a)   What is the probability that a particular arrival time measurement will be in error by less than 0.01 microseconds?

(b)   Find the mean and variance of the measurement errors.

4.30   According to Y. Zimmels [11], the sizes of particles used in sedimentation experiments often have a uniform distribution. In sedimentation with mixtures of various sized particles, the larger particles hinder the movements of the smaller ones. Thus it is important to study both the mean and variance of particle sizes. Suppose that spherical particles have diameters that are uniformly distributed between 0.01 and 0.05 cm. Find the mean and variance of the *volumes* of these particles. (Recall that the volume of a sphere is $(4/3)\pi r^3$.)

4.5 The Normal Probability Distribution

We have talked about bell-shaped, or normal, distributions of data in connection with the empirical rule, and we will subsequently give an argument to justify the common occurrence of bell-shaped distributions of data in nature. Regardless of the reasons for their occurrence, it is a fact that measurements on many random variables appear to have been generated from population frequency distributions that are closely approximated by a normal probability distribution. The normal density function is as follows:

Definition 4.7

A random variable Y is said to have a *normal probability distribution* if and only if the density function of Y is

$$f(y) = \frac{e^{-(y-\mu)^2/2\sigma^2}}{\sigma\sqrt{2\pi}} \qquad \sigma > 0, \; -\infty < \mu < \infty, \; -\infty < y < \infty$$

Note that the normal density function contains two parameters, μ and σ.

Theorem 4.7

If Y is a normally distributed random variable with parameters μ and σ, then

$$E(Y) = \mu \qquad \text{and} \qquad V(Y) = \sigma^2$$

The proof of this theorem will be deferred to Section 4.9, where we derive the moment-generating function of a random variable with a normal distribution. The results contained in Theorem 4.7 imply that the parameter μ locates the center of the distribution and that σ measures its spread. A graph of the normal density function is shown in Figure 4.10.

Figure 4.10
The normal probability density function

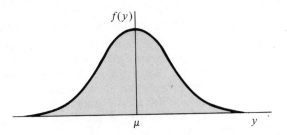

Areas under the normal density function corresponding to $P(a \le Y \le b)$ require evaluation of the integral

$$\int_a^b \frac{1}{\sigma\sqrt{2\pi}} e^{-(y-\mu)^2/2\sigma^2} \, dy$$

Unfortunately, a closed-form expression for the integral does not exist, and hence its evaluation can be obtained only by approximate procedures. As a consequence, areas under the normal density function are presented in Table 4, Appendix III.

The normal density function is symmetric with respect to μ, so areas need be tabulated only on one side of the mean. The tabulated areas are areas to the right of points z, where z is the distance from the mean measured in standard deviations. This area is shaded in Figure 4.11.

Figure 4.11
Tabulated area for the normal density function

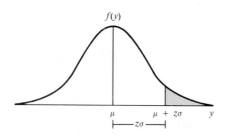

EXAMPLE 4.8   Let Z denote a normal random variable with mean 0 and standard deviation 1.

    (a)   Find $P(Z > 2)$.

    (b)   Find $P(-2 \leq Z \leq 2)$.

    (c)   Find $P(0 \leq Z \leq 1.73)$.

Solution   (a) Proceed down the first (z) column in Table 4, and read the area opposite $z = 2.0$. This area, denoted by the symbol $A(z)$, is $A(z = 2.0) = .0228$. Thus $P(Z > 2) = .0228$.

(b) Refer to Figure 4.12 where we have shaded the area of interest.

Figure 4.12
Desired area for Example 4.8(b).

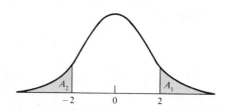

In part (a) we determined that $A_1 = A(z = 2.0) = .0228$. Because the density function is symmetric about the mean, $\mu = 0$, it follows that $A_2 = A_1 = .0228$ and hence that

$$P(-2 \leq Z \leq 2) = 1 - A_1 - A_2 = 1 - 2\,(.0228) = .9544$$

(c) Note that $P(0 \leq Z \leq 1.73) = .5 - A(1.73)$ where $A(1.73)$ is obtained by proceeding down the z column in Table 4 to the entry "1.7" and then across the top of the table to the column labeled ".03" to read $A(1.73) = .0418$. Thus

$$P(0 \leq Z \leq 1.73) = .5 - .0418 = .4582.$$

EXAMPLE 4.9   The achievement scores for a college entrance examination are normally distributed with mean 75 and standard deviation 10. What fraction of the scores lies between 80 and 90?

Solution   Recall that z is the distance from the mean of a normal distribution expressed in units of standard deviation. Thus

$$z = \frac{y - \mu}{\sigma}$$

Then the desired fraction of the population is given by the area between

$$z_1 = \frac{80 - 75}{10} = .5 \quad \text{and} \quad z_2 = \frac{90 - 75}{10} = 1.5$$

This area is shaded in Figure 4.13.

Figure 4.13
Required area for
Example 4.9

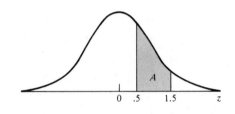

You can see from Figure 4.13 that $A = A(.5) - A(1.5) = .3085 - .0668 = .2417.$

We can always transform a normal random variable to Z, and thereby use Table 4 by the relationship

$$Z = \frac{Y - \mu}{\sigma}$$

We have noted that Z locates a point measured from the mean of a normal random variable with the distance *expressed in units of standard deviation* of the original normal random variable. Thus the mean value of Z must be 0 and its standard deviation must equal 1. Proof that Z, called a *standardized normal random variable*, is normally distributed with mean equal to 0 and standard deviation equal to 1 is given in Chapter 6.

Exercises

4.31   Use Table 4, Appendix III, to find the following probabilites for a standard normal random variable Z.

(a) $P(0 \le Z \le 1.2)$

(b) $P(-.9 \le Z \le 0)$

(c) $P(.3 \le Z \le 1.56)$

(d) $P(-.2 \leq Z \leq .2)$

(e) $P(-1.56 \leq Z \leq -.2)$

4.32   **(a)**   Find the value z_0 such that $P(Z > z_0) = .5$.

       (b)   Find the value z_0 such that $P(Z < z_0) = .8643$.

       (c)   Find the value z_0 such that $P(-z_0 < Z < z_0) = .90$.

       (d)   Find the value z_0 such that $P(-z_0 < Z < z_0) = .99$.

4.33   The weekly amount spent for maintenance and repairs in a certain company was observed, over a long period of time, to be approximately normally distributed with a mean of $400 and a standard deviation of $20. If $450 is budgeted for next week, what is the probability that the actual costs will exceed the budgeted amount?

4.34   In Exercise 4.33, how much should be budgeted for weekly repairs and maintenance in order that the budgeted amount be exceeded with a probability of only .1?

4.35   A machining operation produces bearings with diameters that are normally distributed with a mean of 3.0005 inches and a standard deviation of .0010 inch. Specifications require the bearing diameters to lie in the interval $3.000 \pm .0020$ inches. Those outside the interval are considered scrap and must be remachined. With the existing machine setting, what fraction of total production will be scrap?

4.36   In Exercise 4.35, what should the mean diameter be in order that the fraction of bearings scrapped be minimized?

4.37   The grade point averages of a large population of college students are approximately normally distributed with a mean of 2.4 and a standard deviation of .8. What fraction of the students will possess a grade point average in excess of 3.0?

4.38   Refer to Exercise 4.37. If students possessing a grade point average equal to or less than 1.9 are dropped from college, what percentage of the students will be dropped?

4.39   Refer to Exercise 4.37. Suppose that three students are randomly selected from the student body. What is the probability that all three will possess a grade point average in excess of 3.0?

4.40   Wires manufactured for use in a certain computer system are specified to have resistances between 0.12 and 0.14 ohms. The actual measured resistances of the wires produced by Company A have a normal probability distribution with a mean of 0.13 ohm and a standard deviation of 0.005 ohm.

       (a)   What is the probability that a randomly selected wire from Company A's production will meet the specifications?

       (b)   If four such wires are used in the system and all are selected from Company A, what is the probability that all four will meet the specifications?

4.41   One method of arriving at economic forecasts is to use a consensus approach. A forecast is obtained from each of a large number of analysts; the average of these individual forecasts is the consensus forecast. Suppose that the individual 1985 January prime-interest-rate forecasts of all economic analysts are approximately normally distributed with mean equal to 14 % and a standard deviation of 2.6 %. If a single analyst is randomly selected from among this group, what is the probability that the analyst's forecast of the prime interest rate will

       (a)   exceed 18 %?

       (b)   be less than 16 %?

4.42   The width of bolts of fabric is normally distributed with a mean of 950 millimeters and a standard deviation of 10 millimeters.

    (a)   What is the probability that a randomly chosen bolt has a width between 947 and 958 millimeters?

    (b)   What is the appropriate value for C such that a randomly chosen bolt has a width less than C with probability .8531?

4.43   Scores on an examination are assumed to be normally distributed with a mean of 78 and a variance of 36.

    (a)   What is the probability that a person taking the examination scores higher than 72?

    (b)   Suppose that students scoring in the top 10% of this distribution are to receive an A grade. What is the minimum score a student must achieve to earn an A grade?

    (c)   What must be the cutoff point for passing the examination if the examiner wants only the top 28.1% of all scores to be passing?

    (d)   Find, approximately, what proportion of students have scores 5 or more points above the score that cuts off the lowest 25%.

    (e)   If it is known that a student's score exceeds 72, what is the probability that his or her score exceeds 84?

4.44   A soft drink machine can be regulated so that it discharges an average of μ ounces per cup. If the ounces of fill are normally distributed with standard deviation equal to .3 ounce, give the setting for μ so that 8-ounce cups will overflow only 1% of the time.

4.45   Show that the maximum value of the normal density with parameters μ and σ is $1/(\sigma\sqrt{2\pi})$ and occurs when $y = \mu$.

4.46   Show that the normal density with parameters μ and σ has inflection points at the values $\mu - \sigma$ and $\mu + \sigma$. (Recall that an inflection point is a point where the curve changes direction from concave up to concave down, or vice versa, and occurs when the second derivative changes sign. Such a change in sign may occur when the second derivative equals zero.)

4.47   Assume that Y is normally distributed with a mean μ and a standard deviation σ. After observing a value of Y, a rectangle is constructed with length $L = |Y|$ and width $W = 3|Y|$. Let A denote the area of the resulting rectangle. What is $E(A)$?

4.6 The Gamma-Type Probability Distribution

Some random variables are always nonnegative and for various reasons yield distributions of data that are skewed (nonsymmetric) to the right. That is, most of the area under the density function is located near the origin and the density function drops gradually as y increases. A skewed probability distribution is shown in Figure 4.14.

    The lengths of time between malfunctions for aircraft engines possess a skewed frequency distribution, as do the lengths of time between arrivals at a supermarket checkout queue (that is, the line at the checkout counter). Similarly, the lengths of time to complete a maintenance checkout for an

Figure 4.14
A skewed probability
distribution

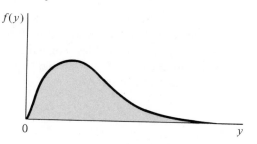

automobile or aircraft engine form a skewed frequency distribution. The populations associated with these random variables frequently possess distributions that are adequately modeled by the gamma-type density function.

Definition 4.8

A random variable Y is said to have a *gamma-type probability distribution with parameters* α and β if and only if the density function of Y is

$$f(y) = \frac{y^{\alpha-1}e^{-y/\beta}}{\beta^{\alpha}\Gamma(\alpha)}, \qquad \alpha, \beta > 0; 0 \leq y \leq \infty$$

$$0, \qquad\qquad \text{elsewhere}$$

where

$$\Gamma(\alpha) = \int_{0}^{\infty} y^{\alpha-1}e^{-y}\, dy$$

The quantity $\Gamma(\alpha)$ is known as the *gamma function*. Direct integration will verify that $\Gamma(1) = 1$. Integration by parts will verify that $\Gamma(\alpha) = (\alpha - 1)\Gamma(\alpha - 1)$ for any $\alpha \geq 1$ and that $\Gamma(n) = (n - 1)!$, provided n is an integer.

The gamma-type density functions for $\alpha = 1$, 2, and 4 and $\beta = 1$ are graphically presented in Figure 4.15.

Figure 4.15
The gamma density
function, $\beta = 1$

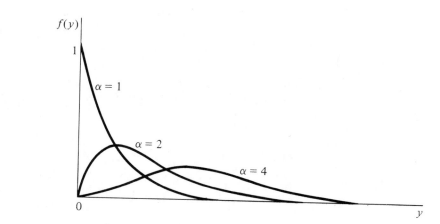

In the special case when α is an integer, it is possible to express the distribution function of a gamma-type random variable as a sum of certain Poisson random variables. You will find this representation in Exercise 4.57. If α is not an integer, it is impossible to give a closed-form expression for

$$\int_c^d \frac{y^{\alpha-1}e^{-y/\beta}}{\beta^\alpha \Gamma(\alpha)}\, dy, \quad \text{where } 0 < c < d < \infty$$

and hence it is impossible to obtain areas under the gamma-type density function by direct integration. Tabulated values of the integral are given in *Tables of the Incomplete Gamma Function* [4].

Theorem 4.8

> If Y is a gamma-type random variable with parameters α and β, then
>
> $$\mu = E(Y) = \alpha\beta \quad \text{and} \quad \sigma^2 = V(Y) = \alpha\beta^2.$$

Proof $$E(Y) = \int_{-\infty}^{\infty} yf(y)\, dy = \int_0^\infty y\, \frac{y^{\alpha-1}e^{-y/\beta}}{\beta^\alpha \Gamma(\alpha)}\, dy$$

By definition, the gamma-type density function is such that

$$\int_0^\infty \frac{y^{\alpha-1}e^{-y/\beta}}{\beta^\alpha \Gamma(\alpha)}\, dy = 1$$

Hence $$\int_0^\infty y^{\alpha-1}e^{-y/\beta}\, dy = \beta^\alpha \Gamma(\alpha)$$

and

$$E(Y) = \int_0^\infty \frac{y^\alpha e^{-y/\beta}\, dy}{\beta^\alpha \Gamma(\alpha)} = \frac{1}{\beta^\alpha \Gamma(\alpha)} \int_0^\infty y^\alpha e^{-y/\beta}\, dy$$

$$= \frac{1}{\beta^\alpha \Gamma(\alpha)}\,[\beta^{\alpha+1}\Gamma(\alpha+1)] = \frac{\beta\alpha\Gamma(\alpha)}{\Gamma(\alpha)}$$

or $E(Y) = \alpha\beta$.

From the continuous analog of Theorem 3.6, $V(Y) = E[Y^2] - [E(Y)]^2$. Further,

$$E(Y^2) = \int_0^\infty y^2\, \frac{y^{\alpha-1}e^{-y/\beta}}{\beta^\alpha \Gamma(\alpha)}\, dy = \frac{1}{\beta^\alpha \Gamma(\alpha)} \int_0^\infty y^{\alpha+1}e^{-y/\beta}\, dy$$

$$= \frac{1}{\beta^\alpha \Gamma(\alpha)}\,[\beta^{\alpha+2}\Gamma(\alpha+2)] = \frac{\beta^2(\alpha+1)\alpha\Gamma(\alpha)}{\Gamma(\alpha)} = \alpha(\alpha+1)\beta^2$$

Then $V(Y) = E[Y^2] - [E(Y)]^2$ where, from the earlier part of the derivation, $E(Y) = \alpha\beta$. Substituting $E[Y^2]$ and $E(Y)$ into the formula for $V(Y)$, we obtain

$$V(Y) = \alpha(\alpha + 1)\beta^2 - (\alpha\beta)^2 = \alpha^2\beta^2 + \alpha\beta^2 - \alpha^2\beta^2 = \alpha\beta^2$$

Two special cases of gamma-type random variables merit particular consideration.

Definition 4.9

> Let v be a positive integer. A random variable Y is said to have a *chi-square distribution with v degrees of freedom* if and only if Y is a gamma-type random variable with parameters $\alpha = v/2$ and $\beta = 2$.

A random variable with a chi-square distribution is called a *chi-square (χ^2) random variable*. Such a random variable occurs often in statistical theory. The motivation behind calling the parameter v the degrees of freedom of the chi-square distribution rests on one of the major applications of the distribution and cannot be explained intuitively at this point. The mean and variance of a χ^2 random variable follow directly from Theorem 4.8.

Theorem 4.9

> If Y is a chi-square random variable with v degrees of freedom, then
>
> $$\mu = E(Y) = v \quad \text{and} \quad \sigma^2 = V(Y) = 2v.$$

Proof   Apply Theorem 4.8 with $\alpha = v/2$ and $\beta = 2$.

The gamma-type density function for the special case $\alpha = 1$ is called the *exponential density function.*

Definition 4.10

> A random variable Y is said to have an *exponential distribution with parameter β* if and only if the density function of Y is
>
> $$f(y) = \begin{cases} \dfrac{1}{\beta}e^{-y/\beta}, & \beta > 0; 0 \le y < \infty \\ 0, & \text{elsewhere} \end{cases}$$

The exponential density function is often useful for modeling the length of life of electronic components. Suppose that the length of time a component already has operated does not affect its chance of operating at least b additional time units. That is, the probability that the component operates for more than $a + b$ time units, given that it has already operated for at least a time units, is the same as the probability that a new component will operate at least b time units if the new component is put into service at time 0. A fuse is an example of a component for which this assumption often is reasonable. We will see in the next example that the exponential distribution provides a model for the distribution of the lifetime of such a component.

Theorem 4.10

If Y is an exponential random variable with parameter β, then

$$\mu = E(Y) = \beta \quad \text{and} \quad \sigma^2 = V(Y) = \beta^2.$$

Proof   Follows directly from Theorem 4.8 with $\alpha = 1$.

EXAMPLE 4.10   Suppose Y has an exponential probability density function. Show that, for numbers a and b with $0 < a$ and $0 < b$.

$$P(Y > a + b | Y > a) = P(Y > b)$$

Solution   From the definition of conditional probability we have that

$$P(Y > a + b | Y > a) = \frac{P(Y > a + b)}{P(Y > a)}$$

because the intersection of the events $(Y > a + b)$ and $(Y > a)$ is the event $(Y > a + b)$. Now

$$P(Y > a + b) = \int_{a+b}^{\infty} \frac{1}{\beta} e^{-y/\beta} \, dy = -e^{-y/\beta} \Big]_{a+b}^{\infty} = e^{-(a+b)/\beta}$$

Similarly,

$$P(Y > a) = \int_{a}^{\infty} \frac{1}{\beta} e^{-y/\beta} \, dy = e^{-a/\beta}$$

Thus

$$P(Y > a + b | Y > a) = \frac{e^{-(a+b)/\beta}}{e^{-a/\beta}} = e^{-b/\beta} = P(Y > b)$$

This property of the exponential distribution is often called the *memoryless property* of the distribution.   ∽

Exercises

4.48 The magnitudes of earthquakes recorded in a region of North America can be modeled as having an exponential distribution with mean 2.4, as measured on the Richter scale. Find the probability that an earthquake striking this region will

(a) exceed 3.0 on the Richter scale,

(b) fall between 2.0 and 3.0 on the Richter scale.

4.49 Refer to Exercise 4.48. Out of the next ten earthquakes to strike this region, what is the probability that at least one will exceed 5.0 on the Richter scale?

4.50 The operator of a pumping station has observed that demand for water during early afternoon hours has an approximate exponential distribution with mean 100 cfs (cubic feet per second).

(a) Find the probability that the demand will exceed 200 cfs during the early afternoon on a randomly selected day.

(b) What water pumping capacity should the station maintain during early afternoons so that demand will exceed capacity with probability only 0.01?

4.51 The length of time Y necessary to complete a key operation in the construction of houses has an exponential distribution with mean ten hours. The cost C of completing this operation is related to the square of the time to completion by the formula $C = 100 + 40Y + 3Y^2$. Find the mean and variance of C.

4.52 Times between accidents for all fatal accidents on scheduled American domestic passenger flights during the years 1948 through 1961 were found to have an approximate exponential distribution with mean forty-four days (Pyke [7], p. 426).

(a) If one of the accidents occurred on July 1, what is the probability that another accident occurred that same month?

(b) What is the variance of the times between accidents for the years just indicated?

4.53 One-hour carbon monoxide concentrations in air samples from a large city have an approximate exponential distribution with mean 3.6 parts per million (Zamurs [10], p. 637).

(a) Find the probability that the carbon monoxide concentration exceeds 9 parts per million during a 1-hour period.

(b) A traffic control strategy reduced the mean to 2.5 parts per million. Now find the probability that the concentration exceeds 9 parts per million.

4.54 Suppose a random variable Y has a probability density function given by

$$f(y) = \begin{cases} ky^3 e^{-y/2}, & y > 0 \\ 0, & \text{elsewhere} \end{cases}$$

Find the value of k that makes $f(y)$ a density function.

4.55 A certain manufacturing plant makes use of a specific bulk product. The amount of product used in 1 day can be modeled by an exponential distribution with $\beta = 4$ (measurements in tons). Find the probability that the plant will use more than 4 tons on a given day.

4.56   Consider the plant of Exercise 4.55. How much of the bulk product should be stocked so that its chance of running out of the product is only .05?

4.57   The relationship between incomplete gamma integrals and sums of Poisson probabilities is given by

$$\frac{1}{\Gamma(\alpha)} \int_{\lambda}^{\infty} y^{\alpha-1} e^{-y} \, dy = \sum_{y=0}^{\alpha-1} \frac{\lambda^y e^{-\lambda}}{y!}$$

for integer values of α. If Y has a gamma distribution with $\alpha = 2$ and $\beta = 1$, find $P(Y > 1)$ by using the equality above and Table 3 of Appendix III.

4.58   The lifetime (in hours), X, of a certain electronic component is a random variable with a density function given by

$$f(x) = \begin{cases} \dfrac{1}{100} e^{-x/100}, & x > 0 \\ 0, & \text{elsewhere} \end{cases}$$

Three of these components operate independently in a piece of equipment. The equipment fails if at least two of the components fail. Find the probability that the equipment operates for at least 200 hours without failure.

4.59   Four-week summer rainfall totals in a section of the Midwest United States have an approximate gamma-type distribution with $\alpha = 1.6$ and $\beta = 2.0$. Find the mean and variance of the four-week rainfall totals.

4.60   The response times on an online computer terminal have an approximate gamma distribution with mean 4 seconds and variance 8 seconds^2. Write the probability density function for the response times.

4.61   Refer to Exercise 4.60. Use Tchebysheff's Theorem to give an interval that contains at least 75% of the response times.

4.62   Annual incomes for household heads in a certain section of a city have approximately a gamma distribution with $\alpha = 1000$ and $\beta = 20$. Find the mean and variance of these incomes. Would you expect to find many incomes in excess of $40,000 in this section of the city?

4.63   The weekly amount of downtime Y (in hours) for a certain industrial machine has approximately a gamma distribution with $\alpha = 3$ and $\beta = 2$. The loss, in dollars, to the industrial operation as a result of this downtime is given by $L = 30Y + 2Y^2$. Find the expected value and variance of L.

4.64   If Y has a probability density function given by

$$f(y) = \begin{cases} 4y^2 e^{-2y}, & y > 0 \\ 0, & \text{elsewhere} \end{cases}$$

obtain $E(Y)$ and $V(Y)$ by inspection.

4.7 The Beta Probability Distribution

The beta density function is a two-parameter density function defined over the closed interval $0 \leq y \leq 1$. As such, it is often used as a model for proportions, such as the proportion of impurities in a chemical product or the proportion of time that a machine is in a state of being repaired.

Definition 4.11

A random variable Y is said to have a *beta probability distribution with parameters α and β* if and only if the density function of Y is

$$f(y) = \begin{cases} \dfrac{y^{\alpha-1}(1-y)^{\beta-1}}{B(\alpha, \beta)}, & \alpha, \beta > 0; 0 \leq y \leq 1 \\ 0, & \text{elsewhere} \end{cases}$$

where

$$B(\alpha, \beta) = \int_0^1 y^{\alpha-1}(1-y)^{\beta-1} \, dy = \frac{\Gamma(\alpha)\Gamma(\beta)}{\Gamma(\alpha+\beta)}$$

The graph of the density function will assume widely differing shapes for various values of the two parameters α and β. Some of these are shown in Figure 4.16.

Figure 4.16
The beta density function

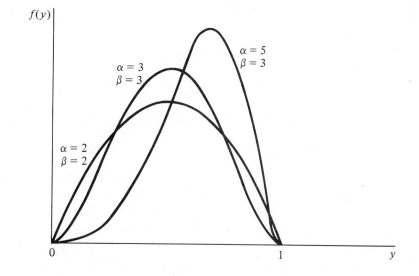

Note that defining y over the interval $0 \leq y \leq 1$ does not restrict its use. If $c \leq y \leq d$, $y^* = (y - c)/(d - c)$ will define a new variable defined on the interval $0 \leq y \leq 1$. Thus the beta density function can be applied to a random variable defined on the interval $c \leq y \leq d$ by translation and a change of scale.

The cumulative distribution function for the beta random variable is commonly called the *incomplete beta function* and is denoted by

$$F(y) = \int_0^y \frac{t^{\alpha - 1}(1 - t)^{\beta - 1}}{B(\alpha, \beta)} \, dt = I_y(\alpha, \beta)$$

A tabulation of $I_y(\alpha, \beta)$ is given in *Tables of the Incomplete Beta Function* [5]. For integral values of α and β, $I_y(\alpha, \beta)$ is related to the binomial probability function. When $y = p$, it can be shown that

$$F(p) = \int_0^p \frac{y^{\alpha - 1}(1 - y)^{\beta - 1}}{B(\alpha, \beta)} \, dy = \sum_{y = \alpha}^{n} \binom{n}{y} p^y (1 - p)^{n - y}$$

where $0 < p < 1$ and $n = \alpha + \beta - 1$. The binomial cumulative distribution function is presented in Table 1, Appendix III, for $n = 5, 10, 15, 20, 25$ ($p = .01$, $.05, .10, .20, .30, .40, .50, .60, .70, .80, .90, .95, .99$). A more extensive tabulation of the binomial probability function is given in *Tables of the Binomial Probability Distribution* [9].

Theorem 4.11

> If Y is a beta distributed random variable with parameters α and β, then
>
> $$\mu = E(Y) = \frac{\alpha}{\alpha + \beta} \qquad \text{and} \qquad \sigma^2 = V(Y) = \frac{\alpha\beta}{(\alpha + \beta)^2(\alpha + \beta + 1)}$$

Proof   By definition,

$$E(Y) = \int_{-\infty}^{\infty} yf(y) \, dy$$

$$= \int_0^1 y \frac{y^{\alpha - 1}(1 - y)^{\beta - 1}}{B(\alpha, \beta)} \, dy$$

$$= \frac{1}{B(\alpha, \beta)} \int_0^1 y^{\alpha}(1 - y)^{\beta - 1} \, dy$$

$$= \frac{B(\alpha + 1, \beta)}{B(\alpha, \beta)}$$

$$= \frac{\Gamma(\alpha + \beta)}{\Gamma(\alpha)\Gamma(\beta)} \times \frac{\Gamma(\alpha + 1)\Gamma(\beta)}{\Gamma(\alpha + \beta + 1)}$$

$$= \frac{\Gamma(\alpha + \beta)}{\Gamma(\alpha)\Gamma(\beta)} \times \frac{\alpha\Gamma(\alpha)\Gamma(\beta)}{(\alpha + \beta)\Gamma(\alpha + \beta)} = \frac{\alpha}{(\alpha + \beta)}$$

The derivation of the variance is left to the reader (see Exercise 4.72).

We will see in the next example that integration of the beta density function can be carried out directly in the case when α and β are both integers.

EXAMPLE 4.11   A gasoline wholesale distributor has bulk storage tanks that hold fixed supplies and are filled every Monday. Of interest to the wholesaler is the proportion of this supply that is sold during the week. Over many weeks of observation it was found that this proportion could be modeled by a beta distribution with $\alpha = 4$ and $\beta = 2$. Find the probability that the wholesaler will sell at least 90% of her stock in a given week.

Solution   If Y denotes the proportion sold during the week, then

$$f(y) = \begin{cases} \dfrac{\Gamma(4 + 2)}{\Gamma(4)\Gamma(2)} y^3(1 - y), & 0 \leq y \leq 1 \\ 0, & \text{elsewhere} \end{cases}$$

and

$$P(Y > .9) = \int_{.9}^{1} f(y)\, dy = \int_{.9}^{1} 20(y^3 - y^4)\, dy$$

$$= 20 \left\{ \frac{y^4}{4} \Big]_{.9}^{1} - \frac{y^5}{5} \Big]_{.9}^{1} \right\} = 20(.004) = .08$$

It is *not* very likely that 90% of the stock will be sold in a given week. ∽

Exercises

4.65   The relative humidity Y, when measured at a certain location, has a probability density function given by

$$f(y) = \begin{cases} ky^3(1 - y)^2, & 0 \leq y \leq 1 \\ 0, & \text{elsewhere} \end{cases}$$

Find the value of k that makes $f(y)$ a density function.

4.66   The percentage of impurities per batch in a certain chemical product is a random variable Y having the density function

$$f(y) = \begin{cases} 12y^2(1 - y), & 0 \leq y \leq 1 \\ 0, & \text{elsewhere} \end{cases}$$

A batch with more than 40% impurities cannot be sold. What is the probability that a randomly selected batch cannot be sold because of excessive impurities?

4.67   Refer to Exercise 4.66. Find the mean and variance of the percentage of impurities in a randomly selected batch of chemical.

4.68   Suppose a random variable Y has a probability density function given by

$$f(y) = \begin{cases} 6y(1-y), & 0 \le y \le 1 \\ 0, & \text{elsewhere} \end{cases}$$

   (a)   Find $F(y)$.

   (b)   Graph $F(y)$ and $f(y)$.

   (c)   Find $P(.5 \le Y \le .8)$.

4.69   Verify that the beta distribution reduces to the uniform distribution over $(0, 1)$ if $\alpha = \beta = 1$.

4.70   The weekly repair cost Y for a certain machine has a probability density function given by

$$f(y) = \begin{cases} 3(1-y)^2, & 0 < y < 1 \\ 0, & \text{elsewhere} \end{cases}$$

with measurements in hundreds of dollars. How much money should be budgeted each week for repair costs so that the actual cost will exceed the budgeted amount only 10% of the time?

4.71   During any 8-hour shift, the proportion of time, Y, that a sheet-metal stamping machine is down for maintenance or repairs has a beta distribution with $\alpha = 1$ and $\beta = 2$. That is,

$$f(y = \begin{cases} 2(1-y), & 0 \le y \le 1, \\ 0, & \text{elsewhere} \end{cases}$$

The cost (in hundreds of dollars) of this downtime, due to lost production and cost of maintenance and repair, is given by

$$C = 10 + 20Y + 4Y^2$$

Find the mean and variance of C.

4.72   Prove that the variance of a beta distributed random variable with parameters α and β is

$$\sigma^2 = \frac{\alpha\beta}{(\alpha + \beta)^2(\alpha + \beta + 1)}$$

4.73   Errors in measuring the time of arrival of a wave front from an acoustic source sometimes have an approximate beta distribution (Perruzzi and Hilliard, [6], p. 197). Suppose that these errors, measured in microseconds, have an approximate beta distribution with $\alpha = 1$ and $\beta = 2$.

   (a)   What is the probability that a measurement error is less than 0.5 microseconds?

   (b)   Give the mean and standard deviation of the measurement errors.

4.74   Proper blending of fine and coarse powders prior to copper sintering is essential for uniformity in the finished product. One way to check the homogeneity of the blend is to select many small samples of the blended powders and measure the proportion of the total weight contributed by the fine powders. These measurements should be relatively stable if a homogeneous blend has been obtained.

(a)   Suppose that the proportion of total weight contributed by the fine powders has a beta distribution with $\alpha = \beta = 3$. Find the mean and variance of the proportion of weight contributed by the fine powders.

(b)   Repeat (a) if $\alpha = \beta = 2$.

(c)   Repeat (a) if $\alpha = \beta = 1$.

(d)   Which of the cases in (a), (b), and (c) yields the most homogeneous blending?

4.75   The proportion of time per day that all checkout counters in a certain supermarket are busy is a random variable, Y, with a density function given by

$$f(y) = \begin{cases} cy^2(1-y)^4, & 0 \leq y \leq 1 \\ 0, & \text{elsewhere} \end{cases}$$

(a)   Find the value of c that makes $f(y)$ a probability density function.

(b)   Find $E(Y)$.

(Use what you have learned about the beta-type distribution. Compare your answers to those obtained in Exercise 4.15.)

4.8 Some General Comments

Keep in mind that density functions are theoretical models for populations of real data that occur in nature. How do we know which to use, and to what extent does it matter if we have the wrong one?

To answer the last question first, it is unlikely that we will ever select a density function that will provide a perfect representation of nature, but goodness of fit is not the criterion for assessing the adequacy of our model. The purpose of a probabilistic model is to provide the mechanism for making inferences about a population based on information contained in a sample. As noted earlier, the probability of the observed sample (or a quantity proportional to it) will be instrumental in making an inference about the population. It follows that a density function that provides a poor fit to the population frequency distribution could (but does not necessarily) yield incorrect probability statements and lead to erroneous inferences about the population. A good model is one that yields good inferences about the population of interest.

A reasonable selection of a model is sometimes implied by theoretical considerations. For example, a situation for which the discrete Poisson random variable is appropriate often will be indicated by the random behavior of events in time. Knowing this, it is possible to show that the interval between any adjacent pair of events follows an exponential distribution. Similarly, we will later encounter a theorem (called the *central limit theorem*) that will outline some conditions that imply normal distributions of data.

A second way to select a model is to form a frequency histogram (Chapter 1) for some data drawn from the population and to choose a density function that would visually appear to give a similar frequency curve. For example, if a set of $n = 100$ sample measurements yields a bell-shaped frequency distribution, we might conclude that the normal density function would adequately model the population frequency distribution.

Not all model selection is completely subjective. Statistical procedures are available to test a hypothesis that a population frequency distribution is of a particular type. We also can calculate a measure of goodness of fit for several distributions and select the best. Studies of many common inferential methods have been made to determine the magnitude of the errors of inference introduced by incorrect population models. It is comforting to know that many statistical methods of inference are insensitive to assumptions about the form of the underlying population frequency distribution.

The uniform, normal, gamma-type, and beta distributions give an assortment of density functions that will fit many population frequency distributions. Another, the Weibull distribution, will appear in the exercises at the end of the chapter.

4.9 Other Expected Values

Moments for continuous random variables have definitions analogous to those given in the discrete case.

Definition 4.12

> If Y is a continuous random variable, then the kth *moment about the origin* is given by
>
> $$\mu'_k = E(Y^k), \ k = 1, 2, \ldots$$
>
> The kth *moment about the mean*, or the kth *central moment* is given by
>
> $$\mu_k = E[(Y - \mu)^k], \ k = 1, 2, \ldots$$

Note that for $k = 1$, $\mu'_1 = \mu$, and for $k = 2$, $\mu_2 = V(Y) = \sigma^2$.

EXAMPLE 4.12   Find μ'_k for the uniform random variable with $\theta_1 = 0$ and $\theta_2 = \theta$.

Solution

By definition,

$$\mu'_k = E(Y^k) = \int_{-\infty}^{\infty} y^k f(y)\, dy = \int_{0}^{\theta} y^k \frac{1}{\theta}\, dy = \frac{y^{k+1}}{\theta(k+1)} \Bigg]_{0}^{\theta} = \frac{\theta^k}{k+1}$$

Thus

$$\mu'_1 = \mu = \frac{\theta}{2}, \qquad \mu'_2 = \frac{\theta^2}{3}, \qquad \mu'_3 = \frac{\theta^3}{4},$$

and so on.

Definition 4.13

> If Y is a continuous random variable, then the *moment-generating function of Y* is given by
>
> $$m(t) = E(e^{tY}).$$
>
> The moment-generating function is said to exist if there exists a positive constant b such that $m(t)$ is finite for $|t| \le b$.

This is simply the continuous analogue of Definition 3.14. That $m(t)$ generates moments is established in exactly the same manner as in Section 3.9. If $m(t)$ exists, then

$$E(e^{tY}) = \int_{-\infty}^{\infty} e^{ty} f(y)\, dy = \int_{-\infty}^{\infty} \left(1 + ty + \frac{t^2 y^2}{2!} + \frac{t^3 y^3}{3!} + \cdots \right) f(y)\, dy$$

$$= \int_{-\infty}^{\infty} f(y)\, dy + t \int_{-\infty}^{\infty} y f(y)\, dy + \frac{t^2}{2!} \int_{-\infty}^{\infty} y^2 f(y)\, dy + \cdots$$

$$= 1 + t\mu'_1 + \frac{t^2}{2!}\mu'_2 + \frac{t^3}{3!}\mu'_3 + \cdots$$

Note that the moment-generating function,

$$m(t) = 1 + t\mu'_1 + \frac{t^2}{2!}\mu'_2 + \cdots$$

takes the same form for both discrete and continuous random variables. Hence Theorem 3.12 holds for continuous random variables and

$$\frac{d^k m(t)}{dt^k}\Bigg]_{t=0} = \mu'_k$$

EXAMPLE 4.13   Find the moment-generating function for the gamma-type random variable.

Solution

$$m(t) = E(e^{tY}) = \int_0^\infty \frac{e^{ty} y^{\alpha-1}}{\beta^\alpha \Gamma(\alpha)} e^{-y/\beta} \, dy$$

$$= \frac{1}{\beta^\alpha \Gamma(\alpha)} \int_0^\infty y^{\alpha-1} \exp\left[-y\left(\frac{1}{\beta} - t\right) \right] dy$$

$$= \frac{1}{\beta^\alpha \Gamma(\alpha)} \int_0^\infty y^{\alpha-1} \exp\left[\frac{-y}{\beta/(1 - \beta t)} \right] dy$$

[exp( ) is simply a more convenient way to write $e^{(\ )}$ when the term in the exponent is long or complex.]

To complete the integration, note that the integral of the variable factor of any density function must equal the reciprocal of the constant factor. That is, if $f(y) = cg(y)$, where c is a constant, then

$$\int_{-\infty}^\infty f(y) \, dy = \int_{-\infty}^\infty cg(y) \, dy = 1$$

Therefore,

$$\int_{-\infty}^\infty g(y) \, dy = \frac{1}{c}$$

Applying this result to the integral in $m(t)$, and noting that $g(y) = y^{\alpha-1} \times \exp\{-y/[\beta/(1 - \beta t)]\}$ is the variable factor of a gamma-type density function with parameters α and $\beta/(1 - \beta t)$, we obtain

$$m(t) = \frac{1}{\beta^\alpha \Gamma(\alpha)} \left[\left(\frac{\beta}{1 - \beta t}\right)^\alpha \Gamma(\alpha) \right] = \frac{1}{(1 - \beta t)^\alpha} \qquad \text{for} \qquad t < \frac{1}{\beta}$$

The moments μ_k' can be extracted from the moment-generating function by differentiating with respect to t, Theorem 3.12, or by expanding the function into a power series in t. We will demonstrate the latter approach.

EXAMPLE 4.14   Expand the moment-generating function of Example 4.13 into a power series in t and thereby obtain μ_k'.

Solution   From Example 4.13, $m(t) = 1/(1 - \beta t)^\alpha = (1 - \beta t)^{-\alpha}$. Using the expansion for a binomial, we have

$$m(t) = (1 - \beta t)^{-\alpha} = 1 + (-\alpha)(1)^{-\alpha-1}(-\beta t)$$

$$+ \frac{(-\alpha)(-\alpha - 1)(1)^{-\alpha+2}(-\beta t)^2}{2!} + \cdots$$

$$= 1 + t(\alpha\beta) + \frac{t^2[\alpha(\alpha + 1)\beta^2]}{2!} + \frac{t^3[\alpha(\alpha + 1)(\alpha + 2)\beta^3]}{3!} + \cdots$$

Because μ'_k will be the coefficient of $t^k/k!$, we find, by inspection,

$$\mu'_1 = \mu = \alpha\beta$$
$$\mu'_2 = \alpha(\alpha + 1)\beta^2$$
$$\mu'_3 = \alpha(\alpha + 1)(\alpha + 2)\beta^3$$

and, in general, $\mu'_k = \alpha(\alpha + 1)(\alpha + 2)\cdots(\alpha + k - 1)\beta^k$. Note that μ'_1 and μ'_2 agree with the results of Theorem 4.8.   ∽

We have already explained the importance of the expected values of Y^k, $(Y - \mu)^k$, and e^{tY}, all of which are functions of the random variable Y and provide numerical descriptive measures for its density function. However, it is important to note that we sometimes are interested in the expected value of a function of a random variable as an end in itself. (We also are interested in the probability distribution of functions of random variables, but we defer discussion of this topic until Chapter 6.)

EXAMPLE 4.15   The kinetic energy k associated with a mass m moving at velocity v is given by the expression

$$k = \frac{mv^2}{2}$$

Consider a device that fires a serrated nail into concrete at a mean velocity of 2000 feet per second, where V, the random velocity, possesses a density function given by

$$f(v) = \frac{v^3 e^{-v/b}}{b^4 \Gamma(4)}, \qquad b = 500, v \geq 0$$

If each nail possesses mass m, find the expected kinetic energy associated with each nail.

Solution

$$E(K) = E\left(\frac{mV^2}{2}\right) = \frac{m}{2} E(V^2)$$

by Theorem 4.5, part 2. The density function for V is a gamma-type function with $\alpha = 4$ and $\beta = 500$. Therefore, $E(V^2) = \mu'_2$ for the random variable V. Referring to Example 4.14, we have $\mu'_2 = \alpha(\alpha + 1)\beta^2 = 4(5)(500)^2 = 5,000,000$. Therefore,

$$E(K) = \frac{m}{2} E(V^2) = \frac{m}{2} (5,000,000) = 2,500,000m$$ ∽

Finding the moments of a function of a random variable is frequently facilitated by using moment-generating functions.

Theorem 4.12

> Let $g(Y)$ be a single-valued function of a random variable Y with density function $f(y)$. Then the moment-generating function for $g(Y)$ is
>
> $$E[e^{tg(Y)}] = \int_{-\infty}^{\infty} e^{tg(y)} f(y) \, dy$$

This theorem follows directly from Definition 4.13 and Theorem 4.4.

EXAMPLE 4.16   Let $g(Y) = Y - \mu$, where Y is a normally distributed random variable with mean μ and variance σ^2. Find the moment-generating function for $(Y - \mu)$.

Solution

$$m(t) = E[e^{t(Y-\mu)}] = \int_{-\infty}^{\infty} e^{t(y-\mu)} \frac{\exp[-(y-\mu)^2/2\sigma^2]}{\sigma\sqrt{2\pi}} \, dy$$

To integrate, let $u = y - \mu$. Then $du = dy$ and

$$m(t) = \frac{1}{\sigma\sqrt{2\pi}} \int_{-\infty}^{\infty} e^{tu} e^{-u^2/2\sigma^2} \, du$$

$$= \frac{1}{\sigma\sqrt{2\pi}} \int_{-\infty}^{\infty} \exp\left[-\left(\frac{1}{2\sigma^2}\right)(u^2 - 2\sigma^2 tu)\right] du$$

Complete the square in the exponent of e by multiplying and dividing by $e^{t^2\sigma^2/2}$. Then

$$m(t) = e^{t^2\sigma^2/2} \int_{-\infty}^{\infty} \frac{\exp[-(1/2\sigma^2)(u^2 - 2\sigma^2 tu + \sigma^4 t^2)]}{\sigma\sqrt{2\pi}} \, du$$

$$= e^{t^2\sigma^2/2} \int_{-\infty}^{\infty} \frac{\exp[-(u - \sigma^2 t)^2/2\sigma^2]}{\sigma\sqrt{2\pi}} \, du$$

The function inside the integral is a normal density function with a mean of $\sigma^2 t$ and a variance of σ^2. (See the equation for the normal density function.) Hence the integral is equal to 1. Then

$$m(t) = e^{(t^2/2)\sigma^2}$$

The moments of $U = Y - \mu$ can be obtained from $m(t)$ by differentiating $m(t)$ in accordance with Theorem 3.12 or by expanding $m(t)$ into a series.   ∽

For your convenience, the probability and density functions, means, variances, and moment-generating functions for some common random variables are presented in Appendix II.

Exercises

4.76   Suppose that the waiting time for the first customer to enter a certain shop after 9:00 A.M. is a random variable Y with an exponential density function given by

$$f(y) = \begin{cases} \left(\dfrac{1}{\theta}\right)e^{-y/\theta}, & y > 0 \\ 0, & \text{elsewhere} \end{cases}$$

  (a) Find the moment-generating function for Y.

  (b) Use the answer for (a) to find $E(Y)$ and $V(Y)$.

4.77   Example 4.16 derives the moment-generating function for $Y - \mu$, where Y is normally distributed with mean μ and variance σ^2.

  (a) Use the answer to Example 4.16 to find the moment-generating function for Y.

  (b) Differentiate the moment-generating function found in (a) to show that $E(Y) = \mu$ and $V(Y) = \sigma^2$.

4.78   The moment-generating function for the gamma random variable is derived in Example 4.13. Differentiate this moment-generating function to find the mean and variance of the gamma distribution.

4.79   Consider a random variable Y with the density function given by

$$f(y) = ke^{-y^2/2}, \quad -\infty < y < \infty$$

  (a) Find k.

  (b) Find the moment-generating function.

  (c) Find $E(Y)$ and $V(Y)$.

4.80   If Y is a random variable with mean μ and variance σ^2 and if

$$U = a + bY$$

for constants a and b, find the moment-generating function for U in terms of the moment-generating function for Y, and use the result to derive the mean and variance of U.

4.81   A random variable Y has the density function

$$f(y) = \begin{cases} e^y, & y < 0 \\ 0, & \text{elsewhere} \end{cases}$$

  (a) Find $E(e^{3Y/2})$.

  (b) Find the moment-generating function for Y.

  (c) Find $V(Y)$.

4.10 Tchebysheff's Theorem

The purpose in presenting the preceding discussion of moments is two-fold. First, as in the case of a finite set of measurements, they can be used as numerical, descriptive measures to describe the data. Second, they can be used in a theoretical sense to prove that a random variable possesses a particular probability distribution. It can be shown that if two random variables Y and Z possess identical moment-generating functions, then Y and Z possess identical probability distributions. This latter application of moments was mentioned in discussing moment-generating functions for discrete random variables in Section 3.9; it applies to continuous random variables as well.

The interpretation of μ and σ for continuous random variables utilizes the empirical rule and Tchebysheff's theorem. The justification for their use is that they provide useful approximations for many distributions.

We would expect Tchebysheff's theorem to hold for probability distributions, both discrete and continuous. We will restate the theorem and give a proof applicable to a continuous random variable.

Theorem 4.13
Tchebysheff's
Theorem

> Let Y be a continuous (discrete) random variable with density function $f(y)$ [or probability function $p(y)$]. Then for any $k > 0$,
>
> $$P(|Y - \mu| < k\sigma) \geq 1 - \frac{1}{k^2} \qquad \text{or} \qquad P(|Y - \mu| \geq k\sigma) \leq \frac{1}{k^2}$$
>
> where $E(Y) = \mu$ and $V(Y) = \sigma^2 < \infty$.

Proof   We will give the proof for a continuous random variable. The proof for the discrete case would proceed in a similar manner.

$$V(Y) = \sigma^2 = \int_{-\infty}^{\infty} (y - \mu)^2 f(y)\, dy$$

$$= \int_{\mu - k\sigma}^{\mu + k\sigma} (y - \mu)^2 f(y)\, dy + \int_{-\infty}^{\mu - k\sigma} (y - \mu)^2 f(y)\, dy + \int_{\mu + k\sigma}^{\infty} (y - \mu)^2 f(y)\, dy$$

The first integral is always greater than or equal to zero and $(y - \mu)^2 \geq k^2\sigma^2$ for the second and third integrals; that is, the region of integration is in the tails of the density function and covers only values of y for which $(y - \mu)^2 \geq k^2\sigma^2$. Replace the first integral by zero and substitute $k^2\sigma^2$ for $(y - \mu)^2$ in the second and third integrals to obtain the inequality

$$V(Y) = \sigma^2 \geq \int_{-\infty}^{\mu - k\sigma} k^2\sigma^2 f(y)\, dy + \int_{\mu + k\sigma}^{\infty} k^2\sigma^2 f(y)\, dy$$

Then
$$\sigma^2 \geq k^2\sigma^2 \left[\int_{-\infty}^{\mu - k\sigma} f(y)\, dy + \int_{\mu + k\sigma}^{+\infty} f(y)\, dy \right]$$

or
$$\sigma^2 \geq k^2\sigma^2 P(|Y - \mu| \geq k\sigma)$$

Dividing by $k^2\sigma^2$, we obtain

$$P(|Y - \mu| \geq k\sigma) \leq \frac{1}{k^2}$$

Hence it follows that

$$P(|Y - \mu| < k\sigma) \geq 1 - \frac{1}{k^2}$$

One real value of Tchebysheff's theorem is that it enables us to find bounds on probabilities that ordinarily would be obtained by tedious mathematical manipulations (integration or summation).

EXAMPLE 4.17   Suppose that the length of time Y (in minutes) to conduct a periodic maintenance check (from previous experience) on a dictating machine follows a gamma distribution with $\alpha = 3$ and $\beta = 2$. Suppose that a new maintenance worker requires 19 minutes to check a machine. Does it appear that this time to perform a maintenance check disagrees with prior experience?

Solution   The mean and variance for the length of maintenance times (prior experience) are (from Theorem 4.8),

$$\mu = \alpha\beta \quad \text{and} \quad \sigma^2 = \alpha\beta^2$$

Then for our example

$$\mu = \alpha\beta = (3)(2) = 6$$
$$\sigma^2 = \alpha\beta^2 = (3)(2)^2 = 12$$
$$\sigma = \sqrt{12} = 3.46$$

and the observed deviation $(Y - \mu) = 19 - 6 = 13$ minutes.

For our example $y = 19$ minutes exceeds the mean $\mu = 6$ minutes by $k = 13/3.46$ standard deviations. Then from Tchebysheff's theorem

$$P(|Y - \mu| \geq k\sigma) \leq \frac{1}{k^2}$$

or

$$P(|Y - 6| \geq 13) \leq \frac{1}{k^2} = \frac{(3.46)^2}{(13)^2} = \frac{12}{169} = .071$$

Note that this probability is based on the assumption that the distribution of maintenance times has not changed from prior experience. Then observing that $P(Y \geq 19$ minutes$)$ is small, we must conclude that either our new maintenance worker has generated a lengthy maintenance time that occurs with low probability or is somewhat slower than preceding ones. Noting the low probability for $P(Y \geq 19)$, we might be inclined to favor the latter view.   ∽

The exact probability, $P(Y \geq 19)$, for Example 4.17 would require evaluation of the integral

$$P(Y \geq 19) = \int_{19}^{\infty} \frac{y^2 e^{-y/2}}{8\Gamma(3)} \, dy$$

(This integral can be shown to equal approximately .004.) Similar integrals are difficult to evaluate for the beta function and for many other density functions. Tchebysheff's theorem often provides a rapid approximate procedure for circumventing laborious integration.

Exercises

4.82   A manufacturer of tires wants to advertise a mileage interval that excludes no more than 10% of the mileages on tires he sells. All he knows is that, for a large number of tires tested, the mean mileage was 25,000 miles, with a standard deviation of 4,000 miles. What interval would you suggest?

4.83   A machine used to fill cereal boxes dispenses, on the average, μ ounces per box. The manufacturer wants the actual ounces dispensed, Y, to be within 1 ounce of μ at least 75% of the time. What is the largest value of σ, the standard deviation of Y, that can be tolerated if the manufacturer's objectives are to be met?

4.84   Find $P(|Y - \mu| \leq 2\sigma)$ for Exercise 4.6. Compare with the comparable probabilistic statements given by Tchebysheff's theorem and the empirical rule.

4.85   Find $P(|Y - \mu| \leq 2\sigma)$ for the uniform random variable. Compare with the comparable probabilistic statements given by Tchebysheff's theorem and the empirical rule.

4.86   Find $P(Y - \mu| \leq 2\sigma)$ for the exponential random variable. Compare with the comparable probabilistic statements given by Tchebysheff's theorem and the empirical rule.

4.87   Refer to Exercise 4.51. Would you expect C to exceed 2000 very often?

4.88   Refer to Exercise 4.63. Find an interval that will contain L for at least 89% of the weeks that the machine is in use.

4.89   Refer to Exercise 4.71. Find an interval in which C will lie with a probability of at least .75.

4.11 Expectations of Discontinuous Functions and Mixed Probability Distributions (Optional)

Problems in probability and statistics sometimes involve functions that are partly continuous and partly discrete in one of two ways. First, we may be interested in the properties, perhaps the expectation, of a random variable $g(Y)$ that is a discontinuous function of a discrete or continuous random variable Y. Second, the random variable of interest itself may have a probability distribution made up of isolated points having discrete probabilities and intervals having continuous probability.

We illustrate the first of these two situations with the following example.

EXAMPLE 4.18   A certain retailer for a petroleum product sells a random amount Y each day. Suppose that Y, measured in hundreds of gallons, has the probability density function

$$f(y) = \begin{cases} (3/8)y^2, & 0 \le y \le 2 \\ 0, & \text{elsewhere} \end{cases}$$

The retailer's profit turns out to be \$5 for each 100 gallons sold (5¢ per gallon) if $Y \le 1$ and \$8 per 100 gallons if $Y > 1$. Find the retailer's expected profit for any given day.

Solution   Let $g(Y)$ denote the retailer's daily profit. Then

$$g(Y) = \begin{cases} 5Y, & 0 \le Y \le 1 \\ 8Y, & 1 < Y \le 2 \end{cases}$$

We want to find expected profit, and, by Theorem 4.4, the expectation is

$$E[g(Y)] = \int_{-\infty}^{\infty} g(y)f(y)\,dy$$

$$= \int_0^1 5y\left[\left(\frac{3}{8}\right)y^2\right]dy + \int_1^2 8y\left[\left(\frac{3}{8}\right)y^2\right]dy$$

$$= \frac{15}{(8)(4)}\,[y^4]_0^1 + \frac{24}{(8)(4)}\,[y^4]_1^2$$

$$= \frac{15}{32}\,(1) + \frac{24}{32}\,(15)$$

$$= \frac{(15)(25)}{32} = 11.72$$

Thus the retailer can expect to profit by \$11.72 on the daily sale of this particular product.

A random variable Y that has some of its probability at discrete points and the remainder spread over intervals is said to have a *mixed distribution*. Let $F(y)$ denote a distribution function representing a mixed distribution. For all practical purposes any mixed distribution function $F(y)$ can be written uniquely as

$$F(y) = c_1 F_1(y) + c_2 F_2(y)$$

where $F_1(y)$ is a step distribution function, $F_2(y)$ is a continuous distribution function, c_1 is the accumulated probability of all discrete points, and $c_2 = 1 - c_1$ is the accumulated probability of all continuous portions.

The following example gives an illustration of a mixed distribution.

EXAMPLE 4.19   Let Y denote the length of life (in hundreds of hours) of a certain type of electronic component. These components frequently fail immediately upon insertion into a system. It has been observed that the probability of immediate failure is 1/4. If a component does not fail immediately, its distribution for length of life has the exponential density function

$$f(y) = \begin{cases} e^{-y}, & y > 0 \\ 0, & \text{elsewhere} \end{cases}$$

Find the distribution function for Y and evaluate $P(Y > 10)$.

Solution   There is only one discrete point, $Y = 0$, and this point has probability 1/4. Hence $c_1 = 1/4$ and $c_2 = 3/4$. It follows that Y is a mixture of two random variables, X_1 and X_2, where X_1 has probability 1 at the point 0 and X_2 has the given exponential density. That is,

$$F_1(y) = \begin{cases} 0, & y < 0 \\ 1, & y \geq 0 \end{cases}$$

$$F_2(y) = \int_0^y e^{-x}\, dx = 1 - e^{-y}, \qquad y > 0$$

Now
$$F(y) = (1/4)F_1(y) + (3/4)F_2(y)$$

and hence

$$P(Y > 10) = 1 - P(Y \leq 10) = 1 - F(10)$$
$$= 1 - [(1/4) + (3/4)(1 - e^{-10})]$$
$$= (3/4)[1 - (1 - e^{-10})] = (3/4)e^{-10}$$

A graphical representation of $F(y)$ is given in Figure 4.17.

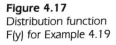

Figure 4.17
Distribution function
F(y) for Example 4.19

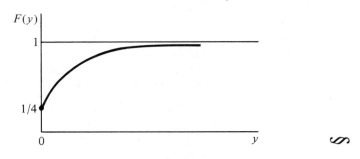

An easy method for finding expectations of random variables having mixed distributions is given in Definition 4.14.

Definition 4.14

Let Y have the mixed distribution function

$$F(y) = c_1 F_1(y) + c_2 F_2(y)$$

and suppose that X_1 is a discrete random variable having distribution function $F_1(y)$ and X_2 is a continuous random variable having distribution function $F_2(y)$. Let $g(Y)$ denote a function of Y. Then

$$E[g(Y)] = c_1 E[g(X_1)] + c_2 E[g(X_2)]$$

EXAMPLE 4.20   Find the mean and variance of the random variable defined in Example 4.19.

Solution   With all definitions as in Example 4.19, it follows that

$$E(X_1) = 0 \quad \text{and} \quad E(X_2) = \int_0^\infty y e^{-y}\, dy = 1$$

Therefore,

$$\mu = E(Y) = (1/4)E(X_1) + (3/4)E(X_2) = 3/4$$

Also, $$E(X_1^2) = 0 \quad \text{and} \quad E(X_2^2) = \int_0^\infty y^2 e^{-y}\, dy = 2$$

Therefore,

$$E(Y^2) = (1/4)E(X_1^2) + (3/4)E(X_2^2) = (1/4)(0) + (3/4)(2) = 3/2$$

Then $$V(Y) = E(Y^2) - \mu^2 = (3/2) - (3/4)^2 = 15/16$$

Exercises

*4.90   A builder of houses has to order some supplies that have a waiting time for delivery, Y, with a continuous uniform distribution over the interval from 1 to 4 days. Because she can get by without them for 2 days, the cost of the delay is fixed at \$100 for any waiting time up to 2 days. However, after 2 days the cost of the delay is \$100 plus \$20 per day for any time beyond 2 days. That is, if the waiting time is 3.5 days, the cost of the delay is \$100 + \$20(1.5) = \$130. Find the expected value of the builder's cost due to waiting for supplies.

*4.91   The duration, Y, of long-distance telephone calls (in minutes) monitored by a certain station is a random variable with the properties that

$$P(Y = 3) = .2 \quad \text{and} \quad P(Y = 6) = .1$$

and otherwise Y has a continuous density function given by

$$f(y) = \begin{cases} (1/4)ye^{-y/2}, & y > 0 \\ 0, & \text{elsewhere} \end{cases}$$

The discrete points at 3 and 6 are due to the fact that the length of call is noted to the caller in 3-minute intervals. Find the expected duration of a randomly selected long-distance call.

4.92   The life length, Y, of a certain component being used in a complex electronic system is known to have an exponential density with a mean of 100 hours. The component is replaced at failure or at age 200 hours, whichever comes first.

(a)   Find the distribution function for X, the length of time that the component is in use.

(b)   Find $E(X)$.

4.12 Summary

This chapter has been related to inference because it presents a probabilistic model for continuous random variables. The density function, which provides a model for a population-frequency distribution associated with a continuous random variable, subsequently will yield the mechanism for inferring character- istics of the population based on measurements contained in a sample. As a consequence, the density function provides a model for a real distribution of data that exists or could be generated by repeated experimentation. Similar distributions for small sets of data (samples from populations) were discussed in Chapter 1.

Four density functions, the uniform, normal, gamma-type, and beta, were presented to provide a wide assortment of models for population frequency

* Exercises preceded by an asterisk are optional.

distributions. Many other density functions could be employed to fit real situations, but the four described adequately suit most situations. A few other density functions are presented in the exercises at the end of the chapter.

The adequacy of a density function to model the frequency distribution for a random variable depends upon the inference-making technique to be employed. If modest disagreement between the model and the real population frequency distribution does not affect the goodness of the inferential procedure, the model is adequate.

The latter part of the chapter concerned expectations, particularly moments and moment-generating functions. It is important to focus attention on the reason for presenting these quantities and avoid excessive concentration on the mathematical aspects of the material. Moments, particularly the mean and variance, are numerical descriptive measures for random variables. Particularly, we subsequently will learn that it is sometimes difficult to find the probability distribution for a random variable, Y, or a function, $g(Y)$, and we already have observed that integration over intervals for many density functions (the normal and gamma-type, for example) is very difficult. When this occurs, we can approximately describe the behavior of the random variable by using its moments along with Tchebysheff's theorem and the empirical rule (Chapter 1).

References and Further Readings

1. Hogg, R. V., and Craig, A. T. *Introduction to Mathematical Statistics.* 4th ed. New York: Macmillan, 1978.

2. Mendenhall, W. *An Introduction to Probability and Statistics.* 7th ed. N. Scituate, Mass.: Duxbury Press, 1987.

3. Parzen, E. *Modern Probability Theory and Its Applications.* New York: Wiley, 1960.

4. Pearson, K., ed. *Tables of the Incomplete Beta Function.* New York: Cambridge University Press, 1956.

5. Pearson, K., ed. *Tables of the Incomplete Gamma Function.* New York: Cambridge University Press, 1956.

6. Perruzzi, J. J., and Hilliard, E. J. "Modeling Time-Delay Measurement Errors Using a Generalized Beta Density Function," *Journal of the Acoustical Society of America* 75, no. 1 (1984): 197–201.

7. Pyke, R. "Spacings," *Journal of the Royal Statistical Society* B, 27, no. 3 (1965): 426.

8. *Standard Mathematical Tables.* 17th ed. Cleveland: Chemical Rubber Company, 1969.

9. *Tables of the Binomial Probability Distribution.* Department of Commerce, National Bureau of Standards, Applied Mathematics Series 6, 1950.

10. Zamurs, J. "Assessing the Effect of Transportation Control Strategies on Urban Carbon Monoxide Concentrations," *Air Pollution Control Association Journal* 34, no. 6 (1984): 637.

11. Zimmels, Y. "Theory of Kindered Sedimentation of Polydisperse Mixtures," *AIChE Journal* 29, no. 4 (1983): 669–676.

Supplementary Exercises

4.93   Let the density function of a random variable Y be given by

$$f(y) = \begin{cases} \dfrac{2}{\pi(1 + y^2)}, & -1 \leq y \leq 1 \\ 0, & \text{elsewhere} \end{cases}$$

(a)   Find the distribution function.

(b)   Find $E(Y)$.

4.94   The length of time required to complete a college achievement test is found to be normally distributed with a mean of 70 minutes and a standard deviation of 12 minutes. When should the test be terminated if we wish to allow sufficient time for 90% of the students to complete the test?

4.95   A manufacturing plant utilizes 3000 electric light bulbs that have a length of life that is normally distributed with a mean of 500 hours and a standard deviation of a 50 hours. To minimize the number of bulbs that burn out during operating hours, all the bulbs are replaced after a given period of operation. How often should the bulbs be replaced if we wish not more than 1% of the bulbs to burn out between replacement periods?

4.96   Refer to Exercise 4.35. Suppose that five bearings are drawn from production. What is the probability that at least one will be defective?

4.97   The length of life of oil drilling bits depends upon the types of rock and soil that the drill encounters, but it is estimated that the mean length of life is 75 hours. If an oil exploration company purchases drill bits that have a length of life that is approximately normally distributed with mean equal to 75 hours and standard deviation equal to 12 hours,

(a)   What proportion of the company's drill bits will fail before 60 hours of use?

(b)   What proportion will last at least 60 hours?

(c)   What proportion will have to be replaced after more than 90 hours of use?

4.98   Let Y have density function

$$f(y) = \begin{cases} cye^{-2y}, & 0 \leq y < \infty \\ 0, & \text{elsewhere} \end{cases}$$

(a)   Find the value of c that makes $f(y)$ a density function.

(b)   Give the mean and variance for Y.

(c)   Give the moment-generating function for Y.

4.99   Use the fact that

$$e^z = 1 + z + \frac{z^2}{2!} + \frac{z^3}{3!} + \frac{z^4}{4!} + \cdots$$

to expand the moment-generating function of Example 4.16 into a series to find μ_1, μ_2, μ_3, and μ_4 for the normal random variable.

4.100   Find an expression for $\mu_k' = E(X^k)$, where the random variable X has a beta distribution.

4.101 The number of arrivals, n, at a supermarket checkout counter in the time interval from 0 to t follows a Poisson probability distribution with mean λt. Let T denote the length of time until the first arrival. Find the density function for T. [Note: $P(T > t_0) = P(n = 0$ at $t = t_0)$.]

4.102 An argument similar to that of Exercise 4.101 can be used to show that if events are occurring in time according to a Poisson distribution with mean λt, then the interarrival times between events have an exponential distribution with mean $1/\lambda$. If calls come into a police emergency center at the rate of 10 per hour, what is the probability that more than 15 minutes will elapse between the next two calls?

4.103 Suppose that plants of a certain species are randomly dispersed over an area, so that the number of plants in a given area follows a Poisson distribution with a mean density of λ plants per unit area. If a plant is randomly selected in this area, find the probability density function of the distance to the *nearest* neighboring plant. [Hint: If R denotes the distance to the nearest neighbor, then $P(R > r)$ is the same as the probability of seeing no plants in a circle of radius r.]

4.104 The time (in hours) it takes a manager to interview a job applicant has an exponential distribution with $\beta = 1/2$. The applicants are scheduled at quarter-hour intervals, beginning at 8:00 A.M., and the applicants arrive exactly on time. When the applicant with an 8:15 A.M. appointment arrives at the manager's office, what is the probability that she has to wait before seeing the manager?

4.105 The median value of a continuous random variable is that value, y, such that $F(y) = .5$. Find the median value of the random variable in Exercise 4.1.

4.106 Graph the beta probability density function for $\alpha = 3$ and $\beta = 2$. If Y has this beta density function, find $P(.1 \leq Y \leq .2)$ by using binomial probabilities to evaluate $F(y)$. (See Section 4.7.)

*4.107 A retail grocer has a daily demand, Y, for a certain food sold by the pound, where Y, measured in hundreds of pounds, has probability density function given by

$$f(y) = \begin{cases} 3y^2, & 0 \leq y \leq 1 \\ 0, & \text{elsewhere} \end{cases}$$

(She cannot stock over 100 pounds.) The grocer wants to order $100k$ pounds of food. She buys the food at 6 cents per pound and sells it at 10 cents per pound. What value of k will maximize her expected daily profit?

4.108 Suppose Y has a gamma distribution with $\alpha = 3$ and $\beta = 1$. Use Poisson probabilities to evaluate $P(Y \leq 4)$. (See Exercise 4.57.)

4.109 Suppose that Y is a normally distributed random variable with a mean of μ and a variance of σ^2. Use the results of Example 4.16 to find the moment-generating function, mean, and variance of

$$Z = \frac{Y - \mu}{\sigma}$$

*4.110 A random variable, Y, is said to have a log-normal distribution if $X = \ln(Y)$ has a normal distribution. (The symbol *ln* denotes natural logarithm.) In this case Y must be nonnegative. The shape of the log-normal probability density function is similar to that of the gamma distribution, with long tails to the right. The equation of the log-normal density function is given by

$$f(y) = \begin{cases} \dfrac{1}{\sqrt{2\pi}\sigma y} e^{-(\ln(y)-\mu)^2/2\sigma^2}, & y > 0 \\ 0, & \text{elsewhere} \end{cases}$$

Because $\ln(Y)$ is a monotonic function of y,

$$P(Y \le y) = P[\ln(Y) \le \ln(y)] = P[X \le \ln(y)]$$

where X has a normal distribution with a mean of μ and a variance of σ^2. Thus probabilities in the log-normal case can be found by transforming them to the normal case.

If Y has a log-normal distribution with $\mu = 4$ and $\sigma^2 = 1$, find

(a) $P(Y \le 4)$.

(b) $P(Y > 8)$.

4.111   If Y has a log-normal distribution with parameters μ and σ^2, then it can be shown that

$$E(Y) = e^{\mu + \sigma^2/2}$$

and

$$V(Y) = e^{2\mu + \sigma^2}(e^{\sigma^2} - 1)$$

The grains composing polycrystalline metals tend to have weights that follow a log-normal distribution. For a certain type of aluminum, gram weights have a log-normal distribution with $\mu = 3$ and $\sigma = 4$ (in units of 10^{-2} gram).

(a)   Find the mean and variance of the grain weights.

(b)   Find an interval in which at least 75% of the grain weights should lie. (Use Tchebysheff's Theorem.)

(c)   Find the probability that a randomly chosen grain weighs less than the mean grain weight.

4.112   Let Y denote a random variable with probability density function given by

$$f(y) = (1/2)e^{-|y|}, \qquad -\infty < y < \infty.$$

Find the moment-generating function of Y and use it to find $E(Y)$.

*4.113   Let $f_1(y)$ and $f_2(y)$ be density functions and a be a constant such that $0 \le a \le 1$. Consider the function $f(y) = af_1(y) + (1 - a)f_2(y)$.

(a)   Show that $f(y)$ is a density function. Such a density function is often referred to as a mixture of two density functions.

(b)   Suppose that Y_1 is a random variable with density function $f_1(y)$, and that $E(Y_1) = \mu_1$ and $Var(Y_1) = \sigma_1^2$ and similarly that Y_2 is a random variable with density function $f_2(y)$, and that $E(Y_2) = \mu_2$ and $\text{Var}(Y_2) = \sigma_2^2$. If Y is a random variable with density that is a mixture of the densities corresponding to Y_1 and Y_2,

(i) show that $E(Y) = a\mu_1 + (1-a)\mu_2$

(ii) show that $\text{Var}(Y) = a\sigma_1^2 + (1-a)\sigma_2^2 + a(1-a)[\mu_1 - \mu_2]^2$ (Hint: $E(Y_i^2) = \mu_i^2 + \sigma_i^2, i = 1, 2$)

*4.114   The random variable Y, with a density function given by

$$f(y) = \frac{m y^{m-1}}{\alpha} e^{-y^m/\alpha}, \qquad 0 \le y < \infty; \alpha, m > 0$$

is called a *Weibull random variable*. The Weibull density function provides a good model for the distribution of length of life for many mechanical devices and biological plants and animals. Find the mean and variance for a Weibull random variable with $m = 2$.

4.115   Resistors used in the construction of an aircraft guidance system have life lengths that follow a Weibull distribution with $m = 2$ and $\alpha = 10$, with measurements in thousands of hours.

(a)   Find the probability that a randomly selected resistor of this type has a life length that exceeds 5000 hours.

(b)   If three resistors of this type are operating independently, find the probability that exactly one of the three burns out prior to 5000 hours of use.

*4.116   A function that is sometimes associated with continuous nonnegative random variables is the failure rate (or hazard rate) function. This function is defined by

$$r(t) = \frac{f(t)}{1 - F(t)}$$

for a density function $f(t)$ with corresponding distribution function $F(t)$. If we think of the random variable in question as the length of life of a component, $r(t)$ is proportional to the probability of failure in a small interval after t, given that the component has survived up to time t.

(a)   Show that for an exponential density function $r(t)$ is constant.

(b)   Show that for a Weibull density function with $m > 1$, $r(t)$ is an increasing function of t. (See Exercise 4.114.)

*4.117   Suppose Y is a continuous random variable with a distribution function given by $F(y)$ and probability density function $f(y)$. We often are interested in conditional probabilities of the form $P(Y \le y | Y \ge c)$ for a constant c.

(a)   Show that for $y \ge c$,

$$P(Y \le y | Y \ge c) = \frac{F(y) - F(c)}{1 - F(c)}$$

(b)   Show that the function in part (a) has all the properties of a distribution function.

(c)   If the length of life Y for a battery has a Weibull distribution with $m = 2$ and $\alpha = 3$ (with measurements in years), find the probability that the battery lasts less than 4 years given that it is now 2 years old.

*4.118   The velocities of gas particles can be modeled by the Maxwell distribution, with a probability density function given by

$$f(v) = 4\pi \left(\frac{m}{2\pi K T} \right)^{3/2} v^2 e^{-v^2(m/2KT)}, \qquad v > 0$$

where m is the mass of the particle, K is Boltzmann's constant, and T is the absolute temperature.

(a)   Find the mean velocity of these particles.

(b)   The kinetic energy of a particle is given by $(1/2)mV^2$. Find the mean kinetic energy for a particle.

*4.119   Because

$$P(Y \le y \,|\, Y \ge c) = \frac{F(y) - F(c)}{1 - F(c)}$$

has the properties of a distribution function, its derivative will have the properties of a probability density function. This derivative is given by

$$\frac{f(y)}{1 - F(c)}, \qquad y \ge c$$

We can then find the expected value of Y, given that Y is greater than c, by using

$$E(Y \,|\, Y \ge c) = \frac{1}{1 - F(c)} \int_c^\infty y f(y)\, dy$$

If Y, the length of life of a certain component, has an exponential distribution with a mean of 100 hours, find the expected value of Y given that this component already has been in use for 50 hours.

*4.120   We can show that the normal density function integrates to unity by showing that

$$\frac{1}{\sqrt{2\pi}} \int_{-\infty}^{\infty} e^{-(1/2)uy^2} dy = \frac{1}{\sqrt{u}}$$

This, in turn, can be shown by considering the product of two such integrals,

$$\frac{1}{2\pi} \left(\int_{-\infty}^{\infty} e^{-(1/2)uy^2}\, dy \right) \left(\int_{-\infty}^{\infty} e^{-(1/2)ux^2}\, dx \right) = \frac{1}{2\pi} \int_{-\infty}^{\infty} \int_{-\infty}^{\infty} e^{-(1/2)u(x^2 + y^2)}\, dx\, dy$$

By transforming to polar coordinates, show that the preceding double integral is equal to $1/u$.

*4.121   The function $\Gamma(u)$ is defined by

$$\Gamma(u) = \int_0^\infty y^{u-1} e^{-y}\, dy$$

Integrate by parts to show that $\Gamma(u) = (u - 1)\Gamma(u - 1)$. Hence if n is a positive integer, it follows that $\Gamma(n) = (n - 1)!$.

*4.122 Show that $\Gamma(1/2) = \sqrt{\pi}$ by writing

$$\Gamma(1/2) = \int_0^\infty y^{-1/2} e^{-y}\, dy$$

making the transformation $y = (1/2)x^2$ and employing the result of Exercise 4.120.

*4.123 The function $B(\alpha, \beta)$ is defined by

$$B(\alpha, \beta) = \int_0^1 y^{\alpha-1}(1 - y)^{\beta-1}\, dy$$

(a)   Letting $y = \sin^2 \theta$, show that

$$B(\alpha, \beta) = 2 \int_0^{\pi/2} \sin^{2\alpha-1} \theta \cos^{2\beta-1} \theta\, d\theta$$

(b)   Write $\Gamma(\alpha)\Gamma(\beta)$ as a double integral, transform to polar coordinates, and conclude that

$$B(\alpha, \beta) = \frac{\Gamma(\alpha)\Gamma(\beta)}{\Gamma(\alpha + \beta)}$$

*4.124 Let $g(y)$ be a function of the random variable Y, with $E[|g(y)|] < \infty$. Show that for every positive constant k,

$$P[|g(y)| \le k] \ge 1 - \frac{E[|g(y)|]}{k}$$

chapter five

MULTIVARIATE PROBABILITY DISTRIBUTIONS

5.1 Introduction

The intersection of two or more events is frequently of interest to an experimenter. For example, the gambler playing blackjack is interested in the event of drawing both an ace and a face card from a fifty-two–card deck. The biologist, observing the number of animals surviving in a litter, is concerned with the intersection of these events:

A: The litter contains n animals

B: y animals survive.

Similarly, the observation of both height and weight on an individual represents the intersection of a specific pair of height-weight measurements.

Most important to statisticians are the intersections that occur when sampling. Suppose that $Y_1, Y_2, \ldots, Y_n$ denote the outcomes on n successive trials of an experiment. For example, this sequence could represent the weights of n people or the measurements of n physical characteristics of a single person. A specific set of outcomes, or sample measurements, may be expressed in terms of the intersection of the n events $(Y_1 = y_1), (Y_2 = y_2), \ldots, (Y_n = y_n)$, which we will denote as $(y_1, y_2, \ldots, y_n)$. Then to make inferences about the population from which the sample was drawn, we will wish to calculate the probability of the intersection $(y_1, y_2, \ldots, y_n)$.

A review of the role probability plays in making inferences, Section 2.2, emphasizes the need for acquiring the probability of the observed sample or, equivalently, the probability of the intersection of a set of numerical events. Knowledge of this probability is fundamental to making an inference about the population from which the sample was drawn. Indeed, this need motivates the discussion of multivariate probability distributions.

5.2 Bivariate and Multivariate Probability Distributions

Many random variables can be defined over the same sample space. For example, consider the experiment of tossing a pair of dice. The sample space contains thirty-six sample points, corresponding to the $mn = (6)(6) = 36$ ways which numbers may appear on the faces of the dice. Any one of the following random variables could be defined over the sample space and might be of interest to the experimenter:

Y_1:   The number of dots appearing on die 1.

Y_2:   The number of dots appearing on die 2.

Y_3:   The sum of the number of dots on the dice.

Y_4:   The product of the number of dots appearing on the dice.

The thirty-six sample points associated with the experiment are equiprobable and correspond to the thirty-six numerical events (y_1, y_2). Thus throwing a pair of 1s would be the simple event $(1, 1)$. Throwing a 2 on die 1 and a 3 on die 2 would be the simple event $(2, 3)$. Because all pairs (y_1, y_2) occur with the same relative frequency, we would assign a probability of 1/36 to each sample point. For this simple example the intersection (y_1, y_2) contains only one sample point. Hence the bivariate probability function is

$$p(y_1, y_2) = 1/36 \qquad y_1 = 1, 2, \ldots, 6; y_2 = 1, 2, \ldots, 6$$

A graphical depiction of the bivariate probability distribution for the die-tossing experiment is shown in Figure 5.1. Note that a nonzero probability is assigned to a point (y_1, y_2) in the plane if and only if $y_1 = 1, 2, \ldots, 6$ and $y_2 = 1, 2, \ldots, 6$. Thus exactly thirty-six points in the plane are assigned nonzero probabilities. Further, the probabilities are assigned in such a way that the sum

Figure 5.1
Bivariate probability
distribution; y_1 =
number of dots on
die 1, y_2 = number of
dots on die 2

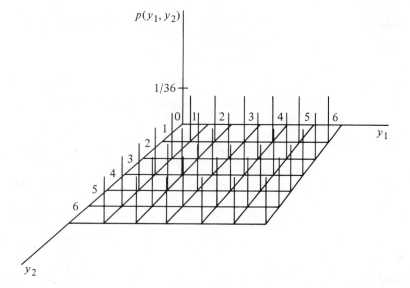

of the nonzero probabilities is 1. In Figure 5.1 the points assigned nonzero probabilities are represented in the (y_1, y_2) plane, whereas the probabilities associated with these points are given by the lengths of the lines above these points. Figure 5.1 may be viewed as a theoretical, three-dimensional relative frequency histogram for the pairs of observations (y_1, y_2). As in the single-variable discrete case, the theoretical histogram provides a model for the sample histogram that would be obtained if the die-tossing experiment was repeated a large number of times.

Definition 5.1

Let Y_1 and Y_2 be discrete random variables. The *joint* (or bivariate) *probability distribution* for Y_1 and Y_2 is given by

$$p(y_1, y_2) = P(Y_1 = y_1, Y_2 = y_2)$$

defined for all real numbers y_1 and y_2. The function $p(y_1, y_2)$ will be referred to as the *joint probability function*.

In the single-variable case discussed in Chapter 3, we saw that the probability function for a discrete random variable Y assigns nonzero probabilities to a countable number of distinct values of Y in such a way that the sum of the probabilities is 1. Similarly, in the bivariate case the joint probability function $p(y_1, y_2)$ will assign nonzero probabilities to only a countable number of pairs of values (y_1, y_2). Further, the nonzero probabilities must sum to 1.

Theorem 5.1

If Y_1 and Y_2 are discrete random variables with joint probability function $p(y_1, y_2)$, then

1. $p(y_1, y_2) \geq 0$ for all y_1, y_2.
2. $\sum\limits_{y_1, y_2} p(y_1, y_2) = 1$, where the sum is over all values (y_1, y_2) that are assigned nonzero probabilities.

Once the joint probability function is determined for discrete random variables Y_1 and Y_2, calculation of joint probabilities involving Y_1 and Y_2 is straightforward. For the die-tossing experiment, $P(2 \leq Y_1 \leq 3, 1 \leq Y_2 \leq 2)$ is

$$P(2 \leq Y_1 \leq 3, 1 \leq Y_2 \leq 2) = p(2, 1) + p(2, 2) + p(3, 1) + p(3, 2)$$
$$= 4/36 = 1/9$$

EXAMPLE 5.1

There are three checkout counters at a local supermarket. Two customers arrive at the counters at different times when the counters are serving no other customers. Each customer chooses a counter at random and independently of the other. Let Y_1 denote the number of customers who choose counter 1 and Y_2 the number who select counter 2. Find the joint distribution of Y_1 and Y_2.

Solution

We might proceed with the derivation in many ways. The most direct is to consider the sample space associated with the experiment. Let the pair $\{i, j\}$ denote the simple event that the first customer chose counter i and the second customer chose counter j, where $i, j = 1, 2, 3$. Using the mn rule, the sample space consists of $3 \times 3 = 9$ sample points. Under the assumptions given earlier, each sample point is equally likely and hence has a probability of 1/9. The sample space associated with the experiment is

$$S = [\{1, 1\}, \{1, 2\}, \{1, 3\}, \{2, 1\}, \{2, 2\}, \{2, 3\}, \{3, 1\}, \{3, 2\} \{3, 3\}].$$

Note that sample point $\{1, 1\}$ is the only sample point corresponding to $(Y_1 = 2, Y_2 = 0)$ and hence $P(Y_1 = 2, Y_2 = 0) = 1/9$. In a similar manner, it follows that $P(Y_1 = 1, Y_2 = 1) = P(\{1, 2\} \text{ or } \{2, 1\}) = 2/9$. Table 5.1 contains the probabili-

Table 5.1
Probability distribution for Y_1 and Y_2, Example 5.1

		Y_1	
Y_2	0	1	2
0	1/9	2/9	1/9
1	2/9	2/9	0
2	1/9	0	0

ties associated with each possible pair of values for Y_1 and Y_2; that is, the joint probability distribution for Y_1 and Y_2. As must always be the case, we note that the results of Theorem 5.1 hold for this example. ∽

As in the case of univariate random variables, the distinction between jointly discrete and jointly continuous random variables may be characterized in terms of their (joint) distribution functions.

Definition 5.2

For any random variables, Y_1 and Y_2, the joint (bivariate) distribution function, $F(a, b)$, is given by

$$F(a, b) = P(Y_1 \leq a, Y_2 \leq b)$$

For two discrete variables Y_1 and Y_2, $F(a, b)$ has the form

$$F(a, b) = \sum_{y_1 = -\infty}^{a} \sum_{y_2 = -\infty}^{b} p(y_1, y_2)$$

For the die-tossing experiment

$$F(2, 3) = P(Y_1 \leq 2, Y_2 \leq 3)$$
$$= p(1, 1) + p(1, 2) + p(1, 3) + p(2, 1) + p(2, 2) + p(2, 3)$$

Because $p(y_1, y_2) = 1/36$ for all pairs of values of y_1 and y_2, $F(2, 3) = 6/36 = 1/6$.

EXAMPLE 5.2   Consider the random variables Y_1 and Y_2 of Example 5.1. Find $F(-1, 2)$, $F(1.5, 2)$ and $F(5, 7)$.

Solution   Using the results in Table 5.1, we see that

$$F(-1, 2) = P(Y_1 \leq -1, Y_2 \leq 2) = P(\phi) = 0.$$

Further,

$$F(1.5, 2) = P(Y_1 \leq 1.5, Y_2 \leq 2) = p(0, 0) + p(0, 1) + p(0, 2) + p(1, 0)$$
$$+ p(1, 1) + p(1, 2) = 8/9.$$

Similarly

$$F(5, 7) = P(Y_1 \leq 5, Y_2 \leq 7) = 1.$$

Note that $F(a, b) = 1$ for all a, b such that min $\{a, b\} \geq 2$. Also, $F(a, b) = 0$ if min $\{a, b\} < 0$. Thus, the only values of a and b for which $0 \leq F(a, b) \leq 1$ are those where both a and b are in the interval $[0, 2]$.   ∽

Two random variables will be said to be jointly continuous if their joint distribution function $F(a, b)$ is continuous in both arguments.

Definition 5.3

Let Y_1 and Y_2 be continuous random variables with joint distribution function $F(a, b)$. If there exists a nonnegative function $f(a, b)$ such that

$$F(a, b) = \int_{-\infty}^{a} \int_{-\infty}^{b} f(y_1, y_2)\, dy_2\, dy_1$$

for any real numbers a and b, then Y_1 and Y_2 are said to be *jointly continuous random variables*. The function $f(y_1, y_2)$ is called the *joint probability density function*.

Bivariate cumulative distribution functions satisfy a set of properties similar to those specified for the univariate cumulative distribution function.

Theorem 5.2

If Y_1 and Y_2 are random variables with joint distribution function $F(a, b)$, then

1. $F(-\infty, -\infty) = F(-\infty, y_2) = F(y_1, -\infty) = 0$.

2. $F(\infty, \infty) = 1$.

3. if $a_2 \geq a_1$ and $b_2 \geq b_1$, then

$$F(a_2, b_2) - F(a_2, b_1) - F(a_1, b_2) + F(a_1, b_1) \geq 0$$

Note that the expression in part 3 is simply $P(a_1 < Y_1 \leq a_2, b_1 < Y_2 \leq b_2)$, which must be nonnegative.

Notice that $F(\infty, \infty) = 1$ implies that the joint density function $f(y_1, y_2)$ must be such that the integral of $f(y_1, y_2)$ over all values of (y_1, y_2) is 1.

Theorem 5.3

If Y_1 and Y_2 are jointly continuous random variables with a joint density function given by $f(y_1, y_2)$ then

1. $f(y_1, y_2) \geq 0$ for all y_1, y_2.

2. $\displaystyle\int_{-\infty}^{\infty} \int_{-\infty}^{\infty} f(y_1, y_2)\, dy_1\, dy_2 = 1$.

As in the univariate continuous case discussed in Chapter 4, the joint density function may be intuitively interpreted as a model for the joint relative frequency histogram for Y_1 and Y_2.

For the univariate case, areas under the probability density over an interval correspond to probabilities. Similarly the bivariate probability density function $f(y_1, y_2)$ traces a probability density surface over the (y_1, y_2) plane (Figure 5.2). Volumes under this surface correspond to probabilities. Thus

Figure 5.2
A bivariate density function f(y₁, y₂)

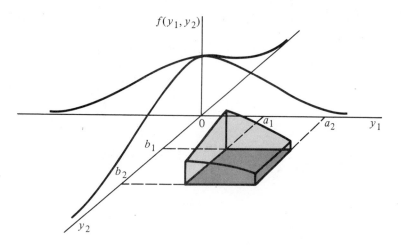

$P(a_1 \leq Y_1 \leq a_2, b_1 \leq Y_2 \leq b_2)$ is the shaded volume shown in Figure 5.2 and is equal to

$$\int_{b_1}^{b_2} \int_{a_1}^{a_2} f(y_1, y_2)\, dy_1\, dy_2$$

EXAMPLE 5.3

Suppose that a radioactive particle is randomly located in a square with sides of unit length. That is, if two regions of equal area are considered, the particle is equally likely to be in either. Let Y_1 and Y_2 denote the coordinates locating the particle. A reasonable model for the relative frequency histogram for Y_1 and Y_2 would be the bivariate analogue of the univariate uniform distribution,

$$f(y_1, y_2) = \begin{cases} 1, & 0 \le y_1 \le 1; 0 \le y_2 \le 1 \\ 0, & \text{elsewhere} \end{cases}$$

(a) Sketch the probability density surface.

(b) Find $F(.2, .4)$.

(c) Find $P(.1 \le Y_1 \le .3, 0 \le Y_2 \le .5)$.

Solution

(a) The sketch is shown in Figure 5.3.

Figure 5.3
Geometric representation of $f(y_1, y_2)$, Example 5.3

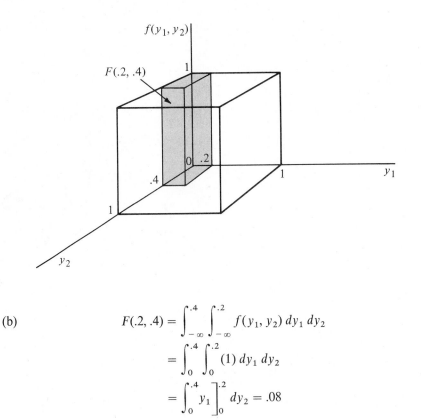

(b)

$$F(.2, .4) = \int_{-\infty}^{.4} \int_{-\infty}^{.2} f(y_1, y_2)\, dy_1\, dy_2$$

$$= \int_{0}^{.4} \int_{0}^{.2} (1)\, dy_1\, dy_2$$

$$= \int_{0}^{.4} y_1 \Big]_{0}^{.2} dy_2 = .08$$

The probability, $F(.2, .4)$, would correspond to the volume under $f(y_1, y_2) = 1$, part (a), over the shaded region of the (y_1, y_2) plane. From geometrical considerations you will observe that the desired probability (volume) is equal to .08.

(c) $$P(.1 \le Y_1 \le .3, 0 \le Y_2 \le .5) = \int_0^{.5} \int_1^{.3} f(y_1, y_2) \, dy_1 \, dy_2$$

$$= \int_0^{.5} \int_1^{.3} dy_1 \, dy_2 = .10$$

This probability would correspond to the volume under $f(y_1, y_2) = 1$ over the region $.1 \le y_1 \le .3, 0 \le y_2 \le .5$. Like the solution for (b), the solution for (c) can be obtained by using the elementary concepts of geometry. The density or height of the probability is equal to 1 and hence the desired probability (volume) is

$$P(.1 \le Y_1 \le .3, 0 \le Y_2 \le .5) = (.2)(.5)(1) = .1$$

We will illustrate a slightly more complicated bivariate model in the following example.

EXAMPLE 5.4   Gasoline is to be stocked in a bulk tank once at the beginning of each week and then sold to individual customers. Let Y_1 denote the proportion of the capacity of the bulk tank that is available after the tank is stocked at the beginning of the week. Because of the limited supplies, Y_1 varies from week to week. Let Y_2 denote the proportion of the capacity of the bulk tank that is sold during the week. Because Y_1 and Y_2 are both proportions, it is obvious that both variables take on values between 0 and 1. Further, it is obvious that the amount sold, y_2, cannot exceed the amount available, y_1. Suppose that a model for the relative frequency histogram for Y_1 and Y_2 is given by

$$f(y_1, y_2) = \begin{cases} 3y_1, & 0 \le y_2 \le y_1 \le 1 \\ 0, & \text{elsewhere} \end{cases}$$

A sketch of this function is given in Figure 5.4.

Figure 5.4
The joint density function for Example 5.4

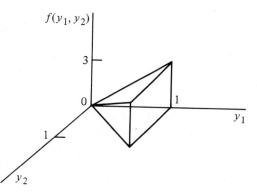

Find the probability that less than 1/2 of the tank will be stocked but more than 1/4 of the tank will be sold.

We want to find $P(0 \le Y_1 \le .5, \ Y_2 > .25)$. For any continuous random variable the probability of observing a value in a region is the volume under the density function above the region of interest. Note that the density function $f(y_1, y_2)$ is positive only in the large triangular portion of the (y_1, y_2) plane shown in Figure 5.5. We are interested only in values of y_1 and y_2 such that $0 \le y_1 \le .5$ and $y_2 > .25$. The intersection of this region and the region where the density function is positive is given by the small (shaded) triangle in Figure 5.5. Thus the probability we desire is the volume under the density function of Figure 5.4 above the shaded region in the (y_1, y_2) plane shown in Figure 5.5.

Figure 5.5
Region of integration
for Example 5.4

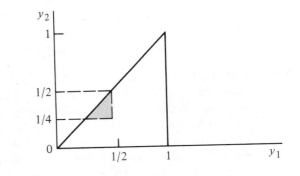

Thus we have

$$P(0 \le Y_1 \le .5, .25 \le Y_2) = \int_{1/4}^{1/2} \int_{1/4}^{y_1} 3y_1 \, dy_2 \, dy_1$$

$$= \int_{1/4}^{1/2} 3y_1 [y_2]_{1/4}^{y_1} \, dy_1$$

$$= \int_{1/4}^{1/2} 3y_1 (y_1 - 1/4) \, dy_1$$

$$= [y_1^3 - (3/8)y_1^2]_{1/4}^{1/2}$$

$$= [(1/8) - (3/8)(1/4)] - [(1/64) - (3/8)(1/16)]$$

$$= 5/128$$

Calculation of the probability specified in Example 5.4 involved integration of the joint density function for Y_1 and Y_2 over the appropriate region. The specification of the limits of integration was made easier by sketching the region of integration in Figure 5.5. Sketching the appropriate region of integration often will facilitate setting up the appropriate integral.

The methods discussed in this section can be used to calculate the probability of the intersection of two events $(Y_1 = y_1, Y_2 = y_2)$. In a similar manner, we can define a probability function (or probability density function) for the intersection of n events $(Y_1 = y_1, Y_2 = y_2, \ldots, Y_n = y_n)$. The probability function corresponding to the discrete case is given by

$$p(y_1, y_2, \ldots, y_n) = P(Y_1 = y_1, Y_2 = y_2, \ldots, Y_n = y_n)$$

The joint density function of $Y_1, Y_2, \ldots, Y_n$ is the function $f(y_1, y_2, \ldots, y_n)$. As in the bivariate case, these functions provide models for the joint relative frequency distributions of the populations of joint observations $(y_1, y_2, \ldots, y_n)$ for the discrete and the continuous case, respectively. In the continuous case

$$P(Y_1 \le y_1, Y_2 \le y_2, \ldots, Y_n \le y_n) = F(y_1, \ldots, y_n)$$

$$= \int_{-\infty}^{y_1} \int_{-\infty}^{y_2} \cdots \int_{-\infty}^{y_n} f(t_1, t_2, \ldots, t_n) \, dt_n \cdots dt_1$$

for every set of real numbers $(y_1, \ldots, y_n)$. Multivariate distribution functions, defined by this equality, will satisfy properties similar to those specified for the bivariate case.

Exercises

5.1 Contracts for two construction jobs are randomly assigned to one or more of three firms, A, B, and C. Let Y_1 denote the number of contracts assigned to firm A and Y_2 the number of contracts assigned to firm B. Recall that each firm can receive 0, 1, or 2 contracts.

(a) Find the joint probability function for Y_1 and Y_2.

(b) Find $F(1, 0)$.

5.2 Three balanced coins are tossed independently. One of the variables of interest is $Y_1 =$ the number of heads. Let Y_2 denote the amount of money won on a side bet in the following manner. If the first head occurs on the first toss, you win $1. If the first head occurs on toss 2 or on toss 3 you win $2 or $3, respectively. If no heads appear, you lose $1 (that is, win $-$1).

(a) Find the joint probability function for Y_1 and Y_2.

(b) What is the probability that less than three heads occur and you win $1 or less? [That is, find $F(2, 1)$].

5.3 In a group of nine executives of a certain business firm, four are married, three have never married, and two are divorced. Three of the executives are to be selected for promotion. Let Y_1 denote the number of married executives and Y_2 the number of never married executives among the three selected for promotion. Assuming that the three are randomly selected from the nine available, find the joint probability distribution of Y_1 and Y_2.

5.4   Let Y_1 and Y_2 have the joint probability density function given by

$$f(y_1, y_2) = \begin{cases} Ky_1y_2, & 0 \le y_1 \le 1; 0 \le y_2 \le 1 \\ 0, & \text{elsewhere} \end{cases}$$

(a)   Find the value of K that makes this a probability density function.

(b)   Find the joint distribution function for Y_1 and Y_2.

(c)   Find $P(Y_1 \le 1/2, Y_2 \le 3/4)$.

5.5   Let Y_1 and Y_2 have the joint probability density function given by

$$f(y_1, y_2) = \begin{cases} K(1 - y_2), & 0 \le y_1 \le y_2 \le 1 \\ 0, & \text{elsewhere} \end{cases}$$

(a)   Find the value of K that makes this a probability density function.

(b)   Find $P(Y_1 \le 3/4, Y_2 \ge 1/2)$.

5.6   An environmental engineer measures the amount (by weight) of particulate pollution in air samples of a certain volume collected over the smokestack of a coal-operated power plant. Let Y_1 denote the amount of pollutant per sample collected when a certain cleaning device on the stack is not operating and let Y_2 denote the amount of pollutant per sample collected under the same environmental conditions when the cleaning device is operating. It is observed that the relative frequency behavior of Y_1 and Y_2 can be modeled by

$$f(y_1, y_2) = \begin{cases} K, & 0 \le y_1 \le 2; 0 \le y_2 \le 1; 2y_2 \le y_1 \\ 0, & \text{elsewhere} \end{cases}$$

(That is, Y_1 and Y_2 are uniformly distributed over the region inside the triangle bounded by $y_1 = 2$, $y_2 = 0$, and $2y_2 = y_1$.)

(a)   Find the value of K that makes this function a probability density function.

(b)   Find $P(Y_1 \ge 3Y_2)$. (That is, find the probability that the cleaning device will reduce the amount of pollutant by one-third or more.)

5.7   Suppose Y_1 and Y_2 are uniformly distributed over the triangle shaded in the accompanying diagram.

(a)   Find $P(Y_1 \le 3/4, Y_2 \le 3/4)$.

(b)   Find $P[(Y_1 - Y_2) \ge 0]$.

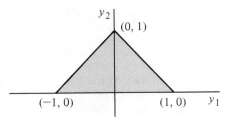

5.8 Let Y_1 and Y_2 denote the proportions of two different types of components in a sample from a mixture of chemicals used as an insecticide. Suppose Y_1 and Y_2 have the joint density function given by

$$f(y_1, y_2) = \begin{cases} 2, & 0 \le y_1 \le 1; 0 \le y_2 \le 1; 0 \le y_1 + y_2 \le 1 \\ 0, & \text{elsewhere} \end{cases}$$

(Note that $Y_1 + Y_2 \le 1$ because the random variables denote proportions within the same sample.)

(a) Find $P(Y_1 \le 3/4, Y_2 \le 3/4)$.

(b) Find $P(Y_1 \le 1/2, Y_2 \le 1/2)$.

5.9 A particular fast-food outlet is interested in the joint behavior of the random variables Y_1, defined as the total time between a customer's arrival at the store and leaving the service window, and Y_2, the time that a customer waits in line before reaching the service window. Because Y_1 contains the time a customer waits in line, we must have $Y_1 \ge Y_2$. The relative frequency distribution of observed values of Y_1 and Y_2 can be modeled by the probability density function

$$f(y_1, y_2) = \begin{cases} e^{-y_1}, & 0 \le y_2 \le y_1 < \infty \\ 0, & \text{elsewhere} \end{cases}$$

with time measured in minutes.

(a) Find $P(Y_1 < 2, Y_2 > 1)$.

(b) Find $P(Y_1 \ge 2Y_2)$.

(c) Find $P(Y_1 - Y_2 \ge 1)$. (Note that $Y_1 - Y_2$ denotes the time spent at the service window.)

5.10 Let Y_1 and Y_2 denote the proportions of time, out of one workday, that employees I and II, respectively, actually spend performing their assigned tasks. The joint relative frequency behavior of Y_1 and Y_2 is modeled by the density function

$$f(y_1, y_2) = \begin{cases} y_1 + y_2, & 0 \le y_1 \le 1; 0 \le y_2 \le 1 \\ 0, & \text{elsewhere} \end{cases}$$

(a) Find $P(Y_1 < 1/2, Y_2 > 1/4)$.

(b) Find $P(Y_1 + Y_2 \le 1)$.

5.11 Let (Y_1, Y_2) denote the coordinates of a point chosen at random inside a unit circle with its center at the origin. That is, Y_1 and Y_2 have a joint density function given by

$$f(y_1, y_2) = \begin{cases} \dfrac{1}{\pi}, & y_1^2 + y_2^2 \le 1 \\ 0, & \text{elsewhere} \end{cases}$$

Find $P(Y_1 \le Y_2)$.

5.12   An electronic system has one of each of two different types of components in joint operation. Let Y_1 and Y_2 denote the random lengths of life of the components of type I and type II, respectively. The joint density function is given by

$$f(y_1, y_2) = \begin{cases} (1/8)y_1 e^{-(y_1 + y_2)/2}, & y_1 > 0; y_2 > 0 \\ 0, & \text{elsewhere} \end{cases}$$

(Measurements are in hundreds of hours.) Find $P(Y_1 > 1, Y_2 > 1)$.

5.3 Marginal and Conditional Probability Distributions

Recall that the values assumed by a discrete random variable represent mutually exclusive events. Similarly all the bivariate events $(Y_1 = y_1, Y_2 = y_2)$, represented by (y_1, y_2), are mutually exclusive events. It then follows that the univariate event $(Y_1 = y_1)$ is the union of bivariate events of the type $(Y_1 = y_1, Y_2 = y_2)$, with the union being taken over all possible values for y_2.

For example, reconsider the die-tossing experiment of Section 5.2, where

$$Y_1 = \text{number of dots on the upper face of die 1}$$
$$Y_2 = \text{number of dots on the upper face of die 2}$$

Then

$$P(Y_1 = 1) = p(1, 1) + p(1, 2) + p(1, 3) + \cdots + p(1, 6)$$
$$= 1/36 + 1/36 + 1/36 + \cdots + 1/36 = 6/36 = 1/6$$
$$P(Y_1 = 2) = p(2, 1) + p(2, 2) + p(2, 3) + \cdots + p(2, 6) = 1/6$$
$$P(Y_1 = 6) = p(6, 1) + p(6, 2) + p(6, 3) + \cdots + p(6, 6) = 1/6$$

Expressed in summation notation, the probabilities are

$$p_1(y_1) = \sum_{y_2 = 1}^{6} p(y_1, y_2)$$

Similarly

$$p_2(y_2) = \sum_{y_1 = 1}^{6} p(y_1, y_2)$$

Summation in the discrete case corresponds to integration in the continuous case, and so we are led to the following definition.

Definition 5.4

(a) Let Y_1 and Y_2 be jointly discrete random variables with probability function $p(y_1, y_2)$. Then the *marginal probability functions* of Y_1 and Y_2, respectively, are given by

$$p_1(y_1) = \sum_{y_2} p(y_1, y_2) \quad \text{and} \quad p_2(y_2) = \sum_{y_1} p(y_1, y_2)$$

(b) Let Y_1 and Y_2 be jointly continuous random variables with joint density function $f(y_1, y_2)$. Then the *marginal density functions* of Y_1 and Y_2, respectively, are given by

$$f_1(y_1) = \int_{-\infty}^{\infty} f(y_1, y_2)\, dy_2 \quad \text{and} \quad f_2(y_2) = \int_{-\infty}^{\infty} f(y_1, y_2)\, dy_1$$

The term *marginal*, as applied to the univariate probability distributions of Y_1 and Y_2, has intuitive meaning. Finding $p_1(y_1)$ implies summing $p(y_1, y_2)$ over all values of y_2 and hence accumulating the probabilities on the y_1 axis (or margin).

We illustrate both the discrete and continuous cases with the following two examples.

EXAMPLE 5.5   From a group of three Republicans, two Democrats, and one Independent, a committee of two people is to be randomly selected. Let Y_1 denote the number of Republicans and Y_2 the number of Democrats on the committee. Find the joint probability distribution of Y_1 and Y_2, and then find the marginal distribution of Y_1.

Solution   The probabilities sought here are similar to the hypergeometric probabilities of Chapter 3. For example,

$$P(Y_1 = 1, Y_2 = 1) = p(1, 1) = \frac{\binom{3}{1}\binom{2}{1}\binom{1}{0}}{\binom{6}{2}} = \frac{3(2)}{15} = \frac{6}{15}$$

because there are fifteen equally likely sample points; and for the event in question we must select one Republican from the three, one Democrat from the two, and zero Independents. Similar calculations will lead to the other probabilities shown in Table 5.2.

Table 5.2
Probability
distribution for Y_1
and Y_2, Example 5.5

		Y_1		
Y_2	0	1	2	*Total*
0	0	3/15	3/15	6/15
1	2/15	6/15	0	8/15
2	1/15	0	0	1/15
Total	3/15	9/15	3/15	1

To find $p_1(y_1)$, we must sum over values of Y_2, as Definition 5.4 indicates. Hence these probabilities are given by the column totals in Table 5.2. That is,

$$p_1(0) = p(0, 0) + p(0, 1) + p(0, 2)$$
$$= 0 + 2/15 + 1/15 = 3/15$$

Similarly,

$$p_1(1) = 9/15, \qquad p_1(2) = 3/15$$

Note that the marginal distribution of Y_2 is given by the row totals. ∽

EXAMPLE 5.6   Let

$$f(y_1, y_2) = \begin{cases} 2y_1, & 0 \le y_1 \le 1; 0 \le y_2 \le 1 \\ 0, & \text{elsewhere} \end{cases}$$

Sketch $f(y_1, y_2)$ and find the marginal density functions for Y_1 and Y_2.

Solution   Viewed geometrically, $f(y_1, y_2)$ traces a surface shaped like a wedge. A sketch is presented in Figure 5.6.

Before applying Definition 5.4 to find $f_1(y_1)$ and $f_2(y_2)$, we will use Figure 5.6 to visualize the result. If the probability represented by the wedge were accumulated on the y_1 axis (accumulating probability along lines parallel to the y_2 axis), the result would be a triangular probability distribution that would look like the side of the wedge in Figure 5.6. If accumulated along the y_2 axis (accumulating along lines parallel to the y_1 axis), the resulting distribution would be uniform. We will confirm the visual solution by applying Definition 5.4. Then

$$f_1(y_1) = \int_{-\infty}^{\infty} f(y_1, y_2)\, dy_2 = \int_0^1 2y_1\, dy_2 = 2y_1 y_2 \Big]_0^1 = 2y_1, \qquad 0 \le y_1 \le 1$$

Figure 5.6
Geometric
representation of
$f(y_1, y_2)$, Example 5.6

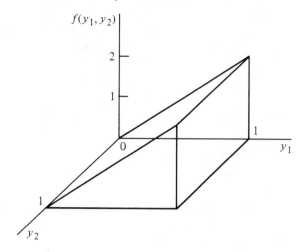

Similarly

$$f_2(y_2) = \int_{-\infty}^{\infty} f(y_1, y_2)\, dy_1 = \int_0^1 2y_1\, dy_1 = y_1^2 \Big]_0^1 = 1, \qquad 0 \le y_2 \le 1$$

Note that graphs of $f_1(y_1)$ and $f_2(y_2)$ trace triangular and uniform probability distributions, respectively, as expected. ∽

We now turn our attention to conditional distributions, looking first at the discrete case.

The multiplicative law (Section 2.8) gives the probability of the intersection $A \cap B$ as

$$P(A \cap B) = P(A)P(B|A)$$

where $P(A)$ is the unconditional probability of A and $P(B|A)$ is the probability of B given A has occurred. Now consider the intersection of the two numerical events, $(Y_1 = y_1) \cap (Y_2 = y_2)$, represented by the bivariate event (y_1, y_2). Then it follows directly from the multiplicative law of probability that the bivariate probability for the intersection (y_1, y_2) is

$$p(y_1, y_2) = p_1(y_1)p(y_2|y_1)$$
$$= p_2(y_2)p(y_1|y_2)$$

The probabilities $p_1(y_1)$ and $p_2(y_2)$ are those associated with the univariate probability distributions for Y_1 and Y_2 (content of Chapter 3). The meaning of the conditional probabilities $p(y_1|y_2)$ and $p(y_2|y_1)$ is given in Chapter 2. Thus $p(y_1|y_2)$ is the probability that the random variable Y_1 takes a specific value y_1, given that Y_2 takes on the value y_2.

Definition 5.5

> If Y_1 and Y_2 are jointly discrete random variables with joint probability function $p(y_1, y_2)$, and marginal probability functions $p_1(y_1)$ and $p_2(y_2)$, respectively, then the *conditional discrete probability function* of Y_1 given Y_2 is
>
> $$p(y_1|y_2) = P(Y_1 = y_1 | Y_2 = y_2) = \frac{P(Y_1 = y_1, Y_2 = y_2)}{P(Y_2 = y_2)} = \frac{p(y_1, y_2)}{p_2(y_2)}$$
>
> provided $p_2(y_2) > 0$.

Thus $P(y_1 = 2 | y_2 = 3)$ is the conditional probability that $Y_1 = 2$ given that $Y_2 = 3$. A similar interpretation can be attached to the conditional probability $p(y_2|y_1)$.

EXAMPLE 5.7   Refer to Example 5.5 and find the conditional distribution of Y_1 given that $Y_2 = 1$. That is, given that one of the two people on the committee is a Democrat, find the conditional distribution for the number of Republicans selected for the committee.

Solution   The joint probabilities are given in Table 5.2. To find $p(y_1 | Y_2 = 1)$, we concentrate on the row headed by $Y_2 = 1$. Then

$$P(Y_1 = 0 | Y_2 = 1) = \frac{p(0, 1)}{p_2(1)} = \frac{2/15}{8/15} = \frac{1}{4}$$

$$P(Y_1 = 1 | Y_2 = 1) = \frac{p(1, 1)}{p_2(1)} = \frac{6/15}{8/15} = \frac{3}{4}$$

and

$$P(Y_1 \geq 2 | Y_2 = 1) = \frac{p(2, 1)}{p_2(1)} = 0$$

In the randomly selected committee, if one person is a Democrat, there is a high probability that the other will be a Republican.   ∽

In the continuous case we can obtain an appropriate analogue of the conditional probability function $p(y_1|y_2)$, but it is not obtained in such a straightforward manner. If Y_1 and Y_2 are continuous, $P(Y_1 = y_1 | Y_2 = y_2)$ cannot be defined as in the discrete case because both events involved have zero probability. The following considerations, however, do lead to a useful and consistent definition for a conditional density function.

Assuming that Y_1 and Y_2 are jointly continuous with density function $f(y_1, y_2)$, we might be interested in a probability of the form

$$P(Y_1 \le y_1 | Y_2 = y_2) = F(y_1 | y_2)$$

which, as a function of y_1 for a fixed y_2, we will call the *conditional distribution function* of Y_1 given $Y_2 = y_2$.

Definition 5.6

> If Y_1 and Y_2 are jointly continuous random variables with joint density function $f(y_1, y_2)$, then the *conditional distribution function* of Y_1 given $Y_2 = y_2$ is
>
> $$F(y_1 | y_2) = P(Y_1 \le y_1 | Y_2 = y_2)$$

Note that $F(y_1 | y_2)$ is a function of y_1 for a fixed value of y_2.

If we could take $F(y_1 | y_2)$, multiply by $P(Y_2 = y_2)$ for each possible value of Y_2, and sum all the resulting probabilities, we would obtain $F(y_1)$. This is not possible because the number of values for y_2 is uncountable and all probabilities $P(Y_2 = y_2)$ are zero. But we can do something analogous by multiplying by $f_2(y_2)$ and then integrating to obtain

$$F(y_1) = \int_{-\infty}^{\infty} F(y_1 | y_2) f_2(y_2) \, dy_2$$

The quantity $f_2(y_2) \, dy_2$ can be thought of as the approximate probability that Y_2 takes on a value in a small interval about y_2, and the integral is a generalized sum.

Now from previous considerations we know that

$$F(y_1) = \int_{-\infty}^{y_1} f_1(t_1) \, dt_1 = \int_{-\infty}^{y_1} \left[\int_{-\infty}^{\infty} f(t_1, y_2) \, dy_2 \right] dt_1$$

$$= \int_{-\infty}^{\infty} \int_{-\infty}^{y_1} f(t_1, y_2) \, dt_1 \, dy_2$$

From the expressions for $F(y_1)$ we must have

$$F(y_1 | y_2) f_2(y_2) = \int_{-\infty}^{y_1} f(t_1, y_2) \, dt_1$$

or

$$F(y_1 | y_2) = \int_{-\infty}^{y_1} \frac{f(t_1, y_2)}{f_2(y_2)} \, dt_1$$

We will call the integrand of this expression the *conditional density function* of Y_1 given $Y_2 = y_2$ and denote it by $f(y_1|y_2)$.

Definition 5.7

Let Y_1 and Y_2 be jointly continuous random variables with joint density $f(y_1, y_2)$ and marginal densities $f_1(y_1)$ and $f_2(y_2)$, respectively. Then the conditional density of Y_1 given $Y_2 = y_2$ is given by

$$f(y_1|y_2) = \begin{cases} \dfrac{f(y_1, y_2)}{f_2(y_2)}, & f_2(y_2) > 0 \\ 0, & \text{elsewhere} \end{cases}$$

and the conditional density of Y_2 given $Y_1 = y_1$ is given by

$$f(y_2|y_1) = \begin{cases} \dfrac{f(y_1, y_2)}{f_1(y_1)}, & f_1(y_1) > 0 \\ 0, & \text{elsewhere} \end{cases}$$

EXAMPLE 5.8   A soft drink machine has a random amount Y_2 in supply at the beginning of a given day and dispenses a random amount Y_1 during the day (with measurements in gallons). It is not resupplied during the day and hence $Y_1 \leq Y_2$. It has been observed that Y_1 and Y_2 have a joint density given by

$$f(y_1, y_2) = \begin{cases} 1/2, & 0 \leq y_1 \leq y_2; 0 \leq y_2 \leq 2 \\ 0, & \text{elsewhere} \end{cases}$$

That is, the points (y_1, y_2) are uniformly distributed over the triangle with the given boundaries. Find the conditional density of Y_1 given $Y_2 = y_2$. Evaluate the probability that less than 1/2 gallon is sold, given that the machine contains 1 gallon at the start of the day.

Solution   The marginal density of Y_2 is given by

$$f_2(y_2) = \int_{-\infty}^{\infty} f(y_1, y_2) \, dy_1$$

or

$$f_2(y_2) = \begin{cases} \displaystyle\int_0^{y_2} (1/2) \, dy_1 = (1/2)y_2, & y_1 \leq y_2 \leq 2 \\ 0, & \text{elsewhere} \end{cases}$$

By Definition 5.7,

$$f(y_1|y_2) = \frac{f(y_1, y_2)}{f_2(y_2)}$$

or

$$f(y_1|y_2) = \begin{cases} \dfrac{1/2}{(1/2)(y_2)} = \dfrac{1}{y_2}, & 0 < y_1 \le y_2 \le 2 \\ 0, & \text{elsewhere} \end{cases}$$

The probability of interest is

$$P(Y_1 \le 1/2 \,|\, Y_2 = 1) = \int_{-\infty}^{1/2} f(y_1|y_2 = 1)\, dy_1 = \int_{0}^{1/2} (1)\, dy_1 = \frac{1}{2}$$

Note that if the machine has contained 2 gallons at the start of the day, then

$$P(Y_1 \le 1/2 \,|\, Y_2 = 2) = \int_{0}^{1/2} \frac{1}{2}\, dy_1 = \frac{1}{4}$$

Thus the conditional probability that $Y_1 \le 1/2$ given $Y_2 = y_2$ changes apprecia-bly depending on the particular choice of y_2.   ᔓ

Exercises

5.13   In Exercise 5.1 we determined that the joint distribution of Y_1, the number of contracts awarded to firm A, and Y_2, the number of contracts awarded to firm B, is given by the entries in the following table.

		Y_1	
Y_2	0	1	2
0	1/9	2/9	1/9
1	2/9	2/9	0
2	1/9	0	0

(a)   Find the marginal probability distribution of Y_1.

(b)   According to results in Chapter 4, Y_1 has a binomial distribution with $n = 2$ and $p = 1/3$. Is there any conflict between this result and the answer you provided in part (a)?

5.14   Refer to Exercise 5.2.

   (a)   Derive the marginal probability distribution for your winnings on the side bet.

   (b)   What is the probability that you obtained three heads given that you won $1 on the side bet?

5.15   In Exercise 5.3, we determined that the joint probability distribution of Y_1, the number of married executives, and Y_2, the number of never married executives, is given by

$$p(y_1, y_2) = \frac{\binom{4}{y_1}\binom{3}{y_2}\binom{2}{3 - y_1 - y_2}}{\binom{9}{3}}$$

where y_1 and y_2 are integers, $0 \le y_1 \le 3, 0 \le y_2 \le 3$, and $0 \le y_1 + y_2 \le 3$.

   (a)   Find the marginal probability distribution of y_1, the number of married executives among the three selected for promotion.

   (b)   Find $P(Y_1 = 1 | Y_2 = 2)$.

   (c)   If we let Y_3 denote the number of divorced executives among the three selected for promotion, then $Y_3 = 3 - Y_1 - Y_2$. Find $P(Y_3 = 1 | Y_2 = 1)$.

   (d)   Compare the marginal distribution derived in part (a) with the hypergeometric distribution with $N = 9, n = 3, r = 4$ encountered in Section 3.7.

5.16   In Exercise 5.4, we derived the fact that

$$f(y_1, y_2) = \begin{cases} 4y_1y_2, & 0 \le y_1 \le 1, 0 \le y_2 \le 1 \\ 0, & \text{elsewhere} \end{cases}$$

is a valid joint probability density function.

   (a)   Find the marginal density functions for Y_1 and Y_2.

   (b)   Find $P(Y_1 \le 1/2 | Y_2 \ge 3/4)$.

   (c)   Find the conditional density function of Y_1 given $Y_2 = y_2$.

   (d)   Find the conditional density function of Y_2 given $Y_1 = y_1$.

   (e)   Find $P(Y_1 \le 3/4 | Y_2 = 1/2)$.

5.17   In Exercise 5.5, we determined that

$$f(y_1, y_2) = \begin{cases} 6(1 - y_2), & 0 \le y_1 \le y_2 \le 1 \\ 0, & \text{elsewhere} \end{cases}$$

is a valid joint probability density function.

   (a)   Find the marginal density functions for Y_1 and Y_2.

   (b)   Find $P(Y_2 \le 1/2 | Y_1 \le 3/4)$.

   (c)   Find the conditional density function of Y_1 given $Y_2 = y_2$.

   (d)   Find the conditional density function of Y_2 given $Y_1 = y_1$.

   (e)   Find $P(Y_2 \ge 3/4 | Y_1 = 1/2)$.

5.18   In Exercise 5.6, we proved that

$$f(y_1, y_2) = \begin{cases} 1, & 0 \le y_1 \le 2, 0 \le y_2 \le 1; 2y_2 \le y_1 \\ 0, & \text{elsewhere} \end{cases}$$

is a valid joint probability density function for Y_1, the amount of pollutant per sample collected when the cleaning device is not operating, and Y_2, the amount collected when the device is operating.

(a)   If the cleaning device is operating, find the probability that the amount of pollutant in a given sample will exceed 0.5.

(b)   Given that the amount of pollutant in a sample taken with the cleaning device operating is observed to be 0.5, find the probability that the amount would have exceeded 1.5 had the cleaning device not been operating.

5.19   Refer to Exercise 5.7.

(a)   Find the marginal density functions for Y_1 and Y_2.

(b)   Find $P(Y_2 > 1/2 | Y_1 = 1/4)$.

5.20   In Exercise 5.8, we were given the following joint probability density function for the random variables Y_1 and Y_2, which were the proportions of two components in a sample from a mixture of insecticide.

$$f(y_1, y_2) = \begin{cases} 2, & 0 \le y_1 \le 1, 0 \le y_2 \le 1; 0 \le y_1 + y_2 \le 1 \\ 0, & \text{elsewhere} \end{cases}$$

(a)   Find $P(Y_1 \ge 1/2 | Y_2 \le 1/4)$.

(b)   Find $P(Y_1 \ge 1/2 | Y_2 = 1/4)$.

5.21   If Y_1 is the total time between a customer's arrival in the store and leaving the service window, and Y_2 is the time spent in line before reaching the window, the joint density of these variables was given in Exercise 5.9 to be

$$f(y_1, y_2) = \begin{cases} e^{-y_1}, & 0 \le y_2 \le y_1 \le \infty \\ 0, & \text{elsewhere} \end{cases}$$

If 2 minutes elapse between a customer's arrival at the store and his departure from the service window, find the probability that he waited in line less than 1 minute to reach the window.

5.22   In Exercise 5.10, Y_1 and Y_2 denoted the proportions of time that employees I and II actually spent on their assigned tasks during a workday. The joint density of Y_1 and Y_2 is given by

$$f(y_1, y_2) = \begin{cases} y_1 + y_2, & 0 \le y_1 \le 1, 0 \le y_2 \le 1 \\ 0, & \text{elsewhere} \end{cases}$$

(a)   Find the marginal density functions for Y_1 and Y_2.

(b)   Find $P(Y_1 \ge 1/2 | Y_2 \ge 1/2)$.

(c)   If employee II spends exactly 50% of the day on assigned duties, find the probability that employee I spends more than 75% of the day on similar duties.

5.23 In Exercise 5.12, Y_1 and Y_2 denoted the lengths of life, in hours, for components of types I and II, respectively, in an electronic system. The joint density of Y_1 and Y_2 is given by

$$f(y_1, y_2) = \begin{cases} (1/8)y_1 e^{-(y_1 + y_2)/2}, & y_1 > 0, y_2 > 0 \\ 0, & \text{elsewhere.} \end{cases}$$

Find the probability that a component of type II will have a life length in excess of 200 hours.

5.24 Let Y_1 denote the weight (in tons) of a certain bulk item stocked by a supplier at the beginning of a week and suppose that Y_1 has a uniform distribution over the interval $0 \le y_1 \le 1$. Let Y_2 denote the weight of this item sold by the supplier during the week and suppose that Y_2 has a uniform distribution over the interval $0 \le y_2 \le y_1$, where y_1 is a specific value of Y_1.

(a) Find the joint density function for y_1 and y_2.

(b) If the supplier stocks a half ton of the item, what is the probability that she sells more than a quarter ton?

(c) If it is known that the supplier sold a quarter ton of the item, what is the probability that she had stocked more than a half ton?

*5.25 A quality control plan calls for randomly selecting three items from the daily production (assumed large) of a certain machine and observing the number of defectives. However, the proportion p of defectives produced by the machine varies from day to day and is assumed to have a uniform distribution on the interval $(0, 1)$. For a randomly chosen day find the unconditional probability that exactly two defectives are observed in the sample.

*5.26 The number of defects per yard, denoted by Y, for a certain fabric is known to have a Poisson distribution with parameter λ. However, λ itself is a random variable with probability density function given by

$$f(\lambda) = \begin{cases} e^{-\lambda}, & \lambda \ge 0 \\ 0, & \text{elsewhere} \end{cases}$$

Find the unconditional probability function for Y.

5.4 Independent Random Variables

In Example 5.8 we saw two random variables that were dependent in the sense that probabilities associated with Y_1 depended on the value of Y_2. We now present a formal definition of independence of random variables.

Two events A and B are independent if $P(AB) = P(A)P(B)$. When discussing random variables, we are often concerned with events of the type

---

*Exercises preceded by an asterisk are optional.

$(a \le Y_1 \le b) \cap (c \le Y_2 \le d)$. To be consistent with the earlier definition of independent events, if Y_1 and Y_2 are independent, we would like to have

$$P(a \le Y_1 \le b, c \le Y_2 \le d) = P(a \le Y_1 \le b)P(c \le Y_2 \le d)$$

for any choice of real numbers a, b, c, and d. That is, if Y_1 and Y_2 are independent, the joint probability can be written as the product of the marginal probabilities. This property will be satisfied if Y_1 and Y_2 are independent in the sense detailed in the following definition.

Definition 5.8

Let Y_1 have distribution function $F_1(y_1)$, Y_2 have distribution function $F_2(y_2)$, and Y_1 and Y_2 have joint distribution function $F(y_1, y_2)$. Then Y_1 and Y_2 are said to be *independent* if and only if

$$F(y_1, y_2) = F_1(y_1)F_2(y_2)$$

for every pair of real numbers (y_1, y_2).
If Y_1 and Y_2 are not independent, they are said to be *dependent*.

It usually is more convenient to establish independence, or the lack of it, by using the result contained in the following theorem. The proof is omitted, see the references at the end of the chapter.

Theorem 5.4

If Y_1 and Y_2 are discrete random variables with joint probability function $p(y_1, y_2)$ and marginal probability functions $p_1(y_1)$ and $p_2(y_2)$, respectively, then Y_1 and Y_2 are independent if and only if

$$p(y_1, y_2) = p_1(y_1)p_2(y_2)$$

for all pairs of real numbers (y_1, y_2).
If Y_1 and Y_2 are continuous random variables with a joint density function of $f(y_1, y_2)$ and marginal density functions of $f_1(y_1)$ and $f_2(y_2)$, respectively, then Y_1 and Y_2 are independent if and only if

$$f(y_1, y_2) = f_1(y_1)f_2(y_2)$$

for all pairs of real numbers (y_1, y_2).

We now illustrate the concept of independence with some examples.

EXAMPLE 5.9   For the die-tossing problem of Section 5.2, show that Y_1 and Y_2 are independent.

Solution   In this problem each of the thirty-six sample points was given an equal probability of 1/36. Consider, for example, the point (1, 2). We know that $p(1,\ 2) = 1/36$. Also, $p_1(1) = P(Y_1 = 1) = 1/6$ and $p_2(2) = P(Y_2 = 2) = 1/6$. Hence

$$p(1, 2) = p_1(1)p_2(2)$$

The same is true for all other points, and it follows that Y_1 and Y_2 are independent.   ∽

EXAMPLE 5.10   Refer to Example 5.5. Is the number of Republicans in the sample independent of the number of Democrats? (Is Y_1 independent of Y_2?)

Solution   Independence of discrete random variables requires that $p(y_1, y_2) = p_1(y_1)p_2(y_2)$ for every choice (y_1, y_2). Thus if this equality is violated for any (y_1, y_2), the random variables are dependent. Looking in the upper left-hand corner of Table 5.2, we see

$$p(0, 0) = 0$$

But $p_1(0) = 3/15$ and $p_2(0) = 6/15$. Hence

$$p(0, 0) \neq p_1(0)p_2(0)$$

and Y_1 and Y_2 are dependent.   ∽

EXAMPLE 5.11   Let

$$f(y_1 y_2) = \begin{cases} 4y_1 y_2, & 0 \le y_1 \le 1; 0 \le y_2 \le 1 \\ 0, & \text{elsewhere} \end{cases}$$

Show that Y_1 and Y_2 are independent.

Solution   We have

$$f_1(y_1) = \int_0^1 f(y_1, y_2)\, dy_2 = \int_0^1 4y_1 y_2\, dy_2$$

$$= 4y_1 \left. \frac{y_2^2}{2} \right]_0^1 = 2y_1, \qquad 0 \le y_1 \le 1$$

Similarly

$$f_2(y_2) = \int_0^1 f(y_1, y_2)\, dy_1 = 2y_2, \qquad 0 \le y_2 \le 1$$

Hence

$$f(y_1, y_2) = f_1(y_1)f_2(y_2)$$

for any real numbers (y_1, y_2), and therefore Y_1 and Y_2 are independent.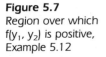

EXAMPLE 5.12   Let

$$f(y_1, y_2) = \begin{cases} 2, & 0 \le y_2 \le y_1; 0 \le y_1 \le 1 \\ 0, & \text{elsewhere} \end{cases}$$

Show that Y_1 and Y_2 are dependent.

Figure 5.7
Region over which
$f(y_1, y_2)$ is positive,
Example 5.12

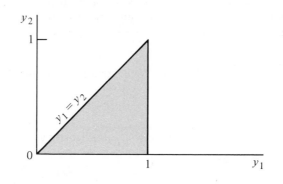

Solution $f(y_1, y_2) = 2$ over the shaded region shown in Figure 5.7. Therefore,

$$f_1(y_1) = \int_0^{y_1} f(y_1, y_2)\, dy_2 = \int_0^{y_1} 2\, dy_2 = 2y_2 \Big]_0^{y_1}$$
$$= 2y_1, \qquad 0 \le y_1 \le 1$$

Similarly,

$$f_2(y_2) = \int_{y_2}^1 f(y_1, y_2)\, dy_1 = \int_{y_2}^1 2\, dy_1 = 2y_1 \Big]_{y_2}^1$$
$$= 2(1 - y_2), \qquad 0 \le y_2 \le 1$$

Hence

$$f(y_1, y_2) \ne f_1(y_1)f_2(y_2)$$

for some real numbers (y_1, y_2), and therefore Y_1 and Y_2 are dependent.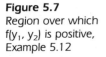

You will note a distinct difference in the limits of integration employed in finding the marginal density functions obtained in Examples 5.11 and 5.12. In finding $f_1(y_1)$, Example 5.12, the limits of integration for y_2 depended on y_1. In contrast, the limits of integration involved in finding the marginal density functions, Example 5.11, were constants. If the limits of integration are constants, the following theorem provides an easy way to show independence of two random variables.

Theorem 5.5

Let Y_1 and Y_2 have a joint density $f(y_1, y_2)$, which is positive if and only if $a \le y_1 \le b$, $c \le y_2 \le d$, for constants a, b, c, and d, and $f(y_1, y_2) = 0$ otherwise. Then Y_1 and Y_2 are independent random variables if and only if

$$f(y_1, y_2) = g(y_1)h(y_2)$$

where $g(y_1)$ is a nonnegative function of y_1 alone and $h(y_2)$ is a nonnegative function of y_2 alone.

The proof of this theorem is omitted. (See the references at the end of the chapter.)

EXAMPLE 5.13   Let Y_1 and Y_2 have a joint density given by

$$f(y_1, y_2) = \begin{cases} 2y_1, & 0 \le y_1 \le 1; 0 \le y_2 \le 1 \\ 0, & \text{elsewhere} \end{cases}$$

Are Y_1 and Y_2 independent variables?

Solution   Note that $f(y_1, y_2)$ is positive if and only if $0 \le y_1 \le 1$ and $0 \le y_2 \le 1$. Further, $f(y_1, y_2) = g(y_1)h(y_2)$, where $g(y_1) = 2y_1$ and $h(y_2) = 1$. Therefore, Y_1 and Y_2 are independent random variables.   ∽

EXAMPLE 5.14   Refer to Example 5.4. Is Y_1, the amount in stock, independent of Y_2, the amount sold?

Solution   Because the density function is positive if and only if $0 \le y_2 \le y_1 \le 1$, there do not exist *constants*, a, b, c, and d, such that the density is positive over the region $a \le y_1 \le b$, $c \le y_2 \le d$. Thus Theorem 5.5 cannot be applied. However, Y_1 and Y_2 can be shown to be dependent random variables because the joint density is not the product of the marginal densities.   ∽

Definition 5.8 easily can be generalized to n dimensions. Suppose we have n random variables, $Y_1, \ldots, Y_n$, with Y_i having distribution function $F_i(y_i)$, $i = 1, \ldots, n$; and $Y_1, \ldots, Y_n$ having joint distribution function $F(y_1, \ldots, y_n)$. Then $Y_1, \ldots, Y_n$ are independent if and only if

$$F(y_1, \ldots, y_n) = F_1(y_1) \cdots F_n(y_n)$$

for all real numbers $(y_1, \ldots, y_n)$, with the obvious equivalent forms for the discrete and continuous cases.

Exercises

5.27   In Exercise 5.1 we determined that the joint distribution of Y_1, the number of contracts awarded to firm A, and Y_2, the number of contracts awarded to firm B, is given by the entries in the following table.

	Y_1		
Y_2	0	1	2
0	1/9	2/9	1/9
1	2/9	2/9	0
2	1/9	0	0

The marginal probability function of Y_1 was derived in Exercise 5.13 to be binomial with $n = 2$ and $p = 1/3$. Are Y_1 and Y_2 independent? Why?

5.28   Refer to Exercise 5.2. The number of heads in three coin tosses is binomially distributed with $n = 3$, $p = 1/2$. Are the total number of heads and your winnings on the side bet independent? [Examine your answer to Exercise 5.14, part (b).]

5.29   In Exercise 5.3, we determined that the joint probability distribution of Y_1, the number of married executives, and Y_2, the number of never married executives, is given by

$$p(y_1, y_2) = \frac{\binom{4}{y_1}\binom{3}{y_2}\binom{2}{3 - y_1 - y_2}}{\binom{9}{3}}$$

where y_1 and y_2 are integers, $0 \le y_1 \le 3$, $0 \le y_2 \le 3$, and $0 \le y_1 + y_2 \le 3$. Are Y_1 and Y_2 independent? [Recall your answer to Exercise 5.15.]

5.30 In Exercise 5.4, we derived the fact that

$$f(y_1, y_2) = \begin{cases} 4y_1y_2, & 0 \le y_1 \le 1, 0 \le y_2 \le 1 \\ 0, & \text{elsewhere} \end{cases}$$

is a valid joint probability density function. Are Y_1 and Y_2 independent?

5.31 In Exercise 5.5, we determined that

$$f(y_1, y_2) = \begin{cases} 6(1 - y_2), & 0 \le y_1 \le y_2 \le 1 \\ 0, & \text{elsewhere} \end{cases}$$

is a valid joint probability density function. Are Y_1 and Y_2 independent?

5.32 In Exercise 5.6, we proved that

$$f(y_1, y_2) = \begin{cases} 1, & 0 \le y_1 \le 2, 0 \le y_2 \le 1; 2y_2 \le y_1 \\ 0, & \text{elsewhere} \end{cases}$$

is a valid joint probability density function for Y_1, the amount of pollutant per sample collected when the cleaning device is not operating, and Y_2, the amount collected when the device is operating. Are the amounts of pollutants per sample collected with and without the cleaning device independent?

5.33 Refer to Exercise 5.7. Are Y_1 and Y_2 independent?

5.34 In Exercise 5.8, we were given the following joint probability density function for the random variables Y_1 and Y_2, which were the proportions of two components in a sample from a mixture of insecticide.

$$f(y_1, y_2) = \begin{cases} 2, & 0 \le y_1 \le 1, 0 \le y_2 \le 1; 0 \le y_1 + y_2 \le 1 \\ 0, & \text{elsewhere} \end{cases}$$

Are Y_1 and Y_2 independent?

5.35 If Y_1 is the total time between a customer's arrival in the store and leaving the service window and Y_2 is the time spent in line before reaching the window, the joint density of these variables was given in Exercise 5.9 to be

$$f(y_1, y_2) = \begin{cases} e^{-y_1}, & 0 \le y_2 \le y_1 \le \infty \\ 0, & \text{elsewhere} \end{cases}$$

Are Y_1 and Y_2 independent?

5.36 In Exercise 5.10, Y_1 and Y_2 denoted the proportions of time that employees I and II actually spent on their assigned tasks during a workday. The joint density of Y_1 and Y_2 is given by

$$f(y_1, y_2) = \begin{cases} y_1 + y_2, & 0 \le y_1 \le 1, 0 \le y_2 \le 1 \\ 0, & \text{elsewhere} \end{cases}$$

Are Y_1 and Y_2 independent?

5.37   In Exercise 5.12, Y_1 and Y_2 denoted the lengths of life, in hours, for components of types I and II, respectively, in an electronic system. The joint density of Y_1 and Y_2 is

$$f(y_1, y_2) = \begin{cases} (1/8)y_1 e^{-(y_1+y_2)/2}, & y_1 > 0, y_2 > 0 \\ 0, & \text{elsewhere} \end{cases}$$

Are Y_1 and Y_2 independent?

5.38   A supermarket has two customers waiting to pay for their purchases at counter I and one customer waiting to pay at counter II. Let Y_1 and Y_2 denote the numbers of customers who purchase more than \$50 of groceries at the respective counters. Suppose Y_1 and Y_2 are independent binomial random variables, with the probability of a customer spending more than \$50 equal to .2 for counter I and .3 for counter II.

(a)   Find the joint probability distribution for Y_1 and Y_2.

(b)   Find the probability that no more than one of the three customers spends in excess of \$50.

5.39   The length of life, Y, for fuses of a certain type is modeled by the exponential distribution, with

$$f(y) = \begin{cases} (1/3)e^{-y/3}, & y > 0 \\ 0, & \text{elsewhere} \end{cases}$$

(The measurements are in hundreds of hours.)

(a)   If two such fuses have independent lengths of life, Y_1 and Y_2, find the joint probability density function for Y_1 and Y_2.

(b)   One fuse in (a) is in a primary system and the other is in a backup system that comes into use only if the primary system fails. The total effective length of life of the two fuses is then $Y_1 + Y_2$. Find $P(Y_1 + Y_2 \le 1)$.

5.40   A bus arrives at a bus stop at a uniformly distributed time over the interval zero to 1 hour. A passenger also arrives at the bus stop at a uniformly distributed time over the interval zero to 1 hour. Assume that the arrival times of the bus and passenger are independent of one another and that the passenger will wait up to 1/4 hour for the bus to arrive. What is the probability that the passenger will catch the bus? [Hint: Let Y_1 denote the bus arrival time and Y_2 the passenger arrival time. Determine the joint density of Y_1 and Y_2, and find $P(Y_2 \le Y_1 \le Y_2 + 1/4)$.]

5.41   Two telephone calls come into a switchboard at random times in a fixed 1-hour period. Assume that the calls are made independently of one another.

(a)   What is the probability that both calls are made in the first half hour?

(b)   What is the probability that the calls are made within 5 minutes of each other?

5.5 The Expected Value of a Function of Random Variables

You need only construct the multivariate analogy to the univariate situation to justify the following definition concerning the expected value of a function of random variables.

Definition 5.9

> Let $g(Y_1, Y_2, \ldots, Y_k)$ be a function of the random variables, Y_1, $Y_2, \ldots, Y_k$, which possess a probability function $p(y_1, y_2, \ldots, y_k)$. Then the expected value of $g(Y_1, Y_2, \ldots, Y_k)$ is
>
> $$E[g(Y_1, Y_2, \ldots, Y_k)] = \sum_{y_k} \cdots \sum_{y_2} \sum_{y_1} g(y_1, y_2, \ldots, y_k) p(y_1, y_2, \ldots, y_k)$$
>
> If $Y_1, Y_2, \ldots, Y_k$ are continuous random variables with a joint density function $f(y_1, y_2, \ldots, y_k)$, then[†]
>
> $$E[g(Y_1, Y_2, \ldots, Y_k)]$$
> $$= \int_{y_k} \cdots \int_{y_2} \int_{y_1} g(y_1, y_2, \ldots, y_k) f(y_1, y_2, \ldots, y_k) \, dy_1 \, dy_2 \cdots dy_k$$

EXAMPLE 5.15   Let Y_1 and Y_2 have a joint density given by

$$f(y_1, y_2) = \begin{cases} 2y_1, & 0 \le y_1 \le 1; 0 \le y_2 \le 1 \\ 0, & \text{elsewhere} \end{cases}$$

Find $E(Y_1 Y_2)$.

Solution   From Definition 5.9 we obtain

$$E(Y_1 Y_2) = \int_0^1 \int_0^1 y_1 y_2 \, f(y_1, y_2) \, dy_1 \, dy_2 = \int_0^1 \int_0^1 y_1 y_2 (2y_1) \, dy_1 \, dy_2$$
$$= \int_0^1 y_2 \frac{2y_1^3}{3} \Big]_0^1 \, dy_2 = \int_0^1 \left(\frac{2}{3}\right) y_2 \, dy_2 = \frac{2}{3} \frac{y_2^2}{2} \Big]_0^1 = \frac{1}{3} \qquad \text{∽}$$

We will show that Definition 5.9 is consistent with Definition 4.4, which defines the expected value of a univariate random variable. Consider two random variables Y_1 and Y_2, with density function $f(y_1, y_2)$. We wish to find the expected value of $g(Y_1, Y_2) = Y_1$.

Then from Definition 5.9 we have

$$E(Y_1) = \int_{-\infty}^{\infty} \int_{-\infty}^{\infty} y_1 f(y_1, y_2) \, dy_1 \, dy_2$$
$$= \int_{-\infty}^{\infty} y_1 \left[\int_{-\infty}^{\infty} f(y_1, y_2) \, dy_2 \right] dy_1$$

[†] Again we say the expectations exist if the corresponding sum or integral of $|g(y_1, \ldots, y_n)|$ is finite.

The quantity within the brackets, by definition, is the marginal density function for Y_1. Therefore, we obtain

$$E(Y_1) = \int_{-\infty}^{\infty} y_1 f_1(y_1) \, dy_1$$

which agrees with Definition 4.4.

EXAMPLE 5.16   Let Y_1 and Y_2 have a joint density given by

$$f(y_1, y_2) = \begin{cases} 2y_1, & 0 \le y_1 \le 1; 0 \le y_2 \le 1 \\ 0, & \text{elsewhere} \end{cases}$$

Find the expected value of Y_1.

Solution

$$E(Y_1) = \int_0^1 \int_0^1 y_1(2y_1) \, dy_1 \, dy_2$$

$$= \int_0^1 \frac{2y_1^3}{3} \bigg]_0^1 \, dy_2 = \int_0^1 \frac{2}{3} \, dy_2 = \frac{2}{3} y_2 \bigg]_0^1 = \frac{2}{3}$$

Refer to Figure 5.6 and estimate the expected value of Y_1. The value $E(Y_1) = 2/3$ would appear to be quite reasonable.
$\backsim$

EXAMPLE 5.17   Examine Figure 5.6 and note that the mean value of Y_2 would appear to equal .5. Let us confirm this visual estimate. Find $E(Y_2)$.

Solution

$$E(Y_2) = \int_0^1 \int_0^1 y_2(2y_1) \, dy_1 \, dy_2 = \int_0^1 y_2 \frac{2y_1^2}{2} \bigg]_0^1 \, dy_2$$

$$= \int_0^1 y_2 \, dy_2 = \frac{y_2^2}{2} \bigg]_0^1 = \frac{1}{2}$$
$\backsim$

EXAMPLE 5.18   Let Y_1 and Y_2 be random variables with density function

$$f(y_1, y_2) = \begin{cases} 2y_1, & 0 \le y_1 \le 1; 0 \le y_2 \le 1 \\ 0, & \text{elsewhere} \end{cases}$$

Find $V(Y_1)$.

Solution

The marginal density for Y_1 was obtained in Example 5.6 as $f_1(y_1) = 2y_1$, $0 \le y_1 \le 1$. Then $V(Y_1) = E(Y_1^2) - [E(Y_1)]^2$ and

$$E(Y_1^k) = \int_0^1 y_1^k f(y_1) \, dy_1 = \int_0^1 y_1^k (2y_1) \, dy_1 = \frac{2y_1^{k+2}}{k+2} \Big]_0^1 = \frac{2}{k+2}$$

Letting $k = 1$ and $k = 2$, $E(Y_1)$ and $E(Y_1^2)$ are 2/3 and 1/2, respectively. Then $V(Y_1) = E(Y_1^2) - [E(Y_1)]^2 = 1/2 - (2/3)^2 = 1/18$.   ༤

EXAMPLE 5.19   A certain process for producing an industrial chemical yields a product containing two types of impurities. For a specified sample from this process, let Y_1 denote the proportion of impurities in the sample and Y_2 the proportion of type I impurity among all impurities found. Suppose the joint distribution of Y_1 and Y_2 can be modeled by the following probability density function:

$$f(y_1, y_2) = \begin{cases} 2(1 - y_1), & 0 \le y_1 \le 1; 0 \le y_2 \le 1 \\ 0, & \text{elsewhere} \end{cases}$$

Find the expected value of the proportion of type I impurities in the sample.

Solution

Because Y_1 is the proportion of impurities in the sample and Y_2 is the proportion of type I impurities among the sample impurities, it follows that $Y_1 Y_2$ is the proportion of type I impurities in the entire sample. Thus we want to find $E(Y_1 Y_2)$.

$$E(Y_1 Y_2) = \int_0^1 \int_0^1 2y_1 y_2 (1 - y_1) \, dy_2 \, dy_1 = 2 \int_0^1 y_1 (1 - y_1) \left(\frac{1}{2} \right) dy_1$$
$$= \int_0^1 (y_1 - y_1^2) \, dy_1 = \frac{y_1^2}{2} - \frac{y_1^3}{3} \Big]_0^1 = \frac{1}{2} - \frac{1}{3} = \frac{1}{6}$$

Therefore, we would expect 1/6 of the sample to be made up of type I impurities.   ༤

5.6 Special Theorems

Theorems concerning the expected value of a constant, the expected value of a constant times a function of random variables, and the expected value of the sum of functions of random variables are similar to those for the univariate case.

Theorem 5.6

Let c be a constant. Then

$$E(c) = c$$

Theorem 5.7

> Let $g(Y_1, Y_2)$ be a function of the random variables Y_1, Y_2, and let c be a constant. Then
>
> $$E[cg(Y_1, Y_2)] = cE[g(Y_1, Y_2)]$$

Theorem 5.8

> Let Y_1 and Y_2 be random variables with a joint density function of $f(y_1, y_2)$, and let $g_1(Y_1, Y_2), g_2(Y_1, Y_2), \ldots, g_k(Y_1, Y_2)$ be functions of Y_1 and Y_2. Then
>
> $$E[g_1(Y_1, Y_2) + g_2(Y_1, Y_2) + \cdots + g_k(Y_1, Y_2)]$$
> $$= E[g_1(Y_1, Y_2)] + E[g_2(Y_1, Y_2)] + \cdots + E[g_k(Y_1, Y_2)]$$

The proofs of these three theorems are analogous to the univariate cases discussed in Chapters 3 and 4.

EXAMPLE 5.20   Refer to Example 5.4. The random variable $Y_1 - Y_2$ denotes the proportional amount of gasoline remaining at the end of the week. Find $E(Y_1 - Y_2)$.

Solution   Employing Theorem 5.8 with $g_1(Y_1, Y_2) = Y_1$ and $g(Y_1, Y_2) = -Y_2$, we see that

$$E(Y_1 - Y_2) = E(Y_1) + E(-Y_2)$$

Theorem 5.7 applies to yield that $E(-Y_2) = -E(Y_2)$, and therefore it follows that

$$E(Y_1 - Y_2) = E(Y_1) - E(Y_2)$$

Also,

$$E(Y_1) = \int_0^1 \int_0^{y_1} y_1(3y_1)\, dy_2\, dy_1 = \int_0^1 3y_1^3\, dy_1 = \frac{3}{4}\{y_1^4]_0^1\} = \frac{3}{4}$$

$$E(Y_2) = \int_0^1 \int_0^{y_1} y_2(3y_1)\, dy_2\, dy_1 = \int_0^1 3y_1\left\{\frac{y_2^2}{2}\right]_0^{y_1}\right\} dy_1 = \int_0^1 \frac{3}{2} y_1^3\, dy_1$$
$$= \frac{3}{8}\{y_1^4]_0^1\} = \frac{3}{8}$$

Thus

$$E(Y_1 - Y_2) = (3/4) - (3/8) = 3/8$$

and we would expect 3/8 of the tank to be filled at the end of the week's sales.

If the random variables under study are independent, we sometimes can simplify the work involved in finding expectations. The following theorem is quite useful in this regard.

Theorem 5.9

Let Y_1 and Y_2 be independent random variables with a joint density of $f(y_1, y_2)$. Let $g(Y_1)$ and $h(Y_2)$ be functions of Y_1 and Y_2, respectively. Then

$$E[g(Y_1)h(Y_2)] = E[g(Y_1)]E[h(Y_2)]$$

provided the expectations exist.

Proof We will give the proof of the result for the continuous case. The product $g(Y_1)h(Y_2)$ is a function of Y_1 and Y_2. Hence by Definition 5.9,

$$E[g(Y_1)h(Y_2)] = \int_{-\infty}^{\infty} \int_{-\infty}^{\infty} g(y_1)h(y_2)f(y_1, y_2) \, dy_2 \, dy_1$$

$$= \int_{-\infty}^{\infty} \int_{-\infty}^{\infty} g(y_1)h(y_2)f_1(y_1)f_2(y_2) \, dy_2 \, dy_1$$

(because Y_1 and Y_2 are independent)

$$= \int_{-\infty}^{\infty} g(y_1)f_1(y_1) \left[\int_{-\infty}^{\infty} h(y_2)f_2(y_2) \, dy_2 \right] dy_1$$

$$= \int_{-\infty}^{\infty} g(y_1)f_1(y_1)E[h(Y_2)] \, dy_1$$

$$= E[h(Y_2)] \int_{-\infty}^{\infty} g(y_1)f_1(y_1) \, dy_1 = E[g(Y_1)]E[h(Y_2)]$$

The proof for the discrete case follows in an analogous manner.

EXAMPLE 5.21 Refer to Example 5.19. In that example we found $E(Y_1 Y_2)$ directly. By investigating the form of the joint density function given there, we can see that Y_1 and Y_2 are independent. Find $E(Y_1 Y_2)$ by using the result that $E(Y_1 Y_2) = E(Y_1)E(Y_2)$ if Y_1 and Y_2 are independent.

Solution The joint density function is given by

$$f(y_1, y_2) = \begin{cases} 2(1 - y_1), & 0 \le y_1 \le 1; 0 \le y_2 \le 1 \\ 0, & \text{elsewhere} \end{cases}$$

Hence

$$f_1(y_1) = 2(1 - y_1), \qquad 0 \leq y_1 \leq 1$$
$$f_2(y_2) = \int_0^1 2(1 - y_1)\, dy_1 = -(1 - y_1)^2]_0^1 = 1, \qquad 0 \leq y_2 \leq 1$$

We then have

$$E(Y_1) = \int_0^1 y_1\{2(1 - y_1)\, dy_1 = 2\left[\frac{y_1^2}{2} - \frac{y_1^3}{3}\right]_0^1 = \frac{1}{3}$$
$$E(Y_2) = 1/2$$

because Y_2 is uniformly distributed over $(0, 1)$.
   It follows that

$$E(Y_1 Y_2) = E(Y_1)E(Y_2) = (1/3)(1/2) = 1/6$$

which agrees with the answer in Example 5.19.   ∽

Exercises

5.42   In Exercise 5.1 we determined that the joint distribution of Y_1, the number of contracts awarded to firm A, and Y_2, the number of contracts awarded to firm B, is given by the entries in the following table.

	Y_1		
Y_2	0	1	2
0	1/9	2/9	1/9
1	2/9	2/9	0
2	1/9	0	0

The marginal probability function of Y_1 was derived in Exercise 5.13 to be binomial with $n = 2$ and $p = 1/3$.
   (a)   Find $E(Y_1)$.
   (b)   Find $V(Y_1)$.
   (c)   Find $E(Y_1 - Y_2)$.

5.43   In Exercise 5.3, we determined that the joint probability distribution of Y_1, the number of married executives, and Y_2, the number of never married executives, is given by

$$p(y_1, y_2) = \frac{\binom{4}{y_1}\binom{3}{y_2}\binom{2}{3 - y_1 - y_2}}{\binom{9}{3}}$$

where y_1 and y_2 are integers, $0 \le y_1 \le 3$, $0 \le y_2 \le 3$, and $0 \le y_1 + y_2 \le 3$. Find the expected number of married executives among the three selected for promotion. [See Exercise 5.15.]

5.44   In Exercise 5.4, we derived the fact that

$$f(y_1, y_2) = \begin{cases} 4y_1 y_2, & 0 \le y_1 \le 1, 0 \le y_2 \le 1 \\ 0, & \text{elsewhere} \end{cases}$$

(a)   Find $E(Y_1)$.
(b)   Find $V(Y_1)$.
(c)   Find $E(Y_1 - Y_2)$.

5.45   In Exercise 5.5, we determined that

$$f(y_1, y_2) = \begin{cases} 6(1 - y_2), & 0 \le y_1 \le y_2 \le 1 \\ 0, & \text{elsewhere} \end{cases}$$

is a valid joint probability density function.
(a)   Find $E(Y_1)$ and $E(Y_2)$.
(b)   Find $V(Y_1)$ and $V(Y_2)$.
(c)   Find $E(Y_1 - 3Y_2)$.

5.46   In Exercise 5.6, we proved that

$$f(y_1, y_2) = \begin{cases} 1, & 0 \le y_1 \le 2, 0 \le y_2 \le 1; 2y_2 \le y_1 \\ 0, & \text{elsewhere} \end{cases}$$

is a valid joint probability density function for Y_1, the amount of pollutant per sample collected when the cleaning device is not operating, and Y_2, the amount collected when the device is operating.
(a)   Find $E(Y_1)$ and $E(Y_2)$.
(b)   Find $V(Y_1)$ and $V(Y_2)$.
(c)   The random variable $Y_1 - Y_2$ represents the amount by which the weight of pollutant can be reduced by using the cleaning device. Find $E(Y_1 - Y_2)$.
(d)   Find $V(Y_1 - Y_2)$. Within what limits would you expect $Y_1 - Y_2$ to fall?

5.47   Refer to Exercise 5.7. Find $E(Y_1 Y_2)$.

5.48   In Exercise 5.10, Y_1 and Y_2 denoted the proportions of time that employees I and II actually spent on their assigned tasks during a workday. The joint density of Y_1 and Y_2 is given by

$$f(y_1, y_2) = \begin{cases} y_1 + y_2, & 0 \le y_1 \le 1, 0 \le y_2 \le 1 \\ 0, & \text{elsewhere} \end{cases}$$

Employee I has a higher productivity rating than employee II and a measure of the total productivity of the pair of employees is $30Y_1 + 25Y_2$. Find the expected value of this measure of productivity.

5.49   In Exercise 5.12, Y_1 and Y_2 denoted the lengths of life, in hours, for components of types I and II, respectively, in an electronic system. The joint density of Y_1 and Y_2 is

$$f(y_1, y_2) = \begin{cases} (1/8)y_1 e^{-(y_1 + y_2)/2}, & y_1 > 0, y_2 > 0 \\ 0, & \text{elsewhere} \end{cases}$$

One way to measure the relative efficiency of the two components is to compute the ratio Y_2/Y_1. Find $E(Y_2/Y_1)$. [Hint: In Exercise 5.37 we proved that Y_1 and Y_2 are independent.]

5.50   In Exercise 5.24 we determined that the joint density function for Y_1, the weight in tons of a bulk item stocked by a supplier, and Y_2, the weight of the item sold by the supplier, has joint density

$$f(y_1, y_2) = \begin{cases} 1/y_1, & 0 \le y_2 \le y_1 \le 1 \\ 0, & \text{elsewhere} \end{cases}$$

In this case, the random variable $Y_1 - Y_2$ measures the amount of stock remaining at the end of the week, a quantity of great importance to the supplier. Find $E(Y_1 - Y_2)$.

5.51   In Example 5.26, we determined that the unconditional probability distribution for Y, the number of defects per yard in a certain fabric, is

$$p(y) = (1/2)^{y-1}, \qquad y = 0, 1, 2, \ldots$$

Find the expected number of defects per yard.

5.7 The Covariance of Two Random Variables

Intuitively, we think of the dependence of two random variables Y_1 and Y_2 as implying that one variable, say, Y_1, either increases or decreases as Y_2 changes. We will confine our attention to two measures of dependence, the covariance and the simple coefficient of linear correlation, and will utilize Figures 5.8(a) and (b) to justify their choice as measures of dependence. Figures 5.8(a) and (b) represent plotted points of two random samples of $n = 10$ experimental units drawn from a population. Measurements on Y_1 and Y_2 were made on each experimental unit. If all the points fall along a straight line, as indicated in

Figure 5.8
Dependent and
independent
observations for
(y_1, y_2)

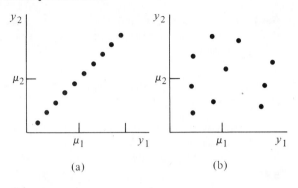

(a)   (b)

Figure 5.8(a), Y_1 and Y_2 are obviously dependent. In contrast, Figure 5.8(b) would indicate little or no dependence between Y_1 and Y_2.

Suppose that we really knew the values of $E(Y_1) = \mu_1$ and $E(Y_2) = \mu_2$ and located this point on the graph, Figure 5.8. Now locate a plotted point on Figure 5.8(a) and measure the deviations $(y_1 - \mu_1)$ and $(y_2 - \mu_2)$. Note that both deviations will assume the same algebraic sign for a particular point and hence that their product, $(y_1 - \mu_1)(y_2 - \mu_2)$, will be positive. This will be true for *all* plotted points on Figure 5.8(a). Points to the right of μ_1 will yield pairs of positive deviations; points to the left will produce pairs of negative deviations; and the average of the product of the deviations, $(y_1 - \mu_1)(y_2 - \mu_2)$, will be large and positive. If the linear relation indicated in Figure 5.8(a) had sloped downward to the right, all corresponding pairs of deviations would have been of the opposite sign, and the average value of $(y_1 - \mu_1)(y_2 - \mu_2)$ would have been a large negative number.

The situation just described will not occur for Figure 5.8(b), where little dependence exists between Y_1 and Y_2. Corresponding deviations, $(y_1 - \mu_1)$ and $(y_2 - \mu_2)$, will assume the same algebraic sign for some points and opposite signs for others. Thus the product $(y_1 - \mu_1)(y_2 - \mu_2)$ will be positive for some points, negative for others, and will average to some value near zero.

It is clear from the foregoing discussion that the average value of $(y_1 - \mu_1)(y_2 - \mu_2)$ will provide a measure of the linear dependence of Y_1 and Y_2. This quantity, defined over the bivariate population associated with Y_1 and Y_2, is called the *covariance* of Y_1 and Y_2.

Definition 5.10

> The *covariance* of Y_1 and Y_2 is defined to be the expected value of $(Y_1 - \mu_1)(Y_2 - \mu_2)$. In the notation of expectation the covariance will equal
>
> $$\text{Cov}(Y_1,\ Y_2) = E[(Y_1 - \mu_1)(Y_2 - \mu_2)]$$
>
> where $\mu_1 = E(Y_1)$ and $\mu_2 = E(Y_2)$.

The larger the absolute value of the covariance of Y_1 and Y_2, the greater the linear dependence between Y_1 and Y_2. Positive values indicate that Y_1 increases as Y_2 increases; negative values indicate that Y_1 decreases as Y_2 increases. A zero value of the covariance would indicate no linear dependence between Y_1 and Y_2.

Unfortunately, it is difficult to employ the covariance as an absolute measure of dependence because its value depends upon the scale of measurement and so it is hard to determine whether a particular covariance is large at first glance. This problem can be eliminated by standardizing its value, using the simple coefficient of linear correlation. The population linear *coefficient of correlation*, ρ, is related to the covariance and is defined as

$$\rho = \frac{\text{Cov}(Y_1, Y_2)}{\sigma_1 \sigma_2}$$

where σ_1 and σ_2 are the standard deviations of Y_1 and Y_2, respectively. Supplemental discussion of correlation may be found in Mendenhall [3].

It can be shown that the coefficient of correlation ρ satisfies the inequality $-1 \le \rho \le 1$. The proof is outlined in Exercise 5.105. Thus -1 or $+1$ imply perfect correlation, with all points falling on a straight line. A value $\rho = 0$ implies a zero covariance and no correlation. The sign of the correlation coefficient is dependent on the sign of the covariance. Thus a positive coefficient of correlation indicates that Y_2 increases as Y_1 increases. A negative coefficient of correlation implies a decrease in Y_2 as Y_1 increases.

We will now derive a convenient computational formula for the covariance.

Theorem 5.10

> Let Y_1 and Y_2 be random variables with a joint density function of $f(y_1, y_2)$. Then
>
> $$\text{Cov}(Y_1, Y_2) = E[(Y_1 - \mu_1)(Y_2 - \mu_2)] = E(Y_1 Y_2) - E(Y_1)E(Y_2)$$

Proof

$$\text{Cov}(Y_1, Y_2) = E[(Y_1 - \mu_1)(Y_2 - \mu_2)] = E(Y_1 Y_2 - \mu_1 Y_2 - \mu_2 Y_1 + \mu_1 \mu_2)$$

From Theorem 5.8, the expected value of a sum is equal to the sum of the expected values, and from Theorem 5.7, the expected value of a constant times a random variable is the constant times the expected value. Thus

$$\text{Cov}(Y_1, Y_2) = E(Y_1 Y_2) - \mu_1 E(Y_2) - \mu_2 E(Y_1) + \mu_1 \mu_2$$

Because $E(Y_1) = \mu_1$ and $E(Y_2) = \mu_2$, it follows that

$$\text{Cov}(Y_1, Y_2) = E(Y_1 Y_2) - E(Y_1)E(Y_2)$$

or

$$\text{Cov}(Y_1, Y_2) = E(Y_1 Y_2) - \mu_1 \mu_2$$

EXAMPLE 5.22   Refer to Example 5.4. Find the covariance between the amount in stock, Y_1, and amount of sales, Y_2.

Solution   Recall that Y_1 and Y_2 have a joint density function given by

$$f(y_1, y_2) = \begin{cases} 3y_1, & 0 \le y_2 \le y_1 \le 1 \\ 0, & \text{elsewhere} \end{cases}$$

Thus

$$E(Y_1 Y_2) = \int_0^1 \int_0^{y_1} y_1 y_2 3y_1 \, dy_2 \, dy_1 = \int_0^1 3y_1^2 \left\{ \frac{y_2^2}{2} \right\}_0^{y_1} \, dy_1$$

$$= \int_0^1 \frac{3}{2} y_1^4 \, dy_1 = \frac{3}{2} \left\{ \frac{y_1^5}{5} \right\}_0^1 = \frac{3}{10}$$

From Example 5.20, we know that $E(Y_1) = 3/4$ and $E(Y_2) = 3/8$. Thus from Theorem 5.10, we have

$$\text{Cov}(Y_1, Y_2) = E(Y_1 Y_2) - E(Y_1)E(Y_2) = (3/10) - (3/4)(3/8) = .30 - .28 = .02$$

Note that, intuitively, the covariance here should be positive because large values of Y_2 can occur only with large values of Y_1, and the larger values of Y_1 tend to have a larger probability mass associated with them. (See Figure 5.4.)

EXAMPLE 5.23   Let Y_1 and Y_2 have a joint density given by

$$f(y_1, y_2) = \begin{cases} 2y_1, & 0 \le y_1 \le 1; 0 \le y_2 \le 1 \\ 0, & \text{elsewhere} \end{cases}$$

Find the covariance of Y_1 and Y_2.

Solution   From Example 5.15, $E(Y_1 Y_2) = 1/3$. Also from Examples 5.16 and 5.17, $E(Y_1)$ and $E(Y_2)$ equal 2/3 and 1/2, respectively. Then

$$\text{Cov}(Y_1, Y_2) = E(Y_1 Y_2) - \mu_1 \mu_2 = (1/3) - (2/3)(1/2) = 0$$

Example 5.23 furnishes an illustration of a general concept given in Theorem 5.11.

Theorem 5.11

If Y_1 and Y_2 are independent random variables, then

$$\text{Cov}(Y_1, Y_2) = 0$$

Proof   We know that

$$\text{Cov}(Y_1, Y_2) = E(Y_1 Y_2) - \mu_1 \mu_2$$

But if Y_1 and Y_2 are independent, then

$$E(Y_1 Y_2) = E(Y_1)E(Y_2) = \mu_1 \mu_2$$

from Theorem 5.9. The desired result follows immediately.

Note that the random variables Y_1 and Y_2 of Example 5.23 are independent, and hence by Theorem 5.11 their covariance must be zero. The converse of Theorem 5.11 is not true, as will be illustrated in the following example.

EXAMPLE 5.24   Let Y_1 and Y_2 be discrete random variables with a joint probability distribution as shown in Table 5.3. Show that Y_1 and Y_2 are dependent but have zero covariance.

Table 5.3
Joint probability
distribution,
Example 5.24

		Y_1	
Y_2	-1	0	$+1$
-1	1/16	3/16	1/16
0	3/16	0	3/16
$+1$	1/16	3/16	1/16

Solution

Calculation of marginal probabilities yields $p_1(-1) = p_2(-1) = 5/16$, $p_1(0) = p_2(0) = 6/16$, and $p_1(1) = p_2(1) = 5/16$. Looking at the upper left-hand cell, we have $p(-1, -1) = 1/16$. Obviously,

$$p(-1, -1) \neq p_1(-1)p_2(-1)$$

and this is sufficient to show that Y_1 and Y_2 are dependent.

Again looking at the marginal probabilities, we see that $E(Y_1) = E(Y_2) = 0$. Also,

$$E(Y_1 Y_2) = \sum_{y_1} \sum_{y_2} y_1 y_2 p(y_1, y_2)$$

$$= (-1)(-1)(1/16) + (0)(-1)(3/16) + \cdots + (1)(1)(1/16) = 0$$

Thus

$$\text{Cov}(Y_1, Y_2) = E(Y_1 Y_2) - E(Y_1)E(Y_2) = 0$$

This example shows that the converse of Theorem 5.11 is not true. That is, if the covariance of two random variables is zero, the variables need not be independent. ∽

Exercises

5.52   In Exercise 5.1 we determined that the joint distribution of Y_1, the number of contracts awarded to firm A, and Y_2, the number of contracts awarded to firm B, is given by the entries in the following table.

		Y_1	
Y_2	0	1	2
0	1/9	2/9	1/9
1	2/9	2/9	0
2	1/9	0	0

Find $\text{Cov}(Y_1, Y_2)$. Does it surprise you that $\text{Cov}(Y_1, Y_2)$ is negative? Why?

5.53   In Exercise 5.3, we determined that the joint probability distribution of Y_1, the number of married executives, and Y_2, the number of never married executives, is given by

$$p(y_1, y_2) = \frac{\binom{4}{y_1} \binom{3}{y_2} \binom{2}{3 - y_1 - y_2}}{\binom{9}{3}}$$

where y_1 and y_2 are integers, $0 \le y_1 \le 3$, $0 \le y_2 \le 3$, and $0 \le y_1 + y_2 \le 3$. Find $\text{Cov}(Y_1, Y_2)$.

5.54   In Exercise 5.4, we derived the fact that

$$f(y_1, y_2) = \begin{cases} 4y_1 y_2, & 0 \le y_1 \le 1, 0 \le y_2 \le 1 \\ 0, & \text{elsewhere} \end{cases}$$

Show that $\text{Cov}(Y_1, Y_2) = 0$. Does it surprise you that $\text{Cov}(Y_1, Y_2)$ is zero? Why?

5.55   In Exercise 5.5, we determined that

$$f(y_1, y_2) = \begin{cases} 6(1 - y_2), & 0 \le y_1 \le y_2 \le 1 \\ 0, & \text{elsewhere} \end{cases}$$

is a valid joint probability density function. Find $\text{Cov}(Y_1, Y_2)$. Are Y_1 and Y_2 independent?

5.56   Refer to Exercises 5.7 and 5.47. Find the covariance and correlation coefficient between Y_1 and Y_2.

5.57   Let Y_1 and Y_2 be uncorrelated random variables. Find the covariance and correlation between $U_1 = Y_1 + Y_2$ and $U_2 = Y_1 - Y_2$ in terms of the variances of Y_1 and Y_2.

5.58   Let the discrete random variables Y_1 and Y_2 have the joint probability function

$$p(y_1, y_2) = 1/3 \quad \text{for} \quad (y_1, y_2) = (-1, 0), (0, 1), (1, 0)$$

Find $\text{Cov}(Y_1, Y_2)$. Note that Y_1 and Y_2 are dependent. (Why?) This is an example of uncorrelated random variables that are not independent.

5.8 The Expected Value and Variance of Linear Functions of Random Variables

We frequently will be concerned with parameter estimators that are linear functions of the sample measurements $Y_1, Y_2, \ldots, Y_n$. Consequently, we will wish to find the expected value and variance of a linear function

$$U_1 = a_1 Y_1 + a_2 Y_2 + a_3 Y_3 + \cdots + a_n Y_n = \sum_{i=1}^{n} a_i Y_i$$

when $a_1, a_2, \ldots, a_n$ are constants and $Y_1, Y_2, \ldots, Y_n$ are random variables.

We also may be interested in the covariance between such linear functions. The necessary results are summarized in the following theorem.

Theorem 5.12

Let $Y_1, \ldots, Y_n$ and $X_1, \ldots, X_m$ be random variables with $E(Y_i) = \mu_i$ and $E(X_i) = \xi_i$. Define

$$U_1 = \sum_{i=1}^{n} a_i Y_i \qquad U_2 = \sum_{j=1}^{m} b_j X_j$$

for constants $a_1, \ldots, a_n, b_1, \ldots, b_m$. Then the following hold:

(a) $E(U_1) = \sum_{i=1}^{n} a_i \mu_i$.

(b) $V(U_1) = \sum_{i=1}^{n} a_i^2 V(Y_i) + 2 \sum\sum_{i<j} a_i a_j \, \mathrm{Cov}(Y_i, Y_j)$, where the double sum is over all pairs (i, j) with $i < j$.

(c) $\mathrm{Cov}(U_1, U_2) = \sum_{i=1}^{n} \sum_{j=1}^{m} a_i b_j \, \mathrm{Cov}(Y_i, X_j)$.

Before proceeding with the proof of Theorem 5.12, we illustrate the use of the theorem with the following example.

EXAMPLE 5.25   Let Y_1, Y_2, and Y_3 be random variables, where $E(Y_1) = 1$, $E(Y_2) = 2$, $E(Y_3) = -1$, $V(Y_1) = 1$, $V(Y_2) = 3$, $V(Y_3) = 5$, $\mathrm{Cov}(Y_1, Y_2) = -0.4$, $\mathrm{Cov}(Y_1, Y_3) = 1/2$, and $\mathrm{Cov}(Y_2, Y_3) = 2$. Find the expected value and variance of $U = Y_1 - 2Y_2 + Y_3$.

Solution $U = a_1 Y_1 + a_2 Y_2 + a_3 Y_3$, where $a_1 = 1$, $a_2 = -2$, and $a_3 = 1$. Then by Theorem 5.12,

$$E(U) = a_1 E(Y_1) + a_2 E(Y_2) + a_3 E(Y_3)$$
$$= (1)(1) + (-2)(2) + (1)(-1) = -4$$

Similarly

$$V(U) = a_1^2 V(Y_1) + a_2^2 V(Y_2) + a_3^2 V(Y_3) + 2a_1 a_2 \, \mathrm{Cov}(Y_1, Y_2)$$
$$+ 2a_1 a_3 \, \mathrm{Cov}(Y_1, Y_3) + 2a_2 a_3 \, \mathrm{Cov}(Y_2, Y_3)$$
$$= (1)^2(1) + (-2)^2(3) + (1)^2(5) + (2)(1)(-2)(-0.4)$$
$$+ (2)(1)(1)(1/2) + (2)(-2)(1)(2)$$
$$= 12.6$$

We now proceed with the proof of Theorem 5.12.

Proof   Part (a) follows directly from Theorems 5.7 and 5.8. To prove part (b), we appeal to the definition of variance and write

$$V(U_1) = E[U_1 - E(U_1)]^2 = E\left[\sum_{i=1}^{n} a_i Y_i - \sum_{i=1}^{n} a_i \mu_i\right]^2$$

$$= E\left[\sum_{i=1}^{n} a_i (Y_i - \mu_i)\right]^2$$

$$= E\left[\sum_{i=1}^{n} a_i^2 (Y_i - \mu_i)^2 + \sum_{i \neq j}\sum a_i a_j (Y_i - \mu_i)(Y_j - \mu_j)\right]$$

$$= \sum_{i=1}^{n} a_i^2 E(Y_i - \mu_i)^2 + \sum_{i \neq j}\sum a_i a_j E[(Y_i - \mu_i)(Y_j - \mu_j)]$$

By definition of variance and covariance, we then have

$$V(U_1) = \sum_{i=1}^{n} a_i^2 V(Y_i) + \sum_{i \neq j}\sum a_i a_j \, \text{Cov}(Y_i, \, Y_j)$$

Note that $\text{Cov}(Y_i, Y_j) = \text{Cov}(Y_j, Y_i)$, and hence we can write

$$V(U_1) = \sum_{i=1}^{n} a_i^2 V(Y_i) + 2 \sum_{i < j}\sum a_i a_j \, \text{Cov}(Y_i, \, Y_j)$$

Part (c) is obtained by similar steps. We have

$$\text{Cov}(U_1, U_2) = E\{[U_1 - E(U_1)][U_2 - E(U_2)]\}$$

$$= E\left[\left(\sum_{i=1}^{n} a_i Y_i - \sum_{i=1}^{n} a_i \mu_i\right)\left(\sum_{j=1}^{m} b_j X_j - \sum_{j=1}^{m} b_j \xi_j\right)\right]$$

$$= E\left\{\left[\sum_{i=1}^{n} a_i (Y_i - \mu_i)\right]\left[\sum_{j=1}^{m} b_j (X_j - \xi_j)\right]\right\}$$

$$= E\left[\sum_{i=1}^{n}\sum_{j=1}^{m} a_i b_j (Y_i - \mu_i)(X_j - \xi_j)\right]$$

$$= \sum_{i=1}^{n}\sum_{j=1}^{m} a_i b_j E(Y_i - \mu_i)(X_j - \xi_j)$$

$$= \sum_{i=1}^{n}\sum_{j=1}^{m} a_i b_j \, \text{Cov}(Y_i, \, X_j)$$

On observing that $\text{Cov}(Y_i, Y_i) = V(Y_i)$, we can see that part (b) is a special case of part (c).

EXAMPLE 5.26   Refer to Examples 5.4 and 5.20. In Example 5.20 we were interested in $Y_1 - Y_2$, the proportional amount of gasoline remaining at the end of a week. Find the variance of $Y_1 - Y_2$.

Solution   Using Theorem 5.12, we have

$$V(Y_1 - Y_2) = V(Y_1) + V(Y_2) - 2 \text{ Cov}(Y_1, Y_2)$$

Because $f_1(y_1) = 3y_1^2$, $0 \le y_1 \le 1$, and $f_2(y_2) = (3/2)(1 - y_2^2)$, $0 \le y_2 \le 1$, it follows that

$$E(Y_1^2) = \int_0^1 3y_1^4 \, dy_1 = \frac{3}{5}$$

$$E(Y_2^2) = \int_0^1 \frac{3}{2} y_2^2(1 - y_2^2) \, dy_2 = \frac{3}{2}\left[\frac{1}{3} - \frac{1}{5}\right] = \frac{1}{5}$$

From Example 5.20, we have

$$E(Y_1) = 3/4 \quad \text{and} \quad E(Y_2) = 3/8$$

Thus

$$V(Y_1) = E(Y_1^2) - [E(Y_1)]^2 = (3/5) - (3/4)^2 = .04$$
$$V(Y_2) = (1/5) - (3/8)^2 = .06$$

In Example 5.22 we determined that $\text{Cov}(Y_1, Y_2) = .02$. Thus

$$V(Y_1 - Y_2) = V(Y_1) + V(Y_2) - 2 \text{ Cov}(Y_1, Y_2)$$
$$= .04 + .06 - 2(.02) = .06$$

The standard deviation of $Y_1 - Y_2$ is then $\sqrt{.06} = .245$.   ∽

EXAMPLE 5.27   Let $Y_1, Y_2, \ldots, Y_n$ be independent random variables with $E(Y_i) = \mu$ and $V(Y_i) = \sigma^2$. (These variables may denote the outcomes on n independent trials of an experiment.) Define

$$\bar{Y} = \frac{1}{n} \sum_{i=1}^{n} Y_i$$

and show that $E(\bar{Y}) = \mu$ and $V(\bar{Y}) = \sigma^2/n$.

Solution

Note that $\bar{Y}$ is a linear function of $Y_1, Y_2, \ldots, Y_n$ with all constants a_i equal to $1/n$. That is,

$$\bar{Y} = \left(\frac{1}{n}\right)Y_1 + \cdots + \left(\frac{1}{n}\right)Y_n$$

By Theorem 5.12, part (a)

$$E(\bar{Y}) = \sum_{i=1}^{n} a_i \mu = \mu \sum_{i=1}^{n} a_i = \mu \sum_{i=1}^{n} \frac{1}{n} = \frac{n\mu}{n} = \mu$$

By Theorem 5.12, part (b)

$$V(\bar{Y}) = \sum_{i=1}^{n} a_i^2 V(Y_1) + 2 \sum \sum_{i<j} a_i a_j \, \text{Cov}(Y_i, \, Y_j)$$

but the covariance terms all are zero because the random variables are independent. Thus

$$V(\bar{Y}) = \sum_{i=1}^{n} \left(\frac{1}{n}\right)^2 \sigma^2 = \frac{1}{n^2} \sum_{i=1}^{n} \sigma^2 = \frac{n\sigma^2}{n^2} = \frac{\sigma^2}{n} \qquad \text{⌇}$$

EXAMPLE 5.28   The number of defectives Y in a sample of $n = 10$ items selected from a manufacturing process follows a binomial probability distribution. An estimator of the fraction defective in the lot is the random variable $\hat{p} = Y/n$. Find the expected value and variance of $\hat{p}$.

Solution

The term $\hat{p}$ is a linear function of a single random variable Y, where $\hat{p} = a_1 Y$ and $a_1 = 1/n$. Then by Theorem 5.12

$$E(\hat{p}) = a_1 E(Y) = \frac{1}{n} E(Y)$$

The expected value and variance of a binomial random variable are np and npq, respectively. Substituting for $E(Y)$, we obtain

$$E(\hat{p}) = \frac{1}{n} np = p$$

Thus the expected value of the number of defectives Y, divided by the sample size, is p. Similarly

$$V(\hat{p}) = a_1^2 V(Y) = \left(\frac{1}{n}\right)^2 npq = \frac{pq}{n} \qquad \text{⌇}$$

EXAMPLE 5.29   Suppose an urn contains r white balls and $(N - r)$ black balls. A random sample of n balls is drawn without replacement and Y, the number of white balls in the sample, is observed. From Chapter 3 we know that Y has a hypergeometric probability distribution. Find the mean and variance of Y.

Solution   We will first observe some characteristics of sampling without replacement. Suppose the sampling is done sequentially and we observe outcomes for X_1, $X_2, \ldots, X_n$, where

$$X_i = \begin{cases} 1, & \text{if the } i\text{th draw results in a white ball} \\ 0, & \text{otherwise.} \end{cases}$$

Unquestionably $P(X_1 = 1) = r/N$. But it is also true that $P(X_2 = 1) = r/N$ because

$$\begin{aligned} P(X_2 = 1) &= P(X_1 = 1, X_2 = 1) + P(X_1 = 0, X_2 = 1) \\ &= P(X_1 = 1)P(X_2 = 1 | X_1 = 1) + P(X_1 = 0)P(X_2 = 1 | X_1 = 0) \\ &= \left(\frac{r}{N}\right)\left(\frac{r - 1}{N - 1}\right) + \left(\frac{N - r}{N}\right)\left(\frac{r}{N - 1}\right) = \frac{r(N - 1)}{N(N - 1)} = \frac{r}{N} \end{aligned}$$

The same is true for X_k; that is,

$$P(X_k = 1) = \frac{r}{N}, \qquad k = 1, \ldots, n$$

Thus the probability of drawing a white ball on any draw, given no knowledge of the outcomes on previous draws, is r/N.

In a similar way it can be shown that

$$P(X_j = 1, X_k = 1) = \frac{r(r - 1)}{N(N - 1)}, \qquad j \neq k$$

Now, observe that $Y = \sum_{i=1}^{n} X_i$, and hence

$$E(Y) = \sum_{i=1}^{n} E(X_i) = n\left(\frac{r}{N}\right)$$

To find $V(Y)$ we need $V(X_i)$ and $\text{Cov}(X_i, X_j)$. Because X_i is 1 with probability r/N and 0 with probability $1 - r/N$, it follows that

$$V(X_i) = \frac{r}{N}\left(1 - \frac{r}{N}\right)$$

Also,

$$\text{Cov}(X_i, X_j) = E(X_i X_j) - E(X_i)E(X_j) = \frac{r(r-1)}{N(N-1)} - \left(\frac{r}{N}\right)^2$$

$$= -\frac{r}{N}\left(1 - \frac{r}{N}\right)\left(\frac{1}{N-1}\right)$$

because $X_i X_j = 1$ if and only if $X_i = 1$ and $X_j = 1$. From Theorem 5.12 we know that

$$V(Y) = \sum_{i=1}^{n} V(X_i) + 2 \sum\sum_{i<j} \text{Cov}(X_i, X_j)$$

$$= n\left(\frac{r}{N}\right)\left(1 - \frac{r}{N}\right) + 2\sum\sum_{i<j}\left[-\frac{r}{N}\left(1 - \frac{r}{N}\right)\left(\frac{1}{N-1}\right)\right]$$

$$= n\left(\frac{r}{N}\right)\left(1 - \frac{r}{N}\right) - n(n-1)\left(\frac{r}{N}\right)\left(1 - \frac{r}{N}\right)\left(\frac{1}{N-1}\right)$$

because there are $n(n-1)/2$ terms in the double summation. A little algebra yields

$$V(Y) = n\left(\frac{r}{N}\right)\left(1 - \frac{r}{N}\right)\left(\frac{N-n}{N-1}\right)$$

An appreciation for the usefulness of Theorem 5.12 can be gained by trying to find the expected value and variance for the hypergeometric random variable by proceeding directly from the definition of an expectation. The necessary summations are exceedingly difficult to obtain.

Exercises

5.59   A firm purchases two types of industrial chemicals. The type I chemical cost $3.00 per gallon, whereas type II costs $5.00 per gallon. The mean and variance for the number of gallons of type I chemical purchased, Y_1, are 40 and 4, respectively. The amount of type II chemical purchased, Y_2, has $E(Y_2) = 65$ gallons and $V(Y_2) = 8$. Assume that Y_1 and Y_2 are independent and find the mean and variance of the total amount of money spent per week on the two chemicals.

5.60   Assume Y_1, Y_2, and Y_3 are random variables, with

$$E(Y_1) = 2 \qquad E(Y_2) = -1 \qquad E(Y_3) = 4$$
$$V(Y_1) = 4 \qquad V(Y_2) = 6 \qquad V(Y_3) = 8$$
$$\text{Cov}(Y_1, Y_2) = 1 \quad \text{Cov}(Y_1, Y_3) = -1 \quad \text{Cov}(Y_2, Y_3) = 0$$

Find $E(3Y_1 + 4Y_2 - 6Y_3)$ and $V(3Y_1 + 4Y_2 - 6Y_3)$.

5.61   In Exercise 5.3, we determined that the joint probability distribution of Y_1, the number of married executives, and Y_2, the number of never married executives, is given by

$$p(y_1, y_2) = \frac{\binom{4}{y_1}\binom{3}{y_2}\binom{2}{3 - y_1 - y_2}}{\binom{9}{3}}$$

where y_1 and y_2 are integers, $0 \le y_1 \le 3, 0 \le y_2 \le 3$, and $0 \le y_1 + y_2 \le 3$.

(a)   Find $E(Y_1 + Y_2)$ and $V(Y_1 + Y_2)$ by first finding the probability distribution of $Y_1 + Y_2$.

(b)   Recall that in Exercise 5.53 we determined that $\text{Cov}(Y_1, Y_2) = -1/3$. Find $E(Y_1 + Y_2)$ and $V(Y_1 + Y_2)$ by using Theorem 5.12.

5.62   In Exercise 5.4 we derived the fact that

$$f(y_1, y_2) = \begin{cases} 4y_1y_2, & 0 \le y_1 \le 1, 0 \le y_2 \le 1 \\ 0, & \text{elsewhere} \end{cases}$$

is a valid joint probability density function. In Exercise 5.44 we derived that $E(Y_1 - Y_2) = 0$. Find $V(Y_1 - Y_2)$.

5.63   In Exercise 5.5 we determined that

$$f(y_1, y_2) = \begin{cases} 6(1 - y_2), & 0 \le y_1 \le y_2 \le 1 \\ 0, & \text{elsewhere} \end{cases}$$

is a valid joint probability density function. In Exercise 5.45 we derived that $E(Y_1 - Y_2) = -5/4$ and in Exercise 5.55 we proved that $\text{Cov}(Y_1, Y_2) = 1/40$. Find $V(Y_1 - Y_2)$.

5.64   In Exercise 5.8, we were given the following joint probability density function for the random variables Y_1 and Y_2, which were the proportions of two components in a sample from a mixture of insecticide.

$$f(y_1, y_2) = \begin{cases} 2, & 0 \le y_1 \le 1, 0 \le y_2 \le 1; 0 \le y_1 + y_2 \le 1 \\ 0, & \text{elsewhere} \end{cases}$$

For the two chemicals under consideration, an important quantity is the total proportion $Y_1 + Y_2$ found in any sample. Find $E(Y_1 + Y_2)$ and $V(Y_1 + Y_2)$.

5.65   If Y_1 is the total time between a customer's arrival in the store and leaving the service window and Y_2 is the time spent in line before reaching the window, the joint density of these variables was given in Exercise 5.9 to be

$$f(y_1, y_2) = \begin{cases} e^{-y_1}, & 0 \le y_2 \le y_1 \le \infty \\ 0, & \text{elsewhere} \end{cases}$$

The random variable $Y_1 - Y_2$ represents the time spent at the service window. Find $E(Y_1 - Y_2)$ and $V(Y_1 - Y_2)$. Would you think it highly likely that a customer would spend more than 2 minutes at the service window?

5.66 In Exercise 5.10, Y_1 and Y_2 denoted the proportions of time that employees I and II actually spent on their assigned tasks during a workday. The joint density of Y_1 and Y_2 is given by

$$f(y_1, y_2) = \begin{cases} y_1 + y_2, & 0 \le y_1 \le 1, 0 \le y_2 \le 1 \\ 0 & \text{elsewhere} \end{cases}$$

In Exercise 5.48, we derived the mean of the productivity measure $30Y_1 + 25Y_2$. Find the variance of this measure of productivity. Give an interval in which you think the total productivity measures of the two employees should lie at least 75% of the days in question.

5.67 In Exercise 5.12, Y_1 and Y_2 denoted the lengths of life, in hours, for components of types I and II, respectively, in an electronic system. The joint density of Y_1 and Y_2 is

$$f(y_1, y_2) = \begin{cases} (1/8)y_1 e^{-(y_1 + y_2)/2}, & y_1 > 0, y_2 > 0 \\ 0, & \text{elsewhere} \end{cases}$$

The cost, C, of replacing the two components depends upon their length of life at failure and is given by $C = 50 + 2Y_1 + 4Y_2$. Find $E(C)$ and $V(C)$.

5.68 A retail grocery merchant figures that her daily gain from sales, X, is a normally distributed random variable with $\mu = 50$ and $\sigma = 3$ (measurements in dollars). X could be negative if she is forced to dispose of perishable goods. Also, she figures daily overhead costs, Y, to have a gamma distribution with $\alpha = 4$ and $\beta = 2$. If X and Y are independent, find the expected value and variance of her net daily *gain*. Would you expect her net gain for tomorrow to go above $70?

5.69 For the daily output of a certain industrial operation, let Y_1 denote the amount of sales and Y_2 the costs, in thousands of dollars. Assume Y_1 has a probability density function of

$$f_1(y) = \begin{cases} (1/6)y^3 e^{-y}, & y > 0 \\ 0, & \text{elsewhere} \end{cases}$$

and Y_2 has a density function of

$$f_2(y) = \begin{cases} (1/2)e^{-y/2}, & y > 0 \\ 0, & \text{elsewhere} \end{cases}$$

The daily profit is given by $U = Y_1 - Y_2$.

(a) Find $E(U)$.

(b) Assuming Y_1 and Y_2 are independent, find $V(U)$.

(c) Would you anticipate that the daily profit would drop below zero very often? Why?

5.70 A population of N alligators is to be sampled in order to obtain an approximate measure of the difference between the proportions of sexually mature males and sexually mature females, because this parameter has important implications for the future of the population. Assume n animals are to be sampled without replacement. Let Y_1 denote the number of mature females and Y_2 the number of mature males in the sample. If the population contains proportions p_1 and p_2 of mature females and males, respectively (with $p_1 + p_2 < 1$), find expressions for

$$E\left(\frac{Y_1}{n} - \frac{Y_2}{n}\right) \quad \text{and} \quad V\left(\frac{Y_1}{n} - \frac{Y_2}{n}\right)$$

5.71   The total sustained load on the concrete footing of a planned building is the sum of the dead load plus the occupancy load. Suppose the dead load, X_1, has a gamma distribution with $\alpha_1 = 50$ and $\beta_1 = 2$ whereas the occupancy load, X_2, has a gamma distribution with $\alpha_2 = 20$ and $\beta_2 = 2$. (Units are in kips.)

(a)   Find the mean and variance of the total sustained load on the footing.

(b)   Find a value for the sustained load that should only be exceeded with probability less than 1/16.

5.9 The Multinomial Probability Distribution

Recall from Chapter 3 that a binomial random variable results from an experiment consisting of n trials with two possible outcomes per trial. Frequently a similar situation arises in which the number of possible outcomes per trial is more than two. For example, experiments that involve blood typing will typically have at least four possible outcomes per trial. Experiments that involve sampling for defectives may categorize the type of defects observed into more than one class.

A *multinomial experiment*, a generalization of the binomial experiment, possesses the characteristics given in the following definition.

Definition 5.11

A *multinomial experiment* possesses the following properties:

1. The experiment consists of n identical trials.

2. The outcome of each trial falls into one of k classes or cells.

3. The probability that the outcome of a single trial will fall in a particular cell, say, cell i, is $p_i (i = 1, 2, \ldots, k)$ and remains the same from trial to trial. Note that

$$p_1 + p_2 + p_3 + \cdots + p_k = 1$$

4. The trials are independent.

5. The random variables of interest are $Y_1, Y_2, \ldots, Y_k$, where $Y_i (i = 1, 2, \ldots, k)$ is equal to the number of trials in which the outcome falls in cell i. Note that $Y_1 + Y_2 + Y_3 + \cdots + Y_k = n$.

The joint probability function for $Y_1, \ldots, Y_k$ is given by

$$p(y_1, y_2, \ldots, y_k) = \frac{n!}{y_1! y_2! \cdots y_k!} \, p_1^{y_1} p_2^{y_2} \cdots p_k^{y_k}$$

where

$$\sum_{i=1}^{k} p_i = 1 \quad \text{and} \quad \sum_{i=1}^{k} y_i = n$$

Finding the probability that the n trials result in $Y_1, Y_2, \ldots, Y_k$, where Y_i is the number of trials falling in cell i, is an excellent application of the probabilistic methods of Chapter 2. We leave this problem as an exercise.

Definition 5.12

Assume that $p_1, p_2, \ldots, p_k$ are such that $\sum_{i=1}^{k} p_i = 1$, and $p_i \geq 0$ for $i = 1, 2, \ldots, k$. The random variables $Y_1, Y_2, \ldots, Y_k$, are said to have a *multinomial distribution* with parameters n and $p_1, p_2, \ldots, p_k$ if the joint probability function of $Y_1, Y_2, \ldots, Y_k$ is given by

$$p(y_1, y_2, \ldots, y_k) = \frac{n!}{y_1! y_2! \cdots y_k!} \, p_1^{y_1} p_2^{y_2} \cdots p_k^{y_k}$$

where $\sum_{i=1}^{k} y_i = n$.

Many experiments involving classification yield multinomial experiments. For example, the classification of people into five income brackets would result in an enumeration or count corresponding to each of five income classes. Or we might be interested in studying the reaction of a mouse to a particular stimulus in a psychological experiment. If a mouse could react in one of three ways when the stimulus is applied and if a large number of mice were subjected to the stimulus, the experiment would yield three counts indicating the number of mice falling in each reaction class. Similarly, a traffic study might require a count and classification of the type of motor vehicles using a section of highway. An industrial process manufactures items that fall into one of three quality classes: acceptable, seconds, and rejects. A student of the arts might classify paintings in one of k categories according to style and period in order to study trends in style over time. We might wish to classify ideas in a philosophical study or style in the field of literature. The result of an advertising campaign would yield count data indicating a classification of consumer reactions. Indeed, many observations in the physical sciences are not amenable to measurement on a continuous scale and hence result in enumerative or classificatory data.

Notice that the binomial experiment is a special case of the multinomial experiment (where $k = 2$ classes).

EXAMPLE 5.30   According to the adjusted 1980 census figures, the proportions of adults (those over 18 years of age) in the United States associated with five age categories were as follows:

Age	Proportion
18–24	.18
25–34	.23
35–44	.16
45–64	.27
65–	.16

If five adults were randomly sampled from this population, find the probability that the sample would contain one person between the ages of 18 and 24, two between the ages of 25 and 34, and two between the ages of 45 and 64.

Solution   We will number the five age classes 1, 2, 3, 4, and 5 from top to bottom and assume that the proportions given are the probabilities associated with each of the classes. Then we wish to find

$$p(y_1, y_2, y_3, y_4, y_5) = \frac{n!}{y_1! y_2! y_3! y_4! y_5!} p_1^{y_1} p_2^{y_2} p_3^{y_3} p_4^{y_4} p_5^{y_5}$$

for $n = 5$ and $y_1 = 1$, $y_2 = 2$, $y_3 = 0$, $y_4 = 2$, and $y_5 = 0$. Substituting these values into the formula for the joint probability function, we obtain

$$p(1, 2, 0, 2, 0) = \frac{5!}{1! 2! 0! 2! 0!} (.18)^1 (.23)^2 (.16)^0 (.27)^2 (.16)^0$$

$$= 30(.18)(.23)^2 (.27)^2 = .0208 \qquad \backsim$$

Theorem 5.13

If $Y_1, Y_2, \ldots, Y_k$ have a multinomial distribution with parameters n and $p_1, p_2, \ldots, p_k$, then

1. $E(Y_i) = np_i$, $V(Y_i) = np_i q_i$,

2. $\text{Cov}(Y_s, Y_t) = -np_s p_t$,   if $s \neq t$.

Proof   The marginal distribution of Y_i can be used to derive the mean and variance. Recall that Y_i may be interpreted as the number of trials falling in cell i. Imagine all of the cells, excluding cell i, combined into a single large cell.

Hence every trial will result in cell i or in a cell other than cell i with probabilities p_i and $1 - p_i$, respectively. Thus Y_i possesses a binomial marginal probability distribution. Consequently

$$E(Y_i) = np_i \qquad V(Y_i) = np_i q_i \qquad \text{where } q_i = 1 - p_i$$

Note: The same results can be obtained by setting up the expectations and evaluating. For example,

$$E(Y_1) = \sum_{y_1} \sum_{y_2} \cdots \sum_{y_k} y_1 \frac{n!}{y_1! y_2! \cdots y_k!} p_1^{y_1} p_2^{y_2} \cdots p_k^{y_k}$$

Because we have already derived the expected value and variance of Y_i, we leave the tedious summation of this expectation to the interested reader.

The proof of part 2 makes use of Theorem 5.12. Think of the multinomial experiment as a sequence of n independent trials and define

$$U_i = \begin{cases} 1, & \text{if trial } i \text{ results in class } s \\ 0, & \text{otherwise} \end{cases}$$

and

$$W_i = \begin{cases} 1, & \text{if trial } i \text{ results in class } t \\ 0, & \text{otherwise} \end{cases}$$

Then

$$Y_s = \sum_{i=1}^{n} U_i \qquad \text{and} \qquad Y_t = \sum_{j=1}^{n} W_j$$

(Note that because $U_i = 1, 0$ depending upon whether the ith trial resulted in class s, Y_s is simply the sum of a series of 0s and 1s. There is a 1 in the sum everytime we observe an item from class s and a 0 everytime we observe any other class. Thus Y_s is simply the number of times class s is observed. A similar interpretation applies to Y_t.)

To evaluate $\text{Cov}(Y_s, Y_t)$, we need the following results:

$$E(U_i) = p_s \qquad E(W_j) = p_t$$
$$\text{Cov}(U_i, W_j) = 0 \qquad \text{if } i \neq j \text{ because the trials are independent}$$
$$\text{Cov}(U_i, W_i) = E(U_i W_i) - E(U_i)E(W_i) = 0 - p_s p_t$$

because $U_i W_i$ always equals zero. From Theorem 5.12 we then have

$$\text{Cov}(Y_s, Y_t) = \sum_{i=1}^{n} \sum_{j=1}^{n} \text{Cov}(U_i, W_j)$$

$$= \sum_{i=1}^{n} \text{Cov}(U_i, W_i) + \sum \sum_{i \neq j} \text{Cov}(U_i, W_j)$$

$$= \sum_{i=1}^{n} (-p_s p_t) + 0 = -n p_s p_t$$

Note that the covariance is negative, which is to be expected because a large number of outcomes in cell s would force the number in cell t to be small.

Inferential problems associated with the multinomial experiment will be discussed later.

Exercises

5.72   An experiment on learning requires a rat to run a maze (a network of pathways) until it locates one of three possible exits. Exit 1 presents a reward of food, but exits 2 and 3 do not. (If the rat eventually selects exit 1 almost every time, learning may have taken place.) Let Y_i denote the number of times exit i is chosen in successive runnings. For the following assume that the rat chooses an exit at random on each run.

(a)   Find the probability that $n = 6$ runs result in $Y_1 = 3$, $Y_2 = 1$, and $Y_3 = 2$.

(b)   For general n, find $E(Y_1)$ and $V(Y_1)$.

(c)   Find $\text{Cov}(Y_2, Y_3)$ for general n.

(d)   To check for the rat's preference between exits 2 and 3, we may look at $Y_2 - Y_3$. Find $E(Y_2 - Y_3)$ and $V(Y_2 - Y_3)$ for general n.

5.73   A sample of size n is selected from a large lot of items of which a proportion p_1 contains exactly one defect and a proportion p_2 contains more than one defect (with $p_1 + p_2 < 1$). The cost of repairing the defective items in the sample is $C = Y_1 + 3Y_2$, where Y_1 denotes the number of items with one defect and Y_2 denotes the number with two or more defects. Find the expected value and variance of C.

5.74   Refer to Exercise 5.70. Suppose the number N of alligators in the population is very large, with $p_1 = .3$ and $p_2 = .1$.

(a)   Find the probability that in a sample of five alligators $Y_1 = 2$ and $Y_2 = 1$.

(b)   If $n = 5$, find

$$E\left(\frac{Y_1}{n} - \frac{Y_2}{n}\right) \quad \text{and} \quad V\left(\frac{Y_1}{n} - \frac{Y_2}{n}\right)$$

5.75   The weights of a population of mice fed on a certain diet since birth are assumed to be normally distributed with $\mu = 100$ and $\sigma = 20$ (measurement in grams). Suppose a random sample of $n = 4$ mice is taken from this population.

   (a)   Find the probability that exactly two weigh between 80 and 100 grams and exactly one weighs over 100 grams.

   (b)   Find the probability that each of the four mice weighs over 100 grams.

5.76   The National Fire Incident Reporting Service stated that, among residential fires, 73% are in family homes, 20% are in apartments, and 7% are in other types of dwellings. If four fires are independently reported on a single day, what is the probability that two are in family homes, one is in an apartment, and one is in another type of dwelling?

5.77   The typical cost of damages for a fire in a family home is $20,000, whereas the typical cost of an apartment fire is $10,000 and a fire in other dwelling types is $2,000. If four fires are independently reported, use the information in Exercise 5.76 to

   (a)   find the expected total damage cost,

   (b)   find the variance of the total damage cost.

5.78   When commercial aircrafts are inspected, wing cracks are reported as nonexistent, detectable, or critical. The history of a particular fleet indicated that 70% of the planes inspected have no wing cracks, 25% have detectable wing cracks, and 5% have critical wing cracks. Five planes are randomly selected. Find the probability that

   (a)   one has a critical crack, two have detectable cracks, and two have no cracks,

   (b)   at least one plane has observable critical cracks.

5.79   A large lot of manufactured items contains 10% with exactly one defect, 5% with more than one defect, and the remainder defect free. Ten items are randomly selected from this lot for sale. If Y_1 denotes the number of items with one defect and Y_2 the number with more than one defect, the repair costs are $Y_1 + 3Y_2$. Find the mean and variance of the repair costs.

5.80   Refer to Exercise 5.79. Let Y denote the number of items among the ten that contain at least one defect. Find the probability that Y

   (a)   equals 2,

   (b)   is at least 1.

5.10 The Bivariate Normal Distribution (Optional)

No discussion of multivariate probability distributions would be complete without reference to the multivariate normal distribution, which is a keystone to much modern statistical theory. In general, the multivariate normal density function would be defined for k continuous random variables, $Y_1, Y_2, \ldots, Y_k$. Because of its complexity, we will present only the bivariate density function $(k = 2)$.

$$f(y_1, y_2) = \frac{e^{-Q/2}}{2\pi\sigma_1\sigma_2\sqrt{1 - \rho^2}} \qquad \begin{array}{l} -\infty < y_1 < \infty \\ -\infty < y_2 < \infty \end{array}$$

where

$$Q = \frac{1}{1 - \rho^2} \left[\frac{(y_1 - \mu_1)^2}{\sigma_1^2} - 2\rho \frac{(y_1 - \mu_1)(y_2 - \mu_2)}{\sigma_1 \sigma_2} + \frac{(y_2 - \mu_2)^2}{\sigma_2^2} \right]$$

The bivariate normal density function is a function of five parameters, μ_1, μ_2, σ_1^2, σ_2^2, and ρ. The choice of symbols employed for these parameters is not coincidental. With a bit of tedious integration we can show that $E(Y_i) = \mu_i$, $V(Y_i) = \sigma_i^2$, $i = 1, 2$, and the covariance of Y_1 and Y_2 is $\rho\sigma_1\sigma_2$.

Note that if $\text{Cov}(Y_1, Y_2) = 0$, or, equivalently, $\rho = 0$, then

$$f(y_1, y_2) = g(y_1)h(y_2)$$

and hence Y_1 and Y_2 are independent by Theorem 5.5. Recall that zero covariance for two random variables does not generally imply independence. However, for normally distributed random variables the variables are independent if and only if the covariance is zero.

The expression for the joint density function, $k > 2$, is most easily expressed by using the matrix algebra. A discussion of the general case can be found in the references at the end of the chapter.

Exercise

*5.81 Let $Y_1, Y_2, \ldots, Y_n$ be independent random variables with $E(Y_i) = \mu$ and $V(Y_i) = \sigma^2$, $i = 1, \ldots, n$. Let

$$U_1 = \sum_{i=1}^{n} a_i Y_i \quad \text{and} \quad U_2 = \sum_{i=1}^{n} b_i Y_i$$

where $a_1, \ldots, a_n, b_1, \ldots, b_n$ are constants. U_1 and U_2 are said to be orthogonal if $\text{Cov}(U_1, U_2) = 0$.

(a) Show that U_1 and U_2 are orthogonal if and only if $\sum_{i=1}^{n} a_i b_i = 0$

(b) Suppose, in addition, that $Y_1, \ldots, Y_n$ are normally distributed. Then U_1 and U_2 will have a bivariate normal distribution. Show that U_1 and U_2 are independent if they are orthogonal.

5.11 Conditional Expectations

Section 5.3 contains a discussion of conditional probability functions and conditional density functions, which we will now relate to conditional expectations. Conditional expectations are defined in the same manner as univariate expectations except that the conditional density is used in place of the marginal density function.

Definition 5.13

> If Y_1 and Y_2 are any two random variables, the *conditional expectation* of Y_1 given that $Y_2 = y_2$ is defined to be
>
> $$E(Y_1 | Y_2 = y_2) = \int_{-\infty}^{\infty} y_1 f(y_1 | y_2) \, dy_1$$
>
> if Y_1 and Y_2 are jointly continuous and
>
> $$E(Y_1 | Y_2 = y_2) = \sum_{y_1} y_1 p(y_1 | y_2)$$
>
> if Y_1 and Y_2 are jointly discrete.

EXAMPLE 5.31   Refer to the random variables Y_1 and Y_2 of Example 5.8, where the joint density function is given by

$$f(y_1, y_2) = \begin{cases} 1/2, & 0 \le y_1 \le y_2; 0 \le y_2 \le 2 \\ 0, & \text{elsewhere} \end{cases}$$

Find the conditional expectation of amount of sales Y_1 given that $Y_2 = 1$.

Solution   In Example 5.8 we found that

$$f(y_1 | y_2) = \begin{cases} 1/y_2, & 0 < y_1 \le y_2 \le 2 \\ 0, & \text{elsewhere} \end{cases}$$

Thus from Definition 5.13

$$E(Y_1 | Y_2 = 1) = \int_{-\infty}^{\infty} y_1 f(y_1 | y_2) \, dy_1 = \int_0^1 y_1(1) \, dy_1 = \left. \frac{y_1^2}{2} \right]_0^1 = \frac{1}{2}$$

That is, if the soft drink machine contains 1 gallon at the start of the day, the expected sales for that day is 1/2 gallon.   ∽

In general, the conditional expectation of Y_1 given $Y_2 = y_2$ is a function of y_2. If we now let Y_2 range over all its possible values, we can think of the conditional expectation as a function of the random variable Y_2 and hence we can find the expected value of this conditional expectation. The result of this type of iterated expectation is given in Theorem 5.14.

Theorem 5.14

> Let Y_1 and Y_2 denote random variables. Then
>
> $$E(Y_1) = E[E(Y_1 | Y_2)]$$
>
> where, on the right-hand side, the inside expectation is with respect to the conditional distribution of Y_1 given Y_2, and the outside expectation is with respect to the distribution of Y_2.

Proof   Let Y_1 and Y_2 have a joint density function of $f(y_1, y_2)$ and marginal densities of $f_1(y_1)$ and $f_2(y_2)$, respectively. Then

$$E(Y_1) = \int_{-\infty}^{\infty} y_1 f_1(y_1)\, dy_1 = \int_{-\infty}^{\infty} \int_{-\infty}^{\infty} y_1 f(y_1, y_2)\, dy_1\, dy_2$$

$$= \int_{-\infty}^{\infty} \int_{-\infty}^{\infty} y_1 f(y_1|y_2) f_2(y_2)\, dy_1\, dy_2$$

$$= \int_{-\infty}^{\infty} \left[\int_{-\infty}^{\infty} y_1 f(y_1|y_2)\, dy_1 \right] f_2(y_2)\, dy_2$$

$$= \int_{-\infty}^{\infty} E(Y_1 | Y_2 = y_2) f_2(y_2)\, dy_2 = E[E(Y_1 | Y_2)]$$

The proof is similar for the discrete case.

EXAMPLE 5.32   A quality control plan for an assembly line involves sampling $n = 10$ finished items per day and counting Y, the number of defectives. If p denotes the probability of observing a defective, then Y has a binomial distribution, assuming that the number of items produced by the line is large. But p varies from day to day and is assumed to have a uniform distribution on the interval from 0 to 1/4. Find the expected value of Y for any given day.

Solution   From Theorem 5.14 we know that $E(Y) = E[E(Y|p)]$. For a given p, Y has a binomial distribution, and hence $E(Y|p) = np$. Thus

$$E(Y) = E(np) = nE(p) = n \int_0^{1/4} 4p\, dp = n\left(\frac{1}{8}\right)$$

and for $n = 10$

$$E(Y) = 10/8 = 5/4$$

This inspection policy should average 5/4 defectives per day in the long run.

These calculations could be checked by actually finding the unconditional distribution of Y and computing $E(Y)$ directly. In doing so, we would need to find the joint distribution of Y and p. From this joint distribution the marginal distribution of Y can be obtained and $E(Y)$ determined by evaluating $\sum_y yp(y)$.

Thus we can see that Theorem 5.14 permits a much more efficient (timewise) calculation of the desired expected value.   ∽

Exercises

5.82   In Exercise 5.5 we determined that

$$f(y_1, y_2) = \begin{cases} 6(1 - y_2), & 0 \le y_1 \le y_2 \le 1 \\ 0, & \text{elsewhere} \end{cases}$$

is a valid joint probability density function.

(a)   Find $E(Y_1 \mid Y_2 = y_2)$.

(b)   Use the answer derived in part (a) to find $E(Y_1)$. (Compare with the answer found in Exercise 5.45.)

5.83   In Exercise 5.25 we considered a quality control plan that calls for randomly selecting three items from the daily production (assumed large) of a certain machine and observing the number of defectives. The proportion, p, of defectives produced by the machine varies from day to day and has a uniform distribution on the interval $(0, 1)$. Find the expected number of defectives observed among the three sampled items.

5.84   In Exercise 5.26 the number of defects per yard in a certain fabric, Y, was known to have a Poisson distribution with parameter λ. The parameter λ was assumed to be a random variable with a density function given by

$$f(\lambda) = \begin{cases} e^{-\lambda}, & \lambda \ge 0 \\ 0, & \text{elsewhere} \end{cases}$$

Find the expected number of defects per yard by first finding the conditional expectation of Y for given λ.

5.85   In Exercise 5.24 we assumed that Y_1, the weight of bulk item stocked by a supplier, has a uniform distribution over the interval $(0, 1)$. The random variable Y_2 denoted the weight of the item sold and was assumed to have a uniform distribution over the interval $(0, y_1)$, where y_1 is a specific value of Y_1. If the supplier stocks three-quarters of a ton, what is the expected amount sold during the week?

5.86   Assume that Y denotes the number of bacteria per cubic centimeter in a particular liquid and that Y has a Poisson distribution with parameter λ. Further assume that λ varies from location to location and has a gamma-type distribution with parameters α and β where α is a positive integer. If we randomly select a location, what is the expected number of bacteria per cubic centimeter?

*5.87   If Y_1 and Y_2 are independent random variables, each having a normal distribution with a mean of 0 and a variance of 1, find the moment-generating function of $U = Y_1 Y_2$. Use this moment-generating function to find $E(U)$ and $V(U)$. Check the result by evaluating $E(U)$ and $V(U)$ directly from the density functions for Y_1 and Y_2.

5.12 Summary

The multinomial experiment, Section 5.9, and its associated multinomial probability distribution convey the theme of this chapter. Most experiments yield sample measurements, $y_1, y_2, \ldots, y_k$, which may be regarded as observations on k random variables. Inferences about the underlying structure that generates the observations, the probabilities of falling in cells $1, 2, \ldots, k$, are based on knowledge of the probabilities associated with various samples $(y_1, y_2, \ldots, y_k)$. Joint, marginal, and conditional distributions are essential concepts in finding the probabilities of various sample outcomes.

Generally we draw from a population a sample of n observations, which are specific values of $Y_1, Y_2, \ldots, Y_n$. Many times the random variables are independent and have the same probability distribution. As a consequence, the concept of independence is useful in finding the probability of observing the given sample.

You should note that the objective of this chapter has been to convey the ideas contained in the two preceding paragraphs. The details contained in the chapter are numerous and are essential in providing a solid background for a study of inference. At the same time, you should be careful to avoid over-emphasis on details; be sure to keep the broader inferential objectives in mind.

References and Further Readings

1.   Hoel, P. G. *Introduction to Mathematical Statistics.* 5th ed. New York: Wiley, 1984.

2.   Hogg, R. V., and Craig, A. T. *Introduction to Mathematical Statistics.* 4th ed. New York: Macmillan, 1978.

3.   Mendenhall, W. *Introduction to Probability and Statistics.* 7th ed. N. Scituate, Mass.: Duxbury Press, 1987.

4.   Mood, A. M.; Graybill, F. A.; and Boes, D. *Introduction to the Theory of Statistics.* 3d ed. New York: McGraw-Hill, 1974.

5.   Parzen, E. *Modern Probability Theory and Its Applications.* New York: Wiley, 1960.

Supplementary Exercises

5.88   A target for a bomb is in the center of a circle with radius of 1 mile. A bomb falls at a randomly selected point inside that circle. If the bomb destroys everything within 1/2 mile of its landing point, what is the probability that the target is destroyed?

5.89   Two friends are to meet at the library. Each independently and randomly selects an arrival time within the same 1-hour period. Each agrees to wait a maximum of 10 minutes for the other to arrive. What is the probability that they will meet?

5.90 A committee of three people is to be randomly selected from a group containing four Republicans, three Democrats, and two Independents. Let Y_1 and Y_2 denote the number of Republicans and Democrats, respectively, on the committee.

    (a) What is the joint probability distribution for Y_1 and Y_2?

    (b) Find the marginal distributions of Y_1 and Y_2.

    (c) Find $P(Y_1 = 1 \mid Y_2 \geq 1)$.

5.91 Let Y_1 and Y_2 have a joint density function given by

$$f(y_1, y_2) = \begin{cases} 3y_1, & 0 \leq y_2 \leq y_1 \leq 1 \\ 0, & \text{elsewhere} \end{cases}$$

    (a) Find the marginal density functions of Y_1 and Y_2.

    (b) Find $P(Y_1 \leq 3/4 \mid Y_2 \leq 1/2)$.

    (c) Find the conditional density function of Y_1 given $Y_2 = y_2$.

    (d) Find $P(Y_1 \leq 3/4 \mid Y_2 = 1/2)$.

5.92 Refer to Exercise 5.91.

    (a) Find $E(Y_2 \mid Y_1 = y_1)$.

    (b) Use Theorem 5.14 to find $E(Y_2)$.

    (c) Find $E(Y_2)$ directly from the marginal density of Y_2.

5.93 The lengths of life, Y, for a type of fuses has an exponential distribution with a density function given by

$$f(y) = \begin{cases} (1/\beta)e^{-y/\beta}, & y \geq 0 \\ 0, & \text{elsewhere} \end{cases}$$

    (a) If two such fuses have independent life lengths Y_1 and Y_2, find their joint probability density function.

    (b) One fuse from part (a) is in a primary system and the other is in a backup system that comes into use only if the primary system fails. The total effective life length of the two fuses therefore is $Y_1 + Y_2$. Find $P(Y_1 + Y_2 \leq a)$ where $a > 0$.

5.94 In the production of a certain type of copper, two types of copper powder (types A and B) are mixed together and sintered (heated) for a certain length of time. For a fixed volume of sintered copper it is then important to measure the proportion Y_1 of the volume due to solid copper (there will be some pores filled with air) and the proportion Y_2 of the solid mass due to type A crystals. Assume that appropriate probability densities for Y_1 and Y_2 are

$$f_1(y_1) = \begin{cases} 6y_1(1 - y_1), & 0 \leq y_1 \leq 1 \\ 0, & \text{elsewhere} \end{cases}$$

$$f_2(y_2) = \begin{cases} 3y_2^2, & 0 \leq y_2 \leq 1 \\ 0, & \text{elsewhere} \end{cases}$$

The proportion of the sample volume due to type A crystals is then $Y_1 Y_2$. Assuming Y_1 and Y_2 are independent, find $P(Y_1 Y_2 \leq .5)$.

5.95   Suppose the number of eggs laid by a certain insect has a Poisson distribution with mean λ. The probability that any one egg hatches is p. Assuming that the eggs hatch independently of one another, find the expected value of Y, the total number of eggs that hatch.

5.96   In a clinical study of a new drug formulated to reduce the effects of rheumatoid arthritis, researchers found that the proportion, p, of patients who respond favorably to the drug is a random variable that varies from batch to batch of the drug. Assume that p has a probability density function given by

$$f(p) = \begin{cases} 12p^2(1-p), & 0 \le p \le 1 \\ 0, & \text{elsewhere} \end{cases}$$

Suppose that n patients are injected with portions of the drug taken from the same batch. Let Y denote the number showing a favorable response.

(a)   Find the unconditional probability distribution of Y for general n.

(b)   Find $E(Y)$ for $n = 2$.

5.97   A forester studying diseased pine trees models the number of diseased trees per acre, Y, as a Poisson random variable with mean λ. However, λ changes from area to area, and its random behavior is modeled by a gamma distribution. That is,

$$f(\lambda) = \begin{cases} \dfrac{1}{\Gamma(\alpha)\beta^{\alpha}} \lambda^{\alpha-1} e^{-\lambda/\beta}, & \lambda > 0 \\ \\ 0, & \text{elsewhere} \end{cases}$$

Find the unconditional probability distribution for Y.

5.98   A coin has probability p of coming up heads when tossed. In n independent tosses of the coin, let $X_i = 1$ if the ith toss results in heads and $X_i = 0$ if the ith toss results in tails. Then Y, the number of heads in the n tosses, has a binomial distribution and can be represented as

$$Y = \sum_{i=1}^{n} X_i$$

Find $E(Y)$ and $V(Y)$ using Theorem 5.12.

5.99   The negative binomial random variable Y was defined in Section 3.6 as the number of the trial on which the rth success occurs, in a sequence of independent trials with constant probability p of success on each trial. Let X_i denote a random variable defined as the number of the trial on which the ith success occurs, for $i = 1, 2, \ldots, r$. Now define

$$W_i = X_i - X_{i-1}, \qquad i = 1, 2, \ldots, r$$

where X_0 is defined to be zero. Then we can write

$$Y = \sum_{i=1}^{r} W_i$$

Note that the random variables $W_1, \ldots, W_r$ have identical geometric distributions and are mutually independent. Use Theorem 5.12 to show that $E(Y) = r/p$ and $V(Y) = r(1 - p)/p^2$.

5.100   A box contains four balls, numbered 1 through 4. One ball is selected at random from this box. Let

$$X_1 = 1 \text{ if ball number 1 or ball number 2 is drawn}$$
$$X_2 = 1 \text{ if ball number 1 or ball number 3 is drawn}$$
$$X_3 = 1 \text{ if ball number 1 or ball number 4 is drawn}$$

and the X_i's are zero otherwise. Show that any two of the random variables X_1, X_2, and X_3 are independent but the three together are not.

5.101   Suppose we are to observe two independent random samples, with $Y_1, \ldots, Y_n$ denoting a random sample from a normal distribution with mean μ_1 and variance σ_1^2 and $X_1, \ldots, X_m$ denoting a random sample from another normal distribution with mean μ_2 and variance σ_2^2. An approximation for $\mu_1 - \mu_2$ is given by $\bar{Y} - \bar{X}$, the difference between the sample means. Find $E(\bar{Y} - \bar{X})$ and $V(\bar{Y} - \bar{X})$.

*5.102   Let X_1, X_2, and X_3 be random variables, either continuous or discrete. The joint moment-generating function of X_1, X_2, and X_3 is defined by

$$m(t_1, t_2, t_3) = E(e^{t_1 X_1 + t_2 X_2 + t_3 X_3})$$

(a)   Show that $m(t, t, t)$ gives the moment-generating function of $X_1 + X_2 + X_3$.

(b)   Show that $m(t, t, 0)$ gives the moment-generating function of $X_1 + X_2$.

(c)   Show that

$$\left. \frac{\partial^{k_1 + k_2 + k_3} m(t_1, t_2, t_3)}{\partial t_1^{k_1} \, \partial t_2^{k_2} \, \partial t_3^{k_3}} \right|_{t_1 = t_2 = t_3 = 0} = E(X_1^{k_1} X_2^{k_2} X_3^{k_3})$$

*5.103   Let X_1, X_2, and X_3 have a multinomial distribution with probability function

$$p(x_1, x_2, x_3) = \frac{n!}{x_1! x_2! x_3!} \, p_1^{x_1} p_2^{x_2} p_3^{x_3}, \qquad \sum_{i=1}^{n} x_i = n$$

Employ the results of Exercise 5.102 to answer the following.

(a)   Find the joint moment-generating function of X_1, X_2, and X_3.

(b)   Use the answer to part (a) to show that the marginal distribution of X_1 is binomial with parameter p_1.

(c)   Use the joint moment-generating function to find $\text{Cov}(X_1, X_2)$.

*5.104   A box contains N_1 white balls, N_2 black balls, and N_3 red balls ($N_1 + N_2 + N_3 = N$). A random sample of n balls is selected from the box (without replacement). Let Y_1, Y_2, and Y_3 denote the number of white, black, and red balls, respectively, observed in the sample. Find the correlation coefficient for Y_1 and Y_2. (Let $p_i = N_i/N$, $i = 1, 2, 3$.)

*5.105   Let Y_1 and Y_2 be jointly distributed random variables with finite variances.

(a)   Show that $[E(Y_1 Y_2)]^2 \le E(Y_1^2)E(Y_2^2)$. Hint: Observe that for any real number t, $E[(t Y_1 - Y_2)^2] \ge 0$ or, equivalently,

$$t^2 E(Y_1^2) - 2t E(Y_1 Y_2) + E(Y_2^2) \ge 0$$

This is a quadratic expression of the form $At^2 + Bt + C$, and because it is nonnegative, we must have $B^2 - 4AC \le 0$. The inequality above follows directly.

(b)   Let ρ denote the correlation coefficient of Y_1 and Y_2. Using the inequality of part (a), show that $\rho^2 \le 1$.

chapter six

FUNCTIONS OF RANDOM VARIABLES

6.1 Introduction

As we indicated in Chapter 1, the objective of statistics is to make an inference about a population based on information contained in a sample and to provide an associated measure of goodness for the inference. Each topic discussed in the preceding chapters plays a role in the story of statistical inference, but none of the topics discussed thus far is as closely related to the objective of statistics as is the study of functions of random variables. This is due to the fact that all quantities used to estimate population parameters or to make decisions about a population are functions of the n random observations that appear in a sample.

To illustrate, consider the problem of estimating a population mean. Intuitively we draw a random sample of n observations, $y_1, y_2, \ldots, y_n$, from the population and employ the sample mean

$$\bar{y} = \frac{y_1 + y_2 + \cdots + y_n}{n} = \frac{\displaystyle\sum_{i=1}^{n} y_i}{n}$$

as an estimate of μ. How good will this estimate be? The answer depends upon the behavior of the random variables $Y_1, Y_2, \ldots, Y_n$ and their effect on $\bar{Y} = (1/n) \sum_{i=1}^{n} Y_i$.

The measure of goodness of an estimate is the *error of estimation*, the difference between the estimate and the parameter estimated (for our example,

$\bar{y}$ and μ). Because $Y_1, Y_2, \ldots, Y_n$ are random variables, in repeated sampling $\bar{Y}$ will also be a random variable (which is indeed a function of the n variables Y_1, $Y_2, \ldots, Y_n$). Therefore, we cannot be certain that the error of estimation will be less than a specific value; say, B. However, if we can determine the probability distribution of the estimator $\bar{Y}$, this probability distribution can be used to determine the probability that the error of estimation is less than or equal to B.

To determine the probability distribution for a function of n random variables, $Y_1, \ldots, Y_n$, we must find the joint probability distribution for the random variables themselves. We generally assume that observations are obtained through random sampling, which was defined in Section 2.12. We saw in Section 3.7 that random sampling from a finite population (sampling without replacement) results in dependent trials but that these trials become essentially independent if the population is large relative to the sample.

We will assume throughout the remainder of this text that populations are large relative to the sample size and consequently that the random variables obtained through a random sample, in fact, are independent of one another. Thus in the discrete case the joint probability function for $Y_1, Y_2, \ldots, Y_n$, all sampled from the same population, is

$$p(y_1, y_2, \ldots, y_n) = p(y_1)p(y_2) \cdots p(y_n)$$

In the continuous case the joint density function is

$$f(y_1, y_2, \ldots, y_n) = f(y_1)f(y_2) \cdots f(y_n)$$

The statement " $Y_1, Y_2, \ldots, Y_n$ is a random sample from $f(y)$" will mean that the random variables are independent with common density function $f(y)$.

6.2 Finding the Probability Distribution of a Function of Random Variables

We will present three methods for finding the probability distribution for a function of random variables. Any one of these may be employed for a given function, but one method usually leads to a simpler derivation than another. Hence an acquaintance with all three is desirable. Although the three methods for finding the probability distribution of a function of random variables will be discussed separately in the next three sections, we will summarize the methods here.

Consider random variables $Y_1, Y_2, \ldots, Y_n$ and a function $U(Y_1, Y_2, \ldots, Y_n)$, which we will denote simply as U. Then the three methods for finding the probability distribution of U are as follows:

1. Method of distribution functions: Find $P(U \leq u) = F(u)$. This is a probability problem. For every joint event $(Y_1 = y_1, Y_2 = y_2, \ldots, Y_n = y_n)$, there corresponds to one and only one value of U. Thus we must

find the region in the $y_1, y_2, \ldots, y_n$ space for which $U \leq u$ and then find $P(U \leq u)$ by integrating $f(y_1, y_2, \ldots, y_n)$ over this region. The density function for U then is obtained by differentiating $F(u)$. A detailed account of this procedure will be presented in Section 6.3.

2. Method of transformations: If we are given the density function of a random variable Y, the method of transformations results in a general expression for the density of $U = h(Y)$ for an increasing or decreasing function $h(y)$. Then if Y_1 and Y_2 are jointly distributed random variables, we can use the univariate result explained earlier to find the joint density of Y_1 and U. By integrating over y_1, we then find the marginal probability density function of U, which is our objective. This method will be illustrated in Section 6.4.

3. Method of moment-generating functions: The moment-generating–function method is based on a uniqueness theorem, Theorem 6.1, which states that, if two moment-generating functions are identical, the two random variables possess the same probability distributions. Hence, we must find the moment-generating function for U and compare it with the moment-generating functions for the common discrete and continuous random variables derived in Chapters 3 and 4. If it is identical to some well-known moment-generating function, the probability distribution of U will be identified because of the uniqueness theorem. The applications of moment-generating functions will be presented in Section 6.5. Probability-generating functions can be employed in a similar way. If you are interested in their use, see the references at the end of the chapter.

6.3 Method of Distribution Functions

We will illustrate the method of distribution functions with a simple univariate example. If Y has a probability density function $f(y)$, and if U is some function of Y, then we can find $F_U(u) = P(U \leq u)$ directly by integrating $f(y)$ over the region for which $U \leq u$. We can find the probability density function for U by differentiating $F_U(u)$. The following example illustrates the method.

EXAMPLE 6.1   A process for refining sugar yields up to 1 ton of pure sugar per day, but the actual amount produced, Y, is a random variable because of machine breakdowns and other slowdowns. Suppose Y has a density function given by

$$f(y) = \begin{cases} 2y, & 0 \leq y \leq 1 \\ 0, & \text{elsewhere} \end{cases}$$

The company is paid at the rate of \$300 per ton for the refined sugar, but it also has a fixed overhead cost of \$100 per day. Thus the daily profit, in hundreds of dollars, is $U = 3Y - 1$. Find the probability density function for U.

Solution   To employ the distribution function approach, we must find

$$F_U(u) = P(U \le u) = P(3Y - 1 \le u) = P\left(Y \le \frac{u+1}{3}\right)$$

If $u < -1$, then $(u+1)/3 < 0$, and therefore $F_U(u) = P[Y \le (u+1)/3] = 0$. Also, if $u > 2$, then $(u+1)/3 > 1$, and $F_U(u) = P[Y \le (u+1)/3] = 1$. However, if $-1 \le u \le 2$, the probability can be written as an integral of $f(y)$, and

$$P\left(Y \le \frac{u+1}{3}\right) = \int_0^{(u+1)/3} f(y)\, dy = \int_0^{(u+1)/3} 2y\, dy = \left(\frac{u+1}{3}\right)^2$$

(Note that as Y ranges from 0 to 1, U ranges from -1 to 2.) Thus

$$F_U(u) = \begin{cases} 0, & u < -1 \\ \left(\dfrac{u+1}{3}\right)^2, & -1 \le u \le 2 \\ 1, & u > 2 \end{cases}$$

and

$$f_U(u) = \frac{dF_U(u)}{du} = \begin{cases} (2/9)(u+1), & -1 \le u \le 2 \\ 0, & \text{elsewhere} \end{cases}$$

∽

In the bivariate situation let Y_1 and Y_2 be random variables with joint density $f(y_1, y_2)$, and let $U = h(Y_1, Y_2)$ be a function of Y_1, Y_2. Then for every point (y_1, y_2) there corresponds one and only one value of U. If we can find the points (y_1, y_2) such that $U \le u$, then the integral of the joint density function $f(y_1, y_2)$ over the region will equal $P(U \le u)$ or, equivalently, $F_U(u)$ (by Definition 4.1). Then the density function for U can be obtained by differentiation.

We will illustrate these ideas with two examples.

EXAMPLE 6.2   In Example 5.4 we considered the random variables Y_1, the proportional amount of gasoline stocked at the beginning of a week, and Y_2, the proportional amount of gasoline sold during the week. The joint density function of Y_1 and Y_2 is given by

$$f(y_1, y_2) = \begin{cases} 3y_1, & 0 \le y_2 \le y_1 \le 1 \\ 0, & \text{elsewhere} \end{cases}$$

Find the probability density function for $U = Y_1 - Y_2$, the proportional amount of gasoline remaining at the end of the week. Use the density function of U to find $E(U)$.

Solution

The region over which $f(y_1, y_2)$ is not zero is sketched in Figure 6.1. Also shown there is the line $y_1 - y_2 = u$, for a value of u between 0 and 1. Notice that any point (y_1, y_2) such that $y_1 - y_2 \leq u$ will be above the line $y_1 - y_2 = u$.

Figure 6.1
Region over which
f(y₁, y₂) is positive,
Example 6.2

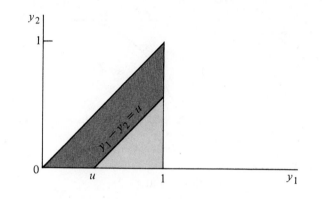

Further, for $u < 0$, $F_U(u) = P(Y_1 - Y_2 \leq u) = 0$, and for $u > 1$, $F_U(u) = 1$. For $0 \leq u \leq 1$, $F_U(u) = P(Y_1 - Y_2 \leq u)$ will be the integral over the shaded region above the line $y_1 - y_2 = u$. Because it will be easier to integrate over the lower triangular region, we can write

$$F_U(u) = P(U \leq u) = 1 - P(U \geq u)$$

$$= 1 - \int_u^1 \int_0^{y_1 - u} 3y_1 \, dy_2 \, dy_1$$

$$= 1 - \int_u^1 3y_1(y_1 - u) \, dy_1$$

$$= 1 - 3\left[\frac{y_1^3}{3} - \frac{uy_1^2}{2}\right]_u^1$$

$$= 1 - \left[1 - \frac{3}{2}(u) + \frac{u^3}{2}\right]$$

$$= \frac{1}{2}(3u - u^3), \qquad 0 \leq u \leq 1$$

Thus

$$F_U(u) = \begin{cases} 0, & u < 0 \\ (1/2)(3u - u^3), & 0 \leq u \leq 1 \\ 1, & u > 1 \end{cases}$$

A graphical depiction of $F_U(u)$ is given in Figure 6.2(a).

Figure 6.2
Distribution and
density functions for
Example 6.2

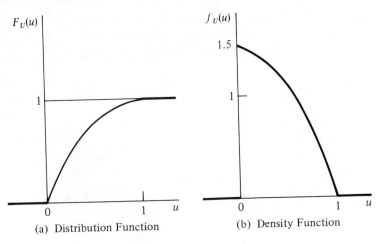

(a) Distribution Function   (b) Density Function

It follows that

$$f_U(u) = \frac{dF_U(u)}{du} = \begin{cases} (3/2)(1 - u^2), & 0 \le u \le 1 \\ 0, & \text{elsewhere} \end{cases}$$

The density function $f_U(u)$ is graphed in Figure 6.2(b).
We can use this derived density function to find $E(U)$, because

$$E(U) = \int_0^1 u\left(\frac{3}{2}\right)(1 - u^2)\, du = \frac{3}{2}\left\{\frac{u^2}{2} - \frac{u^4}{4}\right\}_0^1 = \frac{3}{8}$$

which agrees with $E(Y_1 - Y_2)$ found in Example 5.20 by using the methods
developed in Chapter 5 for finding the expected value of a linear function of
random variables.   ∽

EXAMPLE 6.3   Let (Y_1, Y_2) denote a random sample of size $n = 2$ from the uniform distribution
on the interval (0, 1). Find the probability density function for $U = Y_1 + Y_2$.

Solution   The density function for each Y_i is

$$f(y) = \begin{cases} 1, & 0 \le y \le 1 \\ 0, & \text{elsewhere} \end{cases}$$

Therefore, because we have a random sample, Y_1 and Y_2 are independent, and

$$f(y_1, y_2) = f(y_1)f(y_2) = \begin{cases} 1, & 0 \le y_1 \le 1; 0 \le y_2 \le 1 \\ 0, & \text{elsewhere} \end{cases}$$

The random variables Y_1 and Y_2 are defined over the unit square, as shown
in Figure 6.3. We wish to find $F_U(u) = P(U \le u)$. The first step is to find the

Figure 6.3
The region of
integration for
Example 6.3

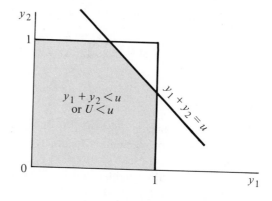

points (y_1, y_2) that imply $y_1 + y_2 < u$. The easiest way to find this region is to locate the points that divide the regions $U \leq u$ and $U > u$. These points lie on the line $y_1 + y_2 = u$.

Graphing this relationship in Figure 6.3 and arbitrarily selecting y_2 as the dependent variable, the line possesses a slope equal to -1 and a y_2 intercept equal to u. The points associated with $U < u$ are either above or below the line and can be obtained by testing. Suppose that $u = 1.5$. Let $y_1 = y_2 = 1/4$; then $y_1 + y_2 = 1/4 + 1/4 = 1/2$ satisfies the inequality $y_1 + y_2 < u$ and falls in the shaded region below the line. You can see intuitively that all points such that $y_1 + y_2 < u$ similarly will lie below the line $y_1 + y_2 = u$. Thus

$$F_U(u) = P(U \leq u) = P(Y_1 + Y_2 \leq u) = \iint\limits_{y_1 + y_2 \leq u} f(y_1, y_2) \, dy_1 \, dy_2$$

Note that u can assume any value in the interval $0 \leq u \leq 2$, and that the limits of integration depend upon u (where u is the y_2 intercept of the line $y_1 + y_2 = u$). Thus the mathematical expression for $F_U(u)$ changes depending on whether $0 \leq u \leq 1$ or $1 \leq u \leq 2$.

The region $y_1 + y_2 \leq u$, $0 \leq u \leq 1$, is the shaded area in Figure 6.4.

Figure 6.4
The region
$y_1 + y_2 \leq u$,
$0 \leq u \leq 1$

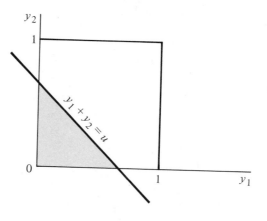

Then for $f(y_1, y_2) = 1$, we have

$$F_U(u) = \int_0^u \int_0^{u-y_2} (1)\, dy_1\, dy_2 = \int_0^u (u - y_2)\, dy_2$$

$$= uy_2 - \frac{y_2^2}{2}\Bigg]_0^u = u^2 - \frac{u^2}{2} = \frac{u^2}{2}, \qquad 0 \leq u \leq 1$$

The solution, $F_U(u), 0 \leq u \leq 1$, could have been acquired directly by using knowledge of elementary geometry. The bivariate density $f(y_1, y_2) = 1$ is uniform over the unit square, $0 \leq y_1 \leq 1, 0 \leq y_2 \leq 1$. Hence $F_U(u)$ is the volume of a solid with height equal to $f(y_1, y_2) = 1$ and triangular cross section as shown in Figure 6.4. Hence

$$F_U(u) = (\text{area of triangle}) \cdot (\text{altitude}) = \frac{u^2}{2}\,(1) = \frac{u^2}{2}$$

The distribution function can be acquired in a similar manner when u is defined over the interval $1 \leq u \leq 2$. Although the geometric solution is easier, we will acquire $F_U(u)$ directly by integration. Thus the region $y_1 + y_2 \leq u$, $1 \leq u \leq 2$, is the shaded area indicated in Figure 6.5.

Figure 6.5
The region
$y_1 + y_2 \leq u,$
$1 \leq u \leq 2$

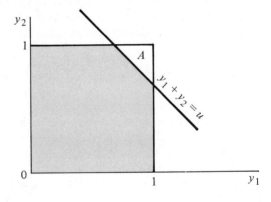

The complement of the event $U \leq u$ is the event that (Y_1, Y_2) falls in the region A of Figure 6.5. Then

$$F_U(u) = 1 - \int_A \int f(y_1, y_2)\, dy_1\, dy_2, \qquad 1 \leq u \leq 2$$

$$= 1 - \int_{u-1}^1 \int_{u-y_2}^1 (1)\, dy_1\, dy_2 = 1 - \int_{u-1}^1 y_1 \Bigg]_{u-y_2}^1 dy_2$$

$$= 1 - \int_{u-1}^1 (1 - u + y_2)\, dy_2 = 1 - \left[(1 - u)y_2 + \frac{y_2^2}{2}\right]_{u-1}^1$$

$$= -\frac{u^2}{2} + 2u - 1, \qquad 1 \leq u \leq 2$$

You will observe that this probability (volume) under $f(y_1, y_2)$ over the region $y_1 + y_2 \le u$, $1 \le u \le 2$, could have been acquired directly and simply by using elementary geometry concepts.

Because there are no values (y_1, y_2) with positive density for which $y_1 + y_2 < 0$, it follows that $F_U(u) = 0$ if $u < 0$. Further, because every pair (y_1, y_2) with positive density is such that $y_1 + y_2 \le 2$, then $F_U(u) = 1$ if $u > 2$. To summarize,

$$
F_U(u) = \begin{cases}
0, & u < 0 \\[2mm]
\dfrac{u^2}{2}, & 0 \le u \le 1 \\[2mm]
\dfrac{-u^2}{2} + 2u - 1, & 1 \le u \le 2 \\[2mm]
1, & u > 2
\end{cases}
$$

The distribution function for U is shown in Figure 6.6(a).

Figure 6.6
Distribution and density functions for Example 6.3

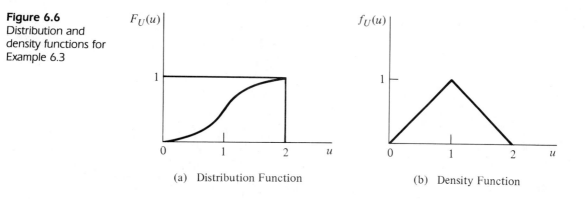

(a) Distribution Function   (b) Density Function

The density function $f_U(u)$ can be obtained by differentiating $F_U(u)$. Thus

$$
f_U(u) = \frac{dF_U(u)}{du} = \frac{d(u^2/2)}{du} = u, \qquad 0 \le u < 1
$$

and

$$
f_U(u) = \frac{d[-(u^2/2) + 2u - 1]}{du} = 2 - u, \qquad 1 \le u \le 2
$$

A graph of $f_U(u)$ is shown in Figure 6.6(b).

Summary of the Distribution Function Method

Let U be a function of the random variables $Y_1, Y_2, \ldots, Y_n$.

1. Find the region $U = u$ in the $(y_1, y_2, \ldots, y_n)$ space.
2. Find the region $U \leq u$.
3. Find $F_U(u) = P(U \leq u)$ by integrating $f(y_1, y_2, \ldots, y_n)$ over the region $U \leq u$.
4. Find the density function $f_U(u)$ by differentiating $F_U(u)$. Thus $f_U(u) = dF_U(u)/du$.

To illustrate, we will consider the case $U = h(Y) = Y^2$, where Y is still continuous with distribution function $F_Y(y)$ and density function $f_Y(y)$. We then have (see Figure 6.7)

$$F_U(u) = P(U \leq u) = P(Y^2 \leq u) = P(-\sqrt{u} \leq Y \leq \sqrt{u})$$

$$= \int_{-\sqrt{u}}^{\sqrt{u}} f(y) \, dy = F_Y(\sqrt{u}) - F_Y(-\sqrt{u})$$

Figure 6.7
The function
$h(y) = y^2$

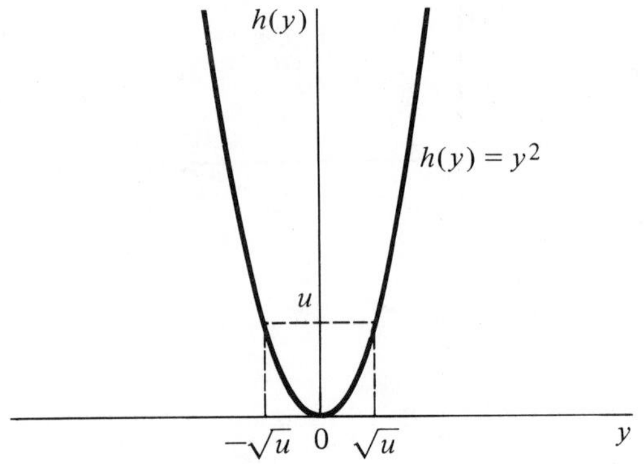

On differentiating with respect to u, we see that

$$f_U(u) = f_Y(\sqrt{u}) \left(\frac{1}{2\sqrt{u}} \right) + f_Y(-\sqrt{u}) \left(\frac{1}{2\sqrt{u}} \right) = \frac{1}{2\sqrt{u}} \left[f_Y(\sqrt{u}) + f_Y(-\sqrt{u}) \right]$$

EXAMPLE 6.4   Let Y have the probability density function given by

$$f_Y(y) = \begin{cases} \dfrac{y + 1}{2}, & -1 \leq y \leq 1 \\ 0, & \text{elsewhere} \end{cases}$$

Find the density function for $U = Y^2$.

Solution

We know that

$$f_U(u) = \frac{1}{2\sqrt{u}} [f_Y(\sqrt{u}) + f_Y(-\sqrt{u})]$$

and on substituting into this equation, we obtain

$$f_U(u) = \frac{1}{2\sqrt{u}} \left(\frac{\sqrt{u}+1}{2} + \frac{-\sqrt{u}+1}{2} \right) = \frac{1}{2\sqrt{u}}, \quad 0 \le u \le 1$$

$$f_U(u) = 0, \quad \text{elsewhere}$$

Note that, because Y has positive density over the interval $-1 \le y \le 1$, then $U = Y^2$ has positive density over the interval $0 \le u \le 1$. $\backsim$

In some instances, it is possible to find a transformation that, when applied to a random variable with a uniform distribution on the interval $[0, 1]$, results in a random variable with some other specified distribution function; say, $F(x)$. The next example illustrates a technique that often can be used to achieve this objective. A brief discussion of one of the practical uses of this transformation follows the example.

EXAMPLE 6.5   Let U be a uniform random variable on the interval $[0, 1]$. Find a transformation $G(U)$ such that $G(U)$ possesses an exponential distribution with mean β.

Solution

If U possesses a uniform distribution on the interval $[0, 1]$, then the distribution function of U (see Exercise 4.21) is given by

$$F_U(u) = \begin{cases} 0, & u < 0 \\ u, & 0 \le u \le 1 \\ 1, & u > 1 \end{cases}$$

Let X denote a random variable that has an exponential distribution with mean β. Then (see Section 4.6) X has distribution function

$$F_X(x) = \begin{cases} 0, & x < 0 \\ 1 - e^{-x/\beta}, & x \ge 0. \end{cases}$$

Note that $F_X(x)$ is strictly increasing on the interval $[0, \infty]$. Let $0 < u < 1$, and observe that there is a unique value x such that $F_X(x) = u$. Thus $F_X^{-1}(u)$, $0 < u < 1$ is well defined. In this case $F_X(x) = 1 - e^{-x/\beta} = u$ if and only if

$x = -\beta \ln(1 - u) = F_X^{-1}(u)$. Consider the random variable $F_X^{-1}(U) = -\beta \ln(1 - U)$, and observe that if $x > 0$

$$
\begin{aligned}
P(F_X^{-1}(U) \le x) &= P(-\beta \ln(1 - U) \le x) \\
&= P(\ln(1 - U) \ge -x/\beta) \\
&= P(U \le 1 - e^{-x/\beta}) \\
&= 1 - e^{-x/\beta}
\end{aligned}
$$

Thus $F_X^{-1}(U) = -\beta \ln(1 - U)$ possesses an exponential distribution with mean β, as desired.

Many times computer simulations are used to evaluate proposed statistical techniques. Typically, these simulations require that we obtain observed values of random variables with a prescribed distribution. As noted in Section 4.4, most computer systems contain a subroutine that provides observed values of a random variable, U, which has a uniform distribution on the interval $[0, 1]$. How can the result of Example 6.5 be used to generate a set of observations from an exponential distribution with mean β? Simply use the computer's random number generator to produce values $u_1, u_2, \ldots, u_n$ from a uniform $[0, 1]$ distribution, and then form $x_i = -\beta \ln(1 - u_i)$, $i = 1, 2, \ldots, n$ to obtain values of random variables with the required exponential distribution.

As long as a prescribed distribution function, $F(y)$, possesses a unique inverse, F^{-1}, the preceding technique can be applied. In instances like that illustrated in Example 6.5, we can readily write down the form of F^{-1} and proceed as earlier. If the form of a distribution function cannot be written in an easily invertible form (recall that the distribution functions of normal, gamma, and beta distributed random variables are given through tables that are derived using numerical integration techniques), our task is more difficult. In these instances we would use the random number subroutine to generate values, say, $u_1, u_2, \ldots, u_n$, of uniform variables, and then use another subroutine for numerically computing the inverse function F^{-1} and the values of $y_i = F^{-1}(u_i)$ for each $i = 1, 2, \ldots, n$.

In the following exercise set you will find problems that can be solved by using the techniques presented in this section. The exercises that involve finding $F^{-1}(U)$ for some specific distribution, F, focus on cases where F^{-1} exists in a closed form.

Exercises

6.1   Let Y be a random variable with a probability density function given by

$$
f(y) = \begin{cases} 2(1 - y), & 0 \le y \le 1 \\ 0, & \text{elsewhere} \end{cases}
$$

(a)   Find the density function of $U_1 = 2Y - 1$.

(b)   Find the density function of $U_2 = 1 - 2Y$.

(c)   Find the density function of $U_3 = Y^2$.

(d)   Find $E(U_1)$, $E(U_2)$, and $E(U_3)$ by using the derived density functions for these random variables.

(e)   Find $E(U_1)$, $E(U_2)$, and $E(U_3)$ by the methods of Chapter 4.

6.2   Let Y be a random variable with a density function given by

$$f(y) = \begin{cases} (3/2)y^2, & -1 \le y \le 1 \\ 0, & \text{elsewhere} \end{cases}$$

(a)   Find the density function of $U_1 = 3Y$.

(b)   Find the density function of $U_2 = 3 - Y$.

(c)   Find the density function of $U_3 = Y^2$.

6.3   A supplier of kerosene has a weekly demand Y possessing a probability density function given by

$$f(y) = \begin{cases} y, & 0 \le y \le 1 \\ 1, & 1 < y \le 1.5 \\ 0, & \text{elsewhere} \end{cases}$$

with measurements in hundreds of gallons. (This problem was introduced in Exercise 4.3.) The supplier's profit is given by $U = 10Y - 4$.

(a)   Find the probability density function for U.

(b)   Use the answer to (a) to find $E(U)$.

(c)   Find $E(U)$ by the methods of Chapter 4.

6.4   The amount of flour used per day by a bakery is a random variable Y having an exponential distribution with mean equal to 4 tons. The cost of the flour is proportional to $U = 3Y + 1$.

(a)   Find the probability density function for U.

(b)   Use the answer in (a) to find $E(U)$.

6.5   The waiting time Y until delivery of a new component for an industrial operation is uniformly distributed over the interval from 1 to 5 days. The cost of this delay is given by $U = 2Y^2 + 3$. Find the probability density function for U.

6.6   The joint distribution of amount of pollutant emitted from a smokestack without a cleaning device (Y_1) and with a cleaning device (Y_2) was given in Exercise 5.6 to be

$$f(y_1, y_2) = \begin{cases} 1, & 0 \le y_1 \le 2; 0 \le y_2 \le 1; 2y_2 \le y_1 \\ 0, & \text{elsewhere} \end{cases}$$

The reduction in amount due to the cleaning device is given by $U = Y_1 - Y_2$.

(a)   Find the probability density function for U.

(b)   Use the answer in (a) to find $E(U)$. (Compare your results with those of Exercise 5.46.)

6.7 Suppose that a unit of mineral ore contains a proportion Y_1 of metal A and a proportion Y_2 of metal B. Experience has shown that the joint probability density function of (Y_1, Y_2) is uniform over the region $0 \le y_1 \le 1, 0 \le y_2 \le 1, 0 \le y_1 + y_2 \le 1$. Let $U = Y_1 + Y_2$, the proportion of metals A and B per unit.

  (a) Find the probability density function for U.

  (b) Find $E(U)$ by using the answer to part (a).

  (c) Find $E(U)$ by using only the marginal densities of Y_1 and Y_2.

6.8 The total time from arrival to completion of service at a fast-food outlet, Y_1, and the time spent waiting in line before arriving at the service window, Y_2, were given in Exercise 5.9 to have joint density function

$$f(y_1, y_2) = \begin{cases} e^{-y_1}, & 0 \le y_2 \le y_1 < \infty \\ 0, & \text{elsewhere} \end{cases}$$

Another random variable of interest is $U = Y_1 - Y_2$, the time spent at the service window.

  (a) Find the probability density function for U.

  (b) Find $E(U)$ and $V(U)$. (Compare your answers with the results of Exercise 5.65.)

6.9 Suppose two electronic components in the guidance system for a missile operate independently, but each has a length of life governed by the exponential distribution with a mean of 1 (with measurements in hundreds of hours).

  (a) Find the probability density function for the average length of life of the two components.

  (b) Find the mean and variance of this average, using the answer in (a). Check your answer by computing the mean and variance, using Theorem 5.12.

6.10 In a process of sintering two types of copper powder (introduced in Exercise 5.94) the density function for Y_1, the volume proportion of solid copper in a sample, was given by

$$f_1(y_1) = \begin{cases} 6y_1(1 - y_1), & 0 \le y_1 \le 1 \\ 0, & \text{elsewhere} \end{cases}$$

The density function for Y_2, the proportion of type A crystals among the solid copper, was given as

$$f_2(y_2) = \begin{cases} 3y_2^2, & 0 \le y_2 \le 1 \\ 0, & \text{elsewhere} \end{cases}$$

The variable $U = Y_1 Y_2$ gives the proportion of the sample volume due to type A crystals. Find the probability density function for U, assuming Y_1 and Y_2 are independent.

6.11 Let Y have a distribution function given by

$$F(y) = \begin{cases} 0, & y < 0 \\ 1 - e^{-y^2}, & y \ge 0 \end{cases}$$

Find a transformation $G(U)$ such that, if U has a uniform distribution on the interval $[0, 1]$, $G(U)$ has the same distribution as Y.

6.12 In Exercise 4.5 we determined that

$$f(y) = \begin{cases} b/y^2, & y \geq b \\ 0, & \text{elsewhere} \end{cases}$$

is a bona fide probability density function for a random variable, Y. Assuming b is a known constant and U has a uniform distribution on the interval $[0, 1]$, transform U to obtain a random variable with the same distribution as Y.

6.13 A member of the power family of distributións has a distribution function given by

$$F(y) = \begin{cases} 0, & y < 0 \\ (y/\theta)^\alpha, & 0 \leq y \leq \theta \\ 1, & y > \theta \end{cases}$$

where $\alpha, \theta > 0$.

(a) Find the density function.

(b) For fixed values of α and θ, find a transformation $G(U)$ so that $G(U)$ has a distribution function of F when U possesses a uniform $[0, 1]$ distribution.

(c) If a random sample of size 5 from a uniform distribution on the interval $[0, 1]$ yielded the values .2700, .6901, .1413, .1523, and .3609, use the transformation derived in part (b) to give values associated with a random variable with a power family distribution with $\alpha = 2$, $\theta = 4$.

6.14 A member of the Pareto family of distributions (often used in economics to model income distributions) has a distribution function given by

$$F(y) = \begin{cases} 0, & y < \beta \\ 1 - (\beta/y)^\alpha, & y \geq \beta \end{cases}$$

where $\alpha, \beta > 0$.

(a) Find the density function.

(b) For fixed values of β and α, find a transformation $G(U)$ such that $G(U)$ has a distribution function of F when U has a uniform distribution on the interval $[0, 1]$.

(c) If a random sample of size 5 from a uniform distribution on the interval $[0, 1]$ yielded the values .0058, .2048, .7692, .2475, .6078, use the transformation derived in part (b) to give values associated with a random variable with a Pareto distribution with $\alpha = 2$, $\beta = 3$.

6.15 Refer to Exercises 6.13 and 6.14. If Y possesses a Pareto distribution with parameters α and β, prove that $X = 1/Y$ has a power family distribution with parameters α and $\theta = \beta^{-1}$.

6.16 Let the random variable Y possess a uniform distribution on the interval $[0, 1]$.

(a) Derive the distribution of the random variable $W = Y^2$.

(b) Derive the distribution of the random variable $W = \sqrt{Y}$.

6.4 Method of Transformations

The transformation method for finding the probability distribution of a function of random variables is an offshoot of the distribution function method of Section 6.3. Through the distribution function approach we can arrive at a simple method of writing down the density function of $U = h(Y)$ provided that $h(y)$ is either decreasing or increasing. [By $h(y)$ increasing we mean that if $y_1 < y_2$, then $h(y_1) < h(y_2)$ for any real numbers y_1 and y_2.] The graph of an increasing function $h(y)$ appears in Figure 6.8.

Figure 6.8
An increasing function

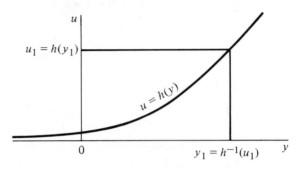

Suppose that $h(y)$ is an increasing function of y and that $U = h(Y)$, where Y has density function $f_Y(y)$. We see from Figure 6.8 that the set of points y such that $h(y) \leq u_1$ is precisely the same as the set of points y such that $y \leq h^{-1}(u_1)$. Therefore (see Figure 6.8),

$$P(U \leq u) = P(Y \leq y) \qquad \text{where } y = h^{-1}(u)$$

or

$$F_U(u) = F_Y(y) \qquad \text{where } y = h^{-1}(u)$$

Then differentiating with respect to u, we have

$$f_U(u) = \frac{dF_U(u)}{du} = \frac{dF_Y(y)}{du} = f_Y(y)\frac{dy}{du}$$

where $y = h^{-1}(u)$. [Note that $dy/du = 1/(du/dy)$.]

Thus we have acquired a new way to find $f_U(u)$ that evolved from the general method of distribution functions. To find $f_U(u)$, solve for y in terms of u; that is, find $y = h^{-1}(y)$ and substitute this expression into $f_Y(y)$. Then multiply this quantity by dy. We will illustrate the procedure with an example.

EXAMPLE 6.6   In Example 6.1 we worked with a random variable Y (amount of sugar produced) with a density function given by

$$f_Y(y) = \begin{cases} 2y, & 0 \le y \le 1 \\ 0, & \text{elsewhere} \end{cases}$$

We were interested in a new random variable (profit) given by $U = 3Y - 1$. Find the probability density function for U by the transformation method.

Solution   The function of interest here is $h(y) = 3y - 1$, which is increasing in y. If $u = 3y - 1$, then

$$y = h^{-1}(u) = \frac{u + 1}{3} \qquad \text{and} \qquad \frac{dy}{du} = \frac{1}{3}$$

Thus

$$f_U(u) = f_Y(y)\frac{dy}{du} = 2y\frac{dy}{du} = 2\left(\frac{u + 1}{3}\right)\left(\frac{1}{3}\right)$$

$$= \frac{2(u + 1)}{9}, \qquad -1 < u < 2$$

$$f_U(u) = 0, \qquad\qquad \text{elsewhere}$$

The range over which $f_U(u)$ is positive is simply the interval $0 < y < 1$ transformed to the u axis by the function $u = 3y - 1$. Note that this answer agrees with that of Example 6.1. ∽

If $h(y)$ is a decreasing function, as in Figure 6.9, then the set of points y such as $h(y) \le u_1$ is the same as the set of points such that $y \ge h^{-1}(u_1)$.

Figure 6.9
A decreasing function

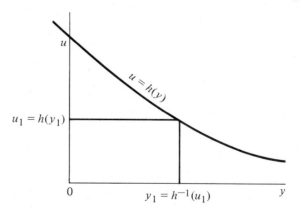

u

$u = h(y)$

$u_1 = h(y_1)$

0

$y_1 = h^{-1}(u_1)$

y

It follows that for $U = h(Y)$ (see Figure 6.9)

$$P(U \leq u) = P(Y \geq y) \qquad \text{where } y = h^{-1}(u)$$

or

$$F_U(u) = 1 - F_Y(y) \qquad \text{where } y = h^{-1}(u)$$

If we differentiate with respect to u, we obtain [with $y = h^{-1}(u)$]

$$f_U(u) = -f_Y(y) \frac{dy}{du}$$

Because dy/du is negative for a decreasing function, then

$$f_U(u) = f_Y(y) \left| \frac{dy}{du} \right|$$

These results are combined in the following statement:

> Let Y have probability density function $f_Y(y)$. If $h(y)$ is either increasing or decreasing in y, then $U = h(Y)$ has density function
>
> $$f_U(u) = f_Y(y) \left| \frac{dy}{du} \right| \qquad \text{where } y = h^{-1}(u)$$

EXAMPLE 6.7   Let Y have the probability density function given by

$$f_Y(y) = \begin{cases} 2y, & 0 < y < 1 \\ 0, & \text{elsewhere} \end{cases}$$

Find the density function of $U = -4Y + 3$.

Solution   The function of interest, $h(y) = -4y + 3$, is decreasing in y. If $u = -4y + 3$, then

$$y = h^{-1}(u) = \frac{3 - u}{4} \qquad \text{and} \qquad \frac{dy}{du} = -\frac{1}{4}$$

Thus

$$f_U(u) = f_Y(y)\left|\frac{dy}{du}\right| = 2y\left|\frac{dy}{du}\right| = 2\left(\frac{3-u}{4}\right)\left(\frac{1}{4}\right)$$

$$= \frac{3-u}{8}, \qquad -1 < u < 3$$

$$f_U(u) = 0, \qquad \text{elsewhere} \qquad \qquad \backsim$$

Direct application of the method of transformation requires that the function $h(y)$ be either increasing or decreasing. If you want to use this method to find the distribution of $U = h(Y)$, you should be very careful to check that the function h is either increasing or decreasing. If it is not, the method of transformations cannot be utilized directly, and you should utilize the method of distribution functions discussed in Section 6.3.

The transformation method also can be used in multivariate situations. The following example gives an illustration for the bivariate case.

EXAMPLE 6.8   Let Y_1 and Y_2 have a joint density function given by

$$f(y_1, y_2) = \begin{cases} e^{-(y_1 + y_2)}, & 0 \le y_1, 0 \le y_2 \\ 0, & \text{elsewhere} \end{cases}$$

Find the density function for $U = Y_1 + Y_2$.

Solution   This problem must be solved in two stages: first, we will find the joint density of Y_1 and U; and second, we will find the marginal density of U. The approach is to let one of the original variables, say, Y_1, be fixed at a value y_1. Then $U = Y_1 + Y_2$ and we can consider the one-dimensional transformation problem in which $U = h(Y_2) = Y_1 + Y_2$. Letting $g(y_1, u)$ denote the joint density of Y_1 and U, we have, with $y_2 = u - y_1$,

$$g(y_1, u) = f(y_1, y_2)\left|\frac{dy_2}{du}\right| = e^{-u}(1), \quad 0 \le u, 0 \le y_1 \le u$$

$$g(y_1, u) = 0, \qquad \qquad \text{otherwise}$$

(Note that $Y_1 \le U$.) The marginal density of U then is given by

$$f_U(u) = \int_{-\infty}^{\infty} g(y_1, u)\, dy_1 = \int_0^u e^{-u}\, dy_1 = ue^{-u}, \qquad 0 \le u$$

$$f_U(u) = 0, \qquad \text{elsewhere} \qquad \qquad \backsim$$

We will use the bivariate transformation case with another example, this one involving a product of random variables.

EXAMPLE 6.9   In Example 5.19 we considered a random variable Y_1, which denoted the proportion of impurities in a chemical sample, and a variable Y_2, which denoted the proportion of type I impurities among all impurities in the sample. The joint density function was given by

$$f(y_1, y_2) = \begin{cases} 2(1 - y_1), & 0 \le y_1 \le 1, 0 \le y_2 \le 1 \\ 0, & \text{elsewhere} \end{cases}$$

We are interested in $U = Y_1 Y_2$, which denotes the proportion of type I impurities in the sample. Find the probability density function for U and use it to find $E(U)$.

Solution   Because we are interested in $U = Y_1 Y_2$, let us first fix one of the random variables, Y_i, say, Y_1, and think in terms of the univariate transformation $U = h(Y_2) = y_1 Y_2$. We can then determine the joint density function for Y_1 and U to be (with $y_2 = u/y_1$)

$$g(y_1, u) = f(y_1, y_2) \left| \frac{dy_2}{du} \right| = 2(1 - y_1)\left(\frac{1}{y_1}\right), \qquad 0 \le u \le y_1 \le 1$$

(Note that U also ranges between 0 and 1 but that Y_1 always must be greater than or equal to U.) Further,

$$f_U(u) = \int_{-\infty}^{\infty} g(y_1, u)\, dy_1 = \int_u^1 2(1 - y_1)\left(\frac{1}{y_1}\right) dy_1$$

$$= 2\int_u^1 \left(\frac{1}{y_1} - 1\right) dy_1 = 2\{\ln y_1]_u^1 - y_1]_u^1\}$$

$$= 2\{-\ln u - 1 + u\} = 2(u - \ln u - 1), \qquad 0 \le u \le 1$$

$$f_U(u) = 0 \qquad \text{elsewhere}$$

(The symbol ln stands for natural logarithm.)
We now find $E(U)$:

$$E(U) = \int_{-\infty}^{\infty} u f_U(u)\, du = \int_0^1 2u(u - \ln u - 1)\, du$$

$$= 2\left\{\int_0^1 u^2\, du - \int_0^1 u(\ln u)\, du - \int_0^1 u\, du\right\}$$

$$= 2\left\{\frac{u^3}{3}\Big]_0^1 - \int_0^1 u(\ln u)\, du - \frac{u^2}{2}\Big]_0^1\right\}$$

The middle integral is most easily solved by using integration by parts, and it becomes

$$\int_0^1 u(\ln u)\, du = \left(\frac{u^2}{2}\right)(\ln u)\Big]_0^1 - \int_0^1 \left(\frac{u^2}{2}\right)\left(\frac{1}{u}\right) du = 0 - \left[\frac{u^2}{4}\right]_0^1 = -\frac{1}{4}$$

Thus,

$$E(U) = 2[(1/3) - (-1/4) - (1/2)] = 2(1/12) = 1/6$$

This answer agrees with the answer to Example 5.21, where $E(U) = E(Y_1 Y_2)$ was found by a different method. ∽

Summary of the Transformation Method

Let U be an increasing or decreasing function of the random variable Y; say, $U = h(Y)$.

1. Find the inverse function, $Y = h^{-1}(U)$.

2. Evaluate dy/du.

3. Find $f_U(u)$ by

$$f_U(u) = f_Y(y)\left|\frac{dy}{du}\right| \qquad \text{where } y = h^{-1}(u)$$

Exercises

6.17   In Exercise 6.1 we considered a random variable, Y, with a probability density function given by

$$f(y) = \begin{cases} 2(1-y), & 0 \le y \le 1 \\ 0, & \text{elsewhere} \end{cases}$$

and used the method of distribution functions to find the density functions of

(a) $U_1 = 2Y - 1$.

(b) $U_2 = 1 - 2Y$.

(c) $U_3 = Y^2$.

Use the method of transformations to find the densities of U_1, U_2, and U_3.

6.18   In Exercise 6.4 we considered a random variable, Y, that possessed an exponential distribution with a mean of 4 and used the method of distribution functions to derive the density function for $U = 3Y + 1$. Use the method transformations to derive the density function for U.

6.19   In Exercise 6.9 we considered two electronic components that operate independently, each with life length governed by the exponential distribution with a mean of 1. We proceeded to use the method of distribution functions to obtain the distribution of the average length of life for the two components. Use the method of transformations to obtain the density function for the average life length of the two components.

6.20   The Weibull density function is given by

$$f(y) = \begin{cases} \dfrac{1}{\alpha} m y^{m-1} e^{-y^m/\alpha}, & y > 0 \\ 0, & \text{elsewhere} \end{cases}$$

where α and m are positive constants. This density function often is used as a model for the lengths of life of physical systems. Suppose Y has the Weibull density just given.

(a)   Find the density function of $U = Y^m$.

(b)   Find $E(Y^k)$ for any positive integer k.

6.21   Let Y have an exponential distribution with a mean of β.

(a)   Prove that $W = \sqrt{Y}$ has a Weibull density with $\alpha = \beta$ and $m = 2$.

(b)   Use the result in Exercise 6.20, part (b), to give $E(Y^{k/2})$ for any positive integer k.

6.22   Let Y have a uniform $[0, 1]$ distribution. Show that $U = -2 \ln Y$ has an exponential distribution with a mean of 2.

6.23   The speed of a molecule in a uniform gas at equilibrium is a random variable, V, whose density function is given by

$$f(v) = a v^2 e^{-bv^2}, \qquad v > 0$$

where $b = m/2kT$ and k, T, and m denote the Boltzmann's constant, the absolute temperature, and the mass of the molecule, respectively.

(a)   Derive the distribution of $W = mV^2/2$, the kinetic energy of the molecule.

(b)   Find $E(W)$.

6.24   A fluctuating electric current, I, may be considered a uniformly distributed random variable over the interval $(9, 11)$. If this current flows through a 2-ohm resistor, find the probability density function of the power $P = 2I^2$.

6.25   The joint distribution for the length of life of two different types of components operating in a system was given in Exercise 5.12 by

$$f(y_1, y_2) = \begin{cases} (1/8) y_1 e^{-(y_1 + y_2)/2}, & y_1 > 0, y_2 > 0 \\ 0, & \text{elsewhere} \end{cases}$$

The relative efficiency of the two types of components is measured by $U = Y_2/Y_1$. Find the probability density function for U.

6.26   In Exercise 6.5 we considered a random variable, Y, that has a uniform distribution on the interval $[1, 5]$. The cost of delay is given by $U = 2Y^2 + 3$. Use the method of transformations to derive the density function of U.

6.27   The proportion of impurities in certain ore samples is a random variable, Y, with a density function given by

$$f(y) = \begin{cases} (3/2)y^2 + y, & 0 \le y \le 1 \\ 0, & \text{elsewhere} \end{cases}$$

The dollar value of such samples is $U = 5 - (Y/2)$. Find the probability density function for U.

6.28   A density function sometimes used by engineers to model lengths of life of electronic components is the Rayleigh density, given by

$$f(y) = \begin{cases} \left(\dfrac{2y}{\theta}\right)e^{-y^2/\theta}, & y > 0 \\ 0, & \text{elsewhere} \end{cases}$$

(a)   If Y has the Rayleigh density, find the probability density function for $U = Y^2$.

(b)   Use the result of (a) to find $E(Y)$ and $V(Y)$.

6.5 Method of Moment-Generating Functions

The moment-generating–function method for finding the probability distribution of a function of random variables $Y_1, Y_2, \ldots, Y_n$ is based on the following uniqueness theorem.

Theorem 6.1

> Suppose that for each of two random variables, X and Y, moment-generating functions exist and are given by $m_X(t)$ and $m_Y(t)$, respectively. If $m_X(t) = m_Y(t)$ for all values of t, then X and Y have the same probability distribution.

(The proof of Theorem 6.1 is beyond the scope of this text.)

The first step in using Theorem 6.1 is to find the moment-generating function of U, where, as before, U is a function of n random variables, $Y_1, Y_2, \ldots, Y_n$. Thus U is a random variable and, by definition, its moment-generating function is

$$m_U(t) = E(e^{tU})$$

Once the moment-generating function for U has been found, it is compared with the moment-generating functions for other well-known random variables. If $m_U(t)$ is identical to one of these, say, the moment-generating function for a random variable V, then by Theorem 6.1 U and V possess identical probability distributions. The density functions, means, variances, and moment-generating functions for some frequently encountered random variables are presented in Appendix II. We will illustrate the procedure with a few examples.

EXAMPLE 6.10   Suppose that Y is a normally distributed random variable with mean μ and variance σ^2. Show that

$$Z = \frac{Y - \mu}{\sigma}$$

is normally distributed with a mean of 0 and a variance of 1.

Solution   We have seen in Example 4.16 that $Y - \mu$ has moment-generating function $e^{t^2\sigma^2/2}$. Hence

$$m_Z(t) = E(e^{tZ}) = E[e^{(t/\sigma)(Y - \mu)}] = m_{(Y - \mu)}\left(\frac{t}{\sigma}\right) = e^{(t/\sigma)^2(\sigma^2/2)} = e^{t^2/2}$$

On comparing $m_Z(t)$ with the moment-generating function of a normal random variable, we see that Z must be normally distributed with $E(Z) = 0$ and $V(Z) = 1$.

EXAMPLE 6.11   Let Z be a normally distributed random variable with a mean of 0 and a variance of 1. Use moment-generating functions to find the probability distribution of Z^2.

Solution   The moment-generating function for Z^2 is

$$m_{Z^2}(t) = E(e^{tZ^2}) = \int_{-\infty}^{\infty} e^{tz^2} f(z)\, dz = \int_{-\infty}^{\infty} e^{tz^2} \frac{e^{-z^2/2}}{\sqrt{2\pi}}\, dz$$

$$= \int_{-\infty}^{\infty} \frac{1}{\sqrt{2\pi}} e^{-(z^2/2)(1 - 2t)}\, dz$$

This integral can be evaluated either by consulting a table of integrals or by noting that the integrand

$$\frac{\exp\left[-\left(\frac{z^2}{2}\right)(1 - 2t)\right]}{\sqrt{2\pi}} = \frac{\exp\left[-\left(\frac{z^2}{2}\right)\Big/(1 - 2t)^{-1}\right]}{\sqrt{2\pi}}$$

is proportional to a normal density function for a random variable with a mean of 0 and a variance of $(1 - 2t)^{-1}$. To make the integrand a normal density function (so that the definite integral is equal to 1), multiply the numerator and denominator by the standard deviation, $(1 - 2t)^{-1/2}$. Then

$$m_{Z^2}(t) = \frac{1}{(1 - 2t)^{1/2}} \int_{-\infty}^{\infty} \frac{1}{\sqrt{2\pi(1 - 2t)^{-1/2}}} \exp\left[-\left(\frac{z^2}{2}\right) \middle/ (1 - 2t)^{-1} \right] dz$$

Because the integral integrates to 1,

$$m_{Z^2}(t) = \frac{1}{(1 - 2t)^{1/2}}$$

A comparison of $m_{Z^2}(t)$ with the moment-generating functions in Appendix II shows $m_{Z^2}(t)$ identical to the moment-generating function for the gamma-type density function with $\alpha = 1/2$ and $\beta = 2$. Thus Z^2 has a chi-square distribution with $v = 1$ degree of freedom. Substituting these values for the chi-square density function (Section 4.6), the density function for $Y = Z^2$ is

$$f_Y(y) = \begin{cases} \dfrac{y^{-1/2}e^{-y/2}}{\Gamma(1/2)2^{1/2}}, & y \geq 0 \\ 0, & \text{elsewhere} \end{cases}$$

A comparison of $m_{Z^2}(t)$ Another very useful application of moment-generating functions comes when dealing with sums of independent random variables.

Theorem 6.2

Let $Y_1, \ldots, Y_n$ be independent random variables with moment-generating functions $m_{Y_1}(t), \ldots, m_{Y_n}(t)$, respectively. If $U = Y_1 + Y_2 + \cdots + Y_n$, then

$$m_U(t) = m_{Y_1}(t)m_{Y_2}(t) \cdots m_{Y_n}(t)$$

Proof   We know that

$$m_U(t) = E[e^{t(Y_1 + \cdots + Y_n)}] = E(e^{tY_1}e^{tY_2} \cdots e^{tY_n})$$
$$= E(e^{tY_1})E(e^{tY_2}) \cdots E(e^{tY_n})$$

because the random variables $Y_1, \ldots, Y_n$ are independent (see Theorem 5.9). Thus by the definition of moment-generating function,

$$m_U(t) = m_{Y_1}(t)m_{Y_2}(t) \cdots m_{Y_n}(t)$$

EXAMPLE 6.12   The number of customer arrivals at a checkout counter in a given interval of time possesses approximately a Poisson probability distribution (see Section 3.8). If Y_1 denotes the time until the first arrival, Y_2 denotes the time between the first and second arrival, ..., Y_n denotes the time between the $(n-1)$st and nth arrival, then it can be shown that $Y_1, Y_2, \ldots, Y_n$ are independent random variables, with the density function for Y_i given by

$$f_{Y_i}(y_i) = \begin{cases} \dfrac{1}{\theta} e^{-y_i/\theta}, & y_i > 0 \\ 0, & \text{otherwise} \end{cases}$$

[Because the Y_i, $i = 1, 2, \ldots, n$, are exponentially distributed, it follows that $E(Y_i) = \theta$; that is, θ is the average time between arrivals.] Find the probability density function for the waiting time from the opening of the counter until the nth customer arrival. (If $Y_1, Y_2, \ldots$ denote successive interarrival times, we want the density function of $U = Y_1 + Y_2 + \cdots + Y_n$.)

Solution   To use Theorem 6.2, we must first find $m_{Y_i}(t)$, where

$$m_{Y_i}(t) = \int_0^\infty \frac{1}{\theta} e^{ty} e^{-y/\theta}\, dy = \frac{1}{\theta} \int_0^\infty e^{-y[(1-\theta t)/\theta]}\, dy = (1 - \theta t)^{-1}$$

Hence

$$m_U(t) = m_{Y_1}(t) \cdots m_{Y_n}(t) = (1 - \theta t)^{-n}$$

and it follows that U has a gamma-type density function with $\alpha = n$ and $\beta = \theta$. That is,

$$f_U(u) = \begin{cases} \dfrac{1}{\Gamma(n)\theta^n} (u^{n-1} e^{-u/\theta}), & u > 0 \\ 0, & \text{elsewhere} \end{cases}$$

∽

The method of moment-generating functions can be used to establish some interesting and useful results about the distributions of functions of normally distributed random variables. Because these results will be used throughout Chapters 7, 8, and 9, we present them in the form of theorems.

Theorem 6.3

> Let $Y_1, \ldots, Y_n$ be independent normally distributed random variables with $E(Y_i) = \mu_i$ and $V(Y_i) = \sigma_i^2$, $i = 1, \ldots, n$. Define U by
>
> $$U = \sum_{i=1}^{n} a_i Y_i = a_1 Y_1 + a_2 Y_2 + \cdots + a_n Y_n$$
>
> where $a_1, \ldots, a_n$ are constants. Then U is a normally distributed random variable with
>
> $$E(U) = \sum_{i=1}^{n} a_i \mu_i = a_1 \mu_1 + a_2 \mu_2 + \cdots + a_n \mu_n$$
>
> $$V(U) = \sum_{i=1}^{n} a_i^2 \sigma_i^2 = a_1^2 \sigma_1^2 + a_2^2 \sigma_2^2 + \cdots + a_n^2 \sigma_n^2$$

Proof   Because Y_i is normally distributed with a mean of μ_i and a variance of σ_i^2, Y_i has the moment-generating function given by

$$m_{Y_i}(t) = \exp\left(\mu_i t + \frac{\sigma_i^2 t^2}{2}\right)$$

Therefore, $a_i Y_i$ has a moment-generating function given by

$$m_{a_i Y_i}(t) = E(e^{t a_i Y_i}) = m_{Y_i}(a_i t) = \exp\left(\mu_i a_i t + \frac{a_i^2 \sigma_i^2 t^2}{2}\right)$$

Because the random variables Y_i are independent, the random variables $a_i Y_i$ are independent, $i = 1, 2, \ldots, n$, and Theorem 6.2 implies that

$$m_U(t) = m_{a_1 Y_1}(t) m_{a_2 Y_2}(t) \cdots m_{a_n Y_n}(t)$$

$$= \exp\left(\mu_1 a_1 t + \frac{a_1^2 \sigma_1^2 t^2}{2}\right) \cdots \exp\left(\mu_n a_n t + \frac{a_n^2 \sigma_n^2 t^2}{2}\right)$$

$$= \exp\left(t \sum_{i=1}^{n} a_i \mu_i + \frac{t^2}{2} \sum_{i=1}^{n} a_i^2 \sigma_i^2\right)$$

[$\exp(\ )$ is simply a more convenient way to write $e^{(\ )}$ when the term in the exponent is long or complex.] Thus U is normal with a mean of $\sum_{i=1}^{n} a_i \mu_i$ and a variance of $\sum_{i=1}^{n} a_i^2 \sigma_i^2$.

Theorem 6.4

> Let $Y_1, \ldots, Y_n$ be defined as in Theorem 6.3 and define Z_i by
>
> $$Z_i = \frac{Y_i - \mu_i}{\sigma_i}, \qquad i = 1, \ldots, n$$
>
> Then $\sum_{i=1}^{n} Z_i^2$ has a χ^2 distribution with n degrees of freedom.

Proof   Because Y_i is normally distributed with a mean of μ_i and a variance of σ_i^2, the result of Example 6.9 implies that Z_i is normally distributed with a mean of 0 and a variance of 1. From Example 6.10 we then have that Z_i^2 is a χ^2 random variable with 1 degree of freedom. Thus

$$m_{Z_i^2}(t) = (1 - 2t)^{-1/2}$$

and from Theorem 6.2, with $V = \sum_{i=1}^{n} Z_i^2$,

$$m_V(t) = (1 - 2t)^{-n/2}$$

or V is distributed as a χ^2 random variable with n degrees of freedom.

> ### Summary of the Moment-Generating–Function Method
>
> Let U be a function of the random variables $Y_1, Y_2, \ldots, Y_n$.
> 1. Find the moment-generating function for U, $m_U(t)$.
> 2. Compare $m_U(t)$ with other well-known moment-generating functions. If $m_U(t) = m_V(t)$ for all values of t, then U and V have identical density functions, by Theorem 6.1.

Exercises

6.29   In Exercises 6.9 and 6.19 we considered two electronic components that operate independently, each with a life length governed by the exponential distribution with a mean of 1. Use the method of moment-generating functions to obtain the density function for the average life length of the two components.

6.30   Suppose that Y_1 and Y_2 are independent, standard normal random variables. Find the density function of $U = Y_1^2 + Y_2^2$.

6.31   Let $Y_1, Y_2, \ldots, Y_n$ be independent normal random variables, each with a mean of μ and a variance of σ^2. Let $a_1, a_2, \ldots, a_n$ denote known constants. Find the density function of the linear combination $U = \sum_{i=1}^{n} a_i Y_i$.

6.32   A certain type of elevator has a maximum weight capacity Y_1, which is normally distributed with a mean of 5000 pounds and a standard deviation of 300 pounds. For a certain building equipped with this type of elevator, the elevator loading Y_2 is a normally distributed random variable with a mean of 4000 pounds and a standard deviation of 400 pounds. For any given time that the elevator is in use, find the probability that it will be overloaded, assuming that Y_1 and Y_2 are independent.

6.33   The manager of a construction job needs to figure prices carefully before submitting a bid. He also needs to account for uncertainty (variability) in the amounts of products he might need. To oversimplify the real situation, suppose a project manager decides that the amount of sand, in yards, needed for a construction project is a random variable Y_1, which is normally distributed with a mean of 10 yards and a standard deviation of .5 yard. The amount of cement mix needed, in hundreds of pounds, also is a random variable, Y_2, which is normally distributed with a mean of 4 and a standard deviation of .2. The sand costs \$7 per yard and the cement mix costs \$3 per hundred pounds. Adding \$100 for other costs, he computes his total cost to be

$$U = 100 + 7Y_1 + 3Y_2$$

If Y_1 and Y_2 are independent, how much should the manager bid in order to ensure that the true costs will exceed the amount bid with a probability of only .01? Is the independence assumption reasonable here?

6.34   In a missile-testing program one random variable of interest is the distance between the point at which the missile lands and the center of the target at which the missile was aimed. If we think of the center of the target as the origin of a coordinate system, then we can let Y_1 denote the north-south distance between the landing point and the target center and Y_2 denote the corresponding east-west distance. (Assume north and east define positive directions.) The distance between the landing point and the target center then is $U = \sqrt{Y_1^2 + Y_2^2}$. If Y_1 and Y_2 are independent, standard normal random variables, find the probability density function for U.

6.35   Let Y_1 be a binomial random variable with n_1 trials and probability of success given by p. Let Y_2 be another binomial random variable with n_2 trials and probability of success given by p. If Y_1 and Y_2 are independent, find the probability function of $Y_1 + Y_2$.

6.36   Let Y_1 and Y_2 be independent Poisson random variables with mean λ_1 and λ_2, respectively.

(a)   Find the probability function of $Y_1 + Y_2$.

(b)   Find the conditional probability function of Y_1, given $Y_1 + Y_2 = m$.

6.37   Customers arrive at a department store checkout counter according to a Poisson distribution with a mean of 7 per hour. In a given 2-hour period, what is the probability that more than twenty customers arrive at the counter?

6.38   The length of time necessary to tune up a car is exponentially distributed with a mean of .5 hour. If two cars are waiting for a tune-up and the service times are independent, what is the probability that the two tune-ups will require longer than 1.5 hours? (Hint: Recall the result of Example 6.12.)

6.39   Let $Y_1, Y_2, \ldots, Y_n$ be independent random variables such that each Y_i has a gamma distribution with parameters α_i and β. Prove that $U = Y_1 + Y_2 + \cdots + Y_n$ has a gamma distribution with parameters $\alpha_1 + \alpha_2 + \cdots + \alpha_n$ and β.

6.40   We saw in Exercise 5.99 that the negative binomial random variable Y can be written as $Y = \sum_{i=1}^{r} W_i$, where $W_1, \ldots, W_r$ are independent geometric random variables with parameter p. Use this fact to derive the moment-generating function for Y. Use the moment-generating function to show that $E(Y) = r/p$ and $V(Y) = r(1-p)/p^2$.

6.41   Show that, if Y_1 has a χ^2 distribution with v_1 degrees of freedom and Y_2 has a χ^2 distribution with v_2 degrees of freedom, then $U = Y_1 + Y_2$ has a χ^2 distribution with $v_1 + v_2$ degrees of freedom, provided Y_1 and Y_2 are independent.

*6.42   Let Y_1 and Y_2 be independent normal random variables, each with a mean of 0 and a variance of σ^2. Define $U_1 = Y_1 + Y_2$ and $U_2 = Y_1 - Y_2$. Show that U_1 and U_2 are independent normal random variables, each with a mean of 0 and a variance of $2\sigma^2$. [Hint: If (U_1, U_2) has a moment-generating function of $m(t_1, t_2)$, then U_1 and U_2 are independent if and only if $m(t_1, t_2) = m_{U_1}(t_1)m_{U_2}(t_2)$.]

6.6 Order Statistics

Many functions of random variables that are of interest in practice depend on the relative magnitudes of the observed variables. For instance, we may be interested in the fastest time in an automobile race or the heaviest mouse among those fed on a certain diet. Thus we often order observed random variables according to their magnitudes. The resulting ordered variables are called *order statistics*.

Formally, let $Y_1, Y_2, \ldots, Y_n$ denote independent continuous random variables with distribution function $F(y)$ and density function $f(y)$. We will denote the ordered random variables Y_i by $Y_{(1)}, Y_{(2)}, \ldots, Y_{(n)}$, where $Y_{(1)} \leq Y_{(2)} \leq \cdots \leq Y_{(n)}$. (Because the random variables are continuous, the equality signs can be ignored.) That is,

$$Y_{(1)} = \min(Y_1, \ldots, Y_n)$$

the minimum of the Y_i's, and

$$Y_{(n)} = \max(Y_1, \ldots, Y_n)$$

the maximum of the Y_i's.

The probability density functions for $Y_{(1)}$ and $Y_{(n)}$ can be found using the method of distribution functions. We will derive the density function of $Y_{(n)}$ first. Because $Y_{(n)}$ is the maximum of $Y_1, Y_2, \ldots, Y_n$, the event $(Y_{(n)} \leq y)$ will occur if and only if the events $(Y_i \leq y)$ occur, for every $i = 1, 2, \ldots, n$. That is,

$$P(Y_{(n)} \leq y) = P(Y_1 \leq y, Y_2 \leq y, \ldots, Y_n \leq y)$$

---

* Exercises preceded by an asterisk are optional.

Because the Y_i are independent and $P(Y_i \leq y) = F(y)$ for $i = 1, 2, \ldots, n$, it follows that

$$P(Y_{(n)} \leq y) = P(Y_1 \leq y)P(Y_2 \leq y) \cdots P(Y_n \leq y) = [F(y)]^n$$

Letting $g_n(y)$ denote the density function of $Y_{(n)}$, we see that, on taking derivatives of both sides,

$$g_n(y) = n[F(y)]^{n-1}f(y)$$

The density function for $Y_{(1)}$ can be found in a similar manner. We have that

$$P(Y_{(1)} \leq y) = 1 - P(Y_{(1)} > y)$$

Because $Y_{(1)}$ is the minimum of $Y_1, Y_2, \ldots, Y_n$, it follows that the event $(Y_{(1)} > y)$ occurs if and only if the events $(Y_i > y)$ occur for $i = 1, 2, \ldots, n$. Because the Y_i are independent and $P(Y_i > y) = 1 - F(y)$ for $i = 1, 2, \ldots, n$, we see that

$$\begin{aligned} P(Y_{(1)} \leq y) &= 1 - P(Y_1 > y, Y_2 > y, \ldots, Y_n > y) \\ &= 1 - [P(Y_1 > y)P(Y_2 > y) \cdots P(Y_n > y)] \\ &= 1 - [1 - F(y)]^n \end{aligned}$$

Thus if $g_1(y)$ denotes the density function of $Y_{(1)}$, differentiation of both sides of the last expression yields

$$g_1(y) = n[1 - F(y)]^{n-1}f(y)$$

Let us now consider the case $n = 2$ and find the joint density function for $Y_{(1)}$ and $Y_{(2)}$. The event $[Y_{(1)} \leq y_1, Y_{(2)} \leq y_2]$ means that either $(Y_1 \leq y_1, Y_2 \leq y_2)$ or $(Y_2 \leq y_1, Y_1 \leq y_2)$. [Note that $Y_{(1)}$ could be either Y_1 or Y_2, whichever is smaller.] Therefore, for $y_1 \leq y_2$, $P(Y_{(1)} \leq y_1, Y_{(2)} \leq y_2)$ is equal to the probability of the union of the two events $(Y_1 \leq y_1, Y_2 \leq y_2)$ and $(Y_2 \leq y_1, Y_1 \leq y_2)$. That is,

$$P(Y_{(1)} \leq y_1, Y_{(2)} \leq y_2) = P[(Y_1 \leq y_1, Y_2 \leq y_2) \cup (Y_2 \leq y_1, Y_1 \leq y_2)]$$

Using the additive law of probability, we see that

$$\begin{aligned} P(Y_{(1)} \leq y_1, Y_{(2)} \leq y_2) &= P(Y_1 \leq y_1, Y_2 \leq y_2) + P(Y_2 \leq y_1, Y_1 \leq y_2) \\ &\quad - P(Y_1 \leq y_1, Y_2 \leq y_1) \end{aligned}$$

Because Y_1 and Y_2 are independent and $P(Y_i \leq w) = F(w)$, $i = 1, 2$, it follows that

$$P(Y_{(1)} \leq y_1, Y_{(2)} \leq y_2) = F(y_1)F(y_2) + F(y_2)F(y_1) - F(y_1)F(y_1)$$
$$= 2F(y_1)F(y_2) - [F(y_1)]^2$$

Letting $g_{12}(y_1, y_2)$ denote the joint density of $Y_{(1)}$ and $Y_{(2)}$, we see that, on differentiating first with respect to y_2 and then with respect to y_1,

$$g_{12}(y_1, y_2) = \begin{cases} 2f(y_1)f(y_2), & y_1 \leq y_2 \\ 0, & \text{elsewhere} \end{cases}$$

The same method can be used to find the joint density of $Y_{(1)}, \ldots, Y_{(n)}$, which turns out to be

$$g_{12\cdots n}(y_1, \ldots, y_n) = \begin{cases} n! \, f(y_1), \ldots, f(y_n), & y_1 \leq y_2 \leq \cdots \leq y_n \\ 0, & \text{elsewhere} \end{cases}$$

The marginal density function for any of the order statistics can be found from this joint density function, but we will not pursue the matter in this text.

EXAMPLE 6.13   Electronic components of a certain type have a length of life Y, with a probability density given by

$$f(y) = \begin{cases} (1/100)e^{-y/100}, & y > 0 \\ 0, & \text{elsewhere} \end{cases}$$

(Length of life is measured in hours.) Suppose that two such components operate independently and in series in a certain system (that is, the system fails when either component fails). Find the density function for X, the length of life of the system.

Solution   Because the system fails at the first component failure, $X = \min(Y_1, Y_2)$, where Y_1 and Y_2 are independent random variables with the given density. Then because $F(y) = 1 - e^{-y/100}$, $y \geq 0$,

$$f_X(y) = g_1(y) = n[1 - F(y)]^{n-1}f(y) = 2e^{-y/100}(1/100)e^{-y/100}$$
$$= (1/50)e^{-y/50}, \quad y > 0$$
$$f_X(y) = 0, \text{ elsewhere}$$

Thus we see that the minimum of two exponentially distributed random variables has an exponential distribution.   ∽

EXAMPLE 6.14 Suppose that in Example 6.13 the components operate in parallel (that is, the system does not fail until both components fail). Find the density function for X, the length of life of the system.

Solution Now $X = \max(Y_1, Y_2)$ and

$$f_X(y) = g_2(y) = n[F(y)]^{n-1}f(y)$$

$$= \begin{cases} 2(1 - e^{-y/100})(1/100)e^{-y/100}, & y > 0 \\ 0, & \text{elsewhere} \end{cases}$$

and therefore

$$f_X(y) = \begin{cases} (1/50)(e^{-y/100} - e^{-y/50}), & y > 0 \\ 0, & \text{elsewhere} \end{cases}$$

We see here that the maximum of two exponential random variables is not an exponential random variable.

Other examples of the use of order statistics will be given in the exercises that follow.

Exercises

6.43 Let Y_1 and Y_2 be independent and uniformly distributed over the interval $(0, 1)$. Find the probability density function of the following:

(a) $U_1 = \min(Y_1, Y_2)$

(b) $U_2 = \max(Y_1, Y_2)$

6.44 Let $Y_1, Y_2, \ldots, Y_n$ be independent uniformly distributed random variables on the interval $[0, \theta]$.

(a) Find the probability distribution function of $Y_{(n)} = \max(Y_1, Y_2, \ldots, Y_n)$.

(b) Find the density function of $Y_{(n)}$.

6.45 Refer to Exercise 6.44. Suppose that the number of minutes that you need to wait for a bus is uniformly distributed on the interval $[0, 15]$. If you take the bus five times, what is the probability that your longest wait is less than 10 minutes?

6.46 Let $Y_1, Y_2, \ldots, Y_n$ be independent random variables, each with a beta distribution with $\alpha = \beta = 2$.

(a) Find the probability distribution function of $Y_{(n)} = \max(Y_1, Y_2, \ldots, Y_n)$.

(b) Find the density function of $Y_{(n)}$.

6.47   Let $Y_1, Y_2, \ldots, Y_n$ be independent exponentially distributed random variables with a mean of β.

   (a)   Show that $Y_{(1)} = \min(Y_1, Y_2, \ldots, Y_n)$ has an exponential distribution with mean β/n.

   (b)   If $n = 5$ and $\beta = 2$, find $P(Y_{(5)} \leq 3.6)$.

6.48   The opening prices per share of two similar stocks, Y_1 and Y_2, are independent random variables, each with a density function given by

$$f(y) = \begin{cases} (1/2)e^{-(1/2)(y-4)}, & y \geq 4 \\ 0, & \text{elsewhere} \end{cases}$$

On a given morning Mr. A is going to buy shares of whichever stock is less expensive. Find the probability density function for the price per share that Mr. A will have to pay.

6.49   Suppose that the length of time Y that it takes a worker to complete a certain task has the probability density function given by

$$f(y) = \begin{cases} e^{-(y-\theta)}, & y > \theta \\ 0, & \text{elsewhere} \end{cases}$$

where θ is a positive constant that represents the minimum time to task completion. Let $Y_1, \ldots, Y_n$ denote a random sample of completion times from this distribution.

   (a)   Find the density function for $Y_{(1)} = \min(Y_1, \ldots, Y_n)$.

   (b)   Find $E(Y_{(1)})$.

*6.50   Let $Y_1, Y_2, \ldots, Y_n$ denote a random sample from the uniform distribution $f(y) = 1$, $0 \leq y \leq 1$. Find the probability density function for the range $R = Y_{(n)} - Y_{(1)}$.

*6.51   Suppose that the number of occurrences of a certain event in time interval $(0, t)$ has a Poisson distribution. If we know that n such events have occurred in $(0, t)$, then the actual times, measured from 0, for the occurrences of the events in question form an ordered set of random variables, which we denote by $W_{(1)} \leq W_{(2)} \leq \cdots \leq W_{(n)}$. ($W_{(i)}$ actually is the waiting time from 0 until the occurrence of the ith event.) It can be shown that the joint density function for $W_{(1)}, \ldots, W_{(n)}$ is given by

$$f(w_1, \ldots, w_n) = \begin{cases} \dfrac{n!}{t^n}, & w_1 \leq w_2 \leq \cdots \leq w_n \\ 0, & \text{elsewhere} \end{cases}$$

[This is the density function for an ordered sample of size n from a uniform distribution on the interval $(0, t)$.] Suppose that telephone calls coming into a switchboard follow a Poisson distribution with a mean of ten calls per minute. A slow period of 2-minutes duration had only four calls.

   (a)   Find the probability that all four calls come in during the first minute [i.e., find $P(W_{(4)} \leq 1)$].

   (b)   Find the expected waiting time, from the start of the 2-minute period, until the fourth call.

*6.52 Suppose that n electronic components, each having an exponentially distributed length of life with mean θ, are put into operation at the same time. The components operate independently and are observed until r have failed ($r \leq n$). Let W_j denote the length of time until the jth failure, with $W_1 \leq W_2 \leq \cdots \leq W_r$. Let $T_j = W_j - W_{j-1}$ for $j \geq 2$ and $T_1 = W_1$. Note that T_j measures the time elapsed between successive failures.

(a) Show that T_j, for $j = 1, 2, \ldots, r$, has an exponential distribution with mean $\theta/(n - j + 1)$.

(b) Show that

$$U_r = \sum_{j=1}^{r} W_j + (n - r)W_r = \sum_{j=1}^{r} (n - j + 1)T_j$$

and hence that $E(U_r) = r\theta$. (U_r is called the *total observed life*, and we can use U_r/r as an approximation to, or "estimator" of, θ.)

6.7 Summary

This chapter has been concerned with finding probability distributions for functions of random variables. This is an important problem in statistics because estimators of population parameters are functions of random variables. Hence it is necessary to know something about the probability distributions of these functions (or estimators) in order to evaluate the goodness of our statistical procedures. A discussion of estimation will be presented in Chapters 8 and 9.

The methods for finding probability distributions for functions of random variables are the distribution function method, Section 6.3; the transformation method, Section 6.4; and the moment-generating-function method, Section 6.5. It should be noted that no one method is always best, because the method of solution depends a great deal on the nature of the function involved. Facility for handling these methods can be achieved only through practice. The exercises at the end of the chapter provide a good starting point.

Some special functions of random variables useful in statistical inference will be considered in Chapter 7.

References and Further Readings

1. Hoel, P. G . *Introduction to Mathematical Statistics*. 5th ed. New York: Wiley, 1984.

2. Hogg, R. V., and Craig, A. T. *Introduction to Mathematical Statistics*. 4th ed. New York: Macmillan, 1978.

3. Mood, A. M.; Graybill, F. A.; and Boes, D. *Introduction to the Theory of Statistics*. 3d ed. New York: McGraw-Hill, 1974.

4. Parzen, E. *Modern Probability Theory and Its Applications*. New York: Wiley, 1960.

Supplementary Exercises

6.53   If Y_1 and Y_2 are independent and identically distributed normal random variables with mean μ and variance σ^2, find the probability density function for $U = (1/2)(Y_1 - 3Y_2)$.

6.54   When current I flows through resistance R, the power generated is given by $W = I^2 R$. Suppose I has a uniform distribution over the interval $(0, 1)$ and R has a density function given by

$$f(r) = \begin{cases} 2r, & 0 \le r \le 1 \\ 0, & \text{elsewhere} \end{cases}$$

Find the probability density function for W. (Assume I is independent of R.)

6.55   Two efficiency experts take independent measurements Y_1 and Y_2 on the length of time it takes workers to complete a certain task. Each measurement is assumed to have the density function given by

$$f(y) = \begin{cases} (1/4)ye^{-y/2}, & y > 0 \\ 0, & \text{elsewhere} \end{cases}$$

Find the density function for the average $U = (1/2)(Y_1 + Y_2)$. (Hint: Use moment-generating functions.)

6.56   Let Y_1 and Y_2 be independent and uniformly distributed over the interval $(0, 1)$. Find the probability density function of the following:

(a) $U_1 = Y_1^2$

(b) $U_2 = Y_1/Y_2$

(c) $U_3 = -\ln(Y_1 Y_2)$

(d) $U_4 = Y_1 Y_2$

*6.57   The length of time that a certain machine operates without failure is denoted by Y_1 and the length of time to repair a failure is denoted by Y_2. After a repair is made, the machine is assumed to operate like a new machine. Y_1 and Y_2 are independent and each has the density function

$$f(y) = \begin{cases} e^{-y}, & y > 0 \\ 0, & \text{elsewhere} \end{cases}$$

Find the probability density function for

$$U = \frac{Y_1}{Y_1 + Y_2}$$

the proportion of time that the machine is in operation during any one operation-repair cycle.

*6.58 A parachutist wants to land at a target T, but she finds that she is equally likely to land at any point on a straight line (A, B) of which T is the midpoint. Find the probability density function of the distance between her landing point and the target. (Hint: Denote A by -1, B by $+1$, and T by 0. Then the parachutist's landing point has a coordinate X, which is uniformly distributed between -1 and $+1$. The distance between X and T is $|X|$.)

6.59 Two sentries are sent to patrol a road 1 mile long. The sentries are sent to points chosen independently and at random along the road. Find the probability that the sentries will be less than 1/2 mile apart when they reach their assigned posts.

6.60 Let Y_1 and Y_2 be independent, standard normal random variables. Find the probability density function of $U = Y_1/Y_2$.

6.61 If Y is a continuous random variable with distribution function $F(y)$, find the probability density function of $U = F(Y)$.

6.62 Let Y be uniformly distributed over the interval $(-1, 3)$. Find the probability density function of $U = Y^2$.

6.63 If Y denotes the length of life of a component and $F(y)$ is the distribution function of Y, then $P(Y > y) = 1 - F(y)$ is called the *reliability* of the component. Suppose that a system consists of four components with the identical reliability functions, $1 - F(y)$, operating as indicated in the accompanying diagram. The system operates correctly if an unbroken chain of components is in operation between A and B. If the four components operate independently, find the reliability of the system, in terms of $F(y)$.

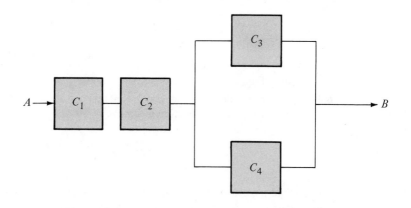

6.64 The percentage of alcohol in a certain compound is a random variable Y, with the following density function:

$$f(y) = \begin{cases} 20y^3(1 - y), & 0 < y < 1 \\ 0, & \text{otherwise} \end{cases}$$

Suppose that the selling price of the compound depends on the alcohol content. Specifically, if $1/3 < y < 2/3$, the compound sells for C_1 dollars per gallon; otherwise it sells for C_2 dollars per gallon. If the production cost is C_3 dollars per gallon, find the probability distribution of the net profit per gallon.

6.65   An engineer has observed that the gap times between vehicles passing a certain point on a highway have an exponential distribution with a mean of 10 seconds.

(a)   Find the probability that the next gap observed will be no longer than 1 minute.

(b)   Find the probability density function for the sum of the next four gap times to be observed. What assumptions are necessary for this answer to be correct?

6.66   If Y is a continuous random variable and m is the median of the distribution, then m is such that $P(Y \le m) = P(Y \ge m) = 1/2$. If Y_1, Y_2, Y_3, and Y_4 are independent random variables with the same continuous distribution, what is the probability that the largest exceeds the median?

6.67   The time until failure of an electronic device has an exponential distribution with mean 15 months. If a random sample of five such devices are put on test, what is the probability that the first failure among the five devices occurs

(a)   after 9 months?

(b)   before 12 months?

*6.68   If a random variable U is normally distributed with mean μ and variance σ^2, and if we write $U = \ln Y$, then Y is said to have a *log-normal* distribution. The log-normal distribution often is used in the biological and physical sciences to model sizes, by volume or weight, of various quantities, such as crushed coal particles, bacteria colonies, and individual animals. Let U and Y be as given above.

(a)   Show that the density function for Y is

$$f(y) = \begin{cases} \left(\dfrac{1}{y\sigma\sqrt{2\pi}}\right)e^{-(\ln y - \mu)^2/2\sigma^2}, & y > 0 \\ 0, & \text{elsewhere} \end{cases}$$

(b)   Find $E(Y)$.

*6.69   A machine produces spherical containers with the radii varying according to the probability density function given by

$$f(r) = \begin{cases} 2r, & 0 \le r \le 1 \\ 0, & \text{elsewhere} \end{cases}$$

Find the probability density function for the volume of the containers.

*6.70   Let v denote the volume of a three-dimensional figure. Let Y denote the number of particles observed in volume v, and assume Y has a Poisson distribution with mean λv. The particles might represent pollution particles in air, bacteria in water, or stars in the heavens.

(a)   If a point is chosen at random within the volume v, show that the distance R to the nearest particle has the probability density function given by

$$f(r) = \begin{cases} 4\lambda\pi r^2 e^{-(4/3)\lambda\pi r^3}, & r > 0 \\ 0, & \text{elsewhere} \end{cases}$$

(b)   If R is as in part (a), show that $U = R^3$ has an exponential distribution.

chapter seven

SAMPLING DISTRIBUTIONS AND THE CENTRAL LIMIT THEOREM

7.1 Introduction

In Chapter 6, we presented methods for finding the distributions of functions of random variables. Throughout this chapter we will be working with functions of the variables $Y_1, Y_2, \ldots, Y_n$ observed in a random sample selected from a population of interest. Thus as discussed in Chapter 6, the variables $Y_1, Y_2, \ldots, Y_n$ are independent with a common distribution. Certain functions of the random variables observed in a sample often are used to estimate or make decisions about unknown population parameters. For example, suppose that we want to estimate a population mean μ. If we obtain a random sample of n observations, $y_1, y_2, \ldots, y_n$, it seems reasonable to estimate μ with the sample mean

$$\bar{y} = \frac{1}{n} \sum_{i=1}^{n} y_i$$

The goodness of the estimate depends on the behavior of the random variables $Y_1, Y_2, \ldots, Y_n$ and the effect that this behavior has on $\bar{Y} = (1/n) \sum_{i=1}^{n} Y_i$. Notice that the random variable $\bar{Y}$ is a function of (only) the random variables $Y_1, Y_2, \ldots, Y_n$ and the (constant) sample size n. The random variable $\bar{Y}$ is therefore an example of a statistic.

Definition 7.1

A *statistic* is a function of the observable random variables in a sample and known constants. Statistics are used to make inferences (estimates or decisions) about unknown population parameters.

Because a statistic is a function of the random variables observed in a sample, the statistic itself is a random variable. Consequently, we will derive its probability distribution, which we will call the *sampling distribution* of the statistic, using the methods of Chapter 6. From a practical point of view, the sampling distribution for a statistic provides a theoretical model for the relative frequency histogram of the possible values of the statistic that we would observe through repeated sampling.

It should be clear to you that the form of the theoretical sampling distribution of a statistic will depend upon the distribution of the observable random variables in the sample. In the next section we will derive the sampling distributions for some statistics used to make inferences about the parameters of a normal distribution.

7.2 Sampling Distributions Related to the Normal Distribution

We already have noted that many phenomena observed in the real world have relative frequency distributions that can be adequately modeled by a normal probability distribution. Thus in many applied problems it is reasonable to assume that the observable random variables in a random sample, $Y_1, Y_2, \ldots, Y_n$, are independent with common normal density function. In such situations the following theorem establishes the sampling distribution of the statistic $\bar{Y} = (1/n)(Y_1 + Y_2 + \cdots + Y_n)$.

Theorem 7.1

Let $Y_1, Y_2, \ldots, Y_n$ be a random sample of size n from a normal distribution with a mean of μ and a variance of σ^2. Then

$$\bar{Y} = \frac{1}{n} \sum_{i=1}^{n} Y_i$$

is normally distributed with a mean of μ and a variance of σ^2/n.

Proof   Because $Y_1, Y_2, \ldots, Y_n$ is a random sample from a normal distribution with a mean of μ and a variance of σ^2, the Y_i are independent normally distributed variables, with $E(Y_i) = \mu$, $V(Y_i) = \sigma^2$, $i = 1, 2, \ldots, n$. Further,

$$\bar{Y} = \frac{1}{n} \sum_{i=1}^{n} Y_i = \frac{1}{n}(Y_i) + \frac{1}{n}(Y_2) + \cdots + \frac{1}{n}(Y_n)$$

and $\bar{Y}$ is a linear combination of $Y_1, Y_2, \ldots, Y_n$, or

$$\bar{Y} = a_1 Y_1 + a_2 Y_2 + \cdots + a_n Y_n$$

with $a_i = 1/n$, $i = 1, 2, \ldots, n$.

Therefore, Theorem 6.3 can be applied to conclude that $\bar{Y}$ that is normally distributed with

$$E(\bar{Y}) = E\left[\frac{1}{n}(Y_1) + \cdots + \frac{1}{n}(Y_n)\right] = \frac{1}{n}(\mu) + \cdots + \frac{1}{n}(\mu) = \mu$$

$$V(\bar{Y}) = V\left[\frac{1}{n}(Y_1) + \cdots + \frac{1}{n}(Y_n)\right] = \frac{1}{n^2}(\sigma^2) + \cdots + \frac{1}{n^2}(\sigma^2)$$

$$= \frac{1}{n^2}(n\sigma^2) = \frac{\sigma^2}{n}$$

That is, the sampling distribution of $\bar{Y}$ is normal with a mean of μ and a variance of σ^2/n.

Notice that the variance of each of the random variables $Y_1, Y_2, \ldots, Y_n$ is σ^2 and the variance of the sampling distribution of the random variable $\bar{Y}$ is σ^2/n. In the discussion that follows we will have occasion to refer to both of these variances. The notation σ^2 will be retained for the variance of the variables $Y_1, Y_2, \ldots, Y_n$ and $\sigma_{\bar{Y}}^2$ will be used to denote the variance of the sampling distribution of the random variable $\bar{Y}$.

Because under the conditions of Theorem 7.1, $\bar{Y}$ is normally distributed with the mean μ and variance $\sigma_{\bar{Y}}^2 = \sigma^2/n$, it follows that

$$Z = \frac{\bar{Y} - \mu}{\sigma_{\bar{Y}}} = \frac{\bar{Y} - \mu}{\sigma/\sqrt{n}} = \sqrt{n}\left(\frac{\bar{Y} - \mu}{\sigma}\right)$$

has a standard normal distribution. We will illustrate the use of Theorem 7.1 in the following example.

EXAMPLE 7.1   A bottling machine can be regulated so that it discharges an average of μ ounces per bottle. It has been observed that the amount of fill dispensed by the machine is normally distributed with $\sigma = 1.0$ ounce. A sample of $n = 9$ filled bottles is randomly selected from the output of the machine on a given day (all bottled with the same machine setting) and the ounces of fill measured for each. Find the probability that the sample mean will be within .3 ounce of the true mean μ for that particular setting.

Solution

If $Y_1, Y_2, \ldots, Y_9$ denote the ounces of fill to be observed, then we know that the Y_i are normally distributed with a mean of μ and a variance of $\sigma^2 = 1$ for $i = 1, 2, \ldots, 9$. Therefore, by Theorem 7.1, $\bar{Y}$ possesses a normal sampling distribution with a mean of μ and $\sigma_{\bar{Y}}^2 = \sigma^2/n = 1/9$. We want to find

$$P(|\bar{Y} - \mu| \leq .3) = P[-.3 \leq (\bar{Y} - \mu) \leq .3]$$

$$= P\left(-\frac{.3}{\sigma/\sqrt{n}} \leq \frac{\bar{Y} - \mu}{\sigma/\sqrt{n}} \leq \frac{.3}{\sigma/\sqrt{n}}\right)$$

$$= P\left(-\frac{.3}{1/\sqrt{9}} \leq Z \leq \frac{.3}{1/\sqrt{9}}\right) = P(-.9 \leq Z \leq .9)$$

because $(\bar{Y} - \mu)/(\sigma/\sqrt{n})$ will have a standard normal distribution. Using Table 4, Appendix III, we find

$$P(-.9 \leq Z \leq .9) = 1 - 2P(Z > .9) = 1 - 2(.1841) = .6318$$

Thus the chance is only .63 that the sample mean will be within .3 ounce of the true population mean.   ∽

EXAMPLE 7.2

Refer to Example 7.1. How many observations should be included in the sample if we wish $\bar{Y}$ to be within .3 ounce of μ with probability .95?

Solution

Now we want

$$P(|\bar{Y} - \mu| \leq .3) = P[-.3 \leq (\bar{Y} - \mu) \leq .3] = .95$$

Multiplying each term of the inequality by $\sqrt{n}/\sigma$ (recall $\sigma = 1$), we have

$$P\left[-.3\sqrt{n} \leq \sqrt{n}\left(\frac{\bar{Y} - \mu}{\sigma}\right) \leq .3\sqrt{n}\right] = P(-.3\sqrt{n} \leq Z \leq .3\sqrt{n}) = .95$$

But using Table 4, Appendix III, we have that

$$P(-1.96 \leq Z \leq 1.96) = .95$$

It must follow that

$$.3\sqrt{n} = 1.96$$

or

$$n = \left(\frac{1.96}{.3}\right)^2 = 42.68$$

If n is taken to be 43, $P(|\bar{Y} - \mu| \leq .3)$ will exceed .95 slightly.   ∽

In succeeding chapters we will be interested in statistics that are functions of the squares of the observations in a random sample from a normal population. Theorem 7.2 establishes the sampling distribution of the sum of the squares of independent standard normal random variables.

Theorem 7.2

> Let $Y_1, Y_2, \ldots, Y_n$ be defined as in Theorem 7.1. Then $Z_i = (Y_i - \mu)/\sigma$ are independent standard normal random variables, $i = 1, 2, \ldots, n$, and
>
> $$\sum_{i=1}^{n} Z_i^2 = \sum_{i=1}^{n} \left(\frac{Y_i - \mu}{\sigma}\right)^2$$
>
> has a χ^2 distribution with n degrees of freedom.

Proof   Because $Y_1, Y_2, \ldots, Y_n$ is a random sample from a normal distribution with mean μ and variance σ^2, Example 6.9 implies that $Z_i = (Y_i - \mu)/\sigma$ has a standard normal distribution for $i = 1, 2, \ldots, n$. Further, the Z_i are independent because the Y_i are independent, $i = 1, 2, \ldots, n$. The fact that $\sum_{i=1}^{n} Z_i^2$ has a χ^2 distribution with n degrees of freedom follows directly from Theorem 6.4.

From Table 6, Appendix III, we can find values χ_α^2 so that

$$P(\chi^2 > \chi_\alpha^2) = \alpha$$

for random variables with χ^2 distributions (see Figure 7.1). For example, if the χ^2 random variable of interest has 10 degrees of freedom (d.f.), Table 6, Appendix III, can be used to find $\chi_{.90}^2$. To do so, look in the row labeled 10 d.f. and the column headed $\chi_{.90}^2$ and read the value 4.86518. Therefore, the probability that a χ^2 random variable with 10 d.f. exceeds 4.86518 is .90.

Figure 7.1
A chi-square distribution showing upper-tail area α

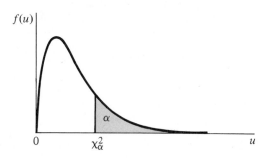

The following example illustrates the combined use of Theorem 7.2 and the χ^2 tables.

EXAMPLE 7.3   If $Z_1, Z_2, \ldots, Z_6$ denotes a random sample from the standard normal distribution, find a number b such that

$$P\left(\sum_{i=1}^{6} Z_i^2 \le b\right) = .95$$

Solution   By Theorem 7.2, $\sum_{i=1}^{6} Z_i^2$ has a χ^2 distribution with 6 degrees of freedom. Looking at Table 6, Appendix III, in the row headed 6 d.f. and the column headed with an upper-tail area of .05, we see the number 12.5916. Thus

$$P\left(\sum_{i=1}^{6} Z_i^2 > 12.5916\right) = .05 \quad \text{or} \quad P\left(\sum_{i=1}^{6} Z_i^2 \le 12.5916\right) = .95$$

and $b = 12.5916$.

The χ^2 distribution plays an important role when we wish to make an inference about the population variance σ^2 based on a random sample $Y_1,$ $Y_2, \ldots, Y_n$ from a normal population. As we will show in Chapter 8, a good estimator of σ^2 is the sample variance[†]

$$S^2 = \frac{1}{n-1} \sum_{i=1}^{n} (Y_i - \bar{Y})^2$$

The following theorem gives the probability distribution for a function of the statistic S^2.

Theorem 7.3

Let $Y_1, Y_2, \ldots, Y_n$ be a random sample from a normal distribution with a mean of μ and a variance of σ^2. Then

$$\frac{1}{\sigma^2} \sum_{i=1}^{n} (Y_i - \bar{Y})^2 = \frac{(n-1)S^2}{\sigma^2}$$

has a χ^2 distribution with $(n-1)$ degrees of freedom. Also, $\bar{Y}$ and S^2 are independent random variables.

[†] Recall that in Chapter 1 we defined the sample variance as

$$S'^2 = \frac{\sum_{i=1}^{n} (\bar{Y}_i - \bar{Y})^2}{n}$$

However, as we will see in Chapter 8, S^2 as defined here (with $n-1$ rather than n) has some convenient properties and hence often is used as the sample variance.

Proof   The complete proof of Theorem 7.3 is outlined in Exercise 13.74. To make the general result more plausible, we will consider the case $n = 2$ and show that $(n - 1)S^2/\sigma^2$ has a χ^2 distribution with 1 degree of freedom. In the case $n = 2$,

$$\bar{Y} = (1/2)(Y_1 + Y_2)$$

and therefore,

$$S^2 = \frac{1}{2-1} \sum_{i=1}^{2} (Y_i - \bar{Y})^2$$

$$= \left[Y_1 - \frac{1}{2}(Y_1 + Y_2) \right]^2 + \left[Y_2 - \frac{1}{2}(Y_1 + Y_2) \right]^2$$

$$= \left[\frac{1}{2}(Y_1 - Y_2) \right]^2 + \left[\frac{1}{2}(Y_2 - Y_1) \right]^2$$

$$= 2 \left[\frac{1}{2}(Y_1 - Y_2) \right]^2 = \frac{(Y_1 - Y_2)^2}{2}$$

It follows that when $n = 2$,

$$\frac{(n-1)S^2}{\sigma^2} = \frac{(Y_1 - Y_2)^2}{2\sigma^2} = \left(\frac{Y_1 - Y_2}{\sqrt{2}\sigma} \right)^2$$

We will show that this quantity is equal to the square of a standard normal random variable, that is, it is a Z^2, which, as we have already shown in Example 6.10, possesses a χ^2 distribution with 1 degree of freedom.

Because $Y_1 - Y_2$ is a linear combination of independent normally distributed random variables $(Y_1 - Y_2 = a_1 Y_1 + a_2 Y_2$ with $a_1 = 1$ and $a_2 = -1)$, Theorem 6.3 tells us that $Y_1 - Y_2$ has a normal distribution with mean $1\mu - 1\mu = 0$ and variance $(1)^2\sigma^2 + (-1)^2\sigma^2 = 2\sigma^2$. Therefore,

$$Z = \frac{Y_1 - Y_2}{\sqrt{2}\sigma}$$

has a standard normal distribution. Because for $n = 2$

$$\frac{(n-1)S^2}{\sigma^2} = Z^2$$

it follows that $(n - 1)S^2/\sigma^2$ has a χ^2 distribution with 1 degree of freedom.

EXAMPLE 7.4   In Example 7.1 the ounces of fill from the bottling machine are assumed to have a normal distribution with $\sigma^2 = 1$. Suppose that we plan to select a random sample of ten bottles and measure the amount of fill in each bottle. If these ten observations are used to calculate S^2, it might be useful to specify an interval of values that will include S^2 with a high probability. Find numbers b_1 and b_2 such that

$$P(b_1 \le S^2 \le b_2) = .90$$

Solution   Notice that

$$P(b_1 \le S^2 \le b_2) = P\left[\frac{(n-1)b_1}{\sigma^2} \le \frac{(n-1)S^2}{\sigma^2} \le \frac{(n-1)b_2}{\sigma^2}\right]$$

Because $\sigma^2 = 1$, it follows that $(n-1)S^2/\sigma^2 = (n-1)S^2$ has a χ^2 distribution with $(n-1)$ degrees of freedom. Therefore, we can use Table 6, Appendix III, to find two numbers, a_1 and a_2, such that

$$P[a_1 \le (n-1)S^2 \le a^2] = .90$$

One method of doing this is to find the a_2 that cuts off an area of .05 in the upper tail and the a_1 that cuts off .05 in the lower tail (.95 in the upper tail). Because there are 9 degrees of freedom, the table gives $a_2 = 16.919$ and $a_1 = 3.325$. Thus we must have

$$a_1 = \frac{(n-1)b_1}{\sigma^2} = (n-1)b_1 = 9b_1$$

$$a_2 = \frac{(n-1)b_2}{\sigma^2} = (n-1)b_2 = 9b_2$$

or

$$b_1 = \frac{3.325}{9} = .369 \quad \text{and} \quad b_2 = \frac{16.919}{9} = 1.880$$

Thus we see that if we wish to have an interval that will include S^2 with probability .90, one such interval is (.369, 1.880). Note that this interval is fairly wide.   ∽

As you will subsequently learn, the result given in Theorem 7.1 will provide the basis for development of inference-making procedures about the mean μ of a normal population with a known variance of σ^2. In that case Theorem 7.1 tells us that $\sqrt{n}(\bar{Y} - \mu)/\sigma$ has a standard normal distribution.

When σ is unknown, it can be estimated by $S = \sqrt{S^2}$, and the quantity

will provide the basis for development of methods for inferences about μ. We will show that the probability distribution of $\sqrt{n}(\bar{Y} - \mu)/S$ is given by a probability density function known as a *Student's t distribution* with $n - 1$ degrees of freedom. The general definition of a random variable that possesses a Student's t distribution (or simply the t distribution) is as follows.

Definition 7.2

> Let Z be a standard normal random variable and let χ^2 be a chi-square random variable with v degrees of freedom. Then if Z and χ^2 are independent,
>
> $$T = \frac{Z}{\sqrt{\chi^2/v}}$$
>
> is said to have a t *distribution* with v degrees of freedom.

If $Y_1, Y_2, \ldots, Y_n$ is a random sample from a normal population with a mean of μ and a variance of σ^2, Theorem 7.1 may be applied to show that $Z = \sqrt{n} \times (\bar{Y} - \mu)/\sigma$ has a standard normal distribution. Theorem 7.3 tells us that $\chi^2 = (n - 1)S^2/\sigma^2$ has a χ^2 distribution with $v = n - 1$ degrees of freedom and that Z and χ^2 are independent (because $\bar{Y}$ and S^2 are independent). Therefore, by Definition 7.2

$$T = \frac{Z}{\sqrt{\chi^2/v}} = \frac{\sqrt{n}(\bar{Y} - \mu)/\sigma}{\sqrt{(n - 1)S^2/\sigma^2(n - 1)}} = \sqrt{n}\left(\frac{\bar{Y} - \mu}{S}\right)$$

had a t distribution with $(n - 1)$ degrees of freedom.

The equation for the t density function will not be given here, but hints about its derivation are given in the exercises at the end of the chapter. Like the standard normal density function, the t density function is symmetric about zero. Further, for $v > 1$, $E(T) = 0$ and for $v > 2$, $V(T) = v/(v - 2)$. Thus we see that a t-distributed random variable has the same expected value as a standard normal random variable. However, a standard normal random variable always has a variance of 1, while the variance of a variable with a t distribution always exceeds 1.

A standard normal density function and a t density function are sketched in Figure 7.2. Note that both density functions are symmetric about the origin but that the t density has more probability mass in the tails.

Figure 7.2
A comparison of the
standard normal and
t density functions

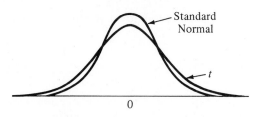

Values of t_α such that $P(T > t_\alpha) = \alpha$ for $\alpha = .100, .050, .025, .010$, and $.005$ are given in Table 5 of Appendix III. For example, if a random variable has a t distribution with 21 degrees of freedom (d.f.), $t_{.100}$ is found by looking in the row labeled 21 d.f. and the column headed $t_{.100}$. Using Table 5, we see that $t_{.100} = 1.323$. Therefore, for 21 d.f. the probability that a t-distributed random variable exceeds 1.323 is .100.

EXAMPLE 7.5   The tensile strength for a certain type of wire is normally distributed with an unknown mean of μ and an unknown variance of σ^2. Six pieces of wire were randomly selected from a large roll and Y_i, the tensile strength for portion i, is measured for $i = 1, 2, \ldots, 6$. The population mean μ and variance σ^2 can be estimated by $\bar{Y}$ and S^2, respectively. Because $\sigma_{\bar{Y}}^2 = \sigma^2/n$, $\sigma_{\bar{Y}}^2$ can be estimated by S^2/n. Find the approximate probability that $\bar{Y}$ will be within $2S/\sqrt{n}$ of the true population mean μ.

Solution   We want to find

$$P\left[-\frac{2S}{\sqrt{n}} \leq (\bar{Y} - \mu) \leq \frac{2S}{\sqrt{n}}\right] = P\left[-2 \leq \sqrt{n}\left(\frac{\bar{Y} - \mu}{S}\right) \leq 2\right]$$

$$= P(-2 \leq T \leq 2)$$

where T has a t distribution with, in this case, $n - 1 = 5$ degrees of freedom. Looking at Table 5, Appendix III, we see that the upper-tail area to the right of 2.015 is .05. Hence

$$P(-2.015 \leq T \leq 2.015) = .90$$

and the probability that $\bar{Y}$ will be within two estimated standard deviations of μ will be just slightly less than .90.

Note that if σ^2 were known, the probability that $\bar{Y}$ would be within $2\sigma_{\bar{Y}}$ of μ is given by

$$P\left[-2\left(\frac{\sigma}{\sqrt{n}}\right) \leq (\bar{Y} - \mu) \leq 2\left(\frac{\sigma}{\sqrt{n}}\right)\right] = P\left[-2 \leq \sqrt{n}\left(\frac{\bar{Y} - \mu}{\sigma}\right) \leq 2\right]$$

$$= P(-2 \leq Z \leq 2) = .9544 \qquad \backsim$$

Suppose that we desire to compare the variances of two normal populations based on information contained in independent random samples from the
two populations. Suppose that one random sample contains n_1 normally
distributed random variables with common variance σ_1^2 and that the other
random sample contains n_2 normally distributed random variables with com-
mon variance σ_2^2. If we calculate S_1^2 from the observations in sample 1, then S_1^2
estimates σ_1^2. Similarly, S_2^2 calculated from the observations in the second sample
estimates σ_2^2. Thus it seems intuitive that the ratio S_1^2/S_2^2 could be used to make
inferences about the relative magnitudes of σ_1^2 and σ_2^2. If we divide each S_i^2 by σ_i^2,
then the resulting ratio

$$\frac{S_1^2/\sigma_1^2}{S_2^2/\sigma_2^2} = \left(\frac{\sigma_2^2}{\sigma_1^2}\right)\left(\frac{S_1^2}{S_2^2}\right)$$

has an *F distribution* with $(n_1 - 1)$ and $(n_2 - 1)$ degrees of freedom. The general
definition of an F distribution is as follows:

Definition 7.3

Let χ_1^2 and χ_2^2 be chi-square random variables with v_1 and v_2 degrees
of freedom, respectively. Then if χ_1^2 and χ_2^2 are independent,

$$F = \frac{\chi_1^2/v_1}{\chi_2^2/v_2}$$

is said to have an *F distribution* with v_1 numerator degrees of
freedom and v_2 denominator degrees of freedom.

The density function for random variables with the F distribution is a
member of the family of beta distributions (see Section 4.7). We omit the formula
for the density of a random variable with the F distribution, but the method of
derivation is outlined in the exercises at the end of the chapter.

Considering once again the independent random sample from normal
distributions, we know that $\chi_1^2 = (n_1 - 1)S_1^2/\sigma_1^2$ and $\chi_2^2 = (n_2 - 1)S_2^2/\sigma_2^2$ have
independent χ^2 distributions with $v_1 = (n_1 - 1)$ and $v_2 = (n_2 - 1)$ degrees of
freedom, respectively. Thus Definition 7.3 implies that

$$F = \frac{\chi_1^2/v_1}{\chi_2^2/v_2} = \frac{(n_1 - 1)S_1^2/\sigma_1^2(n_1 - 1)}{(n_2 - 1)S_2^2/\sigma_2^2(n_2 - 1)} = \frac{S_1^2/\sigma_1^2}{S_2^2/\sigma_2^2}$$

has an F distribution with $(n_1 - 1)$ numerator degrees of freedom and $(n_2 - 1)$
denominator degrees of freedom.

A typical F density function is sketched in Figure 7.3. Values of F_α such
that $P(F > F_\alpha) = \alpha$ are given in Table 7 for values of $\alpha = .100, .050, .025, .010,$

Figure 7.3
A typical F probability
density function

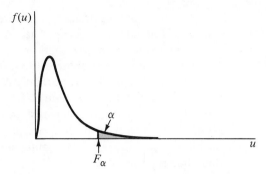

and .005. In Table 7, the column headings are the numerator degrees of freedom, whereas the denominator degrees of freedom are given in the main row headings. Opposite each denominator degrees of freedom (row heading), the values of $\alpha = .100, .050, .025, .010,$ and $.005$ appear. For example, if the F variable of interest has 5 numerator degrees of freedom and 7 denominator degrees of freedom $F_{.100} = 2.88$, $F_{.050} = 3.97$, $F_{.025} = 5.29$, $F_{.010} = 7.46$, and $F_{.005} = 9.52$. Thus, the probability that a random variable with an F distribution with 5 numerator degrees of freedom and 7 denominator degrees of freedom exceeds 7.46 is .01.

EXAMPLE 7.6   If we take independent samples of size $n_1 = 6$ and $n_2 = 10$ from two normal populations with equal population variances, find the number b such that

$$P\left(\frac{S_1^2}{S_2^2} \le b\right) = .95$$

Solution   Because $n_1 = 6$, $n_2 = 10$, and the population variances are equal, then

$$\frac{S_1^2/\sigma_1^2}{S_2^2/\sigma_2^2} = \frac{S_1^2}{S_2^2}$$

has an F distribution with $v_1 = n_1 - 1 = 5$ numerator degrees of freedom and $v_2 = n_2 - 1 = 9$ denominator degrees of freedom. Also,

$$P\left(\frac{S_1^2}{S_2^2} \le b\right) = 1 - P\left(\frac{S_1^2}{S_2^2} > b\right)$$

Therefore, we want to find the number b cutting off an upper-tail area of .05 under the F density function with 5 numerator degrees of freedom and 9 denominator degrees of freedom. Looking in column 5 and row 9 in Table 7, we see that the appropriate value of b is 3.48.

Note that even when the population variances are equal, the probability that the ratio of the sample variances exceeds 3.48 is still .05 (assuming sample sizes of $n_1 = 6$ and $n_2 = 10$).   ∽

This section has been devoted to developing the sampling distributions of various statistics calculated by using the observations in a random sample from a normal population (or independent random samples from two normal populations). In particular, if $Y_1, Y_2, \ldots, Y_n$ represents a random sample from a normal population with a mean of μ and a variance of σ^2, we have seen that $\sqrt{n}(\bar{Y} - \mu)/\sigma$ has a standard normal distribution. Also $(n - 1)S^2/\sigma^2$ has a χ^2 distribution, and $\sqrt{n}(\bar{Y} - \mu)/S$ has a t distribution (both with $n - 1$ degrees of freedom). If we have two independent random samples from normal populations with variances σ_1^2 and σ_2^2, $F = (S_1^2/\sigma_1^2)/(S_2^2/\sigma_2^2)$ has an F distribution. These sampling distributions will enable us to evaluate the properties of some inferential procedures discussed in the following chapters. In the next section we discuss approximations to certain sampling distributions, which can be very useful when the exact form of the sampling distribution is unknown or when probabilities acquired from the sampling distribution are difficult to compute.

Exercises

7.1 A forester studying the effects of fertilization on certain pine forests in the Southeast is interested in estimating the average basal area of pine trees. In studying basal areas of similar trees for many years, he has discovered these measurements (in square inches) to be normally distributed with a standard deviation of approximately 4 square inches. If the forester samples $n = 9$ trees, find the probability that the sample mean will be within 2 square inches of the population mean.

7.2 Suppose the forester in Exercise 7.1 would like the sample mean to be within 1 square inch of the population mean, with probability .90. How many trees must he measure in order to ensure this degree of accuracy?

7.3 The Environmental Protection Agency is concerned with the problem of setting criteria for the amount of certain toxic chemicals to be allowed in freshwater lakes and rivers. A common measure of toxicity for any pollutant is the concentration of the pollutant that will kill half of the test species in a given amount of time (usually 96 hours for fish species). This measure is called the LC50 (lethal concentration killing 50% of the test species). In many studies the ln(LC50) measurements are normally distributed and hence the analysis is based on ln(LC50) data.

Studies of the effects of copper on a certain species of fish (say, species A) show the variance of ln(LC50) measurements to be around .4 with concentration measurements in milligrams per liter. If $n = 10$ studies on LC50 for copper are to be completed, find the probability that the sample mean of ln(LC50) will differ from the true population mean by no more than .5.

7.4 If in Exercise 7.3 it is desired that the sample mean differ from the population mean by no more than .5, with probability .95, how many tests should be run?

7.5 Suppose that $X_1, \ldots, X_m$ and $Y_1, \ldots, Y_n$ are two independent random samples, with the X_i's normally distributed with a mean of μ_1 and a variance of σ_1^2 and the Y_i's normally distributed with a mean of μ_2 and a variance of σ_2^2. The difference between the sample means, $\bar{X} - \bar{Y}$, is then a linear combination of $m + n$ normal random variables and by Theorem 6.3 is itself normally distributed.

(a) Find $E(\bar{X} - \bar{Y})$.

(b) Find $V(\bar{X} - \bar{Y})$.

(c) Suppose $\sigma_1^2 = 2$, $\sigma_2^2 = 2.5$, and $m = n$. Find the sample sizes so that $(\bar{X} - \bar{Y})$ will be within one unit of $(\mu_1 - \mu_2)$ with probability .95.

7.6 Referring to Exercise 7.3, suppose that the effects of copper on a second species (say, species B) of fish show the variance of ln(LC50) measurements to be .8. If the population means of ln(LC50) for the two species are equal, find the probability that, with random samples of ten measurements from each species, the sample mean for species A exceeds the sample mean for species B by at least one unit.

7.7 Ammeters produced by a particular company are marketed under the specification that the standard deviation of gauge readings is no larger than .2 amp. One of these ammeters was used to make ten independent readings on a test circuit with constant current. If the sample variance of these ten measurements is .065, and it is reasonable to assume that the readings are normally distributed, do the results suggest that the ammeter used does not meet the marketing specifications? [Hint: Find the approximate probability that the sample variance will exceed .065 if the true population variance is .04.]

7.8 **(a)** If U has a χ^2 distribution with v degrees of freedom, find $E(U)$ and $V(U)$.

(b) Using the results of Theorem 7.3, find $E(S^2)$ and $V(S^2)$ when $Y_1, \ldots, Y_n$ is a random sample from a normal distribution with mean μ and variance σ^2.

7.9 Refer to Exercise 7.3. If $n = 20$ observations are to be taken on ln(LC50) measurements, with $\sigma^2 = 1.4$, find two numbers a and b such that $P(a \le S^2 \le b) = .95$, where S^2 is the sample variance of the 20 measurements.

7.10 Refer to Exercise 7.1. Suppose that in the forest fertilization problem the population standard deviation of basal areas is not known and must be estimated from the sample. If a random sample of $n = 9$ basal areas is to be measured, find two statistics g_1 and g_2 such that $P[g \le (\bar{Y} - \mu) \le g_2] = .90$.

7.11 If Y is a random variable that has an F distribution with v_1 numerator and v_2 denominator degrees of freedom, show that $U = 1/Y$ has an F distribution with v_2 numerator and v_1 denominator degrees of freedom.

*7.12 Let S_1^2 denote the sample variance for a random sample of ten ln(LC50) values for copper and let S_2^2 denote the sample variance for a random sample of eight ln(LC50) values for lead, both samples using the same species of fish. The population variance for measurements on copper is assumed to be twice the corresponding population variance for measurements on lead. Find two numbers a and b such that

$$P\left(a \le \frac{S_1^2}{S_2^2} \le b\right) = .90$$

assuming S_1^2 to be independent of S_2^2. [Hint: Use the result of Exercise 7.11 and note that $P(U_1/U_2 \le k) = P(U_2/U_1 \ge 1/k).$]

7.13 Let $Y_1, Y_2, \ldots, Y_5$ be a random sample of size 5 from a normal population with a mean of 0 and a variance of 1, and let $\bar{Y} = (1/5) \sum_{i=1}^{5} Y_i$. Let Y_6 be another independent observation from the same population.

* Exercises preceded by an asterisk are optional.

(a)   What is the distribution of $W = \sum_{i=1}^{5} Y_i^2$? Why?

(b)   What is the distribution of $U = \sum_{i=1}^{5} (Y_i - \bar{Y})^2$? Why?

(c)   What is the distribution of $\sum_{i=1}^{5} (Y_i - \bar{Y})^2 + Y_6^2$? Why?

7.14   Suppose $Y_1, Y_2, \ldots, Y_5, Y_6, \bar{Y}, W$, and U are as defined in Exercise 7.13.

(a)   What is the distribution of $\sqrt{5}Y_6/\sqrt{W}$? Why?

(b)   What is the distribution of $2Y_6/\sqrt{U}$? Why?

(c)   What is the distribution of $2(5\bar{Y}^2 + Y_6^2)/U$? Why?

7.3 The Central Limit Theorem

In Chapter 5 we showed that if $Y_1, Y_2, \ldots, Y_n$ represents a random sample from *any* distribution with a mean of μ and a variance of σ^2, then $E(\bar{Y}) = \mu$ and $V(\bar{Y}) = \sigma^2/n$. In this section we will develop an approximation for the sampling distribution of $\bar{Y}$ that can be used whatever the distribution of the population is from which the sample is taken.

If we sample from a normal population, then Theorem 7.1 tells us that $\bar{Y}$ has a sampling distribution that is normal. But what can we say about the sampling distribution of $\bar{Y}$ if the Y_i are not normally distributed? Fortunately, we will see that $\bar{Y}$ will have a sampling distribution that is approximately normal if the sample size is large. The formal statement of this result is called the *central limit theorem*. Before we state this theorem, however, we will look at some empirical investigations that demonstrate the sampling distribution of $\bar{Y}$.

Random samples of size n were generated by computer from an exponential density function with a mean of 10; that is, from

$$f(y) = \begin{cases} (1/10)e^{-y/10}, & y > 0 \\ 0, & \text{elsewhere} \end{cases}$$

A graph of this density function is given in Figure 7.4. The sample mean and sample variance were computed for each sample. The relative frequency

Figure 7.4
An exponential
density function

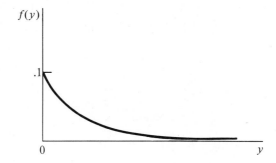

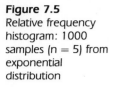

Figure 7.5
Relative frequency
histogram: 1000
samples (n = 5) from
exponential
distribution

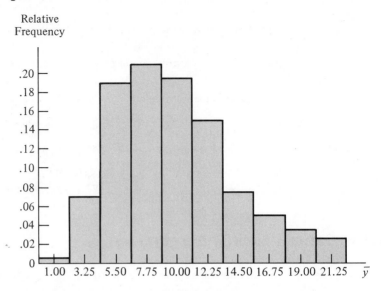

histogram of the mean values for 1000 samples of size $n = 5$ is shown in Figure 7.5. Notice that Figure 7.5 portrays a histogram that is roughly mound-shaped, but the histogram is not symmetric.

    Figure 7.6 is a graphical depiction of a similar relative frequency histogram of the mean values for 1000 samples of size $n = 25$. In this case Figure 7.6 shows a mound-shaped and nearly symmetric histogram, which could be approximated quite closely with a normal density function.

Figure 7.6
Relative frequency
histogram: 1000
samples (n = 25)
from exponential
distribution

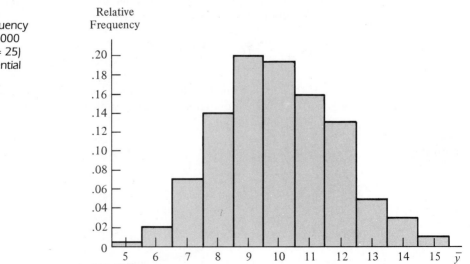

Recall from Chapter 5 that $E(\bar{Y}) = \mu$ and $V(\bar{Y}) = \sigma_{\bar{Y}}^2 = \sigma^2/n$. For the exponential density function used in the simulation, $\mu = E(Y_i) = 10$ and $\sigma^2 = V(Y_i) = (10)^2$. Thus we see that for this situation

$$E(\bar{Y}) = \mu = 10 \quad \text{and} \quad \sigma_{\bar{Y}}^2 = \frac{\sigma^2}{n} = \frac{(10)^2}{n}$$

For each value of n (5 and 25) we calculated the average of the 1000 sample means generated in the study. The observed variance of the 1000 sample means was also calculated for each value of n. The results are shown in Table 7.1.

Table 7.1
Calculations for 1000
sample means

Sample Size	Average of Sample Mean	Population Mean μ	Variance of Sample Means	$\sigma_{\bar{Y}}^2 = \sigma^2/n$
$n = 5$	9.86	10	19.63	20
$n = 25$	9.95	10	3.93	4

Notice that the empirical study gives average sample means and variances of sample means that are quite close to the theoretical expected values.

We now give a formal statement of the central limit theorem.

Theorem 7.4
The Central Limit
Theorem

Let $Y_1, Y_2, \ldots, Y_n$ be independent and identically distributed random variables with $E(Y_i) = \mu$ and $V(Y_i) = \sigma^2 < \infty$. Define

$$U_n = \sqrt{n}\left(\frac{\bar{Y} - \mu}{\sigma}\right) \quad \text{where} \quad \bar{Y} = \frac{1}{n}\sum_{i=1}^{n} Y_i$$

Then the distribution function of U_n converges to a standard normal distribution function as $n \to \infty$.

The conclusion of Theorem 7.4, stated in a slightly different fashion, says that

$$P(a \le U_n \le b) \to \int_a^b \left(\frac{1}{\sqrt{2\pi}}\right) e^{-u^2/2} \, du$$

as $n \to \infty$. That is, probability statements about U_n can be approximated by corresponding probabilities for the standard normal random variable if n is large. (Usually a value of n greater than 30 will ensure that the distribution of U_n can be closely approximated by a normal distribution.)

As a matter of convenience, the conclusion of the central limit theorem often is replaced with the simpler statement that $\bar{Y}$ is asymptotically normal with a mean of μ and a variance of σ^2/n. It is important to notice that the central limit theorem can be applied for a random sample $Y_1, Y_2, \ldots, Y_n$ from *any* distribution, so long as $E(Y_i) = \mu$ and $V(Y_i) = \sigma^2$ are both finite and the sample size is large.

We will give some examples of the use of the central limit theorem but will defer the proof (coverage is optional) until the next section. The proof is not needed for an understanding of the applications of the central limit theorem that appear in the remainder of the text.

EXAMPLE 7.7   Achievement test scores from all high school seniors in a certain state have a mean of 60 and a variance of 64. A specific high school class of $n = 100$ students had a mean score of 58. Is there evidence to suggest that this high school is inferior? (Calculate the probability that the sample mean is at most 58 when $n = 100$.)

Solution   Let $\bar{Y}$ denote the mean of a random sample of $n = 100$ scores from a population with $\mu = 60$ and $\sigma^2 = 64$. We want to approximate $P(\bar{Y} \le 58)$. We know from Theorem 7.4 that $\sqrt{n}(\bar{Y} - \mu)/\sigma$ is approximately a standard normal random variable, which we denote by Z. Hence

$$P(\bar{Y} \le 58) \approx P\left(Z \le \frac{58 - 60}{\sqrt{64/100}}\right) = P(Z \le -2.5) = .0062$$

using Table 4, Appendix III.

Because this probability is so small, it is unlikely that the specific class of interest can be regarded as a random sample from a population with $\mu = 60$ and $\sigma^2 = 64$. There is evidence to suggest that the average score for this high school is lower than the overall average of $\mu = 60$.   ∽

EXAMPLE 7.8   The service times for customers coming through a checkout counter in a retail store are independent random variables with a mean of 1.5 minutes and a variance of 1.0. Approximate the probability that 100 customers can be serviced in less than 2 hours of total service time.

Solution   If we let Y_i denote the service time for the ith customer, then we want

$$P\left(\sum_{i=1}^{100} Y_i \le 120\right) = P\left(\bar{Y} \le \frac{120}{100}\right) = P(\bar{Y} \le 1.20)$$

Because the sample size is large, the central limit theorem tells us that $\bar{Y}$ is approximately normally distributed with a mean of $\mu = 1.5$ and a variance of

$\sigma_{\bar{Y}}^2 = \sigma^2/n = 1.0/100$. Therefore

$$P(\bar{Y} \le 1.20) = P\left(\frac{\bar{Y} - 1.50}{1/\sqrt{100}} \le \frac{1.20 - 1.50}{1/\sqrt{100}}\right)$$

$$\approx P[Z \le (1.2 - 1.5)\sqrt{100}] = P(Z \le -3) = .0013$$

from Table 4, Appendix III.

Thus the probability that 100 customers can be serviced in less than 2 hours is approximately .0013. This small probability indicates that it is virtually impossible to service 100 customers in 2 hours of service time.   ∽

Exercises

7.15   An anthropologist wishes to estimate the average height of men for a certain race of people. If the population standard deviation is assumed to be 2.5 inches and if she randomly samples 100 men, find the probability that the difference between the sample mean and the true population mean will not exceed .5 inch.

7.16   Suppose that the anthropologist of Exercise 7.15 wants the difference between the sample mean and the population mean to be less than .4 inch, with probability .95. How many men should she sample to achieve this objective?

7.17   A large industry has an average wage of $4.00 per hour with a standard deviation of $.50. The industry has 64 workers of a certain ethnic group. These workers have an average wage of $3.90 per hour. Is it reasonable to assume that the ethnic group is a random sample of workers from the industry? (Calculate the probability of obtaining a sample mean less than or equal to $3.90 per hour.)

7.18   The acidity of soils is measured by a quantity called the pH, which may range from 0 (high acidity) to 14 (high alkalinity). A soil scientist wants to estimate the average pH for a large field by randomly selecting n core samples and measuring the pH in each sample. Although the population standard deviation of pH measurements is not known, past experience indicates that most soils have a pH value between 5 and 8. If the scientist selects $n = 40$ samples, find the approximate probability that the sample mean of the 40 pH measurements will be within .2 unit of the true average pH for the field. [Hint: See Exercise 1.11.]

7.19   Suppose the scientist of Exercise 7.18 would like the sample mean to be within .1 of the true mean with probability .90. How many core samples should the scientist take?

7.20   An important aspect of the 1981 federal economic plan was that consumers would save a substantial portion of the money that they received from an income tax reduction. Suppose that early estimates of the portion of total tax saved, based on a random sampling of thirty-five economists, had a mean of 26% and a standard deviation of 12%.

(a)   What is the approximate probability that a sample mean estimate, based on a random sample of $n = 35$ economists, will lie within 1% of the mean of the population of the estimates of all economists?

(b)   Is it necessarily true that the mean of the population of estimates of all economists is equal to the percent tax saving that will actually be achieved?

7.21   The length of time required for the periodic maintenance of an automobile, machine, etc., usually will have a probability distribution that is mound-shaped and, because some occasional long-service times will occur, be skewed to the right. Suppose that the length of time required to run a 5000 mile check and service for a new automobile has a mean equal to 1.4 hours and a standard deviation of .7 hour. Suppose that the service department plans to service fifty automobiles per 8-hour day and that, in order to do so, it must spend no more service time than an average of 1.6 hours per automobile. What proportion of all days will the service department have to work overtime?

7.22   Many bulk products, such as iron ore, coal, and raw sugar, are sampled for quality by a method that requires many small samples to be taken periodically as the material is moving along a conveyor belt. The small samples are then aggregated and mixed to form one composite sample. Let Y_i denote the volume of the ith small sample from a particular lot, and suppose $Y_1, Y_2, \ldots, Y_n$ constitutes a random sample, with each Y_i having mean μ (in cubic inches) and variance σ^2. The average volume of the samples, μ, can be set by adjusting the size of the sampling device. Suppose the variance of sampling volumes, σ^2, is known to be approximately 4 for a particular situation. It is required that the total volume of the composite sample exceed 200 cubic inches with probability approximately .95 when $n = 50$ small samples are selected. Determine a setting for μ that will allow the sampling requirements to be satisfied.

7.23   Twenty-five heat lamps are connected in a greenhouse so that when one lamp fails, another takes over immediately. (Only one lamp is turned on at any time.) The lamps operate independently, and each has a mean life of 50 hours and a standard deviation of 4 hours. If the greenhouse is not checked for 1300 hours after the lamp system is turned on, what is the probability that a lamp will be burning at the end of the 1300-hour period?

7.24   Suppose that $X_1, \ldots, X_n$ are independent random variables, each with a mean of μ_1 and a variance of σ_1^2. Suppose also that $Y_1, \ldots, Y_n$ are independent random variables, each with a mean of μ_2 and a variance of σ_2^2. Show that the random variable

$$U_n = \frac{(\bar{X} - \bar{Y}) - (\mu_1 - \mu_2)}{\sqrt{(\sigma_1^2 + \sigma_2^2)/n}}$$

satisfies the conditions of Theorem 7.4 and thus that the distribution function of U_n converges to a standard normal distribution function as $n \to \infty$.

7.25   An experiment is designed to test whether operator A or operator B gets the job of operating a new machine. Each operator is timed on fifty independent trials involving the performance of a certain task on the machine. If the sample means for the fifty trials differ by more than 1 second, the operator with the smaller mean time gets the job. Otherwise, the experiment is considered to end in a tie. If the standard deviations of times for both operators are assumed to be 2 seconds, what is the probability that operator A gets the job even though both operators have equal ability?

7.26   The result in Exercise 7.24 holds even if the sample sizes differ. That is, if $X_1, \ldots, X_{n_1}$ and $Y_1, \ldots, Y_{n_2}$ constitute independent random samples from populations with means of μ_1 and μ_2 and variances of σ_1^2 and σ_2^2, respectively, then $\bar{X} - \bar{Y}$ will be approximately normally distributed, for large n_1 and n_2, with a mean of $\mu_1 - \mu_2$ and a variance of $(\sigma_1^2/n_1) + (\sigma_2^2/n_2)$.

The flow of water through soil depends on, among other things, the porosity (volume proportion of voids) of the soil. For the comparison of two types of sandy soil,

$n_1 = 50$ measurements are to be taken on the porosity of soil A and $n_2 = 100$ measurements are to be taken on soil B. Assume that $\sigma_1^2 = .01$ and $\sigma_2^2 = .02$. Find the probability that the difference between the sample means will be within .05 unit of the difference between the population means, $\mu_1 - \mu_2$.

7.27   Refer to Exercise 7.26. Suppose $n_1 = n_2 = n$ and find the value of n that will allow the difference between the sample means to be within .04 unit of $\mu_1 - \mu_2$ with a probability of .90.

7.28   The times that a cashier spends processing each person's order are independent random variables with mean 1.5 minutes and standard deviation 1 minute. What is the approximate probability that the orders of 100 persons can be processed in less than 2 hours?

7.29   Refer to Exercise 7.28. Find the number of customers, n, such that the probability that the orders of all n customers can be processed in less than 2 hours is approximately 0.1.

7.4 A Proof of the Central Limit Theorem (Optional)

We will sketch a proof of the central limit theorem for the case in which the moment-generating function exists for the random variables in the sample. The proof depends upon a fundamental result of probability theory, which cannot be proved here but which is stated in Theorem 7.5.

Theorem 7.5

> Let Y_n and Y be random variables with moment-generating functions $m_n(t)$ and $m(t)$, respectively. If
>
> $$\lim_{n \to \infty} m_n(t) = m(t)$$
>
> for all real t, then the distribution function of Y_n converges to the distribution function of Y as $n \to \infty$.

Proof of the Central Limit Theorem (Theorem 7.4)   Define a random variable Z_i by

$$Z_i = \frac{Y_i - \mu}{\sigma}$$

Note that $E(Z_i) = 0$ and $V(Z_i) = 1$. The moment-generating function of Z_i, $m_Z(t)$, can be written as

$$m_Z(t) = 1 + \frac{t^2}{2} + \frac{t^3}{3!} E(Z_i^3) + \cdots$$

Further,

$$U_n = \sqrt{n}\left(\frac{\bar{Y} - \mu}{\sigma}\right) = \frac{1}{\sqrt{n}}\left(\frac{\sum_{i=1}^{n} Y_i - n\mu}{\sigma}\right) = \frac{1}{\sqrt{n}}\sum_{i=1}^{n} Z_i$$

Because the Y_i are independent, it follows that the Z_i are independent, $i = 1, 2, \ldots, n$.

Recall that the moment-generating function of the sum of independent random variables is the product of their individual moment-generating functions. Hence

$$m_n(t) = \left[m_Z\left(\frac{t}{\sqrt{n}}\right)\right]^n = \left(1 + \frac{t^2}{2n} + \frac{t^3}{3!n^{3/2}}k + \cdots\right)^n$$

where $k = E(Z_i^3)$.

Now take the limit of $m_n(t)$ as $n \to \infty$. One way to evaluate the limit is to consider log $m_n(t)$, where

$$\ln m_n(t) = n \ln\left[1 + \left(\frac{t^2}{2n} + \frac{t^3 k}{6n^{3/2}} + \cdots\right)\right]$$

A standard series expansion for $\ln(1 + x)$ is

$$\ln(1 + x) = x - \frac{x^2}{2} + \frac{x^3}{3} - \frac{x^4}{4} + \cdots$$

Letting

$$x = \left(\frac{t^2}{2n} + \frac{t^3 k}{6n^{3/2}} + \cdots\right)$$

we have

$$\ln m_n(t) = n \ln(1 + x) = n\left(x - \frac{x^2}{2} + \cdots\right)$$
$$= n\left[\left(\frac{t^2}{2n} + \frac{t^3 k}{6n^{3/2}} + \cdots\right) - \frac{1}{2}\left(\frac{t^2}{2n} + \frac{t^3 k}{6n^{3/2}} + \cdots\right)^2 + \cdots\right]$$

where the succeeding terms in the expansion involve x^3, x^4, and so on. Multiplying through by n, we see that the first term, $t^2/2$, does not involve n, whereas all other terms will have n to a positive power in the denominator. Thus it can be shown that

$$\lim_{n \to \infty} \ln m_n(t) = \frac{t^2}{2}$$

or

$$\lim_{n \to \infty} m_n(t) = e^{t^2/2}$$

the moment-generating function for a standard normal random variable. Applying Theorem 7.5, we conclude that U_n has a distribution function that converges to that of the standard normal random variable.

7.5 The Normal Approximation to the Binomial Distribution

The central limit theorem also can be used to approximate probabilities for some discrete random variables when the exact probabilities are difficult to calculate for large values of the sample size n. One useful example involves the binomial distribution.

Suppose that Y has a binomial distribution with n trials and a probability of success on any one trial denoted by p. If we want to find $P(Y \leq b)$, then we can use the binomial probability function to compute the probability for each nonnegative integer less than or equal to b and sum these probabilities. Tables are available for some values of the sample size n, but direct calculation is cumbersome for large values of n for which tables are unavailable.

Alternatively, we can view Y, the number of successes in n trials, as a sum of a sample consisting of 0s and 1s; that is,

$$Y = \sum_{i=1}^{n} X_i$$

where

$$X_i = \begin{cases} 1, & \text{if the } i\text{th trial results in success} \\ 0, & \text{otherwise} \end{cases}$$

and X_i, $i = 1, 2, \ldots, n$, are independent (because the trials are independent). Consequently, when n is large, the sample fraction of successes,

$$\frac{Y}{n} = \frac{1}{n} \sum_{i=1}^{n} X_i = \bar{X}$$

will possess approximately a normal sampling distribution with mean $E(X_i) = p$ and variance equal to $V(X_i)/n$; that is, $p(1 - p)/n$.

EXAMPLE 7.9

Candidate A believes that he can win a city election if he can poll at least 55% of the votes in precinct I. He also believes that about 50% of the city's voters favor him. If $n = 100$ voters show up to vote at precinct I, what is the probability that candidate A receives at least 55% of the votes?

Solution

Let Y denote the number of voters at precinct I who vote for candidate A. We must approximate $P(Y/n \geq .55)$, when p, the probability that a randomly selected voter favors candidate A, is .5. If we think of the $n = 100$ voters at precinct I as a random sample from the city, then Y has a binomial distribution with $p = .5$. We have seen that the fraction of voters is

$$\frac{Y}{n} = \frac{1}{n} \sum_{i=1}^{n} X_i$$

where $X_i = 1$ if the ith voter favors candidate A and $X_i = 0$ otherwise.

Because Y/n satisfies the central limit theorem, we know that Y/n is approximately normally distributed with mean $p = .5$ and variance $pq/n = (.5)(.5)/100 = .0025$. Therefore,

$$P\left(\frac{Y}{n} \geq .55\right) = P\left(\frac{Y/n - .5}{\sqrt{.0025}} \geq \frac{.55 - .50}{.05}\right) \approx P(Z \geq 1) = .1587$$

from Table 4, Appendix III.

The normal approximation to binomial probabilities works well for even moderately large n, so long as p is not close to zero or one. A useful rule of thumb is that the normal approximation to the binomial distribution is appropriate when $p \pm 2\sqrt{pq/n}$ lies in the interval $(0, 1)$. In these cases substantial improvement in the approximation can be made by a slight adjustment of the boundaries. If we look at the segment of a binomial distribution graphed in Figure 7.7, we can see what happens when we try to approximate a discrete distribution represented by a histogram with a continuous density function. If we want to find $P(Y \leq 3)$ by using the binomial distribution, we could find the total area in the four rectangles (above 0, 1, 2, 3) illustrated in the binomial histogram (Figure 7.7). Notice that the total area in the rectangles can be approximated by an area under the normal curve. The area under the curve includes some areas not in the histogram and excludes the portion of the histogram that lies above the curve. If we want to approximate $P(Y \leq 3)$ by calculating an area under the density function, you can see that the area under the density function to the left of 3.5 will provide a better approximation than

Figure 7.7
The normal
approximation to the
binomial distribution:
$n = 10$ and $p = .5$

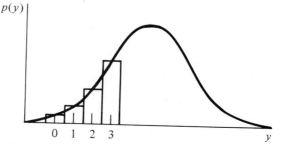

the area to the left of 3.0. The following example illustrates how close the normal approximation is for a case in which some exact binomial probabilities can be found.

EXAMPLE 7.10 Suppose Y has a binomial distribution with $n = 25$ and $p = .4$. Find the exact probabilities that $Y \le 8$ and $Y = 8$, and compare these to the corresponding values found from the normal approximation.

Solution From Table 1, Appendix III, we have that

$$P(Y \le 8) = .274$$

and

$$P(Y = 8) = P(Y \le 8) - P(Y \le 7) = .274 - .154 = .120$$

For the normal approximation we treat Y/n as having approximately the same distribution as U, where U is normally distributed with a mean of p and a variance of $p(1 - p)/n$. Equivalently, we can think of Y as having approximately the same distribution as W, where W is normally distributed with a mean of np and a variance of $np(1 - p)$. Because we want $P(Y \le 8)$, we look at the normal curve area to the left of 8.5. Thus

$$P(Y \le 8) \approx P(W \le 8.5) = P\left(\frac{W - np}{\sqrt{np(1 - p)}} \le \frac{8.5 - 10}{\sqrt{25(.4)(.6)}} \right)$$
$$= P(Z \le -.61) = .2709$$

from Table 4, Appendix III. Note that the approximate value is close to the exact value for $P(Y \le 8) = .274$, which we obtained from the binomial tables.

To find the normal approximation to the binomial probability, $p(8)$, we will find the area under the normal curve between the points 7.5 and 8.5, because this is the interval included in the histogram bar over $y = 8$ (see Figure 7.8).

Figure 7.8
P(Y = 8) for binomial
distribution of
Example 7.10

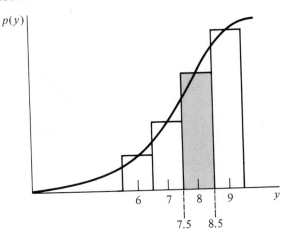

Because Y has approximately the same distribution as W, where W is normally distributed with a mean of $np = 25(.4) = 10$ and a variance of $np(1 - p) = 25(.4)(.6) = 6$, then

$$P(Y = 8) \approx P(7.5 \le W \le 8.5)$$

$$= P\left(\frac{7.5 - 10}{\sqrt{6}} \le \frac{W - 10}{\sqrt{6}} \le \frac{8.5 - 10}{\sqrt{6}}\right)$$

$$= P(-1.02 \le Z \le -.61) = .2709 - .1539 = .1170$$

Again we see that this approximate value is very close to the actual value, $P(Y = 8) = .120$, calculated earlier.

Exercises

7.30 A machine is shut down for repairs if a random sample of 100 items selected from the daily output of the machine reveals at least 15% defectives. (Assume that the daily output is a large number of items.) If the machine, in fact, is producing only 10% defective items, find the probability that it will be shut down on a given day. (Use the continuity correction.)

7.31 An airline finds that 5 percent of the persons making reservations on a certain flight will not show up for the flight. If the airline sells 160 tickets for a flight with only 155 seats, what is the probability that a seat will be available for every person holding a reservation and planning to fly?

7.32 According to a survey reported at the 1981 convention of the American Bar Association, there was one lawyer in every 700 Americans in 1960, one in every 600 in 1970, and one in every 410 Americans in 1981 (*The Philadelphia Inquirer*, August 18, 1981). One in every 64 residents in Washington, D.C., is a lawyer.

(a) If you were to select a random sample of 1500 Americans, what is the approximate probability that the sample would contain at least one lawyer?

(b) If the sample is selected from among the residents of Washington. D.C., what is the approximate probability that the sample would contain more than thirty lawyers?

(c) If you stood on a Washington, D.C., street corner and interviewed the first 1000 persons who walked by and 30 were lawyers, would this suggest that the density of lawyers passing the corner exceeds the density within the city? Explain.

7.33 A pollster believes that 20% of the voters in a certain area favor a bond issue. If 64 voters are randomly sampled from the large number of voters in this area, approximate the probability that the sampled fraction of voters favoring the bond issue will not differ from the true fraction by more than .06.

7.34 **(a)** Show that the variance of Y/n, where Y has a binomial distribution with n trials and a success probability of p, has a maximum at $p = .5$, for fixed n.

(b) A random sample of n items is to be selected from a large lot and the number of defectives, Y, is to be observed. Find a value of n that will guarantee that Y/n will be within .1 of the true fraction of defectives, with probability .95.

7.35 The manager of a supermarket wants to obtain information about the proportion of customers who dislike a new policy on cashing checks. How many customers should he sample if he wants the sample fraction to be within .15 of the true fraction, with probability .98?

7.36 If the supermarket manager (Exercise 7.35) actually samples $n = 50$ customers and if the true fraction of customers who dislike the policy is approximately .9, find the probability that the sample fraction will be within .15 unit of the true fraction.

7.37 Suppose that a random sample of twenty-five items is selected from the machine of Exercise 7.30. If the machine produces 10% defectives, find the probability that the sample will contain at least two defectives in the following ways:

(a) by the normal approximation to the binomial,

(b) by the exact binomial tables.

7.38 Just as the difference between two sample means will tend to be normally distributed for large samples, so will the difference between two sample proportions. That is, if Y_1 and Y_2 are independent binomial random variables with parameters (n_1, p_1) and (n_2, p_2), respectively, then $(Y_1/n_1) - (Y_2/n_2)$ will be approximately normally distributed for large values of n_1 and n_2.

(a) Find $E\left(\dfrac{Y_1}{n_1} - \dfrac{Y_2}{n_2}\right)$.

(b) Find $V\left(\dfrac{Y_1}{n_1} - \dfrac{Y_2}{n_2}\right)$.

7.39 As a check on the relative abundance of certain species of fish in two lakes, $n = 50$ observations are taken on net trap results for each lake. For each observation the experimenter merely records whether or not the desired species was present. Past experience has shown that this species will appear in lake A traps approximately 10% of the time and in lake B traps approximately 20% of the time. Use these results to approximate the probability that the difference between the sample proportions will be within .1 of the difference between the true proportions.

7.40   An auditor samples 100 of a firm's travel vouchers to ascertain how many of the vouchers are improperly documented. What is the approximate probability that more than 30% of the sampled vouchers are improperly documented if, in fact, only 20% of all the vouchers are improperly documented. If you were the auditor and observed more than 30% with improper documentation, what would you conclude about the firm's claim that only 20% suffered from improper documentation? Why?

7.41   The times to process orders at the service counter of a pharmacy are exponentially distributed with a mean of 10 minutes. If 100 customers visit the counter in a 2-day period, what is the probability that at least half of them need to wait more than 10 minutes?

7.6 Summary

To make inferences about population parameters, we need to know the probability distributions for certain functions (statistics) of the observable random variables in the sample (or samples). These probability distributions provide models for the relative frequency behavior of the statistics in repeated sampling and consequently are referred to as *sampling distributions*. We have seen that the normal, χ^2, t, and F distributions provide models for the sampling distributions of statistics used to make inferences about the parameters of normal distributions.

When the sample size is large, the sample mean $\bar{Y}$ will possess approximately a normal distribution if the random sample is taken from *any* distribution with a finite mean of μ and a finite variance of σ^2. This result, known as the *central limit theorem*, also provides the justification for approximating the binomial distribution with a corresponding normal distribution.

The sampling distributions developed in this chapter will be utilized in the inference-making procedures presented in subsequent chapters.

References and Further Readings

1.   Hoel, P. G. *Introduction to Mathematical Statistics*. 5th ed. New York: Wiley, 1984.

2.   Hogg, R. V., and Craig, A. T. *Introduction to Mathematical Statistics*. 4th ed. New York: Macmillan, 1978.

3.   Mood, A. M.; Graybill, F. A.; and Boes, D. *Introduction to the Theory of Statistics*. 3d ed. New York: McGraw-Hill, 1974.

4.   Parzen, E. *Modern Probability Theory and Its Applications*. New York: Wiley, 1960.

Supplementary Exercises

7.42   The efficiency (in lumens per watt) of light bulbs of a certain type have a population mean of 9.5 and standard deviation of .5, according to production specifications. The specifications for a room in which eight of these bulbs are to be installed call for the average efficiency of the eight bulbs to exceed 10. Find the probability that this specification for the room will be met, assuming efficiency measurements are normally distributed.

7.43 Refer to Exercise 7.42. What should the mean efficiency per bulb equal if the specification for the room is to be met with probability approximately .80? (Assume the variance of efficiency measurements remains at .5.)

7.44 Briggs and King developed the technique of nuclear transplantation in which a nucleus of a cell from one of the later stages of the development of an embryo is transplanted into a zygote (a single cell, fertilized egg) to see if the nucleus can support normal development. If the probability that a single transplant from the early gastrula stage will be successful is .65, what is the probability that more than 70 transplants out of 100 will be successful?

7.45 A retail dealer sells three brands of automobiles. For brand A her profit per sale, X, is normally distributed with parameters (μ_1, σ_1^2); for brand B her profit per sale, Y, is normally distributed with parameters (μ_2, σ_2^2); for brand C her profit per sale, W, is normally distributed with parameters (μ_3, σ_3^2). For the year two-fifths of the dealer's sales are in brand A, one-fifth in brand B, and the remaining two-fifths in brand C. If you are given data on profits for n_1, n_2, and n_3 sales of brands A, B, and C, respectively, the quantity $U = .4\bar{X} + .2\bar{Y} + .4\bar{W}$ would provide an approximation to the true average profit per sale for the year. Find the mean, variance, and probability density function for U. Assume X, Y, and W are independent.

7.46 From each of two normal populations with identical means and with standard deviations of 6.40 and 7.20, independent random samples of sixty-four observations are drawn. Find the probability that the difference between the means of the samples exceeds .6 in absolute value.

7.47 If Y has an exponential distribution with a mean of θ, show that $U = 2Y/\theta$ has a χ^2 distribution with 2 degrees of freedom.

7.48 A plant supervisor is interested in budgeting weekly repair costs for a certain type of machine. Records over the past years indicate that these repair costs have an exponential distribution with a mean of 20 for each machine studied. Let $Y_1, Y_2, \ldots, Y_5$ denote the repair costs for five of these machines for the next week. Find a number c such that $P\left(\sum_{i=1}^{5} Y_i > c\right) = .05$, assuming the machines operate independently.

7.49 Suppose T has a t distribution with v degrees of freedom. Show that T^2 has an F distribution with 1 and v numerator and denominator degrees of freedom, respectively.

7.50 The *coefficient of variation* for a sample of values $Y_1, Y_2, \ldots, Y_n$ is defined by

$$\text{C.V.} = \frac{S}{\bar{Y}}$$

This term gives the standard deviation as a proportion of the mean, and it is sometimes an informative quantity. For example, a value of $S = 10$ has little meaning unless we can compare it to something else. If S is observed to be 10 and $\bar{Y}$ is observed to be 1000, then the amount of variation is small relative to the mean. However, if S is observed to be 10 and $\bar{Y}$ is 5, then the variation is quite large relative to the mean. If we were studying the precision (variation in repeated measurements) of a measuring instrument, the first case (C.V. = 10/1000) might give quite acceptable precision but the second case (C.V. = 2) would be quite unacceptable.

Let $Y_1, \ldots, Y_{10}$ denote a random sample of size 10 from a normal distribution with mean 0 and variance σ^2. Use the following steps to find the number c such that

$$P\left(-c \le \frac{S}{\bar{Y}} \le c\right) = .95$$

(a)   Use the result of Exercise 7.49 to find the distribution of $(10)\bar{Y}^2/S^2$.

(b)   Find the distribution of $S^2/(10)\bar{Y}^2$.

(c)   Use the answer to part (b) in finding the constant c.

*7.51   If U_1/U_2 has an F distribution with v_1 and v_2 degrees of freedom in the numerator and denominator, respectively, find $E(U_1/U_2)$.

7.52   Suppose that $Y_1, Y_2, \ldots, Y_{40}$ denotes a random sample of measurements on the proportion of impurities in iron ore samples. Let each Y_i have a probability density function given by

$$f(y) = \begin{cases} 3y^2, & 0 \le y \le 1 \\ 0, & \text{elsewhere} \end{cases}$$

The ore is to be rejected by the potential buyer if $\bar{Y}$ exceeds .7. Find $P(\bar{Y} > .7)$ for the sample of size 40.

*7.53   If Y has a χ^2 distribution with n degrees of freedom, then Y could be represented by

$$Y = \sum_{i=1}^{n} X_i$$

where X_i's are independent, each having a χ^2 distribution with 1 degree of freedom.

(a)   Show that $Z = (Y - n)/\sqrt{2n}$ has an asymptotic standard normal distribution.

(b)   A machine in a heavy-equipment factory produces steel rods of length Y, where Y is a normal random variable with a mean μ of 6 inches and a variance of .2. The cost C of repairing a rod that is not exactly 6 inches in length is proportional to the square of the error and is given, in dollars, by $C = 4(Y - \mu)^2$. If 50 rods with independent lengths are produced in a given day, approximate the probability that the total cost for repairs for that day exceeds $48.

*7.54   Suppose that T is defined as in Definition 7.2.

(a)   If χ^2 is fixed at v, then T is given by Z/c, where $c = \sqrt{v/n}$. Use this idea to find the conditional density of T for a fixed $\chi^2 = v$.

(b)   Find the joint density of T and χ^2, $f(t, v)$, by using $f(t, v) = f(t|v)f(v)$.

(c)   Integrate over v to show that

$$f(t) = \left\{\frac{\Gamma[(n+1)/2]}{\sqrt{\pi n}\,\Gamma(n/2)}\right\}\left(1 + \frac{t^2}{n}\right)^{-(n+1)/2}, \qquad -\infty < t < \infty$$

*7.55 Suppose F is defined as in Definition 7.3.

    (a)   If χ_2^2 is fixed at w, then $F = \chi_1^2/c$, where $c = wn_1/n_2$. Find the conditional density of F for fixed $\chi_2^2 = w$.

    (b)   Find the joint density of F and χ_2^2.

    (c)   Integrate over w to show that the probability density function of F, $g(f)$, is given by

$$g(f) = \frac{\Gamma[(n_1 + n_2)/2](n_1/n_2)^{n_1/2}}{\Gamma(n_1/2)\Gamma(n_2/2)} (f)^{[(n_1/2) - 1]}\left(1 + \frac{n_1 f}{n_2}\right)^{-(n_1 + n_2)/2}, \qquad 0 < f < \infty$$

*7.56 Let X have a Poisson distribution with parameter λ.

    (a)   Show that the moment-generating function of $Y = (X - \lambda)/\sqrt{\lambda}$ is given by

$$m(t) = \exp(\lambda e^{t/\sqrt{\lambda}} - \sqrt{\lambda}t - \lambda)$$

    (b)   Use the expansion

$$e^{t/\sqrt{\lambda}} = \sum_{i=0}^{\infty} \frac{[t/\sqrt{\lambda}]^i}{i!}$$

    to show that

$$\lim_{\lambda \to \infty} m(t) = e^{t^2/2}$$

    (c)   Use Theorem 7.5 to show that the distribution function of Y converges to a standard normal distribution function as $\lambda \to \infty$.

*7.57 In the interest of pollution control an experimenter wants to count the number of bacteria per small volume of water. Let X denote the bacteria count per cubic centimeter of water and assume that X has a Poisson probability distribution with mean $\lambda = 100$. If the allowable pollution in a water supply is a count of 110 per cubic centimeter, approximate the probability that X will be at most 110. [Hint: Use the result in part (c) of Exercise 7.56.]

*7.58 The number of accidents per year at a given intersection, Y, is assumed to have a Poisson distribution. Over the past few years there has been an average of thirty-six accidents per year at this intersection. If the number of accidents per year is at least forty-five, an intersection can qualify to be rebuilt under an emergency program set up by the state. Approximate the probability that the intersection in question will come under the emergency program at the end of next year.

*7.59 An experimenter is comparing two methods of removing bacteria colonies from processed luncheon meats. After treating some samples by method A and other identical samples by method B, a 2-cm^3 subsample is selected from each sample. Bacteria colony counts are made on these subsamples. Let X denote the count for the subsample treated by A and let Y denote the count for the one treated by B. Assume X and Y are independent Poisson random variables with means λ_1 and λ_2, respectively. If X exceeds Y by more than 10, method B will be judged superior to A. Suppose that, in fact, $\lambda_1 = \lambda_2 = 50$. Find the approximate probability that B will be judged superior to A.

*7.60   Let Y_n be a binomial random variable with n trials and succes probability p. Suppose that n tends to infinity and p tends to zero in such a way that np remains fixed so that $np = \lambda$. Use the result in Theorem 7.5 to prove that the distribution of Y_n converges to a Poisson distribution with a mean of λ.

7.61   If the probability that a person will suffer an adverse reaction from a medication is .001, use the result of Exercise 7.60 to approximate the probability that 2 or more will suffer an adverse reaction if the medication is administered to 1000 individuals.

ESTIMATION

8.1 Introduction

As stated in Chapter 1, the objective of statistics is to make an inference about a population based on information contained in a sample. Because populations are characterized by numerical descriptive measures called *parameters*, the objective of many statistical investigations is to make an inference about one or more population parameters. Most statistical inference procedures involve either estimation or hypothesis testing. In this chapter we consider the general topic of estimation of population parameters. As you will subsequently see, the sampling distributions derived in Chapter 7 will play an important role in the development of our estimation procedures.

Estimation has many practical applications. For example, a manufacturer of washing machines might be interested in estimating the proportion, p, of washers that would fail prior to the expiration of a 1-year guarantee time. Other important population parameters are the population mean, variance, and standard deviation. For example, we might wish to estimate the mean waiting time μ at a supermarket checkout station or the standard deviation of the error of measurement, σ, of an electronic instrument. To simplify our terminology, we will call the parameter of interest in the experiment the *target parameter*.

Suppose that we wish to estimate the average amount of mercury, μ, that a particular process can remove from 1 ounce of ore obtained at a particular geographic location. We could give our estimate in two distinct forms. A single number, for instance, .13 ounce, could be given. Our intention is that this number should be close to the unknown population mean μ. This type of estimate is called a *point estimate* because a single value, or point, is given as the estimate of μ. On the other hand, we might say that μ will fall between two numbers, for example, between .07 and .19 ounce. In the second type of estimation procedure we give two values that may be used to construct an interval (.07, .19), which is intended to enclose the parameter of interest. This second type of estimation, wherein an interval of possible values of μ is specified, is called an *interval estimate*.

A point estimation procedure utilizes the information in the sample to arrive at a single number or point that estimates the target parameter. An interval estimation procedure makes use of the sample information to arrive at two numbers that are intended to enclose the parameter of interest. In either case the actual estimation is accomplished by an *estimator*, which is a rule that tells us how to employ the sample data to determine the value (or values) that we use as our point (or interval) estimate.

Definition 8.1

> An *estimator* is a rule that tells how to calculate an estimate based on the measurements contained in a sample.

Most frequently an estimator is expressed as a formula. For example, the sample mean

$$\bar{Y} = \frac{1}{n} \sum_{i=1}^{n} Y_i$$

is one possible point estimator of the population mean μ. Clearly the expression for $\bar{Y}$ is both a rule and a formula. It tells us to sum the sample observations and divide by the sample size n.

An experimenter who wants an interval estimate of a parameter must use the sample data to calculate two points. It is hoped that the interval formed by the two points will possess a high probability of including the target parameter. Examples of interval estimators will be given in subsequent sections.

It is possible to obtain many different estimators (rules for estimating) for the same population parameter. This should not be surprising. Ten engineers, each assigned to estimate the cost of a large construction job, would most likely arrive at different estimates of the total cost. Such engineers, called *estimators* in the construction industry, use certain fixed guidelines plus intuition to achieve their estimates. Each represents a unique human subjective rule for obtaining a single estimate. This brings us to a most important point: some estimators are considered *good*, some *bad*. How would the management of a construction firm define *good* and *bad* as they relate to the estimation of the cost of a job—and how would we establish a criterion of goodness to compare one statistical estimator with another? We will discuss this idea in the following sections.

8.2 Some Properties of Point Estimators

An investigation of the reasoning used in calculating the goodness of a point estimator is facilitated by considering an analogy. Point estimation is similar, in many respects, to firing a revolver at a target. The estimator, generating estimates, is analogous to the revolver, a particular estimate to the bullet, and the parameter of interest to the bull's-eye. Drawing a sample from the population and estimating the value of the parameter is equivalent to firing a single shot at the target.

Suppose that a man fires a single shot at a target and that the shot pierces the bull's-eye. Do we conclude that he is an excellent shot? Would you want to hold the target while a second shot is fired? Obviously, we would not decide that the man is an expert marksman based on such a small amount of evidence. On the other hand, if 1 million shots in succession hit the bull's-eye, we might acquire sufficient confidence in the marksman to hold the target for the next shot, if the compensation was adequate. The point we wish to make is clear. We cannot evaluate the goodness of a point estimation procedure on the basis of a single estimate; rather, we must observe the results when the estimation procedure is used many, many times. Because the estimates are numbers, we would evaluate the goodness of the point estimator by constructing a frequency distribution of the estimates obtained in repeated sampling and note how closely the distribution clusters about the parameter of interest.

Suppose that we wish to specify a point estimate for a population parameter that we will call θ. The estimator of θ will be indicated by the symbol $\hat{\theta}$, where the "hat" indicates that we are estimating the parameter immediately beneath. With the revolver-firing example in mind, we see that the desirable properties of a good estimator are quite clear. We would like the distribution of estimates, or, more properly, the sampling distribution of the estimator, to center about the target parameter as shown in Figure 8.1. In other words, we would like the mean or expected value of the distribution of estimates to equal

Figure 8.1
A distribution of estimates

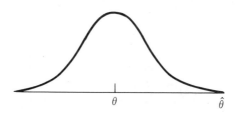

the parameter estimated; that is, $E(\hat{\theta}) = \theta$. Point estimators that satisfy this property are said to be *unbiased*. The sampling distribution for a positively biased point estimator, one for which $E(\hat{\theta}) > \theta$, is shown in Figure 8.2.

Figure 8.2
Sampling distribution
for a positively biased
estimator

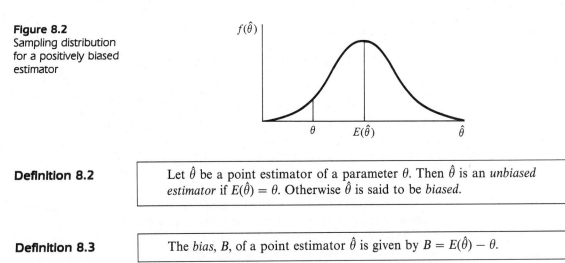

Definition 8.2

Let $\hat{\theta}$ be a point estimator of a parameter θ. Then $\hat{\theta}$ is an *unbiased estimator* if $E(\hat{\theta}) = \theta$. Otherwise $\hat{\theta}$ is said to be *biased*.

Definition 8.3

The *bias*, B, of a point estimator $\hat{\theta}$ is given by $B = E(\hat{\theta}) - \theta$.

Figure 8.3 shows two possible sampling distributions for unbiased point estimators for a target parameter θ. We would prefer that our estimator have the type of distribution indicated in Figure 8.3(b) because the smaller variance guarantees that in repeated sampling a higher fraction of values of $\hat{\theta}_2$ will be "close" to θ. Thus in addition to unbiasedness, we like the spread of a distribution of estimates to be as small as possible. That is, we want $V(\hat{\theta})$ to be a minimum. Given two unbiased estimators of a parameter θ, and all other things being equal, we would select the estimator with the smaller variance.

Figure 8.3
Sampling distributions
for two unbiased
estimators:
(a) estimator with
large variation;
(b) estimator with
small variation

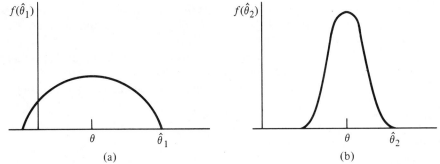

Rather than use the bias and variance to characterize the goodness of a point estimator, we might employ the expected value of $(\hat{\theta} - \theta)^2$, the square of the distance between $\hat{\theta}$ and its target parameter.

Definition 8.4

> The *mean square error* of a point estimator $\hat{\theta}$ is defined to be the expected value of $(\hat{\theta} - \theta)^2$.

The mean square error of an estimator $\hat{\theta}$, denoted by the symbol $\text{MSE}(\hat{\theta})$, is a function of both its variance and its bias.

It can be shown that

$$\text{MSE}(\hat{\theta}) = V(\hat{\theta}) + B^2$$

We will leave the proof of this result to Exercise 8.1.

In this section we have defined properties for point estimations that are sometimes desirable. In particular, we often seek unbiased estimators with relatively small variances. In the next section we will consider some common and useful unbiased point estimators.

Exercises

8.1   Using the identity

$$(\hat{\theta} - \theta) = [\hat{\theta} - E(\hat{\theta})] + [E(\hat{\theta}) - \theta] = [\hat{\theta} - E(\hat{\theta})] + B$$

show that

$$\text{MSE}(\hat{\theta}) = E[(\hat{\theta} - \theta)^2] = V(\hat{\theta}) + B^2$$

8.2   Suppose that $E(\hat{\theta}_1) = E(\hat{\theta}_2) = \theta$, $V(\hat{\theta}_1) = \sigma_1^2$ and $V(\hat{\theta}_2) = \sigma_2^2$. A new unbiased estimator, $\hat{\theta}_3$, is to be formed by

$$\hat{\theta}_3 = a\hat{\theta}_1 + (1 - a)\hat{\theta}_2$$

How should the constant a be chosen in order to minimize the variance of $\hat{\theta}_3$? Assume that $\hat{\theta}_1$ and $\hat{\theta}_2$ are independent.

8.3   Consider the situation described in Exercise 8.2. How should the constant a be chosen to minimize the variance of $\hat{\theta}_3$ if $\hat{\theta}_1$ and $\hat{\theta}_2$ are not independent but are such that $\text{cov}(\hat{\theta}_1, \hat{\theta}_2) = c \neq 0$.

8.4   Suppose Y_1, Y_2, Y_3 denotes a random sample from an exponential distribution with density function

$$f(y) = \begin{cases} \left(\dfrac{1}{\theta}\right)e^{-y/\theta}, & y > 0 \\ 0, & \text{elsewhere} \end{cases}$$

Consider the following five estimators of θ:

$$\hat{\theta}_1 = Y_1 \qquad \hat{\theta}_2 = \frac{Y_1 + Y_2}{2} \qquad \hat{\theta}_3 = \frac{Y_1 + 2Y_2}{3} \qquad \hat{\theta}_4 = \min(Y_1, Y_2, Y_3) \qquad \hat{\theta}_5 = \bar{Y}$$

(a)   Which estimators are unbiased?

(b)   Among the unbiased estimators, which has smallest variance?

8.5   Suppose $Y_1, Y_2, \ldots, Y_n$ is a random sample from a population with probability density function

$$f(y) = \begin{cases} \left(\dfrac{1}{\theta + 1}\right) e^{-y/(\theta + 1)}, & y > 0; \theta > -1 \\ 0, & \text{elsewhere} \end{cases}$$

Suggest a suitable statistic to be used as an unbiased estimator for θ. (Hint: Use $\bar{Y}$.)

8.6   The number of breakdowns per week for a certain type of minicomputer is a random variable Y having the Poisson distribution with mean λ. A random sample $Y_1, Y_2, \ldots, Y_n$ of observations on the weekly number of breakdowns is available.

(a)   Suggest an unbiased estimator for λ.

(b)   The weekly cost of repairing these breakdowns is $C = 3Y + Y^2$. Show that $E(C) = 4\lambda + \lambda^2$.

(c)   Find a function of $Y_1, Y_2, \ldots, Y_n$ that is an unbiased estimator of $E(C)$.

8.7   Let $Y_1, Y_2, \ldots, Y_n$ denote a random sample of size n from a population with a mean of 3. Assume that $\hat{\theta}_2$ is an unbiased estimator of the $E(Y^2)$ and $\hat{\theta}_3$ is an unbiased estimator of $E(Y^3)$. Give an unbiased estimator for the third central moment of the underlying distribution.

8.8   The reading on a voltage meter connected to a test circuit is uniformly distributed over the interval $(0, \theta + 1)$, where θ is the true but unknown voltage of the circuit. Suppose that $Y_1, Y_2, \ldots, Y_n$ denotes a random sample of such readings.

(a)   Show that $\bar{Y}$ is a biased estimator of θ, and compute the bias.

(b)   Find a function of $\bar{Y}$ that is an unbiased estimator of θ.

(c)   Find $\text{MSE}(\bar{Y})$ when $\bar{Y}$ is used as an estimator of θ.

8.9   If Y has a binomial distribution with parameters n and p, then we have seen that Y/n is an unbiased estimator of p. To estimate the variance of Y, we generally use $n(Y/n)(1 - Y/n)$.

(a)   Show that the suggested estimator is a biased estimator of $V(Y)$.

(b)   Modify $n(Y/n)(1 - Y/n)$ slightly to form an unbiased estimator of $V(Y)$.

8.10   Let $Y_1, Y_2, \ldots, Y_n$ denote a random sample of size n from a population with a density given by

$$f(y) = \begin{cases} \alpha y^{\alpha - 1}/\theta^{\alpha}, & 0 \leq y \leq \theta \\ 0, & \text{elsewhere} \end{cases}$$

where $\alpha > 0$ is a known fixed value but θ is unknown. (Note that this is the power family distribution introduced in Exercise 6.13.) Consider the estimator $\hat{\theta} = \max(Y_1, Y_2, \ldots, Y_n)$.

(a) Show that $\hat{\theta}$ is a biased estimator for θ.

(b) Find a multiple of $\hat{\theta}$ that is an unbiased estimator of θ.

(c) Derive MSE($\hat{\theta}$).

8.11 Let $Y_1, Y_2, \ldots, Y_n$ denote a random sample of size n from a population with a density given by

$$f(y) = \begin{cases} 3\beta^3 y^{-4}, & \beta \le y \\ 0, & \text{elsewhere} \end{cases}$$

where $\beta > 0$ is unknown. (Note that this is one of the Pareto family of distributions introduced in Exercise 6.14.) Consider the estimator $\hat{\theta} = \min(Y_1, Y_2, \ldots, Y_n)$.

(a) Derive the bias of the estimator $\hat{\theta}$.

(b) Derive MSE($\hat{\theta}$).

*8.12 Suppose $Y_1, Y_2, \ldots, Y_n$ is a random sample from a normal distribution with parameters μ and σ^2.

(a) Show that $S = \sqrt{S^2}$ is a biased estimator of σ. [Hint: Recall the distribution of $(n-1)S^2/\sigma^2$.]

(b) Adjust S to form an unbiased estimator of σ.

(c) Find an unbiased estimator of $\mu - z_\alpha \sigma$, the point that cuts off a lower-tail area of α under this normal curve.

8.13 If Y has a binomial distribution with parameters n and p, then $\hat{p}_1 = Y/n$ is an unbiased estimator of p. Another estimator of p is $\hat{p}_2 = (Y+1)/(n+2)$.

(a) Derive the bias of $\hat{p}_2$.

(b) Derive MSE($\hat{p}_1$) and MSE($\hat{p}_2$).

(c) For what values of p is MSE($\hat{p}_2$) < MSE($\hat{p}_1$)?

8.14 Let $Y_1, Y_2, \ldots, Y_n$ denote a random sample of size n from a population with a uniform distribution on the interval $(0, \theta)$. Consider $Y_{(1)} = \min(Y_1, Y_2, \ldots, Y_n)$, the smallest order statistic. Use the methods of Section 6.6 to derive $E(Y_{(1)})$. Find a multiple of $Y_{(1)}$ that is an unbiased estimator for θ.

8.15 Suppose that $Y_1, Y_2, \ldots, Y_n$ denotes a random sample of size 4 from a population with an exponential distribution with a density given by

$$f(y) = \begin{cases} (1/\theta)e^{-y/\theta}, & y > 0 \\ 0, & \text{elsewhere} \end{cases}$$

If $Y_{(1)} = \min(Y_1, Y_2, \ldots, Y_n)$ denotes the smallest order statistic, show that $\hat{\theta} = nY_{(1)}$ is an unbiased estimator for θ and find MSE($\hat{\theta}$). (Hint: Recall the results of Exercise 6.47.)

* Exercises preceded by an asterisk are optional.

*8.16   Suppose that Y_1, Y_2, Y_3, Y_4 denotes a random sample of size 4 from a population with an exponential distribution with a density given by

$$f(y) = \begin{cases} (1/\theta)e^{-y/\theta}, & y > 0 \\ 0, & \text{elsewhere} \end{cases}$$

(a)   Let $X = \sqrt{Y_1 Y_2}$. Find a multiple of X that is an unbiased estimator for θ. [Hint: Use your knowledge of the gamma distribution and the fact that $\Gamma(1/2) = \sqrt{\pi}$ to find $E(\sqrt{Y_1})$. Recall that the Y_i's are independent.]

(b)   Let $W = \sqrt{Y_1 Y_2 Y_3 Y_4}$ and find a multiple of W that is an unbiased estimator for θ^2. Recall the hint for part (a).

8.3 Some Common Unbiased Point Estimators

Some formal methods for deriving point estimators for target parameters are derived in Chapter 9. In this section we present some estimators that merit consideration on the basis of intuition. For example, it seems natural to use the sample mean $\bar{Y}$ to estimate the population mean μ, and the sample proportion $\hat{p} = Y/n$ to estimate a binomial parameter p. How, then, would we estimate the difference between corresponding parameters for two different populations, say, the difference in means, $(\mu_1 - \mu_2)$, or the difference in two binomial parameters, $(p_1 - p_2)$, when the inference is to be based on random samples of n_1 and n_2 observations selected independently from the two populations? Again, our intuition suggests the point estimators $(\bar{Y}_1 - \bar{Y}_2)$, the difference in the sample means, for estimating $(\mu_1 - \mu_2)$, and $(\hat{p}_1 - \hat{p}_2)$, the difference in the sample proportions, for estimating $(p_1 - p_2)$.

Because the four estimators, $\bar{Y}$, $\hat{p}$, $(\bar{Y}_1 - \bar{Y}_2)$, and $(\hat{p}_1 - \hat{p}_2)$, are functions of the random sample measurements, we could find their expected values and variances by using the expectation theorems of Sections 5.6, 5.7, and 5.8. Such an effort would show that all four point estimators are unbiased and that they possess the variances shown in Table 8.1 when random sampling has been employed. To facilitate communication we use the notation $\sigma_{\hat{\theta}}^2$ to denote the variance of the sampling distribution of the estimator $\hat{\theta}$.

A survey of Chapter 5 would show that we have done much of the derivation required for Table 8.1. In particular, we found the means and variances of $\bar{Y}$ and $\hat{p}$ in Examples 5.27 and 5.28, respectively. If the random samples are independent, these results and Theorem 5.12 imply that

$$E(\bar{Y}_1 - \bar{Y}_2) = E(\bar{Y}_1) - E(\bar{Y}_2) = \mu_1 - \mu_2$$

$$V(\bar{Y}_1 - \bar{Y}_2) = V(\bar{Y}_1) + V(\bar{Y}_2) = \frac{\sigma_1^2}{n_1} + \frac{\sigma_2^2}{n_2}$$

Table 8.1
Expected values and variances of some common point estimators

Target Parameter θ	Sample Size(s)	Point Estimator $\hat{\theta}$	$E(\hat{\theta})$	$\sigma_{\hat{\theta}}^2$
μ	n	$\bar{Y}$	μ	$\dfrac{\sigma^2}{n}$
p	n	$\hat{p} = \dfrac{Y}{n}$	p	$\dfrac{pq}{n}$
$\mu_1 - \mu_2$	n_1 and n_2	$\bar{Y}_1 - \bar{Y}_2$	$\mu_1 - \mu_2$	$\dfrac{\sigma_1^2}{n_1} + \dfrac{\sigma_2^2}{n_2}$ *
$p_1 - p_2$	n_1 and n_2	$\hat{p}_1 - \hat{p}_2$	$p_1 - p_2$	$\dfrac{p_1 q_1}{n_1} + \dfrac{p_2 q_2}{n_2}$

* σ_1^2 and σ_2^2 are the variances of populations 1 and 2, respectively.

The expected value and variance of $(\hat{p}_1 - \hat{p}_2)$, shown in Table 8.1, are acquired in a similar manner.

Although unbiasedness is often a desirable property for a point estimator, not all estimators are unbiased. Example 8.1 shows that the quantity

$$S'^2 = \frac{\sum_{i=1}^{n}(Y_i - \bar{Y})^2}{n}$$

is a biased estimator of the population variance σ^2 and, particularly, that this bias can be corrected by dividing the sum of squares of deviations of the measurements about $\bar{Y}$ by $(n-1)$ rather than n. Because it is used most often in practice to estimate σ^2, the unbiased estimator

$$S^2 = \frac{\sum_{i=1}^{n}(Y_i - \bar{Y})^2}{n-1}$$

henceforth will be called the *sample variance*.

EXAMPLE 8.1   Let $Y_1, \ldots, Y_n$ be a random sample with $E(Y_i) = \mu$ and $V(Y_i) = \sigma^2$. Show that

$$S'^2 = \frac{1}{n}\sum_{i=1}^{n}(Y_i - \bar{Y})^2$$

is a biased estimator for σ^2 and that

$$S^2 = \frac{1}{n-1} \sum_{i=1}^{n} (Y_i - \bar{Y})^2$$

is an unbiased estimator for σ^2.

Solution   It can be shown (see Exercise 1.5) that

$$\sum_{i=1}^{n} (Y_i - \bar{Y})^2 = \sum_{i=1}^{n} Y_i^2 - n\bar{Y}^2$$

Hence

$$E\left[\sum_{i=1}^{n} (Y_i - \bar{Y})^2 \right] = E\left(\sum_{i=1}^{n} Y_i^2 \right) - nE(\bar{Y}^2) = \sum_{i=1}^{n} E(Y_i^2) - nE(\bar{Y}^2)$$

Now note that $E(Y_i^2)$ is the same for $i = 1, 2, \ldots, n$ and use the fact that the variance of a random variable is given by $V(Y) = E(Y^2) - \mu^2$. Then $E(Y^2) = V(Y) + \mu^2$ and

$$E\left[\sum_{i=1}^{n} (Y_i - \bar{Y})^2 \right] = \sum_{i=1}^{n} (\sigma^2 + \mu^2) - n\left(\frac{\sigma^2}{n} + \mu^2 \right)$$

$$= n(\sigma^2 + \mu^2) - n\left(\frac{\sigma^2}{n} + \mu^2 \right)$$

$$= n\sigma^2 - \sigma^2 = (n - 1)\sigma^2$$

It follows that

$$E(S'^2) = \left(\frac{n-1}{n} \right)\sigma^2$$

and that S'^2 is biased since $E(S'^2) \neq \sigma^2$. However,

$$S^2 = \frac{1}{n-1} \sum_{i=1}^{n} (Y_i - \bar{Y})^2$$

$$= \left(\frac{n}{n-1} \right)\left[\frac{1}{n} \sum_{i=1}^{n} (Y_i - \bar{Y})^2 \right] = \left(\frac{n}{n-1} \right)S'^2$$

Therefore,

$$E(S^2) = \left(\frac{n}{n-1} \right)E(S'^2) = \left(\frac{n}{n-1} \right)\left[\left(\frac{n-1}{n} \right)\sigma^2 \right] = \sigma^2$$

and we see that S^2 is an unbiased estimator for σ^2.

Two final comments can be made concerning the point estimators of Table 8.1. First, the expected values and variances shown in the table are valid regardless of the form of the population probability density functions. Second, all four estimators will possess probability distributions that are approximately normal for large samples. The central limit theorem justifies this statement for $\bar{Y}$ and $\hat{p}$, whereas similar theorems for functions of sample means justify the assertion for $(\bar{Y}_1 - \bar{Y}_2)$ and $(\hat{p}_1 - \hat{p}_2)$. How large is "large"? For most populations the probability distributions of $\bar{Y}$ will be mound-shaped for relatively small samples, as low as $n = 5$, and will tend rapidly to normality as the sample size approaches $n = 30$ or larger. However, you sometimes will need to select larger samples from binomial populations because the required sample size depends on p. The binomial probability distribution is perfectly symmetrical about its mean when $p = 1/2$ and becomes more and more asymmetric as p tends to 0 or 1. As a rough rule you can assume that the distribution of $\hat{p}$ will be mound-shaped and approaching normality for sample sizes such that $p \pm 2\sqrt{pq/n}$ lies in the interval $(0, 1)$.

Because we know that $\bar{Y}$, $\hat{p}$, $(\bar{Y}_1 - \bar{Y}_2)$, and $(\hat{p}_1 - \hat{p}_2)$ are unbiased with near-normal (at least mound-shaped) sampling distributions for moderate-sized samples, let us now see how we can use this information to answer a practical question. If we use an estimator once and acquire a single estimate, how good will this estimate be? How much faith can we place in the validity of our inference? The answers to these questions are provided in the next section.

8.4   Evaluating the Goodness of a Point Estimator

One way to measure the goodness of any point estimation procedure is in terms of the distances between the estimates it generates and the target parameter. This quantity, which varies in a random manner in repeated sampling, is called the *error of estimation*. Naturally we would like the error of estimation to be as small as possible.

Definition 8.5

> The *error of estimation* ε is the distance between an estimator and its target parameter. That is, $\varepsilon = |\hat{\theta} - \theta|$.

Because the error of estimation will be a random quantity, we cannot say how large or small it will be for a particular estimate, but we can make probability statements about it. For example, suppose that $\hat{\theta}$, an unbiased estimator of θ, possesses a sampling distribution as shown in Figure 8.4. If we select two points, $(\theta - b)$ and $(\theta + b)$, located near the tails of the probability

Figure 8.4
Sampling distribution
of a point estimator $\hat{\theta}$

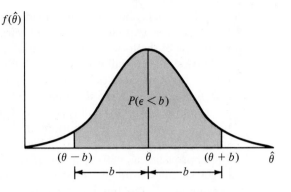

distribution, the probability that the error of estimation ε is less than b is the shaded area in Figure 8.4. That is,

$$P(|\hat{\theta} - \theta| < b) = P[-b < (\hat{\theta} - \theta) < b] = P(\theta - b < \hat{\theta} < \theta + b)$$

We can think of b as a probabilistic bound on the error of estimation. Hence we are not certain that a given error will be less than b, but we may know that the probability of such an event is high. If b can be regarded as small, from a practical point of view, then $P(\varepsilon < b)$ provides a measure of the goodness of a single estimate. This probability gives the fraction of times, in repeated sampling, that the estimator $\hat{\theta}$ will be within b units of θ, the target parameter.

Finding the value of b for a given estimation problem is easy if you know the probability distribution of $\hat{\theta}$. Suppose that you want ε to be less than b with a probability equal to .90. Then you seek a value b such that

$$\int_{\theta - b}^{\theta + b} f(\hat{\theta}) \, d\hat{\theta} = .90$$

Whether you know the probability distribution of $\hat{\theta}$ or not, an approximate bound on ε for unbiased estimators can be found by expressing b as a multiple of the standard deviation of $\hat{\theta}$. For example, if we let $b = k\sigma_{\hat{\theta}}$, $k \geq 1$, then we know from Tchebysheff's theorem that ε will be less than $k\sigma_{\hat{\theta}}$ with probability at least $1 - 1/k^2$. A convenient and often used value of k is $k = 2$. Hence we know that ε will be less than $b = 2\sigma_{\hat{\theta}}$ with probability at least .75.

You will find that most random variables observed in nature lie within two standard deviations of their mean, with a probability in the vicinity of .95. The probability that Y lies in the interval $(\mu \pm 2\sigma)$ is shown in Table 8.2 for the normal, uniform, and exponential probability distributions. The point we make, of course, is that $B = 2\sigma_{\hat{\theta}}$ is a good approximate bound on the error of estimation in a practical situation. The probability that the error of estimation will be less than this bound will be near .95.

Table 8.2
Probability that
$(\mu - 2\sigma) < Y < (\mu + 2\sigma)$

Distribution	Probability
Normal	.9544
Uniform	1.0000
Exponential	.9502

EXAMPLE 8.2   A sample of $n = 1000$ voters, randomly selected from a city, showed $y = 560$ in favor of candidate Jones. Estimate p, the fraction of voters in the population favoring Jones, and place a two-standard-deviation bound on the error of estimation.

Solution   We will use the estimator $\hat{p} = Y/n$ to estimate p. Hence the estimate of p, the fraction of voters favoring candidate Jones, is

$$\hat{p} = \frac{y}{n} = \frac{560}{1000} = .56$$

How much faith can we place in this figure? The probability distribution of $\hat{p}$ will be very accurately approximated by a normal probability distribution for samples as large as $n = 1000$. Then when $b = 2\sigma_{\hat{p}}$, the probability that ε will be less than b is approximately .95.

From Table 8.1, $V(\hat{p}) = pq/n$. Hence

$$b = 2\sigma_{\hat{p}} = 2\sqrt{\frac{pq}{n}}$$

Unfortunately, we need to know p in order to calculate b, and finding p was the objective of our sampling. This apparent stalemate is not a handicap, because $\sigma_{\hat{p}}$ will vary little for small changes in p. Hence the substitution of the estimate $\hat{p}$ for p will produce little error in calculating the exact value of $b = 2\sigma_{\hat{p}}$. Then for our example, we have

$$b = 2\sigma_{\hat{\theta}} = 2\sqrt{\frac{pq}{n}} \approx 2\sqrt{\frac{(.56)(.44)}{1000}} = .03$$

Now what is the significance of our calculations? The probability that the error of estimation is less than .03 is approximately .95. Hence we are reasonably confident that our estimate, .56, is within .03 of the population value of p.

EXAMPLE 8.3   A comparison of the wearing quality of two types of automobile tires was obtained by road testing samples of $n_1 = n_2 = 100$ tires for each type. The number of miles until wear-out was recorded, where wear-out was defined as a specific amount of tire wear. The test results were as follows:

$$\bar{y}_1 = 26{,}400 \text{ miles} \qquad \bar{y}_2 = 25{,}100 \text{ miles}$$
$$s_1^2 = 1{,}440{,}000 \qquad s_2^2 = 1{,}960{,}000$$

Estimate the difference in mean time to wear-out and place a two-standard-deviation bound on the error of estimation.

Solution   The point estimate of $(\mu_1 - \mu_2)$ is

$$(\bar{y}_1 - \bar{y}_2) = 26{,}400 - 25{,}100 = 1{,}300 \text{ miles}$$

and

$$\sigma_{(\bar{Y}_1 - \bar{Y}_2)} = \sqrt{\frac{\sigma_1^2}{n_1} + \frac{\sigma_2^2}{n_2}}$$

You will note that we must know σ_1^2 and σ_2^2, or have good approximate values for them, in order to calculate $\sigma_{(\bar{Y}_1 - \bar{Y}_2)}$. Fairly accurate values of σ_1^2 and σ_2^2 often can be calculated from similar experimental data collected at some prior time or they can be obtained from the current sample data by using the unbiased estimator

$$\hat{\sigma}_i^2 = S_i^2 = \frac{\sum\limits_{j=1}^{n_i} (Y_{ij} - \bar{Y}_i)^2}{n_i - 1} \qquad i = 1, 2$$

These estimates will be adequate if the sample sizes are reasonably large, say, $n_i \geq 30$, $i = 1, 2$. The calculated values of S_1^2 and S_2^2, based on the two wear tests, are $s_1^2 = 1{,}440{,}000$ and $s_2^2 = 1{,}960{,}000$. Substituting these values for σ_1^2 and σ_2^2 in the formula for $\sigma_{(\bar{Y}_1 - \bar{Y}_2)}$, we have

$$\sigma_{(\bar{Y}_1 - \bar{Y}_2)} \approx \sqrt{\frac{s_1^2}{n_1} + \frac{s_2^2}{n_2}} = \sqrt{\frac{1{,}440{,}000}{100} + \frac{1{,}960{,}000}{100}}$$

$$= \sqrt{34{,}000} = 184 \text{ miles}$$

Consequently, we estimate the difference in mean wear to be 1,300 miles, and we expect the error of estimation to be less than $2\sigma_{(Y_1 - Y_2)}$ or 368 miles, with probability approximately .95.   ∽

Exercises

8.17   Geologists are interested in shifts and movements of the earth's surface indicated by fractures (cracks) in the earth's crust. One of the most famous large fractures is the San Andreas fault (moving fracture) in California. A geologist attempting to study the movement of the relative shifts in the earth's crust at a particular location found many fractures in the local rock structure. In an attempt to determine the mean angle of the breaks, she samples $n = 50$ fractures and found the sample mean and standard deviation to be $39.8°$ and $17.2°$, respectively. Estimate the mean angular direction of the fractures and place a bound on the error of estimation.

8.18 An increase in the rate of consumer savings frequently is tied to a lack of confidence in the economy and is said to be an indicator of a recessional tendency in the economy. A random sampling of $n = 200$ savings accounts in a local community showed a mean increase in savings account values of 7.2% over the past 12 months and a standard deviation of 5.6%. Estimate the mean percent increase in savings account values over the past 12 months for depositors in the community. Place a bound on your error of estimation.

8.19 The Environmental Protection Agency and the University of Florida recently cooperated in a large study of the possible effects of trace elements in drinking water on kidney stone disease. Given below are data on age, amount of calcium in home drinking water (measured in parts per million), and smoking activity. These data were obtained from individuals with recurrent kidney stone problems, who all lived in the Carolinas and the Rocky Mountain states.

	Carolinas	Rockies
Sample size	467	191
Mean age	45.1	46.4
Standard deviation of age	10.2	9.8
Mean calcium content (ppm)	11.3	40.1
Standard deviation of calcium	16.6	28.4
Proportion now smoking	.78	.61

(a) Estimate the average calcium concentration in drinking water for kidney stone patients in the Carolinas. Place a bound on the error of estimation.

(b) Estimate the difference in mean ages for kidney stone patients in the Carolinas and in the Rockies. Place a bound on the error of estimation.

(c) Estimate and place a two-standard-deviation bound on the difference in proportions of kidney stone patients from the Carolinas and Rockies who were smokers at the time of the study.

8.20 Results of a public opinion poll reported in a news magazine (*Time*, January 2, 1984)[†] show that 51% of the respondents completely agreed with the statement:
   "The Soviets are just as afraid of nuclear war as we are, and therefore it is in our mutual interest to find ways to negotiate."
The article states that "The findings are based on a telephone survey of 1000 registered voters.... The potential sampling error is plus or minus 3%."
   How was the 3% calculated, and what is its interpretation? Can we conclude that a majority of registered voters completely agreed with the statement?

[†] *Source* Reprinted from *Time*, © 1984, Time-Life Books, Inc. All rights reserved.

8.21   A study was conducted to compare the mean number of police emergency calls per 8-hour shift in two districts of a large city. Samples of 100 8-hour shifts were randomly selected from the police records for each of the two regions, and the number of emergency calls was recorded for each shift. The sample statistics are as follows:

	Region	
	1	2
Sample size	100	100
Sample mean	2.4	3.1
Sample variance	1.44	2.64

Estimate the difference in the mean number of police emergency calls per shift between the two districts in the city. Find a bound for the error of estimation.

8.22   A study by the General Accounting Office (GAO) of Veterans Administration (VA) hospitals suggests that some of the $8.3 billion spent on the treatment of VA patients in 1984 might have been saved. After evaluating 800 cases at six VA hospitals, the GAO concluded that 244, or 31%, of the patients "did not belong in medical and surgical acute care beds and could have been treated in less costly facilities if available" (*Philadelphia Inquirer*, August 8, 1985). If we can regard the 800 patients as a random sample of all patients treated in VA hospitals in 1984, how accurate is the estimate that 31 percent could have been treated in less costly facilities? Find a bound on the error of estimation.

8.23   The *Orlando Sentinel Star* (October 6, 1981), reporting on a September 1981, AP-NBC news poll, states that "even among those who personally believe abortion is wrong, more than 40% say it still should be legal." The poll of 1601 adults in a nationwide "scientific random sampling" showed that 44% of the respondents viewed abortion as not wrong, 49% viewed it as wrong, and 7 percent had no opinion. Of the 49% who regarded abortion as wrong, 40% still believed that it should be legal. The article concludes, "the results [of the poll] are subject to an error margin of 3 percentage points either way because of chance variations. That is, if one could have talked this past week to all Americans with telephones, there is only one chance in twenty that the findings would vary by more than 3 percentage points."

(a)   Do you agree that, based on 1601 respondents, the margin of error for the reported percentages should be less than 3% with probability equal to .95?

(b)   If the sample included only these adults who have telephones, do you think that the 1601 respondents represent a random sample from the population of all United States adults?

8.24   In a study of the relationship between birth order and college success, an investigator found that 126 in a sample of 180 college graduates were first-born or only children; in a sample of 100 nongraduates of comparable age and socioeconomic background the number of first born or only children was 54. Estimate the difference in the proportion of first-born or only children for the two populations from which these samples were drawn. Give a bound for the error of estimation.

8.25 In a study of the relationship between law and the use of lethal force, W. B. Waegel classified 459 police shootings in Philadelphia over the period 1970 to 1978 to see whether new restrictive laws, introduced in 1973, produced a change in the percentage of unjustified uses of lethal force by police (W. B. Waegel, "The Use of Lethal Force by Police: The Effect of Statutory Change," *Crime and Delinquency* 30, [January 1984]. © 1984 Sage Publications, Inc. with permission).

	Total for period 1970–1972	1973	Total for period 1974–1978
Justified	72	30	173
Not justified	11	9	59
Unable to determine	18	7	53
Accidental	10	5	12
Total	111	51	297

Assume that the data in the table represent samples from populations of cases that could have occurred and in which lethal force would have been used, one population corresponding to the legal situation existing prior to 1973 and the other to the legal situation after the new restrictive laws were introduced, the period 1974–1978.

(a) Estimate the difference in the proportions of nonjustified use of lethal force. Find a bound for the error of estimation.

(b) Estimate the difference in the proportions of unable-to-determine cases. Find a bound for the error of estimation.

8.26 An auditor randomly samples 20 accounts receivable from the 500 accounts of a certain firm. The auditor lists the amount of each account and checks to see if the underlying documents are in compliance with stated procedures. The data are as follows (amounts are in dollars, Y = Yes, N = No):

Account	Amount	Compliance	Account	Amount	Compliance
1	278	Y	11	188	N
2	192	Y	12	212	N
3	310	Y	13	92	Y
4	94	N	14	56	Y
5	86	Y	15	142	Y
6	335	Y	16	37	Y
7	310	N	17	186	N
8	290	Y	18	221	Y
9	221	Y	19	219	N
10	168	Y	20	305	Y

Estimate the total accounts receivable for the 500 accounts of the firm, and place a bound on the error of estimation. Do you think that the *average* account receivable for the firm exceeds $250? Why?

8.27   Refer to Exercise 8.26. From the data given on the compliance checks, estimate the proportion of the firm's accounts that fail to comply with stated procedures. Place a bound on the error of estimation. Do you think the proportion of accounts that comply with stated procedures exceeds 80%? Why?

8.28   Eighty-two accountants were interviewed for assessment of their attitudes regarding advertising of their services. When asked about the likelihood of their advertising their services in the future, the following data were obtained.

Percentages of respondents in various categories when asked whether or not they planned to advertise	
Very likely to virtual certainty	16%
Somewhat likely	19%
About 50–50	18%
Somewhat unlikely	6%
Very unlikely	12%
Absolutely not, no response	29%

Source   Reprinted from the *Journal of Advertising Research.* © Copyright 1981, by the Advertising Research Foundation.

Estimate and place bounds on the errors of estimation of the population proportion:

(a)   Very likely to advertise in the future.

(b)   With at least a 50–50 chance of advertising in the future.

8.29   We can place a two-standard-deviation bound on the error of estimation with any estimator for which we can find a reasonable estimate of the standard deviation. Suppose $Y_1, \ldots, Y_n$ represents a random sample from a Poisson distribution with mean λ. We know that $V(Y_i) = \lambda$, and hence $E(\bar{Y}) = \lambda$ and $V(\bar{Y}) = \lambda/n$. Show how you would employ $Y_1, \ldots, Y_n$ to estimate λ, and estimate the standard deviation of your estimator.

8.30   In polycrystalline aluminum, the number of grain nucleation sites per unit volume is modeled as having a Poisson distribution with mean λ. Fifty unit-volume test specimens subjected to annealing under regime A produced an average of 20 sites per unit volume. Fifty independently selected unit-volume test specimens subjected to annealing regime B produced an average of 23 sites per unit volume.

(a)   Estimate the mean number, λ_A, of nucleation sites for regime A, and place a two-standard-deviation bound on the error of estimation.

(b)   Estimate the difference in the mean numbers of nucleation sites, $\lambda_A - \lambda_B$, for regimes A and B. Place a two-standard-deviation bound on the error of estimation. Would you say that regime B tends to produce a larger mean number of nucleation sites? Why?

8.31   If $Y_1, \ldots, Y_n$ denotes a random sample from an exponential distribution with mean θ, then $E(Y_i) = \theta$ and $V(Y_i) = \theta^2$. Thus $E(\bar{Y}) = \theta$ and $V(\bar{Y}) = \theta^2/n$, or $\sqrt{V(\bar{Y})} = \theta/\sqrt{n}$. Suggest an unbiased estimator for θ, and show how you would estimate the standard deviation of your estimator.

8.32 An engineer observes $n = 10$ independent length-of-life measurements on a type of electronic component. The average of these 10 measurements is 1020 hours. If these lengths of life come from an exponential distribution with mean θ, estimate θ and place a two-standard-deviation bound on the error of estimation.

8.33 The number of persons coming through a blood bank until the first person with type A blood is found is a random variable Y with a geometric distribution. If p denotes the probability of any one randomly selected person possessing type A blood, then $E(Y) = 1/p$ and $V(Y) = (1 - p)/p^2$. Find a function of Y that is an unbiased estimator of $V(Y)$. Then suggest how to form a two-standard-deviation bound on the error of estimation when Y is used to estimate $1/p$.

8.5 Confidence Intervals

You will recall that an interval estimator is a rule that specifies the method for using the sample measurements to calculate two numbers that form the endpoints of the interval. Ideally we would like the resulting interval to have two properties. First, we would like the interval to contain the target parameter θ. Second, we would like the interval to be relatively narrow. Note that one or both of the endpoints of the interval, being functions of the sample measurements, will vary in a random manner from sample to sample. Thus the length and location of the interval are random quantities, and we cannot be certain that the (fixed) target parameter θ actually will fall between the endpoints of any single interval calculated from a single sample. This being the case, our objective is to find an interval estimator that generates narrow intervals that enclose θ with a high probability.

Interval estimators are commonly called *confidence intervals.* The upper and lower endpoints of a confidence interval are called the *upper* and *lower confidence limits,* respectively. The probability that a confidence interval will enclose θ is called the *confidence coefficient.* From a practical point of view, the confidence coefficient gives the fraction of the time, in repeated sampling, that the intervals constructed will contain the target parameter θ. If we know that the confidence coefficient associated with our estimator is high, we will be highly confident that one particular confidence interval, constructed in a single sampling situation, will enclose θ.

Suppose that $\hat{\theta}_L$ and $\hat{\theta}_U$ are the lower and upper confidence limits, respectively, for a parameter θ. Then if

$$P(\hat{\theta}_L < \theta < \hat{\theta}_U) = 1 - \alpha$$

the probability, $(1 - \alpha)$, is the confidence coefficient. The resulting random interval defined by $\hat{\theta}_L$ to $\hat{\theta}_U$ is called a *two-sided confidence interval.*

It is also possible to form a *one-sided confidence interval* such that

$$P(\hat{\theta}_L < \theta) = 1 - \alpha$$

Although only one point is random in this case, the confidence interval is $(\hat{\theta}_L, \infty)$. Similarly, we could have an upper one-sided confidence interval such that

$$P(\theta < \hat{\theta}_U) = 1 - \alpha$$

The implied confidence interval is $(-\infty, \hat{\theta}_U)$.

One very useful method for finding confidence intervals is called the *pivotal method*. This method depends upon finding a pivotal quantity that possesses two characteristics:

1. It is a function of the sample measurements and the unknown parameter θ, where θ is the *only* unknown quantity.

2. It has a probability distribution that does not depend upon the parameter θ.

If the probability distribution of the pivotal quantity is known, the following logic can be used to form the desired interval estimate. If Y is any random variable, c is a constant ($c > 0$), and $P(a \leq Y \leq b) = .7$, then certainly $P(ca \leq cY \leq cb) = .7$. Similarly, for any constant d, $P(a + d \leq Y + d \leq b + d) = .7$. That is, the probability of the event $(a < Y < b)$ is unaffected by a change of scale or a translation of Y. Thus if we know the probability distribution of a pivotal quantity, it may be possible to use operations like those just used to form the desired interval estimator. We illustrate this method in the following examples.

EXAMPLE 8.4   Suppose we are to obtain a single observation Y from an exponential distribution with mean θ. Use Y to form a confidence interval for θ with confidence coefficient .90.

Solution   The probability density function for Y is given by

$$f(y) = \begin{cases} \left(\dfrac{1}{\theta}\right)e^{-y/\theta}, & y \geq 0 \\ 0, & \text{elsewhere} \end{cases}$$

By the transformation method of Chapter 6 we can see that $U = Y/\theta$ has the exponential density function given by

$$f_U(u) = \begin{cases} e^{-u}, & u > 0 \\ 0, & \text{elsewhere} \end{cases}$$

The density function for U is graphically depicted in Figure 8.5. Notice that $U = Y/\theta$ is a function of Y (the sample measurement) and θ and the distribution of U does not depend on θ. Thus we can use $U = Y/\theta$ as a pivotal

Figure 8.5
Density function for
U, Example 8.4

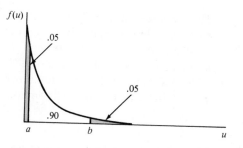

quantity. Because we want an interval estimator with confidence coefficient equal to .90, we find two numbers a and b, such that

$$P(a \le U \le b) = .90$$

One way to do this is to choose a and b to satisfy

$$P(U \le a) = \int_0^a e^{-u} \, du = .05 \quad \text{and} \quad P(U > b) = \int_b^\infty e^{-u} \, du = .05$$

These equations yield

$$1 - e^{-a} = .05 \quad \text{and} \quad e^{-b} = .05$$

or $a = .051$ and $b = 2.996$.
It follows that

$$.90 = P(.051 \le U \le 2.996) = P\left(.051 \le \frac{Y}{\theta} \le 2.996\right)$$

Because we seek an interval estimator for θ, let us manipulate the inequalities describing the event to isolate θ in the middle. Because Y has an exponential distribution, $P(Y > 0) = 1$, and we maintain the direction of the inequalities if we divide through by Y. That is,

$$.90 = P\left(.051 \le \frac{Y}{\theta} \le 2.996\right) = P\left(\frac{.051}{Y} \le \frac{1}{\theta} \le \frac{2.996}{Y}\right)$$

Taking reciprocals (and hence reversing the direction of the inequalities), we obtain

$$.90 = P\left(\frac{Y}{.051} \ge \theta \ge \frac{Y}{2.996}\right) = P\left(\frac{Y}{2.996} \le \theta \le \frac{Y}{.051}\right)$$

Thus we see that $Y/2.996$ and $Y/.051$ form the desired lower and upper confidence limits, respectively. To obtain numerical values for these limits, we must observe an actual value for Y and substitute that value into the given formulas for the confidence limits. We know that limits of the form $(Y/2.996, Y/.051)$ will include the true (unknown) values of θ for 90% of the values of Y we would obtain from repeatedly sampling from this exponential distribution.

EXAMPLE 8.5   Suppose that we take a sample of size $n = 1$ from a uniform distribution defined on the interval $(0, \theta)$, where θ is unknown. Find a 95% lower confidence bound for θ.

Solution   Because Y is uniform on $(0, \theta)$, the methods of Chapter 6 can be used to show that $U = Y/\theta$ will be uniformly distributed over $(0, 1)$. That is,

$$f_U(u) = \begin{cases} 1, & 0 \le u \le 1 \\ 0, & \text{elsewhere} \end{cases}$$

Figure 8.6 contains a graphical depiction of the density function for U. Again we see that U satisfies the requirements of a pivotal quantity. Because we

Figure 8.6
Density function for
U, Example 8.5

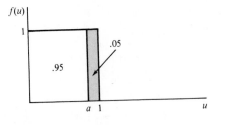

seek a 95% lower confidence limit for θ, let us determine the value for a so that $P(U \le a) = .95$. That is,

$$\int_0^a (1)\, du = .95$$

or $a = .95$. Thus

$$P(U \le .95) = P\left(\frac{Y}{\theta} \le .95\right) = P(Y \le .95\theta) = P\left(\frac{Y}{.95} \le \theta\right) = .95$$

We see that $Y/.95$ will form a lower confidence limit for θ, with confidence coefficient .95. Note that because any observed Y must be less than θ, it is intuitively reasonable to have the lower confidence limit for θ slightly larger than the observed value of Y.

The two preceding examples illustrate the use of the pivotal method for finding confidence limits for unknown parameters. In each instance the interval estimates were developed on the basis of a single observation from the distribution. These examples were introduced primarily for the purpose of illustration. In the remaining sections of this chapter we will utilize the pivotal method in conjunction with the sampling distributions presented in Chapter 7 to develop some interval estimates of greater practical importance.

Exercises

8.34 Suppose the random variable Y has a gamma distribution with parameters $\alpha = 2$ and an unknown β.

(a) Use the method of moment-generating functions to show that $2Y/\beta$ has a chi-square distribution with 4 degrees of freedom.

(b) Use the result of part (a) to form a 90% confidence interval for β.

8.35 Suppose the random variable Y is an observation from a normal distribution with an unknown mean of μ and a variance of 1.

(a) Find a 95% confidence interval for μ.

(b) Find a 95% upper confidence limit for μ.

(c) Find a 95% lower confidence limit for μ.

8.36 Suppose Y is normally distributed with a mean of 0 and an unknown variance of σ^2. Then Y^2/σ^2 has a chi-square distribution with 1 degree of freedom. Use this distribution to find the following:

(a) a 95% confidence interval for σ^2,

(b) a 95% upper confidence limit for σ^2,

(c) a 95% lower confidence limit for σ^2,

8.37 Use the answers from Exercises 8.36 to find the following:

(a) a 95% confidence interval for σ,

(b) a 95% upper confidence limit for σ,

(c) a 95% lower confidence limit for σ,

8.38 Let $Y_1, Y_2, \ldots, Y_n$ denote a random sample of size n from a population with a uniform distribution on the interval $(0, \theta)$. Let $Y_{(n)} = \max(Y_1, Y_2, \ldots, Y_n)$. Let $U = (1/\theta)Y_{(n)}$.

(a) Show that U has a distribution function given by

$$F(u) = \begin{cases} 0, & u < 0 \\ u^n, & 0 \le u \le 1 \\ 1, & u > 1 \end{cases}$$

(b) Note that the distribution of U does not depend on θ. Find a 95% lower confidence bound for θ.

8.6 Large-Sample Confidence Intervals

In Section 8.3 we presented some unbiased point estimators for the parameters μ, p, $\mu_1 - \mu_2$, and $p_1 - p_2$. As we indicated in that section, for large samples all these point estimators have approximately normal sampling distributions. That is, under the conditions of Section 8.3, if the target parameter θ is μ, p, $\mu_1 - \mu_2$, or $p_1 - p_2$, then for large samples

$$Z = \frac{\hat{\theta} - \theta}{\sigma_{\hat{\theta}}}$$

possesses approximately a standard normal distribution. That is, $Z = (\hat{\theta} - \theta)/\sigma_{\hat{\theta}}$ forms a pivotal quantity, and the pivotal method can be employed to develop interval estimators for the target parameter θ.

EXAMPLE 8.6   Let $\hat{\theta}$ be a statistic that is normally distributed with an expected value and a variance equal to θ and $\sigma_{\hat{\theta}}^2$, respectively. Find a confidence interval for θ that possesses a confidence coefficient equal to $(1 - \alpha)$.

Solution   The quantity

$$Z = \frac{\hat{\theta} - \theta}{\sigma_{\hat{\theta}}}$$

has a standard normal distribution. Now select two tail-end values of this distribution, $z_{\alpha/2}$ and $-z_{\alpha/2}$, such that

$$P(-z_{\alpha/2} < Z \le z_{\alpha/2}) = 1 - \alpha$$

See Figure 8.7.

Figure 8.7
Location of $z_{\alpha/2}$
and $-z_{\alpha/2}$

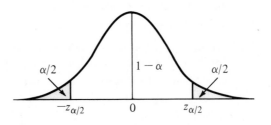

Substituting for Z in the probability statement, we have

$$P\left(-z_{\alpha/2} < \frac{\hat{\theta} - \theta}{\sigma_{\hat{\theta}}} < z_{\alpha/2} \right) = 1 - \alpha$$

Multiplying by $\sigma_{\hat{\theta}}$, we obtain

$$P(-z_{\alpha/2}\sigma_{\hat{\theta}} < \hat{\theta} - \theta < z_{\alpha/2}\sigma_{\hat{\theta}}) = 1 - \alpha$$

and subtracting $\hat{\theta}$ from each term of the inequality, we get

$$P(-\hat{\theta} - z_{\alpha/2}\sigma_{\hat{\theta}} < -\theta < -\hat{\theta} + z_{\alpha/2}\sigma_{\hat{\theta}}) = 1 - \alpha$$

Finally, multiplying each term by -1 and, consequently, changing the direction of the inequalities, we have

$$P(\hat{\theta} - z_{\alpha/2}\sigma_{\hat{\theta}} < \theta < \hat{\theta} + z_{\alpha/2}\sigma_{\hat{\theta}}) = 1 - \alpha$$

Thus the lower and upper confidence limits for θ are

$$\text{lower confidence limit (LCL)} = \hat{\theta} - z_{\alpha/2}\sigma_{\hat{\theta}}$$
$$\text{upper confidence limit (UCL)} = \hat{\theta} + z_{\alpha/2}\sigma_{\hat{\theta}}$$

Example 8.6 can be used to find large-sample confidence intervals for μ, p, $(\mu_1 - \mu_2)$, and $(p_1 - p_2)$, the parameters estimated under the conditions described in Section 8.3. The following examples illustrate the use of the general method developed in Example 8.6.

EXAMPLE 8.7 The shopping times were recorded for $n = 64$ randomly selected customers for a local supermarket. The average and variance of the 64 shopping times were 33 minutes and 256, respectively. Estimate the true average shopping time per customer, μ, with a confidence coefficient of $1 - \alpha = .90$.

Solution In this case we are interested in the parameter $\theta = \mu$. Thus $\hat{\theta} = \bar{y} = 33$ and $s^2 = 256$ for a sample of $n = 64$ shopping times. The population variance σ^2 is unknown, so as in Section 8.3, we use s^2 as its estimated value. The confidence interval

$$\hat{\theta} \pm z_{\alpha/2}\sigma_{\hat{\theta}}$$

will have the form

$$\bar{y} \pm z_{\alpha/2}\left(\frac{\sigma}{\sqrt{n}}\right) \approx \bar{y} \pm z_{\alpha/2}\left(\frac{s}{\sqrt{n}}\right)$$

From Table 4, Appendix III, $z_{\alpha/2} = z_{.05} = 1.645$, and hence the confidence limits are given by

$$\bar{y} - z_{\alpha/2}\left(\frac{s}{\sqrt{n}}\right) = 33 - 1.645\left(\frac{16}{8}\right) = 29.71$$

$$\bar{y} + z_{\alpha/2}\left(\frac{s}{\sqrt{n}}\right) = 33 + 1.645\left(\frac{16}{8}\right) = 36.29$$

Thus our confidence interval for μ is (29.71, 36.29). The chances are good that this interval does include μ because, in repeated sampling, approximately 90% of all intervals of the form $\bar{Y} \pm 1.645(\sigma/\sqrt{n})$ do include μ, the true mean shopping time per customer.   ∽

EXAMPLE 8.8   Two brands of refrigerators, denoted by A and B, are each guaranteed for one year. In a random sample of 50 refrigerators of brand A, 12 were observed to fail before the guarantee period ended. A random sample of 60 brand B refrigerators also revealed 12 failures during the guarantee period. Estimate the true difference between proportions of failures during the guarantee period, $(p_1 - p_2)$, with confidence coefficient .98.

Solution   The confidence interval

$$\hat{\theta} \pm z_{\alpha/2}\sigma_{\hat{\theta}}$$

now has the form

$$(\hat{p}_1 - \hat{p}_2) \pm z_{\alpha/2}\sqrt{\frac{p_1 q_1}{n_1} + \frac{p_2 q_2}{n_2}}$$

Because p_1, q_1, p_2, and q_2 are unknown, the exact value of $\sigma_{\hat{\theta}}$ cannot be evaluated. As indicated in Section 8.3 we can get a good approximation for $\sigma_{\hat{\theta}}$ by substituting $\hat{p}_1$ and $\hat{p}_2$ for p_1 and p_2, respectively.

For this example $\hat{p}_1 = .24$, $\hat{p}_2 = .20$, and $z_{.01} = 2.33$. The confidence interval then becomes

$$(.24 - .20) \pm 2.33\sqrt{\frac{(.24)(.76)}{50} + \frac{(.20)(.80)}{60}}$$

$$.04 \pm 1857$$

Note that this confidence interval overlaps zero, so the true difference, $(p_1 - p_2)$, quite possibly could be positive or negative.   ∽

We close this section with an empirical investigation of the performance of the large-sample interval estimation procedure for a single population mean μ, based on a sample of $Y_1, Y_2, \ldots, Y_n$. In this case $\theta = \mu$, and therefore $\hat{\theta} = \bar{Y}$ and $\sigma_{\hat{\theta}}^2 = \sigma_{\bar{Y}}^2 = \sigma^2/n$, where σ^2 is the population variance. The appropriate confidence limits then are

$$\hat{\theta}_L = \bar{Y} - z_{\alpha/2}\left(\frac{\sigma}{\sqrt{n}}\right) \quad \text{and} \quad \hat{\theta}_U = \bar{Y} + z_{\alpha/2}\left(\frac{\sigma}{\sqrt{n}}\right)$$

In the (large-sample) case when σ is unknown, it can be replaced by the sample estimate s (which is equal to $\sqrt{s^2}$) in these formulas with no serious loss in accuracy.

Independent random samples, each of size $n = 100$, were selected from an exponential probability distribution with mean μ equal to 10. For each sample the sample mean $\bar{y}$ and the sample standard deviation s were calculated and used to form the 95% confidence interval given by $\bar{y} \pm 1.96s/\sqrt{100}$. (Note that $z_{.025} = 1.96$.) The mean, standard deviation, and resulting confidence limits for each sample are shown in Table 8.3 for the 100 samples generated. Among these 100 intervals, 7 do not include the true population mean value of 10. (These 7 intervals are marked with asterisks in Table 8.3.) In other words, 93 out of these 100 intervals do include the true mean value. Note that this fraction is close to the nominal confidence coefficient of .95. Thus if we take only one sample and produce one interval, we should be quite confident that the interval will include the true population mean.

In this section we have used the pivotal method to derive large-sample confidence intervals for the parameters μ, p, $\mu_1 - \mu_2$, and $p_1 - p_2$ under the conditions of Section 8.3. The key formula is

$$\hat{\theta} \pm z_{\alpha/2}\sigma_{\hat{\theta}}$$

where the values of $\hat{\theta}$ and $\sigma_{\hat{\theta}}^2$ are provided in Table 8.1. (Note that $\sigma_{\hat{\theta}} = \sqrt{\sigma_{\hat{\theta}}^2}$.) When $\theta = \mu$ is the target parameter, then $\hat{\theta} = \bar{y}$ and $\sigma_{\hat{\theta}}^2 = \sigma^2/n$, where σ^2 is the population variance. If the true value of σ^2 is known, this value should be used in the calculation of the confidence interval. If σ^2 is not known, there is no serious loss of accuracy if s^2 is substituted for σ^2 in the formula for the confidence interval. Similarly, if σ_1^2 and σ_2^2 are unknown, s_1^2 and s_2^2 can be substituted for these values in the formula for a large-sample confidence interval for $\theta = \mu_1 - \mu_2$. When $\theta = p$ is the target parameter, then $\hat{\theta} = \hat{p}$ and $\sigma_{\hat{\theta}}^2 = pq/n$. Because p is the unknown target parameter, $\sigma_{\hat{p}}^2$ cannot be evaluated. However, if we substitute $\hat{p}$ for p (and $1 - \hat{p}$ for q) in the formula for $\sigma_{\hat{p}}^2$, the resulting confidence interval will have approximately the stated confidence coefficient. Similar statements hold when $\hat{p}_1$ and $\hat{p}_2$ are used to estimate p_1 and p_2, respectively, in the formula for $\sigma_{\hat{p}_1 - \hat{p}_2}^2$. The theoretical justification for these substitutions will be provided in Section 9.3.

Table 8.3
Large-sample
confidence intervals
(n = 100)

Sample	LCL	Mean	UCL	s
1	7.26064	8.6714	10.0822	7.1977
2	7.59523	9.5291	11.4630	9.8668
3	7.68938	9.3871	11.0849	8.6619
4	8.94686	11.1785	13.4100	11.3856
5	7.96793	10.1508	12.3336	11.1369
6	7.61223	10.0101	12.4080	12.2340
7	7.55475	9.2754	10.9961	8.7791
8	7.48558	9.1436	10.8016	8.4593
9	7.71919	9.5219	11.3245	9.1973
10	7.95952	9.7012	11.4428	8.8859
11	7.34959	8.7958	10.2420	7.3785
12	8.71580	10.8458	12.9759	10.8675
*13	6.41955	8.1029	9.7863	8.5885
14	7.27538	8.9598	10.6441	8.5938
15	8.30285	10.2902	12.2776	10.1398
16	9.48735	11.5540	13.6206	10.5441
17	8.35947	10.8521	13.3448	12.7176
18	7.63421	9.9196	12.2049	11.6599
19	7.97129	10.1196	12.2679	10.9608
20	8.86613	10.9223	12.9785	10.4906
21	7.16061	9.2142	11.2679	10.4778
22	6.99569	9.2782	11.5607	11.6454
23	8.13781	10.2701	12.4024	10.8791
24	7.70300	9.4910	11.2790	9.1226
25	9.61738	11.9111	14.2048	11.7026
26	7.23197	9.2257	11.2194	10.1720
27	7.85056	9.9053	11.9601	10.4836
28	8.82373	11.1849	13.5462	12.0470
29	8.36303	10.4008	12.4387	10.3970
30	8.74332	10.4288	12.1144	8.5996
31	7.80013	9.7064	11.6127	9.7259
32	7.88003	10.0706	12.2611	11.1763
33	7.01555	8.6994	10.3833	8.5913
34	8.31036	10.3074	12.3045	10.1891
35	7.91383	9.8344	11.7549	9.7986
36	8.80795	10.5208	12.2337	8.7390
37	7.27322	9.1728	11.0725	9.6919
38	7.47957	9.1333	10.7871	8.4375
39	7.77926	9.8034	11.8275	10.3272
40	7.09624	9.0955	11.0947	10.2001
41	8.32909	10.1639	11.9986	9.3610
42	8.23895	10.2593	12.2797	10.3080
43	8.44813	10.4284	12.4087	10.1034
44	7.11788	8.8708	10.6237	8.9436
45	8.14050	10.0992	12.0580	9.9936
46	7.53466	9.4198	11.3050	9.6183
47	8.93450	11.0165	13.0986	10.6227

Sample	LCL	Mean	UCL	s
48	7.43185	8.9012	10.3705	7.4965
49	7.83028	9.7367	11.6432	9.7268
50	7.61207	9.3698	11.1275	8.9680
51	7.97829	9.7442	11.5101	9.0098
52	8.97313	11.2890	13.6049	11.8158
*53	5.99726	7.7422	9.4871	8.9027
54	7.51538	9.4089	11.3025	9.6609
55	8.59019	10.4373	12.2843	9.4238
56	8.49077	10.3914	12.2920	9.6971
57	8.12970	10.4056	12.6814	11.6115
*58	6.85175	8.3456	9.8395	7.6217
59	8.87405	10.9482	13.0225	10.5827
60	8.70548	10.2597	11.8140	7.9298
61	6.37467	8.3475	10.3203	10.0653
62	8.64835	10.6703	12.6923	10.3161
63	8.37157	10.6196	12.8676	11.4696
64	7.76632	9.6919	11.6174	9.8241
65	9.19745	11.5565	13.9155	12.0359
66	7.54964	9.1818	10.8140	8.3275
67	8.57961	10.8918	13.2039	11.7967
68	8.23986	10.0088	11.7778	9.0254
69	8.08091	9.9399	11.7989	9.4846
70	7.52910	9.2224	10.9156	8.6391
71	7.52705	9.4949	11.4628	10.0401
72	8.61466	10.7728	12.9310	11.0110
73	9.13542	11.2654	13.3953	10.8670
*74	6.73171	8.1351	9.5386	7.1604
75	7.78546	10.1938	12.6021	12.2874
76	8.04925	9.6668	11.2843	8.2526
77	7.61919	9.6068	11.5943	10.1407
*78	5.91167	7.7383	9.5649	9.3194
79	8.36219	10.4003	12.4384	10.3984
80	8.78276	11.0946	13.4065	11.7953
81	7.75108	9.4840	11.2170	8.8415
82	8.41534	10.4168	12.4183	10.2117
83	7.56210	9.1461	10.7300	8.0815
*84	6.51658	8.0757	9.6349	7.9549
85	8.73332	10.7988	12.8643	10.5381
86	7.10357	8.9350	10.7664	9.3439
87	7.46386	9.5521	11.6403	10.6541
88	7.40553	9.0024	10.5993	8.1473
89	7.70464	9.6524	11.6002	9.9376
90	8.16395	10.3755	12.5870	11.2834
91	8.08567	9.8488	11.6120	8.9958
92	8.45424	10.2642	12.0741	9.2344

(continued)

Table 8.3 (cont.)

Sample	LCL	Mean	UCL	s
*93	6.78362	8.1950	9.6064	7.2010
94	8.56442	10.6626	12.7608	10.7050
95	7.49852	9.2474	10.9962	8.9226
96	8.01983	10.9016	13.7833	14.7028
97	7.40014	9.1321	10.8640	8.8363
98	7.80386	9.6444	11.4849	9.3903
99	8.47046	10.3403	12.2101	9.5399
100	8.08183	10.0133	11.9447	9.8542

* The intervals for these samples do not include the true mean value.

Exercises

8.39   A survey conducted by the President's Commission on Pension Policy in the fall of 1979 revealed that a high proportion of Americans are very pessimistic about their prospects when they eventually retire (*Orlando Sentinel Star*, May 3, 1980).* When asked if they expected their retirement to be adequate, 62.9% of the 6100 respondents, full-time workers 18 years old or older, indicated that in their opinion, their income upon retirement definitely (or probably) would not be adequate. Find a 95% confidence interval for the proportion of all workers, 18 or older, who do not believe that, when they retire, their retirement income will be adequate. Interpret the interval.

8.40   A *Wall Street Journal* (July 23, 1981)[†] report on a survey by Warwick, Welsh & Miller, a New York advertisement agency, indicates that matters of taste cannot be ignored in television advertising. Based on a mail survey of 3440 people, 40% indicated that they found TV commercials to be in poor taste; 55% say that they avoid products whose commercials were judged to be in poor taste; and, of this latter group, only 20% ever complained to a TV station or an advertiser about their dissatisfaction.

(a)   Find a 95% confidence interval for the percentage of TV viewers who find TV commercials to be in poor taste.

(b)   Find a 95% confidence interval for the percentage of TV viewers who avoid products which use TV commercials that they consider to be in poor taste.

(c)   Find a 95% confidence interval for the percentage of those who avoid products who have complained to the TV station or the advertiser about poor taste in a TV commercial.

8.41   The administrators for a hospital wished to estimate the average number of days required for treatment of patients between the ages of 25 and 34. A random sample of 500 hospital patients between these ages produced a mean and standard deviation equal to 5.4 and 3.1 days, respectively. Construct a 95% confidence interval for the mean length of stay for the population of patients from which the sample was drawn. Use a confidence coefficient of .95.

8.42   The *Florida Alligator* of May 23, 1979, reported that "law students oppose the death penalty." This statement was made on the basis of a sample survey in which eighty-six law students were randomly selected and interviewed. Of those interviewed, 52% said that they were opposed to the death penalty. Using this information, construct a 95% confidence interval for the true proportion of law students who oppose the death penalty. Is the statement quoted above justified?

8.43   According to *Environment News* (September 1975), acid rain, caused by the reaction of certain air pollutants with rainwater, appears to be a growing problem in the northeastern United States. (Acid rain affects the soil and causes corrosion on exposed metal surfaces.) Pure rain falling through clean air registers a pH value (pH is a measure of acidity; 0 is acid, 14 is alkaline) of 5.7. Suppose that water samples from 40 rainfalls are analyzed for pH and that $\bar{x}$ and s are equal to 3.7 and .5, respectively. Find a 99% confidence interval for the mean pH in rainfalls and interpret the interval. What assumption must be made in order that the confidence interval be valid?

8.44   A small amount of the trace element selenium, from 50 to 200 micrograms (μg) per day, is considered essential to good health (*Prevention*, September 1980). Suppose that random samples of $n_1 = n_2 = 30$ adults were selected from two regions of the United States, and a day's intake of selenium, from both liquids and solids, was recorded for each person. The mean and standard deviation of the selenium daily intakes for the 30 adults from Region 1 were $\bar{y}_1 = 167.1$ and $s_1 = 24.3$ μg, respectively. The corresponding statistics for the 30 adults from Region 2 were $\bar{y}_2 = 140.9$ and $s_2 = 17.6$. Find a 95 percent confidence interval for the difference in the mean selenium intake for the two regions.

8.45   The following statistics are the result of an experiment conducted by P. I. Ward to investigate a theory concerning the moulting behavior of the male *Gammarus pulex*, a small crustacean ("*Gammarus pulex* Control their Moult Timing to Secure Mates," *Animal Behaviour* 32 [1984]). If a male has to moult while paired with a female, he must release her and so loses her. The theory is that the male *Gammarus pulex* is able to postpone the time to moult and thereby reduce the possibility of losing his mate. Ward randomly assigned 100 pairs of males and females to two groups of 50 each. Pairs in the first group were maintained together (Normal); those in the second group were separated (Split). The length of time to moult was recorded for both males and females and the means, standard deviations, and sample sizes are shown in the table. (The number of crustaceans in each of the four samples is less than 50 because some in each group did not survive until moulting time.) Find a 99% confidence interval for the difference in mean moult time for "Normal" males versus those "Split" from their mates. Interpret the interval.

	Time to moult (days)		
	Mean	s	n
Males			
Normal	24.8	7.1	34
Split	21.3	8.1	41
Females			
Normal	8.6	4.8	45
Split	11.6	5.6	48

8.46   Should professional athletes be tested for drugs? Reporting on this question, the *Orlando Sentinel* (October 31, 1985) noted that Donald Fehr, executive director of the Major League Baseball Player's Association, "criticized mandatory testing as being unnecessary."

Baseball commissioner Peter Ueberroth and the public do not seem to agree. A *Washington Post*–ABC News survey found that 73% of 1506 people interviewed favored drug tests for professional athletes. Sixty-eight percent said that professional athletes using drugs for the first time should be banned or suspended from professional sports.

(a)   Find a 95% confidence interval for the percent of the public that favors drug tests for professional athletes.

(b)   Upon what assumptions is your confidence interval in part (a) based?

8.47   A study conducted at the University of Massachusetts Medical School in Worcester indicates that heart attack patients given beta blockers have a far lower fatality rate in hospital than those not given beta blockers. The in-hospital fatality rate for 879 Massachusetts heart attack victims using beta blockers was 9% compared to 27% of 1906 patients who did not use the drug (*USA Today*, March 12, 1985). Find a 99% confidence interval for the difference in in-hospital fatality rates between those heart attack patients on beta blockers and those not using the drug.

8.48   The Florida Opinion Poll of October 25–31, 1984, includes telephone interviews of 497 randomly selected adult residents of Florida. When asked whether the sex of a political candidate was an important consideration, 62% of the men and 49% of the women said that the sex of the candidate made no difference. If 241 men and 256 women were interviewed, construct a 99% confidence interval for the difference in the proportion of men and women in Florida for whom the sex of the candidate would make no difference (*Gainesville Sun*, November 11, 1984).

8.49   According to *Environment News* (April 1975), "The continuing analysis of lead levels in the drinking water of several Boston communities has verified elevated lead levels in the water supplies of Somerville, Brighton and Beacon Hill." Preliminary results of a study carried out in 1974 found that "20 percent of 248 households tested in those communities showed levels exceeding the U.S. Public Health Service standard of 50 parts per million." In contrast, in Cambridge, which adds anticorrosives to its water, "only 5 percent of the 110 households tested showed lead levels exceeding the standard." Find a 95% confidence interval for the difference in the proportions of households which have lead levels exceeding the standard between the communities of Somerville, Brighton, and Beacon Hill and the community of Cambridge.

8.50   For a comparison of the rates of defectives produced by two assembly lines, independent random samples of 100 items are selected from each line. Line *A* yielded 18 defectives in the sample, and line *B* yielded 12 defectives. Find a 98% confidence interval for the true difference in proportions of defectives for the two lines. (Is there evidence here to suggest that one line produces a higher proportion of defectives than the other?)

8.51   The *Research Quarterly* for May 1979 reports the results of a study on the relationship between participation in sports and manual dexterity. A random sample of 37 second graders who participated in sports had manual dexterity scores with a mean of 32.19 and a standard deviation of 4.34. An independent random sample of 37 second graders who did not participate in sports had manual dexterity scores with a mean of 31.68 and a standard deviation of 4.56. Estimate the difference in true mean scores for the two groups with a 90% confidence interval. Does it appear that the average manual dexterity score for those who participate in sports differs from the mean score of those who do not participate in sports?

8.52   One method for solving the electric power shortage uses the construction of floating nuclear power plants located a few miles offshore in the ocean. Because there is concern about the possibility of a ship collision with the floating (but anchored) plant, an estimate of the density of ship traffic in the area is needed. The number of ships passing within 10 miles of the proposed power-plant location per day, recorded for $n = 60$ days during July and August, possessed sample mean and variance equal to

$$\bar{x} = 7.2$$
$$s^2 = 8.8$$

(a)   Find a 95% confidence interval for the mean number of ships passing within 10 miles of the proposed power-plant location during a one-day time period.

(b)   The density of ship traffic was expected to decrease during the winter months. A sample of $n = 90$ daily recordings of ship sightings for December, January, and February gave the following mean and variance:

$$\bar{x} = 4.7$$
$$s^2 = 4.9$$

Find a 90% confidence interval for the difference in mean density of ship traffic between the summer and winter months.

(c)   What is the population associated with your estimate, part (b)? What could be wrong with the sampling procedure, parts (a) and (b)?

*8.53   Suppose (Y_1, Y_2, Y_3, Y_4) has a multinomial distribution with n trials and probabilities for the four cells denoted by p_1, p_2, p_3, and p_4. Just as in the binomial case, any linear combination of Y_1, Y_2, Y_3, and Y_4 will be approximately normally distributed for large n.

(a)   Determine the variance of $Y_1 - Y_2$. (Recall that the Y_i's are dependent.)

(b)   A study of attitudes among residents of Florida with regard to policies for handling nuisance alligators in urban areas showed the following. Among 500 people sampled and presented with four management choices, 6% said the alligators should be completely protected, 16% said they should be destroyed by wildlife officers, 52% said they should be relocated alive, and 26% said that a regulated commercial harvest should be allowed. Estimate the difference between the population proportion favoring complete protection and the population proportion favoring destruction by wildlife officers. Use a confidence coefficient of .95.

*8.54   The *Journal of Communication*, Winter 1978, reports on a study of the viewing of violence on TV. Samples from populations with low viewing rates (10–19 programs per week) and high viewing rates (40–49 programs per week) were divided into age groups, and Y, the number of persons watching a high number of violent programs, was recorded. The data for two age groups are as shown in the accompanying table, with n_i denoting the sample size for each cell. If Y_1, Y_2, Y_3, and Y_4 have independent binomial distributions with parameters p_1, p_2, p_3, and p_4, respectively, find a 95% confidence interval for $(p_3 - p_1) - (p_4 - p_2)$. This function of the p_i's represents a comparison between the change in viewing habits for young adults and the corresponding change for older adults, as we move from those with low viewing rates to those with high viewing rates. (The data suggest that the rate of viewing violence may increase with young adults but decrease with older adults.)

Viewing Rate	Age Group			
	16–34		55 and Over	
Low	$y_1 = 20$	$n_1 = 31$	$y_2 = 13$	$n_2 = 30$
High	$y_3 = 18$	$n_3 = 26$	$y_4 = 7$	$n_4 = 28$

8.7 Selecting the Sample Size

The design of an experiment is essentially a plan for purchasing a quantity of information, which, like any other commodity, may be acquired at varying prices depending upon the manner in which the data are obtained. Some measurements contain a large amount of information concerning the parameter of interest, whereas others may contain little or none. Because the sole product of research is information, we should make its purchase at minimum cost.

The sampling procedure, or *experimental design* as it is usually called, affects the quantity of information per measurement. This, together with the sample size n, controls the total amount of relevant information in a sample. At this point in our study we will be concerned with the simplest sampling situation—random sampling from a relatively large population—and will devote our attention to the selection of the sample size n.

The researcher makes little progress in planning an experiment before encountering the problem of selecting the sample size. Indeed, perhaps one of the most frequent questions asked of the statistician is, How many measurements should be included in the sample? Unfortunately, the statistician cannot answer this question without knowing how much information the experimenter wishes to buy. Certainly the total amount of information in the sample will affect the measure of goodness of the method of inference and must be specified by the experimenter. Referring specifically to estimation, we would like to know how accurate the experimenter wishes the estimate to be. This accuracy may be stated by specifying a bound on the error of estimation.

For instance, suppose that we wish to estimate the average daily yield of a chemical, μ, and we wish the error of estimation to be less than 5 tons with a probability of .95. Because approximately 95% of the sample means will lie within $2\sigma_{\bar{y}}$ of μ in repeated sampling, we are asking that $2\sigma_{\bar{y}}$ equal 5 tons (see Figure 8.8). Then

$$\frac{2\sigma}{\sqrt{n}} = 5$$

Figure 8.8
The approximate
distribution of $\bar{Y}$ for
large samples

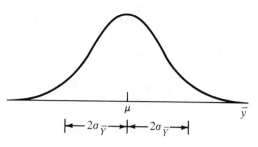

Solving for n, we obtain

$$n = \frac{4\sigma^2}{25}$$

You can see that we cannot obtain a numerical value for n unless the population standard deviation σ is known. This is exactly what we would expect, because the variability of $\bar{Y}$ depends upon the variability of the population from which the sample was drawn.

   Lacking an exact value for σ, we would use the best approximation available, such as an estimate s obtained from a previous sample or knowledge of the range in which the measurements will fall. Because the range is approximately equal to 4σ (the empirical rule), one-fourth of the range will provide an approximate value of σ. For our example suppose we know that the range of the daily yields is approximately 84 tons. Then $\sigma \approx 21$ and

$$n = \frac{4\sigma^2}{25} \approx \frac{(4)(21)^2}{25} = 70.56$$

or

$$n = 71$$

Using a sample size $n = 71$, we would be reasonably certain (with probability approximately equal to .95) that our estimate will lie within $2\sigma_{\bar{Y}} = 5$ tons of the true average daily yield.

   Actually we would expect the error of estimation to be much less than 5 tons. According to the empirical rule, the probability is approximately equal to .68 that the error of estimation would be less than $\sigma_{\bar{Y}} = 2.5$ tons. You will note that the probabilities, .95 and .68, used in these statements will be inexact because σ was approximated. Although this method of choosing the sample sizes is only approximate for a specified desired accuracy of estimation, it is the best available and is certainly better than selecting the sample size on the basis of our intuition.

The method of choosing the sample size for all the large-sample estimation procedures outlined in Table 8.1 is identical to that just described. The experimenter must specify a desired bound on the error of estimation and an associated confidence level, $1 - \alpha$. For example, if the parameter is θ and the desired bound is B, we would equate

$$z_{\alpha/2}\sigma_{\hat{\theta}} = B$$

where $z_{\alpha/2}$ is the z value defined in Section 8.6; that is,

$$P(Z > z_{\alpha/2}) = \frac{\alpha}{2}$$

We will illustrate these ideas with examples.

EXAMPLE 8.9   The reaction of an individual to a stimulus in a psychological experiment may take one of two forms, A or B. If an experimenter wishes to estimate the probability p that a person will react in manner A, how many people must be included in the experiment? Assume that he will be satisfied if the error of estimation is less than .04 with probability equal to .90. Assume also that he expects p to lie somewhere in the neighborhood of .6.

Solution   Since the confidence coefficient is $1 - \alpha = .90$, α must equal .10 and $\alpha/2 = .05$. The z value corresponding to an area equal to .05 in the upper tail of the z distribution is $z_{\alpha/2} = 1.645$. We then require that

$$1.645\sigma_{\hat{p}} = .04 \qquad \text{or} \qquad 1.645\sqrt{\frac{pq}{n}} = .04$$

Because the variability of $\hat{p}$ is dependent upon p, which is unknown, we must use the guessed value of $p = .6$ provided by the experimenter as an approximation. Then

$$1.645\sqrt{\frac{(.6)(.4)}{n}} = .04$$

$$n = 406$$

(Note: If we did not know that $p \approx .6$, we would use $p = .5$, which would yield the maximum possible value for n, $n = 423$.)   ∽

EXAMPLE 8.10 An experimenter wishes to compare the effectiveness of two methods of training industrial employees to perform a certain assembly operation. The selected employees are to be divided into two groups of equal size, the first receiving training method 1 and the second, training method 2. Each employee will perform the assembly operation, and the length of assembly time will be recorded. It is expected that the measurements for both groups will have a range of approximately 8 minutes. If the estimate of the difference in mean time to assemble is desired correct to within 1 minute with a probability equal to .95, how many workers must be included in each training group?

Solution Equating $2\sigma_{(\bar{Y}_1 - \bar{Y}_2)}$ to 1 minute, we obtain

$$2\sqrt{\frac{\sigma_1^2}{n_1} + \frac{\sigma_2^2}{n_2}} = 1$$

Or because we desire n_1 to equal n_2, we may let $n_1 = n_2 = n$ and obtain the equation

$$2\sqrt{\frac{\sigma_1^2}{n} + \frac{\sigma_2^2}{n}} = 1$$

As noted earlier, the variability of each method of assembly is approximately the same, and hence $\sigma_1^2 = \sigma_2^2 = \sigma^2$. Because the range, equal to 8 minutes, is approximately equal to 4σ, then

$$4\sigma \approx 8$$
$$\sigma \approx 2$$

Substituting this value for σ_1 and σ_2 in this equation, we obtain

$$2\sqrt{\frac{(2)^2}{n} + \frac{(2)^2}{n}} = 1$$

Solving, we have $n = 32$, so each group should contain $n = 32$ members.

∽

Exercises

8.55 Let Y be a binomial random variable with parameter p. Find the sample size necessary to estimate p to within .05 with probability .95 in the following situations:

(a) if p is thought to be approximately .9,

(b) if no information about p is known (use $p = .5$ in estimating the variance of $\hat{p}$).

8.56   If a wildlife service wishes to estimate the mean number of days of hunting per hunter for all hunters licensed in the state during a given season, with a bound on the error of estimation equal to two hunting days, how many hunters must be included in the survey? Assume that data collected in earlier surveys have shown σ to be approximately equal to 10.

8.57   A Louis Harris and Associates survey conducted in 1984 for Philip Morris, Inc., showed an increase over the decade in attendance at movies, concerts, and the theater. The poll was based on a telephone survey of 1504 adults and the results were compared with similar polls conducted in 1973, 1975, and 1980 (*Wall Street Journal*, December 1, 1984).

    (a)   If the survey in 1973 was based on an equal number of interviews, how accurately could you estimate the difference in attendance proportions between 1973 and 1984? To answer this question, find the bound on the error of estimation based on a sample size of 1500 interviewees for each sample. Assume that both proportions are near .7 or .8, say .75.

    (b)   Suppose that you were designing the surveys and wished to estimate the difference in a pair of proportions correct to within .02 with probability equal to .9. How many interviewees would have to be included in each sample?

8.58   Refer to Exercise 8.42. How many students should have been interviewed by the newspaper in order to estimate the proportion of students opposing the death penalty, to within .1 with probability .95? Use the proportion from the previous sample in approximating the variance of the estimate.

8.59   Refer to Exercise 8.43 on the measurement of acidity in rainwater and suppose that you want to estimate the mean pH of rainfalls in an area that suffers heavy pollution due to the discharge of smoke from a power plant. Assume you know that σ is in the neighborhood of .5 pH and that you want your estimate to lie within .1 of μ with probability near .95. Approximately how many rainfalls must be included in your sample (one pH reading per rainfall)? Would it be valid to select all of your water specimens from a single rainfall? Explain.

8.60   Refer to Exercise 8.59. Suppose that you wish to estimate the difference between the mean acidity for rainfalls at two different locations, one in a relatively unpolluted area along the ocean and the other in an area subject to heavy air pollution. If you wish your estimate to be correct to the nearest .1 pH with probability near .90, approximately how many rainfalls (pH values) would have to be included in each sample? (Assume that the variance of the pH measurements is approximately .25 at both locations and that the samples will be of equal size.)

8.61   Refer to the comparison of the daily adult intake of selenium in two different regions of the United States in Exercise 8.44. Suppose that you wish to estimate the difference in the mean daily intake between the two regions correct to within 5 μg with probability equal to .90. If you plan to select an equal number of adults from the two regions (that is, $\mu_1 = \mu_2$), how large should n_1 and n_2 be?

8.62   Refer to Exercise 8.24. If the researcher wants to estimate the difference in proportions to within .05 with 90% confidence, how many graduates and nongraduates need to be interviewed? (Assume that an equal number will be interviewed from each group.)

8.63   Refer to Exercise 8.50. How many items should be sampled from each line if a 95% confidence interval for the true difference in proportions is to have width .2? Assume samples of equal size will be taken from each line.

8.64   Refer to Exercise 8.51. If another similar study is to be undertaken, how many observations should be included in each group in order to produce a 90% confidence interval for the true difference in means with a width of 2 units? Assume each group will contain the same number of observations.

8.8 Small-Sample Confidence Intervals for μ and μ₁ − μ₂

The confidence interval we will discuss in this section is based on the assumption that the experimenter's sample has been randomly selected from a normal population. It is appropriate for samples of any size and works satisfactorily even when the population is not normal, as long as the departure from normality is not excessive. That is, we rarely know the form of the population frequency distribution before we sample. So if a confidence interval is to be of any value, it must work reasonably well even when the population is not normal. *Working well* means that the confidence coefficient should not be affected by modest departures from normality. Experimental studies indicate that this particular confidence interval will maintain a confidence coefficient close to the experimenter's specified value for most mound-shaped probability distributions.

We assume that $Y_1, Y_2, \ldots, Y_n$ represents a random sample selected from a normal population, and we let $\bar{Y}$ and S^2 represent the sample mean and variance, respectively. We would like to construct a confidence interval for the population mean when $V(Y_i) = \sigma^2$ is unknown and the sample size is too small to apply the large-sample techniques of the previous section.

Recall from Theorems 7.1 and 7.3 and Definition 7.2 that

$$T = \frac{\bar{Y} - \mu}{S/\sqrt{n}}$$

has a t distribution with $(n - 1)$ degrees of freedom. This T will serve as the pivotal quantity for forming a confidence interval for μ. From Table 5, Appendix III, we can find values $t_{\alpha/2}$ and $-t_{\alpha/2}$ so that

$$P(-t_{\alpha/2} \leq T \leq t_{\alpha/2}) = 1 - \alpha$$

(See Figure 8.9.)

Figure 8.9
Location of $t_{\alpha/2}$
and $-t_{\alpha/2}$

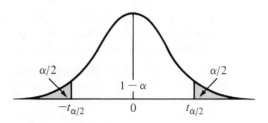

The t distribution has a density function very much like the standard normal except that the tails are thicker (as illustrated in Figure 7.2). Recall that the values of $t_{\alpha/2}$ depend upon the degrees of freedom, $(n - 1)$, as well as the confidence coefficient $(1 - \alpha)$.

The confidence interval for μ is developed just as in Example 8.6. In this case the resulting confidence interval for μ is of the form

$$\bar{Y} \pm t_{\alpha/2}\left(\frac{S}{\sqrt{n}}\right)$$

Note that $\bar{Y} - t_{\alpha/2}(S/\sqrt{n})$ is the lower confidence limit and $\bar{Y} + t_{\alpha/2}(S/\sqrt{n})$ is the upper confidence limit.

EXAMPLE 8.11   A manufacturer of gunpowder has developed a new powder, which was tested in eight shells. The resulting muzzle velocities, in feet per second, were as follows.

3005	2925	2935	2965
2995	3005	2937	2905

Find a confidence interval for the true average velocity μ for shells of this type, with confidence coefficient .95. Assume that muzzle velocities are approximately normally distributed.

Solution   If we assume the velocities Y_i to be normally distributed, the confidence interval for μ is

$$\bar{Y} \pm t_{\alpha/2}\left(\frac{S}{\sqrt{n}}\right)$$

For the given data, $\bar{y} = 2959$ and $s = 39.1$. From Table 5, Appendix III, $t_{\alpha/2} = t_{.025} = 2.365$ because there are $(n - 1) = 7$ degrees of freedom associated with this sample. Thus we obtain

$$2959 \pm 2.365\left(\frac{39.1}{\sqrt{8}}\right)$$

$$2959 \pm 32.7$$

as the observed confidence interval for μ.

Suppose that we are interested in comparing means from two normal populations, one with a mean of μ_1 and a variance of σ_1^2 and the other with a mean of μ_2 and a variance of σ_2^2. A confidence interval for $\mu_1 - \mu_2$ based on a T random variable can be constructed if we assume that $\sigma_1^2 = \sigma_2^2 = \sigma^2$.

If $\bar{Y}_1$ and $\bar{Y}_2$ are the respective sample means obtained from independent random samples, the large-sample confidence interval for $(\mu_1 - \mu_2)$ is developed from the random variable

$$Z = \frac{(\bar{Y}_1 - \bar{Y}_2) - (\mu_1 - \mu_2)}{\sqrt{\dfrac{\sigma_1^2}{n_1} + \dfrac{\sigma_2^2}{n_2}}}$$

which has approximately a standard normal distribution. Under the assumption $\sigma_1^2 = \sigma_2^2 = \sigma^2$, this ratio becomes

$$Z = \frac{(\bar{Y}_1 - \bar{Y}_2) - (\mu_1 - \mu_2)}{\sigma\sqrt{\dfrac{1}{n_1} + \dfrac{1}{n_2}}}$$

Now we need an estimator of the common variance σ^2 in order to construct a quantity with a t distribution.

Let $Y_{11}, Y_{12}, \ldots, Y_{1n_1}$ denote the random sample of size n_1 from the first population and let $Y_{21}, Y_{22}, \ldots, Y_{2n_2}$ denote an independent random sample from the second. Then

$$\bar{Y}_1 = \frac{1}{n_1}\sum_{i=1}^{n_1} Y_{1i} \quad \text{and} \quad \bar{Y}_2 = \frac{1}{n_2}\sum_{i=1}^{n_2} Y_{2i}$$

The usual unbiased estimator of the common variance σ^2 is obtained by pooling the sample data to obtain

$$S^2 = \frac{\displaystyle\sum_{i=1}^{n_1}(Y_{1i} - \bar{Y}_1)^2 + \sum_{i=1}^{n_2}(Y_{2i} - \bar{Y}_2)^2}{n_1 + n_2 - 2} = \frac{(n_1 - 1)S_1^2 + (n_2 - 1)S_2^2}{n_1 + n_2 - 2}$$

where S_i^2 is the sample variance from the ith sample, $i = 1, 2$. Note that

$$\frac{(n_1 + n_2 - 2)S^2}{\sigma^2} = \frac{\sum\limits_{i=1}^{n_1}(Y_{1i} - \bar{Y}_1)^2}{\sigma^2} + \frac{\sum\limits_{i=1}^{n_2}(Y_{2i} - \bar{Y}_2)^2}{\sigma^2}$$

is the sum of two independent χ^2 random variables with $(n_1 - 1)$ and $(n_2 - 1)$ degrees of freedom, respectively. Thus $(n_i + n_2 - 2)S^2/\sigma^2$ has a χ^2 distribution with $v = (n_1 + n_2 - 2)$ degrees of freedom. (See Theorems 7.2 and 7.3.) We now utilize this χ^2 variable and the Z defined in the previous paragraph to form a pivotal quantity. That is,

$$T = \frac{Z}{\sqrt{\dfrac{\chi^2}{v}}} = \left[\frac{(\bar{Y}_1 - \bar{Y}_2) - (\mu_1 - \mu_2)}{\sigma\sqrt{\dfrac{1}{n_1} + \dfrac{1}{n_2}}}\right]\left[\frac{1}{\sqrt{\dfrac{(n_1 + n_2 - 2)S^2}{\sigma^2(n_1 + n_2 - 2)}}}\right]$$

$$= \frac{(\bar{Y}_1 - \bar{Y}_2) - (\mu_1 - \mu_2)}{S\sqrt{\dfrac{1}{n_1} + \dfrac{1}{n_2}}}$$

has a t distribution with $(n_1 + n_2 - 2)$ degrees of freedom.

The confidence interval for $(\mu_1 - \mu_2)$ then has the form

$$(\bar{Y}_1 - \bar{Y}_2) \pm t_{\alpha/2}S\sqrt{\frac{1}{n_1} + \frac{1}{n_2}}$$

where $t_{\alpha/2}$ comes from the t distribution with $(n_1 + n_2 - 2)$ degrees of freedom.

EXAMPLE 8.12 An assembly operation in a manufacturing plant requires approximately a 1-month training period for a new employee to reach maximum efficiency. A new method of training was suggested and a test was conducted to compare the new method with the standard procedure. Two groups of nine new employees were trained for a period of 3 weeks, one group using the new method and the other following standard training procedure. The length of time (in minutes) required for each employee to assemble the device was recorded at the end of the 3-week period. The measurements are as shown in Table 8.4. Estimate the true mean

Table 8.4
Data for
Example 8.12

Procedure	Measurements								
Standard	32	37	35	28	41	44	35	31	34
New	35	31	29	25	34	40	27	32	31

difference, $(\mu_1 - \mu_2)$, with confidence coefficient .95. Assume that the assembly times are approximately normally distributed and that the variances of the assembly times are approximately equal for the two methods.

Solution

For the data in Table 8.4, with sample 1 denoting the standard procedure, we have

$$\bar{y}_1 = 35.22 \qquad\qquad \bar{y}_2 = 31.56$$

$$\sum_{i=1}^{9}(y_{1i} - \bar{y}_1)^2 = 195.56 \qquad \sum_{i=1}^{9}(y_{2i} - \bar{y}_2)^2 = 160.22$$

Hence

$$s^2 = \frac{195.56 + 160.22}{9 + 9 - 2} = 22.24$$

and

$$s = 4.71$$

Also, $t_{.025} = 2.120$ for $(n_1 + n_2 - 2) = 16$ degrees of freedom. The observed confidence interval is then

$$(\bar{y}_1 - \bar{y}_2) \pm t_{\alpha/2} s \sqrt{\frac{1}{n_1} + \frac{1}{n_2}}$$

$$(35.22 - 31.56) \pm (2.120)(4.71)\sqrt{\frac{1}{9} + \frac{1}{9}}$$

$$3.66 \pm 4.71$$

Note that this confidence interval can be written in the form $(-1.05, 8.37)$. Thus we see that the interval is fairly wide and includes both positive and negative values. If $\mu_1 - \mu_2$ really is positive, $\mu_1 > \mu_2$ and the standard procedure has a larger expected assembly time than the new procedure. If $\mu_1 - \mu_2$ is really negative, the reverse is true. Because the interval contains both positive and negative values, neither procedure can be said to differ from the other.

As the sample size n gets large, the number of degrees of freedom for the t distribution will increase, and the t distribution can be approximated quite closely by the standard normal distribution. Thus the small-sample confidence intervals of this section are equivalent to the large-sample confidence intervals of Section 8.6 for large n (or large n_1 and n_2). The intervals are nearly equivalent when the degrees of freedom exceed 30.

Summary of Small-Sample Confidence Intervals for Means of Normal Distributions with Unknown Variance(s)

Parameter	Confidence Interval (v = degrees of freedom)
μ	$\bar{Y} \pm t_{\alpha/2}\left(\dfrac{S}{\sqrt{n}}\right), \qquad v = n-1$
$\mu_1 - \mu_2$	$(\bar{Y}_1 - \bar{Y}_2) \pm t_{\alpha/2} S \sqrt{\dfrac{1}{n_1} + \dfrac{1}{n_2}}$

where

$$S^2 = \frac{(n_1 - 1)S_1^2 + (n_2 - 1)S_2^2}{n_1 + n_2 - 2} \quad \text{and} \quad v = n_1 + n_2 - 2$$

(requires assumption that $\sigma_1^2 = \sigma_2^2$).

Exercises

8.65   The mean and standard deviation of the percent body fat measurements on fourteen elite women volleyball players, members of the Women's University World Games Team, were 17.9 and 3.6%, respectively (J. Puhl, S. Case, S. Fleck, and P. V. Handel, "Physical and Physiological Characteristics of Elite Volleyball Players," *Research Quarterly for Exercise and Sport* 53, no. 3 [1982]). Suppose that these fourteen women can be viewed as a random sample selected from among all elite women volleyball players. Find a 95% confidence interval for the mean percent body fat for this population of women volleyball players.

8.66   The carapace lengths of ten lobsters examined in a study of the infestation of the *T. orientalis* lobster by two types of barnacles, *O. tridens* and *O. lowei* are in the following table. Find a 95% confidence interval for the mean carapace length of *T. orientalis* lobsters caught in the seas in the vicinity of Singapore (W. B. Jeffries, H. K. Voris, and C. M. Yang, "Diversity and Distribution of the Pedunculate Barnacle *Octolasmis* Gray, 1825 Epizoic on the Scyllarid Lobster, *Thenus orientalis* (Lund, 1793)," *Crustaceana* 46, no. 3 [1984]).

Lobster field number	A061	A062	A066	A070	A067	A069	A064	A068	A065	A063
Carapace length (mm)	78	66	65	63	60	60	58	56	52	50

8.67   The scores of a random sample of sixteen people who took the Graduate Record Examination had a mean of 540 and a standard deviation of 50. Construct a 95% confidence interval for the population mean GRE score, assuming the scores are normally distributed.

8.68   The rate of oxygen consumption is a measurement of physiological activity important to runners. *Research Quarterly*, May 1979, reports on the differences between oxygen consumption rates for college men trained by two different methods, one involving continuous training for a period of time each day and the other involving intermittent training of about the same overall duration. The means, standard deviations, and sample sizes are shown in the accompanying table. (The measurements are in milliliters per kilogram-minute.) If the measurements are assumed to come from normally distributed populations with equal variances, estimate the difference between the population means, with confidence coefficient .95.

Continuous Training	Intermittent Training
$n_1 = 9$	$n_2 = 7$
$\bar{y}_1 = 43.71$	$\bar{y}_2 = 39.63$
$s_1 = 5.88$	$s_2 = 7.68$

8.69   Fifteen resistors were randomly selected from the output of a process supposedly producing 10-ohm resistors. The fifteen resistors actually showed a sample mean of 9.8 ohms and a sample standard deviation of 0.5 ohm. Find a 95% confidence interval for the true mean resistance of the resistors produced by this process. Assume resistance measurements are approximately normally distributed.

8.70   Two new drugs were given to patients with heart disease. The first drug lowered the blood pressure of sixteen patients an average of 11 points, with a standard deviation of 6 points. The second drug lowered the blood pressure of another twenty patients an average of 12 points, with a standard deviation of 8 points. Determine a 95% confidence interval for the difference in the mean reductions in blood pressure, assuming the measurements are normally distributed with equal variances.

8.71   The Environmental Protection Agency has collected data on LC50 measurements (concentration killing 50% of test animals) for certain chemicals likely to be found in freshwater rivers and lakes. (See Exercise 7.3 for additional details.) For certain species of fish the LC50 measurements for DDT in twelve experiments were as follows:

$$16, 5, 21, 19, 10, 5, 8, 2, 7, 2, 4, 9$$

(measurements in parts per million). Estimate the true mean LC50 for DDT, with confidence coefficient .90. Assume the LC50 measurements have an approximately normal distribution.

8.72   Refer to Exercise 8.71. Another common insecticide, Diazinon, gave LC50 measurements in three experiments of 7.8, 1.6, and 1.3.

(a)   Estimate the mean LC50 for Diazinon with a 90% confidence interval.

(b)   Estimate the difference between the mean LC50 for DDT and that for Diazinon with a 90% confidence interval. What assumptions are necessary for the method used to be valid?

8.73   *Research Quarterly,* May 1979, reports on a study of impulses applied to the ball by tennis rackets of various construction. Three measurements on ball impulse were taken on each type of racket. For a wood racket the mean was 2.41 and the standard deviation was .02. For a graphite racket the mean was 2.22 and the standard deviation was .07. Estimate the difference between true mean impulses for the two rackets, with confidence coefficient .98. What assumptions are necessary for the method used to be valid?

8.74   Seasonal ranges (in hectares) for alligators were monitored on a lake outside Gainesville, Florida, by biologists from the Florida Game and Fish Commission. Five alligators monitored in the spring showed ranges of 8.0, 12.1, 8.1, 18.2, 31.7. Four different alligators monitored in the summer showed ranges of 102.0, 81.7, 54.7, 50.7. Estimate the difference between mean spring and summer ranges with a 95% confidence interval. What assumptions did you make?

8.75   Copper produced by sintering (heating without melting) a powder under certain conditions is then measured for porosity (the volume fraction due to voids) in a certain laboratory. A sample of $n_1 = 4$ independent porosity measurements shows a mean of $\bar{x}_1 = 0.22$ and a variance of $s_1^2 = 0.0010$. A second laboratory repeats the same process on an identical powder and gets $n_2 = 5$ independent porosity measurements with $\bar{x}_2 = 0.17$ and $s_2^2 = 0.0020$. Estimate the true difference between the population means $(\mu_1 - \mu_2)$ for these two laboratories, with confidence coefficient 0.95.

*8.76   A factory operates with two machines of type A and one machine of type B. The weekly repair costs Y for type A machines are normally distributed with mean μ_1 and variance σ^2. The weekly repair costs X for machines of type B also are normally distributed but with mean μ_2 and variance $3\sigma^2$. The expected repair cost per week for the factory is then $2\mu_1 + \mu_2$. If you are given a random sample $Y_1, \ldots, Y_n$ on costs of type A machines and an independent random sample $X_1, \ldots, X_m$ on costs for type B machines, show how you would construct a 95% confidence interval for $2\mu_1 + \mu_2$. (Assume σ^2 is not known.)

8.9 A Confidence Interval for σ^2

Recall that the population variance σ^2 is a number that quantifies the amount of variability in the population. Many times the actual value of σ^2 will be unknown to an experimentor and he or she will need to estimate σ^2. In Section 8.3 we proved that $S^2 = [1/(n-1)] \sum_{i=1}^{n} (Y_i - \bar{Y})^2$ is an unbiased estimator for σ^2. Throughout our construction of confidence intervals for μ, we used S^2 to estimate σ^2 when σ^2 was unknown.

In addition to requiring information about σ^2 to calculate confidence intervals for μ and $\mu_1 - \mu_2$, we may be interested in forming a confidence interval for σ^2. For example, if we performed a careful chemical analysis of tablets of a particular medication, we would be interested in the mean amount of active ingredient per tablet *and* the amount of tablet-to-tablet variability, as quantified by σ^2. Obviously, for a medication we would desire a small amount of tablet-to-tablet variation and hence a small σ^2.

To proceed with our interval estimation procedure, we require the existence of a pivotal quantity. Again, assume that we have a random sample $Y_1,$

$Y_2, \ldots, Y_n$ from a normal distribution with mean μ and variance σ^2, both unknown. We know from Theorem 7.3 that

$$\frac{\sum_{i=1}^{n} (Y_i - \bar{Y})^2}{\sigma^2} = \frac{(n-1)S^2}{\sigma^2}$$

has a χ^2 distribution with $(n - 1)$ degrees of freedom. We can then proceed, by the pivotal method, to find two numbers χ_L^2 and χ_U^2 such that

$$P\left[\chi_L^2 \leq \frac{(n-1)S^2}{\sigma^2} \leq \chi_U^2\right] = 1 - \alpha$$

for any confidence coefficient $(1 - \alpha)$. (The subscripts L and U stand for *lower* and *upper*, respectively.) The χ^2 density function is not symmetric, and so there is some freedom in the choice of χ_L^2 and χ_U^2. We would like to find the shortest interval that includes probability $(1 - \alpha)$, but this generally is difficult. We compromise by choosing points that cut off equal tail areas, as indicated in Figure 8.10.

Figure 8.10
Location of χ_L^2 and χ_U^2

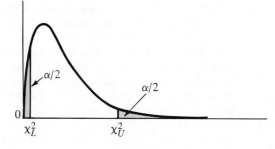

A reordering of the inequality in the probability statement gives

$$P\left[\frac{(n-1)S^2}{\chi_U^2} \leq \sigma^2 \leq \frac{(n-1)S^2}{\chi_L^2}\right] = 1 - \alpha$$

and hence the confidence interval for σ^2 is as follows.

A Confidence Interval for σ^2

$$\left(\frac{(n-1)S^2}{\chi_U^2}, \frac{(n-1)S^2}{\chi_L^2}\right)$$

EXAMPLE 8.13   An experimenter wanted to check the variability of equipment designed to measure the volume of an audio source. Three independent measurements recorded by this equipment for the same sound were 4.1, 5.2, and 10.2. Estimate σ^2 with confidence coefficient .90.

Solution   If normality of the measurements recorded by this equipment can be assumed, the confidence interval just developed applies. For the data given, $s^2 = 10.57$. With $\alpha/2 = .05$ and $(n - 1) = 2$ degrees of freedom, Table 6, Appendix III, gives $\chi_L^2 = .103$ and $\chi_U^2 = 5.991$. Thus the confidence interval for σ^2 is

$$\left(\frac{(n - 1)s^2}{\chi_U^2}, \frac{(n - 1)s^2}{\chi_L^2} \right)$$

$$\left(\frac{(2)(10.57)}{5.991}, \frac{(2)(10.57)}{.103} \right)$$

$$(3.53, 205.24)$$

Note that this interval for σ^2 is very wide, primarily because n is quite small.

Exercises

8.77   In 1978, the EPA set a maximum noise level for heavy trucks at 83 decibels (*Environment News*, October 1976). How this limit is applied will greatly affect the industry and the public. One way to apply the limit would be to require all trucks to conform to the noise limit. A second but less satisfactory method would be to require the truck fleet mean noise level to be less than the limit. If the latter were the rule, variation in the noise level from truck to truck would be important because a large value of σ^2 would imply many trucks exceeding the limit, even if the mean fleet level was 83 decibels. A random sample of six heavy trucks in 1977 produced the following noise levels (in decibels):

$$85.4, \ 86.8, \ 86.1, \ 85.3, \ 84.8, \ 86.0.$$

Use these data to construct a 90% confidence interval for σ^2, the variance of the truck noise emission readings. Interpret your results.

8.78   In Exercise 8.66 we gave the carapace lengths of ten mature *T. orientalis* lobsters caught in the seas in the vicinity of Singapore. The data are reproduced below (W. B. Jeffries, H. K. Voris, and C. M. Yang, "Diversity and Distribution of the Pedunculate Barnacle *Octolasmis* Gray, 1825 Epizoic on the Scyllarid Lobster, *Thenus orientalis* (Lund, 1793)," *Crustaceana* 46, no. 3 [1984]). Suppose that you wished to describe the variability of the carapace lengths of this population of lobsters. Find a 90% confidence interval for the population variance σ^2.

Lobster field number	A061	A062	A066	A070	A067	A069	A064	A068	A065	A063
Carapace length (mm)	78	66	65	63	60	60	58	56	52	50

8.79 A random sample of twenty-one engineers was selected from a large group of engineers employed by an electronics manufacturer. The sample standard deviation of working hours per week was 7 hours. Determine a 90% confidence interval for the population variance of working hours for all engineers employed by the manufacturer, assuming these measurements are normally distributed.

8.80 In laboratory work it is desirable to run careful checks on the variability of readings produced on standard samples. In a study of the amount of calcium in drinking water undertaken as part of a water quality assessment, the same standard was run through the laboratory six times at random intervals. The six readings, in parts per million, were 9.54, 9.61, 9.32, 9.48, 9.70, 9.26. Estimate σ^2, the population variance for readings on this standard, in a 90% confidence interval.

8.81 The ages of a random sample of five university professors are 39, 54, 61, 72, and 59. Using this information, find a 99% confidence interval for the population standard deviation of the ages of all professors at the university, assuming these ages are normally distributed.

8.82 A precision instrument is guaranteed to read accurately to within 2 units. A sample of four instrument readings on the same object yielded the measurements 353, 351, 351, and 355. Find a 90% confidence interval for the population variance. What assumptions are necessary? Does the guarantee seem reasonable?

8.10 Summary

The objective of many statistical investigations is to make inferences about population parameters based on sample data. Often these inferences are in the form of estimates, either point estimates or interval estimates.

In this chapter we have discussed the fact that we like to have unbiased estimators with small variance. The goodness of an estimator $\hat{\theta}$ can be measured by $\sigma_{\hat{\theta}}$ because the error of estimation will generally be smaller than $2\sigma_{\hat{\theta}}$ with high probability.

Interval estimates of many parameters, such as μ and p, can be derived from the normal distribution for large sample sizes because of the central limit theorem. If sample sizes are small, the normality of the population must be assumed and the t distribution is used in deriving confidence intervals.

If sample measurements have been selected from a normal distribution, a confidence interval for σ^2 can be developed through use of the χ^2 distribution.

References and Further Readings

1. Hoel, P. G. *Introduction to Mathematical Statistics.* 5th ed. New York: Wiley, 1984.

2. Hogg, R. V., and Craig, A. T. *Introduction to Mathematical Statistics.* 4th ed. New York: Macmillan, 1978.

3. Mood, A. M.; Graybill, F. A.; and Boes, D. *Introduction to the Theory of Statistics.* 3d ed. New York: McGraw-Hill, 1974.

Supplementary Exercises

8.83 According to an Associated Press–NBC News nationwide telephone poll, 65% of a total of 1601 adults believe that the United States should reduce immigration (*Philadelphia Inquirer*, August 17, 1981). The percentage favoring reduction in immigration varied from one social group to another and tended to be larger in those states most affected by the influx. Suppose that random samples of $n_1 = n_2 = 100$ adults selected from Florida and Texas, respectively, produced $x_1 = 72$ of the Florida respondents favoring reduction of immigration as opposed to $x_2 = 78$ from Texas.

   (a) Find a 95% confidence interval for the difference in the proportions of adults in Florida and Texas who favor a reduction in immigration.

   (b) Find a 95% confidence interval for the difference in the proportion of adults in Florida favoring a reduction in immigration versus the national proportion.

   (c) Find a 95% confidence interval for the difference in the proportion of adults in Texas favoring a reduction in immigration versus the national proportion.

8.84 Refer to Exercise 8.83 and the Associated Press–NBC News nationwide telephone poll of opinions concerning the limits on immigration. Suppose that you want to estimate the difference between the proportions of adults in Florida and Texas favoring a reduction of immigration and that you want your estimate to lie within .08 of the true difference with probability equal to .95. If you plan to select equal sample sizes; that is, $n_1 = n_2$, from the two states, how large should n_1 and n_2 be?

8.85 A chemist has prepared a product designed to kill 60% of a particular type of insect. What sample size should be used if she desires to be 95% confident that she is within .02 of the true fraction of insects killed?

8.86 To estimate the proportion of unemployed workers in Panama, an economist selected at random 400 persons from the working class. Of these, 25 were unemployed.

   (a) Estimate the true proportion of unemployed workers and place bounds on the error of estimation.

   (b) How many persons must be sampled to reduce the bound on error to .02?

8.87 Past experience shows that the standard deviation of the yearly income of textile workers in a certain state is $400. How many textile workers would you need to sample if you wished to estimate the population mean to within $50.00, with a probability of .95?

8.88 How many voters must be included in a sample collected to estimate the fraction of the popular vote favorable to a presidential candidate in a national election if the estimate is desired correct to within .005? Assume that the true fraction will lie somewhere in the neighborhood of .5. Use a confidence coefficient of approximately .95.

8.89 In a poll taken among college students, 300 of 500 fraternity men favored a certain proposition, whereas 64 of 100 nonfraternity men favored it. Estimate the difference in the fractions favoring the proposition and place a two-standard-deviation bound on the error of estimation.

8.90 Refer to Exercise 8.89. How many fraternity and nonfraternity students must be included in a poll if we wish to estimate the difference in the fractions correct to within .05? Assume that the groups will be of equal size and that $p = .6$ will suffice as an approximation of both fractions.

8.91   A chemical process has produced, on the average, 800 tons of chemical per day. The daily yields for the past week are 785, 805, 790, 793, and 802 tons. Estimate the mean daily yield, with confidence coefficient .90, from the data. What assumptions did you make?

8.92   Refer to Exercise 8.91. Find a 90% confidence interval for σ^2, the variance of the daily yields.

8.93   The main stem growth, measured for a sample of seventeen 4-year-old red pine trees, produced a mean of 11.3 inches and a standard deviation of 3.4 inches. Find a 90% confidence interval for the mean main stem growth of a population of 4-year-old red pine trees subjected to similar environmental conditions. Assume the growths are normally distributed.

8.94   Owing to the variability of trade-in allowance, the profit per new car sold by an automobile dealer varies from car to car. The profits per sale (in hundreds of dollars), tabulated for the past week, were 2.1, 3.0, 1.2, 6.2, 4.5, and 5.1. Find a 90% confidence interval for the mean profit per sale. What assumptions are necessary to validly employ the technique you used?

8.95   A mathematics test is given to a class of fifty students randomly selected from high school 1 and also to a class of 45 students randomly selected from high school 2. For the class at high school 1 the sample mean is 75 points and the sample standard deviation is 10 points. For the class at high school 2 the sample mean is 72 points and the sample standard deviation is 8 points. Construct a 95% confidence interval for the difference in the mean scores. What assumptions are necessary?

8.96   Two methods for teaching reading were applied to two randomly selected groups of elementary schoolchildren and compared on the basis of a reading comprehension test given at the end of the learning period. The sample means and variances computed from the test scores are shown in the accompanying table. Find a 95% confidence interval for $(\mu_1 - \mu_2)$. What assumptions are necessary?

Statistic	Method 1	Method 2
No. children in group	11	14
$\bar{y}$	64	69
s^2	52	71

8.97   A comparison of reaction times for two different stimuli in a psychological word association experiment produced the results (in seconds) shown in the accompanying table when applied to a random sample of sixteen people. Obtain a 90% confidence interval for $(\mu_1 - \mu_2)$. What assumptions are necessary?

Stimulus 1		Stimulus 2	
1	2	4	1
3	1	2	2
2	3	3	3
1	2	3	3

8.98   The length of time between the billing and the receipt of payment was recorded for a random sample of 100 of a CPA firm's clients. The sample mean and standard deviation for the 100 accounts were 39.1 days and 17.3 days, respectively. Find a 90% confidence interval for the mean time between billing and receipt of payment for all of the CPA firm's accounts. Interpret the interval.

8.99   Television advertisers may mistakenly believe that most viewers understand most of the advertising that they see and hear. A recent research study used 2300 viewers above age 13. Each viewer looked at 30-second television advertising excerpts. Of these, 1914 of the viewers misunderstood all or part of the excerpt. Find a 95% confidence interval for the proportion of all viewers (of which the sample is representative) that will misunderstand all or part of the television excerpts used in this study.

8.100   A survey of 415 corporate, government, and accounting executives of the Financial Accounting Foundation (*Wall Street Journal*, June 13, 1980) found that 67% rated cash flow (as opposed to earnings per share, etc.) as the most important indicator of a company's financial health. Assume that these 415 executives could be viewed as a random sample from the population of all executives. Use the data to find a 95% confidence interval for the fraction of all corporate executives who would list cash flow as the most important measure of a company's financial health.

*8.101   We noted in Section 8.3 that

$$S'^2 = \frac{\sum\limits_{i=1}^{n}(Y_i - \bar{Y})^2}{n}$$

is a biased estimator of σ^2 and that

$$S^2 = \frac{\sum\limits_{i=1}^{n}(Y_i - \bar{Y})^2}{n-1}$$

is an unbiased estimator of the same parameter. If we sample from a normal population,

(a)   find $V(S'^2)$,

(b)   show that $V(S^2) > V(S'^2)$.

*8.102   Exercise 8.101 suggests that S^2 is superior to S'^2 in regard to bias and that S'^2 is superior to S^2 because it possesses smaller variance. Which is the better estimator? Compare the mean square errors.

*8.103   Suppose that two independent random samples of n_1 and n_2 observations are selected from normal populations. Further, assume that the populations possess a common variance σ^2. Let

$$S_i^2 = \frac{\sum\limits_{j=1}^{n_i}(Y_{ij} - \bar{Y}_i)^2}{n_i - 1} \qquad i = 1, 2$$

(a)   Show that the pooled estimator of σ^2, which follows, is unbiased.

$$S^2 = \frac{(n_1 - 1)S_1^2 + (n_2 - 1)S_2^2}{n_1 + n_2 - 2}$$

(b)   Find $V(S^2)$.

*8.104   Suppose that two independent random samples of n_1 and n_2 observations are selected from normal populations with means μ_i and variances σ_i^2, $i = 1, 2$. We wish to construct a confidence interval for the variance ratio σ_1^2/σ_2^2. Let S_i^2, $i = 1, 2$, be as defined in Exercise 8.103. Use the fact that

$$\frac{(n_i - 1)S_i^2}{\sigma_i^2} \qquad i = 1, 2$$

has a χ^2 distribution with $(n_i - 1) = \nu_i$ degrees of freedom and use the definition of the F random variable, Definition 7.3, to construct a pivotal quantity. Then find a confidence interval for σ_1^2/σ_2^2 with confidence coefficient $(1 - \alpha)$.

*8.105   Note that the small-sample confidence interval for μ, based on Student's t (Section 8.8), possesses a random width [in contrast to the large-sample confidence interval (Section 8.6), where the width is not random]. Find the expected value of the interval width in the small-sample case.

*8.106   A confidence interval is *unbiased* if the expected value of the interval midpoint is equal to the estimated parameter. The expected value of the midpoint of the large-sample confidence interval (Section 8.6) is equal to the estimated parameter, and the same is true for the small-sample confidence intervals for μ and $(\mu_1 - \mu_2)$ (Section 8.8). For example, the midpoint of the interval $\bar{y} \pm ts/\sqrt{n}$ is $\bar{y}$, and $E(\bar{Y}) = \mu$. Now consider the confidence interval for σ^2. Show that the expected value of the midpoint of this confidence interval is not equal to σ^2.

*8.107   As we have noted, the sample mean $\bar{Y}$ is a good point estimator of the population mean μ. It can also be used to predict a future value of Y independently selected from the population. Assume that you have a sample mean and variance, $\bar{Y}$ and S^2, based on a random sample of n measurements from a normal population. Use Student's t to form a pivotal quantity to find a prediction interval for some new value of Y, say Y_p, to be observed in the future. (Hint: Start with the quantity $Y_p - \bar{Y}$.) Note the terminology: parameters are *estimated*; values of random variables are *predicted*.

chapter nine

PROPERTIES OF POINT ESTIMATORS AND METHODS OF ESTIMATION

9.1 Introduction

In Chapter 8 we presented some estimators for parameters, which often are of interest in practical problems. The estimators presented there were estimators that merited consideration on the basis of intuition. Recall that an estimator $\hat{\theta}$ for a target parameter θ is a function of the random variables observed in a sample and therefore itself is a random variable. This being the case, we know that an estimator will have a probability distribution, which we call the *sampling distribution* of the estimator. In Section 8.2 we stated that if $E(\hat{\theta}) = \theta$, then the estimator has the (sometimes) desirable property of being unbiased.

In this chapter we will present a more formal and detailed examination of some of the mathematical properties of point estimators, particularly the notions of efficiency, consistency, and sufficiency. Minimal sufficient statistics will be derived and used to construct minimum-variance unbiased estimators for parameters. We then will consider two other useful methods for deriving estimators, the method of moments and the method of maximum likelihood. Some of the properties of estimators derived from these methods also will be discussed.

9.2 Relative Efficiency

In Section 8.2 (Figure 8.3) we mentioned that it usually is possible to obtain more than one unbiased estimator for the same target parameter θ. If $\hat{\theta}_1$ and $\hat{\theta}_2$ denote two unbiased estimators for the same parameter θ, we would prefer to use the estimator with the smaller variance. That is, if both estimators are unbiased, $\hat{\theta}_1$ is relatively more efficient than $\hat{\theta}_2$ if $V(\hat{\theta}_2) > V(\hat{\theta}_1)$. In fact, we use the ratio $V(\hat{\theta}_2)/V(\hat{\theta}_1)$ to define the *relative efficiency* of two unbiased estimators.

Definition 9.1

Given two unbiased estimators, $\hat{\theta}_1$ and $\hat{\theta}_2$, of a parameter θ, with variances $V(\hat{\theta}_1)$ and $V(\hat{\theta}_2)$, respectively, then the efficiency of $\hat{\theta}_1$ relative to $\hat{\theta}_2$ is defined to be the ratio

$$\text{efficiency} = \frac{V(\hat{\theta}_2)}{V(\hat{\theta}_1)}$$

If $\hat{\theta}_1$ and $\hat{\theta}_2$ are unbiased estimators for θ, the efficiency of $\hat{\theta}_1$ relative to $\hat{\theta}_2$ is greater than one only if $V(\hat{\theta}_2) > V(\hat{\theta}_1)$. In this case $\hat{\theta}_1$ is a better unbiased estimator than $\hat{\theta}_2$. For example, if the efficiency of $\hat{\theta}_1$ relative to $\hat{\theta}_2$ is 1.8, then the amount of variability associated with $\hat{\theta}_2[V(\hat{\theta}_2)]$ is 1.8 times the amount of variability associated with $\hat{\theta}_1$. Similarly, if the efficiency of $\hat{\theta}_1$ relative to $\hat{\theta}_2$ is less than 1, say, .73, then the amount of variability associated with $\hat{\theta}_2$ is only 73% of the variability associated with $\hat{\theta}_1$. In this case $\hat{\theta}_2$ would be preferred to $\hat{\theta}_1$.

Let us consider an example. You know that we can construct many estimators for a population mean. Suppose that we wish to estimate the mean of a normal population. Let $\hat{\theta}_1$ be the sample median, the middle observation when the sample measurements are ordered according to magnitude (n odd) or the average of the two middle observations (n even). Let $\hat{\theta}_2$ be the sample mean. Although proof is omitted, it can be shown that the variance of the sample median, for large n, is $V(\hat{\theta}_1) = (1.2533)^2(\sigma^2/n)$. Then the efficiency of the sample median relative to the sample mean is

$$\text{efficiency} = \frac{V(\hat{\theta}_2)}{V(\hat{\theta}_1)} = \frac{\sigma^2/n}{(1.2533)^2\sigma^2/n} = \frac{1}{(1.2533)^2} = .6366$$

Thus we see that the amount of variability associated with the sample mean is approximately 63% of the variability associated with the sample median. Therefore, we would prefer to use the sample mean as the estimator for the population mean.

EXAMPLE 9.1 Let $Y_1, \ldots, Y_n$ denote a random sample from the uniform distribution on the interval $(0, \theta)$. Two unbiased estimators for θ are

$$\hat{\theta}_1 = 2\bar{Y} \quad \text{and} \quad \hat{\theta}_2 = \left(\frac{n+1}{n}\right)Y_{(n)}$$

where $Y_{(n)} = \max(Y_1, \ldots, Y_n)$. Find the efficiency of $\hat{\theta}_1$ relative to $\hat{\theta}_2$.

Solution

Because each Y_i has a uniform distribution on the interval $(0, \theta)$, $\mu = E(Y_i) = \theta/2$ and $\sigma^2 = V(Y_i) = \theta^2/12$. Therefore,

$$E(\hat{\theta}_1) = E(2\bar{Y}) = 2E(\bar{Y}) = 2(\mu) = 2\left(\frac{\theta}{2}\right) = \theta$$

and $\hat{\theta}_1$ is unbiased, as claimed. Further,

$$V(\hat{\theta}_1) = V(2\bar{Y}) = 4V(\bar{Y}) = 4\left[\frac{V(Y_i)}{n}\right] = \left(\frac{4}{n}\right)\left(\frac{\theta^2}{12}\right) = \frac{\theta^2}{3n}$$

To find the mean and variance of $\hat{\theta}_2$, we must first recall that the density function of $Y_{(n)}$ is given by

$$g_n(y) = n[F_Y(y)]^{n-1}f_Y(y) = n\left(\frac{y}{\theta}\right)^{n-1}\left(\frac{1}{\theta}\right)$$

$$= \frac{ny^{n-1}}{\theta^n}, \quad 0 \le y \le \theta$$

$$g_n(y) = 0, \quad \text{elsewhere}$$

because

$$P(Y_i \le y) = F_Y(y) = \frac{y}{\theta}, \quad 0 \le y \le \theta$$

Thus

$$E(Y_{(n)}) = \frac{n}{\theta^n}\int_0^\theta y^n \, dy = \left(\frac{n}{n+1}\right)\theta$$

and it follows that $E\{[(n+1)/n]Y_{(n)}\} = \theta$; that is $\hat{\theta}_2$ is unbiased. Because

$$E(Y_{(n)}^2) = \frac{n}{\theta^n}\int_0^\theta y^{n+1} \, dy = \left(\frac{n}{n+2}\right)\theta^2$$

it follows that

$$V(Y_{(n)}) = E(Y_{(n)}^2) - [E(Y_{(n)})]^2 = \left[\frac{n}{n+2} - \left(\frac{n}{n+1}\right)^2\right]\theta^2$$

and

$$V(\hat{\theta}_2) = V\left[\left(\frac{n+1}{n}\right)Y_{(n)}\right] = \left(\frac{n+1}{n}\right)^2 V(Y_{(n)})$$

$$= \left[\frac{(n+1)^2}{n(n+2)} - 1\right]\theta^2 = \frac{\theta^2}{n(n+2)}$$

Therefore, the efficiency of $\hat{\theta}_1$ relative to $\hat{\theta}_2$ is given by

$$\text{efficiency} = \frac{V(\hat{\theta}_2)}{V(\hat{\theta}_1)} = \frac{\theta^2/n(n+2)}{\theta^2/3n} = \frac{3}{n+2}$$

Note that this efficiency is less than unity for $n > 1$. That is, $\hat{\theta}_2$ has a smaller variance than $\hat{\theta}_1$, and therefore $\hat{\theta}_2$ would generally be preferred over $\hat{\theta}_1$ as an estimator of θ.

∽

We will present some methods for finding estimators with small variance later in this chapter. For now we only wish to make the point that relative efficiency is one important criterion for comparing estimators.

Exercises

9.1   In Exercise 8.4 we considered a random sample of size three from an exponential distribution with a density function given by

$$f(y) = \begin{cases} (1/\theta)e^{-y/\theta}, & 0 < y \\ 0, & \text{elsewhere} \end{cases}$$

and determined that $\hat{\theta}_1 = Y_1$, $\hat{\theta}_2 = (Y_1 + Y_2)/2$, $\hat{\theta}_3 = (Y_1 + 2Y_2)/3$ and $\hat{\theta}_5 = \bar{Y}$ all are unbiased estimators for θ. Find the efficiency of $\hat{\theta}_1$ relative to $\hat{\theta}_5$, of $\hat{\theta}_2$ relative to $\hat{\theta}_5$, of $\hat{\theta}_3$ relative to $\hat{\theta}_5$.

9.2   Let $Y_1, Y_2, \ldots, Y_n$ denote a random sample from a population with a mean of μ and a variance σ^2. Consider the following three estimators for μ:

$$\hat{\mu}_1 = \frac{1}{2}(Y_1 + Y_2) \qquad \hat{\mu}_2 = \frac{1}{4}Y_1 + \frac{Y_2 + \cdots + Y_{n-1}}{2(n-2)} + \frac{1}{4}Y_n \qquad \hat{\mu}_3 = \bar{Y}$$

(a)   Show that each of the three estimators is unbiased.

(b)   Find the efficiency of $\hat{\mu}_3$ relative to $\hat{\mu}_2$ and $\hat{\mu}_1$, respectively.

9.3   Let $Y_1, \ldots, Y_n$ denote a random sample from the uniform distribution on the interval $(\theta, \theta + 1)$. Let

$$\hat{\theta}_1 = \bar{Y} - \frac{1}{2} \qquad \hat{\theta}_2 = Y_{(n)} - \frac{n}{n+1}$$

(a)   Show that both $\hat{\theta}_1$ and $\hat{\theta}_2$ are unbiased estimators of θ.

(b)   Find the efficiency of $\hat{\theta}_1$ relative to $\hat{\theta}_2$.

9.4   Let $Y_1, Y_2, \ldots, Y_n$ denote a random sample of size n from a uniform distribution on the interval $(0, \theta)$. If $Y_1 = \min(Y_1, Y_2, \ldots, Y_n)$, the result of Exercise 8.14 is that $\hat{\theta}_1 = (n+1)Y_{(1)}$ is an unbiased estimator for θ. If $Y_{(n)} = \min(Y_1, Y_2, \ldots, Y_n)$, the results of Example 9.1 imply that $\hat{\theta}_2 = [(n+1)/n]Y_{(n)}$ also is an unbiased estimator for θ. Show that the efficiency of $\hat{\theta}_1$ to $\hat{\theta}_2$ is $1/n^2$. Note that this implies that $\hat{\theta}_2$ is a markedly superior estimator.

9.5   Suppose $Y_1, \ldots, Y_n$ is a random sample from a normal distribution with mean μ and variance σ^2. Two unbiased estimators of σ^2 are

$$\hat{\sigma}_1^2 = S^2 = \frac{1}{n-1} \sum_{i=1}^{n} (Y_i - \bar{Y})^2 \qquad \text{and} \qquad \hat{\sigma}_2^2 = \frac{1}{2}(Y_1 - Y_2)^2$$

Find the efficiency of $\hat{\sigma}_1^2$ relative to $\hat{\sigma}_2^2$.

9.6   Suppose that $Y_1, Y_2, \ldots, Y_n$ denote a random sample of size n from a Poisson distribution with a mean of λ. Consider $\hat{\lambda}_1 = (Y_1 + Y_2)/2$ and $\hat{\lambda}_2 = \bar{Y}$. Derive the efficiency of $\hat{\lambda}_1$ relative to $\hat{\lambda}_2$.

9.7   Suppose that $Y_1, Y_2, \ldots, Y_n$ denote a random sample of size n from an exponential distribution with a density function given by

$$f(y) = \begin{cases} (1/\theta)e^{-y/\theta}, & 0 < y \\ 0, & \text{elsewhere} \end{cases}$$

In Exercise 8.15 we determined that $\hat{\theta}_1 = nY_{(1)}$ is an unbiased estimator of θ with $\text{MSE}(\hat{\theta}_1) = \theta^2$. Consider the estimator $\hat{\theta}_2 = \bar{Y}$ and find the efficiency of $\hat{\theta}_1$ relative to $\hat{\theta}_2$.

*9.8   Let $Y_1, \ldots, Y_n$ denote a random sample from a probability density function $f(y)$, which has unknown parameter θ. If $\hat{\theta}$ is an unbiased estimator of θ, then under very general conditions

$$V(\hat{\theta}) \geq I(\theta) \qquad \text{where} \qquad I(\theta) = 1 \bigg/ nE\left[-\frac{\partial^2 \ln f(y)}{\partial \theta^2} \right]$$

(This is known as the Cramer-Rao inequality.) If $V(\hat{\theta}) = I(\theta)$, the estimator $\hat{\theta}$ is said to be *efficient*.

(a)   Suppose $f(y)$ is the normal distribution with mean μ and variance σ^2. Show that $\bar{Y}$ is an efficient estimator of μ.

(b)   This inequality also holds for discrete probability functions $p(y)$. Suppose $p(y)$ is the Poisson distribution with mean λ. Show that $\bar{Y}$ is an efficient estimator of λ.

---

* Exercises preceded by an asterisk are optional.

9.3 Consistency

Suppose that a coin, which has probability p of resulting in heads, is tossed n times. If the tosses are independent, then Y, the number of heads among the n tosses, has a binomial distribution. If the true value of p is unknown, the sample proportion Y/n is an estimator of p. What happens to this sample proportion as the number of tosses n increases? Our intuition leads us to believe that as n gets larger, Y/n should get closer to the true value of p. That is, our estimator should get closer to the quantity being estimated as the amount of information in the sample is increased.

Because Y/n is a random variable, we may express this "closeness" to p in probabilistic terms. In particular, let us examine the probability that the distance between the estimator and the target parameter, $|(Y/n) - p|$, will be less than some arbitrary, positive real number ε. This probability,

$$P\left(\left|\frac{Y}{n} - p\right| \le \varepsilon\right)$$

should be close to unity for a large n, if our intuition is correct. If the probability just shown, in fact, does tend to unity as $n \to \infty$, we then say that (Y/n) is a consistent estimator of p or that (Y/n) "converges in probability to p."

Definition 9.2

The estimator $\hat{\theta}_n$ is said to be a *consistent estimator* of θ if, for any positive number ε,

$$\lim_{n \to \infty} P(|\hat{\theta}_n - \theta| \le \varepsilon) = 1$$

or, equivalently,

$$\lim_{n \to \infty} P(|\hat{\theta}_n - \theta| > \varepsilon) = 0$$

The notation $\hat{\theta}_n$ is used to express the fact that the estimator for θ is calculated by using a sample of size n. For example, $\bar{Y}_2$ is the average of two observations, whereas $\bar{Y}_{100}$ is the average of the 100 observations contained in a sample of size $n = 100$. If $\hat{\theta}_n$ is an unbiased estimator, the following theorem often can be used to prove that the estimator is consistent.

Theorem 9.1

An unbiased estimator $\hat{\theta}_n$ for θ is a consistent estimator of θ if $\lim_{n \to \infty} V(\hat{\theta}_n) = 0$.

Proof   If Y is any random variable with $E(Y) = \mu$ and $V(Y) = \sigma^2 < \infty$, and if k is any nonnegative constant, Tchebysheff's theorem (see Theorem 4.13) implies that

$$P(|Y - \mu| > k\sigma) \leq \frac{1}{k^2}$$

Because $\hat{\theta}_n$ is an unbiased estimator for θ, $E(\hat{\theta}_n) = \theta$. If we apply Tchebysheff's theorem for the random variable $\hat{\theta}_n$, we obtain that

$$P[|\hat{\theta}_n - \theta| > k\sqrt{V(\hat{\theta}_n)}] \leq \frac{1}{k^2}$$

For any positive number ε,

$$k = \frac{\varepsilon}{\sqrt{V(\hat{\theta}_n)}}$$

is a positive number. Application of Tchebysheff's theorem for this choice of k shows that

$$P\left\{|\hat{\theta}_n - \theta| > \left[\frac{\varepsilon}{\sqrt{V(\hat{\theta}_n)}}\right]\sqrt{V(\hat{\theta}_n)}\right\} \leq \frac{V(\hat{\theta}_n)}{\varepsilon^2}$$

or

$$P(|\hat{\theta}_n - \theta| > \varepsilon) \leq \frac{V(\hat{\theta}_n)}{\varepsilon^2}$$

If $\lim_{n \to \infty} V(\hat{\theta}_n) = 0$, then

$$\lim_{n \to \infty} P(|\hat{\theta}_n - \theta| > \varepsilon) = 0$$

which means that $\hat{\theta}_n$ is a consistent estimator for θ.

The consistency property given in Definition 9.2 and discussed in Theorem 9.1 involves a certain type of convergence of $\hat{\theta}_n$ to θ. For this reason the statement "$\hat{\theta}_n$ is a consistent estimator for $\hat{\theta}$" is sometimes replaced by the equivalent statement "$\hat{\theta}_n$ converges in probability to θ."

EXAMPLE 9.2   Let $Y_1, \ldots, Y_n$ denote a random sample from a distribution with mean μ and variance $\sigma^2 < \infty$. Show that $\bar{Y}$ is a consistent estimator of μ.

Solution   We know from earlier chapters that $E(\bar{Y}) = \mu$ and $V(\bar{Y}) = \sigma^2/n$. Because $\bar{Y}$ is unbiased for μ and $V(\bar{Y}) \to 0$ as $n \to \infty$, Theorem 9.1 applies to show that $\bar{Y}$ is a consistent estimator of μ. Equivalently, we may say that $\bar{Y}$ converges in probability to μ.

The fact that $\bar{Y}$ is consistent for μ, or converges in probability to μ, is sometimes referred to as the *law of large numbers*. It is the theoretical justification for the averaging process employed by many experimenters to obtain precision in measurements. For example, an experimenter may take the average of several measurements of the weight of an animal to obtain a more precise estimate of an animal's weight. His feeling, a feeling confirmed by Theorem 9.1, is that the average of many independently selected weights should be quite close to the true weight, with high probability.   ∽

In Section 8.3 we considered an intuitive estimator for $\mu_1 - \mu_2$, the difference in the means of two populations. The estimator discussed at that time was $\bar{Y}_1 - \bar{Y}_2$, the difference in the means of independent random samples selected from the two populations. The results of Theorem 9.2 will be very useful in establishing the consistency of such estimators.

Theorem 9.2

> Suppose that $\hat{\theta}_n$ converges in probability to θ and that $\hat{\theta}'_n$ converges in probability to θ'.
>
> (a) $\hat{\theta}_n + \hat{\theta}'_n$ converges in probability to $\theta + \theta'$.
> (b) $\hat{\theta}_n \hat{\theta}'_n$ converges in probability to $\theta\theta'$.
> (c) $\hat{\theta}_n / \hat{\theta}'_n$ converges in probability to θ/θ', provided that $\theta' \neq 0$.
> (d) $\sqrt{\hat{\theta}_n}$ converges in probability to $\sqrt{\theta}$ if $P(\hat{\theta}_n \geq 0) = 1$.

EXAMPLE 9.3   Suppose that $Y_1, \ldots, Y_n$ represents a random sample such that $E(Y_i) = \mu$, $E(Y_i^2) = \mu'_2$, and $E(Y_i^4) = \mu'_4$ are all finite. Show that

$$S^2 = \frac{1}{n-1} \sum_{i=1}^{n} (Y_i - \bar{Y})^2$$

is a consistent estimator of $\sigma^2 = V(Y_i)$.

Solution   We have seen in earlier chapters that S^2 can be written as

$$S^2 = \frac{1}{n-1} \left(\sum_{i=1}^{n} Y_i^2 - n\bar{Y}^2 \right) = \left(\frac{n}{n-1} \right) \left(\frac{1}{n} \sum_{i=1}^{n} Y_i^2 - \bar{Y}^2 \right)$$

The statistic $(1/n) \sum_{i=1}^{n} Y_i^2$ is the average of n independent and identically distributed random variables, with $E(Y_i^2) = \mu_2'$ and $V(Y_i^2) = \mu_4' - (\mu_2')^2$. By the law of large numbers, Example 9.2, $(1/n) \sum_{i=1}^{n} Y_i^2$ converges in probability to μ_2'.

Example 9.2 also implies that $\bar{Y}$ converges in probability to μ. Thus by part (b) of Theorem 9.2, $\bar{Y}^2 = \bar{Y} \cdot \bar{Y}$ converges in probability to μ^2. It then follows from part (a) of Theorem 9.2 that

$$\frac{1}{n} \sum_{i=1}^{n} Y_i^2 - \bar{Y}^2$$

converges in probability to $\mu_2' - \mu^2 = \sigma^2$. Because $n/(n-1)$ is a sequence of constants converging to unity as $n \to \infty$, we can conclude that S^2 converges in probability to $\sigma^2 = \mu_2' - \mu^2$. Thus S^2 is a consistent estimator for σ^2.   ∽

In Section 8.6 we considered large-sample confidence intervals for some parameters of practical interest. In particular, if $Y_1, \ldots, Y_n$ is a random sample from any distribution with mean μ and variance σ^2, we established that

$$\bar{Y} \pm z_{\alpha/2} \left(\frac{\sigma}{\sqrt{n}} \right)$$

is a valid large-sample confidence interval with confidence coefficient approximately equal to $(1 - \alpha)$. If σ^2 is known, this interval can and should be calculated. However, if σ^2 is not known but the sample size is large, we recommended substituting S for σ in the calculation and indicated that there would be no significant loss of accuracy. As we will subsequently see, the following theorem provides the theoretical justification for these claims.

Theorem 9.3

> Suppose that U_n has a distribution function that converges to a standard normal distribution function as $n \to \infty$. If W_n converges in probability to one, then the distribution function of U_n/W_n converges to a standard normal distribution function.

The proof of this result is beyond the scope of this text. However, the usefulness of the result is illustrated in the following example.

EXAMPLE 9.4   Suppose that $Y_1, \ldots, Y_n$ is a random sample of size n from a distribution with $E(Y_i) = \mu$ and $V(Y_i) = \sigma^2$. Define S^2 as

$$S^2 = \frac{1}{n-1} \sum_{i=1}^{n} (Y_i - \bar{Y})^2$$

Show that the distribution function of

$$\sqrt{n}\left(\frac{\bar{Y} - \mu}{S}\right)$$

converges to a standard normal distribution function.

Solution

In Example 9.3 we showed that S^2 converges in probability to σ^2. Hence it follows from Theorem 9.2, parts (c) and (d), that S^2/σ^2 (and hence S/σ) converges in probability to one. We also know from Theorem 7.4 that the distribution function of

$$U_n = \sqrt{n}\left(\frac{\bar{Y} - \mu}{\sigma}\right)$$

converges to a standard normal distribution function. Therefore, Theorem 9.3 implies that the distribution function of

$$\sqrt{n}\left(\frac{\bar{Y} - \mu}{\sigma}\right)\bigg/(S/\sigma) = \sqrt{n}\left(\frac{\bar{Y} - \mu}{S}\right)$$

converges to a standard normal distribution function.   ∽

The result of Example 9.4 tells us that when n is large, $\sqrt{n}(\bar{Y} - \mu)/S$ has approximately a standard normal distribution *whatever* the form of the distribution from which the sample is taken. If the sample is taken from a *normal distribution*, the results of Chapter 7 imply that $t = \sqrt{n}(\bar{Y} - \mu)/S$ has a t distribution with $n - 1$ degrees of freedom. Combining this information, we see that if a large sample is taken from a normal distribution, then the distribution function of $t = \sqrt{n}(\bar{Y} - \mu)/S$ can be approximated by a standard normal distribution function. That is, as n gets large, and hence as the number of degrees of freedom gets large, the t distribution function converges to the standard normal distribution function.

If we obtain a large sample from any distribution, we know from Example 9.4 that $\sqrt{n}(\bar{Y} - \mu)/S$ has approximately a standard normal distribution. Therefore, it follows that

$$P\left[-z_{\alpha/2} \leq \sqrt{n}\left(\frac{\bar{Y} - \mu}{S}\right) \leq z_{\alpha/2}\right] \approx 1 - \alpha$$

If we manipulate the inequalities in the probability statement to isolate μ in the middle, we obtain

$$P\left[\bar{Y} - z_{\alpha/2}\left(\frac{S}{\sqrt{n}}\right) \leq \mu \leq \bar{Y} + z_{\alpha/2}\left(\frac{S}{\sqrt{n}}\right)\right] \approx 1 - \alpha$$

Thus $\bar{Y} \pm z_{\alpha/2}(S/\sqrt{n})$ forms a valid large-sample confidence interval for μ, with confidence coefficient approximately equal to $1 - \alpha$. Similarly, Theorem 9.3 can be applied to show that

$$\hat{p} \pm z_{\alpha/2}\sqrt{\frac{\hat{p}\hat{q}}{n}}$$

is a valid large-sample confidence interval for p with confidence coefficient approximately equal to $1 - \alpha$.

In this section we have seen that consistency is a property that tells us something about the distance between an estimator and the quantity being estimated. We have seen that $\bar{Y}$ should be close to μ and S^2 should be close to σ^2, with high probability, when the sample size is large. We will see other examples of consistent estimators in the exercises and later in the chapter.

Exercises

9.9   Refer to Exercise 9.3. Show that both $\hat{\theta}_1$ and $\hat{\theta}_2$ are consistent estimators for θ.

9.10   Refer to Exercise 9.5. Is $\hat{\sigma}_2^2$ a consistent estimator of σ^2?

9.11   Suppose $X_1, \ldots, X_n$ and $Y_1, \ldots, Y_n$ are independent random samples from populations with means μ_1 and μ_2 and variances σ_1^2 and σ_2^2, respectively. Show that $\bar{X} - \bar{Y}$ is a consistent estimator of $\mu_1 - \mu_2$.

9.12   In Exercise 9.11 suppose that the populations are normally distributed with $\sigma_1^2 = \sigma_2^2 = \sigma^2$. Show that

$$\frac{\sum_{i=1}^{n}(X_i - \bar{X})^2 + \sum_{i=1}^{n}(Y_i - \bar{Y})^2}{2n - 2}$$

is a consistent estimator of σ^2.

9.13   Let $Y_1, \ldots, Y_n$ denote a random sample from the probability density function

$$f(y) = \begin{cases} \theta y^{\theta-1}, & 0 < y < 1; \theta > 0 \\ 0, & \text{elsewhere} \end{cases}$$

Show that $\bar{Y}$ is a consistent estimator of $\theta/(\theta + 1)$.

9.14   If Y has a binomial distribution with n trials and success probability p, show that Y/n is a consistent estimator of p.

9.15   Suppose that $Y_1, Y_2, \ldots, Y_n$ denote a random sample of size n from a normal distribution with mean μ and variance 1. Consider the first observation Y_1 as an estimator for μ.

(a)   Show that Y_1 is an unbiased estimator for μ.

(b)   Find $P(|Y_1 - \mu| \leq 1)$.

(c)   Look at the basic definition of consistency given in Definition 9.2. Using the result of part (b), is Y_1 as a consistent estimator for μ?

*9.16 It sometimes is relatively easy to establish consistency or lack of consistency by appealing directly to Definition 9.2, evaluating $P(|\hat{\theta}_n - \theta| \le \varepsilon)$ directly and then showing that $\lim_{n \to \infty} P(|\hat{\theta}_n - \theta| \le \varepsilon) = 1$. Let $Y_1, Y_2, \ldots, Y_n$ denote a random sample of size n from a uniform distribution on the interval $(0, \theta)$. If $Y_{(n)} = \max(Y_1, Y_2, \ldots, Y_n)$, we showed in Exercise 6.44 that the probability distribution function of $Y_{(n)}$ is given by

$$F_n(y) = \begin{cases} 0, & y < 0 \\ (y/\theta)^n, & 0 \le y \le \theta \\ 1, & y > \theta \end{cases}$$

(a) For each $n \ge 1$ and every $\varepsilon > 0$, $P(|Y_{(n)} - \theta| \le \varepsilon) = P(\theta - \varepsilon \le Y_{(n)} \le \theta + \varepsilon)$. If $\varepsilon > \theta$, verify that $P(\theta - \varepsilon \le Y_{(n)} \le \theta + \varepsilon) = 1$ and that, for every positive $\varepsilon < \theta$, $P(\theta - \varepsilon \le Y_{(n)} \le \theta + \varepsilon) = 1 - [(\theta - \varepsilon)/\theta]^n$.

(b) Using the result from part (a), show that for every $\varepsilon > 0$, $\lim_{n \to \infty} P(|Y_{(n)} - \theta| \le \varepsilon) = 1$, and hence that $Y_{(n)}$ is a consistent estimator for θ.

*9.17 Use the method described in Exercise 9.16 to show that if $Y_{(1)} = \min(Y_1, Y_2, \ldots, Y_n)$ when $Y_1, Y_2, \ldots, Y_n$ are independent uniform random variables on the interval $(0, \theta)$, then $Y_{(1)}$ is not a consistent estimator for θ. (Hint: Using the methods of Section 6.6, $Y_{(1)}$ has distribution function $F(y) = 1 - (1 - y/\theta)^n$ for $0 \le y \le \theta$.)

*9.18 Let $Y_1, Y_2, \ldots, Y_n$ denote a random sample of size n from a Pareto distribution (see Exercise 6.15). Then the methods of Section 6.6 imply that $Y_{(1)} = \min(Y_1, Y_2, \ldots, Y_n)$ has distribution function given by

$$F_n(y) = \begin{cases} 0, & y \le \beta \\ 1 - (\beta/y)^{\alpha n}, & y > \beta \end{cases}$$

Use the method described in Exercise 9.16 to show that $Y_{(1)}$ is a consistent estimator of β.

*9.19 Let $Y_1, Y_2, \ldots, Y_n$ denote a random sample of size n from a power family distribution (see Exercise 6.14). Then the methods of Section 6.6 imply that $Y_{(n)} = \max(Y_1, Y_2, \ldots, Y_n)$ has distribution function given by

$$F_n(y) = \begin{cases} 0, & y < 0 \\ (y/\theta)^{\alpha n}, & 0 \le y \le \theta \\ 1, & y > \theta \end{cases}$$

Use the method described in Example 9.16 to show that $Y_{(n)}$ is a consistent estimator of θ.

9.20 Let $Y_1, Y_2, \ldots, Y_n$ be independent random variables, each with probability density function

$$f(y) = \begin{cases} 3y^2, & 0 \le y \le 1 \\ 0, & \text{elsewhere} \end{cases}$$

Show that $\bar{Y}$ converges in probability to some constant and find the constant.

9.21 If $Y_1, Y_2, \ldots, Y_n$ denotes a random sample from a gamma-type distribution with parameters α and β, show that $\bar{Y}$ converges in probability to some constant and find the constant.

9.22 Let $Y_1, \ldots, Y_n$ denote a random sample from the probability density function

$$f(y) = \begin{cases} \dfrac{2}{y^2}, & y \geq 2 \\[2mm] 0, & \text{elsewhere} \end{cases}$$

Does the law of large numbers apply to $\bar{Y}$ in this case? Why or why not?

9.23 An experimenter wishes to compare the numbers of bacteria of types A and B in samples of water. A total of n independent water samples are taken, and counts are made for each sample. Let X_i denote the number of type A bacteria and let Y_i denote the number of type B bacteria for sample i. Assume the two bacteria types are sparsely distributed within a water sample, so that $X_1, \ldots, X_n$ and $Y_1, \ldots, Y_n$ can be considered independent random samples from Poisson distributions with means λ_1 and λ_2, respectively. Suggest an estimator of $\lambda_1/(\lambda_1 + \lambda_2)$. What properties does your estimator have?

9.24 Use Theorem 9.3 to prove that the distribution of $(\hat{p} - p)/\sqrt{\hat{p}\hat{q}/n}$ converges to a standard normal distribution.

9.4 Sufficiency

Up to this point we have chosen estimators on the basis of intuition. Thus we chose $\bar{Y}$ and S^2 as the estimators of the mean and variance, respectively, of the normal distribution. (It *seems* as though these estimators should be good estimators of the population parameters.) We have seen that it is sometimes desirable to use estimators that are unbiased. Indeed, $\bar{Y}$ and S^2 have been shown to be unbiased estimators of the population mean μ and variance σ^2, respectively. Notice that we have utilized the information in a sample of size n to calculate the value of two statistics that are used as estimators for the parameters of interest. At this stage the actual sample values no longer are important; rather, we summarize the information in the sample about the parameters of interest in the values of $\bar{Y}$ and S^2. Has this process of summarizing, or reducing the data to the two values of $\bar{Y}$ and S^2, retained all the information about μ and σ^2 in the n sample observations, or has some information about these parameters been lost or obscured through the process of reducing the data? In this section we will present methods for finding statistics that, in a sense, summarize *all* the information in a sample about a target parameter. Such statistics are said to have the property of *sufficiency* or, more simply, are called *sufficient statistics*. As we will see in the next section, sufficient statistics often can be used to develop estimators that have minimum variance among all unbiased estimators.

To illustrate the notion of a sufficient statistic, let us consider the outcomes of n trials of a binomial experiment, $X_1, \ldots, X_n$. That is, $X_i = 1$ with probability p and $X_i = 0$ with probability $(1 - p)$, where p is the probability of success on any one trial. Suppose we are given a value of $Y = \sum_{i=1}^{n} X_i$, the number of successes on the n trials. If we know the value of Y, can we gain any further information on p by looking at other functions of $X_1, \ldots, X_n$? One way to answer this question is to look at the conditional distribution of $X_1, \ldots, X_n$ given Y.

$$P(X_1 = x_1, \ldots, X_n = x_n \mid Y = y) = \frac{P(X_1 = x_1, \ldots, X_n = x_n, Y = y)}{P(Y = y)}$$

$$= \frac{p^y(1 - p)^{n-y}}{\binom{n}{y} p^y (1 - p)^{n-y}} = \frac{1}{\binom{n}{y}}$$

because the numerator is the probability of an independent sequence of 0s and 1s, with y 1s, and the denominator is the binomial probability of y successes out of n trials.

It is important to note that the conditional distribution of $X_1, \ldots, X_n$ given Y *does not* depend upon p. That is, once Y is known, no other function of $X_1, \ldots, X_n$ will shed additional light on the possible value of p. In this sense Y contains all the information about p. Therefore, the statistic Y is said to be *sufficient* for p. We generalize this idea in the following definition.

Definition 9.3

> Let $Y_1, \ldots, Y_n$ denote a random sample from a probability distribution with unknown parameter θ. Then the statistic $U = g(Y_1, \ldots, Y_n)$ is said to be *sufficient* for θ if the conditional distribution $Y_1, \ldots, Y_n$ given U does not depend on θ.

Definition 9.3 tells us how to check whether a statistic is sufficient, but it really doesn't tell us how to find a sufficient statistic. Recall that in the discrete case the joint distribution of discrete random variables $Y_1, \ldots, Y_n$ is given by a probability function $p(y_1, \ldots, y_n)$, which gives the probability or *likelihood* of observing the event $(Y_1 = y_1, \ldots, Y_n = y_n)$. In the continuous case the joint distribution of $Y_1, \ldots, Y_n$ is given by a joint density function $f(y_1, \ldots, y_n)$. Henceforth, it will be convenient to have a single name for the function that defines the joint distribution of the variables $Y_1, \ldots, Y_n$ observed in a sample.

Definition 9.4

> Let $y_1, y_2, \ldots, y_n$ be sample observations taken on corresponding random variables $Y_1, Y_2, \ldots, Y_n$. Then if $Y_1, Y_2, \ldots, Y_n$ are discrete random variables, the *likelihood of the sample, $L = L(y_1, y_2, \ldots, y_n)$,* is defined to be the joint probability of $y_1, y_2, \ldots, y_n$. If $Y_1, Y_2, \ldots, Y_n$ are continuous random variables, the likelihood $L(y_1, y_2, \ldots, y_n)$ is defined to be the joint density evaluated at $y_1, y_2, \ldots, y_n$.

If the set of observations $Y_1, Y_2, \ldots, Y_n$ is a random sample from a discrete distribution with probability function $p(y)$, then

$$L(y_1, y_2, \ldots, y_n) = P(Y_1 = y_1, \ldots, Y_n = y_n) = p(y_1) \cdots p(y_n)$$

Whereas if $Y_1, Y_2, \ldots, Y_n$ have a continuous distribution with density function $f(y)$, then

$$L(y_1, y_2, \ldots, y_n) = f(y_1, y_2, \ldots, y_n) = f(y_1) \cdots f(y_n)$$

To simplify notation, we will sometimes denote the likelihood by L instead of $L(y_1, y_2, \ldots, y_n)$.

The following theorem relates the property of sufficiency to the likelihood L.

Theorem 9.4

> Let U be a statistic based on the random sample $Y_1, Y_2, \ldots, Y_n$. Then U is a *sufficient statistic* for the estimation of a parameter θ if and only if the likelihood L can be factored into two nonnegative functions,
>
> $$L(y_1, y_2, \ldots, y_n) = g(u, \theta) \, h(y_1, y_2, \ldots, y_n)$$
>
> where $g(u, \theta)$ is a function only of u and θ and $h(y_1, y_2, \ldots, y_n)$ is not a function of θ.

Although the proof of Theorem 9.4, also known as the *factorization criterion*, is beyond the scope of this book, we illustrate the usefulness of the theorem in the following example.

EXAMPLE 9.5   Let $Y_1, Y_2, \ldots, Y_n$ be a random sample in which Y_i possesses the probability density function $f(y_i) = (1/\alpha)e^{-y_i/\alpha}, 0 \le y_i \le \infty, i = 1, 2, \ldots, n$. Show that $\bar{Y}$ is a sufficient statistic for the estimation of α.

Solution   The likelihood L of the sample is the joint density

$$L = f(y_1, y_2, \ldots, y_n) = f(y_1)f(y_2) \cdots f(y_n)$$

$$= \left(\frac{e^{-y_1/\alpha}}{\alpha} \right)\left(\frac{e^{-y_2/\alpha}}{\alpha} \right) \cdots \left(\frac{e^{-y_n/\alpha}}{\alpha} \right) = \frac{e^{-\Sigma y_i/\alpha}}{\alpha^n} = \frac{e^{-n\bar{y}/\alpha}}{\alpha^n}$$

Note that L is a function only of α and $\bar{y}$ and that

$$g(\bar{y}, \alpha) = \frac{e^{-n\bar{y}/\alpha}}{\alpha^n} \qquad \text{and} \qquad h(y_1, y_2, \ldots, y_n) = 1$$

Hence $\bar{Y}$ is a sufficient statistic for the estimation of α. ∽

Theorem 9.4 can be used to show that there are many possible sufficient statistics for any one population parameter. First of all, according to Definition 9.3 or the factorization criterion (Theorem 9.4), the random sample itself is a sufficient statistic. Second, if $Y_1, \ldots, Y_n$ denotes a random sample from a distribution with a density function with parameter θ [e.g., if Y_i is uniform on $(0, \theta)$], then the set of order statistics $Y_{(1)} \leq Y_{(2)} \leq \cdots \leq Y_{(n)}$, which is a function of $Y_1, \ldots, Y_n$, is sufficient for θ.

In our initial example of this section, involving the number of successes in n trials, $Y = \sum_{i=1}^{n} X_i$ reduces the data $X_1, \ldots, X_n$ to a single value that remains sufficient for p. Generally we would like to find a sufficient statistic that reduces the data in the sample as much as possible. Statistics that achieve this objective are called *minimal sufficient statistics*.

In the next section we present a method for finding minimal sufficient statistics. We will also show how these statistics can be used to develop unbiased estimators with minimum variance.

Exercises

9.25   Let $X_1, \ldots, X_n$ denote n independent and identically distributed random variables such that

$$P(X_i = 1) = p \qquad \text{and} \qquad P(X_i = 0) = 1 - p$$

for each $i = 1, 2, \ldots, n$. Show that $\sum_{i=1}^{n} X_i$ is sufficient for p.

9.26   Let $Y_1, \ldots, Y_n$ denote a random sample from a normal distribution with a mean of μ and a variance of σ^2.

(a)   If μ is unknown and σ^2 is known, show that $\bar{Y}$ is sufficient for μ.

(b)   If μ is known and σ^2 is unknown, show that $\sum_{i=1}^{n} (Y_i - \mu)^2$ is sufficient for σ^2.

(c)   If μ and σ^2 are both unknown, show that $\sum_{i=1}^{n} Y_i$ and $\sum_{i=1}^{n} Y_i^2$ are jointly sufficient for μ and σ^2. (Thus it follows that $\bar{Y}$ and $\sum_{i=1}^{n} (Y_i - \bar{Y})^2$ or $\bar{Y}$ and S^2 also are jointly sufficient for μ and σ^2.)

9.27   Let $Y_1, \ldots, Y_n$ denote a random sample from a Poisson distribution with parameter λ. Show by conditioning that $\sum_{i=1}^{n} Y_i$ is sufficient for λ.

9.28   Let $Y_1, \ldots, Y_n$ denote a random sample from a Rayleigh distribution with parameter θ. (Refer to Exercise 6.28.) Show that $\sum_{i=1}^{n} Y_i^2$ is sufficient for θ.

9.29   Let $Y_1, \ldots, Y_n$ denote a random sample from a Weibull distribution with known m and unknown α. (Refer to Exercise 6.20.) Show that $\sum_{i=1}^{n} Y_i^m$ is sufficient for α.

9.30   If $Y_1, Y_2, \ldots, Y_n$ dnotes a random sample from a geometric distribution with parameter p, show that $\bar{Y}$ is sufficient for p.

9.31   Let $Y_1, Y_2, \ldots, Y_n$ denote independent and identically distributed random variables from a power family distribution with parameters α and θ. Then, by the result in Exercise 6.13, if $\alpha, \theta > 0$,

$$f(y) = \begin{cases} \alpha y^{\alpha - 1}/\theta^{\alpha}, & 0 \le y \le \theta \\ 0, & \text{elsewhere} \end{cases}$$

If θ is known, show that $\prod\limits_{i=1}^{n} Y_i$ is sufficient for α.

9.32   Let $Y_1, Y_2, \ldots, Y_n$ denote independent and identically distributed random variables from a Pareto distribution with parameters α and β. Then, by the result in Exercise 6.14, if $\alpha, \beta > 0$,

$$f(y) = \begin{cases} \alpha \beta^{\alpha} y^{-(\alpha + 1)}, & y \ge \beta \\ 0, & \text{elsewhere} \end{cases}$$

If β is known, show that $\prod\limits_{i=1}^{n} Y_i$ is sufficient for α.

9.33   Suppose that $Y_1, Y_2, \ldots, Y_n$ is a random sample from a probability density function that is in the (one parameter) exponential family so that

$$f(y) = \begin{cases} a(\theta)b(y)e^{-[c(\theta)d(y)]}, & g \le y \le h \\ 0, & \text{elsewhere} \end{cases}$$

where g and h are constants that do not depend on θ. Show that $\sum\limits_{i=1}^{n} d(Y_i)$ is sufficient for θ.

9.34   If $Y_1, Y_2, \ldots, Y_n$ denotes a random sample from an exponential distribution with a mean of β, show that $f(y)$ is in the exponential family and that $\bar{Y}$ is sufficient for β.

9.35   Refer to Exercise 9.31. If θ is known, show that the power family of distributions is in the exponential family. What is a sufficient statistic for α? Does this contradict your answer to Exercise 9.31?

9.36   Refer to Exercise 9.32. If β is known, show that the Pareto distribution is in the exponential family. What is a sufficient statistic for α? Argue that there is no contradiction between your answer to this exercise and the answer found in Exercise 9.32.

*9.37   Let $Y_1, \ldots, Y_n$ denote a random sample from the uniform distribution over the interval $(0, \theta)$. Show that $Y_{(n)} = \max(Y_1, \ldots, Y_n)$ is sufficient for θ.

*9.38   Let $Y_1, \ldots, Y_n$ denote a random sample from the uniform distribution over the interval (θ_1, θ_2). Show that $Y_{(1)} = \min(Y_1, \ldots, Y_n)$ and $Y_{(n)}$ are jointly sufficient for θ_1 and θ_2.

*9.39   Let $Y_1, \ldots, Y_n$ denote a random sample from the probability density function

$$f(y) = \begin{cases} e^{-(y - \theta)}, & y \ge \theta \\ 0, & \text{elsewhere} \end{cases}$$

Show that $Y_{(1)} = \min(Y_1, \ldots, Y_n)$ is sufficient for θ.

*9.40   Let $Y_1, Y_2, \ldots, Y_n$ denote independent and identically distributed random variables from a power family distribution with parameters α and θ. Then, as in Exercise 9.31, if $\alpha, \theta > 0$,

$$f(y) = \begin{cases} \alpha y^{\alpha-1}/\theta^{\alpha}, & 0 \le y \le \theta \\ 0, & \text{elsewhere} \end{cases}$$

Show that $\max(Y_1, Y_2, \ldots, Y_n)$ and $\prod\limits_{i=1}^{n} Y_i$ are jointly sufficient for α and θ.

*9.41   Let $Y_1, Y_2, \ldots, Y_n$ denote independent and identically distributed random variables from a Pareto distribution with parameters α and β. Then, as in Exercise 9.32, if $\alpha, \beta > 0$,

$$f(y) = \begin{cases} \alpha \beta^{\alpha} y^{-(\alpha+1)}, & y \ge \beta \\ 0, & \text{elsewhere} \end{cases}$$

Show that $\prod\limits_{i=1}^{n} Y_i$ and $\min(Y_1, Y_2, \ldots, Y_n)$ are jointly sufficient for α and β.

9.5 Minimal Sufficiency and Minimum-Variance Unbiased Estimation

In the previous section we defined the property of sufficiency. We saw that in a certain sense a sufficient statistic is a statistic that summarizes all the information in a sample about a target parameter. In this section we present a method for finding a sufficient statistic that reduces or summarizes the data as much as possible. That is, we will find *minimal sufficient* statistics and use these statistics to develop minimum-variance unbiased estimators (MVUEs) for target parameters.

How do we find these minimal sufficient statistics? We will present and illustrate the use of a method for finding such statistics, which was developed by Lehmann and Scheffé. The proof that the method actually achieves a minimal sufficient statistic is not given. You can find the proof in the texts listed in the references at the end of the chapter.

Suppose $Y_1, \ldots, Y_n$ denotes a random sample from a probability function $p(y)$, or a density function $f(y)$, with unknown parameter θ. The set of variables $(Y_1, \ldots, Y_n)$ can take on many different values. Let us suppose that one such set of values is $(x_1, \ldots, x_n)$ and another is $(y_1, \ldots, y_n)$. The method of Lehmann and Scheffé makes use of the ratio of the likelihoods evaluated at these two points,

$$\frac{L(x_1, \ldots, x_n)}{L(y_1, \ldots, y_n)}$$

Many times it is possible to find a funtion $g(x_1, \ldots, x_n)$ such that this ratio will be free of the unknown parameter θ if and only if $g(x_1, \ldots, x_n) = g(y_1, \ldots, y_n)$. If such a function g can be found, then $g(Y_1, \ldots, Y_n)$ is the minimal sufficient statistic for θ (proof omitted).

EXAMPLE 9.6   Let $Y_1, \ldots, Y_n$ be a random sample from a distribution where $P(Y_i = 1) = p$ and $P(Y_i = 0) = 1 - p$, with p unknown. Find a minimal sufficient statistic for p.

Solution   In this example the ratio of the likelihoods is

$$\frac{L(x_1, \ldots, x_n)}{L(y_1, \ldots, y_n)} = \frac{p(x_1, \ldots, x_n)}{p(y_1, \ldots, y_n)} = \frac{p^{\Sigma x_i}(1 - p)^{n - \Sigma x_i}}{p^{\Sigma y_i}(1 - p)^{n - \Sigma y_i}} = \left(\frac{p}{1 - p}\right)^{\Sigma x_i - \Sigma y_i}$$

For this ratio to be independent of p, we must have

$$\sum_{i=1}^{n} x_i - \sum_{i=1}^{n} y_i = 0 \qquad \text{or} \qquad \sum_{i=1}^{n} x_i = \sum_{i=1}^{n} y_i$$

Thus the method of Lehmann and Scheffé tells us that $\sum_{i=1}^{n} Y_i$ is a minimal sufficient statistic for p.   ∽

For the probability functions and density functions we usually consider, any unbiased estimator that is a function of a minimal sufficient statistic will be a *minimum-variance unbiased estimator* (MVUE). That is, it will possess the smallest variance possible among unbiased estimators. In the binomial case, Y/n is an unbiased estimator of p and a function of the minimal sufficient statistic for p. Thus Y/n is an MVUE of p.

We will illustrate this concept with other examples.

EXAMPLE 9.7   Suppose $Y_1, \ldots, Y_n$ denotes a random sample from the Weibull density function, given by

$$f(y) = \begin{cases} \left(\dfrac{2y}{\theta}\right) e^{-y^2/\theta}, & y > 0 \\ 0, & \text{elsewhere} \end{cases}$$

Find an MVUE for θ.

Solution   Using the Lehmann-Scheffé criterion, we look at

$$\frac{L(x_1, \ldots, x_n)}{L(y_1, \ldots, y_n)} = \frac{f(x_1, \ldots, x_n)}{f(y_1, \ldots, y_n)} = \frac{\left[2^n(x_1 \cdots x_n) \exp\left(-\dfrac{1}{\theta} \sum_{i=1}^{n} x_i^2\right) \Big/ \theta^n\right]}{\left[2^n(y_1 \cdots y_n) \exp\left(-\dfrac{1}{\theta} \sum_{i=1}^{n} y_i^2\right) \Big/ \theta^n\right]}$$

$$= \left(\frac{x_1 \cdots x_n}{y_1 \cdots y_n}\right) \exp\left[-\frac{1}{\theta}\left(\sum_{i=1}^{n} x_i^2 - \sum_{i=1}^{n} y_i^2\right)\right]$$

For this ratio to be free of θ we must have $\sum_{i=1}^{n} x_i^2 = \sum_{i=1}^{n} y_i^2$, and so $\sum_{i=1}^{n} Y_i^2$ is the minimal sufficient statistic for θ.

We now must find a function of this statistic that is unbiased for θ. Letting $U = Y_i^2$, we have

$$f_U(u) = f(\sqrt{u}) \frac{d\sqrt{u}}{du} = \left(\frac{2}{\theta}\right)(\sqrt{u} e^{-u/\theta})\left(\frac{1}{2\sqrt{u}}\right) = \left(\frac{1}{\theta}\right)e^{-u/\theta}, \qquad u > 0$$

That is, Y_i^2 has an exponential distribution with parameter θ. Because

$$E(Y_i^2) = E(U) = \theta \qquad \text{and} \qquad E\left(\sum_{i=1}^{n} Y_i^2\right) = n\theta$$

it follows that

$$\frac{1}{n} \sum_{i=1}^{n} Y_i^2$$

is an unbiased estimator of θ. Hence $(1/n) \sum_{i=1}^{n} Y_i^2$ is an MVUE of the Weibull parameter θ.   ∽

The following example illustrates the use of this technique for estimating two unknown parameters.

EXAMPLE 9.8   Suppose $Y_1, \ldots, Y_n$ denotes a random sample from a normal distribution with unknown mean μ and variance σ^2. Find the MVUEs for μ and σ^2.

Solution   Again, looking at the ratio of the joint density function at two sample points, we have

$$\frac{L(x_1, \ldots, x_n)}{L(y_1, \ldots, y_n)} = \frac{f(x_1, \ldots, x_n)}{f(y_1, \ldots, y_n)} = \frac{\left(\dfrac{1}{\sigma\sqrt{2\pi}}\right)^n \exp\left[-\dfrac{1}{2\sigma^2} \sum_{i=1}^{n} (x_i - \mu)^2\right]}{\left(\dfrac{1}{\sigma\sqrt{2\pi}}\right)^n \exp\left[-\dfrac{1}{2\sigma^2} \sum_{i=1}^{n} (y_i - \mu)^2\right]}$$

$$= \exp\left\{-\frac{1}{2\sigma^2}\left[\sum_{i=1}^{n} (x_i - \mu)^2 - \sum_{i=1}^{n} (y_i - \mu)^2\right]\right\}$$

$$= \exp\left\{-\frac{1}{2\sigma^2}\left[\left(\sum_{i=1}^{n} x_i^2 - \sum_{i=1}^{n} y_i^2\right)\right.\right.$$

$$\left.\left. - 2\mu\left(\sum_{i=1}^{n} x_i - \sum_{i=1}^{n} y_i\right)\right]\right\}$$

For this ratio to be free of μ and σ^2, we must have $\sum_{i=1}^{n} x_i = \sum_{i=1}^{n} y_i$ and $\sum_{i=1}^{n} x_i^2 = \sum_{i=1}^{n} y_i^2$. Thus $\sum_{i=1}^{n} Y_i$ and $\sum_{i=1}^{n} Y_i^2$, jointly, form the minimal sufficient statistics for μ and σ^2.

We know from past work that $\bar{Y}$ is unbiased for μ and

$$S^2 = \frac{1}{n-1} \sum_{i=1}^{n} (Y_i - \bar{Y})^2 = \frac{1}{n-1} \left[\sum_{i=1}^{n} Y_i^2 - n\bar{Y}^2 \right]$$

is unbiased for σ^2. Because these estimators are functions of the minimal sufficient statistics, they are MVUEs for μ and σ^2.

The minimal sufficient statistics for a set of parameters can be used to find minimum-variance unbiased estimators for functions of those parameters. We illustrate the technique in the following example.

EXAMPLE 9.9 Let $Y_1, \ldots, Y_n$ denote a random sample from the exponential density function given by

$$f(y) \begin{cases} \left(\dfrac{1}{\theta}\right) e^{-y/\theta}, & y > 0 \\ 0, & \text{elsewhere} \end{cases}$$

Find an MVUE of $V(Y_i)$.

Solution In Chapter 4 we determined that $E(Y_i) = \theta$ and that $V(Y_i) = \theta^2$. The criterion of Lehmann and Scheffé will show that $\sum_{i=1}^{n} Y_i$ is the minimal sufficient statistic for θ. In fact, $\bar{Y}$ is the MVUE of θ. Therefore, it seems tempting to use $\bar{Y}^2$ as an estimator of θ^2. But

$$E(\bar{Y}^2) = V(\bar{Y}) + [E(\bar{Y})]^2 = \frac{\theta^2}{n} + \theta^2 = \left(\frac{n+1}{n}\right)\theta^2$$

It follows that $\bar{Y}^2$ is a biased estimate for θ^2. However,

$$\left(\frac{n}{n+1}\right)\bar{Y}^2$$

is an MVUE of θ^2 because it is an unbiased estimator for θ^2 that is a function of the minimal sufficient statistic. No other unbiased estimator of θ^2 will have a smaller variance than this one.

The minimal sufficient statistic for a parameter θ often can be used to construct an exact confidence interval for θ, if the probability distribution of the statistic can be found. The resulting intervals generally will be the shortest that can be found with a specified confidence coefficient. We illustrate the technique with an example involving the Weibull distribution.

EXAMPLE 9.10 The following data, with measurements in hundreds of hours, represent the lengths of life of ten identical electronic components operating in a guidance control system for missiles.

.637	1.531	.733	2.256	2.364
1.601	.152	1.826	1.868	1.126

The length of life of a component of this type is assumed to follow a Weibull distribution with density function given by

$$f(y) = \begin{cases} \left(\dfrac{2y}{\theta}\right) e^{-y^2/\theta}, & y > 0 \\ 0, & \text{elsewhere} \end{cases}$$

Use the data to construct a 95% confidence interval for θ.

Solution We saw in Example 9.7 that the minimal sufficient statistic for θ is $\sum_{i=1}^{n} Y_i^2$. We will use this statistic to form a pivotal quantity that will be used to construct the desired confidence interval.

Recall from Example 9.7 that $U_i = Y_i^2$ has an exponential distribution with mean θ. Now consider the transformation $W_i = 2U_i/\theta$. Then

$$f_W(w) = f_U\left(\frac{\theta w}{2}\right) \frac{d(\theta w/2)}{dw} = \left(\frac{1}{\theta}\right) e^{-(\theta w/2)/\theta} \left(\frac{\theta}{2}\right) = \left(\frac{1}{2}\right) e^{-w/2}, \qquad w > 0$$

Thus for each $i = 1, 2, \ldots, n$, W_i has a χ^2 distribution with 2 degrees of freedom. Further, because the variables Y_i are independent, the variables W_i are independent, $i = 1, 2, \ldots, n$. And because the sum of independent χ^2 random variables has a χ^2 distribution, with degrees of freedom equal to the sum of the degrees of freedom of the variables in the sum, the quantity

$$\sum_{i=1}^{10} W_i = \frac{2}{\theta} \sum_{i=1}^{10} U_i = \frac{2}{\theta} \sum_{i=1}^{10} Y_i^2$$

has a χ^2 distribution with 20 degrees of freedom. Thus

$$\frac{2}{\theta}\sum_{i=1}^{10}Y_i^2$$

is a pivotal statistic and we can use the pivotal method (Section 8.5) to construct the desired confidence interval.

From Table 6, Appendix III, we can find two numbers a and b such that

$$P\left(a\le\frac{2}{\theta}\sum_{i=1}^{10}Y_i^2\le b\right)=.95$$

Manipulating the inequality to isolate θ in the middle, we have

$$.95=P\left(a\le\frac{2}{\theta}\sum_{i=1}^{10}Y_i^2\le b\right)=P\left(\frac{1}{b}\le\frac{\theta}{2\sum_{i=1}^{10}Y_i^2}\le\frac{1}{a}\right)$$

$$=P\left(\frac{2\sum_{i=1}^{10}Y_i^2}{b}\le\theta\le\frac{2\sum_{i=1}^{10}Y_i^2}{a}\right)$$

From Table 6 the value that cuts off an area of .025 in the lower tail of the χ^2 distribution with 20 degrees of freedom is $a = 9.591$. The value that cuts off an area of .025 in the upper tail of the same distribution is $b = 34.170$. For the data given above, $\sum_{i=1}^{10}Y_i^2 = 24.643$. Therefore, the 95% confidence interval for the Weibull parameter θ is

$$\left(\frac{2(24.643)}{34.170},\frac{2(24.643)}{9.591}\right)$$

or $(1.442, 5.139)$

This is a fairly wide interval for θ, but keep in mind that the interval is based on only ten observations.   ∽

In this section we have considered the topics of minimal sufficient statistics and minimum-variance unbiased estimation. We have seen how to use the method of Lehmann and Scheffé to derive minimal sufficient statistics. For the distributions we consider in this text, a minimum-variance unbiased estimator (MVUE) for a target parameter θ can be found if we find some function of a minimal sufficient statistic that is an unbiased estimator for θ. This method often works well. However, sometimes a minimal sufficient statistic is a fairly complicated function of the observable random variables in the sample. In cases like these it may be difficult to find a function of the minimal sufficient statistic that is an unbiased estimator for the target parameter. For this reason two additional methods of finding estimators, the method of moments and the method of maximum likelihood, are presented in the next two sections. A third important method for estimation, the method of least squares, is the topic of Chapter 11.

Exercises

9.42 Let $Y_1, \ldots, Y_n$ denote a random sample from the probability density function

$$ f(y) = \begin{cases} \theta y^{\theta-1}, & 0 < y < 1; \theta > 0 \\ 0, & \text{elsewhere} \end{cases} $$

Find the minimal sufficient statistic for θ.

9.43 Refer to Exercise 9.26(b). Show that $\sum\limits_{i=1}^{n} (Y_i - \mu)^2$ is the minimal sufficient statistic for σ^2. Use this statistic to find an MVUE of σ^2.

9.44 Refer to Exercise 9.12. Is the estimator of σ^2 given there an MVUE of σ^2?

9.45 Refer to Exercise 9.28. Use $\sum\limits_{i=1}^{n} Y_i^2$ to find an MVUE of θ.

9.46 It is observed that the number of breakdowns per day, Y, for a certain machine is a Poisson random variable with mean λ. The daily cost of repairing these breakdowns is given by $C = 3Y^2$. If $Y_1, \ldots, Y_n$ denotes the observed number of breakdowns for n independently selected days, find an MVUE for $E(C)$.

*9.47 Refer to Exercise 9.37. Show that $Y_{(n)}$ is the minimal sufficient statistic for θ. Use $Y_{(n)}$ to find an MVUE of θ. (See Example 9.1.)

9.48 Refer to Exercise 9.39.

(a) Show that $Y_{(1)}$ is the minimal sufficient statistic for θ.

(b) Find a function of $Y_{(1)}$ that is an unbiased estimator for θ.

*9.49 An important theorem in statistics, known as the Rao-Blackwell theorem, states the following. If U is a sufficient statistic for θ and T is an unbiased estimator of θ, but not a function of U alone, then $g(U) = E(T|U)$ has the properties that

$$ E[g(U)] = \theta \quad \text{and} \quad V[g(U)] < V(T) $$

This theorem aids in the search for MVUEs because it implies that a minimum-variance unbiased estimator *must* be a function of a sufficient statistic. The theorem also gives us a method for finding the unbiased estimator $g(U)$, because $g(U)$ must equal $E(T|U)$ for some unbiased estimator T.

Let $Y_1, \ldots, Y_n$ be independent random variables with $P(Y_i = 1) = p$ and $P(Y_i = 0) = 1 - p$. Find the MVUE of $p(1 - p)$ (a term in the variance of Y_i or $Y = \sum\limits_{i=1}^{n} Y_i$) by the following steps.

(a) Let $T = 1$ if $Y_1 = 1$ and $Y_2 = 0$ and let $T = 0$ otherwise. Show that $E(T) = p(1 - p)$.

(b)   Show that

$$P(T = 1 | Y = y) = \frac{y(n - y)}{n(n - 1)}$$

(c)   Show that

$$E(T | Y) = \frac{n}{n - 1}\left[\frac{Y}{n}\left(1 - \frac{Y}{n}\right)\right]$$

and hence that this is the MVUE of $p(1 - p)$.

*9.50   Suppose a statistic U has a probability density function that is positive over the interval $a \leq U \leq b$, and suppose the density depends on a parameter θ that can range over the interval $\alpha_1 \leq \theta \leq \alpha_2$. Suppose also that $g(u)$ is continuous for u in the interval $[a, b]$. If $E[g(U)] = 0$ for all θ in the interval $[\alpha_1, \alpha_2]$ implies that $g(u)$ is identically zero, then the density $f_U(u)$ is said to be *complete*. (Most of the statistics we employ in common practical problems have complete density functions.) Suppose U is a minimal sufficient statistic for θ, and $g_1(U)$ and $g_2(U)$ are both unbiased estimators of θ. Show that if the density of U is complete, $g_1(U)$ must equal $g_2(U)$, and thus there is a *unique* function of U that is an unbiased estimator of θ.

Coupled with the Rao-Blackwell theorem of Exercise 9.49 the property of completeness of $f_U(u)$, along with the sufficiency of U, assures us that there is a unique minimum-variance unbiased estimator of θ.

9.6 The Method of Moments

We indicated in the previous section that the method we presented for finding a minimum-variance unbiased estimator is not always applicable. For this reason we present two additional methods for deriving estimators for population parameters. In this section we will discuss one of the oldest methods for deriving point estimators, the method of moments. A more sophisticated method, the method of maximum likelihood, is the topic of Section 9.7.

The method of moments is a very simple procedure for finding an estimator for one or more population parameters. Recall that the kth moment of a random variable, taken about the origin, is

$$\mu'_k = E(Y^k)$$

The corresponding kth sample moment is the average

$$m'_k = \frac{1}{n}\sum_{i=1}^{n} Y_i^k$$

The method of moments is based on the assumption that sample moments should provide good estimates of the corresponding population moments. That is, m_k' should be a good estimator of μ_k', $k = 1, 2, \ldots$. Then because the population moments $\mu_1', \mu_2', \ldots, \mu_k'$ will be functions of the population parameters, we will equate corresponding population and sample moments and solve for the desired parameters. Hence the method of moments can be stated as follows.

Method of Moments:

Choose as estimates those values of the parameters that are solutions of the equations $\mu_k' = m_k'$, $k = 1, 2, \ldots, t$, where t equals the number of parameters.

EXAMPLE 9.11   A random sample of n observations, $Y_1, Y_2, \ldots, Y_n$, is selected from a population in which Y_i, $i = 1, 2, \ldots, n$, possesses a uniform probability density function over the interval $(0, \theta)$ where θ is unknown. Use the method of moments to estimate the parameter θ.

Solution   The value of μ_1' for a uniform random variable is

$$\mu_1' = \mu = \frac{\theta}{2}$$

(This is easy to derive; or see Appendix II.) The corresponding first sample moment is

$$m_1' = \frac{\sum\limits_{i=1}^{n} Y_i}{n} = \bar{Y}$$

Equating corresponding moments and solving for the unknown parameter θ, we have

$$\mu_1' = \frac{\theta}{2} = \bar{Y} \quad \text{or} \quad \hat{\theta} = 2\bar{Y}$$

Thus $2\bar{Y}$ is the moment estimator of θ.   ∽

For the distributions we consider in this text, the methods of Section 9.3 can be used to show that sample moments are consistent estimators of the corresponding population moments. Because the estimators obtained from the method of moments obviously are functions of the sample moments, the estimators are usually consistent.

EXAMPLE 9.12   Show that the estimator $\hat{\theta} = 2\bar{Y}$, derived in Example 9.11, is a consistent estimator for θ.

Solution   In Example 9.1 we showed that $\hat{\theta} = 2\bar{Y}$ is an unbiased estimator for θ and that $V(\hat{\theta}) = \theta^2/3n$. Because $\lim_{n \to \infty} V(\hat{\theta}) = 0$, Theorem 9.1 implies that $\hat{\theta} = 2\bar{Y}$ is a consistent estimator for θ.   ∽

Although the estimator $\hat{\theta}$ derived in Example 9.11 is consistent, it is not necessarily the best estimator for θ. Indeed, the minimal sufficient statistic for θ can be shown to be $Y_{(n)} = \max(Y_1, \ldots, Y_n)$. Thus the moment estimator will have larger variance than an unbiased estimator based on $Y_{(n)}$. This, in fact, was shown to be the case in Example 9.1.

EXAMPLE 9.13   A random sample of n observations, $Y_1, Y_2, \ldots, Y_n$, is selected from a population where Y_i, $i = 1, 2, \ldots, n$, possesses a gamma probability density function with parameters α and β (see Section 4.6 for the gamma probability density function). Find moment estimators for the unknown parameters α and β.

Solution   Because we seek estimators for two parameters α and β, it will be necessary to equate two pairs of population and sample moments.

You can verify (either by deriving or by consulting Appendix II) that the first two moments of the gamma distribution are

$$\mu_1' = \mu = \alpha\beta$$
$$\mu_2' = \sigma^2 + \mu^2 = \alpha\beta^2 + \alpha^2\beta^2$$

Now equate these quantities to their corresponding sample moments and solve for α and β. Thus

$$\mu_1' = \alpha\beta = m_1' = \bar{Y}$$

$$\mu_2' = \alpha\beta^2 + \alpha^2\beta^2 = m_2' = \frac{\sum_{i=1}^{n} Y_i^2}{n}$$

from the first equation $\beta = m_1'/\alpha$. Substituting into the second equation and solving for α, we obtain

$$\alpha = \frac{m_1'^2}{m_2' - m_1'^2} = \frac{\bar{Y}^2}{(\sum Y_i^2/n) - \bar{Y}^2} = \frac{n\bar{Y}^2}{\sum_{i=1}^{n}(Y_i - \bar{Y})^2}$$

Substituting α into the first equation, we obtain

$$\beta = \frac{m_1'}{\alpha} = \frac{\bar{Y}}{\alpha} = \frac{\sum\limits_{i=1}^{n}(Y_i - \bar{Y})^2}{n\bar{Y}}$$

Hence the moment estimators of the parameters α and β are

$$\hat{\alpha} = \frac{n\bar{Y}^2}{\sum\limits_{i=1}^{n}(Y_i - \bar{Y})^2} \quad \text{and} \quad \hat{\beta} = \frac{\sum\limits_{i=1}^{n}(Y_i - \bar{Y})^2}{n\bar{Y}}$$

The moment estimators, $\hat{\alpha}$ and $\hat{\beta}$, in Example 9.13 are consistent. $\bar{Y}$ converges in probability to $E(Y_i) = \alpha\beta$ and $(1/n) \sum\limits_{i=1}^{n} Y_i^2$ converges in probability to $E(Y_i^2) = \alpha\beta^2 + \alpha^2\beta^2$. Thus

$$\hat{\alpha} = \frac{\bar{Y}^2}{\dfrac{1}{n}\sum\limits_{i=1}^{n} Y_i^2 - \bar{Y}^2}$$

is a consistent estimator of

$$\frac{(\alpha\beta)^2}{\alpha\beta^2 + \alpha^2\beta^2 - (\alpha\beta)^2} = \alpha$$

and

$$\hat{\beta} = \frac{\bar{Y}}{\hat{\alpha}}$$

is a consistent estimator of $\alpha\beta/\alpha = \beta$. The minimal sufficient statistics for the gamma density function can be shown to be $\sum\limits_{i=1}^{n} Y_i$ and the product $(Y_1)\cdots(Y_n)$.

Because the moment estimators, $\hat{\alpha}$ and $\hat{\beta}$, are not functions of the minimal sufficient statistics, it is possible to find more efficient estimators for the parameters α and β. However, it is considerably more difficult to apply other methods to find estimators for these parameters.

To summarize, the method of moments finds estimators of unknown parameters by equating corresponding sample and population moments. It is easy to employ and it provides estimators that are consistent. However, the estimators derived by this method generally are not functions of minimal sufficient statistics and hence are not very efficient. In many cases the moment estimators will not be unbiased. The primary value of this method is that it often provides estimators when other methods fail to do so.

Exercises

9.51 Let $Y_1, \ldots, Y_n$ denote a random sample from the probability density function

$$f(y) = \begin{cases} (\theta + 1)y^\theta, & 0 < y < 1; \theta > -1 \\ 0, & \text{elsewhere} \end{cases}$$

Find an estimator for θ by the method of moments. Show that the estimator is consistent. Is the estimator a function of the minimal sufficient statistic for θ?

9.52 Suppose $Y_1, \ldots, Y_n$ is a random sample from a Poisson distribution with a mean of λ. Find the moment estimator of λ.

9.53 If $Y_1, \ldots, Y_n$ denotes a random sample from the normal distribution with known mean $\mu = 0$ and unknown variance σ^2, find the moment estimator of σ^2.

9.54 If $Y_1, \ldots, Y_n$ denotes a random sample from the normal distribution with mean μ and variance σ^2, find the moment estimators of μ and σ^2.

9.55 An urn contains θ black balls and $N - \theta$ white balls. A sample of n balls is to be selected without replacement. Let Y denote the number of black balls in the sample. Show that $(N/n)Y$ is the moment estimator of θ.

9.56 If $Y_1, \ldots, Y_n$ is a random sample from the probability density function given by

$$f(y) = \begin{cases} \left(\dfrac{2}{\theta^2}\right)(\theta - y), & 0 \le y \le \theta \\ 0, & \text{elsewhere} \end{cases}$$

find an estimator for θ by the method of moments. Is this estimator a function of a sufficient statistic?

9.57 Let $Y_1, \ldots, Y_n$ be a random sample from the probability density function given by

$$f(y) = \begin{cases} \dfrac{\Gamma(2\theta)}{[\Gamma(\theta)]^2} (y^{\theta-1})(1 - y)^{\theta-1}, & 0 \le y \le 1 \\ 0, & \text{elsewhere} \end{cases}$$

Find the moment estimator for θ.

9.58 Let $X_1, X_2, X_3, \ldots$ be independent random variables such that $P(X_i = 1) = p$ and $P(X_i = 0) = 1 - p$ for each $i = 1, 2, 3, \ldots$. Let the random variable Y denote the number of trials necessary to obtain the first success, that is, the value of i for which $X_i = 1$ first occurs. Then Y has a geometric distribution with $P(Y = y) = (1 - p)^{y-1}p$, $y = 1, 2, 3, \ldots$. Find the moment estimator of p based on this single observation Y.

9.59 Let $Y_1, Y_2, \ldots, Y_n$ denote independent and identically distributed uniform random variables on the interval $(0, 3\theta)$. Derive the method of moments estimator for θ.

9.60   Let $Y_1, Y_2, \ldots, Y_n$ denote independent and identically distributed random variables from a power family distribution with parameters α and $\theta = 3$. Then, as in Exercise 9.31, if $\alpha > 0$,

$$f(y) = \begin{cases} \alpha y^{\alpha-1}/3^{\alpha}, & 0 \leq y \leq 3 \\ 0, & \text{elsewhere} \end{cases}$$

Show that $E(Y_1) = 3\alpha/(\alpha + 1)$ and derive the method of moments estimator for α.

9.61   Let $Y_1, Y_2, \ldots, Y_n$ denote independent and identically distributed random variables from a Pareto distribution with parameters α and β, where β is known. Then, if $\alpha > 0$,

$$f(y) = \begin{cases} \alpha \beta^{\alpha} y^{-(\alpha+1)}, & y \geq \beta \\ 0, & \text{elsewhere} \end{cases}$$

Show that $E(Y_1) = \alpha\beta/(\alpha - 1)$ if $\alpha > 1$ and $E(Y_1)$ is undefined if $0 < \alpha < 1$. Thus, the method of moments estimator for α is undefined.

9.7 The Method of Maximum Likelihood

In Section 9.5 we presented a method for deriving a minimum-variance unbiased estimator for a target parameter. The method requires that we find some function of a minimal sufficient statistic that is an unbiased estimator for the target parameter. Although we have a method for finding a minimal sufficient statistic, the determination of the function of the minimal sufficient statistic that gives us an unbiased estimator is largely a matter of hit or miss. The previous section contained a discussion of the method of moments. The method of moments is intuitive and easy to apply but does not usually lead to the best estimators. In this section we present a method, the method of maximum likelihood, that often leads to minimum-variance unbiased estimators.

We will use an example to illustrate the logic upon which the method of maximum likelihood is based. Suppose that we are confronted with a box that contains three balls. We know that some of the balls are white and some are red, but we do not know the number of either color. However, we are allowed to randomly sample two of the balls. If our random sample yields two red balls, what would be a good estimate of the total number of red balls in the box? Obviously the number of red balls in the box must be two or three. If there are two red balls and one white ball in the box, the probability of sampling two red balls is

$$\binom{2}{2}\binom{1}{0}\bigg/\binom{3}{2} = \frac{1}{3}$$

On the other hand, if there are three red balls in the box, the probability of sampling two red balls is

$$\binom{3}{2}\bigg/\binom{3}{2} = 1$$

It should seem reasonable to choose three as the estimate of the number of red balls in the box, because this estimate maximizes the probability of the observed sample. Of course, it is possible for the box to contain only two red balls, but the observed outcome gives more credence to there being three red balls in the box.

This example illustrates a method for finding an estimator that can be applied to any situation. The technique, called the *method of maximum likelihood*, selects as estimates those values of the parameters that maximize the likelihood (the probability function or joint density function) of the observed sample. (See Definition 9.4.)

Method of Maximum Likelihood:

Choose as estimates those values of the parameters that maximize the likelihood $L(y_1, \ldots, y_n)$.

We will illustrate the method with an example.

EXAMPLE 9.14 A binomial experiment consisting of n trials resulted in observations y_1, $y_2, \ldots, y_n$, where $y_i = 1$ if the ith trial was a success, $y_i = 0$ otherwise. Find the maximum-likelihood estimator of p, the probability of a success.

Solution The likelihood of the observed sample is the probability of observing y_1, $y_2, \ldots, y_n$. Hence

$$L(y_1, y_2, \ldots, y_n) = p^y(1 - p)^{n-y} \qquad \text{where} \qquad y = \sum_{i=1}^{n} y_i$$

We now wish to find the value of p that maximizes L. Because L is zero at $p = 0$ and $p = 1$ and is continuous for values of p between 0 and 1, we can find the maximum point by setting the derivative dL/dp equal to zero and solving for p.

You will note that $\ln L$ is a monotonically increasing function of L and hence both $\ln L$ and L will be maximized for the same value of p. Because L is a product of functions of p, and finding the derivative of products is tedious, it is easier to find the value of p that maximizes $\ln L$. We have

$$\ln L = y \ln p + (n - y) \ln(1 - p)$$

If we take the derivative of ln L with respect to p, we obtain

$$\frac{d \ln L}{dp} = y\left(\frac{1}{p}\right) + (n-y)\left(\frac{-1}{1-p}\right)$$

Then the value of p that maximizes (or minimizes) ln L is the solution of the equation

$$\frac{y}{\hat{p}} - \frac{n-y}{1-\hat{p}} = 0$$

Solving, we obtain the estimate $\hat{p} = y/n$. You can easily verify that this solution occurs when ln L (and hence L) achieves a maximum.

Note that the estimator $\hat{p} = Y/n$ is the fraction of successes in the total number of trials, n. Hence the maximum-likelihood estimator of p is one that is intuitively appealing. ∽

EXAMPLE 9.15   Let $Y_1, Y_2, \ldots, Y_n$ be a random sample from a normal distribution with mean μ and variance σ^2. Find the maximum-likelihood estimators of μ and σ^2.

Solution   Because $Y_1, Y_2, \ldots, Y_n$ are continuous random variables, L is the joint density of the sample. Thus $L = f(y_1, y_2, \ldots, y_n)$. In this case

$$L = f(y_1, y_2, \ldots, y_n) = f(y_1)f(y_2) \cdots f(y_n)$$

$$= \left\{\frac{\exp\left[\dfrac{-(y_1 - \mu)^2}{2\sigma^2}\right]}{\sigma\sqrt{2\pi}}\right\}\left\{\frac{\exp\left[\dfrac{-(y_2 - \mu)^2}{2\sigma^2}\right]}{\sigma\sqrt{2\pi}}\right\} \cdots \left\{\frac{\exp\left[\dfrac{-(y_n - \mu)^2}{2\sigma^2}\right]}{\sigma\sqrt{2\pi}}\right\}$$

$$= \left(\frac{1}{\sigma^n(2\pi)^{n/2}}\right)\exp\left[\frac{-\displaystyle\sum_{i=1}^{n}(y_i - \mu)^2}{2\sigma^2}\right]$$

[recall that exp( ) is just another way of writing $e^{(\ )}$] and

$$\ln L = -\frac{n}{2}\ln \sigma^2 - \frac{n}{2}\ln 2\pi - \frac{\displaystyle\sum_{i=1}^{n}(y_i - \mu)^2}{2\sigma^2}$$

The maximum-likelihood estimators of μ and σ^2 are those values that make ln L a maximum. Taking derivatives with respect to μ and σ^2, we obtain

$$\frac{d(\ln L)}{d\mu} = \frac{\displaystyle\sum_{i=1}^{n}(y_i - \mu)}{\sigma^2}$$

and

$$\frac{d \ln L}{d\sigma^2} = -\left(\frac{n}{2}\right)\left(\frac{1}{\sigma^2}\right) + \frac{\sum_{i=1}^{n}(y_i - \mu)^2}{2\sigma^4}$$

Setting the derivatives equal to zero and solving simultaneously, we obtain from the first equation

$$\frac{\sum_{i=1}^{n}(y_i - \hat{\mu})}{\hat{\sigma}^2} = 0$$

$$\sum_{i=1}^{n} y_i - n\hat{\mu} = 0$$

or

$$\hat{\mu} = \frac{\sum_{i=1}^{n} y_i}{n} = \bar{y}$$

Substituting $\bar{y}$ for $\hat{\mu}$ in the second equation and solving for $\hat{\sigma}^2$, we have

$$\frac{-n}{\hat{\sigma}^2} + \frac{\sum_{i=1}^{n}(y_i - \bar{y})^2}{\hat{\sigma}^4} = 0$$

or

$$\hat{\sigma}^2 = \frac{\sum_{i=1}^{n}(y_i - \bar{y})^2}{n} = s'^2$$

Thus $\bar{Y}$ and S'^2 are the maximum-likelihood estimators of μ and σ^2, respectively. Note that $\bar{Y}$ is unbiased for μ. Although S'^2 is not unbiased for σ^2, it can be easily adjusted to the unbiased estimator S^2 (see Example 8.1).

EXAMPLE 9.16   Let $Y_1, Y_2, \ldots, Y_n$ be a random sample of observations from a uniform distribution with probability density function $f(y_i) = 1/\theta$, $0 \le y_i \le \theta$, $i = 1, 2, \ldots, n$. Find the maximum-likelihood estimator of θ.

Solution   In this case the likelihood is given by

$$L = f(y_1, y_2, \ldots, y_n) = f(y_1)f(y_2)\cdots f(y_n) = \frac{1}{\theta}\frac{1}{\theta}\cdots\frac{1}{\theta} = \frac{1}{\theta^n}$$

You will note that L is a monotonically decreasing function of θ and hence that nowhere in the interval $0 < \theta < \infty$ is $dL/d\theta$ equal to zero. However, note that L increases as θ decreases and that θ must be equal to or greater than the maximum observation in the set $y_1, y_2, \ldots, y_n$. Hence the value of θ that maximizes L is the largest observation in the sample. That is, $\hat{\theta} = Y_{(n)} = \max(Y_1, \ldots, Y_n)$. The estimator $\hat{\theta} = Y_{(n)}$ is not an unbiased estimator of θ, but it can be adjusted to be unbiased, as shown in Example 9.1.   ∽

We have seen that minimal sufficient statistics have desirable properties and often can be used to find an MVUE for parameters of interest. If U is *any* sufficient statistic (including the case that U is a minimal sufficient statistic) for the estimation of a parameter θ, the maximum-likelihood estimator always is some function of U. That is, the maximum-likelihood estimator depends on the sample observations only through the value of a sufficient statistic. To show this, we need observe only that if U is a sufficient statistic for θ, the factorization criterion (Theorem 9.4) implies that the likelihood can be factored as

$$L(y_1, \ldots, y_n) = g(u, \theta)h(y_1, \ldots, y_n)$$

where $g(u, \theta)$ is a function of only u and θ and $h(y_1, \ldots, y_n)$ *does not depend on* θ. Therefore, it follows that

$$\ln L(y_1, \ldots, y_n) = \ln g(u, \theta) + \ln h(y_1, \ldots, y_n)$$

Notice that $\ln h(y_1, \ldots, y_n)$ does not depend on θ and therefore that maximizing $\ln L(y_1, \ldots, y_n)$ relative to θ is equivalent to maximizing $\ln [g(u, \theta)]$ relative to θ. Because $\ln [g(u, \theta)]$ depends on the data only through the value of the sufficient statistic U, it follows that $\hat{\theta}$ depends on the data only through the value of U. That is, if U is *any* sufficient statistic for estimating θ, the maximum-likelihood estimator always is some function of U. Consequently, if a maximum-likelihood estimator for a parameter can be found and then adjusted by a constant to be unbiased, the resulting estimator generally will be an MVUE of the parameter in question. This situation makes the method of maximum likelihood a very useful tool in finding estimators with good properties.

Exercises

9.62   Suppose $Y_1, \ldots, Y_n$ denotes a random sample from the Poisson distribution with a mean of λ.

   (a)   Find the maximum-likelihood estimator $\hat{\lambda}$ for λ.

   (b)   Find the expected value and variance of $\hat{\lambda}$.

   (c)   Show that the estimator of part (a) is consistent for λ.

9.63   Let $Y_1, \ldots, Y_n$ denote a random sample from the density function given by

$$f(y) = \begin{cases} \left(\dfrac{1}{\theta}\right) r y^{r-1} e^{-y^r/\theta}, & \theta > 0, \, y > 0 \\ 0, & \text{elsewhere} \end{cases}$$

where r is a known positive constant.

(a)   Find a sufficient statistic for θ.

(b)   Find the maximum-likelihood estimator of θ.

(c)   Is the estimator in part (b) an MVUE for θ?

9.64   Suppose $Y_1, Y_2, \ldots, Y_n$ is a random sample from a uniform distribution with probability density function

$$f(y) = \begin{cases} \dfrac{1}{2\theta + 1}, & 0 \le y \le 2\theta + 1 \\ 0, & \text{otherwise} \end{cases}$$

Obtain the maximum-likelihood estimator of θ.

9.65   A certain type of electronic component has a lifetime Y (in hours) with probability density function given by

$$f(y) = \begin{cases} \left(\dfrac{1}{\theta^2}\right) y e^{-y/\theta}, & y > 0 \\ 0, & \text{otherwise} \end{cases}$$

That is, Y has a gamma distribution with parameter $\alpha = 2$ and θ. Let $\hat{\theta}$ denote the maximum-likelihood estimator of θ. Suppose three such components, tested independently, gave lifetimes of 120, 130, and 128 hours.

(a)   Find the maximum-likelihood estimate of θ.

(b)   Find $E(\hat{\theta})$ and $V(\hat{\theta})$.

(c)   Suppose that θ actually equals 130. Give an approximate bound that you might expect for the error of estimation.

9.66   Let $Y_1, \ldots, Y_n$ denote a random sample from the density function given by

$$f(y) = \begin{cases} \left(\dfrac{1}{\Gamma(\alpha)\theta^\alpha}\right) y^{\alpha-1} e^{-y/\theta}, & y > 0 \\ 0, & \text{elsewhere} \end{cases}$$

(a)   Find the maximum-likelihood estimator $\hat{\theta}$ of θ if α is known.

(b)   Find the expected value and variance of $\hat{\theta}$.

(c)   Show that $\hat{\theta}$ is consistent for θ.

(d)   What is the minimal sufficient statistic for θ in this problem?

(e)   Suppose that $n = 5$ and $\alpha = 2$. Use the minimal sufficient statistic to construct a 90% confidence interval for θ. (Hint: Transform to a χ^2 distribution.)

9.67 Suppose that $X_1, \ldots, X_m$, representing yields per acre for corn variety A, is a random sample from a normal distribution with a mean of μ_1 and a variance of σ^2. Also, $Y_1, \ldots, Y_n$, representing yields for corn variety B, is a random sample from a normal distribution with a mean of μ_2 and a variance of σ^2. If the X's and Y's are independent, find the maximum-likelihood estimator for the common variance σ^2. Assume that μ_1 and μ_2 are unknown.

9.68 A random sample of 100 voters selected from a large population revealed 30 favoring candidate A, 38 favoring candidate B, and 32 favoring candidate C. Find maximum-likelihood estimates for the proportions of voters in the population favoring candidates A, B, and C, respectively. Estimate the difference between the fractions favoring A and B and place a two-standard-deviation bound on the error of estimation.

9.69 Let $Y_1, \ldots, Y_n$ denote a random sample from the probability density function

$$f(y) = \begin{cases} (\theta + 1)y^\theta, & 0 < y < 1; \theta > -1 \\ 0, & \text{elsewhere} \end{cases}$$

Find the maximum-likelihood estimator for θ. Compare your answer to the moment estimator found in Exercise 9.51.

9.70 It is known that the probability, p, of tossing heads on an unbalanced coin is either 1/4 or 3/4. The coin is tossed twice and a value for Y, the number of heads, is observed. For each possible value of Y, which of the two values for p (1/4 or 3/4) maximizes the probability that $Y = y$? Depending on the value of y actually observed, what is the maximum-likelihood estimator of p?

9.71 A random sample of 100 men produced a total of 25 who were in favor of a controversial local issue. An independent random sample of 100 women produced a total of 30 who were in favor of the issue. Assume that p_M is the true underlying proportion of men in favor of the issue, and p_W is the true underlying proportion of women in favor of the issue. If it actually is true that $p_W = p_M = p$, find the maximum-likelihood estimator of the common proportion, p.

*9.72 Find the maximum-likelihood estimator of θ based on a random sample of size n from a uniform distribution on the integral $(0, 2\theta)$.

*9.73 Suppose that $\hat{\theta}$ is the maximum-likelihood estimator for a parameter θ. Let $g(\theta)$ be a function of θ that possesses a unique inverse [i.e., if $\beta = g(\theta)$, then $\theta = g^{-1}(\beta)$]. Show that $g(\hat{\theta})$ is the maximum-likelihood estimator of $g(\theta)$.

*9.74 A random sample of n items is selected from the large number of items produced by a certain production line in one day. Find the maximum-likelihood estimator of the ratio R, the number of defective items to the number of good items.

9.75 Consider a random sample of size n from a normal population with a mean of μ and a variance of σ^2, both unknown. Derive the maximum-likelihood estimator of σ.

9.8 Summary

In this chapter we continued and extended the discussion of estimation begun in Chapter 8. Good estimators are consistent and efficient when compared to other estimators. The most efficient estimators, those with the smallest variances, are functions of minimal sufficient statistics.

Two methods of finding estimators, the method of moments and the method of maximum likelihood, were presented. Moment estimators are consistent but generally not very efficient. Maximum-likelihood estimators, on the other hand, are consistent and, if adjusted to be unbiased, very often lead to minimum-variance unbiased estimators.

References and Further Readings

1.   Cramer, H. *Mathematical Statistics*. Princeton, N.J.: Princeton University Press, 1946.

2.   Hogg, R. V., and Craig, A. T. *Introduction to Mathematical Statistics*. 4th ed. New York: Macmillan, 1978.

3.   Mood, A. M.; Graybill, F. A.; and Boes, D. *Introduction to the Theory of Statistics*. 3d ed. New York: McGraw-Hill, 1974.

4.   Wilks, S. S. *Mathematical Statistics*. New York: Wiley, 1962.

Supplementary Exercises

9.76   Suppose that $Y_1, \ldots, Y_n$ is a random sample from the density function

$$f(y) = \begin{cases} e^{-(y-\theta)}, & y > \theta \\ 0, & \text{elsewhere} \end{cases}$$

where θ is an unknown, positive constant.

(a)   Find an estimator $\hat{\theta}_1$ for θ by the method of moments.

(b)   Find an estimator $\hat{\theta}_2$ for θ by the method of maximum likelihood.

(c)   Adjust $\hat{\theta}_1$ and $\hat{\theta}_2$ so that they are unbiased. Find the efficiency of the adjusted $\hat{\theta}_1$ relative to the adjusted $\hat{\theta}_2$.

9.77   Refer to Exercise 9.26(b). Under the conditions outlined there, find the maximum-likelihood estimator of σ^2.

*9.78   Suppose $Y_1, \ldots, Y_n$ denotes a random sample from a Poisson distribution with a mean of λ. Find the MVUE of $P(Y_i = 0) = e^{-\lambda}$. (Hint: Make use of the Rao-Blackwell theorem, Exercise 9.49.)

*9.79   Suppose that a random sample of length-of-life measurements, $Y_1, \ldots, Y_n$, is to be taken on components whose length of life has an exponential distribution with mean θ. It is frequently of interest to estimate

$$\bar{F}(t) = 1 - F(t) = e^{-t/\theta}$$

the *reliability* at time t of a component of this type. Find the maximum-likelihood estimator of $\bar{F}(t)$.

*9.80   The maximum-likelihood estimator obtained in Exercise 9.79 is a function of the minimal sufficient statistic for θ, but it is not unbiased. Use the Rao-Blackwell theorem to find the MVUE of $e^{-t/\theta}$ by the following steps.

(a)   Let

$$V = \begin{cases} 1, & Y_1 > t \\ 0, & \text{otherwise} \end{cases}$$

Show that V is an unbiased estimator of $e^{-t/\theta}$.

(b)   Because $U = \sum\limits_{i=1}^{n} Y_i$ is the minimal sufficient statistic for θ, show that the conditional density function for Y_1 given $U = u$ is

$$f_{Y_1 | U}(y_1) = \begin{cases} \left(\dfrac{n-1}{u^{n-1}}\right)(u - y_1)^{n-2}, & 0 < y_1 < u \\ 0, & \text{elsewhere} \end{cases}$$

(c)   Show that

$$E(V|U) = P(Y_1 > t|U) = \left(1 - \frac{t}{U}\right)^{n-1}$$

This is the MVUE of $e^{-t/\theta}$ by the Rao-Blackwell theorem and by the fact that the density function for U is complete.

*9.81   Suppose that n integers are drawn at random and *with replacement* from the integers 1, 2, ..., N. That is, each sampled integer has probability $1/N$ of taking on any of the values 1, 2, ..., N, and the sampled values are independent.

(a)   Find the moment estimator $\hat{N}_1$ of N.

(b)   Find $E(\hat{N}_1)$ and $V(\hat{N}_1)$.

*9.82   Refer to Exercise 9.81.

(a)   Find the maximum-likelihood estimator $\hat{N}_2$ of N.

(b)   Show that $E(\hat{N}_2)$ is approximately $[n/(n+1)]N$. Adjust $\hat{N}_2$ to form an estimator $\hat{N}_3$ that is approximately unbiased for N.

(c)   Find an approximate variance for $\hat{N}_3$ by using the fact that for large N the variance of the largest sampled integer is approximately

$$\frac{nN^2}{(n+1)^2(n+2)}$$

(d)   Show that for large N and $n > 1$, $V(\hat{N}_3) < V(\hat{N}_1)$.

*9.83   Suppose that enemy tanks have serial numbers 1, 2, ..., N. A spy randomly observes five tanks (with replacement) with serial numbers 97, 64, 118, 210, and 57. Estimate N and place a bound on the error of estimation.

chapter ten

HYPOTHESIS TESTING

10.1 Introduction

Recall that often the objective of statistics is to make inferences about unknown population parameters based on information contained in sample data. These inferences are phrased in one of two ways, as estimates of the respective parameters or as tests of hypotheses about their values. Chapters 8 and 9 dealt with estimation. In this chapter we deal with the general topic of hypothesis testing.

In many ways the formal procedure for hypothesis testing is similar to the scientific method. The scientist observes nature, formulates a theory, and then tests this theory against observation. In our context the scientist poses a theory concerning one or more population parameters—that they equal specified values. He or she then samples the population and compares observation with theory. If the observations disagree with the theory, the scientist rejects the hypothesis. If not, the scientist concludes either that the theory is true or that the sample did not detect the difference between the real and hypothesized values of the population parameters.

For example, a medical researcher may hypothesize that a new drug is more effective than another in combating a disease. To test her hypothesis, she randomly selects patients infected with the disease and randomly divides them into two groups. The new drug A is applied to the first group of patients and the other drug, B, is applied to the second. Then based on the number of patients in each group who recover from the disease, the researcher must decide whether or not the new drug is more effective than the old.

Hypothesis tests are conducted in all fields in which theory can be tested against observation. A quality control engineer may hypothesize that a new assembly method produces only 5% defective items. An educator may claim that two methods of teaching reading are equally effective, or a political candidate may claim that a plurality of voters favor her election. All these hypotheses can be subjected to statistical verification by comparing the hypotheses with observed sample data.

What is the role of statistics in testing hypotheses? Putting it more bluntly, of what value is statistics in this hypothesis-testing procedure? Note that testing a hypothesis requires a decision when comparing the observed sample with theory. How do we decide whether the sample disagrees with the scientist's hypothesis? When should we reject the hypothesis, when should we accept it, and when should we withhold judgment? What is the probability that we will make the wrong decision and be led to a consequential loss? And, particularly, what function of the sample measurements should be employed to reach a decision? The answers to these questions are contained in a study of statistical hypothesis testing.

You will recall that Chapter 8 introduced the general topic of estimation and presented some intuitive estimation procedures. In Chapter 9 we presented some properties of estimators and some formal methods for deriving estimators. We will use the same approach in our discussion of hypothesis testing. That is, we will introduce the topic, present some intuitive testing procedures, and then consider some formal methods for deriving statistical testing procedures.

10.2 Elements of a Statistical Test

Many times the objective of a statistical test is to test a hypothesis concerning the values of one or more population parameters. We generally will have a theory, a *research hypothesis*, about the parameter(s) that we wish to support. For example, suppose that a political candidate, Jones, claims that he will gain more than 50% of the votes in a city election and thereby emerge as the winner. If we do not believe Jones's claim, we might seek to support the research hypothesis that Jones is *not* favored by more than 50% of the electorate. Support for this research hypothesis, also called the *alternative hypothesis*, is obtained by showing (using the sample data as evidence) that the converse of the alternative hypothesis, the *null hypothesis*, is false. Thus support for one theory is obtained by showing lack of support for its converse, in a sense, of proof by contradiction. Because we seek support for the alternative hypothesis that Jones's claim is false, our alternative hypothesis is that p, the probability of selecting a voter favoring Jones, is less than .5. If we can show that the data support the rejection of the null hypothesis, $p = .5$ (the minimum value needed for a plurality), in favor of the alternative hypothesis, $p < .5$, we have achieved our research objective. Although it is common to speak of testing a null hypothesis, keep in mind that the research objective usually is to show support for the alternative hypothesis, if support is warranted.

How do we utilize that data to decide between the null hypothesis and the alternative hypothesis? Suppose that $n = 15$ voters are randomly selected from the city and Y, the number favoring Jones, is recorded. If none in the sample favor Jones ($Y = 0$), what would you conclude about Jones's claim? If Jones is actually favored by more than 50% of the electorate, it is not *impossible* to observe $Y = 0$ favoring Jones in a sample of size $n = 15$, but it is highly *improbable*. It is much more likely that we would observe $Y = 0$ if the alternative hypothesis were true. Thus we would reject the null hypothesis ($p = .5$) in favor of the alternative hypothesis ($p < .5$). If we observed $Y = 1$ (or any small value of Y), the same type of reasoning would lead us to the same conclusion.

Any statistical test of hypotheses works in exactly the same way and is composed of the same essential elements.

The Elements of a Statistical Test

1. null hypothesis, H_0

2. alternative hypothesis, H_a

3. test statistic

4. rejection region

For our example the hypothesis to be tested, called the *null hypothesis* and denoted by H_0, is $p = .5$. The alternative (or research) hypothesis, denoted as H_a, is the hypothesis to be accepted in case H_0 is rejected. The alternative hypothesis usually is the hypothesis we seek to support on the basis of the information contained in the sample; thus in our example H_a is $p < .5$.

The functioning parts of a statistical test are the test statistic and an associated rejection region. The *test statistic* (like an estimator) is a function of the sample measurements (Y in our example) upon which the statistical decision will be based. The *rejection region*, which will henceforth be denoted by RR, specifies the values of the test statistic for which the null hypothesis is rejected. If for a particular sample the computed value of the test statistic falls in the rejection region, we reject the null hypothesis H_0 and accept the alternative hypothesis H_a. If the value of the test statistic does not fall into the rejection region RR, we accept H_0. As previously indicated, for our example small values of Y would lead us to reject H_0. Therefore, one rejection region we might want to consider is the set of all values of Y that are less than or equal to 2. We will use the notation $RR = \{y:y \leq 2\}$, or, more simply, $RR = \{y \leq 2\}$, to denote this rejection region.

Finding a good rejection region for a statistical test is an interesting problem and one that bears further attention. It is clear that small values of Y, say, $y \leq k$ (see Figure 10.1), are contradictory to the hypothesis $H_0: p = .5$ but favorable to the alternative $H_a: p < .5$. So we intuitively choose the rejection region as $RR = \{y \leq k\}$. But what value should we choose for k? More generally, we seek some objective criteria for deciding which value of k would specify a good rejection region of the form $\{y \leq k\}$.

Figure 10.1
Rejection region,
$RR = \{y \leq k\}$, for a
test of the hypothesis
$H_0: p = .5$ against
the alternative
$H_a: p < .5$

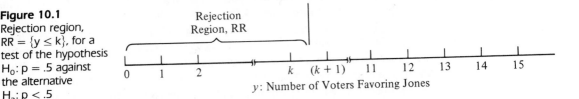

Note that for any fixed rejection region (determined by a particular value of k), two types of errors can be made in reaching a decision. We can decide in favor of H_a when H_0 is true (called a type I error), or we can decide in favor of H_0 when H_a is true (called a type II error).

Definition 10.1

> A *type I error* is made if H_0 is rejected when H_0 is true. The *probability of a type I error* is denoted by α.
> A *type II error* is made if H_0 is accepted when H_a is true. The *probability of a type II error* is denoted by β.

For Jones's political poll a type I error, rejecting $H_0: p = .5$ when in fact H_0 is true, means that we conclude that Jones will lose when, in fact, he is going to win. In contrast, making a type II error means that we accept $H_0: p = .5$ when $p < .5$ and conclude that Jones will win when, in fact, he will lose. For most real situations incorrect decisions cost money, prestige, and so on, and imply a loss. Thus α and β, the probabilities of making these two types of errors, measure the risks associated with the two possible erroneous decisions that might result from a statistical test. As such, they provide a very practical way to measure the goodness of a test.

EXAMPLE 10.1   For Jones's political poll $n = 15$ voters are sampled. We wish to test $H_0: p = .5$ against the alternative $H_a: p < .5$. The test statistic is Y, the number of sampled voters favoring Jones. Calculate α if we select $RR = \{y \leq 2\}$ as the rejection region.

Solution

By definition,

$$\alpha = P(\text{type I error}) = P(\text{rejecting } H_0 \text{ when } H_0 \text{ is true})$$
$$= P(\text{value of test statistic is in RR when } H_0 \text{ is true})$$
$$= P(Y \leq 2 \text{ when } p = .5)$$

Noting that Y is a binomial random variable with $n = 15$ and $p = .5$, we obtain

$$\alpha = \sum_{y=0}^{2} \binom{15}{y}(.5)^{15} = \binom{15}{0}(.5)^{15} + \binom{15}{1}(.5)^{15} + \binom{15}{2}(.5)^{15}$$

Using Table I, Appendix III, to circumvent this computation, we find $\alpha = .004$. Thus we see that if we decide to use the rejection region RR $= \{y \leq 2\}$, we subject ourselves to a very small risk ($\alpha = .004$) of concluding that Jones will lose if, in fact, he is a winner. ↶

EXAMPLE 10.2 Refer to Example 10.1. Is our test equally good in protecting us from concluding that Jones is a winner if, in fact, he will lose? Suppose that he really will win 30% of the vote ($p = .30$). What is the probability β that the sample will erroneously lead us to conclude that H_0 is true and that Jones is going to win?

Solution

By definition,

$$\beta = P(\text{type II error}) = P(\text{accepting } H_0 \text{ when } H_a \text{ is true})$$
$$= P(\text{value of the test statistic is } not \text{ in RR when } H_a \text{ is true})$$

Because we want to calculate β when $p = .3$ (a particular value of p that is in H_a),

$$\beta = P(Y > 2 \text{ when } p = .30) = \sum_{y=3}^{15} \binom{15}{y}(.30)^y(.70)^{15-y}$$

Again consulting Table 1, Appendix III, we find that $\beta = .873$. In other words, if we use RR $= \{y \leq 2\}$, our test will almost always lead us to conclude that Jones is a winner (with probability $\beta = .873$), even if p is as low as $p = .3$. ↶

Note that β depends on the true value of p. The larger the difference between p and the (null) hypothesized value of $p = .5$, the less is the likelihood that we will fail to reject the null hypothesis.

EXAMPLE 10.3 Refer to Examples 10.1 and 10.2. Calculate the value of β if Jones really has only 10% of the votes ($p = .1$).

Solution
In this case we want to calculate β when $p = .1$ (another particular value of p in H_a).

$$\beta = P(\text{type II error}) = P(\text{accepting } H_0 \text{ when } p = .1)$$
$$= P(\text{value of test statistic is } not \text{ in RR when } p = .1)$$
$$= P(Y > 2 \text{ when } p = .1) = \sum_{y=3}^{15} \binom{15}{y}(.1)^y(.9)^{15-y} = .184$$

Consequently, if we use $\{y \le 2\}$ as the rejection region, there is a fairly large probability that we will claim that Jones is a winner if, in fact, he will receive only 10% of the votes.   ∽

Examples 10.1 through 10.3 show that the test using $RR = \{y \le 2\}$ guarantees low risk in making a type I error ($\alpha = .004$) but it does not offer adequate protection against a type II error. How can we improve our test? One way is to balance α and β by changing the rejection region. If we enlarge RR to a new rejection region RR* (that is $RR \subset RR^*$), then the test using RR* will reject H_0 more often. If α^* and α denote the probabilities of type I errors when we use RR* and RR as the rejection regions, respectively, then

$$\alpha^* = P(\text{test statistic is in RR*}) \ge P(\text{test statistic is in RR}) = \alpha$$

Likewise, if we use the enlarged rejection region RR*, then the test procedure will accept H_0 less often. If β^* and β denote the probabilities of type II errors for the tests using RR* and RR, respectively, then $\beta^* \le \beta$. Hence if we change the rejection region to increase α, then β will decrease. Similarly, if the change in rejection region results in a decrease in α, then β will increase. Thus α and β are inversely related.

EXAMPLE 10.4
Refer to the test discussed in Example 10.1. Now assume that $RR = \{y \le 5\}$. Calculate α and calculate β if $p = .3$. Compare the results with the values obtained in Examples 10.1 and 10.2 (where we used $RR = \{y \le 2\}$).

Solution
In this case

$$\alpha = P(\text{test statistic is in RR when } H_0 \text{ is true})$$
$$= P(Y \le 5 \text{ when } p = .5) = \sum_{y=0}^{5} \binom{15}{y}(.5)^{15} = .151$$

When $p = .3$,

$$\beta = P(\text{test statistic not in RR when } H_a \text{ is true and } p = .3)$$
$$= P(Y > 5 \text{ when } p = .3) = \sum_{y=6}^{15} \binom{15}{y}(.3)^y(.7)^{15-y} = .278$$

A comparison of the α and β calculated here with the results of Examples 10.1 and 10.2 shows that enlarging the rejection region from $RR = \{y \le 2\}$ to $RR^* = (y \le 5)$ increased α and decreased β (see Table 10.1). Hence we have achieved a better balance between the risks of type I and type II errors, but both α and β are uncomfortably large. *How, then, can we reduce both α and β?* The answer is intuitively clear—shed more light on the true nature of the population by increasing the sample size. For almost all statistical tests α and β will both decrease as the sample size increases.

Table 10.1
A comparison of α and β for two different rejection regions

Probabilities of error	RR	
	$(0, 1, 2)$	$(0, 1, 2, \ldots, 5)$
α	.004	.151
β for $p = .3$	.873	.278

In this section we have defined the essential elements of any statistical test. We have seen that two possible types of error can be made when testing hypotheses, type I and type II errors. The probabilities of these errors serve as criteria for evaluating a testing procedure. In the next few sections we will utilize the sampling distributions derived in Chapter 7 to develop methods for testing hypotheses about parameters of frequent practical interest.

Exercises

10.1 Define α and β for a statistical test of hypotheses.

10.2 An experimenter has prepared a drug dosage level that she claims will induce sleep for at least 80% of those people suffering from insomnia. After examining the dosage, we feel that her claims regarding the effectiveness of the dosage are inflated. In an attempt to disprove her claim, we administer her prescribed dosage to twenty insomniacs, and we observe Y, the number having sleep induced by the drug dose. We wish to test the hypothesis $H_0: p = .8$ versus the alternative $H_a: p < .8$. Assume the rejection region $\{y \le 12\}$ is used.

(a) Find α.

(b) Find β for $p = .6$.

(c) Find β for $p = .4$.

10.3 Refer to Exercise 10.2.

(a) Find the rejection region of the form $\{y \le c\}$ so that $\alpha = .01$.

(b) For the rejection region in (a) find β for $p = .6$.

(c) For the rejection region in (a) find β for $p = .4$.

10.4 Suppose that we wish to test the null hypothesis, H_0, that the proportion of ledger sheets with errors, p, is equal to .05 versus the alternative, H_a, that the proportion is larger than .05 by using the following scheme. Two ledger sheets are selected at random. If both are error free, we reject H_0. If one or more contains an error, we look at a third sheet. If the third sheet is error free, we reject H_0. In all other cases we accept H_0.

(a) What is the value of α associated with this test?

(b) Calculate $\beta = P(\text{type II error})$ as a function of p.

10.3 Common Large-Sample Tests

Suppose that we want to test a hypothesis concerning a parameter, θ, based on a random sample $Y_1, \ldots, Y_n$. In this section we will develop a hypothesis-testing procedure based on an estimator $\hat{\theta}$, which has an (approximately) normal sampling distribution with a mean of θ and a variance of $\sigma_{\hat{\theta}}^2$. The large-sample estimators of Chapter 8 (Table 8.1), such as $\overline{Y}$ and $\hat{p}$ used for estimating a population mean, μ, and proportion, p, respectively, satisfy these requirements. So also do the estimators for the comparison of two means, $(\mu_1 - \mu_2)$, and the comparison of two binomial parameters, $(p_1 - p_2)$.

If θ_0 is a specific value of θ, we may wish to test $H_0: \theta = \theta_0$ versus $H_a: \theta > \theta_0$. Figure 10.2 contains a graphical depiction of the sampling distributions of $\hat{\theta}$ for various values of θ. If $\hat{\theta}$ is close to θ_0, it seems reasonable to accept

Figure 10.2
Sampling distributions of the estimator $\hat{\theta}$ for various values of θ

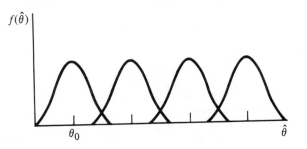

H_0. However, if $\theta > \theta_0$, it is more likely that $\hat{\theta}$ will be large. Consequently, large values of $\hat{\theta}$ (larger than θ_0) favor the rejection of $H_0: \theta = \theta_0$ and acceptance of $H_a: \theta > \theta_0$. That is, the null and alternative hypotheses, the test statistic, and the rejection region are as follows:

H_0: $\theta = \theta_0$.

H_a: $\theta > \theta_0$.

Test Statistic: $\hat{\theta}$.

Rejection Region: RR $= \{\hat{\theta} > k\}$ for some choice of k.

Figure 10.3
Large-sample
rejection region for
$H_0: \theta = \theta_0$ versus
$H_a: \theta > \theta_0$

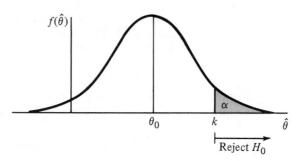

The actual value of k in RR is determined by fixing the type-I-error probability α and choosing k accordingly (see Figure 10.3). If H_0 is true, $\hat\theta$ has an approximately normal distribution with a mean of θ_0 and a variance $\sigma_{\hat\theta}^2$. Therefore, if we desire $P(\text{type I error}) = \alpha$, then

$$k = \theta_0 + z_\alpha \sigma_{\hat\theta}$$

is the appropriate choice for k [where z_α is such that $P(Z > z_\alpha) = \alpha$ when Z has a standard normal distribution]. Because

$$\text{RR} = \{\hat\theta: \hat\theta > \theta_0 + z_\alpha \sigma_{\hat\theta}\} = \left\{\hat\theta: \frac{\hat\theta - \theta_0}{\sigma_{\hat\theta}} > z_\alpha\right\}$$

if $Z = (\hat\theta - \theta_0)/\sigma_{\hat\theta}$ is used as the test statistic, the rejection region also may be written as $\text{RR} = \{Z > z_\alpha\}$. Note that Z measures the number of standard deviations between $\hat\theta$ and θ_0. Thus an equivalent form of the test of hypothesis, with $\alpha = P(\text{type I error})$, is as follows:

$H_0: \quad \theta = \theta_0$.

$H_a: \quad \theta > \theta_0$.

Test Statistic: $Z = \dfrac{\hat\theta - \theta_0}{\sigma_{\hat\theta}}$.

Rejection Region: $z > z_\alpha$.

Note that H_0 is rejected if Z falls far enough into the upper tail of the standard normal distribution. The alternative hypothesis $H_a: \theta > \theta_0$ is called an *upper-tail* alternative, and $\text{RR} = \{z > z_\alpha\}$ is referred to as an *upper-tail rejection region*.

EXAMPLE 10.5   A vice-president in charge of sales for a large corporation claims that salespeople are averaging no more than fifteen sales contacts per week. (He would like to increase this figure.) As check on his claim, $n = 36$ salespeople are selected at random, and the number of contacts is recorded for a single randomly selected week. The sample reveals a mean of seventeen contacts and a variance of 9. Does the evidence contradict the vice-president's claim? Use $\alpha = .05$.

Solution

We are interesed in testing a hypothesis about the mean number of sales contacts per week, μ. Specifically, we are testing

$$H_0: \mu = 15 \quad \text{against} \quad H_a: \mu > 15$$

We know that the sample mean $\bar{Y}$ is a point estimator of μ that satisfies the assumptions just described. Hence our test statistic is

$$Z = \frac{\bar{Y} - \mu_0}{\sigma_{\bar{Y}}} = \frac{\bar{Y} - \mu_0}{\sigma/\sqrt{n}}$$

The rejection region, with $\alpha = .05$, is given by $z > 1.645$ (see Table 4, Appendix III). The population variance σ^2 is not known, but it can be estimated very accurately (because $n = 36$ is sufficiently large) by the sample variance $s^2 = 9$. Thus the observed value of the test statistic is approximately

$$z = \frac{\bar{y} - \mu}{s/\sqrt{n}} = \frac{17 - 15}{3/\sqrt{36}} = 4$$

Because the observed value of Z lies in the rejection region (exceeds $z_\alpha = 1.645$), we reject $H_0: \mu = 15$. Thus it appears that the vice-president's claim is incorrect and that the average number of sales contacts per week exceeds 15. ∽

EXAMPLE 10.6

A machine in a certain factory must be repaired if it produces more than 10% defectives among the large lot of items it produces in a day. A random sample of 100 items from the day's production contains 15 defectives, and the foreman says that the machine must be repaired. Does the sample evidence support his decision? Use $\alpha = .01$.

Solution

If Y denotes the number of observed defectives, then Y is a binomial random variable, with p denoting the probability that a randomly selected item is defective. Hence we want to test the null hypothesis

$$H_0: p = .10$$

against the alternative

$$H_a: p > .10$$

The test statistic is based on $\hat{p} = Y/n$, the unbiased point estimator of p, and is given by

$$Z = \frac{\hat{p} - p_0}{\sigma_{\hat{p}}} = \frac{\hat{p} - p_0}{\sqrt{\dfrac{p_0(1 - p_0)}{n}}}$$

[We could have used $\hat{p}(1 - \hat{p})/n$ to approximate $\sigma_{\hat{p}}^2$, but because we are considering the distribution of Z under H_0, it is more appropriate to use $p_0(1 - p_0)/n$, the true value of $\sigma_{\hat{p}}^2$ when H_0 is true.]

From Table 4, Appendix III, we see that $P(Z > 2.33) = .01$. Hence we take $z > 2.33$ as the rejection region. The observed value of the test statistic is given by

$$z = \frac{\hat{p} - p_0}{\sqrt{\dfrac{p_0(1 - p_0)}{n}}} = \frac{.15 - .10}{\sqrt{\dfrac{(.1)(.9)}{100}}} = \frac{5}{3}$$

Since the observed value of Z is not in the rejection region, we conclude that the evidence does not support the foreman's decision. Is the foreman wrong? We would not reach this conclusion until we have evaluated the probability of accepting H_0 when H_a is true, that is, until we have calculated β. The method for calculating β is presented in Section 10.4.   ∽

A test of $H_0: \theta = \theta_0$ against $H_a: \theta < \theta_0$ would be carried out in an analogous manner, except that we now reject H_0 for values of $\hat{\theta}$ that are much smaller than θ_0. The test statistic remains as

$$Z = \frac{\hat{\theta} - \theta_0}{\sigma_{\hat{\theta}}}$$

but for fixed α we reject the null hypothesis when $z < -z_\alpha$. Because we reject H_0 in favor of H_a when z falls far enough into the lower tail of the standard normal distribution, we call $H_a: \theta < \theta_0$ a *lower-tail* alternative and RR: $z < -z_\alpha$ a *lower-tail rejection region*.

If we wish to test $H_0: \theta = \theta_0$ against $H_a: \theta \neq \theta_0$, we would want to reject H_0 if $\hat{\theta}$ is either much smaller or much larger than θ_0. The test statistic is still Z, as earlier, but the rejection region is located symmetrically in the two tails of the probability distribution for Z. Thus we reject H_0 if either $z < -z_{\alpha/2}$ or $z > z_{\alpha/2}$. Equivalently, we reject H_0 if $|z| > z_{\alpha/2}$. This test is called a *two-tailed test*, as opposed to the *one-tailed tests* used for the alternatives $\theta < \theta_0$ and $\theta > \theta_0$. The rejection regions for the lower-tail alternative, $H_a: \theta < \theta_0$, and the two-sided alternative, $H_a: \theta \neq \theta_0$, are displayed in Figure 10.4.

Figure 10.4
Rejection regions for
testing $H_0: \theta = \theta_0$
versus (a) $H_a: \theta < \theta_0$,
and (b) $H_a: \theta \neq \theta_0$,
based on $Z = \dfrac{\hat{\theta} - \theta_0}{\sigma_{\hat{\theta}}}$

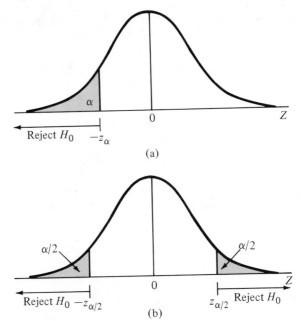

(a)

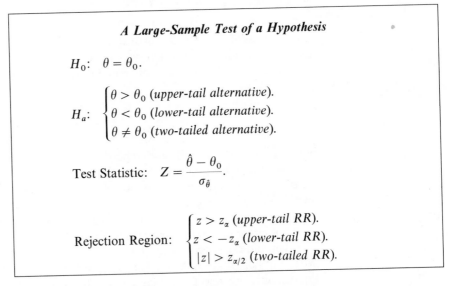

(b)

A summary of the large-sample hypothesis tests developed this far is as follows.

A Large-Sample Test of a Hypothesis

H_0: $\theta = \theta_0$.

H_a: $\begin{cases} \theta > \theta_0 \ (upper\text{-}tail\ alternative). \\ \theta < \theta_0 \ (lower\text{-}tail\ alternative). \\ \theta \neq \theta_0 \ (two\text{-}tailed\ alternative). \end{cases}$

Test Statistic: $Z = \dfrac{\hat{\theta} - \theta_0}{\sigma_{\hat{\theta}}}$.

Rejection Region: $\begin{cases} z > z_\alpha \ (upper\text{-}tail\ RR). \\ z < -z_\alpha \ (lower\text{-}tail\ RR). \\ |z| > z_{\alpha/2} \ (two\text{-}tailed\ RR). \end{cases}$

Note that in any particular test only one of the alternatives H_a just given will be appropriate. Whatever alternative hypothesis we choose, we must be sure to use the corresponding rejection region.

How do we decide which alternative we should use for a test? The answer depends on the hypothesis we seek to support. If we are interested only in detecting an increase in the percentage of defectives (Example 10.6), we should locate the rejection region in the upper tail of the Z distribution. On the other hand, if it is practically important for us to detect a change in p either above or below $p = .10$, then locate the rejection region in both tails of the Z distribution and employ a two-tailed test. The following example illustrates a situation in which a two-tailed test is appropriate.

EXAMPLE 10.7   A psychological study was conducted to compare the reaction times of men and women to a certain stimulus. Independent random samples of 50 men and 50 women were employed in the experiment. The results are shown in Table 10.2. Do the data present sufficient evidence to suggest a difference between true mean reaction times for men and women? Use $\alpha = .05$.

Table 10.2
Data for
Example 10.7

Men	Women
$n_1 = 50$	$n_2 = 50$
$\bar{y}_1 = 3.6$ seconds	$\bar{y}_2 = 3.8$ seconds
$s_1^2 = .18$	$s_2^2 = .14$

Solution   Let μ_1 and μ_2 denote the true mean reaction times for men and women, respectively. Then if we wish to test the hypothesis that the means are equal, we will test $H_0: (\mu_1 - \mu_2) = 0$ against $H_a: (\mu_1 - \mu_2) \neq 0$. Note that we use the two-sided alternative to detect either the case $\mu_1 > \mu_2$ or the reverse $\mu_2 > \mu_1$, in case H_0 is false.

The point estimator of $(\mu_1 - \mu_2)$ is $(\bar{Y}_1 - \bar{Y}_2)$ and it satisfies the assumptions of our large-sample test. Hence if we desire to test $H_0: \mu_1 - \mu_2 = D_0$ (D_0 fixed) versus any alternative, the test statistic is given by

$$Z = \frac{(\bar{Y}_1 - \bar{Y}_2) - D_0}{\sqrt{\dfrac{\sigma_1^2}{n_1} + \dfrac{\sigma_2^2}{n_2}}}$$

where σ_1^2 and σ_2^2 are the respective population variances. In this application we desire a two-tailed test. Thus for $\alpha = .05$ we reject H_0 for $|z| > z_{\alpha/2} = z_{.025} = 1.96$.

For large samples (say, $n > 30$) the sample variances provide good estimates of their corresponding population variances. Substituting these values, along with $\bar{y}_1, \bar{y}_2, n_1, n_2,$ and $D_0 = 0$, into the formula for the test statistic, we have

$$z = \frac{\bar{y}_1 - \bar{y}_2 - 0}{\sqrt{\dfrac{\sigma_1^2}{n_1} + \dfrac{\sigma_2^2}{n_2}}} \approx \frac{.36 - .38}{\sqrt{\dfrac{.18}{50} + \dfrac{.14}{50}}} = -2.5$$

This value is less than $-z_{\alpha/2} = -1.96$ and therefore falls in the rejection region. Hence we reject the hypothesis of no difference in mean reaction times for men and women.

In this section we have described the general procedure to implement a large-sample test of hypotheses for some parameters of frequent practical interest. We will show you in Section 10.4 how to calculate β, the probability of a type II error, for these large-sample tests.

Exercises

10.5   The daily wages in a particular industry are normally distributed with a mean of $13.20 and a standard deviation of $2.50. If a company in this industry employing forty workers pays these workers, on the average, $12.20, can this company be accused of paying inferior wages? Use $\alpha = .01$.

10.6   The output voltage for a certain electric circuit is specified to be 130. A sample of forty independent readings on the voltage for this circuit gave a sample mean of 128.6 and a standard deviation of 2.1. Test the hypothesis that the average output voltage is 130 against the alternative that it is less than 130. Use a 5% significance level.

10.7   The Rockwell hardness index for steel is determined by pressing a diamond point into the steel and measuring the depth of penetration. For fifty specimens of a certain type of steel, the Rockwell hardness index averaged 62 with a standard deviation of 8. The manufacturer claims that this steel has an average hardness index of at least 64. Is there sufficient evidence to refute the manufacturer's claim at the 1% significance level?

10.8   Shear strength measurements derived from unconfined compression tests for two types of soils gave the following results (measurements in tons per square foot).

Soil Type I	Soil Type II
$n_1 = 30$	$n_2 = 35$
$\bar{y}_1 = 1.65$	$\bar{y}_2 = 1.43$
$s_1 = 0.26$	$s_2 = 0.22$

Do the soils appear to differ with respect to average shear strength, at the 1% significance level?

10.9   Do soldiers who reenlist enjoy greater job satisfaction than those who do not reenlist or do they reenlist because of other factors, such as reenlistment bonuses, a lack of opportunity in civilian life, and so on? A study by Chisholm, Gauntner, and Munzenrider [2] of the attitudes of a sample of soldiers stationed in the United States in 1976 addresses this and other questions relating to army life. Each soldier included in the study completed a job satisfaction questionnaire and received a job satisfaction score and, in addition, indicated his or her reenlistment intentions. The sample sizes, means, and standard deviations for the two groups are shown below.

	Intention	
	To reenlist	Not to reenlist
Sample size	30	297
Sample mean	136.9	108.8
Sample standard deviation	29.8	31.3
Population mean	μ_1	μ_2

(a)   Suppose that you had a preconceived theory that the job satisfaction mean score μ_1 for soldiers intending to reenlist is higher than the mean score μ_2 for soldiers who do not intend to reenlist. Based on this theory, what alternative hypothesis would you choose for a statistical test of the null hypothesis $H_0: \mu_1 = \mu_2$?

(b)   Does your alternative hypothesis in part (a) imply a one-tailed or a two-tailed statistical test? Locate the rejection region for the test.

(c)   Conduct the test and state your conclusions.

10.10   In "A Strategy for Big Bucks," Charles Dickey [4] discusses studies of the habits of white-tailed deer that indicate that they live and feed within very limited ranges, approximately 150 to 205 acres. To determine whether there was a difference in the ranges of deer located in two different geographical areas, forty deer were caught, tagged, and fitted with small radio transmitters. Several months later, the deer were tracked and identified and the distance y from the release point was recorded. The mean and standard deviation of the distances from the release point were as follows:

	Location	
	1	2
Sample size	40	40
Sample mean	2980 ft	3205 ft
Sample standard deviation	1140 ft	963 ft
Population mean	μ_1	μ_2

(a)   If you have no preconceived reason for believing one population mean to be larger than another, what would you choose for your alternative hypothesis? Your null hypothesis?

(b)   Would your alternative hypothesis in part (a) imply a one- or a two-tailed test? Explain.

(c)   Do the data provide sufficient evidence to indicate that the mean distances differ for the two geographical locations? Test using $\alpha = .10$.

10.11   Many in Britain believe that to spare the rod is to spoil the child. Writing on this subject, the *New York Times* (August 18, 1985) gives the results of a national poll in Britain conducted in February by Marketing and Opinion Research International for the *Times of London*. Of 604 parents questioned, 63 % were in favor of corporal punishment in the schools. Does this provide sufficient evidence to indicate that the majority of the British populace favor corporal punishment in the schools? Test using $\alpha = .01$.

10.12   Human skin can be removed from a dead person, stored in a skin bank, and then grafted onto burn victims. Only skin from flat areas, such as the chest, back, and thighs, is acceptable and, because persons with infections, cancer, and so on must be excluded, only approximately 7 % of all potential donors are acceptable. The *New York Times* (June 14, 1981) notes that a burn research center at one large city hospital screens approximately 8000 deaths annually and, from these, gets only 80 donors, or, 1 %. Do the data provide sufficient evidence to indicate that the burn research center is selecting its donors from a population containing a lower percentage of acceptables than found in the population at large (7 %)?

10.13   "Minorities leave their newspaper jobs at a rate three times higher than whites because of what they consider a lack of advancement," concluded a study conducted by the Institute for Journalism Education (*Orlando Sentinel*, May 7, 1985). This conclusion is based on the interviews of 175 minority journalists and 125 white journalists. Twenty-three (or 13 %) of the minority journalists left journalism compared with 6 (or 5 %) of the white journalists.

(a)   Do these data provide sufficient evidence to indicate differences in the proportions of minorities and whites leaving the journalism profession? Test using $\alpha = .05$.

(b)   If there is evidence of a difference (part a), does it follow that the difference is due to a perceived lack of advancement opportunities for minorities?

(c)   What assumptions must you make for the test in part (a) to be valid?

10.14   A surprising event occurred during the 1976 presidential race. The Gallup and the Harris polls published conflicting results. The Gallup poll of 1078 registered voters showed Ford beating Humphrey by 51 to 39 %. The Harris poll (taken slightly later) of 950 people gave Humphrey a 52 to 41 % margin over Ford. If you were to compare the percentages in the two samples favoring Ford and if the two samples were randomly selected from the same population, what is the probability that the sample percentages favoring Ford would differ by as much as 10 % (51 versus 41 %)? Do the data provide sufficient evidence to indicate that the two samples were selected from different populations? Or could the disparity in the conclusions of the pollsters be due to the methods of selecting the samples?

10.15   A manufacturer of automatic washers provides a particular model in one of three colors, A, B, or C. Of the first 1000 washers sold, it is noticed that 400 of the washers were of color A. Would you conclude that customers have a preference for color A? Justify your answer.

10.16 A manufacturer claimed that at least 20% of the public preferred her product. A sample of 100 persons is taken to check her claim. With $\alpha = .05$, how small would the sample percentage need to be before the claim could be rightfully refuted? (Note that this would require a one-tailed test of a hypothesis.)

10.17 What conditions must be met in order that the Z test be used to test a hypothesis concerning a population mean μ?

10.18 The Florida Poll of February–March, 1984 interviewed 871 adults from around the state. On one question, 53% of the respondents favored strong support of Israel. Would you conclude that a majority of adults in Florida favor strong support of Israel? (Source: *Gainesville Sun*, April 1, 1984.)

10.19 A political researcher believes that the fraction p_1 of Republicans in favor of the death penalty is greater than the fraction p_2 of Democrats in favor of the death penalty. He acquired independent random samples of 200 Republicans and 200 Democrats and found 46 Republicans and 34 Democrats favoring the death penalty. Does this evidence provide statistical support for the researcher's belief? Use $\alpha = .05$.

10.20 Exercise 8.41 stated that a random sample of 500 measurements on the length of stay in hospitals gave a sample mean of 5.4 days and a sample standard deviation of 3.1 days. It is hypothesized by a federal regulatory agency that the average length of stay is in excess of 5 days. Do the data support this hypothesis? Use $\alpha = .05$.

10.21 Refer to Exercise 8.51. Using the data given there, test the hypothesis that there is no difference between mean manual dexterity scores ($H_0: \mu_1 = \mu_2$) versus the alternative that those who participate in sports have a higher average score ($H_a: \mu_1 > \mu_2$). Use $\alpha = .05$.

*10.22 Refer to Exercise 8.53(b). Is there evidence of a difference between the proportion of residents favoring complete protection of alligators and the proportion favoring their destruction? Use $\alpha = .01$.

10.4 Calculating Type-II-Error Probabilities and Finding the Sample Size for the Z Test

Calculating β can be very difficult for some statistical tests, but it is easy for the test of Section 10.3. Consequently, we can use the Z test to demonstrate both the calculation of β and the logic employed in selecting the sample size for a test.

For the test of $H_0: \theta = \theta_0$ against $H_a: \theta > \theta_0$, it is possible to calculate type-II-error probabilities for only specific points in H_a. Suppose that the experimenter has a specific alternative, say, $\theta = \theta_a(\theta_a > \theta_0)$, in mind. Because the rejection region is of the form

$$\text{RR} = \{\hat{\theta}: \hat{\theta} > k\}$$

* Exercises preceded by an asterisk are optional.

the probability of a type II error, β, is

$$\beta = P(\hat{\theta} \text{ not in RR when } H_a \text{ is true})$$

$$\beta = P(\hat{\theta} \le k \text{ when } \theta = \theta_a) = P\left(\frac{\hat{\theta} - \theta_a}{\sigma_{\hat{\theta}}} \le \frac{k - \theta_a}{\sigma_{\hat{\theta}}} \text{ when } \theta = \theta_a\right)$$

If θ_a is the true value of θ, then $(\hat{\theta} - \theta_a)/\sigma_{\hat{\theta}}$ has approximately a standard normal distribution, and the probability β can be determined (approximately) by finding an area under a standard normal curve.

For a fixed sample size n the size of β will depend upon the distance between θ_a and θ_0. If θ_a is close to θ_0, it will be difficult to detect which is the true value of θ, and the probability of accepting H_0 when H_a is true will tend to be large. If θ_a is far from θ_0, it will be relatively easy to see which is the true value, and β will be considerably smaller. As we saw in Section 10.2, α and β both can be made small by choosing a large sample size n.

EXAMPLE 10.8   Suppose that the vice-president of Example 10.5 wants to be able to detect a difference equal to one call in the mean number of customer calls per week. That is, he is interested in testing $H_0 : \mu = 15$ against $H_a : \mu = 16$. With the data as given in Example 10.5, find β for this test.

Solution   In Example 10.5 we had $n = 36$, $\bar{y} = 17$, and $s^2 = 9$. The rejection region was given by ($\alpha = .05$.)

$$z = \frac{\bar{y} - \mu_0}{\sigma/\sqrt{n}} > 1.645$$

which is equivalent to

$$\bar{y} - \mu_0 > 1.645\left(\frac{\sigma}{\sqrt{n}}\right)$$

or

$$\bar{y} > \mu_0 + 1.645\left(\frac{\sigma}{\sqrt{n}}\right)$$

Substituting $\mu_0 = 15$ and $n = 36$, and using s to approximate σ, we find the rejection region to be

$$\bar{y} > 15 + 1.645\left(\frac{3}{\sqrt{36}}\right)$$

or

$$\bar{y} > 15.8225$$

This rejection region is shown in Figure 10.5. Then by definition, $\beta = P(\bar{Y} \leq 15.8225$ when $\mu = 16)$ is the shaded area under the dashed curve to the left of $k = 15.8225$ in Figure 10.5. Thus for $\mu_a = 16$

$$\beta = P\left(\frac{\bar{Y} - \mu_a}{\sigma/\sqrt{n}} \leq \frac{15.8225 - 16}{3/\sqrt{36}}\right) = P(Z \leq -.36) = .3594$$

Figure 10.5
Rejection region for
Example 10.8
$(k = 15.8225)$

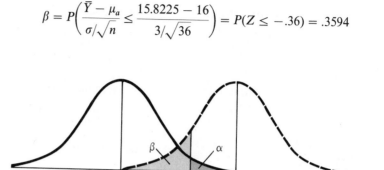

The large value of β tells us that samples of size $n = 36$ frequently will not detect a difference of one unit from the hypothesized mean. We can reduce the value of β by increasing the sample size n.

The preceding example suggests the procedure an experimenter employs when choosing the sample size(s) for an experiment. Suppose that you want to test $H_0: \mu = \mu_0$ versus $H_a: \mu > \mu_0$. If you specify the desired value of α and β (where β is evaluated for a particular value of μ, $\mu_a > \mu_0$), the test depends upon two remaining quantities that must be determined. These are n, the sample size, and k, the point at which the rejection region begins. Because α and β can be written as probabilities involving n and k, we have two equations in two unknowns, which can be solved simultaneously for n. Thus

$$\alpha = P(\bar{Y} > k \text{ when } \mu = \mu_0)$$

$$= P\left(\frac{\bar{Y} - \mu_0}{\sigma/\sqrt{n}} > \frac{k - \mu_0}{\sigma/\sqrt{n}} \text{ when } \mu = \mu_0\right) = P(Z > z_\alpha)$$

$$\beta = P(\bar{Y} \leq k \text{ when } \mu = \mu_a)$$

$$= P\left(\frac{\bar{Y} - \mu_a}{\sigma/\sqrt{n}} \leq \frac{k - \mu_a}{\sigma/\sqrt{n}} \text{ when } \mu = \mu_a\right) = P(Z \leq -z_\beta)$$

(See Figure 10.5.)

From the equation for α we have

$$\frac{k - \mu_0}{\sigma/\sqrt{n}} = z_\alpha$$

and from the equation for β we have

$$\frac{k - \mu_a}{\sigma/\sqrt{n}} = -z_\beta$$

Eliminating k from these two equations gives

$$\mu_0 + z_\alpha\left(\frac{\sigma}{\sqrt{n}}\right) = \mu_a - z_\beta\left(\frac{\sigma}{\sqrt{n}}\right)$$

Thus

$$(z_\alpha + z_\beta)\left(\frac{\sigma}{\sqrt{n}}\right) = \mu_a - \mu_0$$

$$\sqrt{n} = \frac{(z_\alpha + z_\beta)\sigma}{(\mu_a - \mu_0)}$$

Sample Size for an Upper-Tail Alternative Hypothesis

$$n = \frac{(z_\alpha + z_\beta)^2\sigma^2}{(\mu_a - \mu_0)^2}$$

Exactly the same solution would be obtained for a one-tailed alternative, say, $H_a: \mu = \mu_a$ with $\mu_a < \mu_0$. The method just employed can be used to develop a similar formula for the sample size for any one-tailed, hypothesis-testing problem satisfying the conditions of Section 10.3.

EXAMPLE 10.9   Suppose that the vice-president of Example 10.5 wants to test $H_0: \mu = 15$ against $H_a: \mu = 16$ with $\alpha = \beta = .05$. Find the sample size that will ensure this accuracy. Assume that σ^2 is approximately 9.

Solution   Because $\alpha = \beta = .05$, it follows that $z_\alpha = z_\beta = z_{.05} = 1.645$. Then

$$n = \frac{(z_\alpha + z_\beta)^2\sigma^2}{(\mu_a - \mu_0)^2} = \frac{(1.645 + 1.645)^2(9)}{(16 - 15)^2} = 97.4$$

Hence $n = 98$ observations should be used to ensure that $\alpha = \beta = .05$ for this test.

Exercises

10.23   Refer to Exercise 10.6. If the voltage falls as low as 128 serious consequences may result. For testing $H_0: \mu = 130$ versus $H_a: \mu = 128$ find the probability of a type II error, β, for the rejection region used in Exercise 10.6.

10.24   Refer to Exercise 10.7. The steel is sufficiently hard for a certain use so long as the mean Rockwell hardness measure does not drop below 60. Using the rejection region found in Exercise 10.7, find β for the specific alternative $\mu = 60$.

10.25   Refer to Exercise 10.16. Calculate the value of β for the alternative $p_a = .15$.

10.26   Refer to Exercise 10.19. The political researcher should have designed a test for which β is tolerably low when p_1 exceeds p_2 by a meaningful amount. For example, find a common sample size n for a test with $\alpha = .05$ and $\beta \leq .20$ when, in fact, p_1 exceeds p_2 by .1. [Hint: The maximum value of $p(1 - p)$ is .25.]

10.27   Refer to Exercise 10.20. Using the rejection region found there, calculate β for testing $H_0: \mu = 5.0$ versus the alternative $H_a: \mu = 5.5$.

10.28   In Exercise 10.20 and 10.27 how large should the sample size be for testing $H_0: \mu = 5$ versus $H_a: \mu = 5.5$, in order that $\alpha = .01$ and $\beta = .05$.

10.29   Refer to Exercise 10.21. For the rejection region used there calculate β for the alternative $H_a: \mu_1 - \mu_2 = 3$.

10.30   Refer to Exercise 10.21 and 10.29. For $H_0: \mu_1 - \mu_2 = 0$ and $H_a: \mu_1 - \mu_2 = 3$, find the sample size that gives $\alpha = \beta = .05$. (Assume equal-sized samples for each group.)

10.5 Another Way to Report the Results of a Statistical Test: Attained Significance Levels or p-Values

The probability of a type I error, α, is often called the *significance level* associated with a test, a term that originated in the following way. The probability of the observed value of the test statistic, or some value even more contradictory to the null hypothesis, measures, in a sense, the weight of evidence favoring rejection of the null hypothesis. Although small values of α are recommended, the actual choice of α to be used in an analysis is somewhat arbitrary. One experimenter might choose to implement a test with $\alpha = .05$, whereas another experimenter may prefer $\alpha = .01$. It therefore is possible that two persons could analyze the same data, one concluding that the null hypothesis should be rejected at the $\alpha = .05$ significance level, and the other deciding that the null hypothesis cannot be rejected with $\alpha = .01$. Further, α-values of .05 or .01 often are used because of habit rather than through careful consideration of the ramifications of making a type I error.

Once a test statistic (Y in our polling example, or one of the Z's of Section 10.3) is decided upon, it often is possible to report the *p*-value or attained significance level associated with a test. This quantity is a statistic that represents the smallest value of α for which the null hypothesis is rejected.

Definition 10.2

> If W is a test statistic, the *p-value* or *attained significance level* is the smallest level of significance, α, for which the observed data indicates that the null hypothesis should be rejected.

If an experimenter's choice of α is greater than or equal to the p-value, the null hypothesis is rejected. Otherwise, if α is less than the p-value, the null hypothesis cannot be rejected. Many scientific journals require that researchers report p-values associated with statistical tests because these values provide the reader with *more information* than simply a statement that the null hypothesis is rejected or not rejected for some value of α chosen by the writer of the report. In a sense, the p-value allows the reader of published research to evaluate the extent to which the observed data disagrees with the null hypothesis. Particularly, the p-value permits each reader to use his or her own choice for α in deciding whether the observed data lead to a rejection of the null hypothesis.

The procedures for finding p-values for tests we have discussed thus far are presented in the following examples.

EXAMPLE 10.10   Recall our discussion of the political poll (see Examples 10.1 through 10.4) where $n = 15$ voters were sampled. If we wish to test $H_0: p = .5$ versus $H_a: p < .5$ using $Y =$ the number of voters favoring Jones as our test statistic, what is the p-value if $Y = 3$? Interpret the result.

Solution   In previous discussions we indicated that H_0 should be rejected for small values of Y. Thus the p-value for this test is given by p-value $= P\{Y \leq 3\}$ where Y has a binomial distribution with $n = 15$ and $p = .5$ (the shaded area in the binomial distribution of Figure 10.6). Using Table 1, Appendix III, we find that the p-value is .018.

Figure 10.6
Illustration of p-value
for Example 10.10

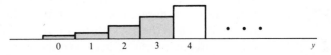

Because the p-value $= .018$ represents the smallest value of α for which the null hypothesis is rejected, an experimenter who specifies any value of $\alpha \geq .018$ would be led to the rejection of H_0 and the conclusion that Jones does *not* have a plurality of the vote. If one were to choose a value of α less than .018, the null hypothesis could *not* be rejected.

This example illustrates that the reporting of p-values is particularly beneficial where the appropriate test statistic possesses a *discrete* distribution. In cases like these, often one cannot find any rejection region that yields an α-value of a particular magnitude. For example, in this instance, no rejection region of the form $\{y \leq a\}$ can be found for which α is exactly equal to .05. In cases like this, reporting the p-value could be preferable to limiting oneself to values of α

that actually can be obtained on the basis of the discrete distribution of the test statistic.

Example 10.10 also indicates the general method of computing p-values. If one were to reject H_0 in favor of H_a for small values of a test statistic W, the p-value associated with an observed value of W, say w_0, is given by

$$p\text{-value} = P(W \leq w_0, \text{ when } H_0 \text{ is true})$$

Analogously, if H_0 should be rejected in favor of H_a for large values of W, the p-value associated with the observed value, w_0, is

$$p\text{-value} = P(W \geq w_0, \text{ when } H_0 \text{ is true})$$

We illustrate the calculation of a p-value for a two-tailed alternative in the following example.

EXAMPLE 10.11   Find the p-value for the statistical test of Example 10.7.

Solution   Example 10.7 presents a test of the null hypothesis $H_0: \mu_1 - \mu_2 = 0$ versus the alternative hypothesis $H_a: \mu_1 - \mu_2 \neq 0$. The value of the test statistic, computed from the observed data, was $z = -2.5$. Because this test is two-tailed, the p-value is the probability that $Z \leq -2.5$ or $Z \geq 2.5$, the shaded area in Figure 10.7. From Table 4, Appendix III, $P(Z \geq 2.5) = P(Z \leq -2.5) = .0062$. Because this is a two-tailed test, the p-value $= 2(.0062) = .0124$. Thus if $\alpha = .05$ (a value larger than .0124) we reject H_0 in favor of H_a and, in agreement with the conclusion of Example 10.7, conclude that there is evidence of a difference in mean reaction time for men and women. However, if $\alpha = .01$ (or any value of $\alpha < .0124$) were chosen, we could *not* claim a difference in mean reaction times for the two sexes.   ⌇

Figure 10.7
Shaded area gives
p-value for
Example 10.11

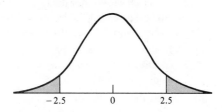

For the statistical tests we have developed thus far, it is possible to compute exact p-values using the binomial and Z tables in Appendix III. Tables of distributions for other test statistics that we encounter in later sections give critical values only for largely differential values of α (for example, .01, .025, .05, .001, etc.). Consequently, exact p-values cannot be calculated. In these cases, we can make a statement only about a region of values, inside which the p-value is known to lie. For example, if a test result is statistically significant for $\alpha = .05$ but *not* for $\alpha = .025$, we will report that $.025 \leq p\text{-value} \leq .05$. Thus, for any $\alpha > .05$ we reject the null hypothesis, for $\alpha < .025$ we do *not* reject the null hypothesis. For values of α between .025 and .05 we would need to seek more complete tables of the appropriate distribution before reaching a conclusion. Nevertheless, we have very useful information about the p-value available to us.

The recommendation that a researcher report the p-value for a test and leave its interpretation to a reader does not violate the traditional (decision theoretic) statistical testing procedures described in the preceding sections. The reporting of a p-value simply leaves the decision regarding whether to reject the null hypothesis (with the associated potential of committing type I or type II errors) to the reader. Thus, the responsibility of choosing α, and possibly the problem of evaluating the probability β of making a type I error, are shifted to the reader.

Exercises

10.31   High airline occupancy rates on scheduled flights are essential to profitability. Suppose that a scheduled flight must average at least 60% occupancy to be profitable and that an examination of the occupancy rates for 120 10:00 A.M. flights from Atlanta to Dallas showed a mean occupancy rate per flight of 58% and a standard deviation of 11%. Test to see if there is sufficient evidence to claim that the flight is unprofitable. Find the p-value associated with the test. What would you conclude if you wished to implement the test at the $\alpha = .10$ level?

10.32   Two sets of elementary schoolchildren were taught to read by using different methods, fifty by each method. At the conclusion of the instructional period, a reading test gave the results $\bar{y}_1 = 74$, $\bar{y}_2 = 71$, $s_1 = 9$, and $s_2 = 10$. What is the attained significance level if you wish to see if there is evidence of a real difference between the two population means? What would you conclude if you desired an α value of .05?

10.33   Does auto safety education increase the use of auto safety devices? Researchers at the University of California at San Francisco gave 78 of a group of 136 expectant couples a special 30-minute lecture on the value of child safety seats (*USA Today*, March 12, 1985). Contacted 4 to 6 weeks after the birth of their children, the researchers found that 96% of those who had attended the lecture claimed that they were using the safety seats versus 78% for those who did not attend the lecture.

   (a)   Is this difference in percentages large enough to imply that the lecture was effective in increasing the percentage of couples using the child safety seats? Test using $\alpha = .05$.

   (b)   Find the p-value for the test.

10.34   How would you like to live to be 200 years old? For centuries, humankind has sought the key to the mystery of aging. What causes it? How can it be slowed? Recent studies focus on biomarkers, physical or biological changes that occur at a predictable time in a person's life. The theory is that if ways can be found to delay the occurrence of these biomarkers, human life can be extended. A key biomarker, according to scientists, is forced vital capacity (FVC), the volume of air that you can expel after taking a deep breath. A study of 5209 men and women aged 30 to 62 showed that FVC declined on the average 3.8 deciliters per decade for men and 3.1 deciliters per decade for women [1]. Suppose that you wished to determine whether a physical fitness program for men and women aged 50 to 60 would delay aging and that you measured the FVC for 30 men and 30 women at the beginning and end of the 50-to-60-year age interval and recorded the drop in FVC for each person. The data follow.

	Men	*Women*
Sample size	30	30
Sample average drop in FVC	3.6	2.7
Sample standard deviation	1.1	1.2
Population mean drop in FVC	μ_1	μ_2

(a)   Do the data provide sufficient evidence to indicate that the decrease in the mean FVC over the decade for the men on the physical fitness program is less than 3.8 deciliters? Find the observed significance level for the test.

(b)   Refer to part (a). If you choose $\alpha = .05$, do the data support the contention that the mean decrease in FVC is less than 3.8 deciliters?

(c)   Test to determine whether the FVC drop for women on the physical fitness program was less than 3.1 deciliters for the decade. Find the observed significance level for the test.

(d)   Refer to part (c). If you choose $\alpha = .05$, do the data support the contention that the mean decrease in FVC is less than 3.1 deciliters?

10.35   Do you believe that an exceptionally high percentage of the executives of large corporations are right-handed? Although 85% of the general public is right-handed, a survey of 300 chief executive officers of large corporations found that 96% were right-handed (*Orlando Sentinel*, December 25, 1982).

(a)   Is this difference in percentages statistically significant? Test using $\alpha = .01$.

(b)   Find the p-value for the test and explain what it means.

10.36   A pharmaceutical company conducted an experiment to compare the mean times (in days) necessary to recover from the effects and complications that follow the onset of the common cold. This experiment compared persons on a daily dose of 500 mg. of vitamin C to those who were not given a vitamin supplement. Thirty-five adults were randomly selected for each treatment category, and the mean recovery times and standard deviations for the two groups follow. Do the data indicate that the use of vitamin C reduces the mean time required to recover? Find the attained significance level. What would the company conclude at the $\alpha = .05$ level?

	Treatment	
	No Supplement	*500 mg. Vitamin C*
Sample size	35	35
Sample mean	6.9	5.8
Sample standard deviation	2.9	1.2

10.37   A check-cashing service found that approximately 5% of all checks submitted to the service were bad. After instituting a check verification system to reduce its losses, the service found that only 45 were bad in a random sample of 1124 that were cashed. Is there sufficient evidence that the check verification system reduced the proportion of bad checks? What is the attained significance level associated with the test? What would you conclude at the $\alpha = .01$ level?

10.38   A publisher of a news magazine has found through past experience that 60% of subscribers renew their subscriptions. In a random sample of 200 subscribers, 108 indicated that they planned to renew their subscriptions. What is the p-value associated with the test that the current proportion of renewals differs from that previously experienced?

10.39   In a study to assess various effects of using a female model in automobile advertising, each of 100 male subjects was shown photographs of two automobiles matched for price, color, and size, but of different makes. One of the automobiles was shown with a female model to 50 of the subjects (group A), and both automobiles were shown without the model to the other 50 subjects (group B). In group A, the automobile shown with the model was judged to be more expensive by 37 subjects. In group B, the same automobile was judged as more expensive by 23 subjects. Do these results indicate that using a female model increases the perceived cost of an automobile? Find the associated p-value and indicate your conclusion for an $\alpha = .05$ level test.

10.6 Some Comments on the Theory of Hypothesis Testing

As previously indicated, we have a choice between implementing a one- or two-tailed test for a given situation. We emphasize that this choice is dictated by the practical aspects of the problem and will depend upon the alternative value of the parameter, say, θ, the experimenter is trying to detect. Thus if we were to sustain a large financial loss if θ were greater than θ_0 but not if it were less, we would concentrate our attention on the detection of values of θ greater than θ_0. Hence we would reject in the upper tail of the distribution for the test statistics previously discussed. On the other hand, if we are equally interested in detecting values of θ that are either less than or greater than θ_0, we would employ a two-tailed test.

The theory of a statistical test of hypothesis outlined in Section 10.2 and utilized in Section 10.3 is a very clear-cut procedure that enables the researcher to either reject or accept the null hypothesis with measured risks α and β. Unfortunately, this theoretical framework does not suffice for all practical situations.

For any statistical test the probability of a type I error, α, depends upon the value of the parameter specified in the null hypothesis. This probability can be calculated, at least approximately, for each of the testing procedures discussed in this text. For the procedures discussed thus far the probability of a type II error, β, can be calculated only after a *specific* value of the parameter of interest is singled out for detection. The selection of a practically meaningful value for this parameter is often difficult. Even if a meaningful alternative can be identified, the actual calculation of β can be quite tedious. Specification of a meaningful alternative hypothesis is even more difficult for some of the testing procedures we will present in subsequent chapters.

Of course, we do not want to ignore the possibility of committing a type II error. Later in the chapter we will determine methods for selecting tests with the smallest possible value of β for tests where the probability of a type I error is a fixed value α. Even in these situations the smallest possible value of β still could be quite large.

These obstacles do not invalidate the use of statistical tests. Rather, these considerations urge us to be cautious about drawing conclusions where insufficient evidence is available to reject the null hypothesis. If a truly meaningful (and believable) value for β can be calculated, we should feel justified in accepting H_0 if the value of β is small. In the more typical situation where a truly meaningful value for β is unavailable, we will modify our procedure as follows. When the value of the test statistic is not in the rejection region, we will "fail to reject" rather than "accept" the null hypothesis. In the polling example discussed in Example 10.1, we tested $H_0: p = .5$ versus $H_a: p < .5$. If our observed value of Y falls into the rejection region, then we reject H_0 and say that the evidence supports the research hypothesis that Jones will lose. In this situation we will have demonstrated support for the hypothesis we wanted to support, the research hypothesis. If, however, Y does not fall in the rejection region, and we can determine no specific value of p in H_a that is of specific interest, we simply state that we will *not* reject H_0 and seek additional information before reaching a conclusion. Or, instead, we could report the p-value associated with the statistical test and leave the interpretation to the reader.

Finally, it should be noted that if H_0 is rejected for a "small" value of α (or for a small p-value), this occurrence does *not* imply that the null hypothesis is "wrong by a large amount." This occurrence *does mean* that the null hypothesis can be rejected based on a procedure that incorrectly rejects the null hypothesis (when H_0 is true) with a small probability (i.e., with a small probability of a type I error). We also must refrain from equating *statistical* and *practical* significance. If we reconsider the experiment described and analyzed in Examples 10.7 and 10.11, the p-value of .0124 is "small" and the result is statistically significant for any choice of $\alpha \geq .0124$. However, the difference in the samples' mean reaction times is only .2 seconds, a result that may or may not be *practically* significant. In order to assess the practical significance of such a difference, you may wish to form a confidence interval for $\mu_1 - \mu_2$ by using the methods of Section 8.6.

10.7 Two Tests Based on Statistics that Possess a Student's t Distribution

In Section 10.3 we presented large-sample, hypothesis-testing procedures that, like the interval estimation procedures developed in Section 8.6, are useful for large samples. We require that the sample size be large enough so that $Z = (\hat{\theta} - \theta_0)/\sigma_{\hat{\theta}}$ has approximately a standard normal distribution. Section 8.8 contains procedures based on the t distribution for construction of confidence intervals for μ (the mean of a single normal population) and $\mu_1 - \mu_2$ (the difference in the means of two normal populations with equal variances). In this section we will develop formal procedures for testing hypotheses about μ and

$\mu_1 - \mu_2$, procedures that are appropriate for small samples from normal populations.

We assume that $Y_1, \ldots, Y_n$ denotes a random sample of size n from a normal distribution with unknown mean μ and unknown variance σ^2. If $\bar{Y}$ and S denote the sample mean and standard deviation, respectively, and if $H_0: \mu = \mu_0$ is true, then

$$T = \frac{\bar{Y} - \mu_0}{S/\sqrt{n}}$$

has a t distribution with $n - 1$ degrees of freedom (see Section 8.8).

Because the t distribution is symmetric and mound-shaped, it is clear that the rejection region for a small-sample test of the hypothesis $H_0: \mu = \mu_0$ would be located in the tails of the t distribution and would be determined in exactly the same way as for the large-sample z statistic. By analogy with the Z test developed in Section 10.3, the proper rejection region for the upper-tail alternative $H_a: \mu > \mu_0$ is given by

$$\mathrm{RR} = \{t > t_\alpha\}$$

where t_α is such that $P\{T > t_\alpha\} = \alpha$ for a t distribution with $n - 1$ degrees of freedom (see Table 5, Appendix III).

A summary of the tests for μ based on the t distribution, known as t tests, is as follows.

A Small-Sample Test for μ

Assumptions: $Y_1, \ldots, Y_n$ is a random sample from a normal distribution with $E(Y_i) = \mu$.

$H_0: \quad \mu = \mu_0$.

$H_a: \quad \begin{cases} \mu > \mu_0 \text{(upper-tail alternative)}. \\ \mu < \mu_0 \text{(lower-tail alternative)}. \\ \mu \neq \mu_0 \text{(two-tailed alternative)}. \end{cases}$

Test Statistic: $\quad T = \dfrac{\bar{Y} - \mu_0}{S/\sqrt{n}}$.

Rejection Region: $\quad \begin{cases} t > t_\alpha \text{(upper-tail RR)}. \\ t < -t_\alpha \text{(lower-tail RR)}. \\ |t| > t_{\alpha/2} \text{(two-tailed RR)}. \end{cases}$

(See Table 5, Appendix III, for values of t_α.)

EXAMPLE 10.12   Example 8.11 gives muzzle velocities of eight shells tested with a new gunpowder along with the sample mean and standard deviation, $\bar{y} = 2959$ and $s = 39.1$. The manufacturer claims that the new gunpowder produces an average velocity of no less than 3000 feet per second. Do the sample data provide sufficient evidence to contradict the manufacturer's claim at the .025 level of significance?

Solution   Assuming that the observed velocities come from a normal distribution, we can use the test just outlined. We want to test $H_0: \mu = 3000$ versus the alternative $H_a: \mu < 3000$. The rejection region is given by $t < -t_{.025} = -2.365$, where t possesses $\nu = (n-1) = 7$ degrees of freedom. Computing, we find that the observed value of the test statistic is

$$t = \frac{\bar{y} - \mu_0}{s/\sqrt{n}} = \frac{2959 - 3000}{39.1/\sqrt{8}} = -2.966$$

This value falls in the rejection region (that is, t is less than -2.365), hence the null hypothesis is rejected at the $\alpha = .025$ level of significance. So, we conclude that the sample data do not support the manufacturer's claim.   ∽

EXAMPLE 10.13   If you planned to report the results of the statistical test, Example 10.12, what is the associated p-value?

Solution   Because the null hypothesis should be rejected if t is "small," the smallest value of α for which the null hypothesis is rejected is p-value $= P(T < -2.966)$ where T has a t distribution with $n - 1 = 7$ degrees of freedom.
   Unlike the table of areas under the normal curve (Table 4, Appendix III), Table 5 in Appendix III does not give areas corresponding to many values of t. Rather, it gives the values of t corresponding to upper tail areas equal to .10, .05, .025, .010, and .005. Because the t distribution is symmetric about 0, we can use these upper tail areas to provide corresponding lower tail areas. In this instance, the t statistic is based on 7 degrees of freedom, hence we consult the d.f. = 7 row of Table 5 and find that -2.966 falls between $-t_{.025} = -2.365$ and $-t_{.01} = -2.998$. These values are indicated in Figure 10.8. Because the observed value of

Figure 10.8
Determination of
p-value for
Example 10.13

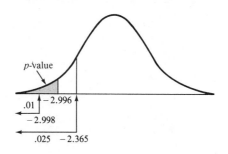

T, -2.966, is less than $-t_{.025} = -2.376$ but not less than $-t_{.01} = -2.998$, we reject H_0 for $\alpha = .025$ but not for $\alpha = .01$. Thus the p-value for the test satisfies $.01 < p < .025$.

A second application of the t distribution is its use in constructing a small-sample test to compare the means of two normal populations that possess equal variances. Suppose that independent random samples are selected from each of two normal populations, $Y_{11}, Y_{12}, \ldots, Y_{1n_1}$ from the first and $Y_{21}, Y_{22}, \ldots, Y_{2n_2}$ from the second, where the mean and variance of the ith population are μ_i and σ^2, $i = 1, 2$. Further, assume that $\bar{Y}_i$ and S_i^2, $i = 1, 2$, are the corresponding sample means and variances. When these assumptions are satisfied, we showed in Section 8.8 that if

$$S^2 = \frac{(n_1 - 1)S_1^2 + (n_2 - 1)S_2^2}{n_1 + n_2 - 2}$$

is the pooled estimator for σ^2, then

$$T = \frac{(\bar{Y}_1 - \bar{Y}_2) - (\mu_1 - \mu_2)}{S\sqrt{\dfrac{1}{n_1} + \dfrac{1}{n_2}}}$$

has a Student's t distribution with $n_1 + n_2 - 2$ degrees of freedom. If we want to test the null hypothesis $H_0: \mu_1 - \mu_2 = D_0$ for some fixed value D_0, it follows that if H_0 is true, then

$$T = \frac{\bar{Y}_1 - \bar{Y}_2 - D_0}{S\sqrt{\dfrac{1}{n_1} + \dfrac{1}{n_2}}}$$

has a Student's t distribution with $n_1 + n_2 - 2$ degrees of freedom. Note that this small-sample test statistic is similar in form to its large-sample counterpart, the Z statistic of Section 10.3. Tests of the hypothesis $H_0: \mu_1 - \mu_2 = D_0$ versus upper-tail, lower-tail, and two-tailed alternatives are conducted in the same

Table 10.3
Data for
Example 10.14

Standard procedure	New procedure
$n_1 = 9$	$n_2 = 9$
$\bar{y}_1 = 35.22$ seconds	$\bar{y}_2 = 31.56$ seconds
$\sum_{i=1}^{9}(y_{1i} - \bar{y}_1)^2 = 195.56$	$\sum_{i=1}^{9}(y_{2i} - \bar{y}_2)^2 = 160.22$

manner as in the large-sample test, except that we employ the t statistic. A summary of the small-sample testing procedures for $\mu_1 - \mu_2$ follows.

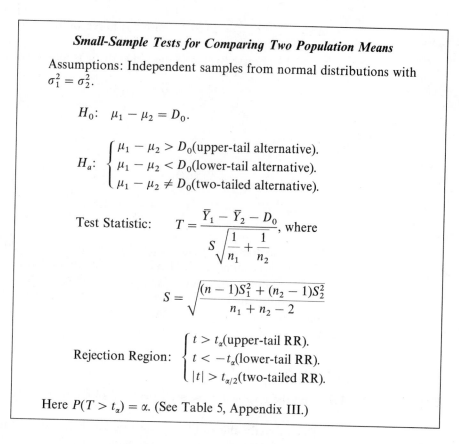

Small-Sample Tests for Comparing Two Population Means

Assumptions: Independent samples from normal distributions with $\sigma_1^2 = \sigma_2^2$.

H_0: $\mu_1 - \mu_2 = D_0$.

H_a:
$\begin{cases} \mu_1 - \mu_2 > D_0 \text{(upper-tail alternative)}. \\ \mu_1 - \mu_2 < D_0 \text{(lower-tail alternative)}. \\ \mu_1 - \mu_2 \ne D_0 \text{(two-tailed alternative)}. \end{cases}$

Test Statistic: $T = \dfrac{\bar{Y}_1 - \bar{Y}_2 - D_0}{S\sqrt{\dfrac{1}{n_1} + \dfrac{1}{n_2}}}$, where

$$S = \sqrt{\frac{(n-1)S_1^2 + (n_2-1)S_2^2}{n_1 + n_2 - 2}}$$

Rejection Region:
$\begin{cases} t > t_\alpha \text{(upper-tail RR)}. \\ t < -t_\alpha \text{(lower-tail RR)}. \\ |t| > t_{\alpha/2} \text{(two-tailed RR)}. \end{cases}$

Here $P(T > t_\alpha) = \alpha$. (See Table 5, Appendix III.)

EXAMPLE 10.14   Example 8.12 gives data on length of time to complete an assembly procedure for two different training methods. The sample data are as shown in Table 10.3. Is there sufficient evidence to indicate a difference in true mean times for the two methods? Test at the $\alpha = .05$ level of significance.

Solution

We are testing H_0: $(\mu_1 - \mu_2) = 0$ against the alternative H_a: $(\mu_1 - \mu_2) \ne 0$. Consequently, we will require a two-tailed test. The test statistic is

$$T = \frac{(\bar{Y}_1 - \bar{Y}_2) - D_0}{S\sqrt{\dfrac{1}{n_1} + \dfrac{1}{n_2}}}$$

with $D_0 = 0$, and the rejection region for $\alpha = .05$ is $|t| > t_{\alpha/2}$, where

$$P[|T| > t_{\alpha/2}] = .05$$

In this case $t_{.025} = 2.120$ because t is based on $(n_1 + n_2 - 2) = 16$ degrees of freedom.

The observed value of the test statistic is found by first computing

$$s^2 = \frac{195.56 + 160.22}{9 + 9 - 2} = 22.24$$

Then

$$t = \frac{\bar{y}_1 - \bar{y}_2}{s\sqrt{\dfrac{1}{n_1} + \dfrac{1}{n_2}}} = \frac{35.22 - 31.56}{4.71\sqrt{\dfrac{1}{9} + \dfrac{1}{9}}} = 1.65$$

This value does not fall in the rejection region ($|t| > 2.120$), and hence the null hypothesis is not rejected. There is not sufficient evidence to indicate a difference in the mean assembly times for the two training methods.

Notice that, in line with the comments of Section 10.6, we have not accepted $H_0: \mu_1 - \mu_2 = 0$. Rather, we stated that we do not have sufficient evidence to accept $H_a: \mu_1 - \mu_2 \neq 0$.

∽

EXAMPLE 10.15 Find the p-value for the statistical test of Example 10.14.

Solution   The oberved value of the test statistic for this two-tailed test was $t = 1.65$. The p-value for this test is thus the probability that $T > 1.65$ or $T < -1.65$, the area shaded in Figure 10.9, that is, $A_1 + A_2$.

Because this test statistic is based on $n_1 + n_2 - 2 = 16$ d.f., we consult Table 5, Appenix III to find $t_{.05} = 1.746$ and $t_{.10} = 1.337$, Thus, the probability

Figure 10.9
Shaded area is
p-value for
Example 10.15

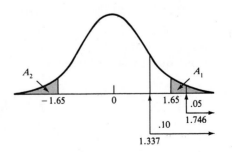

that $T > 1.65$, A_1, lies between .05 and .10, that is, $.05 < A_1 < .1$. Similarly, $.05 < A_2 < .1$. Because the p-value $= A_1 + A_2$, it follows that $.1 < p$-value $< .2$.

In this case, we see that if we select $\alpha = .05$ we cannot reject the null hypothesis, the same conclusion that we reached in Example 10.14. ∽

The test of Example 10.12 is based on the assumption that the muzzle velocity measurements have been randomly selected from a normal population. In most cases it is impossible to verify this assumption. We might ask how this predicament affects the validity of our conclusions.

Empirical studies of the test statistic

$$\frac{\bar{Y} - \mu}{S/\sqrt{n}}$$

have been conducted for sampling from certain nonnormal populations. Such investigations have shown that moderate departures from normality in the probability distribution for Y have little effect on the probability distribution of the test statistic. This result, coupled with the common occurrence of near-normal distributions of data in nature, make the t test of a population mean extremely useful. Statistical tests that lack sensitivity to departures from the assumptions upon which they are based possess wide applicability. Because of their insensitivity to assumptions, they have been called *robust statistical tests*.

Like the t test for a single mean, the t test for comparing two population means (often called the *two-sample* t *test*) is robust relative to the assumption of normality. It also is robust relative to the assumption that $\sigma_1^2 = \sigma_2^2$ when n_1 and n_2 are equal (or nearly equal).

Exercises

10.40 Why is the z test usually inappropriate as a test procedure when the sample size is small?

10.41 What assumptions are made when a Student's t test is employed to test a hypothesis concerning a population mean?

10.42 A chemical process has produced, on the average, 800 tons of chemical per day. The daily yields for the past week are 785, 805, 790, 793, and 802 tons. Do these data indicate that the average yield is less than 800 tons, and hence that something is wrong with the process? Test at the 5 % level of significance. What assumptions are required for the valid use of the procedure you used to analyze these data? Give bounds for the associated p-value.

10.43 A coin-operated soft drink machine was designed to discharge, on the average, 7 ounces of beverage per cup. In a test of the machine ten cupfuls of beverage were drawn from the machine and measured. The mean and standard deviation of the ten measurements were 7.1 ounces and .12 ounce, respectively. Do these data present sufficient evidence to indicate that the mean discharge differs from 7 ounces? What can be said about the attained significance level for this test? What is the appropriate decision if $\alpha = .10$?

10.44   A survey by *Industrial Research and Development* magazine found that salaries for research and development personnel rose 9.6% in 1981 to an average of $31,221 per year (*Orlando Sentinel Star*, April 26, 1981). The mean salaries in specific areas of application tended to be higher or lower than this overall mean. If a random sample of the salaries of twenty physicists possesses a mean $\bar{y} = \$33,120$, and standard deviation $s = \$2,140$, would you conclude that physicists are paid, on the average, substantially more than the overall mean of $31,221 per year? Test using $\alpha = .05$. Give bounds for the p-value associated with this test.

10.45   What assumptions are made about the populations from which independent random samples are obtained when utilizing the t distribution in making small-sample inferences concerning the differences in population means?

10.46   Two methods for teaching reading were applied to two randomly selected groups of elementary schoolchildren and compared on the basis of a reading comprehension test given at the end of the learning period. The sample means and variances computed from the test scores are shown in the accompanying table. Do the data present sufficient evidence to indicate a difference in the mean scores for the populations associated with the two teaching methods? What can be said about the attained significance level? What assumptions are required? What would you conclude at the $\alpha = .05$ level of significance?

	Method 1	Method 2
No. of children in group	11	14
$\bar{y}$	64	69
s^2	52	71

10.47   A study was conducted by the Florida Game and Fish Commission to assess the amounts of chemical residues found in the brain tissue of brown pelicans. In a test for DDT, random samples of $n_1 = 10$ juveniles and $n_2 = 13$ nestlings gave the results shown in the accompanying table (measurements in parts per million). Test the hypothesis that there is no difference between mean amounts of DDT found in juveniles and nestlings versus the alternative that the juveniles have a larger mean. Use $\alpha = .05$. (This test has important implications regarding the accumulation of DDT over time.)

Juveniles	Nestlings
$n_1 = 10$	$n_2 = 13$
$\bar{y}_1 = .041$	$\bar{y}_2 = .026$
$s_1 = .017$	$s_2 = .006$

10.48   The strength of concrete depends, to some extent, on the method used for drying. Two different drying methods showed the following results for independently tested specimens (measurements in psi):

Method I	Method II
$n_1 = 7$	$n_2 = 10$
$\bar{y}_1 = 3250$	$\bar{y}_2 = 3240$
$s_1 = 210$	$s_2 = 190$

Do the methods appear to produce concrete with different mean strengths? Use $\alpha = 0.05$. What is the attained significance level?

10.49   Refer to Exercise 10.47. Is there evidence that the mean for juveniles exceeds that for nestlings by more than .01 parts per million? Give the p-value.

10.50   Refer to Exercise 8.68. Is there evidence of a difference between the mean oxygen consumption rates for the two training methods? Use $\alpha = .05$.

10.51   Refer to Exercise 8.71. A certain testing laboratory claims that for these species of fish the average LC50 measurement is 6 parts per million. Use the data of Exercise 8.71 to determine if there is sufficient evidence to indicate that the average LC50 measurement is less than 6 parts per million. Use $\alpha = .05$.

10.52   The tremendous growth of the Florida lobster (called *spiny lobster*) industry over the past 20 years has made it the state's second most valuable fishery industry. A recent declaration by the Bahamian government that prohibits U.S. lostermen from fishing on the Bahamian portion of the continental shelf is expected to produce a dramatic reduction in the landings in pounds per lobster per trap. According to the records, the mean landings per trap is 30.31 pounds. A random sampling of twenty lobster traps since the Bahamian fishing restriction went into effect gave the following results (in pounds):

17.4	18.9	39.6	34.4	19.6
33.7	37.2	43.4	41.7	27.5
24.1	39.6	12.2	25.5	22.1
29.3	21.1	23.8	43.2	24.4

Do these landings provide sufficient evidence to support the contention that the mean landings per trap has decreased since imposition of the Bahamian restrictions? Test using $\alpha = .05$.

10.53   Jan Lindhe, D. M. D. [8], conducted a study on the effect of an oral antiplaque rinse on plaque buildup on teeth. Fourteen subjects, whose teeth were thoroughly cleaned and polished, were randomly assigned to two groups of seven subjects each. Both groups were assigned to use oral rinses (no brushing) for a 2-week period. Group 1 used a rinse that contained an antiplaque agent. Group 2, the control group, received a similar rinse except that, unknown to the subjects, the rinse contained no antiplaque agent. A plaque index, *y*, a measure of plaque buildup, was recorded at 4, 7, and 14 days. The mean and standard deviation for the 14-day plaque measurements for the two groups follows.

	Control Group	Antiplaque Group
Sample size	7	7
Mean	1.26	.78
Standard deviation	.32	.32

(a)   State the null and alternative hypotheses that should be used to test the effectiveness of the antiplaque oral rinse.

(b)   Do the data provide sufficient evidence to indicate that the oral antiplaque rinse is effective? Test using $\alpha = .05$.

(c)   Find the *p*-value for the test.

10.54   J. S. Diana [3] conducted an experiment to investigate the effect of water temperature on the growth of largemouth bass. Sixty fish were randomly partitioned into six groups of ten fish each. The groups were then randomly assigned, one to each of six different water-temperature conditions. The mean of the growth rates of the ten fish after 15 days of feeding and the standard error of the mean are shown for each of the six water-temperature conditions. (Weight gain was determined by a calorific analysis.)

	Growth (calories per day)	
Water condition	Mean	Standard Error of Mean
Constant warm	40.8	12.5
Constant cool	23.6	3.4
Start warm	46.3	8.6
Midwarm	38.5	4.4
Start cool	58.8	24.4
Midcool	23.4	8.0

(a)   Do the data provide sufficient evidence to indicate a difference in mean growth rate between bass feeding in the constant warm and the start warm conditions? Test using $\alpha = .05$.

(b)   Find the approximate *p*-value for the test and interpret it.

(c)   Find a 95% confidence interval for the difference in mean growth rates for bass feeding in the constant warm water condition versus the mean growth for those feeding in the start warm water condition. Interpret the interval.

10.8 Testing Hypotheses Concerning Variances

We again assume that we have a random sample $Y_1, \ldots, Y_n$ from a normal distribution with an unknown mean of μ and an unknown variance of σ^2. In Section 8.9 we used the pivotal method to construct a confidence interval for the parameter σ^2. In this section we consider the problem of testing $H_0: \sigma^2 = \sigma_0^2$ for some fixed value σ_0^2 versus various alternative hypotheses. Note that if H_0 is true ($\sigma^2 = \sigma_0^2$), Theorem 7.3 implies that

$$\chi^2 = \frac{(n-1)S^2}{\sigma_0^2}$$

has a χ^2 distribution with $n - 1$ degrees of freedom. If we desire to test $H_0: \sigma^2 = \sigma_0^2$ versus $H_a: \sigma^2 > \sigma_0^2$, we can use $\chi^2 = (n-1)S^2/\sigma_0^2$ as our test statistic, but how should we select the rejection region, RR?

If H_a is true and the actual value of σ^2 is larger than σ_0^2, then we would expect S^2 (which estimates the true value of σ^2) to be larger than σ_0^2. The larger S^2 is relative to σ_0^2, the stronger will be the evidence to support $H_a: \sigma^2 > \sigma_0^2$. Note that S^2 will be large relative to σ_0^2 if and only if $\chi^2 = (n-1)S^2/\sigma_0^2$ is large. Thus we see that a rejection region of the form $\text{RR} = \{\chi^2 > k\}$ for some constant k would be appropriate for testing $H_0: \sigma^2 = \sigma_0^2$ versus $H_a: \sigma^2 > \sigma_0^2$. If we desire a test for which the probability of a type I error is α, then we use the rejection region

$$\text{RR} = \{\chi^2 > \chi_\alpha^2\}$$

where $P(\chi^2 > \chi_\alpha^2) = \alpha$. (Values of χ_α^2 can be found in Table 6, Appendix III.) A graphical depiction of this rejection region is found in Figure 10.10(a).

Figure 10.10
Rejection regions, RR, for testing
$H_0: \sigma^2 = \sigma_0^2$ versus
(a) $H_a: \sigma^2 > \sigma_0^2$;
(b) $H_a: \sigma^2 < \sigma_0^2$;
and
(c) $H_a: \sigma^2 \neq \sigma_0^2$

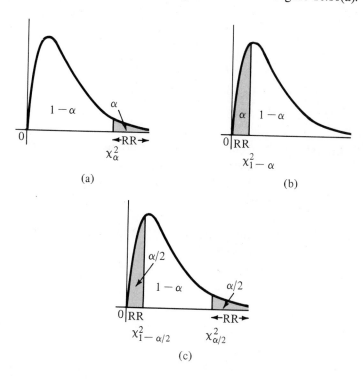

If we want to test $H_0: \sigma^2 = \sigma_0^2$ versus $H_a: \sigma^2 < \sigma_0^2$ (a lower-tail alternative), analogous reasoning will lead to a rejection region located in the lower tail of the χ^2 distribution. We could test $H_0: \sigma^2 = \sigma_0^2$ versus $H_a: \sigma^2 \neq \sigma_0^2$ (a two-tailed test) by using a two-tailed rejection region. Graphical depictions of the appropriate rejection regions are given in Figure 10.10.

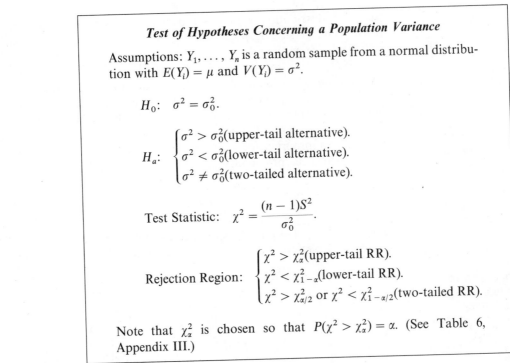

Test of Hypotheses Concerning a Population Variance

Assumptions: $Y_1, \ldots, Y_n$ is a random sample from a normal distribution with $E(Y_i) = \mu$ and $V(Y_i) = \sigma^2$.

$H_0: \quad \sigma^2 = \sigma_0^2.$

$H_a: \begin{cases} \sigma^2 > \sigma_0^2 \text{(upper-tail alternative)}. \\ \sigma^2 < \sigma_0^2 \text{(lower-tail alternative)}. \\ \sigma^2 \neq \sigma_0^2 \text{(two-tailed alternative)}. \end{cases}$

Test Statistic: $\quad \chi^2 = \dfrac{(n-1)S^2}{\sigma_0^2}.$

Rejection Region: $\begin{cases} \chi^2 > \chi_\alpha^2 \text{(upper-tail RR)}. \\ \chi^2 < \chi_{1-\alpha}^2 \text{(lower-tail RR)}. \\ \chi^2 > \chi_{\alpha/2}^2 \text{ or } \chi^2 < \chi_{1-\alpha/2}^2 \text{(two-tailed RR)}. \end{cases}$

Note that χ_α^2 is chosen so that $P(\chi^2 > \chi_\alpha^2) = \alpha$. (See Table 6, Appendix III.)

EXAMPLE 10.16 A machined engine part produced by a company is claimed to have diameter variance no larger than .0002 (diameters measured in inches). A random sample of 10 parts gave a sample variance of .0003. Test, at the 5% level, $H_0: \sigma^2 = .0002$ against $H_a: \sigma^2 > .0002$.

Solution

We must assume that the measured diameters are normally distributed. Then the test statistic is $\chi^2 = (n-1)S^2/\sigma_0^2$, and we reject H_0 for values of this statistic larger than $\chi_{.05}^2 = 16.919$ (based on 9 degrees of freedom). The observed value of the test statistic is

$$\frac{(n-1)s^2}{\sigma_0^2} = \frac{(9)(.0003)}{.0002} = 13.5$$

Thus H_0 is not rejected. There is not sufficient evidence to indicate that σ^2 exceeds .0002.

∽

EXAMPLE 10.17 Determine the p-value associated with the statistical test of Example 10.16.

Solution

The p-value is the probability that a χ^2 random variable with 9 degrees of freedom is larger than the observed value of 13.5. The area corresponding to this probability is shaded in Figure 10.11. Examination of the row corresponding to 9 degrees of freedom in Table 6, Appendix III, yields that $\chi^2_{.1} = 14.6837$. You can see from Figure 10.11 that the shaded area exceeds .1 and thus p-value > .1. That is, for any value of $\alpha < .1$, the null hypothesis cannot be rejected. Again, note the agreement with the conclusion of Example 10.16.

Figure 10.11
Illustration of p-value
for Example 10.17
(χ^2 density with 9
degrees of freedom)

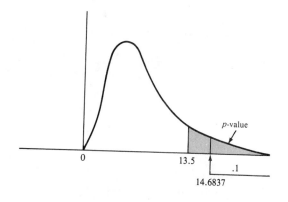

EXAMPLE 10.18 An experimenter was convinced that his measuring equipment possessed variability, which resulted in a standard deviation of 2. Sixteen measurements resulted in a value of $s^2 = 6.1$. Do the data disagree with his claim? Determine the p-value for the test. What would you conclude if you chose $\alpha = .05$?

Solution

We require a test of $H_0: \sigma^2 = 4$ versus $H_a: \sigma^2 \neq 4$. The value of the test statistic is $\chi^2 = 15(6.1)/4 = 22.875$. Note that we require a two-tailed test. Referring to Table 6 we see that for 15 degrees of freedom $\chi^2_{.05} = 24.9958$ and $\chi^2_{.10} = 22.3072$. Thus, the portion of the p-value that falls in the upper tail is between .05 and .10. Since we need to account for a corresponding equal area in the lower tail (this area is also between .05 and .10), it follows that $.1 < p\text{-value} < .2$. The chosen value of $\alpha = .05$ is smaller than even the smallest possible value for the p-value, and therefore we cannot reject the experimenter's claim at the $\alpha = .05$ level.

∽

Sometimes we wish to compare the variances of two normal distributions, particularly by testing to determine whether or not they are equal. These problems are encountered in comparing the precision of two measuring instruments, the variation in quality characteristics of a manufactured product, or the variation in scores for two testing procedures. For example, suppose that $Y_{11}, \ldots, Y_{1n_1}$, and $Y_{21}, \ldots, Y_{2n_2}$ are independent random samples from normal distributions with unknown means and that $V(Y_{1i}) = \sigma_1^2$ and $V(Y_{2i}) = \sigma_2^2$, where σ_1^2 and σ_2^2 are unknown. Suppose that we want to test the null hypothesis $H_0: \sigma_1^2 = \sigma_2^2$ against the alternative $H_a: \sigma_1^2 > \sigma_2^2$.

Because the sample variances S_1^2 and S_2^2 estimate the respective population variances, σ_1^2 and σ_2^2, we would reject H_0 in favor of H_a if S_1^2 is much larger than S_2^2. That is, we use a rejection region RR of the form

$$RR = \left\{ \frac{S_1^2}{S_2^2} > k \right\}$$

where k is chosen so that the probability of a type I error is α. The appropriate value of k depends upon the probability distribution of the statistic S_1^2/S_2^2. Note that $(n_1 - 1)S_1^2/\sigma_1^2$ and $(n_2 - 1)S_2^2/\sigma_2^2$ are independent chi-square random variables, and from Definition 7.3 it follows that

$$F = \frac{(n_1 - 1)S_1^2}{\sigma_1^2(n_1 - 1)} \bigg/ \frac{(n_2 - 1)S_2^2}{\sigma_2^2(n_2 - 1)} = \frac{S_1^2 \sigma_2^2}{S_2^2 \sigma_1^2}$$

has an F distribution with $(n_1 - 1)$ numerator degrees of freedom and $(n_2 - 1)$ denominator degrees of freedom. Under the null hypothesis that $\sigma_1^2 = \sigma_2^2$, then

Figure 10.12
Rejection region RR
for testing
$H_0: \sigma_1^2 = \sigma_2^2$ versus
$H_a: \sigma_1^2 > \sigma_2^2$

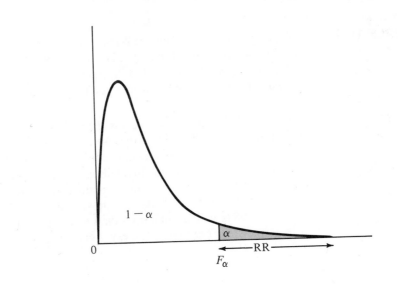

$1 - \alpha$

α

0

F_α

—RR—

$F = S_1^2/S_2^2$ and the rejection region RR given earlier is equivalent to RR $= \{F > k\} = \{F > F_\alpha\}$, where $k = F_\alpha$ is the value of the F distribution with $v_1 = (n_1 - 1)$ and $v_2 = (n_2 - 1)$ such that $P(F > F_\alpha) = \alpha$. Values of F_α are given in Table 7, Appendix III. This rejection region is graphically depicted in Figure 10.12.

EXAMPLE 10.19 Suppose that we wish to compare the variation in diameters for the company in Example 10.16 with the variation of its competitor. Recall that the sample variance for our company, based on $n = 10$ diameters, was $s_1^2 = .0003$. In contrast, the variance of the diameter measurements for a sample of 20 of the competitor's parts was $s_2^2 = .0001$. Do the data provide sufficient information to indicate a smaller variation in diameters for the competitor? Test with $\alpha = .05$.

Solution We are testing $H_0: \sigma_1^2 = \sigma_2^2$ against the alternative $H_a: \sigma_1^2 > \sigma_2^2$. The test statistic, $F = (S_1^2/S_2^2)$, is based on $v_1 = 9$ and $v_2 = 19$ degrees of freedom, and we reject H_0 for values of F larger than $F_{.05} = 2.42$. (See Table 7, Appendix III.) Because the observed value of the test statistic is

$$\frac{s_1^2}{s_2^2} = \frac{.0003}{.0001} = 3$$

$F > F_\alpha$, and we reject $H_0: \sigma_1^2 = \sigma_2^2$. It appears that the variation of the parts diameters is less for the new company.

EXAMPLE 10.20 Give bounds for the p-value associated with the data of Example 10.19.

Solution The calculated F value for this upper-tail test is $F = 3$. Because this value is based on $v_1 = 9$ and $v_2 = 19$ numerator and denominator degrees of freedom, respectively, Table 7, Appendix III, can be used to determine that $F_{.025} = 2.88$, whereas $F_{.01} = 3.52$. Thus, the observed value, $F = 3$, would lead to a rejection of the null hypothesis for $\alpha = .025$ but not for $\alpha = .01$. Hence, $.01 < p\text{-value} < .025$.

Suppose that for Example 10.19 our research hypothesis was $H_a: \sigma_1^2 < \sigma_2^2$. How would we proceed? Note that we are at liberty to identify either population as population 1. Therefore, if we simply interchange the arbitrary labels of 1 and 2 on the two populations (and the corresponding identifiers on sample sizes, sample variances, etc.), our alternative hypothesis becomes $H_a: \sigma_1^2 > \sigma_2^2$ and we can proceed as before. That is, if the research hypothesis is that the variance of one population is larger than that of another population, identify the population with the *hypothesized larger* variance as "population 1" and proceed as previously indicated.

Test of the Hypothesis $\sigma_1^2 = \sigma_2^2$

Assumptions: Independent samples from normal populations.

H_0: $\sigma_1^2 = \sigma_2^2$.

H_a: $\sigma_1^2 > \sigma_2^2$.

Test Statistic: $F = \dfrac{S_1^2}{S_2^2}$.

Rejection Region: $F > F_\alpha$, where F_α, is chosen so that $P(F > F_\alpha) = \alpha$ when F has $v_1 = n_1 - 1$ numerator degrees of freedom and $v_2 = n_2 - 1$ denominator degrees of freedom. (See Table 7, Appendix III.)

If we wish to test $H_0: \sigma_1^2 = \sigma_2^2$ versus $H_a: \sigma_1^2 \neq \sigma_2^2$, with type-I-error probability α, we could employ $F = S_1^2/S_2^2$ as a test statistic and reject H_0 in favor of H_a if the calculated F value is in either the upper or lower $\alpha/2$ tail of the F distribution. The upper-tail critical values can be determined directly from Table 7, Appendix III. How do we determine the lower-tail critical values?

Note that $F = S_1^2/S_2^2$ and $F^{-1} = S_2^2/S_1^2$ both have F distributions, but that the numerator and denominator degrees of freedom are interchanged (the process of inversion switches the roles of numerator and denominator). Let F_b^a denote a random variable with an F distribution with a and b numerator and denominator degrees of freedom, respectively, and let $F_{b,\alpha/2}^a$ be such that

$$P(F_b^a > F_{b,\alpha/2}^a) = \frac{\alpha}{2}$$

Then

$$P[(F_b^a)^{-1} < (F_{b,\alpha/2}^a)^{-1}] = \frac{\alpha}{2}$$

and therefore

$$P[F_a^b < (F_{b,\alpha/2}^a)^{-1}] = \frac{\alpha}{2}$$

That is, the value that cuts off a lower-tail area of $\alpha/2$ for an F_a^b distribution can be found by inverting $F_{b,\alpha/2}^a$. Thus if we use $F = S_1^2/S_2^2$ as a test statistic for testing $H_0: \sigma_1^2 = \sigma_2^2$ versus $H_a: \sigma_1^2 \neq \sigma_2^2$, the appropriate rejection region is

$$\text{RR: } \{F > F_{n_2-1, \alpha/2}^{n_1-1} \quad \text{or} \quad F < (F_{n_1-1, \alpha/2}^{n_2-1})^{-1}\}$$

An equivalent test (see Exercise 10.58) is obtained as follows. Let n_L and n_S denote the sample sizes associated with the larger and smaller sample variances, respectively. Place the large sample variance in the numerator and the smaller sample variance in the denominator of the F statistic and reject $H_0: \sigma_1^2 = \sigma_2^2$ in favor of $H_a: \sigma_1^2 \neq \sigma_2^2$ if $F > F_{\alpha/2}$, where $F_{\alpha/2}$ is determined for $v_1 = n_L - 1$ and $v_2 = n_S - 1$ numerator and denominator degrees of freedom, respectively.

EXAMPLE 10.21   An experiment to explore the pain thresholds to electrical shocks for males and females resulted in the data summary given in Table 10.4. Do the data provide sufficient evidence to indicate a significant difference in the variability of pain thresholds for men and women? Use $\alpha = .10$. What can be said about the p-value?

Table 10.4
Data for
Example 10.21

	Males	Females
n	14	10
$\bar{y}$	16.2	14.9
s^2	12.7	26.4

Solution

Let us assume that the pain thresholds for men and women are approximately normally distributed. We desire to test $H_0: \sigma_M^2 = \sigma_F^2$ versus $H_a: \sigma_M^2 \neq \sigma_F^2$, where σ_M^2 and σ_F^2 are the variances of pain thresholds for men and women, respectively. The larger S^2 is 26.4 (the S^2 for women), and the sample size associated with the larger S^2 is $n_L = 10$. The smaller S^2 is 12.7 (the S^2 for men), and $n_S = 14$ (the number of men in the sample). Therefore, we compute

$$F = \frac{26.4}{12.7} = 2.079$$

and we compare this value to $F_{\alpha/2} = F_{.05}$ with $v_1 = 10 - 1 = 9$ and $v_2 = 14 - 1 = 13$ numerator and denominator degrees of freedom, respectively. Because $F_{.05} = 2.71$ and because 2.079 is not larger than the critical value (2.71), there is not enough evidence to claim a difference in the variability of pain thresholds for men and women.

The p-value associated with the observed value of F for this two-tailed test can be found as follows. Referring to Table 7 with $v_1 = 9$, $v_2 = 13$ numerator and denominator degrees of freedom, respectively, we find $F_{.10} = 2.16$. Thus, p-value $> 2(.10) = .20$. Unless one were willing to work with a very large value of α (some value greater than .2) these results would *not* allow one to conclude that there is a difference in the variances of pain thresholds for men and women.

Exercises

10.55   A manufacturer of hard safety hats for construction workers is concerned about the mean and the variation of the forces helmets transmit to wearers when subjected to a standard external force. The manufacturer desires the mean force transmitted by helmets to be 800 pounds (or less), well under the legal 1000-pound limit, and σ to be less than 40. A random sample of $n = 40$ helmets was tested and the sample mean and variance were found to be equal to 825 pounds and 2350 pounds^2, respectively.

   (a)   If $\mu = 800$ and $\sigma = 40$, is it likely that any helmet, subjected to the standard external force, will transmit a force to a wearer in excess of 1000 pounds? Explain.

   (b)   Do the data provide sufficient evidence to indicate that when subjected to the standard external force, the mean force transmitted by the helmets exceeds 800 pounds?

   (c)   Do the data provide sufficient evidence to indicate that σ exceeds 40?

10.56   A manufacturer of a machine to package soap powder claimed that her machine could load cartons at a given weight with a range of no more than .4 ounce. The mean and variance of a sample of eight 3-pound boxes were found to equal 3.1 and .018, respectively. Test the hypothesis that the variance of the population of weight measurements is $\sigma^2 = .01$ against the alternative $\sigma^2 > .01$. Use an $\alpha = .05$ level of significance. What assumptions are required for this test? What can be said about the attained significance level?

10.57   Under what assumptions may the F distribution be used in making inferences about the ratio of population variances?

10.58   From two normal populations with respective variances σ_1^2 and σ_2^2, we observe independent sample variances S_1^2 and S_2^2, with corresponding degrees of freedom $v_1 = n_1 - 1$ and $v_2 = n_2 - 1$. We wish to test $H_0: \sigma_1^2 = \sigma_2^2$ versus $H_a: \sigma_1^2 \neq \sigma_2^2$.

   (a)   Show that the rejection region given by $\{F > F_{v_2, \alpha/2}^{v_1} \text{ or } F < (F_{v_1, \alpha/2}^{v_2})^{-1}\}$, where $F = S_1^2/S_2^2$, is the same as the rejection region given by $\{S_1^2/S_2^2 > F_{v_2, \alpha/2}^{v_1} \text{ or } S_2^2/S_1^2 > F_{v_1, \alpha/2}^{v_2}\}$.

   (b)   Let S_L^2 denote the larger of (S_1^2, S_2^2) and S_S^2 denote the smaller of (S_1^2, S_2^2). Let v_L and v_S denote the degrees of freedom of S_L^2 and S_S^2, respectively. Use part (a) to show that under H_0

$$P(S_L^2/S_S^2 > F_{v_S, \alpha/2}^{v_L}) = \alpha$$

   Note that this gives an equivalent method of testing equality of two variances.

10.59   A dairy is in the market for a new bottle-filling machine and is considering models A and B manufactured by companies X and Y, respectively. If ruggedness, cost, and convenience are comparable in the two models, the deciding factor is the variability of fills (the model producing fills with the smaller variance being preferred). Let σ_1^2 and σ_2^2 be the fill variances for models A and B, respectively, and consider various tests of the null hypothesis $H_0: \sigma_1^2 = \sigma_2^2$. Obtaining samples of fills from the two machines and utilizing the test statistic s_1^2/s_2^2, one could set up as the rejection region an upper-tail area, a lower-tail area, or a two-tailed area of the F distribution depending on his point of view.

Name the type of rejection region that would be most favored by the following persons, and explain why.

(a)   the manager of the dairy,

(b)   a salesman for company X,

(c)   a salesman for company Y.

10.60   The closing prices of two common stocks were recorded for a period of 16 days. The means and variances were

$$\bar{y}_1 = 40.33 \qquad \bar{y}_2 = 42.54$$
$$s_1^2 = 1.54 \qquad s_2^2 = \ 2.96$$

Do these data present sufficient evidence to indicate a difference in variability of closing prices of the two stocks for the populations associated with the two samples? Give bounds for the attained significance level. What would you conclude with $\alpha = .02$?

10.61   A precision instrument is guaranteed to be accurate to within 2 units. A sample of four instrument readings on the same object yielded the measurements 353, 351, 351, and 355. Give the attained significance level for testing the null hypothesis that $\sigma = 7$ versus the alternative that $\sigma > 7$.

10.62   Aptitude tests should produce scores with a large amount of variation so that an administrator can distinguish between persons with low aptitude and those with high aptitude. The standard test used by a certain industry has been producing scores with a standard deviation of 10 points. A new test is given to twenty prospective employees and produces a sample standard deviation of 12 points. Are scores from the new test significantly more variable than scores from the standard? Use $\alpha = .01$.

10.63   Refer to Exercise 10.47. Is there sufficient evidence, at the 5% significance level, to say that the variance in measurements of DDT levels is greater for juveniles than it is for nestlings?

10.9 Power of Tests: The Neyman-Pearson Lemma

In the remaining sections of this chapter we move from practical examples of statistical tests to a theoretical discussion of their properties. We have suggested specific tests for a number of practical hypothesis-testing situations, but you may be asking why we chose those particular tests. How did we decide on the test statistics that were presented, and how did we know that we had selected the best rejection regions?

Recall that the goodness of a test is measured by α and β, the probabilities of type I and type II errors, where α is chosen in advance and determines the location of the rejection region. A related but more useful concept for evaluating the performance of a test is called the *power* of the test. Basically, the power of a test is the probability that the test will reject the null hypothesis.

Definition 10.3

> Suppose that W is the test statistic and RR is the rejection region for a test of a hypothesis concerning the value of a parameter θ. Then the *power* of the test, noted by power(θ), is the probability that the test rejects H_0 when the actual parameter value is θ. That is,
>
> power(θ) = $P(W$ in RR when the parameter value is θ)

Suppose that we want to test the null hypothesis $H_0: \theta = \theta_0$ and that θ_a is a particular value for θ chosen from H_a. The power of the test at $\theta = \theta_0$, power(θ_0), is equal to the probability of rejecting H_0 when H_0 is true. That is, power(θ_0) = α, the probability of a type I error. For any value of θ from H_a, the power of a test measures the ability of the test to detect that the null hypothesis is false. That is, for $\theta = \theta_a$

$$\text{power}(\theta_a) = P(\text{rejecting } H_0 \text{ when } \theta = \theta_a)$$

Because

$$\beta = P(\text{accepting } H_0 \text{ when } \theta = \theta_a)$$

it follows that the power of the test at θ_a and the probability of a type II error are related as follows:

$$\text{power}(\theta_a) = 1 - \beta$$

A typical *power curve*, a graph of power(θ), is shown in Figure 10.13.

Figure 10.13
A typical power curve for the test $H_0: \theta = \theta_0$ against the alternative $H_a: \theta \neq \theta_0$

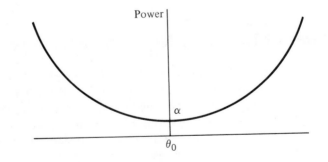

Ideally we would like a test to detect a departure from $H_0: \theta = \theta_0$ with certainty; that is, we would like power(θ_a) to be 1 for all θ_a in H_a (see Figure 10.14). Because for a fixed sample size, α and β both cannot be made arbitrarily small, this is clearly not possible. Therefore, we adopt the procedure of selecting

Figure 10.14
Ideal power curve
for the Test
$H_0: \theta = \theta_0$ versus
$H_a: \theta \neq \theta_0$

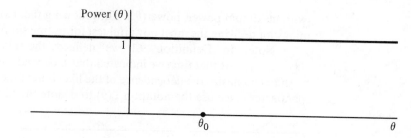

a (small) value for α and finding a rejection region RR to minimize β for the fixed sample size n. That is, we choose RR to maximize power(θ) for θ in H_a. From among all tests with a significance level of α, we seek the test that has the power function closest to the ideal power function, Figure 10.14, if such a test exists. How do we find such a testing procedure?

Before we proceed, it is necessary to define what we mean by *simple* and *composite* hypotheses. Suppose that $Y_1, \ldots, Y_n$ is a random sample from an exponential distribution with parameter λ, that is, $f(y) = (e^{-y/\lambda})/\lambda$, $y > 0$. Then the hypothesis $H: \lambda = 2$ uniquely specifies the distribution from which the sample is taken as having density function $f(y) = (1/2)e^{-y/2}$, $y > 0$. The hypothesis $H: \lambda = 2$ therefore is an example of a *simple* hypothesis. The hypothesis $H: \lambda > 2$ is a *composite* hypothesis because under H the density function $f(y)$ is not uniquely determined. The form of the density is exponential but the parameter λ could be 3 or 15 or any value greater than 2.

Definition 10.4

> If a random sample is taken from a distribution with parameter θ, a hypothesis is said to be a *simple hypothesis* if that hypothesis uniquely specifies the distribution from which the sample is taken. Any hypothesis that is not a simple hypothesis is called a *composite hypothesis*.

If $Y_1, \ldots, Y_n$ is a random sample from a normal distribution with a known variance of, say, $\sigma^2 = 1$, then $H: \mu = 5$ is a simple hypothesis because, if H is true, the density function is uniquely specified to be a normal density function with $\mu = 5$ and $\sigma^2 = 1$. If, on the other hand, σ^2 is not known, the hypothesis $H: \mu = 5$ determines the mean of the normal distribution but does not determine the value of the variance. Therefore, if σ^2 is not known, $H: \mu = 5$ is a composite hypothesis.

Suppose that we would like to test a *simple* null hypothesis $H_0: \theta = \theta_0$ versus a *simple* alternative hypothesis $H_a: \theta = \theta_a$. Because we are concerned only with two particular values of θ (θ_0 and θ_a), we would like to choose a rejection region RR so that $\alpha = $ power (θ_0) is a fixed value and $\beta = 1 - $ power(θ_a) is as small as possible. Equivalently, we seek a *most powerful* test; that is, one with maximum power, power(θ_a). The following theorem provides the methodology for deriving the most powerful test for testing simple H_0 versus simple H_a.

Note: In Definition 9.1 we defined the likelihood function $L = L(y_1, \ldots, y_n)$. At that time we indicated that L depends on $y_1, \ldots, y_n$ and θ. If we wish to emphasize the dependence of the likelihood function on the value of the parameter θ, we use the notation $L(\theta)$ to denote the likelihood function.

Theorem 10.1
Neyman-Pearson
Lemma

Suppose that we wish to test the simple null hypothesis $H_0: \theta = \theta_0$ versus the simple alternative hypothesis $H_a: \theta = \theta_a$ based on a random sample $Y_1, \ldots, Y_n$ from a distribution with parameter θ. Let $L(\theta)$ denote the likelihood of the sample when the value of the parameter is θ. Then for a given α, the test that maximizes the power at θ_a has a rejection region RR determined by

$$\frac{L(\theta_0)}{L(\theta_a)} < k .$$

Such a test will be a most powerful test for H_0 versus H_a.

The proof of Theorem 10.1 will not be given here but it can be found in most of the texts listed in the references at the end of the chapter. We illustrate the application of the theorem with the following example.

EXAMPLE 10.22   Suppose Y represents a single observation from the probability density function given by

$$f(y|\theta) = \begin{cases} \theta y^{\theta-1}, & 0 < y < 1 \\ 0, & \text{elsewhere} \end{cases}$$

Find the most powerful test with significance level $\alpha = .05$ to test $H_0: \theta = 1$ versus $H_a: \theta = 2$.

Solution   Because both of the hypotheses are simple, Theorem 10.1 can be applied to derive the required test. In this case

$$\frac{L(\theta_0)}{L(\theta_a)} = \frac{f(y|\theta_0)}{f(y|\theta_a)} = \frac{(1)y^0}{2y} = \frac{1}{2y}$$

and the form of the rejection region for the most powerful test is

$$\frac{1}{2y} < k$$

Equivalently, the rejection region RR is $y > (1/2k)$. Or because $1/2k = k'$, a constant, the rejection region is $\text{RR}: y > k'$.

Because $\alpha = .05$ is specified, the value of k' is determined by

$$.05 = P(Y \text{ in RR when } \theta = 1) = P(Y > k' \text{ when } \theta = 1) = \int_{k'}^{1} (1)\, dy = 1 - k'$$

Therefore, $k' = 1 - .05 = .95$, and the rejection region is

$$\text{RR}: \{y > .95\}$$

Among all tests for H_0 versus H_a based on a sample of size one and α fixed at .05, this test has the smallest type-II-error probability. That is, it is the most powerful test with significance level $\alpha = .05$. $\qquad\qquad\qquad\qquad \backsim$

Note that the form of the test statistic and rejection region depends upon both H_0 and H_a. If the alternative is changed to $H_a: \theta = 3$, the most powerful test is based upon Y^2, and we reject H_0 in favor of H_a if $Y^2 > k'$, for some constant k'. Also note that the Neyman-Pearson Lemma gives the *form* of the rejection region; the actual rejection region depends upon the specified value for α.

For discrete distributions it is not always possible to find a test that has significance level exactly equal to some predetermined value α. In cases like this we specify the test to be the one for which the probability of a type I error is closest to the predetermined value of α.

Suppose that we sample from a population with a distribution that is completely specified except for the value of a single parameter, θ. If we desire to test $H_0: \theta = \theta_0$ (simple) versus $H_a: \theta > \theta_0$ (composite), there is no general theorem, like Theorem 10.1, that can be applied if either hypothesis is composite. However, Theorem 10.1 can be applied to obtain a most powerful test for $H_0: \theta = \theta_0$ versus $H_a: \theta = \theta_a$ for any single value θ_a, where $\theta_a > \theta_0$. In many situations the form of the rejection region for the most powerful test does not depend upon the particular choice of θ_a. When a test obtained by Theorem 10.1 actually maximizes the power for every value of θ greater than θ_0, it is said to be a *uniformly most powerful* test for $H_0: \theta = \theta_0$ versus $H_a: \theta > \theta_0$. Analogous remarks apply to the derivation of tests for $H_0: \theta = \theta_0$ versus $H_a: \theta < \theta_0$. We illustrate these ideas with the following example.

EXAMPLE 10.23 Suppose that $Y_1, \ldots, Y_n$ is a random sample from a normal distribution with an unknown mean of μ and a known variance of σ^2. We wish to test $H_0: \mu = \mu_0$ against $H_a: \mu > \mu_0$ for a specified constant μ_0. Find the uniformly most powerful test with significance level α.

Solution

We begin by looking for the most powerful α-level test of $H_0: \mu = \mu_0$ versus $H_a: \mu = \mu_a$ for some $\mu_a > \mu_0$. Because

$$f(y: \mu) = \left(\frac{1}{\sigma\sqrt{2\pi}}\right) \exp\left[\frac{-(y - \mu)^2}{2\sigma^2}\right] \qquad -\infty < y < \infty$$

we have

$$L(\mu) = f(y_1: \mu)f(y_2: \mu) \cdots f(y_n: \mu) = \left(\frac{1}{\sigma\sqrt{2\pi}}\right)^n \exp\left[-\sum_{i=1}^{n} \frac{(y_i - \mu)^2}{2\sigma^2}\right]$$

[recall that exp( ) is simply $e^{()}$ in another form]. Appealing to Theorem 10.1, we see that the most powerful test is given by

$$\frac{L(\mu_0)}{L(\mu_a)} < k$$

which in this case is

$$\frac{\left(\frac{1}{\sqrt{2\pi}\sigma}\right)^n \exp\left[-\sum_{i=1}^{n} \frac{(y_i - \mu_0)^2}{2\sigma^2}\right]}{\left(\frac{1}{\sqrt{2\pi}\sigma}\right)^n \exp\left[-\sum_{i=1}^{n} \frac{(y_i - \mu_a)^2}{2\sigma^2}\right]} < k$$

This inequality can be rearranged as follows:

$$\exp\left\{-\frac{1}{2\sigma^2}\left[\sum_{i=1}^{n}(y_i - \mu_0)^2 - \sum_{i=1}^{n}(y_i - \mu_a)^2\right]\right\} < k$$

Taking natural logarithms and simplifying, we have

$$-\frac{1}{2\sigma^2}\left[\sum_{i=1}^{n}(y_i - \mu_0)^2 - \sum_{i=1}^{n}(y_i - \mu_a)^2\right] < \ln k$$

$$\sum_{i=1}^{n}(y_i - \mu_0)^2 - \sum_{i=1}^{n}(y_i - \mu_a)^2 > -2\sigma^2 \ln k$$

$$\sum_{i=1}^{n} y_i^2 - 2n\bar{y}\mu_0 + n\mu_0^2 - \sum_{i=1}^{n} y_i^2 + 2n\bar{y}\mu_a - n\mu_a^2 > -2\sigma^2 \ln k$$

$$\bar{y}(\mu_a - \mu_0) > \frac{-2\sigma^2 \ln k - n\mu_0^2 + n\mu_a^2}{2n}$$

or

$$\bar{y} > \frac{-2\sigma^2 \ln k - n\mu_0^2 + n\mu_a^2}{2n(\mu_a - \mu_0)}$$

Because the quantity on the right-hand side of this inequality is a constant (call it k'), it follows that $\bar{y}$ is the test statistic that yields the most powerful test and that the rejection region is $\bar{y} > k'$. That is,

$$\text{RR} = \{\bar{y} > k'\}$$

The precise value of k' is determined by fixing α and noting that

$$\alpha = P(\bar{Y} \text{ in RR when } \mu = \mu_0)$$

We now observe that the form of the rejection region does not depend upon the particular value assigned to μ_a. That is, any value of μ greater than μ_0 would lead to exactly the same rejection region. Thus we have found the uniformly most powerful test for $H_0: \mu = \mu_0$ versus $H_a: \mu > \mu_0$. It is the Z test of Section 10.3.   ∽

Again consider the situation where the random sample is taken from a distribution that is completely specified except for the value of a single parameter θ. If we wish to derive a test for $H_0: \theta \leq \theta_0$ versus $H_a: \theta > \theta_0$ (note both H_0 and H_a are composite hypotheses), how do we proceed? Suppose that we use the method illustrated in Example 10.23 to find a uniformly most powerful test for $H'_0: \theta = \theta_0$ versus $H_a: \theta > \theta_0$. If θ_1 is less than θ_0 and we use the same test for $H''_0: \theta = \theta_1$ versus H_a, typically both α and β will decrease. In other words, if we have a good test for discriminating between H'_0 and H_a, the same test will be even better for discriminating between H''_0 and H_a. For tests with composite null hypotheses of the form $H_0: \theta \leq \theta_0$ (or $H_0: \theta \geq \theta_0$), we define the significance level α to be the probability of a type I error when $\theta = \theta_0$, that is, $\alpha = \text{power}(\theta_0)$. Generally this value for α is the maximum value of the power function for $\theta \leq \theta_0$ (or $\theta \geq \theta_0$). With this methodology it can be shown that the test derived in Example 10.23 for testing $H_0: \theta = \theta_0$ versus $H_a: \theta > \theta_0$ also is the uniformly most powerful test for testing $H_0: \theta \leq \theta_0$ versus $H_a: \theta > \theta_0$.

In Example 10.23 we derived the uniformly most powerful test for $H_0: \mu = \mu_0$ versus $H_a: \mu > \mu_0$ and found it to have rejection region $\{\bar{y} > k'\}$. That is, for any other test with the same α value and any value of μ_a such that $\mu_a > \mu_0$, the test based on this rejection region gives the smallest value for β. If we wished to test $H_0: \mu = \mu_0$ versus $H_a: \mu < \mu_0$, analogous calculations would lead us to $\{\bar{y} < k^*\}$ as the rejection region for the test that is uniformly most powerful for all $\mu_a < \mu_0$. Therefore, if we wish to test $H_0: \mu = \mu_0$ versus $H_a: \mu \neq \mu_0$, there is no single rejection region that yields the most powerful test for all values of $\mu_a \neq \mu_0$. Thus, there are null and alternative hypotheses for which a uniformly most powerful test does not exist.

The Neyman-Pearson Lemma is useless if we wish to test a hypothesis about a single parameter θ when the sampled distribution contains other unspecified parameters. For example, we might want to test $H_0: \mu = \mu_0$ when the sample is taken from a normal distribution with an unknown variance of σ^2. In this case $H_0: \mu = \mu_0$ does not uniquely determine the form of the distribution (σ^2 could be any nonnegative number) and is therefore *not* a simple hypothesis. The next section will present a general method for developing hypothesis tests when unspecified parameters (called *nuisance parameters*) are present.

Exercises

10.64   Refer to Exercise 10.2. Find the power of the test for each of the following alternatives:

(a) $p = .4$,

(b) $p = .5$,

(c) $p = .6$,

(d) $p = .7$.

Sketch a graph of the power function.

10.65   Let $Y_1, \ldots, Y_{20}$ be a random sample from a normal distribution with an unknown mean of μ and a known variance of $\sigma^2 = 5$. We wish to test $H_0: \mu = 7$ versus $H_a: \mu > 7$.

(a)   Find the uniformly most powerful test with significance level .05.

(b)   For the test in part (a), find the power against each of the following alternatives: $\mu = 7.5$, $\mu = 8.0$, $\mu = 8.5$, and $\mu = 9.0$.

(c)   Sketch a graph of the power function.

10.66   Consider the situation described in Exercise 10.65. What is the smallest sample size such that an $\alpha = .05$ level test has power at least .80 when $\mu = 8$?

10.67   For a normal distribution with a mean of μ and a variance of $\sigma^2 = 25$, one wishes to test $H_0: \mu = 10$ versus $H_a: \mu = 5$. Find the sample size n so that the most powerful test will have $\alpha = \beta = .025$.

10.68   Suppose $Y_1, \ldots, Y_n$ is a random sample from a normal distribution with a *known* mean of μ and an unknown variance of σ^2. Find the most powerful α-level test of $H_0: \sigma^2 = \sigma_0^2$ versus $H_a: \sigma^2 = \sigma_1^2 (\sigma_1^2 > \sigma_0^2)$. Show that this test is equivalent to a χ^2 test. If the test uniformly most powerful for $H_a: \sigma^2 > \sigma_0^2$?

10.69   Suppose we have a random sample of four observations from the density function

$$f(y) = \begin{cases} \left(\dfrac{1}{2\theta^3}\right) y^2 e^{-y/\theta}, & y > 0 \\ 0, & \text{elsewhere} \end{cases}$$

(a)   Find the critical region for the most powerful test of $H_0: \theta = \theta_0$ against $H_a: \theta = \theta_a$, assuming that $\theta_a > \theta_0$. (Hint: Make use of the χ^2 distribution.)

(b)   Is the test given in part (a) uniformly most powerful for the alternative $\theta > \theta_0$?

10.70 Let $Y_1, \ldots, Y_n$ be a random sample from the probability density function given by

$$
f(y) = \begin{cases} \left(\dfrac{1}{\theta}\right) m y^{m-1} e^{-y^m/\theta}, & y > 0 \\ 0, & \text{elsewhere} \end{cases}
$$

with m denoting a known constant.

(a) Find the uniformly most powerful test for testing $H_0: \theta = \theta_0$ against $H_a: \theta > \theta_0$.

(b) If the test in part (a) is to have $\theta_0 = 100$, $\theta_a = 400$, and $\alpha = \beta = .05$, find the appropriate sample size and critical region.

10.71 Let $Y_1, Y_2, \ldots, Y_n$ denote a random sample from a population with a Poisson distribution with mean λ_1. Let $X_1, X_2, \ldots, X_m$ denote an independent random sample from a population with a Poisson distribution with mean λ_2. Derive the most powerful test for testing $H_0: \lambda_1 = \lambda_2 = 2$ versus $H_a: \lambda_1 = 1/2, \lambda_2 = 3$.

*10.72 Let $Y_1, \ldots, Y_n$ denote a random sample from a uniform distribution over the interval $(0, \theta)$.

(a) Find a most powerful α-level test for testing $H_0: \theta = \theta_0$ against $H_a: \theta = \theta_a$, where $\theta_a < \theta_0$.

(b) Is the test in part (a) uniformly most powerful for testing $H_0: \theta = \theta_0$ against $H_a: \theta < \theta_0$?

*10.73 Refer to the random sample of Exercise 10.72.

(a) Find a most powerful α-level test for testing $H_0: \theta = \theta_0$ against $H_a: \theta = \theta_a$ where $\theta_a > \theta_0$.

(b) Is the test in part (a) uniformly most powerful for testing $H_0: \theta = \theta_0$ against $H_a: \theta > \theta_0$?

(c) Is the most powerful α-level test found in part (a) unique?

10.10 Likelihood Ratio Tests

Theorem 10.1 provides a method for constructing most powerful tests for simple hypotheses when the distribution of the observations is known except for the value of a single unknown parameter. In many cases the distribution of concern, in fact, will have more than one unknown parameter. In this section we present a very general method that can be used to derive tests of hypotheses. As you will subsequently see, the procedure works for simple or composite hypotheses and whether or not there are nuisance parameters with unknown values.

Suppose that a random sample is selected from a distribution and that $L(\theta)$ is a function of both unknown nuisance parameters and θ. The null hypothesis specifies that θ lies in a particular set of possible values, say, Ω_0; the alternative hypothesis specifies that θ lies in another set of possible values, say, Ω_a, which does not overlap Ω_0. For example, if we want to test $H_0: \theta = \theta_0$ versus

$H_a: \theta > \theta_0$, then Ω_0 is the single value θ_0, and Ω_a is the set of all values for θ for which $\theta > \theta_0$. Denote the union of the two sets, Ω_0 and Ω_a, by Ω, that is, $\Omega = \Omega_0 \cup \Omega_a$. Many times we cannot completely specify the likelihood of the sample under either hypothesis, because the hypothesis may be composite or other unknown parameters may be present. However, we can employ the observed data to estimate all unknown parameters by the maximum-likelihood method, under either hypothesis. Under H_0 this estimation would be accomplished by using the methods of Section 9.7 to find the maximum-likelihood estimators of all unknown parameters, subject to the restriction that $\theta \in \Omega_0$.

Let $L(\hat{\Omega}_0)$ denote the likelihood function with all unknown parameters replaced by their maximum-likelihood estimators, subject to the restriction that $\theta \in \Omega_0$. Similarly, let $L(\hat{\Omega})$ be obtained the same say, but with the restriction that $\theta \in \Omega$. A likelihood ratio test is then based on the ratio $L(\hat{\Omega}_0)/L(\hat{\Omega})$.

A Likelihood Ratio Test:

Define λ by

$$\lambda = \frac{L(\hat{\Omega}_0)}{L(\hat{\Omega})}$$

A likelihood ratio test of $H_0: \theta \in \Omega_0$ versus $H_a: \theta \in \Omega_a$ employs λ as a test statistic, and the rejection region is determined by $\lambda \leq k$.

It can be shown that $0 \leq \lambda \leq 1$. A value of λ close to zero indicates that the likelihood of the sample appears to be very small under H_0 in comparison with its value under H_a. Equivalently, the data suggest favoring H_a over H_0. The actual value of k is chosen so that α remains at a predetermined level. We will illustrate the mechanics of this method with the following example.

EXAMPLE 10.24   Suppose that $Y_1, \ldots, Y_n$ is a random sample from a normal distribution with an unknown mean of μ and a variance of σ^2. We want to test $H_0: \mu = \mu_0$ versus $H_a: \mu > \mu_0$. Find the appropriate likelihood ratio test.

Solution   Note that Ω_0 is the set $\{\mu_0\}$, $\Omega_a = \{\mu: \mu > \mu_0\}$, and hence that $\Omega = \Omega_0 \cup \Omega_a = \{\mu \geq \mu_0\}$. The constant σ^2 is completely unspecified. We must now find $L(\hat{\Omega}_0)$ and $L(\hat{\Omega})$.

Restricting μ to Ω_0 implies that $\mu = \mu_0$, and hence it need not be estimated. However, we must estimate σ^2 for the case where $\mu = \mu_0$. From Example 9.15 we see that when $\mu = \mu_0$, the maximum-likelihood estimate of σ^2 is

$$\hat{\sigma}_0^2 = \frac{\sum_{i=1}^{n}(Y_i - \mu_0)^2}{n}$$

   The unrestricted maximum-likelihood estimator of μ (see Example 9.15) is $\bar{Y}$. Therefore, for μ restricted to Ω, the maximum-likelihood estimator of μ is $\hat{\mu} = \max(\bar{Y}, \mu_0)$. (If the actual maximum of L is outside the region Ω, the maximum within Ω occurs at the boundary point, μ_0.) Just as earlier, the maximum-likelihood estimator of σ^2 in Ω is

$$\hat{\sigma}^2 = \frac{\sum\limits_{i=1}^{n}(Y_i - \hat{\mu})^2}{n}$$

For the normal density function we have

$$L = \left(\frac{1}{\sqrt{2\pi}}\right)^n\left(\frac{1}{\sigma^2}\right)^{n/2}\exp\left[-\sum_{i=1}^{n}\frac{(y_i - \mu)^2}{2\sigma^2}\right]$$

$L(\hat{\Omega}_0)$ is obtained by replacing μ by μ_0 and σ^2 by $\hat{\sigma}_0^2$, which gives

$$L(\hat{\Omega}_0) = \left(\frac{1}{\sqrt{2\pi}}\right)^n\left(\frac{1}{\hat{\sigma}_0^2}\right)^{n/2}e^{-n/2}$$

$L(\hat{\Omega})$ is obtained by replacing μ by $\hat{\mu}$ and σ^2 by $\hat{\sigma}^2$, which yields

$$L(\hat{\Omega}) = \left(\frac{1}{\sqrt{2\pi}}\right)^n\left(\frac{1}{\hat{\sigma}^2}\right)^{n/2}e^{-n/2}$$

Thus

$$\lambda = \frac{L(\hat{\Omega}_0)}{L(\hat{\Omega})} = \left(\frac{\hat{\sigma}^2}{\hat{\sigma}_0^2}\right)^{n/2}$$

$$= \left[\frac{\sum\limits_{i=1}^{n}(Y_i - \bar{Y})^2}{\sum\limits_{i=1}^{n}(Y_i - \mu_0)^2}\right]^{n/2} \qquad \text{if } \bar{Y} > \mu_0$$

$$= 1 \qquad\qquad\qquad\qquad \text{if } \bar{Y} \leq \mu_0$$

Because

$$\sum_{i=1}^{n}(Y_i - \mu_0)^2 = \sum_{i=1}^{n}[(Y_i - \bar{Y}) + (\bar{Y} - \mu_0)]^2$$

$$= \sum_{i=1}^{n}(Y_i - \bar{Y})^2 + n(\bar{Y} - \mu_0)^2$$

the rejection region, $\lambda \le k$, is equivalent to

$$\frac{\sum_{i=1}^{n}(Y_i - \bar{Y})^2}{\sum_{i=1}^{n}(Y_i - \mu_0)^2} < k^{2/n} = k'$$

$$\frac{\sum_{i=1}^{n}(Y_i - \bar{Y})^2}{\sum_{i=1}^{n}(Y_i - \bar{Y})^2 + n(\bar{Y} - \mu_0)^2} < k'$$

or

$$\frac{1}{1 + \dfrac{n(\bar{Y} - \mu_0)^2}{\sum_{i=1}^{n}(Y_i - \bar{Y})^2}} < k'$$

This inequality, in turn, is equivalent to

$$\frac{n(\bar{Y} - \mu_0)^2}{\sum_{i=1}^{n}(Y_i - \bar{Y})^2} > \frac{1}{k'} - 1 = k''$$

$$\frac{n(\bar{Y} - \mu_0)^2}{\sum_{i=1}^{n}\dfrac{(Y_i - \bar{Y})^2}{n-1}} > (n-1)k''$$

or

$$\frac{\sqrt{n}(\bar{Y} - \mu_0)}{S} > \sqrt{(n-1)k''}$$

where

$$S^2 = \frac{\sum_{i=1}^{n}(Y_i - \bar{Y})^2}{n-1}$$

The last inequality follows because $\bar{Y} > \mu_0$ in the cases for which $\lambda \le k < 1$. Note that $\sqrt{n}(\bar{Y} - \mu_0)/S$ is the t statistic employed in previous sections. Consequently the likelihood ratio test is equivalent to the t test of Section 10.7.

Situations in which the likelihood ratio test assumes a well-known form are not uncommon. In fact, all the tests of Sections 10.7 and 10.8 can be obtained by the likelihood ratio method. For most practical problems the likelihood ratio method produces the best possible test, in the sense of power.

Unfortunately, the likelihood ratio method does not always produce a test statistic with a known probability distribution, such as the t statistic of Example 10.24. However, if the sample size is large, we can obtain an approximation to the distribution of λ.

Theorem 10.2

> Let $Y_1, Y_2, \ldots, Y_n$ have joint likelihood function $L(y_1, y_2, \ldots, y_n)$ where the likelihood function depends on k parameters $\theta_1, \theta_2, \ldots, \theta_k$. Suppose that $r < k$ and that λ is the likelihood ratio statistic for testing $H_0: \theta_1 = \theta_1^0, \theta_2 = \theta_2^0, \ldots, \theta_r = \theta_r^0$ versus $H_a: \theta_1 \ne \theta_1^0, \theta_2 \ne \theta_2^0, \ldots, \theta_r \ne \theta_r^0$ where $\theta_1^0, \theta_2^0, \ldots, \theta_r^0$ are fixed particular values of $\theta_1, \theta_2, \ldots, \theta_r$ and $\theta_{r+1}, \ldots, \theta_k$ are unspecified. Then, for a large n, $-2 \ln \lambda$ has approximately a χ^2 distribution with r degrees of freedom.

Theorem 10.2 allows us to use the χ^2 table for finding rejection regions with fixed α, when n is large. The size of the sample necessary for a "good" approximation varies from application to application.

EXAMPLE 10.25 Suppose that an engineer wishes to compare the number of complaints per week filed by union stewards for two different shifts at a manufacturing plant. One hundred independent observations on the number of complaints gave means $\bar{x} = 20$ for shift 1 and $\bar{y} = 22$ for shift 2. Assume that the number of complaints per week on the ith shift has a Poisson distribution with a mean of θ_i, $i = 1, 2$. Test $H_0: \theta_1 = \theta_2$ versus $H_a: \theta_1 \ne \theta_2$ by the likelihood ratio method, with $\alpha = .01.$

Solution The likelihood of the sample is now the joint density of all X's and Y's and is given by

$$L = \left(\frac{1}{k}\right)\theta_1^{\sum x_i} e^{-n\theta_1} \theta_2^{\sum y_i} e^{-n\theta_2}$$

where $k = x_1! \cdots x_n! \, y_1! \cdots y_n!$, and $n = 100$. Now Ω_0 is the set of parameter values in which $\theta_1 = \theta_2 = \theta$. Hence

$$L = \left(\frac{1}{k}\right) \theta^{\Sigma x_i + \Sigma y_i} e^{-2n\theta}$$

Solving for the maximum-likelihood estimator of θ in Ω_0, we find

$$\hat{\theta} = \frac{1}{2n} \left(\sum_{i=1}^{n} X_i + \sum_{i=1}^{n} Y_i \right) = \frac{1}{2}(\bar{X} + \bar{Y})$$

Separate maximum-likelihood estimators are needed for θ_1 and θ_2 in Ω. Using the likelihood for $\theta_1 \ne \theta_2$ and solving, we find the estimators to be $\hat{\theta}_1 = \bar{X}$ and $\hat{\theta}_2 = \bar{Y}$. Thus

$$\lambda = \frac{L(\hat{\Omega}_0)}{L(\hat{\Omega})} = \frac{(\hat{\theta})^{n\bar{x}+n\bar{y}} e^{-2n\hat{\theta}}}{(\hat{\theta}_1)^{n\bar{x}}(\hat{\theta}_2)^{n\bar{y}} e^{-n\hat{\theta}_1 - n\hat{\theta}_2}} = \frac{(\hat{\theta})^{n\bar{x}+n\bar{y}}}{(\bar{x})^{n\bar{x}}(\bar{y})^{n\bar{y}}}$$

The observed value of $\hat{\theta}$ is $(1/2)(\bar{x} + \bar{y}) = (1/2)(20 + 22) = 21$. The observed value of λ is

$$\lambda = \frac{21^{(100)(20+22)}}{20^{(100)(20)} 22^{(100)(22)}}$$

and hence

$$-2 \ln \lambda = -(2)[4200 \ln (21) - 2000 \ln (20) - 2200 \ln (22)] = 9.53$$

Note that in this application, the number of parameters is $k = 2$. Although H_0 initially does not appear to be of the form described in Theorem 10.2, we can define new parameters $\theta_1' = \theta_1 - \theta_2$, and $\theta_2' = \theta_2$, so that $H_0: \theta_1 = \theta_2$ can be restated as $H_0: \theta_1' = 0$. Thus, $r = 1$ parameter has a fixed specified value. Theorem 10.2 implies that $-2 \ln \lambda$ has an approximate χ^2 distribution with 1 degree of freedom. Small values of λ correspond to large values of $-2 \ln \lambda$ and so the rejection region contains those values of $-2 \ln \lambda$ larger than $\chi^2_{.01} = 6.635$, the value that cuts off an area of .01 in the right-hand tail of a χ^2 density with 1 degree of freedom.

Because the observed value of $-2 \ln \lambda$ is larger than $\chi^2_{.01}$, we reject $H_0: \theta_1 = \theta_2$. It appears that the mean numbers of complaints filed by the union stewards do differ.

Exercises

10.74 Let $Y_1, \ldots, Y_n$ denote a random sample from a normal distribution with a mean of μ (unknown) and a variance of σ^2. For testing $H_0: \sigma^2 = \sigma_0^2$ against $H_a: \sigma^2 > \sigma_0^2$, show that the likelihood ratio test is equivalent to the χ^2 test given in Section 10.8.

10.75 A survey of voter sentiment was conducted in four midcity political wards to compare the fraction of voters favoring candidate A. Random samples of 200 voters were polled in each of the four wards, with the results as shown in the accompanying table. The numbers of voters favoring A in the four samples can be regarded as four independent binomial random variables. Construct a likelihood ratio test of the hypothesis that the fractions of voters favoring candidate A are the same in all four wards. Use $\alpha = .05$.

		Ward			
Opinion	1	2	3	4	Total
Favor A	76	53	59	48	236
Do not favor A	124	147	141	152	564
Total	200	200	200	200	800

10.76 Let S_1^2 and S_2^2 denote, respectively, the variances of independent random samples of sizes n and m selected from normal distributions with means μ_1 and μ_2 and common variance σ^2. If μ_1 and μ_2 are unknown, construct a likelihood ratio test of $H_0: \sigma^2 = \sigma_0^2$ against $H_a: \sigma^2 = \sigma_a^2$, assuming that $\sigma_a^2 > \sigma_0^2$.

10.77 Suppose that $X_1, \ldots, X_{n_1}, Y_1, \ldots, Y_{n_2}$, and $W_1, \ldots, W_{n_3}$ are independent random samples from normal distributions with respective unknown means μ_1, μ_2, and μ_3 and variances σ_1^2, σ_2^2, and σ_3^2.

(a) Find the likelihood ratio test for $H_0: \sigma_1^2 = \sigma_2^2 = \sigma_3^2$ against the alternative of at least one inequality.

(b) Find an approximate critical region for the test in (a) if n_1, n_2, and n_3 are large and $\alpha = .05$.

*10.78 Let $X_1, \ldots, X_m$ denote a random sample from the exponential density with mean θ_1 and let $Y_1, \ldots, Y_n$ denote an independent random sample from an exponential density with mean θ_2. The exponential density is given by

$$f(x) = \left(\frac{1}{\theta_i}\right)e^{-x/\theta_i} \qquad \theta_i > 0; \; x > 0$$

(a)   Find the likelihood ratio criterion for testing $H_0: \theta_1 = \theta_2$ versus $H_a: \theta_1 \neq \theta_2$.

(b)   Show that the test in (a) is equivalent to an exact F test (Hint: Transform $\sum X_i$ and $\sum Y_i$ to chi-square random variables.)

10.79   Suppose independent random samples of sizes n_1 and n_2 are to be selected from normal populations with means of μ_1 and μ_2, respectively, and a common variance of σ^2. For testing $H_0: \mu_1 = \mu_2$ versus $H_a: \mu_1 - \mu_2 > 0$ (σ^2 unknown), show that the likelihood ratio test reduces to the two-sample t test presented in Section 10.7.

10.80   Refer to Exercise 10.79. Show that in testing $H_0: \mu_1 = \mu_2$ versus $H_a: \mu_1 \neq \mu_2$ (σ^2 unknown) the likelihood ratio test reduces to the two-sample t test.

*10.81   Refer to Exercise 10.80. Suppose another independent random sample of size n_3 is selected from a third normal population with a mean of μ_3 and a variance of σ^2. Find the likelihood ratio test for testing $H_0: \mu_1 = \mu_2 = \mu_3$ versus the alternative that there is at least one inequality. Show that this test is equivalent to an exact F test.

10.11 Summary

In Chapters 8, 9, and 10 we have presented the basic concepts associated with the two methods for making inferences, estimation and tests of hypotheses. Philosophically, estimation (Chapters 8 and 9) answers the question, What is the numerical value of a parameter θ? In contrast, a test of a hypothesis answers the question, Is θ equal to a specific numerical value θ_0? The inferential method you employ for a given situation often depends on how you, the experimenter, prefer to phrase your inference. Sometimes this decision is taken out of your hands. That is, the practical question clearly implies that either an estimation or hypothesis-testing procedure be used. For example, the acceptance or rejection of incoming supplies of outgoing products in a manufacturing process clearly requires a decision, or a statistical test.

Associated with both methods for making inferences are measures of their goodness. Thus the expected width of a confidence interval and the confidence coefficient measure the goodness of the estimation procedure. Likewise, the goodness of a statistical test is measured by the probabilities α and β of type I and type II errors. These measures of goodness enable us to compare one statistical test with another and to develop a theory for acquiring statistical tests with desirable properties. The ability to evaluate the goodness of an inference is one of the major contributions of statistics to the analysis of experimental data. Of what value is an inference if you have no measure of its validity?

In this chapter we have presented the elements of a statistical test and explained how a test works. Some useful tests are given to show how they can be used in practical situations, and you will see other interesting applications in the chapters that follow.

Many of the testing procedures developed in this chapter were presented from an intuitive perspective. However, we have also illustrated the use of the Neyman-Pearson Lemma in deriving most powerful procedures for testing a simple null hypothesis versus a simple alternative hypothesis. In addition, we have seen how the Neyman-Pearson method can sometimes be used to find uniformly most powerful tests for composite null and alternative hypotheses if the underlying distribution is specified except for the value of a single parameter. The likelihood ratio procedure provides a general method for developing a statistical test. The likelihood ratio tests can be derived whether or not nuisance parameters are present. In general, the likelihood ratio tests possess desirable properties.

References and Further Readings

1. Boddé, T. "Biomarkers of Aging: Key to a Younger Life," *Bioscience* 31, no. 8 (1981): 566–567.
2. Chisholm, R. F.; Gauther, D. E.; Munzenrider, R. F. "Pre-enlistment Expectation/ Perceptions of Army Life, Satisfaction, and Re-enlistment of Volunteers," *Journal of Political and Military Sociology* 8 (1980): 31–42.
3. Diana, J. S. "The Growth of Largemouth Bass, *Micropterus salmoides* (Lacepede), under Constant and Fluctuating Temperatures," *Journal of Fish Biology* 24 (1984).
4. Dickey, C. "A Strategy for Big Bucks," *Field and Stream* (October 1980).
5. Hoel, P. G. *Introduction to Mathematical Statistics.* 5th ed. New York: Wiley, 1984.
6. Hogg, R. V., and Craig, A. T. *Introduction to Mathematical Statistics.* 4th ed. New York: Macmillan, 1978.
7. Lehmann, E. L. *Testing Statistical Hypotheses.* 2d ed. New York: Wiley, 1986.
8. Lindhe, J. "Clinical Assessment of Antiplaque Agents," *Compendium of Continuing Education in Dentistry*, supl. no. 5 (1984).
9. Mood, A. M.; Graybill, F. A.; and Boes, D. *Introduction to the Theory of Statistics.* 3d ed. New York: McGraw-Hill, 1974.
10. Wilks, S. S. *Mathematical Statistics.* New York: Wiley, 1962.

Supplementary Exercises

10.82 Presently 20% of potential customers buy a certain brand of soap, say, brand *A*. To increase sales, the company will conduct an extensive advertising campaign. At the end of the campaign a sample of 400 potential customers will be interviewed to determine if the campaign was successful.

(a) State H_0 and H_a in terms of p, the probability that a customer prefers soap brand *A*.

(b)   It is decided to conclude that the advertising campaign was a success if at least 92 of the 400 customers interviewed prefer brand A. Find α. (Use the normal approximation to the binomial distribution to evaluate the desired probability.)

10.83   In the past a chemical plant has produced an average of 1100 pounds of chemical per day. The records for the past year, based on 260 operating days, show the following:

$$\bar{y} = 1060 \text{ pounds/day} \qquad s = 340 \text{ pounds/day}$$

We wish to test whether the average daily production has dropped significantly over the past year.

(a)   Give the appropriate null and alternative hypotheses.

(b)   If z is used as a test statistic, determine the rejection region corresponding to a level of significance of $\alpha = .05$.

(c)   Do the data provide sufficient evidence to indicate a drop in average daily production?

10.84   The braking ability was compared for two types of automobiles. Random samples of sixty-four automobiles were tested for each type. The recorded measurement was the distance required to stop when the brakes were applied at 40 miles per hour. The computed sample means and variances were as follows.

$$\begin{array}{cc} \bar{y}_1 = 118 & \bar{y}_2 = 109 \\ s_1^2 = 102 & s_2^2 = 87 \end{array}$$

Do the data provide sufficient evidence to indicate a difference in the mean stopping distance for the two types of automobiles? Give the attained significance level.

10.85   The stability of measurements of the characteristics of a manufactured product is important in maintaining product quality. In fact, it is sometimes better to possess small variation in the measured value of some important characteristic of a product and have the process mean slightly off target than to suffer wide variation with a mean value that perfectly fits requirements. The latter situation may produce a higher percentage of defective product than the former. A manufacturer of light bulbs suspected that one of his production lines was producing bulbs with a high variation in length of life. To test this theory, he compared the lengths of life on $n = 50$ bulbs randomly sampled from the suspect line and $n = 50$ from a line that seemed to be "in control." The sample means and variances for the two samples were as follows.

"Suspect line"	Line "in control"
$\bar{y}_1 = 1{,}520$	$\bar{y}_2 = 1{,}476$
$s_1^2 = 92{,}000$	$s_2^2 = 37{,}000$

(a) Do the data provide sufficient evidence to indicate that bulbs produced by the "suspect line" possess a larger variance in length of life than those produced by the line that is assumed to be in control? Use $\alpha = .05$.

(b) Find the approximate observed significance level for the test and interpret its value.

10.86 A pharmaceutical manufacturer purchases a particular material from two different suppliers. The mean level of impurities in the raw material is approximately the same for both suppliers but the manufacturer is concerned about the variability of the impurities from shipment to shipment. If the level of impurities tends to vary excessively for one source of supply, it could affect the quality of the pharmaceutical product. To compare the variation in percentage impurities for the two suppliers, the manufacturer selects ten shipments from each of the two suppliers and measures the percentage of impurities in the raw material for each shipment. The sample means and variances are shown in the table.

Supplier A	Supplier B
$\bar{y}_1 = 1.89$	$\bar{y}_2 = 1.85$
$s_1^2 = .273$	$s_2^2 = .094$
$n_1 = 10$	$n_2 = 10$

(a) Do the data provide sufficient evidence to indicate a difference in the variability of the shipment impurity levels for the two suppliers? Test using $\alpha = .10$. Based on the results of your test, what recommendation would you make to the pharmaceutical manufacturer?

(b) Find a 90% confidence interval for σ_B^2 and interpret your results.

10.87 The following data give readings in foot-pounds of the impact strength of two kinds of packaging material. Determine whether there is evidence of a difference in mean strength between the two kinds of material. Test at the $\alpha = .10$ level of significance.

A	B
1.25	.89
1.16	1.01
1.33	.97
1.15	.95
1.23	.94
1.20	1.02
1.32	.98
1.28	1.06
1.21	.98
$\sum y_i = 11.13$	$\sum y_i = 8.80$
$\bar{y} = 1.237$	$\bar{y} = .978$
$\sum y_i^2 = 13.7973$	$\sum y_i^2 = 8.6240$

10.88   How much combustion efficiency should a homeowner expect from an oil furnace? The EPA (*Environment News*, January 1977) states that 80% or above is excellent, 75 to 79% is good, 70 to 74% is fair, and below 70% is poor. A home heating contractor who sells two makes of oil heaters (call them *A* and *B*) decided to compare their mean efficiencies. An analysis was made of the efficiencies for eight heaters of type *A* and six of type *B*. The efficiency ratings in percentages, for the fourteen heaters are shown in the table.

Type A	Type B
72	78
78	76
73	81
69	74
75	82
74	75
69	
75	

(a)   Do the data provide sufficient evidence to indicate a difference in mean efficiencies for the two makes of home heaters? Find the approximate *p*-value for the test and interpret its value.

(b)   Find a 90% confidence interval for $(\mu_A - \mu_B)$ and interpret the result.

10.89   A merchant figures his weekly profit to be a function of three variables: retail sales denoted by X, wholesale sales denoted by Y, and overhead costs denoted by W. The variables X, Y, and W are regarded as independent, normally distributed random variables with means μ_1, μ_2, and μ_3 and variances σ^2, $a\sigma^2$, and $b\sigma^2$, respectively, for known constants a and b but unknown σ^2. The merchant's expected profit per week is $\mu_1 + \mu_2 - \mu_3$. If the merchant has independent observations on X, Y, and W for the past n weeks, construct a test of $H_0: \mu_1 + \mu_2 - \mu_3 = k$ against the alternative $H_a: \mu_1 + \mu_2 - \mu_3 \neq k$, for a given constant k. You may specify $\alpha = .05$.

10.90   A reading exam is given to the sixth grades at three large elementary schools. The scores on the exam at each school are regarded as having normal distributions with unknown means μ_1, μ_2, and μ_3, respectively, and unknown common variance σ^2 ($\sigma_1^2 = \sigma_2^2 = \sigma_3^2 = \sigma^2$). Using the accompanying data on independent random samples from each school, test to see if there is evidence of a difference between μ_1 and μ_2. Use $\alpha = .05$.

School I	School II	School III
$n_1 = 10$	$n_2 = 10$	$n_3 = 10$
$\sum x_i^2 = 36{,}950$	$\sum y_i^2 = 25{,}850$	$\sum w_i^2 = 49{,}900$
$\bar{x} = 60$	$\bar{y} = 50$	$\bar{w} = 70$

*10.91 Suppose $Y_1, \ldots, Y_n$ denotes a random sample from the probability density function given by

$$f(y) = \begin{cases} \left(\dfrac{1}{\theta_1}\right) e^{-(y-\theta_2)/\theta_1}, & y > \theta_2 \\ 0, & \text{elsewhere} \end{cases}$$

Find the likelihood ratio test for testing $H_0: \theta_1 = \theta_{1,0}$ versus $H_a: \theta_1 > \theta_{1,0}$, with θ_2 unknown.

*10.92 Refer to Exercise 10.91. Find the likelihood ratio test for testing $H_0: \theta_2 = \theta_{2,0}$ versus $H_a: \theta > \theta_{2,0}$, with θ_1 unknown.

LINEAR MODELS AND ESTIMATION BY LEAST SQUARES

11.1 Introduction

In Chapter 9 we considered several methods for finding estimators of parameters, including the methods of moments and maximum likelihood as well as methods based upon sufficient statistics. Another method of estimation, the method of least squares, is the topic of this chapter.

In all our previous discussions of statistical inference, we assumed that the observable random variables $Y_1, Y_2, \ldots, Y_n$ were independent and identically distributed. One of the implications of this assumption is that the expected value of Y_i, $E(Y_i)$, is constant (if it exists). That is, $E(Y_i) = \mu$ does not depend on any other variable. Obviously this assumption is unrealistic in many inferential problems. For example, the mean stopping distance for a particular type of automobile will depend upon the speed the automobile is traveling; the mean potency of an antibiotic depends upon the amount of time that the antibiotic has been stored; the mean amount of elongation observed in a particular metal alloy depends upon the force applied and the temperature of the alloy. In this chapter we undertake a study of inferential procedures that can be used when a random variable Y, called the *dependent variable,* has a mean that is a function of one or more nonrandom variables, $x_1, x_2, \ldots, x_k$, called *independent variables.* (In this context the terms *independent* and *dependent* are used in their mathematical sense. There is no relationship with the probabilistic concept of independent random variables.)

Many different types of mathematical functions can be used to model a response that is a function of one or more independent variables. These can be classified into two categories, deterministic and probabilistic models. For example, suppose that you are interested in relating a response y to a variable x and that knowledge of the scientific field suggests that y and x are related according to the equation

$$y = \beta_0 + \beta_1 x$$

(where β_0 and β_1 are unknown parameters). This model is called a *deterministic* mathematical model because it does not allow for any error in predicting y as a function of x. We imply that y always takes the value $\beta_0 + \beta_1(20)$ whenever $x = 20$.

Suppose that we collect a sample of n values of y corresponding to n different settings of the independent variable x, and that a plot of the data is as shown in Figure 11.1. It is quite clear from the figure that the expected value of y may increase as a linear function of x but that a deterministic model is far from an adequate description of reality. Repeated experiments when $x = 20$ would find y bobbing about in a random manner. This tells us that the deterministic model is not an exact representation of the relationship between the two variables. Further, if the model were used to predict y when $x = 20$, the prediction would be subject to some unknown error. This, of course, leads us to the use of statistical methods. Predicting y for a given value of x is an inferential process, and we need to know the properties of the error of prediction if the prediction is to be of value in real life.

Figure 11.1
Plot of data

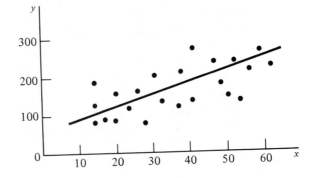

In contrast to the deterministic model, statisticians use *probabilistic* models. For example, we might represent the responses of Figure 11.1 by the model

$$E(Y) = \beta_0 + \beta_1 x$$

or, equivalently,

$$Y = \beta_0 + \beta_1 x + \varepsilon$$

where ε is a random variable having a specified probability distribution with mean zero. We think of Y as having a deterministic component, $E(Y)$, plus a random component, ε. This model accounts for the random behavior of Y exhibited in Figure 11.1 and provides a more accurate description of reality than the deterministic model. Further, the properties of the error of prediction for Y can be derived for many probabilistic models.

Scientific and mathematical textbooks are filled with deterministic models of reality. Indeed, the mathematical functions that appear in your calculus and physics books often are used as deterministic mathematical models of nature. For example, Newton's law relating the force of a moving body to its mass and acceleration,

$$F = ma$$

is a deterministic model that, for practical purposes, predicts with little error. In contrast, many other models—functions graphically represented in scientific journals and texts—are very poor. The spatter of points that would give graphic evidence of their inadequacies, similar to the random behavior of the points in Figure 11.1, has been de-emphasized, leading many budding scientists to accept the "laws" and theories encountered in their science as an exact description of nature.

If deterministic models can be used to predict with negligible error, for all practical purposes, we use them. If not, we seek a probabilistic model, which will not be an exact characterization of nature but will enable us to assess the validity of our inferences.

11.2 Linear Statistical Models

Although infinitely many different functions can be used to model the mean value of the response variable Y as a function of one or more independent variables, we will concentrate on a set of models called *linear statistical models*. If Y is the response variable and x a single independent variable, it may be reasonable to use the model $E(Y) = \beta_0 + \beta_1 x$, for unknown parameter values β_0 and β_1. Note that for this model $E(Y)$ is a linear function of x (for a given β_0 and β_1) and also a linear function of β_0 and β_1 [because $E(Y) = c\beta_0 + d\beta_1$ with $c = 1$ and $d = x$]. In the model $E(Y) = \beta_0 + \beta_1 x^2$, $E(Y)$ is not a linear function of x, but it is a linear function of β_0 and β_1 [because $E(Y) = c\beta_0 + d\beta_1$ with $c = 1$ and $d = x^2$]. When we say we have a linear statistical model for Y, we mean that $E(Y)$ is a linear function of the unknown parameters β_0 and β_1 and *not* necessarily a linear function of x. Thus $Y = \beta_0 + \beta_1 (\ln x) + \varepsilon$ is a linear model (since $\ln x$ is a known constant).

If the model relates $E(Y)$ as a linear function of β_0 and β_1 only, the model is called a *simple* linear regression model. If more than one independent variable, say $x_1, x_2, \ldots, x_k$, is of interest and we model $E(Y)$ by

$$E(Y) = \beta_0 + \beta_1 x_1 + \cdots + \beta_k x_k$$

the model is called a *multiple* linear regression model. Because $x_1, x_2, \ldots, x_k$ are regarded as known constants, they are assumed to be measured without error in an experiment. For example, if you think that the yield y is a function of the variable T, the temperature of a chemical process, you might let $x_1 = T$ and $x_2 = e^T$ and model $E(Y)$ as $E(Y) = \beta_0 + \beta_1 x_1 + \beta_2 x_2$, or, equivalently, $E(Y) = \beta_0 + \beta_1 T + \beta_2 e^T$. Or if y is a function of two variables x_1 and x_2, you might choose a planar approximation to the true mean response, using the linear model $E(Y) = \beta_0 + \beta_1 x_1 + \beta_2 x_2$. Thus $E(Y)$ is a linear function of β_0, β_1, and β_2 and represents a plane in the y, x_1, x_2 space. Similarly,

$$E(Y) = \beta_0 + \beta_1 x + \beta_2 x^2$$

is a linear statistical model, where $E(Y)$ is a second-order polynomial function of the independent variable x, with $x_1 = x$ and $x_2 = x^2$. This model would be appropriate for a response that traces a segment of a parabola over the experimental region.

The expected percentage of water, $E(Y)$, in paper during its manufacture could be represented as a second-order function of the temperature of the dryer, x_1, and the speed of the paper machine, x_2. Thus

$$E(Y) = \beta_0 + \beta_1 x_1 + \beta_2 x_2 + \beta_3 x_1 x_2 + \beta_4 x_1^2 + \beta_5 x_2^2$$

where $\beta_0, \beta_1, \ldots, \beta_5$ are unknown parameters in the model. Geometrically, $E(Y)$ traces a second-order (conic) surface over the x_1, x_2 plane.

Definition 11.1

A *linear statistical* model relating a random response Y to a set of independent variables $x_1, x_2, \ldots, x_k$ is of the form

$$Y = \beta_0 + \beta_1 x_1 + \beta_2 x_2 + \cdots + \beta_k x_k + \varepsilon$$

where $\beta_0, \beta_1, \ldots, \beta_k$ are unknown parameters, ε is a random variable, and $x_1, x_2, \ldots, x_k$ are known constants. We will assume that $E(\varepsilon) = 0$ and hence that

$$E(Y) = \beta_0 + \beta_1 x_1 + \beta_2 x_2 + \cdots + \beta_k x_k$$

Note the physical interpretation of the linear model Y. It says that Y is equal to an expected value, $\beta_0 + \beta_1 x_1 + \beta_2 x_2 + \cdots + \beta_k x_k$ (a function of the independent variables $x_1, x_2, \ldots, x_k$), plus a random error ε. From a practical point of view, ε acknowledges our inability to provide an exact model for nature. In repeated experimentation Y bobs about $E(Y)$ in a random manner because we have failed to include in our model all the many variables that may affect Y. Fortunately, the net effect of these unmeasured, and most often unknown, variables is to cause Y to vary in a manner that may be adequately approximated by an assumption of random behavior.

In this chapter we will use the *method of least squares* to derive estimators for the parameters $\beta_0, \beta_1, \ldots, \beta_k$ in a linear regression model. In many applications one or more of these parameters will have meaningful interpretations. For this reason we develop inferential methods for an individual β parameter and sets of β parameters. If we estimate the parameters $\beta_0, \beta_1, \ldots, \beta_5$ in the model expressing the expected percentage of water, $E(Y)$, in paper as a second-order polynomial in x_1, the dryer temperature, and x_2, the dryer speed, we will develop methods for estimating and forming confidence intervals for the value of $E(Y)$ when x_1 and x_2 take on specific values. Similarly, we develop methods for predicting a future value of Y when the independent variables are set at particular values.

11.3 The Method of Least Squares

A procedure for estimating the parameters of any linear model, the method of least squares, can be illustrated simply by employing it to fit a straight line to a set of data points. Suppose that we wish to fit the model

$$E(Y) = \beta_0 + \beta_1 x$$

to the set of data points shown in Figure 11.2. (Note that the independent variable x could be w^2 or $(w)^{1/2}$ or $\ln w$, and so on, for some other independent variable w.) That is, we postulate that $Y = \beta_0 + \beta_1 x + \varepsilon$, where ε possesses some

Figure 11.2
Fitting a straight line through a set of data points

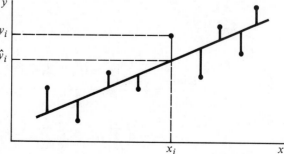
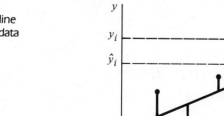

probability distribution with $E(\varepsilon) = 0$. If $\hat{\beta}_0$ and $\hat{\beta}_1$ are estimators of the parameters β_0 and β_1, then $\hat{Y} = \hat{\beta}_0 + \hat{\beta}_1 x$ is clearly an estimator of $E(Y)$.

The least squares procedure for fitting a line through a set of n data points is similar to the method we might use if we fit a line by eye; that is, we want the deviations to be "small" in some sense. A convenient way to accomplish this, and one that yields estimators with good properties, is to minimize the sum of squares of the vertical deviations from the fitted line (see the deviations indicated in Figure 11.2). Thus if

$$\hat{y}_i = \hat{\beta}_0 + \hat{\beta}_1 x_i$$

is the predicted value of the ith y value (when $x = x_i$), then the deviation of the observed value of y from the $\hat{y}$ line (sometimes called the *error*) is

$$y_i - \hat{y}_i$$

and the sum of squares of deviations to be minimized is

$$\text{SSE} = \sum_{i=1}^{n} (y_i - \hat{y}_i)^2 = \sum_{i=1}^{n} [y_i - (\hat{\beta}_0 + \hat{\beta}_1 x_i)]^2$$

The quantity SSE is also called the *sum of squares for error*, for reasons that will subsequently become apparent.

If SSE possesses a minimum, it will occur for values of β_0 and β_1 that satisfy the equations, $\partial \text{SSE}/\partial \hat{\beta}_0 = 0$ and $\partial \text{SSE}/\partial \hat{\beta}_1 = 0$. Taking the partial derivatives of SSE with respect to $\hat{\beta}_0$ and $\hat{\beta}_1$, and setting them equal to zero, we obtain

$$\frac{\partial \text{SSE}}{\partial \hat{\beta}_0} = \frac{\partial \left\{ \sum_{i=1}^{n} [y_i - (\hat{\beta}_0 + \hat{\beta}_1 x_i)]^2 \right\}}{\partial \hat{\beta}_0} = -\sum_{i=1}^{n} 2[y_i - (\hat{\beta}_0 + \hat{\beta}_1 x_i)]$$

$$= -2 \left(\sum_{i=1}^{n} y_i - n\hat{\beta}_0 - \hat{\beta}_1 \sum_{i=1}^{n} x_i \right) = 0$$

and

$$\frac{\partial \text{SSE}}{\partial \hat{\beta}_1} = \frac{\partial \left\{ \sum_{i=1}^{n} [y_i - (\hat{\beta}_0 + \hat{\beta}_1 x_i)]^2 \right\}}{\partial \hat{\beta}_1} = -\sum_{i=1}^{n} 2[y_i - (\hat{\beta}_0 + \hat{\beta}_1 x_i)]x_i$$

$$= -2 \left(\sum_{i=1}^{n} x_i y_i - \hat{\beta}_0 \sum_{i=1}^{n} x_i - \hat{\beta}_1 \sum_{i=1}^{n} x_i^2 \right) = 0$$

The equations $\partial SSE/\partial \hat{\beta}_0 = 0$ and $\partial SSE/\partial \hat{\beta}_1 = 0$ are called the *least squares equations* for estimating the parameters of a line.

Note that the least squares equations are linear in $\hat{\beta}_0$ and $\hat{\beta}_1$ and hence can be solved simultaneously. You can verify that the solutions are

$$\hat{\beta}_1 = \frac{\sum_{i=1}^{n}(x_i - \bar{x})(y_i - \bar{y})}{\sum_{i=1}^{n}(x_i - \bar{x})^2} = \frac{n\sum_{i=1}^{n}x_iy_i - \sum_{i=1}^{n}x_i\sum_{i=1}^{n}y_i}{n\sum_{i=1}^{n}x_i^2 - \left(\sum_{i=1}^{n}x_i\right)^2}$$

$$\hat{\beta}_0 = \bar{y} - \hat{\beta}_1\bar{x}$$

Further, it can be shown that the simultaneous solution for the two least squares equations yields values of $\hat{\beta}_0$ and $\hat{\beta}_1$ that minimize SSE. We leave this for you to prove as an exercise.

We will illustrate the use of the equations above with a simple example.

EXAMPLE 11.1   Use the method of least squares of fit a straight line to the $n = 5$ data points given in Table 11.1.

Table 11.1 Data, Example 11.1

x	y
-2	0
-1	0
0	1
1	1
2	3

Solution   We commence by constructing Table 11.2 to compute the coefficients in the least squares equations. Then we have

$$\hat{\beta}_1 = \frac{n\sum_{i=1}^{n}x_iy_i - \sum_{i=1}^{n}x_i\sum_{i=1}^{n}y_i}{n\sum_{i=1}^{n}x_i^2 - \left(\sum_{i=1}^{n}x_i\right)^2} = \frac{(5)(7) - (0)(5)}{(5)(10) - (0)^2} = .7$$

$$\hat{\beta}_0 = \bar{y} - \hat{\beta}_1\bar{x} = \frac{5}{5} - (.7)(0) = 1$$

and the fitted line is

$$\hat{y} = 1 + .7x$$

The five points and the fitted line are shown in Figure 11.3.

Table 11.2
Calculations for
finding the
coefficients

x_i	y_i	$x_i y_i$	x_i^2
-2	0	0	4
-1	0	0	1
0	1	0	0
1	1	1	1
2	3	6	4
$\sum_{i=1}^{n} x_i = 0$	$\sum_{i=1}^{n} y_i = 5$	$\sum_{i=1}^{n} x_i y_i = 7$	$\sum_{i=1}^{n} x_i^2 = 10$

Figure 11.3
Plot of data points
and least squares line
for Example 11.1

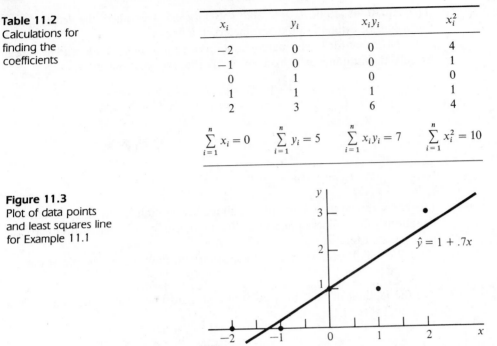

$\hat{y} = 1 + .7x$

In this section we have determined the least squares estimators for the parameters β_0 and β_1 in the model $E(Y) = \beta_0 + \beta_1 x$. A simple example was used to illustrate the use of the formulas developed. We will continue to use this simple example to illustrate calculations in future sections. Some exercises of a more realistic nature are presented at the ends of the sections, and a set of examples involving data from actual experiments is presented and analyzed in Section 11.12. In the next section we will use the methods of matrix algebra to determine the least squares estimators. The most immediate benefit of the matrix algebra approach is that it can be used to provide estimators for the parameters $\beta_0, \beta_1, \ldots, \beta_k$ in the multiple linear regression model $Y = \beta_0 + \beta_1 x_1 + \cdots + \beta_k x_k + \varepsilon$.

Exercises

11.1   Fit a straight line to the accompanying five data points. Give the estimates of β_0 and β_1. Plot the points and sketch the fitted line as a check on the calculations.

y	x
3	-2
2	-1
1	0
1	1
.5	2

11.2   Auditors often are required to compare the audited (or current) value of an inventory item with the book (or listed) value. If a company is keeping its inventory and books up to date, there should be a strong linear relationship between the audited and book values. A sample of ten inventory items from a certain company gave the accompanying data on audited and book values. Fit the model $Y = \beta_0 + \beta_1 x + \varepsilon$ to these data. What is your estimate for the expected change in audited value for a one-unit change in book value? If the book value is $x = 100$, what would you use to estimate the audited value?

Item	Audit Value, y_i	Book Value, x_i
1	9	10
2	14	12
3	7	9
4	29	27
5	45	47
6	109	112
7	40	36
8	238	241
9	60	59
10	170	167

11.3   The median sales prices for new single-family houses over an 8-year period are given in the accompanying table. Letting Y denote the median sales price and x the year (using integers $1, \ldots, 8$), fit the model $Y = \beta_0 + \beta_1 x + \varepsilon$. What can you conclude from the results?

Year	Median Sales Price ($\times$ 1000)
1972(1)	27.6
1973(2)	32.5
1974(3)	35.9
1975(4)	39.3
1976(5)	44.2
1977(6)	48.8
1978(7)	55.7
1979(8)	62.9

Source Adapted from *Time* (July 23, 1979), p. 67; the 1979 figure is for only the first quarter.

11.4   Laboratory experiments designed to measure LC50 values for the effect of certain toxicants on fish are run by two different methods. One method has water continuously flowing through laboratory tanks, and the other method has static water conditions. For purposes of establishing criteria for toxicants, the Environmental Protection Agency (EPA) wants to adjust all results to the flow-through condition. Thus a model is needed to relate the two types of observations. Observations on certain toxicants examined under both static and flow-through conditions yielded the accompanying data (measurements in parts per million). Fit the model $Y = \beta_0 + \beta_1 x + \varepsilon$. What interpretation can you give to the results? Estimate the flow-through value for a toxicant with an LC50 static value of $x = 12$ parts per million.

Toxicant	LC50 Flow-Through, y	LC50 Static, x
1	23.00	39.00
2	22.30	37.50
3	9.40	22.20
4	9.70	17.50
5	.15	.64
6	.28	.45
7	.75	2.62
8	.51	2.36
9	28.00	32.00
10	.39	.77

11.5 The EPA 1980 49-state (all except California) combined mileage rating and engine volume are tabulated for ten standard-transmission, four-cylinder, gasoline-fueled, subcompact cars. The engine sizes are in total cubic inches of cylinder volume.

Car	Cylinder Volume x	mpg (combined) y
VW Rabbit	97	24
Datsun 210	85	29
Chevette	98	26
Dodge Omni	105	24
Mazda 626	120	24
Oldsmobile Starfire	151	22
Mercury Capri	140	23
Toyota Celica	134	23
Datsun 810	146	21

(a) Plot the data points on graph paper.

(b) Find the least-squares line for the data.

(c) Graph the least-squares line to see how well it fits the data.

(d) Use the least-squares line to estimate the mean miles per gallon for a subcompact automobile which has 125 cubic inches of engine volume.

11.6 Suppose that we have postulated the model

$$Y_i = \beta_1 x_i + \varepsilon_i \qquad i = 1, 2, \ldots, n$$

where the ε_i are independent and identically distributed random variables with $E(\varepsilon_i) = 0$. Then $\hat{y}_i = \hat{\beta}_1 x_i$ is the predicted value of the value of y when $x = x_i$ and SSE $= \sum_{i=1}^{n} [y_i - \hat{\beta}_1 x_i]^2$. Find the least squares estimator of β_1. (Note that the equation $y = \beta x$ describes a straight line passing through the origin. The model just described often is called the *no intercept* model.)

11.7 Some data obtained by Carlos E. Macellari [4] on the height, x, and diameter, y, of shells follow. If we consider the model

$$E(Y) = \beta_1 x$$

then the slope, β_1, is a ratio of the mean diameter to the height. Use the following data and the result of Exercise 11.6 to obtain the least squares estimate of the mean diameter to height ratio.

Specimen	Diameter y	Height x
OSU 36651	185	78
OSU 36652	194	65
OSU 36653	173	77
OSU 36654	200	76
OSU 36655	179	72
OSU 36656	213	76
OSU 36657	134	75
OSU 36658	191	77
OSU 36659	177	69
OSU 36660	199	65

11.4 Fitting the Linear Model by Using Matrices

A convenient way to manipulate the linear equations introduced in the previous sections is through matrices. Suppose that we have the linear model

$$Y = \beta_0 + \beta_1 x_1 + \cdots + \beta_k x_k + \varepsilon$$

and we make n independent observations, $y_1, \ldots, y_n$, on Y. We can write the observation y_i as

$$y_i = \beta_0 + \beta_1 x_{i1} + \beta_2 x_{i2} + \cdots + \beta_k x_{ik} + \varepsilon_i$$

where x_{ij} is the setting of the jth independent variable for the ith observation, $i = 1, \ldots, n$. We now define the following matrices, with $x_0 = 1$:

$$\mathbf{Y} = \begin{bmatrix} y_1 \\ y_2 \\ \vdots \\ y_n \end{bmatrix} \qquad \mathbf{X} = \begin{bmatrix} x_0 & x_{11} & x_{12} & \cdots & x_{1k} \\ x_0 & x_{21} & x_{22} & \cdots & x_{2k} \\ \vdots & \vdots & \vdots & & \vdots \\ x_0 & x_{n1} & x_{n2} & \cdots & x_{nk} \end{bmatrix}$$

$$\boldsymbol{\beta} = \begin{bmatrix} \beta_0 \\ \beta_1 \\ \vdots \\ \beta_k \end{bmatrix} \qquad \boldsymbol{\varepsilon} = \begin{bmatrix} \varepsilon_1 \\ \varepsilon_2 \\ \vdots \\ \varepsilon_n \end{bmatrix}$$

Thus the n equations representing y_i as a function of the x's, β's and ε's can be simultaneously written as

$$\mathbf{Y} = \mathbf{X}\boldsymbol{\beta} + \boldsymbol{\varepsilon}$$

(See Appendix I for a discussion of matrix operations.)
For n observations from a simple linear model of the form

$$Y = \beta_0 + \beta_1 x + \varepsilon$$

we have

$$\mathbf{Y} = \begin{bmatrix} y_1 \\ y_2 \\ \vdots \\ y_n \end{bmatrix} \qquad \mathbf{X} = \begin{bmatrix} 1 & x_1 \\ 1 & x_2 \\ \vdots & \vdots \\ 1 & x_n \end{bmatrix} \qquad \boldsymbol{\varepsilon} = \begin{bmatrix} \varepsilon_1 \\ \varepsilon_2 \\ \vdots \\ \varepsilon_n \end{bmatrix} \qquad \boldsymbol{\beta} = \begin{bmatrix} \beta_0 \\ \beta_1 \end{bmatrix}$$

(We suppress the first subscript on x because only one x variable is involved.) The least squares equations for β_0 and β_1 were given in Section 11.3 as

$$n\hat{\beta}_0 + \hat{\beta}_1 \sum_{i=1}^n x_i = \sum_{i=1}^n y_i$$

$$\hat{\beta}_0 \sum_{i=1}^n x_i + \hat{\beta}_1 \sum_{i=1}^n x_i^2 = \sum_{i=1}^n x_i y_i$$

Because

$$\mathbf{X}'\mathbf{X} = \begin{bmatrix} 1 & 1 & \cdots & 1 \\ x_1 & x_2 & \cdots & x_n \end{bmatrix} \begin{bmatrix} 1 & x_1 \\ 1 & x_2 \\ \vdots & \vdots \\ 1 & x_n \end{bmatrix} = \begin{bmatrix} n & \sum_{i=1}^n x_i \\ \sum_{i=1}^n x_i & \sum_{i=1}^n x_i^2 \end{bmatrix}$$

and

$$\mathbf{X}'\mathbf{Y} = \begin{bmatrix} \sum_{i=1}^n y_i \\ \sum_{i=1}^n x_i y_i \end{bmatrix}$$

we see that the least squares equations are given by

$$(\mathbf{X}'\mathbf{X})\hat{\boldsymbol{\beta}} = \mathbf{X}'\mathbf{Y}$$

where

$$\hat{\boldsymbol{\beta}} = \begin{bmatrix} \hat{\beta}_0 \\ \hat{\beta}_1 \end{bmatrix}$$

Hence

$$\hat{\boldsymbol{\beta}} = (\mathbf{X'X})^{-1}\mathbf{X'Y}$$

Although we have shown only that this result holds for a simple case, it can be shown that in general the least squares equations and solutions presented in matrix notation are as follows.

Least Squares Equations and Solutions for a General Linear Model

Equations: $(\mathbf{X'X})\hat{\boldsymbol{\beta}} = \mathbf{X'Y}$.

Solutions: $\hat{\boldsymbol{\beta}} = (\mathbf{X'X})^{-1}\mathbf{X'Y}$.

EXAMPLE 11.2   Solve Example 11.1 using matrix operations.

Solution   From the data given in Example 11.1 we see that

$$\mathbf{Y} = \begin{bmatrix} 0 \\ 0 \\ 1 \\ 1 \\ 3 \end{bmatrix} \quad \text{and} \quad \mathbf{X} = \begin{array}{cc} \begin{array}{cc} x_0 & x_1 \end{array} \\ \begin{bmatrix} 1 & -2 \\ 1 & -1 \\ 1 & 0 \\ 1 & 1 \\ 1 & 2 \end{bmatrix} \end{array}$$

It follows that

$$\mathbf{X'X} = \begin{bmatrix} 5 & 0 \\ 0 & 10 \end{bmatrix} \quad \mathbf{X'Y} = \begin{bmatrix} 5 \\ 7 \end{bmatrix} \quad (\mathbf{X'X})^{-1} = \begin{bmatrix} 1/5 & 0 \\ 0 & 1/10 \end{bmatrix}$$

Thus

$$\hat{\boldsymbol{\beta}} = (\mathbf{X'X})^{-1}\mathbf{X'Y} = \begin{bmatrix} 1/5 & 0 \\ 0 & 1/10 \end{bmatrix}\begin{bmatrix} 5 \\ 7 \end{bmatrix} = \begin{bmatrix} 1 \\ .7 \end{bmatrix}$$

or $\hat{\beta}_0 = 1$ and $\hat{\beta}_1 = .7$. Thus

$$\hat{y} = 1 + .7x$$

just as in Example 11.1.   ∽

EXAMPLE 11.3   Fit a parabola to the data of Example 11.1 using the model

$$Y = \beta_0 + \beta_1 x + \beta_2 x^2 + \varepsilon$$

Solution   The **X** matrix for this example will differ from that of Example 11.1 only by the addition of a third column corresponding to x^2. (Note that $x_1 = x$, $x_2 = x^2$, and $k = 2$ in the notation of the general linear model.) Thus

$$\mathbf{Y} = \begin{bmatrix} 0 \\ 0 \\ 1 \\ 1 \\ 3 \end{bmatrix} \qquad \mathbf{X} = \begin{matrix} x_0 \quad x \quad x^2 \\ \begin{bmatrix} 1 & -2 & 4 \\ 1 & -1 & 1 \\ 1 & 0 & 0 \\ 1 & 1 & 1 \\ 1 & 2 & 4 \end{bmatrix} \end{matrix}$$

The three variables, x_0, x, and x^2, are shown above their respective columns in the **X** matrix. Thus for the first measurement $y = 0$, $x_0 = 1$, $x = -2$, and $x^2 = 4$. For the second measurement $y = 0$, $x_0 = 1$, $x = -1$, and $x^2 = 1$. Succeeding rows of the **Y** and **X** matrices are obtained in a similar manner.

The matrix products, **X'X** and **X'Y**, are

$$(\mathbf{X'X}) = \begin{bmatrix} 1 & 1 & 1 & 1 & 1 \\ -2 & -1 & 0 & 1 & 2 \\ 4 & 1 & 0 & 1 & 4 \end{bmatrix} \begin{bmatrix} 1 & -2 & 4 \\ 1 & -1 & 1 \\ 1 & 0 & 0 \\ 1 & 1 & 1 \\ 1 & 2 & 4 \end{bmatrix} = \begin{bmatrix} 5 & 0 & 10 \\ 0 & 10 & 0 \\ 10 & 0 & 34 \end{bmatrix}$$

$$(\mathbf{X'Y}) = \begin{bmatrix} 1 & 1 & 1 & 1 & 1 \\ -2 & -1 & 0 & 1 & 2 \\ 4 & 1 & 0 & 1 & 4 \end{bmatrix} \begin{bmatrix} 0 \\ 0 \\ 1 \\ 1 \\ 3 \end{bmatrix} = \begin{bmatrix} 5 \\ 7 \\ 13 \end{bmatrix}$$

We omit the process of inverting $(\mathbf{X'X})$ and simply state that it is equal to

$$(\mathbf{X'X})^{-1} = \begin{bmatrix} 17/35 & 0 & -1/7 \\ 0 & 1/10 & 0 \\ -1/7 & 0 & 1/14 \end{bmatrix}$$

[You may verify that $(\mathbf{X'X})^{-1}(\mathbf{X'X}) = \mathbf{I}$.]

Finally,

$$\hat{\beta} = (\mathbf{X'X})^{-1}\mathbf{X'Y}$$

$$= \begin{bmatrix} 17/35 & 0 & -1/7 \\ 0 & 1/10 & 0 \\ -1/7 & 0 & 1/14 \end{bmatrix} \begin{bmatrix} 5 \\ 7 \\ 13 \end{bmatrix} = \begin{bmatrix} 4/7 \\ 7/10 \\ 3/14 \end{bmatrix} \approx \begin{bmatrix} .571 \\ .700 \\ .214 \end{bmatrix}$$

Hence $\hat{\beta}_0 = .571$, $\hat{\beta}_1 = .7$, and $\hat{\beta}_2 = .214$, and the prediction equation is

$$\hat{y} = .571 + .7x + .214x^2$$

A graph of this parabola on Figure 11.3 will indicate a good fit to the data points. ∽

Exercises

11.8   Refer to Exercise 11.1. Fit the model suggested there by use of matrices.

11.9   Fit a straight line to the accompanying data, plot the points, and then sketch the fitted line as a check on the calculations. Note that the data points are the same as for Exercise 11.1 except that they are translated one unit in the positive direction along the x axis. What effect does symmetric spacing of the x values about $x = 0$ have on the form of the $(\mathbf{X'X})$ matrix and the resulting calculations?

y	x
3	−1
2	0
1	1
1	2
.5	3

11.10   Fit the quadratic model $Y = \beta_0 + \beta_1 x + \beta_2 x^2 + \varepsilon$ to the accompanying data points. Plot the points and sketch the fitted parabola as a check on the calculations.

y	x
1	−3
0	−2
0	−1
−1	0
−1	1
0	2
0	3

11.11   The Florida morbidity statistics for the decade ending in 1976 show that infectious hepatitis had the incidence rates shown in the accompanying table (in cases per 100,000 population).

x	y
1967	10.5
1968	18.5
1969	22.6
1970	27.2
1971	31.2
1972	33.0
1973	44.9
1974	49.4
1975	35.0
1976	27.6

(a)   Letting Y denote incidence rate and x denote the coded year (-9 for 1967, -7 for 1968, through 9 for 1976), fit the model $Y = \beta_0 + \beta_1 x + \varepsilon$.

(b)   For the same data fit the model $Y = \beta_0 + \beta_1 x + \beta_2 x^2 + \varepsilon$.

11.12   Cucumbers are usually preserved by fermenting them in a low-salt brine (6 to 9% sodium chloride) and then storing them in a high-salt brine until they are used by processors to produce various types of pickles. The high-salt brine is needed to retard softening of the pickles and to prevent freezing when stored outside in northern climates. Data showing the reduction in firmness of pickles stored over time in a low-salt brine (2 to 3 percent) follow.

	Weeks in Storage at 72°F				
	0	*4*	*14*	*32*	*52*
Firmness in pounds y	19.8	16.5	12.8	8.1	7.5

Source   Buescher, R. W., J. M. Hudson, J. R. Adams, and D. H. Wallace, "Calcium Makes It Possible to Store Cucumber Pickles in Low-Salt Brine," *Arkansas Farm Research*, 30, no. 4 (July–August 1981).

(a)   Fit a least-squares line to the data.

(b)   As a check on your calculations, plot the five data points and graph the line. Does the line appear to provide a good fit to the data points?

(c)   Use the least-squares line to estimate the mean firmness of pickles stored for 20 weeks.

11.13   The table shown below gives the catches in Peruvian anchovies (in millions of metric tons) and the prices of fish meal (in current dollars per ton) for the years 1965 to 1978.

	1965	1966	1967	1968	1969	1970	1971	1972	1973	1974	1975	1976	1977	1978
Price, Fish Meal, y	190	160	134	129	172	197	167	239	542	372	245	376	454	410
Anchovy Catch, x	7.23	8.53	9.82	10.26	8.96	12.27	10.28	4.45	1.78	4.0	3.3	4.3	0.8	0.5

Source   Bardach, John E., and Regina M. Santerre, "Climate and the Fish in the Sea," *BioScience* 31, no. 3 (March 1981), pp 206 ff. Copyright © 1981 by the American Institute of Biological Sciences.

(a)   Find the least-squares line appropriate for these data.

(b)   Plot the points and graph the line as a check on your calculations.

11.14   Matis and Wehrly [5] report the accompanying data on the proportion of green sunfish that survive a fixed level of thermal pollution for varying lengths of time.

(a)   Fit the linear model $Y = \beta_0 + \beta_1 x + \varepsilon$. Give your interpretation.

(b)   Plot the points and graph the result of (a). Does the line fit through the points?

Proportion of Survivors, y	Scaled Time, x
1.00	.10
.95	.15
.95	.20
.90	.25
.85	.30
.70	.35
.65	.40
.60	.45
.55	.50
.40	.55

11.5 Properties of the Least Squares Estimators for the Model $Y = \beta_0 + \beta_1 x + \varepsilon$

In order to utilize the least squares estimators for making statistical inferences, we need to determine their properties. In this section we will show that the least squares estimators $\hat{\beta}_0$ and $\hat{\beta}_1$ for the parameters in the simple linear model

$$Y = \beta_0 + \beta_1 x + \varepsilon$$

are unbiased estimators of their respective parameter values. We also will derive the variances of these estimators and, under the assumption that the error term ε is normally distributed, show that $\hat{\beta}_0$ and $\hat{\beta}_1$ have normal sampling distributions. Results applicable to the general linear model will be presented without proof in Section 11.6.

Recall that ε previously was assumed to be a random variable with $E(\varepsilon) = 0$. We now add the assumption that $V(\varepsilon) = \sigma^2$. That is, we are assuming that the difference between the random variable Y and $E(Y) = \beta_0 + \beta_1 x$ is distributed about zero, with a variance that does not depend on x. Note that $V(Y) = V(\varepsilon) = \sigma^2$, because the other terms in the linear model are constants.

Assume that n independent observations are to be made on this model so that, before sampling, we have n independent random variables of the form

$$Y_i = \beta_0 + \beta_1 x_i + \varepsilon_i$$

From Section 11.3 we know that

$$\hat{\beta}_1 = \frac{\sum_{i=1}^{n} (x_i - \bar{x})(Y_i - \bar{Y})}{\sum_{i=1}^{n} (x_i - \bar{x})^2}$$

which can be written as

$$\hat{\beta}_1 = \frac{\sum_{i=1}^{n} (x_i - \bar{x})Y_i - \bar{Y} \sum_{i=1}^{n} (x_i - \bar{x})}{\sum_{i=1}^{n} (x_i - \bar{x})^2}$$

Then because $\sum_{i=1}^{n} (x_i - \bar{x}) = 0$, we have

$$\hat{\beta}_1 = \frac{\sum_{i=1}^{n} (x_i - \bar{x})Y_i}{\sum_{i=1}^{n} (x_i - \bar{x})^2}$$

Because all summations in the following discussion will be summed from $i = 1$ to n, we will simplify our notation by omitting the variable of summation and its index. Now let us find the expected value and variance of $\hat{\beta}_1$.

From the expectation theorems, Section 5.8, we have

$$E(\hat{\beta}_1) = E\left[\frac{\sum (x_i - \bar{x})Y_i}{\sum (x_i - \bar{x})^2}\right] = \frac{\sum (x_i - \bar{x})E(Y_i)}{\sum (x_i - \bar{x})^2}$$

$$= \frac{\sum (x_i - \bar{x})(\beta_0 + \beta_1 x_i)}{\sum (x_i - \bar{x})^2}$$

$$= \beta_0 \frac{\sum (x_i - \bar{x})}{\sum (x_i - \bar{x})^2} + \beta_1 \frac{\sum (x_i - \bar{x})x_i}{\sum (x_i - \bar{x})^2}$$

Then because $\sum (x_i - \bar{x}) = 0$ and $\sum (x_i - \bar{x})^2 = \sum (x_i - \bar{x})x_i$, we have

$$E(\hat{\beta}_1) = 0 + \beta_1 \frac{\sum (x_i - \bar{x})^2}{\sum (x_i - \bar{x})^2} = \beta_1$$

Thus $\hat{\beta}_1$ is an unbiased estimator of β_1.

To find $V(\hat{\beta}_1)$, we use Theorem 5.12. Then because $Y_1, Y_2, \ldots, Y_n$ are independent, we have

$$V(\hat{\beta}_1) = V\left[\frac{\sum (x_i - \bar{x})Y_i}{\sum (x_i - \bar{x})^2}\right] = \left[\frac{1}{\sum (x_i - \bar{x})^2}\right]^2 \sum V[(x_i - \bar{x})Y_i]$$

$$= \left[\frac{1}{\sum (x_i - \bar{x})^2}\right]^2 \sum (x_i - \bar{x})^2 V(Y_i)$$

Because $V(Y_i) = \sigma^2$, $i = 1, 2, \ldots, n$,

$$V(\hat{\beta}_1) = \frac{\sigma^2}{\sum (x_i - \bar{x})^2}$$

Now let us find the expected value and variance of $\hat{\beta}_0$, where $\hat{\beta}_0 = \bar{Y} - \hat{\beta}_1 \bar{x}$. From Theorem 5.12 we have

$$V(\hat{\beta}_0) = V(\bar{Y}) + \bar{x}^2 V(\hat{\beta}_1) - 2\bar{x} \operatorname{cov}(\bar{Y}, \hat{\beta}_1)$$

Consequently, we must find $V(\bar{Y})$ and $\operatorname{cov}(\bar{Y}, \hat{\beta}_1)$ in order to obtain $V(\hat{\beta}_0)$. Because $Y_i = \beta_0 + \beta_1 x_i + \varepsilon_i$, we see that

$$\bar{Y} = \frac{1}{n}\sum Y_i = \beta_0 + \beta_1 \bar{x} + \bar{\varepsilon}$$

Thus

$$E(\bar{Y}) = \beta_0 + \beta_1 \bar{x} + E(\bar{\varepsilon}) = \beta_0 + \beta_1 \bar{x}$$

and

$$V(\bar{Y}) = V(\bar{\varepsilon}) = \left(\frac{1}{n}\right) V(\varepsilon_i) = \frac{\sigma^2}{n}$$

To find $\mathrm{cov}\,(\bar{Y}, \hat{\beta}_1)$, rewrite the expression for $\hat{\beta}_1$ as

$$\hat{\beta}_1 = \sum c_i Y_i$$

where

$$c_i = \frac{x_i - \bar{x}}{\sum (x_i - \bar{x})^2}$$

[Note that $\sum c_i = 0$.] Then

$$\mathrm{cov}\,(\bar{Y}, \hat{\beta}_1) = \mathrm{cov}\left[\sum \left(\frac{1}{n}\right) Y_i, \sum c_i Y_i\right]$$

and, using Theorem 5.12,

$$\mathrm{cov}\,(\bar{Y}, \hat{\beta}_1) = \sum \left(\frac{c_i}{n}\right) V(Y_i) + 2 \sum \sum_{i<j} \left(\frac{c_j}{n}\right) \mathrm{cov}\,(Y_i, Y_j)$$

Because Y_i and Y_j, $i \neq j$, are independent, $\mathrm{cov}\,(Y_i, Y_j) = 0$. Also, $V(Y_i) = \sigma^2$, and hence

$$\mathrm{cov}\,(\bar{Y}, \hat{\beta}_1) = \frac{\sigma^2}{n} \sum c_i = 0$$

Returning to our original task of finding the expected value and variance of $\hat{\beta}_0 = \bar{Y} - \hat{\beta}_1 \bar{x}$, we apply expectation theorems to obtain

$$E(\hat{\beta}_0) = E(\bar{Y}) - E(\hat{\beta}_1)\bar{x} = \beta_0 + \beta_1 \bar{x} - \beta_1 \bar{x} = \beta_0$$

We have shown that both $\hat{\beta}_0$ and β_1 are unbiased estimators of their respective parameters.

Because we now have $V(\bar{Y})$, $V(\hat{\beta}_1)$, and $\mathrm{cov}\,(\bar{Y}, \hat{\beta}_1)$, we are ready to find $V(\hat{\beta}_0)$. Thus

$$V(\hat{\beta}_0) = V(\bar{Y}) + \bar{x}^2 V(\beta_1) - 2\bar{x}\,\mathrm{cov}\,(\bar{Y}, \hat{\beta}_1)$$

Substituting the values for $V(\bar{Y})$, $V(\hat{\beta}_1)$, and $\text{cov}(\bar{Y}, \hat{\beta}_1)$, we obtain

$$V(\hat{\beta}_0) = \frac{\sigma^2}{n} + \bar{x}^2 \left[\frac{\sigma^2}{\sum (x_i - \bar{x})^2} \right] + 0$$

$$= \sigma^2 \left[\frac{1}{n} + \frac{\bar{x}^2}{\sum (x_i - \bar{x})^2} \right] = \frac{\sigma^2 \sum x_i^2}{n \sum (x_i - \bar{x})^2}$$

Further (see Exercise 11.22), Theorem 5.12 can be employed to show that

$$\text{cov}(\hat{\beta}_0, \hat{\beta}_1) = \frac{-\bar{x}\sigma^2}{\sum (x_i - \bar{x})^2}$$

Note that $\hat{\beta}_0$ and $\hat{\beta}_1$ and correlated (and therefore dependent) unless $\bar{x} = 0$.

The expressions for $V(\hat{\beta}_0)$, $V(\hat{\beta}_1)$, and $\text{cov}(\hat{\beta}_0, \hat{\beta}_1)$ can be conveniently expressed in terms of matrices. We saw in Section 11.4 that $\mathbf{X'X}$ for the linear model $Y = \beta_0 + \beta_1 x + \varepsilon$ is given by

$$\mathbf{X'X} = \begin{bmatrix} n & \sum x_i \\ \sum x_i & \sum x_i^2 \end{bmatrix}$$

It can be shown that

$$(\mathbf{X'X})^{-1} = \begin{bmatrix} \dfrac{\sum x_i^2}{n \sum (x_i - \bar{x})^2} & -\dfrac{\sum x_i}{n \sum (x_i - \bar{x})^2} \\ -\dfrac{\sum x_i}{n \sum (x_i - \bar{x})^2} & \dfrac{1}{\sum (x_i - \bar{x})^2} \end{bmatrix} = \begin{bmatrix} c_{00} & c_{01} \\ c_{10} & c_{11} \end{bmatrix}$$

By checking the variances and covariances derived above, you can see that

$$V(\hat{\beta}_i) = c_{ii}\sigma^2 \qquad i = 0, 1$$

and

$$\text{cov}(\hat{\beta}_0, \hat{\beta}_1) = c_{01}\sigma^2 = c_{10}\sigma^2$$

EXAMPLE 11.4   Find the variances of the estimators $\hat{\beta}_0$ and $\hat{\beta}_1$ for Example 11.2.

Solution   In Example 11.2 we found that

$$(\mathbf{X'X})^{-1} = \begin{bmatrix} 1/5 & 0 \\ 0 & 1/10 \end{bmatrix}$$

Hence

$$V(\hat{\beta}_0) = c_{00}\sigma^2 = (1/5)\sigma^2$$

and

$$V(\hat{\beta}_1) = c_{11}\sigma^2 = (1/10)\sigma^2$$

Note that cov $(\hat{\beta}_0, \hat{\beta}_1) = 0$ in this case since $\sum x_i = 0$.

The expressions above give the variances for the least squares estimators in terms of σ^2, the variance of the error term ε. Usually the value of σ^2 will be unknown, and we will need to make use of the sample observations to estimate σ^2. If $\bar{Y}$ is used to estimate the mean, then

$$S^2 = \left(\frac{1}{n-1}\right)\sum_{i=1}^{n}(Y_i - \bar{Y})^2$$

is an unbiased estimator for the population variance σ^2. Because we are now using $\hat{Y}_i$ to estimate $E(Y_i)$, it seems natural to base an estimate of σ^2 upon SSE $= \sum_{i=1}^{n}(Y_i - \hat{Y}_i)^2$. Indeed, we will show that

$$S^2 = \left(\frac{1}{n-2}\right)\sum_{i=1}^{n}(Y_i - \hat{Y}_i)^2 = \left(\frac{1}{n-2}\right)\text{SSE}$$

provides an unbiased estimator for σ^2. Note that the 2 occurring in the denominator of S^2 corresponds to the number of β parameters estimated in the model.

Because

$$E(S^2) = E\left[\left(\frac{1}{n-2}\right)\text{SSE}\right] = \left(\frac{1}{n-2}\right)E(\text{SSE})$$

it is necessary to find $E(\text{SSE})$ in order to verify that $E(S^2) = \sigma^2$.

Note that

$$
\begin{aligned}
E(\text{SSE}) &= E[\sum(Y_i - \hat{Y}_i)^2] = E[\sum(Y_i - \hat{\beta}_0 - \hat{\beta}_1 x_i)^2] \\
&= E[\sum(Y_i - \bar{Y} + \hat{\beta}_1\bar{x} - \hat{\beta}_1 x_i)^2] \\
&= E[\sum[(Y_i - \bar{Y}) - \hat{\beta}_1(x_i - \bar{x})]^2] \\
&= E[\sum(Y_i - \bar{Y})^2 + \hat{\beta}_1^2\sum(x_i - \bar{x})^2 - 2\hat{\beta}_1\sum(x_i - \bar{x})(Y_i - \bar{Y})]
\end{aligned}
$$

Because $\sum(x_i - \bar{x})(Y_i - \bar{Y}) = \sum(x_i - \bar{x})^2\hat{\beta}_1$, the last two terms in the expectation collapse to $-\hat{\beta}_1^2\sum(x_i - \bar{x})^2$. Also,

$$\sum(Y_i - \bar{Y})^2 = \sum Y_i^2 - n\bar{Y}^2$$

and therefore

$$E[\sum (Y_i - \hat{Y}_i)^2] = E[\sum Y_i^2 - n\bar{Y}^2 - \hat{\beta}_1^2 \sum (x_i - \bar{x})^2]$$
$$= \sum E(Y_i^2) - nE(\bar{Y}^2) - \sum (x_i - \bar{x})^2 E(\hat{\beta}_1^2)$$

Noting that for any random variable U, $E(U^2) = V(U) + [E(U)]^2$, we see that

$$E[\sum (Y_i - \hat{Y}_i)^2] = \sum \{V(Y_i) + [E(Y_i)]^2\} - n\{V(\bar{Y}) + [E(\bar{Y})]^2\}$$
$$- \sum (x_i - \bar{x})^2 \{V(\hat{\beta}_1) + [E(\hat{\beta}_1)]^2\}$$
$$= n\sigma^2 + \sum (\beta_0 + \beta_1 x_i)^2 - n\left[\frac{\sigma^2}{n} + (\beta_0 + \beta_1 \bar{x})^2\right]$$
$$- \sum (x_i - \bar{x})^2 \left[\frac{\sigma^2}{\sum (x_i - \bar{x})^2} + \beta_1^2\right]$$

This expression simplifies to $(n - 2)\sigma^2$. Thus we find that an unbiased estimator of σ^2 is given by

$$S^2 = \left(\frac{1}{n-2}\right) \sum (Y_i - \hat{Y}_i)^2 = \left(\frac{1}{n-2}\right) \text{SSE}$$

One task remains, finding an easy way to calculate $\sum (y_i - \hat{y}_i)^2 = \text{SSE}$. A bit of matrix algebra will show that

$$\boxed{\text{SSE} = \mathbf{Y}'\mathbf{Y} - \hat{\boldsymbol{\beta}}'\mathbf{X}'\mathbf{Y}}$$

(Note that $\mathbf{Y}'\mathbf{Y} = \sum Y_i^2$.)

EXAMPLE 11.5   Estimate σ^2 from the data given in Example 11.1.

Solution   For these data

$$\mathbf{Y} = \begin{bmatrix} 0 \\ 0 \\ 1 \\ 1 \\ 3 \end{bmatrix} \qquad \mathbf{X} = \begin{bmatrix} 1 & -2 \\ 1 & -1 \\ 1 & 0 \\ 1 & 1 \\ 1 & 2 \end{bmatrix} \qquad \hat{\boldsymbol{\beta}} = \begin{bmatrix} 1 \\ .7 \end{bmatrix}$$

Hence

$$SSE = \mathbf{Y'Y} - \boldsymbol{\hat{\beta}'X'Y}$$

$$= \begin{bmatrix} 0 & 0 & 1 & 1 & 3 \end{bmatrix} \begin{bmatrix} 0 \\ 0 \\ 1 \\ 1 \\ 3 \end{bmatrix} - \begin{bmatrix} 1 & .7 \end{bmatrix} \begin{bmatrix} 1 & 1 & 1 & 1 & 1 \\ -2 & -1 & 0 & 1 & 2 \end{bmatrix} \begin{bmatrix} 0 \\ 0 \\ 1 \\ 1 \\ 3 \end{bmatrix}$$

$$= 11 - \begin{bmatrix} 1 & .7 \end{bmatrix} \begin{bmatrix} 5 \\ 7 \end{bmatrix} = 11 - 9.9 = 1.1$$

Then

$$s^2 = \frac{SSE}{n-2} = \frac{1.1}{5-2} = \frac{1.1}{3} = .367$$

The derivations above establish the means and variances of the estimators $\hat{\beta}_0$ and $\hat{\beta}_1$ and show that $S^2 = SSE/(n-2)$ is an unbiased estimator for the parameter σ^2. The only assumptions we thus far have made about the error term ε in the model $Y = \beta_0 + \beta_1 x + \varepsilon$ is that $E(\varepsilon) = 0$ and that $V(\varepsilon) = \sigma^2$, independent of x. The form of the sampling distributions for $\hat{\beta}_0$ and $\hat{\beta}_1$ depends upon the distribution of the error term ε. Because of the common occurrence of the normal distribution in nature, it often is reasonable to assume that ε is normally distributed with a mean of zero and a variance of σ^2. If this assumption of normality is warranted, it follows that Y_i is normally distributed with a mean of $\beta_0 + \beta_1 x_i$ and a variance of σ^2. Because both $\hat{\beta}_0$ and $\hat{\beta}_1$ are linear functions of $Y_1, \ldots, Y_n$, the estimators will be normally distributed, with means and variances as previously derived. As you will subsequently see, the assumption of normality of the distribution of the error term ε and the resulting normal distributions for $\hat{\beta}_0$ and $\hat{\beta}_1$ will allow us to develop tests and confidence intervals based upon the t distribution.

Exercises

11.15   Derive the following identity:

$$SSE = \sum_{i=1}^{n} (y_i - \hat{y}_i)^2 = \sum_{i=1}^{n} (y_i - \hat{\beta}_0 - \hat{\beta}_1 x_i)^2$$

$$= \sum_{i=1}^{n} (y_i - \bar{y})^2 - \hat{\beta}_1 \sum_{i=1}^{n} (x_i - \bar{x})(y_i - \bar{y})$$

Note that this gives an easier computational method of finding SSE.

11.16   An experiment was conducted to observe the effect of an increase in temperature on the potency of an antibiotic. Three 1-ounce portions of the antibiotic were stored for equal lengths of time at each of the following temperatures: $30°$, $50°$, $70°$, and $90°$. The potency readings observed at the end of the experimental period were

Potency readings, y	38, 43, 29	32, 26, 33	19, 27, 23	14, 19, 21
Temperature, x	30°	50°	70°	90°

(a)   Find the least-squares line appropriate for this data.

(b)   Plot the points and graph the line as a check on your calculations.

(c)   Calculate S^2.

11.17 (a)   Calculate SSE and S^2 for Exercise 11.3.

(b)   We see from previous exercises that it sometimes is convenient, for computational purposes, to have equally spaced x values symmetrically spaced about zero. The x values can be rescaled (or coded) in any convenient manner with no loss of information in the statistical analysis. Refer to Exercise 11.3. Code the x values (originally given on a scale of 1 to 8) by use of the formula

$$x^* = \frac{x - 4.5}{.5}$$

Then fit the model $Y = \beta_0^* + \beta_1^* x^* + \varepsilon$. Calculate SSE. (Note that the x^* values are integers symmetrically spaced about zero.) Compare the SSE with the value obtained in part (a).

11.18 (a)   Calculate SSE and S^2 for Exercise 11.2.

(b)   Refer to Exercise 11.2. Fit the model suggested there for the relationship between audit values and book values by using matrices. Note that we can simplify the computations by defining

$$x_i^* = x_i - \bar{x}$$

and fitting the model $Y = \beta_0^* + \beta_1^* x^* + \varepsilon$. Fit this latter model and calculate SSE. Compare your answer with the SSE calculation in part (a).

11.19 (a)   Calculate SSE and S^2 for Exercise 11.4.

(b)   Refer to Exercise 11.4. Code the x values in a convenient manner and fit a simple linear model to the LC50 measurements presented there. Compute SSE and compare your answer to the result of part (a).

11.20 A study was conducted to determine the effects of sleep deprivation on subjects' ability to solve simple problems. The amount of sleep deprivation varied over 8, 12, 16, 20, and 24 hours without sleep. A total of ten subjects participated in the study, two at each sleep-deprivation level. After his or her specified sleep-deprivation period, each subject was administered a set of simple addition problems and the number of errors recorded. The following results were obtained.

Number of errors, y	8, 6	6, 10	8, 14	14, 12	16, 12
Number of hours without sleep, x	8	12	16	20	24

(a)   Find the least-squares line appropriate to these data.

(b)   Plot the points and graph the least-squares line as a check on your calculations.

(c)   Calculate S^2.

11.21   Suppose that $Y_1, \ldots, Y_n$ are independent normal random variables with $E(Y_i) = \beta_0 + \beta_1 x_i$ and $V(Y_i) = \sigma^2$, $i = 1, \ldots, n$. Show that the maximum-likelihood estimators of β_0 and β_1 are the same as the least squares estimators of Section 11.3.

11.22   Under the assumptions of Exercise 11.21 find cov $(\hat{\beta}_0, \hat{\beta}_1)$. Use this answer to show that $\hat{\beta}_0$ and $\hat{\beta}_1$ are independent if $\sum\limits_{i=1}^{n} x_i = 0$. [Hint: cov $(\hat{\beta}_0, \hat{\beta}_1) = $ cov $(\bar{Y} - \hat{\beta}_1 \bar{X}, \hat{\beta}_1)$. Use Theorem 5.12 and the results of this section.]

11.23   Under the assumptions of Exercise 11.21, find the maximum-likelihood estimator of σ^2.

11.6 Properties of the Least Squares Estimators for the Multiple Linear Regression Model

All the theoretical results of the previous section can be extended to the multiple linear regression model,

$$Y_i = \beta_0 + \beta_1 x_{i1} + \cdots + \beta_k x_{ik} + \varepsilon_i, \qquad i = 1, \ldots, n$$

Suppose that $\varepsilon_1, \ldots, \varepsilon_n$ are independent random variables with $E(\varepsilon_i) = 0$ and $V(\varepsilon_i) = \sigma^2$. Then the least squares estimators are given by

$$\hat{\boldsymbol{\beta}} = (\mathbf{X}'\mathbf{X})^{-1} \mathbf{X}'\mathbf{Y}$$

provided that $(\mathbf{X}'\mathbf{X})^{-1}$ exists. The properties of these estimators are as follows (proof omitted).

Properties of the Least Squares Estimators

1. $E(\hat{\beta}_i) = \beta_i$, $i = 0, 1, \ldots, k$.
2. $V(\hat{\beta}_i) = c_{ii}\sigma^2$, where c_{ij} is the element in row i and column j of $(\mathbf{X}'\mathbf{X})^{-1}$. (Recall that this matrix has a row and column numbered 0.)
3. Cov $(\hat{\beta}_i, \hat{\beta}_j) = c_{ij}\sigma^2$.
4. An unbiased estimator of σ^2 is $S^2 = \text{SSE}/[n - (k + 1)]$, where $\text{SSE} = \mathbf{Y}'\mathbf{Y} - \hat{\boldsymbol{\beta}}'\mathbf{X}'\mathbf{Y}$. (Note that there are $k + 1$ unknown β_i's in the model.)

If, in addition, the ε_i, $i = 1, 2, \ldots, n$ are normally distributed:

5. Each $\hat{\beta}_i$ is normally distributed.
6. The random variable

$$\frac{[n - (k + 1)]S^2}{\sigma^2}$$

has a χ^2 distribution with $n - (k + 1)$ degrees of freedom.
7. The statistics S^2 and $\hat{\beta}_i$, $i = 0, 1, 2, \ldots, k$ are independent.

11.7 Inferences Concerning the Parameters β_i

Suppose that an engineer has fit the model

$$Y = \beta_0 + \beta_1 x + \varepsilon$$

where Y is the strength of concrete after 28 days and x is the water/cement ratio used in the concrete. If, in reality, the strength of concrete does not change with the water/cement ratio, then $\beta_1 = 0$. Thus the engineer may desire a test of $H_0: \beta_1 = 0$ versus $H_a: \beta_1 \neq 0$ in order to assess whether the independent variable has an influence on the dependent variable. Or the engineer may wish to estimate the mean rate β_1 of change in y for a one-unit change in the water/cement ratio x.

In general, for any linear regression model, if the random error ε is normally distributed, we have established that $\hat{\beta}_i$ is an unbiased, normally distributed estimator of β_i with $V(\hat{\beta}_i) = c_{ii}\sigma^2$. Using this information, we can construct a test of the hypothesis $H_0: \beta_i = \beta_{i0}$ (β_{i0} is a specified value of β_i), using the test statistic

$$Z = \frac{\hat{\beta}_i - \beta_{i0}}{\sigma\sqrt{c_{ii}}}$$

The rejection region for a two-tailed test would be

$$|z| \geq z_{\alpha/2}$$

As in the case of the simple normal-deviate tests studied in Chapter 10, we must either know σ or possess a good estimate based upon an adequate number of degrees of freedom. (What would be adequate is a debatable point. We suggest that the estimate be based upon 30 or more degrees of freedom.) When this estimate is unavailable (which usually is the case), an estimate of σ may be calculated from the experimental data (in accordance with the procedure of Section 11.5) and substituted for σ in the z statistic. The resulting quantity,

$$T = \frac{\hat{\beta}_i - \beta_{i0}}{S\sqrt{c_{ii}}}$$

can be shown to possess a Student's t distribution with $[n - (k + 1)]$ degrees of freedom.

Test of Hypothesis for β_i

H_0: $\beta_i = \beta_{i0}$.

H_a: $\begin{array}{l} \beta_i > \beta_{i0} \text{ (upper-tail alternative)}. \\ \beta_i < \beta_{i0} \text{ (lower-tail alternative)}. \\ \beta_i \neq \beta_{i0} \text{ (two-tailed alternative)}. \end{array}$

Test Statistic: $T = \dfrac{\hat{\beta}_i - \beta_{i0}}{S\sqrt{c_{ii}}}.$

Rejection Region: $\begin{array}{l} t > t_\alpha \text{ (upper-tail rejection region)}. \\ t < -t_\alpha \text{ (lower-tail rejection region)}. \\ |t| > t_{\alpha/2} \text{ (two-tailed rejection region)}. \end{array}$

Note that t_α is based on $[n - (k + 1)]$ degrees of freedom.

EXAMPLE 11.6   Do the data of Example 11.1 present sufficient evidence to indicate curvature in the response function? Test using $\alpha = .05$ and give bounds to the attained significance level.

Solution   The verbal question just stated assumes that the probabilistic model is a realistic description of the true response and implies a test of the hypothesis $H_0: \beta_2 = 0$ versus $H_a: \beta_2 \neq 0$ in the linear model $Y = \beta_0 + \beta_1 x + \beta_2 x^2 + \varepsilon$. (If $\beta_2 = 0$, the quadratic term will not appear and the expected value of Y will represent a straight-line function of x.) The first step in the solution is the calculation of SSE and s^2.

$$ \text{SSE} = \mathbf{Y'Y} - \hat{\boldsymbol{\beta}}'\mathbf{X'Y} = 11 - [.571 \quad .700 \quad .214] \begin{bmatrix} 5 \\ 7 \\ 13 \end{bmatrix} $$

$$ = 11 - 10.537 = .463 $$

Then

$$ s^2 = \frac{\text{SSE}}{n-3} = \frac{.463}{2} = .232 \quad \text{and} \quad s = .48 $$

[Note: The model contains three parameters and hence SSE is based upon $n - 3 = 2$ degrees of freedom.]

   The estimate of β_2 obtained from Example 11.3 is $\hat{\beta}_2 = 3/14 \approx .214$. Then

$$ t = \frac{\hat{\beta}_2 - 0}{s\sqrt{c_{22}}} = \frac{.214}{.48\sqrt{1/14}} = 1.67 $$

If we take $\alpha = .05$, the value of $t_{\alpha/2} = t_{.025}$ for 2 degrees of freedom is 4.303, and the rejection region would be

$$\text{reject if } |t| \geq 4.303$$

Because the absolute value of the calculated value of t is less than 4.303, we cannot reject the null hypothesis that $\beta_2 = 0$. Note that we do not accept $H_0: \beta_2 = 0$. We would need to know the probability of making a type II error—that is, the probability of falsely accepting H_0 for a specified alternative value of β_2—before we would accept. Because the test is two-tailed, p-value = $2P(t > 1.67)$ where t has a t distribution with 2 degrees of freedom. Using Table 5, Appendix III, we find that $P(t > 1.67) > .10$. Thus we conclude that p-value $> .2$. Unless we are willing to work with a relatively large value of α (greater than .2), we cannot reject H_0. Again we note the agreement between the conclusions reached by the formal (fixed α) test procedure and the proper interpretation of the attained significance level.

As a further step in the analysis, we could look at the width of a confidence interval for β_2 to see whether it is short enough to detect a departure from zero that would be of practical significance. We will show that the confidence interval for β_2 is quite large, suggesting that the experimenter collect more data before reaching a decision.   ∽

Based on the t statistic given earlier, we can follow the procedures of Chapter 10 to show that a confidence interval for β_i, with confidence coefficient $1 - \alpha$, is as follows.

A $(1 - \alpha)$ 100% Confidence Interval for β_i

$$\hat{\beta}_i \pm t_{\alpha/2} S \sqrt{c_{ii}}$$

EXAMPLE 11.7   Calculate a 95% confidence interval for the parameter β_2 of Example 11.6.

Solution   The tabulated value for $t_{.025}$, based upon 2 degrees of freedom, is 4.303. Then the 95% confidence interval for β_2 is

$$\hat{\beta}_2 \pm t_{.025} S \sqrt{c_{22}}$$

Substituting, we get

$$.214 \pm (4.303)(.48)\sqrt{1/14} \qquad \text{or} \qquad .214 \pm .552$$

For example, if we wish to esimate β_2 correct to within .15 unit, it is obvious that the confidence interval is too wide and that the sample size must be increased.   ∽

Exercises

11.24   Refer to Exercise 11.1.

(a)   Do the data present sufficient evidence to indicate that the slope β_1 differs from zero? (Test at the 5% significance level.)

(b)   Find a 95% confidence interval for β_1.

11.25   Refer to Exercise 11.13. Do the data present sufficient evidence to indicate that the size of the anchovy catch, x, contributes information for the prediction of price, y, of fish meal? Give bounds on the attained significance level. What would you conclude at the $\alpha = .10$ level of significance?

11.26   Do the data in Exercise 11.20 present sufficient evidence to indicate that the number of errors is linearly related to the number of hours without sleep?

(a)   Give bounds on the attained significance level.

(b)   What would you conclude at the $\alpha = .05$ level of significance?

(c)   Would you expect the relationship between y and x to be linear if x were varied over a wider range, say from $x = 4$ to $x = 48$?

(d)   Give a 95% confidence interval for the slope. Provide a practical interpretation for this interval estimate.

11.27   Refer to Exercise 11.11.

(a)   Is there evidence of a quadratic effect in the relationship between Y and x? (Test $H_0: \beta_2 = 0$.) Use $\alpha = .10$.

(b)   Find a 90% confidence interval for β_2.

11.28   Most sophomore physics students are required to conduct an experiment verifying Hooke's Law. Hooke's Law states that when a force is applied to a body that is long in comparison to its cross-sectional area, the change y in its length is proportional to the force x; that is,

$$y = \beta_1 x,$$

where β_1 is a constant of proportionality. The results of an actual physics student's laboratory experiment are shown in the table. Six lengths of steel wire, .34 millimeters (mm) in diameter and 2 meters (m) long, were used to obtain the six force–length change measurements.

Force, x (kg)	Change in Length, y (mm)
29.4	4.25
39.2	5.25
49.0	6.50
58.8	7.85
68.6	8.75
78.4	10.00

(a)   Fit the model, $Y = \beta_0 + \beta_1 x + \varepsilon$, to the data using the method of least squares.

(b)   Find a 95% confidence interval for the slope of the line.

(c)   According to Hooke's Law, the line should pass through the point $(0, 0)$; that is, β_0 should equal 0. Test the hypothesis that $E(Y) = 0$ when $x = 0$. Give bounds for the attained significance level. What would you conclude at the $\alpha = .05$ level?

11.29 Use the properties of the least squares estimators, given in Section 11.6, to derive the confidence interval for β_i given in this section.

11.30 Suppose that $Y_1, \ldots, Y_n$ are independent, normally distributed random variables with $E(Y_i) = \beta_0 + \beta_1 x_i$ and $V(Y_i) = \sigma^2$, $i = 1, \ldots, n$. Show that the likelihood ratio test of $H_0: \beta_1 = 0$ versus $H_a: \beta_1 \neq 0$ is equivalent to the t test given in this section.

*11.31 Let $Y_1, \ldots, Y_n$ be as given in Exercise 11.30. Suppose that we have an additional set of independent random variables $W_1, \ldots, W_m$, where W_i is normally distributed with $E(W_i) = \gamma_0 + \gamma_1 c_i$ and $V(W_i) = \sigma^2$, $i = 1, \ldots, m$. Construct a test of $H_0: \beta_1 = \gamma_1$ against $H_a: \beta_1 \neq \gamma_1$.

11.32 The octane number Y of refined petroleum is related to the temperature x of the refining process, but it is also related to the particle size of the catalyst. An experiment with a small-particle catalyst gave a fitted least squares line of

$$\hat{y} = 9.360 + .155x$$

with $n = 31$, $V(\hat{\beta}_1) = (.0202)^2$, and SSE $= 2.04$. An independent experiment with a large-particle catalyst gave

$$\hat{y} = 4.265 + .190x$$

with $n = 11$, $V(\hat{\beta}_1) = (.0193)^2$, and SSE $= 1.86.$[†]

(a)   Test the hypotheses that the slopes are significantly different from zero, with each test at a significance level of .05.

*(b)   Test, at the .05 significance level, that the two types of catalyst produce the same slope in the relationship between octane number and temperature. (Use the test in Exercise 11.31.)

11.33 The Florida morbidity statistics for the decade ending in 1976 show the accompanying data on the number of cases of tuberculosis (per 100,000 population) for the years 1967 through 1976. Is there sufficient evidence to state that the rate of tuberculosis is decreasing over this period? Use $\alpha = .05$. (The years may be coded in any convenient manner.)

Year	Cases
1967	26.3
1968	26.1
1969	24.7
1970	22.8
1971	22.1
1972	20.4
1973	19.0
1974	17.7
1975	19.3
1976	17.5

* Exercises preceded by an asterisk are optional.

† *Source*   Gweyson and Cheasley, *Petroleum Refiner* (August 1959): 135.

11.34   Refer to Exercises 11.3 and 11.17.

   (a)   Is there sufficient evidence to state that the median sales price for new single-family houses has been increasing over the period 1972 through 1979, at the .01 level of significance?

   (b)   Estimate the expected yearly increase in the median sales price by constructing a 99% interval.

11.35   Refer to Exercise 11.4 and 11.19. Is there evidence of a linear relationship between flow-through and static LC50's? Test at the .05 significance level.

11.36   Refer to Exercise 11.4 and 11.19. Is there evidence of a linear relationship between flow-through and static LC50's? Give bounds for the attained significance level.

11.37   An experiment was conducted to investigate the effect of four factors—temperature T_1, pressure P, catalyst C, and temperature T_2—on the yield Y of a chemical.

   (a)   The values (or levels) of the four factors used in the experiment are shown in the accompanying table. If each of the four factors is coded to produce the four variables x_1, x_2, x_3, and x_4, respectively, give the transformation relating each coded variable to its corresponding original.

T_1	x_1	P	x_2	C	x_3	T_2	x_4
50	−1	10	−1	1	−1	100	−1
70	1	20	1	2	1	200	1

   (b)   Fit the linear model

$$Y = \beta_0 + \beta_1 x_1 + \beta_2 x_2 + \beta_3 x_3 + \beta_4 x_4 + \varepsilon$$

to the accompanying data.

				x_4			
				+1		−1	
				x_3		x_3	
				−1	1	−1	1
x_1	−1	x_2	−1	22.2	24.5	24.4	25.9
			1	19.4	24.1	25.2	28.4
	+1	x_2	−1	22.1	19.6	23.5	16.5
			1	14.2	12.7	19.3	16.0

   (c)   Do the data present sufficient evidence to indicate that T_1 contributes information for the estimation of Y? Does P? Does C? Does T_2? (Test the hypotheses, respectively, that $\beta_1 = 0$, $\beta_2 = 0$, $\beta_3 = 0$, and $\beta_4 = 0$.) Give bounds for the p-value associated with each test. What would you conclude if you used $\alpha = .01$ in each case?

11.8 Inferences Concerning Linear Functions of the Model Parameters

In addition to inferences about a single β_i, we frequently are interested in making inferences concerning linear functions of the model parameters β_0, $\beta_1, \ldots, \beta_k$. For example, we might wish to estimate $E(Y)$, given by

$$E(Y) = \beta_0 + \beta_1 x_1 + \cdots + \beta_k x_k$$

where $E(Y)$ might represent the mean yield of a chemical process for settings of controlled process variables $x_1, x_2, \ldots, x_k$, or the mean profit of a corporation for various investment expenditures $x_1, x_2, \ldots, x_k$. Properties of estimators of such linear functions will be given in this section.

Suppose that we wish to make an inference about the linear function

$$a_0 \beta_0 + a_1 \beta_1 + a_2 \beta_2 + \cdots + a_k \beta_k$$

where $a_0, a_1, a_2, \ldots, a_k$ are constants (some of which may equal zero). Then it is easy to see that the same linear function of the parameter estimators,

$$U = a_0 \hat{\beta}_0 + a_1 \hat{\beta}_1 + a_2 \hat{\beta}_2 + \cdots + a_k \hat{\beta}_k$$

is an unbiased estimator because, by Theorem 5.12

$$E(U) = a_0 E(\hat{\beta}_0) + a_1 E(\hat{\beta}_1) + \cdots + a_k E(\hat{\beta}_k)$$
$$= a_0 \beta_0 + a_1 \beta_1 + a_2 \beta_2 + \cdots + a_k \beta_k$$

Applying the same theorem, we find the variance of U:

$$V(U) = a_0^2 V(\hat{\beta}_0) + a_1^2 V(\hat{\beta}_1) + a_2^2 V(\hat{\beta}_2) + \cdots + a_k^2 V(\hat{\beta}_k)$$
$$+ 2a_0 a_1 \text{ cov}(\hat{\beta}_0, \hat{\beta}_1) + 2a_0 a_2 \text{ cov}(\hat{\beta}_0, \hat{\beta}_2)$$
$$+ \cdots + 2a_1 a_2 \text{ cov}(\hat{\beta}_1, \hat{\beta}_2) + \cdots + 2a_{k-1} a_k \text{ cov}(\hat{\beta}_{k-1}, \hat{\beta}_k)$$

where $V(\hat{\beta}_i) = c_{ii} \sigma^2$ and $\text{cov}(\hat{\beta}_i, \hat{\beta}_j) = c_{ij} \sigma^2$. Defining the $(k+1) \times 1$ matrix,

$$\mathbf{a} = \begin{bmatrix} a_0 \\ a_1 \\ a_2 \\ \vdots \\ a_k \end{bmatrix}$$

you may verify that $V(U)$ is given by

$$V(U) = [\mathbf{a}'(\mathbf{X}'\mathbf{X})^{-1}\mathbf{a}]\sigma^2$$

Finally, recalling that $\hat{\beta}_0, \hat{\beta}_1, \hat{\beta}_2, \ldots, \hat{\beta}_k$ are normally distributed in repeated sampling (Section 11.5), it is clear that U is a linear function of normally distributed random variables and hence itself will be normally distributed in repeated sampling.

Because U is normally distributed with

$$E(U) = \beta_0 a_0 + \beta_1 a_1 + \beta_2 a_2 + \cdots + \beta_k a_k$$

and $V(U) = [\mathbf{a}'(\mathbf{X}'\mathbf{X})^{-1}\mathbf{a}]\sigma^2$, we conclude that

$$Z = \frac{U - E(U)}{\sqrt{V(U)}} = \frac{U - E(U)}{\sigma\sqrt{\mathbf{a}'(\mathbf{X}'\mathbf{X})^{-1}\mathbf{a}}}$$

is a standard normal deviate that could be employed to test a hypothesis

$$H_0: E(U) = E_0(U)$$

when $E_0(U)$ is some specified value. Likewise, a large-sample $(1 - \alpha)100\%$ confidence interval for $E(U)$ would be

$$U \pm z_{\alpha/2}\sigma\sqrt{\mathbf{a}'(\mathbf{X}'\mathbf{X})^{-1}\mathbf{a}}$$

Furthermore, as we might suspect, if we substitute S for σ, the quantity

$$T = \frac{U - E(U)}{S\sqrt{\mathbf{a}'(\mathbf{X}'\mathbf{X})^{-1}\mathbf{a}}}$$

possesses a Student's t distribution in repeated sampling, with $[n - (k + 1)]$ degrees of freedom, and hence provides a test statistic to test the hypothesis

$$H_0: E(U) = E_0(U)$$

A Test for E(U)

H_0: $E(U) = E_0(U)$.

H_a: $\begin{cases} E(U) > E_0(U). \\ E(U) < E_0(U). \\ E(U) \neq E_0(U). \end{cases}$

Test Statistic: $T = \dfrac{U - E_0(U)}{S\sqrt{\mathbf{a}'(\mathbf{X}'\mathbf{X})^{-1}\mathbf{a}}}.$

$$\text{Rejection Region:} \quad \begin{cases} t > t_\alpha. \\ t < -t_\alpha. \\ |t| > t_{\alpha/2}. \end{cases}$$

Here t_α is based upon $[n - (k + 1)]$ degrees of freedom.

The corresponding $(1 - \alpha)100\%$ confidence interval for $E(U)$ is as follows.

A $(1 - \alpha)100\%$ confidence interval for $E(U)$

$$U \pm t_{\alpha/2} S \sqrt{\mathbf{a}'(\mathbf{X}'\mathbf{X})^{-1}\mathbf{a}}$$

As earlier, in this formula the tabulated $t_{\alpha/2}$ is based upon $[n - (k + 1)]$ degrees of freedom.

As previously indicated, one useful application of the hypothesis-testing and confidence interval techniques just presented is in the problem of estimating the mean value of Y, $E(Y)$, for fixed values of the independent variables x_1, $x_2, \ldots, x_k$. In particular, if $x_{i,0}$ denotes a particular value of x_i, for $i = 1, \ldots, k$, then

$$E(Y) = \beta_0 + \beta_1 x_{1,0} + \beta_2 x_{2,0} + \cdots + \beta_k x_{k,0}$$

Note that $E(Y)$ is a special case of $a_0 \beta_0 + a_1 \beta_1 + \cdots + a_k \beta_k$ with $a_0 = 1$ and $a_i = x_{i,0}$ for $i = 1, \ldots, k$. Thus an inference about $E(Y)$ when $x_i = x_{i,0}$, for $i = 1, \ldots, k$, can be made by using the techniques developed earlier for a general linear combination of the β's.

In the special case that we are interested in estimating the mean value for Y, $E(Y)$, for a *simple* linear regression model when the single independent variable x takes on the value x_0, it can be shown (see Exercise 11.39) that, with $\mathbf{a}' = [1, x_0]$,

$$\mathbf{a}'(\mathbf{X}'\mathbf{X})^{-1}\mathbf{a} = \frac{1}{n} + \frac{(x_0 - \bar{x})^2}{\sum (x_i - \bar{x})^2}$$

Thus for the simple linear regression model, $\mathbf{a}'(\mathbf{X}'\mathbf{X})^{-1}\mathbf{a}$ can be calculated through matrix operations or through the formula just given. Note that this representation makes it easy to see that $\mathbf{a}'(\mathbf{X}'\mathbf{X})^{-1}\mathbf{a}$ is smallest when $x_0 = \bar{x}$, the average of the x values used in the experiment.

EXAMPLE 11.8   For the data of Example 11.1, find a 90% confidence interval for $E(Y)$ when $x = 1$.

Solution   For the model of Example 11.1,

$$E(Y) = \beta_0 + \beta_1 x$$

To estimate $E(Y)$ for any fixed value $x = x_0$, we use the unbiased estimator $\hat{Y}$. Then

$$u = \hat{y} = \hat{\beta}_0 + \hat{\beta}_1 x_0 = 1 + .7 x_0$$

Hence

$$\mathbf{a} = \begin{bmatrix} a_0 \\ a_1 \end{bmatrix} = \begin{bmatrix} 1 \\ x_0 \end{bmatrix}$$

and for the case $x_0 = 1$,

$$\mathbf{a} = \begin{bmatrix} 1 \\ 1 \end{bmatrix}$$

Thus

$$V(\hat{Y}) = \mathbf{a}'(\mathbf{X}'\mathbf{X})^{-1}\mathbf{a}\sigma^2 = \begin{bmatrix} 1 & 1 \end{bmatrix} \begin{bmatrix} 1/5 & 0 \\ 0 & 1/10 \end{bmatrix} \begin{bmatrix} 1 \\ 1 \end{bmatrix} \sigma^2 = .3\sigma^2$$

In Example 11.5 we found s^2 to be .367 or $s = .606$ for these data. The value of $t_{.05}$ with $n - 2 = 3$ degrees of freedom is 2.353.
The confidence interval for $E(Y)$ is

$$\hat{y} \pm t_{\alpha/2} s \sqrt{\mathbf{a}'(\mathbf{X}'\mathbf{X})^{-1}\mathbf{a}}$$
$$[1 + (.7)(1)] \pm (2.353)(.606)\sqrt{.3}$$
$$1.7 \pm .781$$

That is, we are 90% confident that when the independent variable takes on the value $x = 1$, the mean value of the dependent variable, $E(Y)$, is between .919 and 2.481. This interval obviously is very wide, but remember that it is based on only five data points and was used solely for purposes of illustration. We will show you some practical applications of regression analysis in Section 11.12.

Exercises

11.38   Consider the general linear model

$$Y = \beta_0 + \beta_1 x_1 + \beta_2 x_2 + \cdots + \beta_k x_k + \varepsilon$$

where $E(\varepsilon) = 0$ and $V(\varepsilon) = \sigma^2$. Note that $\beta_i = \mathbf{a}'\boldsymbol{\beta}$, where the vector $\mathbf{a}$ is defined by

$$a_j = \begin{cases} 1, & \text{if } j = i \\ 0, & \text{if } j \neq i \end{cases}$$

Use this to verify that $E(\hat{\beta}_i) = \beta_i$ and $V(\hat{\beta}_i) = c_{ii}\sigma^2$, where c_{ii} is the element in row i and column i of $(\mathbf{X}'\mathbf{X})^{-1}$.

11.39   For the simple linear model $Y = \beta_0 + \beta_1 x + \varepsilon$, with $E(\varepsilon) = 0$ and $V(\varepsilon) = \sigma^2$, show that

$$V(\hat{\beta}_0 + \hat{\beta}_1 x_0) = [\mathbf{a}'(\mathbf{X}'\mathbf{X})^{-1}\mathbf{a}]\sigma^2 = \left[\frac{1}{n} + \frac{(x_0 - \bar{x})^2}{\sum (x_i - \bar{x})^2} \right]\sigma^2$$

Hint: Simplify $\mathbf{a}'(\mathbf{X}'\mathbf{X})^{-1}\mathbf{a}\sigma^2$ with

$$\mathbf{a} = \begin{bmatrix} 1 \\ x_0 \end{bmatrix}$$

For what value of x_0 will the confidence interval for $E(Y)$ attain its minimum length?

11.40   Refer to Exercise 11.13 and 11.25. Find the 90% confidence interval for the mean price per ton of fish meal if the anchovy catch is 5 million metric tons.

11.41   Using the model fit to the data of Exercise 11.4, construct a 95% confidence interval for the mean value of flow-through LC50 measurements for a toxicant that has a static LC50 of 12 parts per million. (Also see Exercise 11.19.)

11.42   Refer to Exercise 11.37. Find a 90% confidence interval for the expected yield, given that $T_1 = 50$, $P = 20$, $C = 1$, and $T_2 = 200$.

11.43   Refer to Exercise 11.1. Find a 90% confidence interval for $E(Y)$ when $x_0 = 0$. Then find 90% confidence intervals for $E(Y)$ when $x_0 = -2$ and $x_0 = +2$. Compare the lengths of these intervals. Plot these confidence limits on the graph you constructed for Exercise 11.1.

11.44   The experimenter who collected the data in Exercise 11.10 claims that the *minimum* value of $E(Y)$ occurs at $x = 1$. Test this claim at the 5% significance level. [Hint: $E(Y) = \beta_0 + \beta_1 x + \beta_2 x^2$ has its minimum at the point x_0, which satisfies the equation $\beta_1 + 2\beta_2 x_0 = 0$.]

11.45   Refer to Exercise 11.16. Find a 95% confidence interval for the mean potency of a 1-ounce portion of antibiotic stored at 65°.

11.46   Refer to Exercise 11.14. Find a 90% confidence interval for the expected proportion of survivors at time period .30.

*11.47   Refer to Exercise 11.2. Suppose the sample given there came from a large, but finite, population of inventory items. We wish to estimate the population mean of the audit values, using the fact that book values are known for every item on inventory. If the population contains N items and

$$E(Y_i) = \mu_i = \beta_0 + \beta_1 x_i$$

then the population mean is given by

$$\mu_Y = \frac{1}{N} \sum_{i=1}^{N} \mu_i = \beta_0 + \beta_1 \left(\frac{1}{N}\right) \sum_{i=1}^{N} x_i = \beta_0 + \beta_1 \mu_x$$

(a)   Using the least squares estimators of β_0 and β_1, show that μ_Y can be estimated by

$$\hat{\mu}_Y = \bar{y} + \hat{\beta}_1 (\mu_x - \bar{x})$$

(Note that $\bar{y}$ is adjusted up or down depending upon whether $\bar{x}$ is larger or smaller than μ_x.)

(b)   Using the data of Exercise 11.2 and the fact that $\mu_x = 74.0$, estimate μ_Y, the mean of the audit values, and place a two-standard-deviation bound on the error of estimation. (Regard the x_i's as constants when computing the variance of $\hat{\mu}_Y$.)

11.9 Predicting a Particular Value of Y

Suppose that the yield Y for a chemical experiment is a function of the temperature x_1 and the pressure x_2 at which the experiment is run. Assume that a linear model of the form

$$Y = \beta_0 + \beta_1 x_1 + \beta_2 x_2 + \varepsilon$$

is an adequate representation of the response surface traced by Y over the experimental region of interest. In Section 11.8 we discussed methods for estimating $E(Y)$ for a given temperature, say $x_{1,0}$, and pressure, say $x_{2,0}$. That is, we now know how to estimate the mean yield of the process, $E(Y)$, at the settings $x_{1,0}$ and $x_{2,0}$.

Now consider a different problem. Instead of estimating the mean yield at $(x_{1,0}, x_{2,0})$, we wish to *predict* the particular response Y that we will observe if the experiment is run at some time in the future (like next Monday). This situation would occur if, for some reason, the response next Monday held a special significance to us. Prediction problems frequently occur in business in which we may be interested in a particular gain associated with the investment we intend to make next month, rather than the mean gain over a long series of investments.

Note that Y is a random variable, not a parameter, and predicting its value therefore presents a departure from the stated objective of making inferences concerning population parameters. Yet if the distribution of a random variable Y is known and a single value of Y is selected at random from the population, what would you use to predict the observed value? We contend that you would select a value of Y near the *center* of the distribution, in particular, the expected value of Y. We could employ $\hat{Y}$ as a predictor of a particular value of Y as well as of $E(Y)$.

The error of predicting a particular value of Y, using $\hat{Y}$ as the predictor, is the difference between the observed value of Y and the predicted:

$$\text{error} = Y - \hat{Y}$$

Let us now investigate the properties of this error in repeated sampling.

First note that both Y and $\hat{Y}$ are normally distributed random variables and that the error is a linear function of Y and $\hat{Y}$. Then we conclude that the error is normally distributed because it is a linear function of normally distributed random variables.

Applying Theorem 5.12, which gives the formulas for the expected value and variance of a linear function of random variables, we obtain

$$E(\text{error}) = E(Y - \hat{Y}) = E(Y) - E(\hat{Y})$$

and because $E(\hat{Y}) = E(Y)$,

$$E(\text{error}) = 0$$

Likewise,

$$V(\text{error}) = V(Y - \hat{Y}) = V(Y) + V(\hat{Y}) - 2\,\text{Cov}\,(Y,\ \hat{Y})$$

Because we are predicting a future value, Y, not employed in the computation of $\hat{Y}$, it follows that Y and $\hat{Y}$ are independent and therefore that the covariance of Y and $\hat{Y}$ is equal to zero. Then

$$V(\text{error}) = V(Y) + V(\hat{Y}) = \sigma^2 + [\mathbf{a}'(\mathbf{X}'\mathbf{X})^{-1}\mathbf{a}]\sigma^2$$
$$= \sigma^2[1 + \mathbf{a}'(\mathbf{X}'\mathbf{X})^{-1}\mathbf{a}]$$

We have shown that the error of predicting a particular value of Y is normally distributed with a mean and a variance equal to zero and $\sigma^2[1 + \mathbf{a}'(\mathbf{X}'\mathbf{X})^{-1}\mathbf{a}]$, respectively. It follows that

$$Z = \frac{Y - \hat{Y}}{\sigma\sqrt{1 + \mathbf{a}'(\mathbf{X}'\mathbf{X})^{-1}\mathbf{a}}}$$

has a standard normal distribution. Furthermore, if S is substituted for σ, it can be shown that

$$T = \frac{Y - \hat{Y}}{S\sqrt{1 + \mathbf{a}'(\mathbf{X}'\mathbf{X})^{-1}\mathbf{a}}}$$

possesses a Student's t distribution with $[n - (k + 1)]$ degrees of freedom. We will use this result to place a bound on the error of prediction and, in doing so, will construct a prediction interval for the random variable Y. The procedure employed will be similar to that used to construct all the confidence intervals presented in the preceding chapters.

We begin by observing that

$$P(-t_{\alpha/2} < T < t_{\alpha/2}) = 1 - \alpha$$

Substituting for T, we obtain

$$P\left[-t_{\alpha/2} < \frac{Y - \hat{Y}}{S\sqrt{1 + \mathbf{a}'(\mathbf{X}'\mathbf{X})^{-1}\mathbf{a}}} < t_{\alpha/2}\right] = 1 - \alpha$$

In other words, in repeated sampling the inequality within the brackets will hold with probability equal to $(1 - \alpha)$. Furthermore, the inequality will continue to hold with the same probability if each term is multiplied by the same factor or if the same quantity is added to each term of the inequality. Hence multiply each term by $S\sqrt{1 + \mathbf{a}'(\mathbf{X}'\mathbf{X})^{-1}\mathbf{a}}$ and then add $\hat{Y}$ to each. The result,

$$P[\hat{Y} - t_{\alpha/2}S\sqrt{1 + \mathbf{a}'(\mathbf{X}'\mathbf{X})^{-1}\mathbf{a}} < Y < \hat{Y} + t_{\alpha/2}S\sqrt{1 + \mathbf{a}'(\mathbf{X}'\mathbf{X})^{-1}\mathbf{a}}] = 1 - \alpha$$

places an interval about $\hat{Y}$ that constitutes a $(1 - \alpha)100\%$ prediction interval for Y.

A $(1 - \alpha)100\%$ Prediction Interval for Y

$$\hat{y} \pm t_{\alpha/2}S\sqrt{1 + \mathbf{a}'(\mathbf{X}'\mathbf{X})^{-1}\mathbf{a}}$$

Thinking in terms of a bound on the error of predicting Y, we would expect the error to be less in absolute value than

$$t_{\alpha/2}S\sqrt{1 + \mathbf{a}'(\mathbf{X}'\mathbf{X})^{-1}\mathbf{a}}$$

with probability equal to $(1 - \alpha)$. We will illustrate with an example.

EXAMPLE 11.9 Suppose the experiment that generated the data of Example 11.1 is to be run again with $x = 2$. Predict the particular value of Y with $1 - \alpha = .90$.

Solution   From Example 11.1 we have that

$$\hat{y} = 1 + .7x$$

and the predicted value of Y with $x = 2$ then is

$$\hat{y} = 1 + (.7)(2) = 2.4$$

In this case $\mathbf{a} = \begin{bmatrix} 1 \\ 2 \end{bmatrix}$ and so

$$\mathbf{a}'(\mathbf{X}'\mathbf{X})^{-1}\mathbf{a} = \begin{bmatrix} 1 & 2 \end{bmatrix} \begin{bmatrix} 1/5 & 0 \\ 0 & 1/10 \end{bmatrix} \begin{bmatrix} 1 \\ 2 \end{bmatrix} = .6$$

From Example 11.5, $s = .606$. The $t_{.05}$ value with 3 degrees of freedom is 2.353. Thus the prediction interval becomes

$$\hat{y} \pm t_{\alpha/2}s\sqrt{1 + \mathbf{a}'(\mathbf{X}'\mathbf{X})^{-1}\mathbf{a}}$$
$$2.4 \pm (2.353)(.606)\sqrt{1 + .6}$$
$$2.4 \pm 1.804$$

§

Exercises

11.48 Suppose the model $Y = \beta_0 + \beta_1 x + \varepsilon$ is fit to the n data points $(y_1, x_1), \ldots, (y_n, x_n)$. At what value of x will the length of the prediction interval for Y be minimized? [See the form of $\mathbf{a}'(\mathbf{X}'\mathbf{X})^{-1}\mathbf{a}$ given in Exercise 11.39.]

11.49 Refer to Exercises 11.3 and 11.17. Use the data and model given there to construct a 95% prediction interval for the median sales price in 1980.

11.50 Refer to Exercise 11.49. Find a 95% prediction interval for the median sales price for the year 1981. Repeat for 1982. Would you feel confortable in using this model and the data of Exercise 11.3 to predict the median sales price for the year 1988?

11.51 Refer to Exercises 11.4 and 11.19. Find a 95% prediction interval for a flow-through LC50 if the static LC50 is observed to be 12 parts per million. Compare the length of this interval to the interval found in Exercise 11.41.

11.52 Refer to Exercise 11.11. Find a 98% prediction interval for the incidence rate of infectious hepatitis in 1977. Use the quadratic model.

11.53 Refer to Exercise 11.16. Find a 95% prediction interval for the potency of a 1-ounce portion of antibiotic stored at 65°. Compare this interval to that calculated in Exercise 11.45.

11.54   Refer to Exercise 11.14. Find a 95% prediction interval for the proportion of survivors at time $x = .60$.

11.55   Refer to Exercise 11.37. Find a 90% prediction interval for Y, given that $T_1 = 50$, $P = 20$, $C = 1$, and $T_2 = 200$.

11.10 A Test Statistic to Test H_0: $\beta_{g+1} = \beta_{g+2} = \cdots = \beta_k = 0$

In seeking an intuitively appealing test statistic to test a hypothesis concerning a set of parameters of the linear model, we are led to a consideration of the sum of squares of deviations, SSE. Suppose, for example, that we were to fit a model,

$$\text{Model 1: } Y = \beta_0 + \beta_1 x_1 + \beta_2 x_2 + \cdots + \beta_g x_g + \varepsilon$$

to the data and then calculate the sum of squares of deviations of the observed values of Y, SSE_1. Having done this, we then fit the linear model,

$$\text{Model 2: } Y = \beta_0 + \beta_1 x_1 + \beta_2 x_2 + \cdots + \beta_g x_g$$
$$+ \beta_{g+1} x_{g+1} + \cdots + \beta_k x_k + \varepsilon$$

which contains all the terms of Model 1 plus the terms involving x_{g+1}, $x_{g+2}, \ldots, x_k$ (note that $k > g$). Then we calculate the sum of squares of deviations for this model, SSE_2. Finally, let us suppose that $x_{g+1}, x_{g+2}, \ldots, x_k$ really contribute a substantial quantity of information for the prediction of Y not contained in the variables $x_1, x_2, \ldots, x_g$ (that is, at least one of the parameters $\beta_{g+1}, \beta_{g+2}, \ldots, \beta_k$ actually differs from zero). What would be the relation between SSE_1 and SSE_2? Intuitively, we see that if $x_{g+1}, x_{g+2}, \ldots, x_k$ are important information-contributing variables, Model 2, which contains all the variables of Model 1 plus the additions $x_{g+1}, x_{g+2}, \ldots, x_k$, should predict with a *smaller* error of prediction than Model 1 and hence SSE_2 should be less then SSE_1. The greater the difference, $(\text{SSE}_1 - \text{SSE}_2)$, the stronger will be the evidence to support the alternative hypothesis that $x_{g+1}, x_{g+2}, \ldots, x_k$ contribute information for the prediction of Y and to reject the null hypothesis,

$$H_0: \beta_{g+1} = \beta_{g+2} = \cdots = \beta_k = 0$$

Models 1 and 2 are known as the *reduced* and *complete* models, respectively, for a test of the null hypothesis indicated earlier. The decrease in the sum of squares of deviations, $(\text{SSE}_1 - \text{SSE}_2)$, is called the *sum of squares associated with the variables* $x_{g+1}, x_{g+2}, \ldots, x_k$, *adjusted for the variables* $x_1, x_2, x_3, \ldots, x_g$.
Thus large values of $(\text{SSE}_1 - \text{SSE}_2)$ would lead us to reject the hypothesis

$$H_0: \beta_{g+1} = \beta_{g+2} = \cdots = \beta_k = 0$$

How large is "large"? We will develop a test statistic that is a function of $(SSE_1 - SSE_2)$ for which we know the distribution when H_0 is true.

To acquire this test statistic, let us *assume* that the null hypothesis is true and then examine the quantities that we have calculated. Particularly, note that

$$SSE_1 \doteq SSE_2 + (SSE_1 - SSE_2)$$

In other words, we have partitioned SSE_1 into two parts, SSE_2 and the difference $(SSE_1 - SSE_2)$, as indicated in Figure 11.4. Further, if H_0 is true $(\beta_{g+1} = \beta_{g+2} = \cdots = \beta_k = 0)$, then

$$S_1^2 = \frac{SSE_1}{n - (g + 1)}$$

Figure 11.4
Partitioning SSE_1

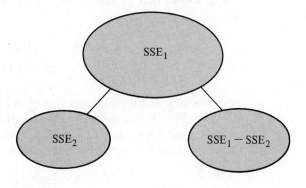

is an unbiased estimator of σ^2, the variance of the random error ε. Also, it can be shown that

$$S_2^2 = \frac{SSE_2}{n - (k + 1)} \quad \text{and} \quad S_3^2 = \frac{SSE_1 - SSE_2}{k - g}$$

are unbiased estimators of σ^2 that are statistically independent.

Consider the ratio

$$F = \frac{S_3^2}{S_2^2}$$

If $H_0: \beta_{g+1} = \beta_{g+2} = \cdots = \beta_k = 0$ is true, then S_3^2 and S_2^2 will be the same relative magnitude and F will assume a value near 1. If H_0 is false, then S_2^2 will still provide an unbiased estimator of σ^2, but S_3^2 will be inflated and possess an expected value equal to σ^2 plus a positive quantity involving the squares of the parameters $\beta_{g+1}, \beta_{g+2}, \ldots, \beta_k$. The larger the value of $SSE_1 - SSE_2$, the more inflated S_3^2 will be in relation to S_2^2 and the greater the evidence favoring rejection of H_0. Thus large values of F favor a rejection of H_0.

The definition of a random variable with an F distribution is given in Definition 7.3. Although we omit the proof, if H_0 is true, then

$$\chi_3^2 = \frac{[n - (g + 1)]S_1^2}{\sigma^2} = \frac{\text{SSE}_1}{\sigma^2}$$

$$\chi_2^2 = \frac{[n - (k + 1)]S_2^2}{\sigma^2} = \frac{\text{SSE}_2}{\sigma^2}$$

$$\chi_1^2 = \frac{(k - g)S_3^2}{\sigma^2} = \frac{\text{SSE}_1 - \text{SSE}_2}{\sigma^2}$$

possess chi-square probability distributions in repeated sampling, with $[n - (g + 1)]$, $[n - (k + 1)]$, and $(k - g)$ degrees of freedom, respectively. Further, it can be shown that χ_2^2 and χ_1^2 are statistically independent. Thus the resulting ratio

$$\frac{S_3^2}{S_2^2} = \frac{\chi_1^2/(k - g)}{\chi_2^2/(n - [k + 1])}$$

possesses an F distribution with $v_1 = (k - g)$ numerator degrees of freedom and $v_2 = (n - [k + 1])$ denominator degrees of freedom.

Because large values of F indicate that H_0 should be rejected, if we desire a test with a type-I-error probability equal to α, it follows that

$$F > F_\alpha$$

is the appropriate rejection region. (See Table 7, Appendix III.)

EXAMPLE 11.10   Do the data of Example 11.3 provide sufficient evidence to indicate that the second-order model

$$Y = \beta_0 + \beta_1 x + \beta_2 x^2 + \varepsilon$$

contributes information for the prediction of y? That is, test the hypothesis $H_0: \beta_1 = \beta_2 = 0$ against the alternative hypothesis H_a: at least one of the parameters, β_1, β_2, differs from 0. Use $\alpha = .05$. Give bounds for the attained significance level.

Solution   For the complete model, we determined in Example 11.6 that $\text{SSE}_2 = .463$. Because we want to test $H_0: \beta_1 = \beta_2 = 0$, the appropriate reduced model is

$$Y = \beta_0 + \varepsilon$$

for which

$$\mathbf{Y} = \begin{bmatrix} 0 \\ 0 \\ 1 \\ 1 \\ 3 \end{bmatrix} \quad \text{and} \quad \mathbf{X} = \overset{x_0}{\begin{bmatrix} 1 \\ 1 \\ 1 \\ 1 \\ 1 \end{bmatrix}}$$

Because $\mathbf{X'X} = 5$, then $(\mathbf{X'X})^{-1} = 1/5$, and $\hat{\beta} = (\mathbf{X'X})^{-1}\mathbf{X'Y} = (1/5)\sum_{i=1}^{5} y_i = \bar{y} = 5/5 = 1$. Thus

$$\text{SSE}_1 = \mathbf{Y'Y} - \hat{\beta}'\mathbf{X'Y}$$

$$= \sum_{i=1}^{5} y_i^2 - \bar{y}\left(\sum_{i=1}^{n} y_1\right) = \sum_{i=1}^{5} y_i^2 - \frac{1}{n}\left(\sum_{i=1}^{5} y_i\right)^2$$

$$= 11 - (1/5)(5)^2 = 11 - 5 = 6$$

Note that in this case the number of independent variables in the complete model is $k = 2$, whereas the number of independent variables in the reduced model is $g = 0$. Thus

$$S_2^2 = \frac{\text{SSE}_2}{n - (k+1)} = \frac{.463}{5 - (3)} = .232$$

and

$$S_3^2 = \frac{\text{SSE}_1 - \text{SSE}_2}{k - g} = \frac{6 - .463}{2 - 0} = 2.768$$

Finally,

$$F = \frac{S_3^2}{S_2^2} = \frac{2.768}{.232} = 11.931$$

The tabulated F value for $\alpha = .05$ with $v_1 = k - g = 2$ numerator degrees of freedom and $v_2 = n - (k+1) = 2$ denominator degrees of freedom is 19.00. Hence the observed value of the test statistic does not fall in the rejection region and we conclude that, at the $\alpha = .05$ level, there is not enough evidence to claim that either β_1 and β_2 differs from zero. Because the proper form of the rejection region is $F > F_\alpha$, the p-value is given by $P(F > 11.931)$ when F is based on 2 numerator and 2 denominator degrees of freedom. Using Table 7, Appendix III, it can be seen that $.05 < p\text{-value} < .10$. Thus if we chose $\alpha = .05$ (in agreement with the previous discussion) there is not enough evidence to claim that either β_1

of β_2 differs from zero. However, if an α value of .10 were selected, we can claim either $\beta_1 \neq 0$ or $\beta_2 \neq 0$. Note the little additional effort required to place bounds on the p-value provides a considerable amount of additional information.

∽

Another application of the method developed in this section is found in Section 11.12.

Exercises

11.56   Refer to Exercise 11.33. Answer the question on the decreasing rate of tuberculosis by constructing an F test.

11.57   Refer to Exercise 11.11.

(a)   For the quadratic model carry out an F test of $H_0: \beta_2 = 0$, using $\alpha = .05$. Compare to the test in Exercise 11.27.

(b)   Test $H_0: \beta_1 = \beta_2 = 0$ at the 5% significance level.

11.58   Refer to Exercise 11.37. Test the hypothesis, at the 5% level of significance, that neither T_1 nor T_2 affects the yield.

11.59   Utility companies, which must plan the operation and expansion of electricity generation, are vitally interested in predicting customer demand over both short and long periods of time. A short-term study was conducted to investigate the effect of mean monthly daily temperature x_1 and cost per kilowatt x_2 on the mean daily consumption (KWH) per household. The company expected the demand for electricity to rise in cold weather (due to heating), fall when the weather was moderate, and rise again when the temperature rose and there was a need for air conditioning. They expected demand to decrease as the cost per kilowatt hour increased, reflecting greater attention to conservation. Data were available for 2 years, a period in which the cost per kilowatt hour, x_2, increased due to the increasing costs of fuel. The company fitted the model,

$$Y = \beta_0 + \beta_1 x_1 + \beta_2 x_1^2 + \beta_3 x_2 + \beta_4 x_1 x_2 + \beta_5 x_1^2 x_2 + \varepsilon$$

to the data that follows and obtained $\hat{Y} = 325.606 - 11.383 x_1 + .113 x_1^2 - 21.699 x_2 + .873 x_1 x_2 - .009 x_1^2 x_2$ with SSE = 152.177.

x_2, Price per KWH	Mean Daily Consumption (KWH) per Household											
8¢ Mean daily temperature (°F), x_1	31	34	39	42	47	56	62	66	68	71	75	78
Mean daily consumption, y	55	49	46	47	40	43	41	46	44	51	62	73
10¢ Mean daily temperature, x_1	32	36	39	42	48	56	62	66	68	72	75	79
Mean daily consumption, y	50	44	42	42	38	40	39	44	40	44	50	55

When the model $Y = \beta_0 + \beta_1 x_1 + \beta_2 x_1^2 + \varepsilon$ was fit, the prediction equation was $Y = 130.009 - 3.302 x_1 + .033 x_1^2$ with SSE = 465.134. Test whether the terms involving x_2 $(x_2, x_1 x_2, x_1^2 x_2)$ contribute to a significantly better fit of the model to the data. Give bounds for the attained significance level.

11.11 Correlation

The previous sections of this chapter dealt with modeling a response Y as a linear function of a nonrandom variable x, so that appropriate inferences could be made concerning the expected value of Y, or a future value of Y, for a given value of x. These models are useful in two quite different practical situations.

First, the variable x may be completely controlled by the experimenter. This occurs, for example, if x is the temperature setting and Y the yield in a chemical experiment. Then x is merely the point at which the temperature dial is set when the experiment is run. Of course, x could vary from experiment to experiment, but it is under the complete control, practically speaking, of the experimenter. The linear model

$$Y = \beta_0 + \beta_1 x + \varepsilon$$

then implies that

$$E(Y) = \beta_0 + \beta_1 x$$

or the average yield is a linear function of the temperature setting.

Second, the variable x may be an observed value of a random variable X. For example, we may want to relate the volume of usable timber in a tree Y to the circumference of the base X. If a functional relationship could be established, then in the future we could predict the amount of timber in any tree by simply measuring the circumference of the base. For this situation we use the model

$$Y = \beta_0 + \beta_1 x + \varepsilon$$

to imply that

$$E(Y \mid X = x) = \beta_0 + \beta_1 x$$

That is, we are assuming that the conditional expectation of Y for a fixed value of X is a linear function of the x value. We generally assume that the vector random variable, (X, Y), has a bivariate normal distribution (see Section 5.10), in which case it can be shown that

$$E(Y \mid X = x) = \beta_0 + \beta_1 x$$

The statistical theory for making inferences about the parameters β_0 and β_1 is exactly the same for both of these cases, but the differences in model interpretation should be kept in mind.

For the case where (X, Y) has a bivariate distribution, the experimenter may not always be interested in the linear relationship defining $E(Y|X)$. He or she may want to know only whether X and Y are *independent* random variables. If (X, Y) has a bivariate normal distribution, then testing for independence is equivalent to testing that the correlation coefficient, ρ, is equal to zero. Recall from Section 5.7 that ρ is positive if X and Y tend to increase together and ρ is negative if Y decreases as X increases.

Let $(X_1, Y_1), (X_2, Y_2), \ldots, (X_n, Y_n)$ denote a random sample from a bivariate normal distribution. The maximum-likelihood estimator of ρ is given by the sample correlation coefficient

$$r = \frac{\sum_{i=1}^{n}(X_i - \bar{X})(Y_i - \bar{Y})}{\sqrt{\sum_{i=1}^{n}(X_i - \bar{X})^2 \sum_{i=1}^{n}(Y_i - \bar{Y})^2}}$$

or an equivalent expression

$$r = \frac{n\sum_{i=1}^{n} X_i Y_i - \sum_{i=1}^{n} X_i \sum_{i=1}^{n} Y_i}{\sqrt{\left[n\sum_{i=1}^{n} X_i^2 - \left(\sum_{i=1}^{n} X_i\right)^2\right]\left[n\sum_{i=1}^{n} Y_i^2 - \left(\sum_{i=1}^{n} Y_i\right)^2\right]}}$$

Note that the numerator of r is exactly the same as the numerator of the estimator for β_1 derived in Section 11.3. Because the denominators of r and $\hat{\beta}_1$ are both nonnegative, it follows that r and $\hat{\beta}_1$ have the same sign.

It would seem natural to use r as a test statistic to test hypotheses about ρ, but difficulties arise because the probability distribution for r is difficult to obtain. This difficulty can be overcome, for moderately large samples, by using the fact that $(1/2) \ln [(1 + r)/(1 - r)]$ is approximately normally distributed with mean $(1/2) \ln [(1 + \rho)/(1 - \rho)]$ and variance $1/(n - 3)$. Thus for testing the hypothesis $H_0: \rho = \rho_0$, we can employ a z test in which

$$Z = \frac{(1/2) \ln\left(\dfrac{1 + r}{1 - r}\right) - (1/2) \ln\left(\dfrac{1 + \rho_0}{1 - \rho_0}\right)}{\dfrac{1}{\sqrt{n - 3}}}$$

If α is the desired probability of a type I error, the form of the rejection region depends upon the alternative hypothesis. The various alternatives of most frequent interest and the corresponding rejection regions are as follows:

$$H_a: \rho > \rho_0. \qquad RR: z > z_\alpha.$$
$$H_a: \rho < \rho_0. \qquad RR: z < -z_\alpha.$$
$$H_a: \rho \neq \rho_0. \qquad RR: |z| > z_{\alpha/2}.$$

We illustrate with an example.

EXAMPLE 11.11   The data in Table 11.3 represent a sample of mathematics achievement test scores and calculus grades for ten independently selected college freshmen. From this evidence, would you say the achievement test scores and calculus grades are independent? Use $\alpha = .05$. Give the corresponding attained significance level.

Table 11.3
Data for
Example 11.11

Student	Mathematics Achievement Test Score	Final Calculus Grade
1	39	65
2	43	78
3	21	52
4	64	82
5	57	92
6	47	89
7	28	73
8	75	98
9	34	56
10	52	75

Solution   Denoting achievement test scores by x and calculus grades by y, we calculate

$$\sum_{i=1}^{10} x_i = 460 \qquad \sum_{i=1}^{10} y_i = 760$$

$$\sum_{i=1}^{10} x_i^2 = 23,634 \qquad \sum_{i=1}^{10} y_i^2 = 59,816$$

$$\sum_{i=1}^{10} x_i y_i = 36,854$$

Thus

$$r = \frac{(10)(36,854) - (460)(760)}{\sqrt{[(10)(23,634) - (460)^2][(10)(59,816) - (760)^2]}} = .84$$

We state as the null hypothesis that X and Y are independent, or, assuming (X, Y) has a bivariate normal distribution, we test $H_0: \rho = 0$ versus $H_a: \rho \neq 0$. The value of the test statistic is

$$z = \frac{(1/2) \ln \left(\dfrac{1 + r}{1 - r}\right) - (1/2) \ln \left(\dfrac{1 + \rho_0}{1 - \rho_0}\right)}{1/\sqrt{n - 3}} = \frac{(1/2) \ln \left(\dfrac{1 + .84}{1 - .84}\right) - 0}{1/\sqrt{7}}$$

$$= 3.231$$

Because $z_{\alpha/2} = z_{.025} = 1.96$, our observed value of the test statistic lies in the rejection region. Thus the evidence strongly suggests that achievement test scores and calculus grades are dependent. Note that $\alpha = .05$ is the probability that our test statistic will fall in the rejection region when H_0 is true. Hence we are fairly confident that we have made a correct decision.

Because we are implementing a two-tailed test, p-value $= 2P(Z > 3.231)$. Based on the values contained in Table 4, Appendix III, it follows that $P(Z > 3.231) < P(Z > 3.00) = .001$. Thus p-value $< 2(.001) = .002$ and for any value of α greater than .002 (including $\alpha = .05$, as used in the initial part of this analysis) we would conclude that $\rho \neq 0$. ∽

Exercises

11.60 Given in the table are the peak power load for a power plant and the high temperature for a random sample of 10 days. Test the hypothesis that the population correlation coefficient ρ between peak power load and high temperature is zero versus the alternative that it is positive. Use $\alpha = .05$. Give the attained significance level.

Day	High Temperature	Peak Load
1	95	214
2	82	152
3	90	156
4	81	129
5	99	254
6	100	266
7	93	210
8	95	204
9	93	213
10	87	150

11.61 In Exercise 11.4 both the flow-through and static LC50 values could be considered random variables. Using the data of Exercise 11.4, test to see if the correlation between static and flow-through values is significantly different from zero. Use $\alpha = .01$. Give the associated p-value.

*11.62   Refer to Exercise 11.61. Suppose that independent tests, with the same toxicants and species but in a different laboratory, showed $r = .85$ with $n = 20$. Test the hypothesis that the two correlation coefficients between static and flow-through measurements are equal. Use $\alpha = .05$.

11.12 Some Practical Examples

In this section we present three examples to illustrate the applicability of the previously developed techniques to real data. Most of the methods will be illustrated somewhere in the course of the discussions. We make no attempt to implement every method for each example.

EXAMPLE 11.12   In his Ph.D. thesis H. Behbahani examined the effect of varying the water/cement ratio on the strength of concrete that had been aged 28 days. For concrete with a cement content of 200 pounds per cubic yard, he obtained the data in Table 11.4. Let Y denote the strength and x denote the water/cement ratio.

Table 11.4
Data for
Example 11.12

Water/Cement Ratio	Strength (100 feet/pound)
1.21	1.302
1.29	1.231
1.37	1.061
1.46	1.040
1.62	.803
1.79	.711

Source Data adapted from Hamid Behbahani, "Econo-crete—Design and Properties" (Ph.D. thesis, University of Florida, 1977), p. 95.

(a)   Fit the model $E(Y) = \beta_0 + \beta_1 x$.

(b)   Test $H_0: \beta_1 = 0$ versus $H_a: \beta_1 < 0$ with $\alpha = .05$. (Note that if H_0 is rejected, we conclude that $\beta_1 < 0$ and that the strength tends to decrease with an increase in water/cement ratio.) Give the corresponding attained significance level.

(c)   Find a 90% confidence interval for the expected strength of concrete when the water/cement ratio is 1.5. What will happen to the confidence interval if we try to estimate mean strengths for water/cement ratios of .3 or 2.7?

Solution   (a)   (Throughout this example, all calculations are carried out to 3 decimal places.) Using the formulas developed in Section 11.3, we have

$$\hat{\beta}_1 = \frac{n \sum_{i=1}^{n} x_i y_i - \sum_{i=1}^{n} x_i \sum_{i=1}^{n} y_i}{n \sum_{i=1}^{n} x_i^2 - \left(\sum_{i=1}^{n} x_i\right)^2} = \frac{6(8.709) - (8.74)(6.148)}{6(12.965) - (8.74)^2} = -1.056$$

and

$$\hat{\beta}_0 = \bar{y} - \hat{\beta}_1 \bar{x} = \frac{6.148}{6} - (-1.056)\left(\frac{8.74}{6}\right) = 2.563$$

Thus the straight-line model that best fits the data is

$$\hat{y} = 2.563 - 1.056x$$

(b)   Because we desire to test whether there is evidence that $\beta_1 < 0$ with $\alpha = .05$, the appropriate test statistic is

$$t = \frac{\hat{\beta}_1 - 0}{S\sqrt{c_{11}}} \quad \text{or} \quad t = \frac{\hat{\beta}_1 - 0}{S\sqrt{\dfrac{1}{\sum_{i=1}^{n}(x_i - \bar{x})^2}}}$$

For a simple linear regression model it can be shown (see Exercise 11.15) that

$$\text{SSE} = \sum_{i=1}^{n}(y_i - \hat{y}_i)^2 = \sum_{i=1}^{n}(y_i - \bar{y})^2 - \hat{\beta}_1 \sum_{i=1}^{n}(x_i - \bar{x})(y_i - \bar{y})$$

For the data in this example,

$$\sum_{i=1}^{n}(y_i - \bar{y})^2 = \sum_{i=1}^{n} y_i^2 - \frac{1}{n}\left(\sum_{i=1}^{n} y_i\right)^2$$

$$= 6.569 - \left(\frac{1}{6}\right)(6.148)^2 = .269$$

and

$$\sum_{i=1}^{n}(x_i - \bar{x})(y_i - \bar{y}) = \sum_{i=1}^{n} x_i y_i - \frac{1}{n}\left(\sum_{i=1}^{n} x_i\right)\left(\sum_{i=1}^{n} y_i\right)$$

$$= 8.709 - \left(\frac{1}{6}\right)(8.74)(6.148) = -.247$$

Therefore,

$$SSE = .269 - (-1.056)(-.247) = .008$$

and hence

$$s = \sqrt{s^2} = \sqrt{\frac{SSE}{n-2}} = \sqrt{\frac{.008}{4}} = .045$$

Also

$$\sum_{i=1}^{n} (x_i - \bar{x})^2 = \sum_{i=1}^{n} x_i^2 - \frac{1}{n}\left(\sum_{i=1}^{n} x_i\right)^2 = .234$$

Thus the value of the appropriate test statistic for testing $H_0: \beta_1 = 0$ versus $H_a: \beta_1 < 0$ is

$$t = \frac{-1.056 - 0}{.045\sqrt{1/.234}} = -11.355$$

Because this statistic is based on $n - 2 = 4$ degrees of freedom, and the appropriate rejection region is $t < -t_{.05} = -2.132$, we reject H_0 in favor of H_a at the $\alpha = .05$ level of significance. Because the appropriate test is a lower-tail test, p-value $= P(t < -11.355)$ where t has a t distribution with 4 degrees of freedom. Thus, p-value $< .005$. In fact, more extensive tables of the t-distribution indicate that the p-value is considerably less than .005. That is, for most commonly used values of α we conclude that there is evidence to indicate that strength decreases with an increase in the water/cement ratio *on the region where the experiment was conducted.* From a practical point of view, the water/cement ratio must be large enough to moisten the cement, sand, and so on that make up concrete. If the water/cement ratio gets too large, the concrete will be useless.

(c) Because the model we are using is a simple linear regression model, the confidence interval can be obtained from the formula

$$\hat{y} \pm t_{\alpha/2} S \sqrt{\frac{1}{n} + \frac{(x_0 - \bar{x})^2}{\sum_{i=1}^{n} (x_i - \bar{x})^2}}$$

We want a confidence interval when $x = 1.5$; therefore, $x_0 = 1.5$ and

$$\hat{y} = 2.563 - (1.056)(1.5) = .979$$

Using calculations from parts (a) and (b), we obtain the 90% confidence interval of

$$.979 \pm (2.132)(.045)\sqrt{\frac{1}{6} + \frac{(1.5 - 1.457)^2}{.234}}$$

or

$$(.938, 1.020)$$

Thus we would estimate the mean strength of concrete with a water/cement ratio of 1.5 to be between .938 and 1.020.

We can see from the variance expression that the confidence interval gets wider as x_0 gets farther from $\bar{x} = 1.457$. Also, the values $x_0 = .3$ and $x_0 = 2.7$ are far from the values that were used in the experiment. Considerable caution should be used before constructing a confidence interval for $E(Y)$ when the values of x_0 are far removed from the experimental region. Water/cement ratios of .3 and 2.7 would probably yield concrete that is completely useless!   ∽

The next example we consider involves the use of a multiple linear regression model.

EXAMPLE 11.13   It is desired to relate abrasion resistance of rubber (Y) to the amount of silica filler x_1' and the amount of coupling agent x_2'. Fine-particle silica fibers are added to rubber to increase strength and resistance to abrasion. The coupling agent chemically bonds the filler to the rubber polymer chains and thus increases the efficiency of the filler. The unit of measurement for x_1' and x_2' is parts per 100 parts of rubber, which is denoted phr. For computational simplicity the actual amounts of silica filler and coupling agent are rescaled by the equations

$$x_1 = \frac{x_1' - 50}{6.7} \quad \text{and} \quad x_2 = \frac{x_2' - 4}{2}$$

(Such rescaling of the independent variables does not affect the analysis or conclusions, but it does simplify computations.)

The data from an actual experiment are given in Table 11.5. Notice that five levels of both x_1 and x_2 are used, with the $(x_1 = 0, x_2 = 0)$ point repeated three times. Let us fit the second-order model

$$Y = \beta_0 + \beta_1 x_1 + \beta_2 x_2 + \beta_3 x_1^2 + \beta_4 x_2^2 + \beta_5 x_1 x_2 + \varepsilon$$

Table 11.5
Data for
Example 11.13

y	x_1	x_2
83	1	-1
113	1	1
92	-1	1
82	-1	-1
100	0	0
96	0	0
98	0	0
95	0	1.5
80	0	-1.5
100	1.5	0
92	-1.5	0

Source Ronald   Suich
and   G.   C.   Derringer,
Technometrics, 19, no. 2
(May 1977): 214.

to these data. This model represents a conic surface over the (x_1, x_2) plane. Fit the second-order model and test $H_0: \beta_3 = \beta_4 = \beta_5 = 0$. (We are testing that the surface is actually a plane versus the alternative that it is a conic surface.) Give bounds for the attained significance level and indicate the proper conclusion if we choose $\alpha = .05$.

Solution

We will first use matrix equations to fit the complete model, as indicated earlier. (With models of this size it is best to use a computer to implement the computations.) We have, for the data in Table 11.5,

$$
Y = \begin{bmatrix} 83 \\ 113 \\ 92 \\ 82 \\ 100 \\ 96 \\ 98 \\ 95 \\ 80 \\ 100 \\ 92 \end{bmatrix}
\qquad
X = \begin{array}{c} \begin{array}{cccccc} & x_1 & x_2 & x_1^2 & x_2^2 & x_1 x_2 \end{array} \\ \begin{bmatrix} 1 & 1 & -1 & 1 & 1 & -1 \\ 1 & 1 & 1 & 1 & 1 & 1 \\ 1 & -1 & 1 & 1 & 1 & -1 \\ 1 & -1 & -1 & 1 & 1 & 1 \\ 1 & 0 & 0 & 0 & 0 & 0 \\ 1 & 0 & 0 & 0 & 0 & 0 \\ 1 & 0 & 0 & 0 & 0 & 0 \\ 1 & 0 & 1.5 & 0 & 2.25 & 0 \\ 1 & 0 & -1.5 & 0 & 2.25 & 0 \\ 1 & 1.5 & 0 & 2.25 & 0 & 0 \\ 1 & -1.5 & 0 & 2.25 & 0 & 0 \end{bmatrix} \end{array}
$$

and

$$(\mathbf{X'X})^{-1} = \begin{bmatrix} 0.33 & 0 & 0 & -0.15 & -0.15 & 0 \\ 0 & 0.12 & 0 & 0 & 0 & 0 \\ 0 & 0 & 0.12 & 0 & 0 & 0 \\ -0.15 & 0 & 0 & 0.15 & 0.05 & 0 \\ -0.15 & 0 & 0 & 0.05 & 0.15 & 0 \\ 0 & 0 & 0 & 0 & 0 & 0.25 \end{bmatrix}$$

These matrices yield

$$\hat{\boldsymbol{\beta}} = (\mathbf{X'X})^{-1}\mathbf{X'Y} = \begin{bmatrix} 98.00 \\ 4.00 \\ 7.35 \\ -0.88 \\ -4.66 \\ 5.00 \end{bmatrix}$$

or the fitted second-order model,

$$\hat{y} = 98.00 + 4.00x + 7.35x_2 - .88x_1^2 - 4.66x_2^2 + 5.00x_1x_2$$

For this model $SSE_2 = \mathbf{Y'Y} - \hat{\boldsymbol{\beta}}'\mathbf{X'Y} = 77.948$.

To test the hypothesis of interest $(H_0: \beta_3 = \beta_4 = \beta_5 = 0)$, we must fit the reduced model

$$Y = \beta_0 + \beta_1 x_1 + \beta_2 x_2 + \varepsilon$$

By deleting the columns for x_1^2, x_2^2, and $x_1 x_2$ in the $\mathbf{X}$ matrix, we have

$$\hat{\boldsymbol{\beta}} = (\mathbf{X'X})^{-1}\mathbf{X'Y} = \begin{bmatrix} 93.73 \\ 4.00 \\ 7.35 \end{bmatrix}$$

and the fitted planar model is

$$\hat{y} = 93.73 + 4.00x_1 + 7.35x_2$$

(Notice that we cannot simply set $\hat{\beta}_3$, $\hat{\beta}_4$, and $\hat{\beta}_5$ equal to zero to produce the fitted model in the reduced case.) For the reduced model $SSE_1 = 326.623$.

The hypothesis $H_0: \beta_3 = \beta_4 = \beta_5 = 0$ is now tested by calculating F (note that $k = 5$, $g = 2$, and $n = 11$):

$$F = \frac{(\text{SSE}_1 - \text{SSE}_2)/(k - g)}{\text{SSE}_2/[n - (k + 1)]} = \frac{(326.623 - 77.948)/3}{77.948/5} = 5.32$$

Because this statistic is based on $v_1 = (k - g) = 3$ numerator degrees of freedom and $v_2 = n - (k + 1) = 5$ denominator degrees of freedom, the p-value is given by $P(F > 5.32)$. Thus, $.05 < p\text{-value} < .10$. If we chose $\alpha = .05$, there is insufficient evidence to claim that the second-order model fits the data significantly better than the planar model. Note that we have tested whether the *group* of variables x_1^2, x_2^2, $x_1 x_2$ contribute to a significantly better fit of the model to the data.

∽

There are many real world situations for which the most appropriate deterministic component of a model is not linear. For example, many populations of plants or animals tend to grow exponentially fast. If Y_t denotes the size of the population at time t, we might employ the model

$$E(Y_t) = \alpha_0 e^{\alpha_1 t}$$

Although this is not linear in the parameters α_0 and α_1, it can be linearized by taking natural logarithms. If Y_t can be observed for various values of t, we could write the model as

$$\ln Y_t = \ln \alpha_0 + \alpha_1 t + \varepsilon$$

and estimate $\ln \alpha_0$ and α_1 by least squares.

Other basic models also can be linearized. In the biological sciences it sometimes is possible to relate the weight (or volume) of an organism to some linear measurement such as length (or weight). If W denotes weight and l length, the model

$$E(W) = \alpha_0 l^{\alpha_1}$$

for unknown α_0 and α_1 is often applicable. (This model is known as an *allometric equation*.) If we want to relate the random weight of organisms to observable fixed lengths, we could take logarithms and obtain the linear model

$$\ln W = \ln \alpha_0 + \alpha_1 \ln l + \varepsilon = \beta_0 + \beta_1 x + \varepsilon$$

with $x = \ln l$. Then $\beta_0 = \ln \alpha_0$ and $\beta_1 = \alpha_1$ can be estimated by the method of least squares. The following example illustrates the use of such a model.

EXAMPLE 11.14   In the data set of Table 11.6 w denotes the weight (in pounds) and l the snout vent length (distance in inches from back of head to end of nose) for 15 alligators captured in central Florida. Because l is easier to observe than w for alligators in their natural habitat, we want to construct a model relating weight to snout vent length. Such a model can be used to predict the weight of alligators with particular snout vent lengths. Fit the model

$$\ln W = \ln \alpha_0 + \alpha_1 \ln l + \varepsilon = \beta_0 + \beta_1 x + \varepsilon$$

to the data. Find a 90% prediction interval for W if $\ln l$ is observed to be 4.00.

Table 11.6
Data for
Example 11.14

Alligator	$x = \ln l$	$y = \ln w$
1	3.87	4.87
2	3.61	3.93
3	4.33	6.46
4	3.43	3.33
5	3.81	4.38
6	3.83	4.70
7	3.46	3.50
8	3.76	4.50
9	3.50	3.58
10	3.58	3.64
11	4.19	5.90
12	3.78	4.43
13	3.71	4.38
14	3.73	4.42
15	3.78	4.25

Solution   We will use the direct calculation formulas of Section 11.3. (The matrix approach could also be used.) Then we have

$$\hat{\beta}_1 = \frac{n \sum\limits_{i=1}^{n} x_i y_i - \sum\limits_{i=1}^{n} x_i \sum\limits_{i=1}^{n} y_i}{n \sum\limits_{i=1}^{n} x_i^2 - \left(\sum\limits_{i=1}^{n} x_i \right)^2} = \frac{15(251.9757) - (56.37)(66.27)}{15(212.6933) - (56.37)^2} = 3.4311$$

and

$$\hat{\beta}_0 = \bar{y} - \hat{\beta}_1 \bar{x} = \frac{66.27}{15} - (3.4311)\left(\frac{56.37}{15} \right) = -8.476$$

We can now estimate α_0 by

$$\hat{\alpha}_0 = e^{\hat{\beta}_0} = e^{-8.476} = .0002$$

and α_1 by $\hat{\alpha}_1 = \hat{\beta}_1$ to arrive at an estimated model of

$$\hat{w} = \hat{\alpha}_0 l^{\hat{\alpha}_1} = (.0002)l^{3.4311}$$

(In many cases α_1 will be close to 3 because a weight or volume generally is proportional to the cube of a lineal measurement.)

For these data SSE $= .1966$, $n = 15$, and $s = \sqrt{\text{SSE}/(n-2)} = .123$. The calculations leading to these numerical values are completely analogous to the calculations of Example 11.12.

To find a prediction interval for W where $x = \ln l = 4$, we must first form a prediction interval for $Y = \ln W$. Because

$$\hat{Y} = -8.456 + 3.4311(4) = 5.2484$$

the prediction interval is

$$\hat{y} \pm t_{.05}s \sqrt{1 + \frac{1}{n} + \frac{(x_0 - \bar{x})^2}{\displaystyle\sum_{i=1}^{n}(x_i - \bar{x})^2}}$$

where $t_{.05}$ is based on $n - 2 = 13$ degrees of freedom. Therefore, $t_{.05} = 1.771$ and the 90% prediction interval for $Y = \ln W$ is

$$5.2484 \pm 1.771(.123)\sqrt{1 + \frac{1}{15} + \frac{(4 - 3.758)^2}{.8548}}$$

$$5.2484 \pm .2321$$

or

$$(5.0164, 5.4805)$$

Because $\hat{Y} = \ln \hat{W}$, we can predict W by $e^{\hat{Y}} = e^{5.2484} = 190.2616$. The observed 90% prediction interval for W is

$$(e^{5.0164}, e^{5.4805}) \qquad \text{or} \qquad (150.867, 239.967)$$

When $x = \ln l = 4$, then $l = e^4 = 54.598$. Thus for an alligator of length 54.598 inches, we predict the alligator to weigh between 150.867 and 239.967 pounds. Note that the rather narrow interval on the natural logarithm scale becomes a rather wide interval when transformed to the original scale.   ∽

The data presented and analyzed in this section are examples of data from real experiments where the methods developed in the previous sections were applied to produce answers of actual concern for experimenters in various areas. Through Example 11.14 we have demonstrated how the theory of linear models sometimes can be applied after transformation of the scale of the original variables. Of course, not all models can be linearized. Numerous techniques for nonlinear least squares estimation are available.

Exercises

11.63   Refer to Example 11.12. Find a 90% prediction interval for the strength of concrete when the water/cement ratio is 1.5.

11.64   Refer to Example 11.13. Using the reduced model, construct a 95% confidence interval for the expected abrasion resistance of rubber when $x_1 = 1$ and $x_2 = -1$.

11.65   Refer to Example 11.13. Construct individual tests of the three hypotheses $H_0: \beta_3 = 0$, $H_0: \beta_4 = 0$, and $H_0: \beta_5 = 0$. Use a 1% level of significance on each test. (If multiple tests are to be conducted on the same set of data, it is wise to use a very small α level on each test.)

11.66   Refer to Example 11.14. Calculate the correlation coefficient r between the variables $\ln W$ and $\ln l$.

*11.67   It is well known that large bodies of water have a mitigating effect on the temperature of the surrounding land masses. On a cold night in central Florida, temperatures were recorded at equal distances along a transect running in the downwind direction from a large lake. The data are given in the accompanying table. Notice that the temperatures drop rapidly and then level off as we move away from the lake. The suggested model for these data is

$$E(Y) = \alpha_0 e^{-\alpha_1 x}$$

(a)   Linearize the model and estimate the parameters by the method of least squares.

(b)   Find a 90% confidence interval for α_0. Give an interpretation of the result.

Site, x	Temperature, y
1	37.00
2	36.25
3	35.41
4	34.92
5	34.52
6	34.45
7	34.40
8	34.00
9	33.62
10	33.90

*11.68   Refer to Exercise 11.14. One model proposed for these data on proportion of survivors of thermal pollution is

$$E(Y) = \exp(-\alpha_0 x^{\alpha_1})$$

Linearize this model and estimate the parameters by using the method of least squares and the data of Exercise 11.14. (Omit the observation with $y = 1.00$.)

*11.69   In the biological and physical sciences a common model for proportional growth over time is

$$E(Y) = 1 - e^{-\beta t}$$

where Y denotes a proportion and t denotes time. Y might represent the proportion of eggs that hatch, the proportion of an organism filled with diseased cells, the proportion of patients reacting to a drug, or the proportion of a liquid that has passed through a porous medium. With n observations of the form (y_i, t_i), outline how you would estimate and then form a confidence interval for β.

11.13 Summary

In this chapter we have been concerned with use of the method of least squares to fit a linear model to an experimental response. We assume the expected value of Y to be a function of a set of variables $x_1, x_2, \ldots, x_k$, where the function is linear in a set of unknown parameters. We used the expression

$$Y = \beta_0 + \beta_1 x_1 + \beta_2 x_2 + \cdots + \beta_k x_k + \varepsilon$$

to denote a linear statistical model.

Inferential problems associated with the linear statistical model are estimation and tests of hypotheses concerning the model parameters β_0, $\beta_1, \ldots, \beta_k$ and even more important, estimation of $E(Y)$, the expected response for a particular setting, and the prediction of some future value of Y. Experiments for which the least squares theory is appropriate include both controlled experiments and those where $x_1, x_2, \ldots, x_k$ are observed values of random variables.

Why use the method of least squares to fit a linear model to a set of data? Where the assumptions on the random errors ε hold [normality, independence, $V(\varepsilon) = \sigma^2$ for all values of $x_1, x_2, \ldots, x_k$], it can be shown that the least squares procedure gives the best *linear* unbiased estimators for $\beta_0, \beta_1, \ldots, \beta_k$. That is, if we estimate the parameters $\beta_0, \beta_1, \ldots, \beta_k$, using linear functions of $y_1, y_2, \ldots, y_k$, the least squares estimators have minimum variance. Some other nonlinear estimators for the parameters may possess a smaller variance than the least squares estimators, but if such estimators exist, they are not known at this time. Again, why use least squares estimators? They are easy to use, we know their properties, and we know that they possess very good properties for many situations.

As you might imagine, the methodology presented in this chapter is employed widely in business and all the sciences for exploring the relationship between a response and a set of independent variables. Estimation of $E(Y)$ or prediction of Y usually is the experimental objective.

References and Further Readings

1. Draper, N. R., and Smith, H. *Applied Regression Analysis*. 2d ed. New York: Wiley, 1981.
2. Graybill, F. *Theory and Application of the Linear Model*. N. Scituate, Mass.: Duxbury Press, 1976.
3. Li, J. C. R. *Introduction to Statistical Inference*. Ann Arbor, Mich.: Edwards, 1961.
4. Marcellari, C. E. "Revision of Serpulids of the Genus Rotularia (Annelida) at Seymour Island (Antarctic Peninsula) and Their Value in Stratigraphy," *Journal of Paleontology* 58, no. 4 (July 1984).
5. Matis, J. H., and Wehrly, T. E. "Stochastic Models of Compartmental Systems," *Biometrics* 35, no. 1 (March 1979): 199–220.
6. Mendenhall, W., *An Introduction to Linear Models and the Design and Analysis of Experiments*. Belmont, Calif: Wadsworth, 1968.

Supplementary Exercises

11.70   At temperatures approaching absolute zero (273° below zero Celsius), helium exhibits traits that defy many laws of conventional physics. An experiment has been conducted with helium in solid form at various temperatures near absolute zero. The solid helium is placed in a dilution refrigerator along with a solid impure substance, and the fraction (in weight) of the impurity passing through the solid helium is recorded. (The phenomenon of solids passing directly through solids is known as *quantum tunneling*.) The data are given in the table.

Temperature, x °C	Proportion of Impurity Passing through Helium, y
−262.0	.315
−265.0	.202
−256.0	.204
−267.0	.620
−270.0	.715
−272.0	.935
−272.4	.957
−272.7	.906
−272.8	.985
−272.9	.987

(a)   Fit a least squares line to the data.

(b)   Test the null hypothesis, $H_0: \beta_1 = 0$, against the alternative hypothesis, $H_a: \beta_1 < 0$, at the $\alpha = 0.01$ level of significance.

(c)   Find a 95% prediction interval for the percentage of the solid impurity passing through solid helium at $-273°C$. (Note that this value of x is outside the experimental region, where use of the model for prediction may be dangerous.)

11.71   A study was conducted to determine whether there is a linear relationship between the breaking strength, y, of wooden beams and the specific gravity, x, of the wood. Ten randomly selected beams of the same cross-sectional dimensions were stressed until they broke. The breaking strengths and the density of the wood are shown for each of the ten beams.

Beam	Specific Gravity, x	Strength, y
1	0.499	11.14
2	0.558	12.74
3	0.604	13.13
4	0.441	11.51
5	0.550	12.38
6	0.528	12.60
7	0.418	11.13
8	0.480	11.70
9	0.406	11.02
10	0.467	11.41

(a)   Fit the model $Y = \beta_0 + \beta_1 x + \varepsilon$.

(b)   Test $H_0: \beta_1 = 0$ against the alternative hypothesis, $H_a: \beta_1 \neq 0$.

(c)   Estimate the mean strength for beams with specific gravity 0.590 using a 90% confidence interval.

11.72   A response y is a function of three independent variables, x_1, x_2, and x_3, that are related as follows.

$$Y = \beta_0 + \beta_1 x_1 + \beta_2 x_2 + \beta_3 x_3 + \varepsilon$$

(a)   Fit this model to the accompanying $n = 7$ data points.

y	x_1	x_2	x_3
1	-3	5	-1
0	-2	0	1
0	-1	-3	1
1	0	-4	0
2	1	-3	-1
3	2	0	-1
3	3	5	1

(b) Predict y when $x_1 = 1$, $x_2 = -3$, $x_3 = -1$. Compare with the observed response in the original data. Why are these two not equal?

(c) Do the data present sufficient evidence to indicate that x_3 contributes information for the prediction of Y? (Test the hypothesis $\beta_3 = 0$, using $\alpha = .05$.)

(d) Find a 95% confidence interval for the expected value of Y, given $x_1 = 1$, $x_2 = -3$, $x_3 = -1$.

(e) Find a 95% prediction interval for Y, given $x_1 = 1$, $x_2 = -3$, $x_3 = -1$.

11.73 If values of independent variables are equally spaced, what is the advantage of coding to new variables that represent symmetric spacing about the origin?

11.74 Suppose that you wish to fit a straight line to a set of n data points, where n is an even integer, and that you can select the n values of x in the interval $-9 \leq x \leq 9$. How should you select the values of x so as to minimize $V(\hat{\beta}_1)$?

11.75 Refer to Exercise 11.74. It is common to employ equal spacing in the selection of the values of x. Suppose that $n = 10$. Find the relative efficiency of the estimator $\hat{\beta}_1$ based on equal spacing versus the same estimator based on the spacing of Exercise 11.74. Assume that $-9 \leq x \leq 9$.

11.76 The accompanying data come from the comparison of the growth rates for bacteria types A and B. The growth Y recorded at five equally spaced (and coded) points of time is shown in the table.

Bacteria Type	\multicolumn{5}{c}{Time}
	-2	-1	0	1	2
A	8.0	9.0	9.1	10.2	10.4
B	10.0	10.3	12.2	12.6	13.9

(a) Fit the linear model

$$Y = \beta_0 + \beta_1 x_1 + \beta_2 x_2 + \beta_3 x_1 x_2 + \varepsilon$$

to the $n = 10$ data points. Let $x_1 = 1$ if the point refers to bacteria type B and $x_1 = 0$ if the point refers to type A. Let $x_2 = $ coded time.

(b) Plot the data points and graph the two growth lines. Note that β_3 is the difference between the slopes of the two lines and represents time–bacteria interaction.

(c) Predict the growth of bacteria type A at time $x_2 = 0$ and compare the answer with the graph. Repeat the process for bacteria type B.

(d) Do the data present sufficient evidence to indicate a difference in the rates of growth for the two types of bacteria?

(e) Find a 90% confidence interval for the expected growth for bacteria type B at time $x_2 = 1$.

(f) Find a 90% prediction interval for the growth Y of bacteria type B at time $x_2 = 1$.

11.77 The following model was proposed for testing whether there was evidence of salary discrimination against women in a state university system:

$$Y = \beta_0 + \beta_1 x_1 + \beta_2 x_2 + \beta_3 x_1 x_2 + \beta_4 x_2^2 + \varepsilon$$

where Y = annual salary (in thousands of dolllars)

$$x_1 = \begin{cases} 1 & \text{if female} \\ 0 & \text{if male} \end{cases}$$

x_2 = amount of experience (in years)

When this model was fit to data obtained from the records of 200 faculty members, SSE = 783.90. The reduced model $Y = \beta_0 + \beta_1 x_2 + \beta_2 x_2^2 + \varepsilon$ also was fit and produced a value of SSE = 795.23. Do the data provide sufficient evidence to support the claim that the mean salary is dependent upon the sexes of the faculty members? Use $\alpha = .05$.

11.78 Show that the least squares prediction equation

$$\hat{Y} = \hat{\beta}_0 + \hat{\beta}_1 x_1 + \cdots + \hat{\beta}_k x_k$$

passes through the point $(\bar{x}_1, \bar{x}_2, \ldots, \bar{x}_k, \bar{Y})$.

11.79 An experiment was conducted to determine the effect of pressure and temperature on the yield of a chemical. Two levels of pressure (pounds per square inch) and three of temperature were used:

Pressure (psi)	Temperature (°F)
50	100
80	200
	300

One run of the experiment at each temperature-pressure combination gave the following data:

Yield	Pressure (psi)	Temperature (°F)
21	50	100
23	50	200
26	50	300
22	80	100
23	80	200
28	80	300

(a) Fit the model $Y = \beta_0 + \beta_1 x_1 + \beta_2 x_2 + \beta_3 x_2^2 + \varepsilon$, where x_1 = pressure and x_2 = temperature.

(b) Test to see if β_3 is significantly different from zero, with $\alpha = .05$.

(c) Test the hypothesis that temperature does not affect the yield, with $\alpha = .05$.

*11.80   Let $(X_1, Y_1), \ldots, (X_n, Y_n)$ denote a random sample from the bivariate normal distribution. A test of $H_0: \rho = 0$ against $H_a: \rho \neq 0$ can be derived as follows.

(a)   Let $S_{yy} = \sum\limits_{i=1}^{n} (y_i - \bar{y})^2$ and $S_{xx} = \sum\limits_{i=1}^{n} (x_i - \bar{x})^2$. Show that

$$\hat{\beta}_1 = r \sqrt{\frac{S_{yy}}{S_{xx}}}$$

(b)   Conditional on $X_i = x_i$, $i = 1, \ldots, n$, show that, under $H_0: \rho = 0$,

$$\frac{\hat{\beta}_1 \sqrt{(n-2)S_{xx}}}{\sqrt{S_{yy}(1 - r^2)}}$$

has a t distribution with $(n - 2)$ degrees of freedom.

(c)   Conditional on $X_i = x_i$, $i = 1, \ldots, n$, conclude that

$$T = \frac{r\sqrt{n-2}}{\sqrt{1 - r^2}}$$

has a t distribution with $(n - 2)$ degrees of freedom, under $H_0: \rho = 0$. Hence conclude that T has the same distribution unconditionally.

11.81   Labor and material costs are two basic components in analyzing the cost of construction. Changes in the component costs, of course, will lead to changes in total construction costs.

Month	Construction Cost*, y	Index of All Construction Materials†, x
January	193.2	180.0
February	193.1	181.7
March	193.6	184.1
April	195.1	185.3
May	195.6	185.7
June	198.1	185.9
July	200.9	187.7
August	202.7	189.6

*Source   United States Department of Commerce, Bureau of the Census.
†Source   United States Department of Labor, Bureau of Labor Statistics. Tables were given in Tables E-1 (p. 43) and E-2 (p. 44), respectively, in *Construction Review*. United States Department of Commerce, Oct. 1976, 22 (8).

Do the data provide sufficient evidence to indicate a nonzero correlation between the monthly construction costs and indexes of all construction materials? Give the attained significance level.

11.82   The data in the table give the mileages per gallon obtained by a test automobile when using gasolines of varying levels of octane.

Mileage, y (miles per gallon)	Octane, x
13.0	89
13.2	93
13.0	87
13.6	90
13.3	89
13.8	95
14.1	100
14.0	98

(a)   Calculate the value of r.

(b)   Do the data provide sufficient evidence to indicate that octane level and miles per gallon are dependent? Give the attained significance level, and indicate your conclusion if you wish to implement an $\alpha = .05$ level test.

chapter twelve

CONSIDERATIONS IN DESIGNING EXPERIMENTS

12.1 The Elements Affecting the Information in a Sample

The information in a sample that is available to make an inference about a population parameter can be measured by the width (or half width) of the confidence interval that could be constructed from the sample data. Recall that a 95% large-sample confidence interval for a population mean is

$$\bar{y} \pm 1.96\left(\frac{\sigma}{\sqrt{n}}\right)$$

The widths of almost all the commonly employed confidence intervals, like the confidence interval for a population mean, are dependent on the population variance σ^2 and the sample size n. The less variation in the population, measured by σ^2, the shorter the confidence interval will be. Similarly, the width of the confidence interval will decrease as n increases. This interesting phenomenon would lead us to believe that two factors affect the quantity of information in a sample pertinent to a parameter; namely, the variation of the data and the sample size n. We will find this deduction to be slightly oversimplified but essentially true.

A strong similarity exists between the audio theory of communication and the theory of statistics. Both are concerned with the transmission of a message (signal) from one point to another and, consequently, both are theories of information. For example, the telephone engineer is responsible for transmitting a verbal message that might originate in New York City and be received in New Orleans. Similarly, a speaker may wish to communicate with a large and noisy audience. If static or background noise is sizable for either example, the receiver may acquire only a sample of the complete signal, and from this partial information must infer the nature of the complete message. Analogously, scientific experimentation is conducted to verify certain theories about natural phenomena, or simply to explore some aspects of nature and hopefully to deduce—either exactly or with a good approximation—the relationships among certain natural variables. Thus one might think of experimentation as the communication between nature and a scientist. The message about the natural phenomenon is contained, in garbled form, in the experimenter's sample data. Imperfections in the measuring instruments, nonhomogeneity of experimental material, and many other factors contribute background noise (or static) that tends to obscure nature's signal and cause the observed response to vary in a random manner. For both the communications engineer and the statistician, two elements affect the quantity of information in an experiment, the magnitude of the background noise (or variation) and the volume of the signal. The greater the noise or, equivalently, the variation, the less information will be contained in the sample. Likewise, the louder the signal, the greater the amplification will be, and hence it is more likely that the message will penetrate the noise and be received.

The design of experiments is a very broad subject concerned with methods of sampling to reduce the variation in an experiment, to amplify nature's signal, and thereby to acquire a *specified quantity* of information at minimum cost. Despite the complexity of the subject, some of the important considerations in the design of good experiments can be easily understood and should be presented to the beginner. We take these considerations as our objective in the succeeding discussion.

12.2 The Physical Process of Designing an Experiment

We commence our discussion of experimental design by clarifying terminology and then, through examples, by identifying the steps that one must take in designing an experiment.

Definition 12.1

> The objects upon which measurements are taken are called *experimental units*.

If an experimenter subjects a set of $n = 10$ rats to a stimulus and measures the response of each, an individual rat is an experimental unit. The collection of $n = 10$ measurements is a sample. Similarly, if a set of $n = 10$ items is selected from a list of hospital supplies in an inventory audit, each item is an experimental unit. The observation made on each experimental unit is the dollar value of the item actually in stock, and the set of ten measurements constitutes a sample.

What one does to the experimental units that makes them differ from one population to another is called a *treatment*. One might wish to study the density of a specific kind of cake when baked at $x = 350°F$, $x = 400°F$, and $x = 450°F$ in a given oven. An experimental unit would be a single mix of batter in the oven at a given point in time. The three temperatures, $x = 350, 400,$ and $450°F$, would represent three treatments. The millions and millions of cakes that conceptually *could* be baked at $350°F$ would generate a population of densities, and one could similarly generate populations corresponding to 400 and $450°F$. The objective of the experiment would be to compare the cake density Y for the three populations. Or we might wish to study the effect of temperature of baking, x, on cake density by fitting a linear or curvilinear model to the data points, using the methods described in Chapter 11.

In another experiment we might wish to compare tire wear for two manufacturers, A and B. Each tire–wheel combination tested at a particular time would represent an experimental unit, and each of the two manufacturers would represent a treatment. Note that one does not physically treat the tires to make them different. They receive two different treatments by the very fact that they are manufactured by two different companies in different locations and in different factories.

As a third example, consider an experiment conducted to investigate the effect of various amounts of nitrogen and phosphate on the yield of a variety of corn. An experimental unit would be a specified acreage, say 1 acre, of corn. A treatment would be a fixed number of pounds of nitrogen, x_1, and phosphate, x_2, applied to a given acre of corn. For example, one treatment might be to use $x_1 = 100$ pounds of nitrogen per acre and $x_2 = 200$ pounds of phosphate. A second treatment could be $x_1 = 150$ and $x_2 = 100$. Note that the experimenter could use different amounts (x_1, x_2) of nitrogen and phosphate and that each conbination would represent a different treatment.

Those variables that are completely controlled by the experimenter, such as the temperature of the oven or the amount of nitrogen per acre, are called *independent* variables. (This usage is consistent with the use of the term *independent variables* in Chapter 11.) Most experiments involve a study of the effect of one or more independent variables upon a response.

Definition 12.2

> Independent experimental variables are called *factors*.

Factors can be *quantitative* or *qualitative*.

Definition 12.3

> A *quantitative factor* is one that can take values corresponding to points on a real line. Factors that are not quantitative are said to be *qualitative*.

Oven temperature, pounds of nitrogen, and pounds of phosphate are examples of quantitative factors. In contrast, manufacturers, types of drugs, or physical locations are factors that cannot be quantified and are called *qualitative*.

Definition 12.4

> The intensity setting of a factor is called a *level*.

The three temperatures, 350, 400, and 450°F, represent three levels of the quantitative factor "oven temperature." Similarly, the two treatments, manufacturer *A* and manufacturer *B*, represent two levels of the qualitative factor "manufacturer." Note that a third or fourth tire manufacturer could have been included in the tire wear experiment, which would have resulted in either three or four levels of the factor "manufacturer."

We noted previously that what one does to experimental units that makes them differ from one population to another is called a *treatment*. Because every treatment implies a combination of one or more factor levels, we have a more precise definition for the term.

Definition 12.5

> A *treatment* is a specific combination of factor levels.

The experiment may involve only a single factor such as temperature in the baking experiment. Or it could be composed of combinations of levels of two (or more) factors as for the corn-fertilizing experiment. Each combination would represent a treatment. One of the early steps in the design of an experiment is the selection of factors to be studied and a decision regarding the combinations of levels (treatments) to be employed in the experiment.

The design of an experiment implies one final problem. After selecting the factor combinations (treatments) to be employed in an experiment, one must decide how the treatments should be assigned to the experimental units. Should the treatments be randomly assigned to the experimental units or should a semirandom pattern be employed? For example, should the tires corresponding to manufacturers *A* and *B* be randomly assigned to all the automobile wheels, or should one each of tire types *A* and *B* be assigned to the rear wheels of each car?

The foregoing discussion suggests that the design of an experiment involves four steps.

Steps Employed in Designing an Experiment

1. Select the factors to be included in the experiment and specify the population parameter(s) of interest.
2. Decide how much information is desired pertinent to the parameter(s) of interest. (For example, how accurately do you wish to estimate the parameters?)
3. Select the treatments (combination of factor levels) to be employed in the experiment and decide on the number of experimental units to be assigned to each.
4. Decide on the manner in which the treatments should be applied to the experimental units.

Steps 3 and 4 correspond to the two elements that affect the quantity of information in an experiment. First, how one selects the treatments (combinations of factor levels) and the number of experimental units assigned to each affects the intensity of nature's signal pertinent to the population parameter(s) of interest to the experimenter. Second, the method of assigning the treatments to the experimental units affects the background noise or, equivalently, the variation of the experimental units. We will examine each of these assertions in detail in Sections 12.4 and 12.5 after digressing briefly in Section 12.3 to consider the implications of random sampling and how to draw a random sample.

12.3 Random Sampling and the Completely Randomized Design

Random sampling—that is, giving every possible sample of fixed size in the population an equal probability of selection—has two purposes. First, it avoids the possibility of bias introduced by a nonrandom selection of sample elements. For example, a sample selection of voters from telephone directories in 1936 indicated a clear win for Landon in the presidential election of that year. However, the sample did not represent a random selection from the whole population of eligible voters because the majority of telephone users in 1936 were Republicans. As another example, suppose that we sample to determine the precentage of home owners favoring the construction of a new city park and we modify our original random sample by ignoring owners who are not at home. The result may yield a biased response because those at home will likely have children and may be more inclined to favor the new park. The second purpose of random sampling is to provide a probabilistic basis for the selection of a sample. That is, it treats the selection of a sample as an experiment (Chapter 2), enabling the statistician to calculate the probability of an observed sample and to use this probability in making inferences. We learned that under fairly general conditions, the mean $\bar{Y}$ of a random sample of n elements will possess a probability distribution that is approximately normal when n is large (the central limit theorem). Fundamental to the proof of the central limit theorem is the assumption of random sampling. Similarly, the confidence intervals and tests of hypotheses based on Student's t (Chapters 8, 9, and 10) required the assumption of random sampling.

The random selection of independent samples to compare two or more populations is the simplest type of experimental design. The populations may differ because we have applied different treatments, and we now wish to consider step 4 in the design of an experiment—deciding how to apply the treatments to the experimental units or, equivalently, how to select the samples.

Definition 12.6

> The selection of independent random samples from k populations is called a *completely randomized design*.

The comparison of five brands of aspirin, A, B, C, D, and E, by randomly selecting 100 pills from the production of each manufacturer would be a completely randomized design with $n_A = n_B = \cdots = n_E = 100$. Similarly, we might wish to compare five teaching techniques, A, B, C, D, and E, using 25 students in each class ($n_A = n_B = \cdots = n_E = 25$). The populations corresponding to A, B, C, D, and E are nonexistent (that is, they are conceptual), because students taught by the five techniques either do not exist or they are unavailable. Consequently, random samples are obtained for the five conceptual populations by randomly selecting 125 students of the type envisioned for the study and then *randomly* assigning 25 students to each of the five teaching techniques. This scheme will yield independent random samples of 25 students subjected to each technique and will result in a completely randomized design.

As you might surmise, not all designs are completely randomized, but all good experiments utilize randomization to some extent. Examples of designs that restrict randomization will be given in Section 12.5.

Before concluding this section on randomization, let us consider a method for selecting a random sample or randomly assigning experimental units to a set of treatments.

A simple and reliable way to select a random sample of n elements from a large population is to employ a table of random numbers, such as that shown in Table 12, Appendix III. The use of this table for selecting random samples was discussed in Section 2.12, but we will give a brief review here. Random number tables are constructed so that integers occur randomly and with equal frequency. For example, suppose that a random sample is to be selected from a population containing $N = 1000$ elements. Number the elements in sequence, from 000 to 999. Then turn to a table of random numbers such as the excerpt shown in Table 12.1.

Select n of the random numbers in order from the table. The population elements to be included in the random sample will be given by the first three

Table 12.1
Portion of a table of
random numbers

15574	35026	98924
45045	36933	28630
03225	78812	50856
88292	26053	21121

digits of the random numbers. If $n = 5$, we would include elements numbered 155, 450, 32, 882, and 350. So as not to use the same sequence of random numbers over and over again, the experimenter should select different starting points in Table 12 to begin the selection of random numbers for different samples.

The random assignment of 40 experimental units, 10 to each of four treatments, A, B, C, and D, is equally easy using the random number table. Number the experimental units 1 to 40. Select a sequence of random numbers and refer only to the first two digits because $n = 40$. Discard random numbers greater than 40. Then assign the experimental units associated with the first 10 numbers that appear to A, those associated with the second 10 to B, and so on. To illustrate, we refer to Table 12.1. Starting in the first column and moving top to bottom, we acquire the numbers 15 and 3 (discard 45 and 88 because no experimental units possess those numbers). This procedure would be continued until all integers 1 to 40 were selected. The resulting numbers would arrange the 40 experimental units in random order.

The analysis of data for a completely randomized design is treated in Section 13.3.

Exercises

12.1   A political analyst wishes to select a sample of $n = 20$ people from a population of 2000. Use the random number table to identify the people to be included in the sample.

12.2   Two drugs, A and B, are to be applied to five rats each. Suppose that the rats are numbered from 1 to 10. Use the random number table to randomly assign the rats to the two treatments.

12.3   Refer to Exercise 12.2. Suppose that the experiment involved three drugs, A, B, and C, with five rats assigned to each. Use the random number table to randomly assign the fifteen rats to the three treatments.

12.4   A population contains 10,000 voters, listed alphabetically on a computer printout. Use the random number table to identify the voters to be included in a random sample of $n = 15$.

12.5   Suppose $n = 50$ names are to be randomly selected from the telephone directory for a small city. Discuss how a random number table could be used in the selection process.

12.6   A chemical engineer has two catalysts and three temperature settings that she wishes to use in a series of experiments. Each experiment makes use of one catalyst-temperature combination. Show how you would use a random number table to randomize the order of the experiments.

12.4 Volume-Increasing Experimental Designs

You will recall that designing to increase the volume of a signal—that is, the information pertinent to one or more population parameters—depends on the selection of treatments and the number of experimental units assigned to each (step 3 in designing an experiment). The treatments, or combinations of factor levels, identify points at which one or more response measurements will be made and indicate the general location in which the experimenter is focusing his or her attention. As we will see, for the same number of observations, some designs contain more information concerning specific population parameters than others. Other very costly experiments contain no information concerning certain population parameters. No single design is best in acquiring information concerning all types of population parameters. Indeed, the problem of finding the best design for focusing information on a specific population parameter has been solved in only a few specific cases.

The purpose of this section is not to present a general theory or find the best selection of factor level combinations for a given experiment but rather to present a few examples to illustrate the principles involved. The optimal design providing the maximum amount of information pertinent to the parameter(s) of interest (for a fixed sample size n) will be given for the following two examples.

The simplest example of an information-focusing experiment is the problem of estimating the difference between a pair of population means, $\mu_1 - \mu_2$, based on independent random samples. In this instance the two treatments have already been selected and the question concerns the allocation of experimental units to the two samples. If the experimenter plans to invest money sufficient to sample a total of n experimental units, how many units should he select from populations 1 and 2, say n_1 and n_2 ($n_1 + n_2 = n$), respectively, so as to maximize the information in the data pertinent to ($\mu_1 - \mu_2$)? If $n = 10$, should he select $n_1 = n_2 = 5$ observations from each population, or would an allocation of $n_1 = 4$ and $n_2 = 6$ be better?

Recall that the estimator of ($\mu_1 - \mu_2$), ($\bar{Y}_1 - \bar{Y}_2$), has standard deviation

$$\sigma_{(\bar{Y}_1 - \bar{Y}_2)} = \sqrt{\frac{\sigma_1^2}{n_1} + \frac{\sigma_2^2}{n_2}}$$

The smaller $\sigma_{(\bar{Y}_1 - \bar{Y}_2)}$, the smaller will be the corresponding error of estimation and the greater will be the quantity of information in the sample pertinent to ($\mu_1 - \mu_2$). If, as we frequently assume, $\sigma_1^2 = \sigma_2^2 = \sigma^2$, then

$$\sigma_{(\bar{Y}_1 - \bar{Y}_2)} = \sigma \sqrt{\frac{1}{n_1} + \frac{1}{n_2}}$$

You can verify that this quantity is a minimum when $n_1 = n_2$ and consequently that the sample contains a maximum of information on ($\mu_1 - \mu_2$) when the n

experimental units are equally divided between the two treatments. A more general case is considered in Example 12.1.

EXAMPLE 12.1   If n observations are to be used to estimate $(\mu_1 - \mu_2)$, based upon independent random samples, find n_1 and n_2 so that $V(\bar{Y}_1 - \bar{Y}_2)$ is minimized (assume that $n_1 + n_2 = n$).

Solution   Let b denote the fraction of the n observations assigned to the sample from population 1, that is, $n_1 = bn$ and $n_2 = (1 - b)n$. Then

$$V(\bar{Y}_1 - \bar{Y}_2) = \frac{\sigma_1^2}{bn} + \frac{\sigma_2^2}{(1 - b)n}$$

To find the fraction b that minimizes this variance, we set the first derivative, with respect to b, equal to zero. This process yields

$$-\frac{\sigma_1^2}{n}\left(\frac{1}{b^2}\right) + \frac{\sigma_2^2}{n}\left(\frac{1}{1 - b}\right)^2 = 0$$

$$\frac{\sigma_1^2}{n}\left(\frac{1}{b^2}\right) = \frac{\sigma_2^2}{n}\left(\frac{1}{1 - b}\right)^2$$

$$\frac{b^2}{(1 - b)^2} = \frac{\sigma_1^2}{\sigma_2^2}$$

$$\frac{b}{1 - b} = \frac{\sigma_1}{\sigma_2}$$

Solving for b, we obtain

$$b = \frac{\sigma_1}{\sigma_1 + \sigma_2}$$

and

$$1 - b = \frac{\sigma_2}{\sigma_1 + \sigma_2}$$

Thus $V(\bar{Y}_1 - \bar{Y}_2)$ is minimized when

$$n_1 = \left(\frac{\sigma_1}{\sigma_1 + \sigma_2}\right)n$$

that is, when sample sizes are allocated proportional to the standard deviations. Note that $n_1 = n/2$ if $\sigma_1 = \sigma_2$.

As a second example, consider the problem of fitting a straight line through a set of n points by using the least squares method of Chapter 11 (see Figure 12.1). Further, suppose that we are primarily interested in the slope of the line, β_1, in the linear model

$$Y = \beta_0 + \beta_1 x + \varepsilon$$

Figure 12.1
Fitting a straight line by the method of least squares

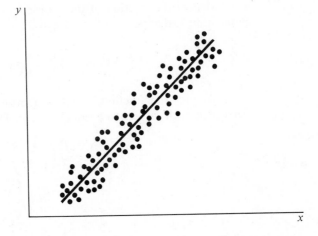

If we have the option of selecting the n values of x for which y will be observed, which values of x will maximize the quantity of information on β_1? We have one single quantitative factor, x, and we have the problem of deciding on the levels to employ, $x_1, x_2, \ldots, x_n$, as well as the number of observations to be taken at each.

A strong lead to the best design for fitting a straight line can be achieved by viewing Figures 12.2(a) and (b). Suppose that y was linearly related to x and

Figure 12.2
Two different level selections for fitting a straight line

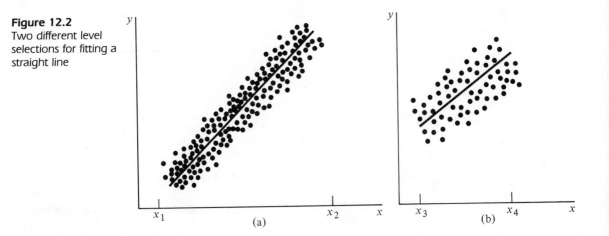

(a)

(b)

generated data similar to that shown in Figure 12.2(a) for the interval $x_1 < x < x_2$. Note the approximate range of variation for a given value of x. Now suppose that instead of the wide range for x employed in Figure 12.2(a), the experimenter selected data from the same population but over the very narrow range $x_3 < x < x_4$, as shown in Figure 12.2(b). The variation in Y, given x, is the same as for Figure 12.2(a). Which distribution of data points would provide the greater amount of information concerning the slope of the line β_1? You might guess (correctly) that the best estimate of slope will occur when the levels of x are selected farther apart, as shown in Figure 12.2(a). The data for Figure 12.2(b) could yield a very inaccurate estimate of the slope and, as a matter of fact, might leave a question as to whether the slope is positive or negative.

The best design for estimating the slope β_1 can be determined by considering the standard deviation of $\hat{\beta}_1$,

$$\sigma_{\hat{\beta}_1} = \frac{\sigma}{\sqrt{\sum_{i=1}^{n} (x_i - \bar{x})^2}}$$

The larger the sum of squares of deviations of $x_1, x_2, \dots, x_n$ about their mean, the smaller will be the standard deviation of $\hat{\beta}_1$. The experimenter usually will have some experimental region, say $x_1 < x < x_2$, over which he or she wishes to observe Y, and this range will frequently be selected prior to experimentation. Then the smallest value for $\sigma_{\hat{\beta}_1}$ will occur when the n data points are equally divided, with half located at the lower boundary of the region, x_1, and half at the upper boundary, x_2. (The proof is omitted.) An experimenter who wished to fit a line by using $n = 10$ data points in the interval $2 < x < 6$ would select five data points at $x = 2$ and five at $x = 6$. Before concluding discussion of this example, you should note that observing all values of y at only two values of x will not provide information on curvature of the response curve in case the assumption of linearity in the relation of y and x is incorrect. It is frequently safer to select a few points (as few as one or two) somewhere near the middle of the experimental region to detect curvature if it should be present (see Figure 12.3).

Figure 12.3
A good design for
fitting a straight line
$(n = 10)$

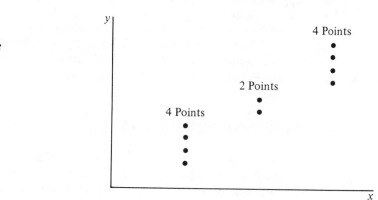

To summarize, we have given optimal designs (factor level combinations and the allocation of experimental units per combination) for comparing a pair of means and fitting a straight line. These two simple designs illustrate the manner in which information in an experiment can be increased or decreased by the selection of the factor level combinations that represent treatments and by changing the allocation of observations to treatments. We have demonstrated that factor level selection and the allocation of experimental units to treatments can greatly affect the information in an experiment pertinent to a particular population parameter and thereby can amplify nature's signal. Thus step 3, page 565, is an important consideration in the design of good experiments.

Exercises

12.7   Suppose that you wish to compare the means for two populations and that $\sigma_1^2 = 9$, $\sigma_2^2 = 25$, and $n = 90$. What allocation of $n = 90$ to the two samples will result in the maximum amount of information about $(\mu_1 - \mu_2)$?

12.8   Refer to Exercise 12.7. Suppose that you allocate $n_1 = n_2$ observations to each sample. How large must n_1 and n_2 be in order to obtain the same amount of information as that implied by the solution to Exercise 12.7?

12.9   Suppose, as in Exercise 12.7, that two populations have respective variances $\sigma_1^2 = 9$ and $\sigma_2^2 = 25$. Find the smallest sample size and the corresponding sample allocation that will yield a 95% confidence interval for $\mu_1 - \mu_2$ that is two units in length.

12.10   Refer to Exercise 12.9. How many observations are needed for a 95% confidence interval to be two units in length if $n_1 = n_2$?

12.11   Suppose that we wish to study the effect of the stimulant digitalis on the blood pressure (Y) of rats over a dosage range of $x = 2$ to $x = 5$ units. The response is expected to be linear over the region; that is, $Y = \beta_0 + \beta_1 x + \varepsilon$. Six rats are available for the experiment and each rat can receive only one dose. What dosages of digitalis should be employed in the experiment, and how many rats should be run at each dosage to maximize the quantity of information in the experiment relative to the slope β_1?

12.12   Refer to Exercise 12.11. Consider two methods for selecting the dosages. Method 1 assigns three rats to the dosage $x = 2$ and three rats to $x = 5$. Method 2 equally spaces the dosages between $x = 2$ and $x = 5$ ($x = 2, 2.6, 3.2, 3.8, 4.4, 5.0$). Suppose that σ is known and that the relationship between $E(Y)$ and x is truly linear (see Chapter 11). How much larger will be the confidence interval for the slope (β_1) for method 2 in comparison with method 1? Approximately how many observations would be required to obtain the same size of confidence interval as obtained by the optimal assignment of method 1?

12.13   Refer to Exercise 12.11. Why might it be advisable to assign one or two points at $x = 3.5$?

12.5 Noise-Reducing Experimental Designs

Noise-reducing experimental designs increase the information in an experiment by decreasing the background noise (variation) caused by uncontrolled nuisance variables. By serving as filters to screen out undesirable noise, they permit nature's signal to be received more clearly. Reduction of noise can be accomplished in step 4 of the design of an experiment, in the method for assigning treatments to the experimental units.

Designing for noise reduction is based on the single principle of making all comparisons of treatments within relatively homogeneous groups of experimental units. The more homogeneous the experimental units, the easier it is to detect a difference in treatments. Because noise-reducing, or filtering, designs work with blocks of relatively homogeneous experimental units, they are called *block* designs. We will illustrate with an example.

A manufacturer would like to compare two procedures, *A* and *B*, for measuring enzyme growth in a fermentation process. The precision of the two methods of measurement appears to be the same, but there is some question as to whether one method produces higher readings than the other.

The experiment could be performed as follows. Samples could be selected at random from the well-mixed vat of fermenting substance. Half of the samples could be measured by method *A* and half by method *B*. One could then estimate the difference between the means, $(\mu_A - \mu_B)$, or could test a hypothesis concerning their equality. Assuming near normality for the two populations of measurements, one would use estimation and test procedures based on the *t* distribution.

A second design for the experiment would choose only half as many samples. Each sample would be divided into two parts, one randomly assigned for analysis by method *A* and the other by method *B*.

Although both of the designs described above result in the same number of analyses and approximately the same cost, the second design could be preferable because it might yield a greater amount of information. The variability or noise for the first procedure would be composed of variability between samples of the fermenting substance in the vat and measurement error introduced by methods *A* and *B*. Admittedly, we expect the within-vat variability of the measurements to be small, because the substance is mixed, but lack of homogeneity between samples, however small, will exist and may be considerably larger than the variability within samples and the measurement error. This within-vat variability would then contribute to the variability of the measurements.

We would like to protect against this contingency. If nonhomogeneity exists, we would expect samples of vat contents near each other to be more nearly homogeneous than those far apart. Hence we would divide each sample and compare *A* and *B* within this relatively homogeneous block of experimental material (that is, the substance in the sample). Comparisons between *A* and *B* would be made in the presence of the reduced noise of the more homogeneous sample.

The experiment, conducted according to the second design, produced the data listed in Table 12.2. Since both methods of measurement are used on each sample, the differences between the responses can be used to compare *A* and *B*. The *i*th paired difference is given by

$$d_i = Y_{Ai} - Y_{Bi}, \quad i = 1, 2, 3, \ldots, 9$$

Table 12.2
Data for a
paired-difference
experiment

Sample	Method A	Method B
1	327.6	327.6
2	327.7	327.7
3	327.7	327.6
4	327.9	327.8
5	327.4	327.4
6	327.7	327.6
7	327.8	327.8
8	327.8	327.7
9	327.4	327.3
	$\bar{y}_A = 327.667$	$\bar{y}_B = 327.611$

The expected value of the ith paired difference may be found using Theorem 5.12:

$$\mu_d = E(d_i) = E(Y_{Ai} - Y_{Bi}) = E(Y_{Ai}) - E(Y_{Bi})$$

or

$$\mu_d = \mu_A - \mu_B$$

In other words, we may make inferences regarding the difference between the means of the methods of measurement, $(\mu_A - \mu_B)$, by making inferences regarding the mean of the differences, μ_d.

The analysis of the paired-difference experiment utilizes the nine paired differences d_i, $i = 1, 2, \ldots, 9$, shown in Table 12.3.

The sample variance of the $n = 9$ differences is

$$s_d^2 = \sum_{i=1}^{n} \frac{(d_i - \bar{d})^2}{n-1} = .002778$$

Table 12.3
The paired
differences for the
data of Table 12.2

Sample	d_i
1	0
2	0
3	.1
4	.1
5	0
6	.1
7	0
8	.1
9	.1
$\bar{d}$	.056

If the differences are normally distributed, the null hypothesis

$$H_0: \mu_A - \mu_B = \mu_d = 0$$

may be tested by using Student's t statistic,

$$t = \frac{\bar{d} - \mu_d}{s_d/\sqrt{n}}$$

For our example we have

$$t = \frac{.056}{.053/\sqrt{9}} = 3.17$$

Because we wish to test that the methods differ, we will employ a two-tailed test. The critical value of t (Table 5, Appendix III), based upon $n - 1 = 8$ degrees of freedom and $\alpha = .05$, is $t = 2.306$. Because the calculated t exceeds this value, we reject the hypothesis that $(\mu_A - \mu_B) = 0$ and conclude that the two methods have different mean responses. The attained significance level is given by p-value $= 2P(t > 3.17)$ and it follows that $.01 < p$-value $< .02$. Thus, for any value of α greater than or equal to .02 we would reject the null hypothesis and conclude that $\mu_A \neq \mu_B$. If we chose $\alpha < .01$, we cannot reject H_0. Before reaching a conclusion for a choice of α between .01 and .02 we would need to seek a more complete table of values of the t distribution.

An indication of the difference in the quantity of information in the unpaired design versus the paired design can be obtained by comparing the relative variability involved in the two resulting methods of analysis. This can be seen most easily by comparing confidence intervals for the two procedures.

The $(1 - \alpha)100\%$ confidence interval for $(\mu_A - \mu_B)$, using the paired differences, is

$$\bar{d} \pm t_{\alpha/2}\left(\frac{s_d}{\sqrt{n}}\right)$$

Substituting, we obtain a 95% confidence interval,

$$.056 \pm \frac{(2.306)(.053)}{\sqrt{9}}$$

or

$$.056 \pm .041$$

Unfortunately, we cannot calculate the $(1 - \alpha)100\%$ confidence interval for $(\mu_A - \mu_B)$ when using the unpaired design because the experiment was not conducted in an unpaired manner. Although it is difficult to state exactly what might have happened if an independent sampling design had been employed, we can approximate the confidence interval by using the variability of the paired-difference data to obtain a pooled estimate of the variance of the experimental error. Specifically, we would assume the population variances to be approximately equal,

$$\sigma_A^2 = \sigma_B^2 = \sigma^2$$

and estimate σ^2 by a pooled estimate of variance,

$$s^2 = \frac{\sum\limits_{i=1}^{n_1}(y_{Ai} - \bar{y}_A)^2 + \sum\limits_{i=1}^{n_2}(y_{Bi} - \bar{y}_B)^2}{n_1 + n_2 - 2}$$

Substituting into this formula, we obtain

$$s^2 = \frac{.2400 + .2289}{16} = .02931$$

and

$$s = .171$$

The 95% confidence interval for $(\mu_A - \mu_B)$ from this unpaired analysis of the paired data is

$$(\bar{y}_A - \bar{y}_B) \pm t_{\alpha/2}s\sqrt{\frac{1}{n_1} + \frac{1}{n_2}}$$

$$(327.667 - 327.611) \pm (2.12)(.171)\sqrt{\frac{1}{9} + \frac{1}{9}}$$

or

$$.056 \pm .171$$

We may now compare the 95% confidence intervals for $(\mu_A - \mu_B)$ for the paired and the unpaired analyses of the data for the paired design.

Paired Analysis: $.056 \pm .041$.
Unpaired Analysis: $.056 \pm .171$.

A comparison of the two intervals reveals the substantial difference in the quantity of information concerning $(\mu_A - \mu_B)$ contained in the two designs. Specifically, the paired analysis produces an interval, and hence a bound on the error of estimation, of length less than one-fourth that obtained by the unpaired analysis. Because the standard deviation of $(\bar{Y}_A - \bar{Y}_B)$ is inversely proportional to the square root of the sample size, it would require approximately $(4)^2 = 16$ times as many observations in the unpaired analysis to reduce the width of the unpaired confidence interval to that obtained from the paired design. In other words, the reduction in noise obtained by the paired design obtains the necessary information at appproximately one-sixteenth the cost required for the unpaired design.

We reiterate that a paired design must utilize the paired-difference analysis and that an unpaired analysis of the same data simply is a procedure for calculating a confidence interval that would be a good approximation to what might have been obtained had the experiment been conducted in an unpaired manner.

The paired design just described, a simple example of a randomized block design, illustrates the principle of noise reduction in increasing the information in an experiment.

Definition 12.7

A *randomized block design* containing b blocks and p treatments consists of b blocks of p experimental units each. The treatments are randomly assigned to the units in each block, with each treatment appearing exactly once in every block.

The difference between a randomized block design and the completely randomized design can be demonstrated by considering an experiment designed to compare subject reaction to a set of four stimuli (treatments) in a stimulus–response psychological experiment. We will denote the treatments as T_1, T_2, T_3, and T_4.

Suppose that eight subjects are to be randomly assigned to each of the four treatments. Random assignment of subjects to treatments (or vice versa) randomly distributes errors due to person-to-person variability to the four treatments and yields four samples that, for all practical purposes, are random and independent. This would be a completely randomized experimental design.

The experimental error associated with a completely randomized design has a number of components. Some of these are due to the difference between subjects, to the failure of repeated measurements within a subject to be identical (due to the variations in physical and psychological conditions), to the failure of the experimenter to administer a given stimulus with exactly the same intensity in repeated measurements, and, finally, to errors of measurement. Reduction of any of these causes of error will increase the information in the experiment.

The subject-to-subject variation in the experiment above can be eliminated by using subjects as blocks. Each subject would receive each of the four treatments assigned in a random sequence. The resulting randomized block design would appear as in Figure 12.4. Now only eight subjects are required to obtain eight response measurements per treatment. Note that each treatment occurs exactly once in each block.

Figure 12.4
A randomized block design

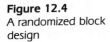

Subjects

1	2	3	4	$\cdots$	8
T_2	T_4	T_1	T_1 $\cdots$		T_2
T_1	T_2	T_3	T_4		T_3
T_4	T_1	T_2	T_3		T_4
T_3	T_3	T_4	T_2 $\cdots$		T_1

The word *randomization* in the name of the design implies that the treatments are randomly assigned within a block. For our experiment, position in the block would pertain to the position in the sequence when assigning the stimuli to a given subject over time. The purpose of the randomization (that is, position in the block) is to eliminate bias caused by fatigue or learning.

Blocks may represent time, location, or experimental material. If three treatments are to be compared and there is a suspected trend in the mean response over time, a substantial part of the time-trend variation may be removed by blocking. All three treatments would be randomly applied to experimental units in one small block of time. This procedure would be repeated in succeeding blocks of time until the required amount of data is collected. A comparison of the sale of competitive products in supermarkets should be made within supermarkets, thus using the supermarkets as blocks and removing market-to-market variability. Animal experiments in agriculture and medicine often utilize animal litters as blocks, applying all the treatments, one each, to animals within a litter. Because of heredity, animals within a litter are more homogeneous than those between litters. This type of blocking removes the litter-to-litter variation. The analysis of data generated by a randomized block design is discussed in Section 13.9.

The randomized block design is only one of many types of block designs. Blocking in two directions can be accomplished by using a Latin square design. Suppose that the subjects of the preceding example became fatigued as the stimuli were applied so that the last stimulus always produced a lower response than the first. If this trend (and consequent lack of homogeneity of the experimental units in a block) were true for all subjects, a Latin square design would be appropriate. The design would be constructed as shown in Figure 12.5. Each stimulus is applied once to each subject and occurs exactly once in each position of the order of presentation. All four stimuli occur in each row and in

Figure 12.5
A Latin square design

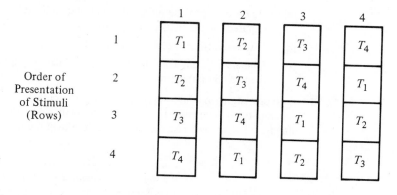

each column of the 4 × 4 configuration. The resulting design is a 4 × 4 Latin square. A Latin square design for three treatments will require a 3 × 3 configuration and, in general, p treatments will require a $p \times p$ array of experimental units. If more observations are desired per treatment, the experimenter would utilize several Latin square configurations in one experiment. In the preceding example, it would be necessary to run two Latin squares to obtain eight observations per treatment. The experiment would then contain the same number of observations per treatment as the randomized block design, Figure 12.4.

A comparison of means for any pair of stimuli would eliminate the effect of subject-to-subject variation but, in addition, would eliminate the effect of the fatigue trend within each stimulus, because each treatment was applied in each position of the stimuli-time administering sequence. Consequently, the effect of the trend would be canceled in comparing the means.

A more extensive discussion of block designs and their analyses is contained in the texts listed in the references at the end of the chapter. The objective of this section is to make you aware of the existence of block designs, how they work, and how they can produce substantial increases in the quantity of information in an experiment by reducing nuisance variation.

Exercises

12.14 Two computers often are compared by running a collection of different "benchmark" programs and recording the difference in CPU time required to complete the same program. Six benchmark programs, run on two computers, produced the following CPU times (in minutes):

Computer	*Benchmark Program*					
	1	*2*	*3*	*4*	*5*	*6*
1	1.12	1.73	1.04	1.86	1.47	2.10
2	1.15	1.72	1.10	1.87	1.46	2.15

(a) Do the data provide sufficient evidence to indicate a difference in mean CPU times required for the two computers to complete a job? Test using $\alpha = .05$.

(b) Give bounds for the associated p-value.

(c) Find a 95% confidence interval for the difference in mean CPU time required for the two computers to complete a job.

12.15 Two procedures for sintering copper are to be compared by testing each procedure on six different types of powder. The measurement of interest is the porosity (volume percentage due to voids) of each test specimen. The results of the tests were as follows:

Powder	Procedure I	Procedure II
1	21	23
2	27	26
3	18	21
4	22	24
5	26	25
6	19	16

Is there sufficient evidence to claim that Procedure II produces higher mean porosity values? Give bounds for the p-value. What would you conclude at the $\alpha = .05$ level?

12.16 An industry, in deciding whether to purchase a machine of design A or design B, checks the times for completing a certain task on each machine. Eight technicians were used in the experiment, with each technican using both machine A and machine B in a randomized order. The times (in seconds) to completion of the task are given in the accompanying table.

Technician	A	B
1	32	30
2	40	39
3	42	42
4	26	23
5	35	36
6	29	27
7	45	41
8	22	21

(a) Test to see if there is a significant difference between mean completion times, at the 5% significance level.

(b) Do you think blocking on technicians was worthwhile in this case? Explain.

(c) What assumptions are necessary for the test in part (a)?

12.17 "Muck" is the rich, highly organic type of soil that serves as the primary growth medium for vegetation in the Florida Everglades. Because of the high concentration of organic material, muck can be destroyed by a variety of natural and manmade causes. Members of the Florida Game and Fresh Water Fish Commission, under the supervision of James Schortemeyer, staked out several plots in the Everglades in May 1972. The depth of muck at each location was measured when each plot was marked, and again in October 1978. The following data (given in inches) is a portion of the data obtained.

Plot	1972 Reading	1978 Reading
1	34.5	31.5
2	44.0	37.9
3	37.5	35.5
4	27.0	23.0
5	37.0	34.5
6	40.0	31.1
7	47.2	46.0
8	35.2	31.0
9	44.0	35.2
10	40.5	37.2
11	27.0	24.7
12	29.5	25.8
13	31.5	29.0
14	35.0	36.8
15	44.0	36.5

(a) Test to see if there is sufficient evidence to indicate that there was a significant loss in average muck depth between 1972 and 1978. Give bounds on the associated p-value. What would you conclude if you desired to implement an $\alpha = .01$ level test? (Although you are free to take the necessary differences in any order that you prefer, the answer provided at the back of the book assumes that the differences were formed by taking 1978 readings minus 1972 readings.)

(b) Give a 95% confidence interval for the difference in mean muck depths for 1978 and 1972. Interpret this interval. (See remark following part (a).)

(c) Give a 95% confidence interval for the 1972 mean muck depth in the portion of the Everglades in which the study was conducted.

(d) Repeat the instructions of part (c) for the year 1978.

(e) What assumptions are necessary to apply the techniques that you used in answering parts (a) and (b)? Parts (c) and (d)?

12.18 An experiment is to be conducted to compare the effect of digitalis on the contraction of the heart muscle of a rat. The experiment is conducted by removing the heart from a live rat, slicing the heart into thin layers, and treating the layers with a dosage of digitalis. The muscle contraction is then measured. If four dosages (A, B, C, and D) are to be employed, what advantage might be derived by applying A, B, C, and D to a slice of tissue from each rat heart? What principle of design is illustrated by this example?

12.19   Complete the assignment of treatments for the following 3×3 Latin square design.

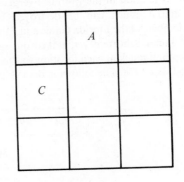

12.6 Summary

The object of this chapter has been to identify the factors that affect the quantity of information in an experiment and to use this knowledge to design better experiments. The subject, design of experiments, is very broad and certainly is not susceptible to condensation into a single chapter in an introductory text. In contrast, the philosophy underlying design, the methods for varying information in an experiment, and desirable strategies for design are easily explained and constitute the objective of this chapter.

Two factors affect the quantity of information in an experiment—the volume of nature's signal and the magnitude of variation caused by uncontrolled variables. The volume of information pertinent to a parameter of interest depends on the selection of factor level combinations (treatments) to be included in the experiment and on the allocation of the total number of experimental units to the treatments. This choice determines the focus of attention of the experimenter.

The second method for increasing the information in an experiment concerns the method for assigning treatments to the experimental units. Blocking, comparing treatments within relatively homogeneous blocks of experimental material, can be used to eliminate block-to-block variation when comparing treatments. As such, it serves as a filter to reduce the unwanted variation that tends to obscure nature's signal.

The selection of factors and the selection of factor levels are important considerations in shifting information in an experiment to amplify the information on a population parameter. The use of blocking in assigning treatments to experimental units reduces the noise created by uncontrolled variables and, consequently, increases the information in an experiment.

The analysis of some elementary experimental designs is given in Chapter 13. A more extensive treatment of the design and analysis of experiments is a course in itself. If you are interested in exploring this subject, consult the texts listed in the references at the end of the chapter.

References and Further Readings

1.   Cochran, W. G., and Cox, G. *Experimental Designs*, 2d ed. New York: Wiley, 1957.

2.   Hicks, C. R. *Fundamental Concepts in the Design of Experiments*, 3d ed. New York: Holt, Rinehart and Winston, 1982.

3.   Mendenhall, W., *An Introduction to Linear Models and the Design and Analysis of Experiments*. Belmont, Calif: Wadsworth Publishing Company, 1968.

4.   Scheaffer, R. L.; Mendenhall, W.; and Ott, L. *Elementary Survey Sampling*, 2d ed. N. Scituate, Mass.: Duxbury Press, 1979.

Supplementary Exercises

12.20   How can one measure the information in a sample pertinent to a specific population parameter?

12.21   What is a *random sample*?

12.22   Give two reasons for the use of random samples.

12.23   What is a *factor*?

12.24   If you were to design an experiment, what part of the design procedure would result in signal amplification? What part of the design procedure would result in noise reduction?

12.25   State the steps involved in designing an experiment.

12.26   Could an independent variable be a factor in one experiment and a nuisance variable (a noise contributor) in another?

12.27   Describe the factors that affect the quantity of information in an experiment and the design procedures that control these factors.

12.28   Refer to the paired-difference experiment of Section 12.5 and assume that the measurement receiving treatment i, $i = 1, 2$, in the jth pair, $j = 1, 2, \ldots, n$, is

$$Y_{ij} = \mu_i + P_j + \varepsilon_{ij}$$

where

μ_i = expected response for treatment i, $i = 1, 2$

P_j = additive random effect (positive or negative) contribution by the jth pair of experimental units, $j = 1, 2, \ldots, n$

ε_{ij} = random error associated with the experimental unit in the jth pair that receives treament i

Assume that ε_{ij} are independent normal random variables with $E(\varepsilon_{ij}) = 0$, $V(\varepsilon_{ij}) = \sigma^2$, and that P_j are independent normal random variables with $E(P_j) = 0$, $V(P_j) = \sigma_p^2$. Also, assume that P_j and ε_{ij} are independent.

(a)   Find $E(Y_{ij})$.

(b)   Find $E(\bar{Y}_i)$ and $V(\bar{Y}_i)$, where $\bar{Y}_i$ is the mean of the n observations receiving treatment i, $i = 1, 2$.

(c)   Let $\bar{d} = \bar{Y}_1 - \bar{Y}_2$. Find $E(\bar{d})$, $V(\bar{d})$, and the probability distribution for $\bar{d}$.

12.29 Refer to Exercise 12.28. Prove that

$$\frac{\bar{d}\sqrt{n}}{S_d}$$

is a Student's t statistic, under H_0: $(\mu_1 - \mu_2) = 0$.

12.30 Refer to Exercise 12.28. Suppose that a completely randomized design is employed for the comparison of the two treatment means. Then a response still could be modeled by the expression

$$Y_{ij} = \mu_i + P_j + \varepsilon_{ij}$$

but the "pair effect" (which will still affect an experimental unit), P_j, will be randomly selected and will likely differ from one of the $2n$ observations to another. Further note that, in contrast to the paired-difference experiment, the pair effects will not cancel when you calculate $(\bar{Y}_1 - \bar{Y}_2)$. Compare $V(\bar{Y}_1 - \bar{Y}_2) = V(\bar{d})$ for this design with the paired-difference design of Exercise 12.28. Why is the variance for the completely randomized design usually larger?

12.31 Persons submitting computing jobs to a computer center usually are required to estimate the amount of computer time required to complete the job. This time is measured in CPUs, the amount of time that a job will occupy a portion of the computer's central processing unit's memory. A computer center decided to perform a comparison of the estimated versus actual CPU times for a particular customer. The corresponding times were available for eleven jobs. The sample data follow.

CPU time (minutes)						Job number					
	1	*2*	*3*	*4*	*5*	*6*	*7*	*8*	*9*	*10*	*11*
Estimated	.50	1.40	.95	.45	.75	1.20	1.60	2.6	1.30	.85	.60
Actual	.46	1.52	.99	.53	.71	1.31	1.49	2.9	1.41	.83	.74

(a) Why would you expect these pairs of data to be correlated?

(b) Do the data provide sufficient evidence to indicate that, *on the average*, the customer tends to underestimate CPU time required for computing jobs? Test using $\alpha = .10$.

(c) Find the observed significance level for the test and interpret its value.

(d) Find a 90% confidence interval for the difference in mean estimated CPU time versus mean actual CPU time.

12.32 The data that follow, gathered by Runzheimer and Co., Inc., give the annual operating cost comparisons for automatic and stick-shift versions of four different makes of automobiles (*USA Today*, January 10, 1985). If the data represent a random sampling of the comparative costs of operating automobiles, find a 90% confidence interval for the mean difference in operating costs between automatic and stick-shift versions of the same type of automobile. (Operating costs include oil, maintenance, tires, and fuel.)

	Annual operating cost	
	Automatic	Manual
Ford Tempo	$1520	$1440
Plymouth Reliant	1240	1150
AMC Encore S	1380	1150
Chevrolet Cavalier	1150	1100

12.33   An experiment was conducted to compare mean reaction time to two types of traffic signs, prohibitive (No Left Turn) and permissive (Left Turn Only). Ten subjects were included in the experiment. Each subject was presented forty traffic signs, twenty prohibitive and twenty permissive, in random order. The mean time to reaction and the number of correct actions were recorded for each subject. The mean reaction times to the twenty prohibitive and twenty permissive traffic signs for each of the ten subjects follow.

Subject	Mean Reaction Times (MS) for 20 Traffic Signs	
	Prohibitive	Permissive
1	824	702
2	866	725
3	841	744
4	770	663
5	829	792
6	764	708
7	857	747
8	831	685
9	846	742
10	759	610

(a)   Explain why this is a paired-difference experiment and give reasons why the pairing should be useful in increasing information on the difference between the mean reaction times to prohibitive and permissive traffic signs.

(b)   Do the data present sufficient evidence to indicate a difference in mean reaction times to prohibitive and permissive traffic signs? Test using $\alpha = .05$.

(c)   Find and interpret the approximate p-value for the test in part (b).

(d)   Find a 95% confidence interval for the difference in mean reaction times to prohibitive and permissive traffic signs.

*12.34   Suppose that you wish to fit a model

$$Y = \beta_0 + \beta_1 x + \beta_2 x^2 + \varepsilon$$

to a set of n data points. If the n points are to be allocated at the design points $x = -1, 0,$ 1, what fraction should be assigned to each value of x so as to minimize $V(\hat{\beta}_2)$? (Assume that n is large and that k_1, k_2, and k_3, $k_1 + k_2 + k_3 = 1$, are the fractions of the total number of observations to be assigned at $x = -1, 0,$ and 1, respectively.)

THE ANALYSIS OF VARIANCE

13.1 Introduction

Most experiments involve a study of the effect of one or more independent variables on a response. In Chapter 12 we learned that a response Y can be affected by two types of independent variables, quantitative and qualitative. Those independent variables that can be controlled in an experiment are called *factors*.

The analysis of data generated by a multivariable experiment requires identification of the independent variables in the experiment. These not only will be factors (controlled independent variables) but also could be directions of blocking. If one studies the wear for three types of tires, A, B, and C, on each of four automobiles, "tire types" is a factor representing a single qualitative variable at three levels. Automobiles are blocks and represent a single qualitative variable at four levels. The response for a Latin square design depends upon the factors that represent treatments, but it is also affected by two qualitative independent block variables, "rows" and "columns."

It is not possible to present a comprehensive treatment of the analysis of multivariable experiments in a single chapter of an introductory text. However, it is possible to introduce the reasoning upon which one method of analysis, the analysis of variance, is based and to show how the technique is applied to a few common experimental designs.

The application of noise reduction and signal amplification to the design of experiments was illustrated in Chapter 12. In particular, the completely randomized and the randomized block designs were shown to be generalizations of simple designs for the unpaired and paired comparisons of means discussed in Chapters 8, 10, and 12. Treatments correspond to combinations of factor levels and identify the different populations of interest to the experimenter. This chapter presents an introduction to the analysis of variance and gives methods for the analysis of the completely randomized and randomized block designs. The analogous methods of analysis for the Latin square design are not presented in this chapter, but they can be found in the texts listed in the references at the end of the chapter.

13.2 The Analysis of Variance Procedure

The method of analysis for experiments involving several independent variables can be explained by intuitively developing the procedure or, more rigorously, by the linear model approach developed in Chapter 11. We begin by presenting an intuitive discussion of a procedure known as the *analysis of variance*. An outline of the linear model approach is presented in Section 13.13.

As the name implies, the analysis of variance procedure attempts to analyze the variation of a response and assign portions of this variation to each of a set of independent variables. The reasoning is that response variables vary only because of variation in a set of unknown independent variables. Because the experimenter rarely, if ever, will include all the variables affecting the response in an experiment, random variation in the response is observed even though all independent variables considered are held constant. The objective of the analysis of variance is to locate important independent variables in a study and determine how they interact and affect the response.

The rationale underlying the analysis of variance can be indicated best with a symbolic discussion. The actual analysis of variance—that is, how to do it—can be illustrated with an example.

You will recall that the variability of a set of n measurements is proportional to the sum of squares of deviations, $\sum_{i=1}^{n} (y_i - \bar{y})^2$, and that this quantity is used to calculate the sample variance. The analysis of variance partitions the sum of squares of deviations, called the *total sum of squares of deviations*, into parts, each of which is attributed to one of the independent variables in the experiment, plus a remainder that is associated with random error. This can be shown diagrammatically, as indicated in Figure 13.1 for three independent variables. If a multivariable linear model were written for the response, as suggested in Chapter 11, the portion of the total sum of squares of deviations assigned to error would be labeled SSE.

Figure 13.1
Partitioning of the total sum of squares of deviations

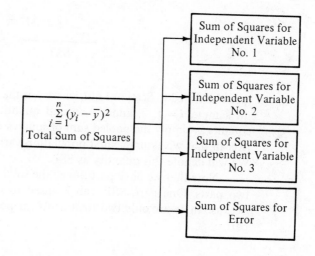

For the cases that we consider, and under the hypothesis that the independent variables are unrelated to the response, it can be shown that each of the pieces of the total sum of squares of deviations, divided by an appropriate constant, provides an independent and unbiased estimator of σ^2, the variance of the experimental error. When a variable is highly related to the response, its portion (called the *sum of squares* for that variable) will be inflated. This condition can be detected by comparing the estimate of σ^2 for a particular independent variable with that obtained from SSE using an F test (see Section 10.8). If the estimate for the independent variable is significantly larger, the F test will reject a hypothesis of no effect for the independent variable and produce evidence to indicate a relation to the response.

The mechanism involved in an analysis of variance can best be illustrated by considering a familiar example; say, the comparison of means for the unpaired experiment for the special case where $n_1 = n_2$. Assume that independent random samples have been drawn from two normal populations with means μ_1 and μ_2 and equal variances, $\sigma_1^2 = \sigma_2^2 = \sigma^2$. This experiment, formerly analyzed by the use of Student's t, will now be approached from another point of view. The total variation of the response measurements about their mean for the two samples is

$$\text{Total SS} = \sum_{i=1}^{2} \sum_{j=1}^{n_1} (y_{ij} - \bar{y})^2$$

where y_{ij} denotes the jth observation in the ith sample and $\bar{y}$ is the mean of all $2n_1 = n$ observations. This quantity can be partitioned into two parts. That is,

$$\text{Total SS} = \sum_{i=1}^{2} \sum_{j=1}^{n_1} (y_{ij} - \bar{y})^2$$

$$= \underbrace{n_1 \sum_{i=1}^{2} (\bar{y}_i - \bar{y})^2}_{\text{SST}} + \underbrace{\sum_{i=1}^{2} \sum_{j=1}^{n_1} (y_{ij} - \bar{y}_i)^2}_{\text{SSE}}$$

(proof deferred to Section 13.6), where $\bar{y}_i$ is the average of the observations in the ith sample, $i = 1, 2$, and the second quantity on the right-hand side of the equality is simply the pooled sum of squares of deviations used to calculate s^2. (Recall that we assume equal population variances for this statistical method.) We will denote this quantity as SSE.

Note that we have partitioned the total sum of squares of deviations into two parts. One part, SSE, can be used to obtain a pooled estimator of σ^2. Because there are only two treatments (or populations) and $n_1 = n_2$, the other part,

$$n_1 \sum_{i=1}^{2} (\bar{y}_i - \bar{y})^2 = \frac{n_1}{2} (\bar{y}_1 - \bar{y}_2)^2 = \text{SST}$$

which we will call the *sum of squares for treatments*, will increase as $(\bar{y}_1 - \bar{y}_2)$ increases. Hence the larger SST, the greater will be the weight of evidence to indicate a difference in $(\mu_1 - \mu_2)$.

How large is "large"? When will SST be large enough to indicate a real difference between μ_1 and μ_2?

Let Y_{ij} denote the random variable that generates the observed value y_{ij}. Because we assume that Y_{ij} is normally distributed with $E(Y_{ij}) = \mu_i$, $i = 1, 2$, and $V(Y_{ij}) = \sigma^2$, and because $\text{SSE}/(2n_1 - 2)$ is identical to the pooled estimator of σ^2 used in Chapters 8 and 10, it follows that

$$E\left(\frac{\text{SSE}}{2n_1 - 2}\right) = \sigma^2$$

and that

$$\frac{\text{SSE}}{\sigma^2} = \sum_{j=1}^{n_1} \frac{(Y_{1j} - \bar{Y}_1)^2}{\sigma^2} + \sum_{j=1}^{n_1} \frac{(Y_{2j} - \bar{Y}_2)^2}{\sigma^2}$$

has a χ^2 distribution with $(2n_1 - 2)$ degrees of freedom (see Section 8.8).

In Section 13.6 we will derive a result that implies that

$$E(\text{SST}) = \sigma^2 + \frac{n_1}{2}(\mu_1 - \mu_2)^2$$

Note that SST estimates σ^2 if $\mu_1 = \mu_2$ and a quantity larger than σ^2 if $\mu_1 \neq \mu_2$. Under the hypothesis that $\mu_1 = \mu_2$,

$$Z = \frac{\bar{Y}_1 - \bar{Y}_2}{\sqrt{2\sigma^2/n_1}}$$

has a standard normal distribution, and hence

$$Z^2 = \left(\frac{n_1}{2}\right)\left[\frac{(\bar{Y}_1 - \bar{Y}_2)^2}{\sigma^2}\right] = \frac{\text{SST}}{\sigma^2}$$

has a χ^2 distribution with 1 degree of freedom.

Because SST and SSE are independent (a fact we will not prove here), it follows from Definition 7.3 that, under the hypothesis $\mu_1 = \mu_2$,

$$\frac{\text{SST}/(1)\sigma^2}{\text{SSE}/(2n_1 - 2)\sigma^2} = \frac{\text{SST}/(1)}{\text{SSE}/(2n_1 - 2)}$$

has an F distribution with $\nu_1 = 1$ numerator degree of freedom and $\nu_2 = (2n_1 - 2)$ denominator degrees of freedom.

Sums of squares divided by their respective degrees of freedom are called *mean squares*. Thus in this case the mean square for error and the mean square for treatments are given by

$$\text{MSE} = \frac{\text{SSE}}{2n_1 - 2} \quad \text{and} \quad \text{MST} = \frac{\text{SST}}{1}$$

Under $H_0: \mu_1 = \mu_2$, both MST and MSE estimate σ^2. However, when H_0 is false and $\mu_1 \neq \mu_2$, MST estimates something larger than σ^2 and will tend to be larger than MSE. To test $H_0: \mu_1 = \mu_2$ versus $H_a: \mu_1 \neq \mu_2$, we utilize

$$F = \frac{\text{MST}}{\text{MSE}}$$

as the test statistic.

Disagreement with the null hypothesis is indicated by a large value of F, and hence the rejection region for a given α is

$$F > F_\alpha$$

Thus the analysis of variance test results in a one-tailed F test. The degrees of freedom for F are those associated with MST and MSE. As previously indicated, in the present instance F is based upon $v_1 = 1$ and $v_2 = 2n_1 - 2$ numerator and denominator degrees of freedom, respectively.

For the two-sample problem under consideration the F test just described is equivalent to the two-tailed t test of Chapter 10. As we will see in Section 13.3, the F test readily generalizes to allow comparison of any number of treatments.

EXAMPLE 13.1   The coded values for the measure of elasticity in plastic, prepared by two different processes, for samples of six drawn randomly from each of the two processes, are given in Table 13.1. Do the data present sufficient evidence to indicate a difference in mean elasticity for the two processes?

Table 13.1
Data for
Example 13.1

A	B
6.1	9.1
7.1	8.2
7.8	8.6
6.9	6.9
7.6	7.5
8.2	7.9

Solution   Although the Student's t could be used as the test statistic for this example, we will use our analysis of variance F test because it is more general and can be used to compare more than two means.

The three desired sums of squares of deviations are

$$\text{Total SS} = \sum_{i=1}^{2} \sum_{j=1}^{6} (y_{ij} - \bar{y})^2 = \sum_{i=1}^{2} \sum_{j=1}^{6} y_{ij}^2 - \frac{\left(\sum_{i=1}^{2} \sum_{j=1}^{6} y_{ij} \right)^2}{12}$$

$$= 711.35 - \frac{(91.9)^2}{12} = 7.5492$$

$$\text{SST} = n_1 \sum_{i=1}^{2} (\bar{y}_i - \bar{y})^2 = 6 \sum_{i=1}^{2} (\bar{y}_i - \bar{y})^2 = 1.6875$$

$$\text{SSE} = \sum_{i=1}^{2} \sum_{j=1}^{6} (y_{ij} - \bar{y}_i)^2 = 5.8617$$

(You may verify that SSE is the pooled sum of squares of the deviations for the two samples. Also, note that Total SS = SST + SSE.) The mean squares for treatment and error, respectively, are

$$MST = \frac{SST}{1} = 1.6875$$

$$MSE = \frac{SSE}{2n_1 - 2} = \frac{5.8617}{10} = .58617$$

To test the null hypothesis $\mu_1 = \mu_2$, we compute the test statistic

$$F = \frac{MST}{MSE} = \frac{1.6875}{.58617} = 2.88$$

The critical value of the F statistic for $\alpha = .05$ is 4.96. Although the mean square for treatments is almost three times as large as the mean square for error, it is not large enough to reject the null hypothesis. Consequently, at the $\alpha = .05$ level of significance, there is not sufficient evidence to indicate a difference between μ_1 and μ_2. The attained significance level is given by p-value $= P(F > 2.88)$ that, according to Table 7, Appendix III, is such that p-value $> .10$.

The purpose of this example was to illustrate the computations involved in a simple analysis of variance. The F test for comparing two means is equivalent to a Student's t test, because an F statistic with 1 degree of freedom in the numerator is equal to t^2. You can easily verify that the square of $t_{.025} = 2.228$ (used for the two-tailed test with $\alpha = .05$ and $v = 10$ degrees of freedom) is equal to $F_{.05} = 4.96$. Had the t test been used for Example 13.1, we would have obtained $t = -1.6967$, which satisfies the relationship $t^2 = (-1.6967)^2 = 2.88 = F$. ∽

Exercises

13.1 The reaction times for two different stimuli in a psychological word association experiment were compared by using each stimulus on independent random samples of size eight. Thus, a total of sixteen people were used in the experiment. Do the following data present sufficient evidence to indicate that there is a difference in the mean reaction times for the two stimuli?

Stimulus 1:	1	3	2	1	2	1	3	2
Stimulus 2:	4	2	3	3	1	2	3	3

(a) Use the analysis of variance approach to test the appropriate hypotheses. Test at the $\alpha = .05$ level of significance.

(b) Test the appropriate hypotheses by using the two-sample t test for comparing population means, which we developed in Section 10.7. Compare the value of the t statistic to the value of the F statistic calculated in part (a).

(c) What assumptions are necessary for the test in part (a)?

13.2 Refer to Exercise 8.73.

(a) Is there evidence of a difference between the mean ball impulses for the two types of rackets? Give bounds for the associated p-value. What would you conclude at the $\alpha = .05$ level of significance?

(b) How does the value of the F statistic obtained in part (a) compare to the value of the t statistic that would be used for testing the equality of the two means?

(c) What assumptions are necessary for the analyses performed in parts (a) and (b)?

13.3 Comparison of More than Two Means: Analysis of Variance for the Completely Randomized Design

An analysis of variance to detect a difference in a set of more than two population means is a simple generalization of the analysis of variance of Section 13.2. The random selection of independent samples from k populations is known as a *completely randomized experimental design*.

Assume that independent random samples have been drawn from k normal populations with means $\mu_1, \mu_2, \ldots, \mu_k$, respectively, and variance σ^2. All populations are assumed to possess equal variances. To be completely general, we will allow the sample sizes to be unequal and let n_i, $i = 1, 2, \ldots, k$, be the number in the sample drawn from the ith population. The total number of observations in the experiment will be $n = n_1 + n_2 + \cdots + n_k$.

Let y_{ij} denote the measured response on the jth experimental unit in the ith sample and let T_i and $\bar{T}_i$ represent the total and mean, respectively, for the observations in the ith sample. (The modification in the symbols for sample totals and averages will simplify the computing formulas for the sums of squares.) Then, as in the analysis of variance involving two means, we have

$$\text{Total SS} = \text{SST} + \text{SSE}$$

(proof deferred to Section 13.6), where

$$\text{Total SS} = \sum_{i=1}^{k} \sum_{j=1}^{n_i} (y_{ij} - \bar{y})^2 = \sum_{i=1}^{k} \sum_{j=1}^{n_i} y_{ij}^2 - \text{CM}$$

$$\text{CM} = \frac{(\text{total of all observations})^2}{n} = \frac{\left(\sum_{i=1}^{k} \sum_{j=1}^{n_i} y_{ij} \right)^2}{n} = n\bar{y}^2$$

(the symbol CM denotes *correction for the mean*),

$$SST = \sum_{i=1}^{k} n_i(\bar{T}_i - \bar{y})^2 = \sum_{i=1}^{k} \frac{T_i^2}{n_i} - CM$$

$$SSE = Total\ SS - SST$$

Although the easy way to compute SSE is by subtraction as shown earlier, it is interesting to note that SSE is the pooled sum of squares for all p samples and is equal to

$$SSE = \sum_{i=1}^{k} \sum_{j=1}^{n_i} (y_{ij} - \bar{T}_i)^2$$

$$= \sum_{i=1}^{k} (n_i - 1)S_i^2 \quad \text{where } S_i^2 = \frac{1}{n_i - 1} \sum_{j=1}^{n_i} (y_{ij} - \bar{T}_i)^2$$

The unbiased estimator of σ^2 based on $(n_1 + n_2 + \cdots + n_k - k)$ degrees of freedom is

$$S^2 = MSE = \frac{SSE}{n_1 + n_2 + \cdots + n_k - k}$$

The mean square for treatments will possess $(k - 1)$ degrees of freedom, that is, one less than the number of means, and is

$$MST = \frac{SST}{k - 1}$$

To test the null hypothesis

$$H_0: \mu_1 = \mu_2 = \cdots = \mu_k$$

against the alternative that at least one of the equalities does not hold, MST is compared with MSE, using the F statistic based upon $v_1 = (k - 1)$ and

$v_2 = \left(\sum_{i=1}^{k} n_i - k\right) = (n - k)$ degrees of freedom. The null hypothesis will be rejected if

$$F = \frac{\text{MST}}{\text{MSE}} > F_\alpha$$

where F_α is the critical value of F for the probability of a type I error of α.

Intuitively, the greater the differences among the observed treatment means, $\bar{T}_1, \bar{T}_2, \ldots, \bar{T}_k$, the greater will be the evidence to indicate a difference among their corresponding population means. It can be seen from the expression above that SST $= 0$ when all the observed treatment means are identical, because then $\bar{T}_1 = \bar{T}_2 = \cdots = \bar{T}_k = \bar{y}$ and the deviations appearing in SST, $(\bar{T}_i - \bar{y}), i = 1, 2, \ldots, k$, will equal zero. As the treatment means get farther apart, the deviations $(\bar{T}_i - \bar{y})$ will increase in absolute value and SST will increase in magnitude. Consequently, the larger the value of SST, the greater will be the weight of evidence favoring a rejection of the null hypothesis. This same line of reasoning will apply to the F tests employed in the analyses of variance for all designed experiments.

The assumptions underlying the analysis of variance F tests should receive particular attention. The samples are assumed to have been randomly selected from the k populations in an independent manner. The populations are assumed to be normally distributed with equal variances σ^2 and means $\mu_1, \mu_2, \ldots, \mu_k$. Moderate departures from these assumptions will not seriously affect the properties of the test. This is particularly true of the normality assumption.

EXAMPLE 13.2   Four groups of students were subjected to different teaching techniques and tested at the end of a specified period of time. As a result of dropouts from the experimental groups (due to sickness, transfer, and so on), the number of students varied from group to group. Do the data shown in Table 13.2 present sufficient evidence to indicate a difference in the mean achievement for the four teaching techniques?

Solution

$$\text{CM} = \frac{\left(\sum_{i=1}^{4} \sum_{j=1}^{n_i} y_{ij}\right)^2}{n} = \frac{(1779)^2}{23} = 137{,}601.8$$

$$\text{Total SS} = \sum_{i=1}^{4} \sum_{j=1}^{n_i} y_{ij}^2 - \text{CM} = 139{,}511 - 137{,}601.8 = 1909.2$$

$$\text{SST} = \sum_{i=1}^{4} \frac{T_i^2}{n_i} - \text{CM} = 138{,}314.4 - 137{,}601.8 = 712.6$$

$$\text{SSE} = \text{Total SS} - \text{SST} = 1196.6$$

Table 13.2
Data for
Example 13.2

	1	2	3	4
	65	75	59	94
	87	69	78	89
	73	83	67	80
	79	81	62	88
	81	72	83	
	69	79	76	
		90		
T_i	454	549	425	351
n_i	6	7	6	4
$\bar{T}_i$	75.67	78.43	70.83	87.75

The mean squares for treatment and error are

$$MST = \frac{SST}{k-1} = \frac{712.6}{3} = 237.5$$

$$MSE = \frac{SSE}{n_1 + n_2 + \cdots + n_k - k} = \frac{SSE}{n-k} = \frac{1196.6}{19} = 63.0$$

The test statistic for testing the hypothesis $\mu_1 = \mu_2 = \mu_3 = \mu_4$ is

$$F = \frac{MST}{MSE} = \frac{237.5}{63.0} = 3.77$$

where

$$\nu_1 = k - 1 = 3 \quad \text{and} \quad \nu_2 = \sum_{i=1}^{k} n_i - 4 = 19$$

The attained significance level is given by p-value $= P(F > 3.77)$. Using Table 7, Appendix III, we see that $.025 < p\text{-value} < .05$. Thus, if we choose $\alpha = .05$ (or any larger value), we reject the null hypothesis and conclude that there is sufficient evidence to indicate a difference in mean achievement among the four teaching procedures.

You may feel that this conclusion could have been made on the basis of visual observation of the treatment means. However, it is not difficult to construct a set of data that will lead the visual decision maker to erroneous results.

13.4 An Analysis of Variance Table for a Completely Randomized Design

The calculations of the analysis of variance are usually displayed in an analysis of variance (ANOVA or AOV) table. The table for the design of Section 13.3 involving k treatment means is shown in Table 13.3. The first column shows the source of each sum of squares of deviations; the second column gives the respective degrees of freedom; the third and forth columns give the corresponding sums of squares and mean squares, respectively. A calculated value of F, comparing MST and MSE, is usually shown in the fifth column. Note that the degrees of freedom and sums of squares add to their respective totals.

Table 13.3
ANOVA table for a completely randomized design

Source	d.f.	SS	MS	F
Treatments	$k-1$	SST	$MST = \dfrac{SST}{k-1}$	$\dfrac{MST}{MSE}$
Error	$n-k$	SSE	$MSE = \dfrac{SSE}{n-k}$	
Total	$n-1$	$\sum\limits_{i=1}^{k}\sum\limits_{j=1}^{n_i}(y_{ij}-\bar{y})^2$		

The ANOVA table for Example 13.2, shown in Table 13.4, gives a compact presentation of the appropriate computed quantities for the analysis of variance.

Table 13.4
ANOVA table for Example 13.2

Source	d.f.	SS	MS	F
Treatments	3	712.6	237.5	3.77
Error	19	1196.6	63.0	
Total	22	1909.2		

Exercises

13.3 State the assumptions underlying the analysis of variance of a completely randomized design.

13.4 Refer to Example 13.2. Calculate SSE by pooling the sums of squares of deviations within each of the four samples, and compare the answer with the value obtained by subtraction. Note that this is an extension of the pooling procedure used in the two-sample case discussed in Section 13.2.

13.5   In a comparison of the strengths of concrete produced by four experimental mixes, three specimens were prepared from each type of mix. Each of the twelve specimens was subjected to increasing compressive loads until breakdown. The accompanying table gives the compressive loads, in tons per square inch, attained at breakdown. Specimen numbers 1–12 are indicated in parentheses for identification purposes. Assuming that the requirements for a completely randomized design are met, analyze the data. State whether there is statistical support at the $\alpha = .05$ level of significance for the conclusion that at least one of the concretes differs in average strength from the others.

Mix A	Mix B	Mix C	Mix D
(1) 2.30	(2) 2.20	(3) 2.15	(4) 2.25
(5) 2.20	(6) 2.10	(7) 2.15	(8) 2.15
(9) 2.25	(10) 2.20	(11) 2.20	(12) 2.25

13.6   A clinical psychologist wished to compare three methods for reducing hostility levels in university students. A certain psychological test (HLT) was used to measure the degree of hostility. High scores on this test indicate great hostility. Eleven students obtaining high and nearly equal scores were used in the experiment. Five were selected at random from among the eleven problem cases and treated by method A. Three were taken at random from the remaining six students and treated by method B. The other three students were treated by method C. All treatments continued throughout a semester. Each student was given the HLT test again at the end of the semester, with the results shown in the accompanying table. Do the data provide sufficient evidence to indicate that at least one of the methods of treatment produces a mean student response different from the other methods? Give bounds for the attained significance level. What would you conclude at the $\alpha = .05$ level of significance?

Method A	Method B	Method C
73	54	79
83	74	95
76	71	87
68		
80		

13.7   It is believed that women in the postmenopausal phase of life suffer from a calcium deficiency. This phenomenon is associated with the relatively high proportion of bone fractures for women in that age group. Is this calcium deficiency caused by an estrogen deficiency, a condition that occurs after menopause? To investigate this theory, Richelson, Wahner, Melton, and Riggs [7] compared the bone mineral density in three groups of women.

    The first group of fourteen women had undergone oophorectomy during young adult womanhood and had lived for a period of 15 to 25 years with an estrogen deficiency. A second group, identified as perimenopausal, were approximately the same age (approximately 50 years) as the oophorectomy group except that the women had never suffered a period of estrogen deficiency. The third group of fourteen women were postmenopausal and had suffered an estrogen deficiency for an average of 20 years. The mean and standard error of the mean for the three samples of lumbar spine bone-density measurements—fourteen measurements in each sample, one for each subject—follow.

Oophorectomized Group 1		Perimenopausal Group 2		Postmenopausal Group 3	
Mean	Standard Error	Mean	Standard Error	Mean	Standard Error
0.93	0.04	1.21	0.03	0.92	0.04

Is there sufficient evidence to conclude that the mean bone-density measurements differ for the three groups of women? What is the p-value associated with your test? What would you conclude at the $\alpha = .05$ level?

13.8 If vegetables intended for human consumption contain any pesticides at all, these pesticides should occur in minute quantities. Detection of pesticides in vegetables sent to market is accomplished by using solvents to extract the pesticides from the vegetables, and then performing tests on this extract to isolate and quantify the pesticides present. The extraction process is thought to be adequate because if known amounts of pesticides are added to "clean" vegetables in a laboratory environment, essentially all of the pesticide can be recovered from the artificially contaminated extract.

The following data were obtained from a study by Wheeler, et al. who sought to determine whether the extraction process also is effective when used in the more realistic situation where pesticides are applied to vegetable crops. Dieldrin (a commonly used pesticide) labeled with (radioactive) Carbon 14 was applied to growing radishes. Fourteen days later the extraction process was used, and these radishes were analyzed for pesticide content. A liquid scintillation counter was used to determine the amount of Carbon 14 present in the extract and also the amount left behind in the vegetable pulp. Because the vegetable pulp typically is discarded when analyzing for pesticides, if an appreciable proportion of pesticide remains in this pulp, a serious underassessment of the amount of pesticide could result. The pesticide was the only source of Carbon 14, thus the proportion of Carbon 14 in the pulp is likely to be indicative of the proportion of pesticide in the pulp. The following table shows a portion of the data that the researchers obtained when low, medium, and high concentrations of the solvent, acetonitrile, were used in the extraction process.

(a) Is there sufficient evidence that the mean percentage of Carbon 14 remaining in the vegetable pulp differs for the different concentrations of acetonitrile used in the extraction process? Give bounds for the attained significance level. What would you conclude at the $\alpha = .01$ level of significance?

(b) What assumptions are necessary to validly employ the analysis that you performed in part (a)? Relate the necessary assumptions to the particular application represented in this exercise.

	Percentage of Carbon 14 in Vegetable Pulp	
Low Concentration of Acetonitrile	Medium Concentration of Acetonitrile	High Concentration of Acetonitrile
23.37	20.39	18.87
25.13	20.87	19.69
23.78	20.78	19.29
27.74	20.19	18.10
25.30	20.01	18.42
25.21	20.23	19.33
22.12	20.73	17.26
20.96	19.53	18.09
23.11	18.87	18.69
22.57	18.17	18.82
24.59	23.34	18.72
23.70	22.45	18.75
Total 287.58	245.56	224.03

Source Wheeler, W. B.; Thompson, N. P.; Edelstein, R. L.; Littel, R. C.; and Krause, R. T.; "Influence of Various Solvent-Water Mixtures on the Extraction of Dieldrin and Methomyl Residues from Radishes," *Journal of the Association of Official Analytical Chemists*, 65, no. 5 (1982): 1112–1117.

13.9 The Institute of Transportation Engineers National Student Paper Award for 1984 was given to Yean-Jye Lu for his report entitled "A Study of Left-Turn Maneuver Time for Signalized Intersections." One portion of the research described in this paper involved an evaluation of maneuver times for vehicles of various sizes that were involved in making a left turn at an intersection with a separate left turn lane, but without a separate left turn phase on the traffic light governing the intersection (an "unprotected" left-turn maneuver). The maneuver time was measured from the instant a vehicle entered the opposing lanes of traffic until it completely cleared the intersection. Four-cylinder automobiles were classified as "small cars," and six- or eight-cylinder automobiles as "large cars." Trucks and buses were combined to form a third category identified as "truck or bus." Other motorized vehicles (motorcycles, etc.) were ignored in the study. A portion of the data, which gives maneuver times (in seconds) for vehicles that attempted the left-turn maneuver from a standing stop, follows:

Vehicle Type	Sample Size	Mean	Standard Deviation
Small car	45	4.59	0.70
Large car	102	4.88	0.64
Truck or bus	18	6.24	0.90

Source Reprinted from *ITE Journal* (Vol. 54, October 1984, pp. 42–47) by permission of the Institute of Transportation Engineers, Washington, D.C. © 1984 *I.T.E.* All rights reserved.

Is there sufficient evidence to claim that the mean maneuver times differ for the three vehicle types? Give bounds for the attained significance level, and indicate the appropriate conclusion for an $\alpha = .05$ level test.

13.10   The Florida Game and Fish Commission desires to compare the amounts of residue from three chemicals found in the brain tissue of brown pelicans. Independent random samples of ten pelicans each yielded the accompanying results (measurements in parts per million). Is there evidence of sufficient differences among the mean residue amounts, at the 5% level of significance?

	Chemical		
	DDE	DDD	DDT
Mean	.032	.022	.041
Standard deviation	.014	.008	.017

13.11   Water samples were taken at four different locations in a river to determine whether the quantity of dissolved oxygen, a measure of water pollution, varied from one location to another. Locations 1 and 2 were selected above an industrial plant, one near the shore and the other in midstream; location 3 was adjacent to the industrial water discharge for the plant; and location 4 was slightly downriver in midstream. Five water specimens were randomly selected at each location, but one specimen, corresponding to location 4, was lost in the laboratory. The data are shown in the accompanying table (the greater the pollution, the lower will be the dissolved oxygen readings). Do the data provide sufficient evidence to indicate a difference in mean dissolved oxygen content for the four locations? Give bounds for the attained significance level.

Location	Mean Dissolved Oxygen Content				
1	5.9	6.1	6.3	6.1	6.0
2	6.3	6.6	6.4	6.4	6.5
3	4.8	4.3	5.0	4.7	5.1
4	6.0	6.2	6.1	5.8	

13.12   An experiment was conducted to examine the effect of age on heart rate when a subject is subjected to a specific amount of exercise. Ten male subjects were randomly selected from four age groups, 10–19, 20–39, 40–59, and 60–69. Each subject walked a treadmill at a fixed grade for a period of 12 minutes and the increase in heart rate, the difference before and after exercise, was recorded (in beats per minute). These data are shown in the table. Do the data provide sufficient evidence to indicate a difference in mean increase in heart rate among the four age groups? Test by using $\alpha = .05$.

	Age			
	10–19	*20–39*	*40–59*	*60–69*
	29	24	37	28
	33	27	25	29
	26	33	22	34
	27	31	33	36
	39	21	28	21
	35	28	26	20
	33	24	30	25
	29	34	34	24
	36	21	27	33
	22	32	33	32
Total	309	275	295	282

13.5 A Statistical Model for the Completely Randomized Design

As earlier, we let Y_{ij} denote the random variables that generate the observed values y_{ij}, $i = 1, 2, \ldots, n_i$, $j = 1, 2, \ldots, k$. The Y_{ij}'s correspond to independent random samples from normal populations with $E(Y_{ij}) = \mu_i$ and $V(Y_{ij}) = \sigma^2$, $i = 1, 2, \ldots, k$, $j = 1, 2, \ldots, n_i$. Let us consider the random sample drawn from population 1 and write

$$Y_{1j} = \mu_1 + \varepsilon_{1j}, \qquad j = 1, 2, \ldots, n_1$$

Equivalently,

$$\varepsilon_{1j} = Y_{1j} - \mu_1, \qquad j = 1, 2, \ldots, n_1$$

Because ε_{1j} simply is the difference between a normally distributed random variable and its mean, it follows that ε_{1j} is normally distributed with $E(\varepsilon_{1j}) = 0$ and $V(\varepsilon_{1j}) = V(Y_{1j}) = \sigma^2$. Further, the independence of Y_{1j}, $j = 1, 2, \ldots, n_1$ implies that $\varepsilon_{1j}, j = 1, 2, \ldots, n_1$ are mutually independent random variables. For each $i = 1, 2, \ldots, k$, we can proceed in an analogous manner to write

$$Y_{ij} = \mu_i + \varepsilon_{ij}, \qquad j = 1, 2, \ldots, n_i.$$

where the "error terms," ε_{ij} are independent, normally distributed random variables with $E(\varepsilon_{ij}) = 0$ and $V(\varepsilon_{ij}) = \sigma^2$, $i = 1, 2, \ldots, k$, $j = 1, 2, \ldots, n_i$. Note that the error terms simply represent the difference between the observations in each sample and the corresponding population mean.

One more set of considerations will lead to the classical model for the completely randomized design. Consider the means μ_i, $i = 1, 2, \ldots, k$, and write

$$\mu_i = \mu + \tau_i$$

where

$$\tau_1 + \tau_2 + \cdots + \tau_k = 0$$

Note that $\sum_{i=1}^{k} \mu_i = k\mu + \sum_{i=1}^{k} \tau_i = k\mu$, and hence $\mu = k^{-1} \sum_{i=1}^{k} \mu_i$ is just the average of the k population means (the μ_i's). For this reason, μ is generally referred to as the *overall mean*. Since for $i = 1, 2, \ldots, k$

$$\tau_i = \mu_i - \mu$$

quantifies the difference between the mean for population i and the overall mean, τ_i usually is referred to as the *effect of population* (or treatment) i. Finally, we present the classical model for the completely randomized design.

Statistical Model For a Completely Randomized Design

For $i = 1, 2, \ldots, k$ and $j = 1, 2, \ldots, n_i$

$$Y_{ij} = \mu + \tau_i + \varepsilon_{ij}$$

where

Y_{ij} = the jth observation from population (treatment) i
μ = the overall mean
τ_i = the nonrandom effect of treatment i, $\left(\sum_{i=1}^{k} \tau_i = 0 \right)$
ε_{ij} = random error terms such that ε_{ij} are independent normally distributed random variables with $E(\varepsilon_{ij}) = 0$ and $V(\varepsilon_{ij}) = \sigma^2$.

The advantage of this model is that it very clearly summarizes all of the assumptions made in the analysis of the data obtained from a completely randomized design. It also gives us a basis for presenting a precise statistical model for the randomized block design. (See Section 13.8.)

Note that $H_0: \mu_1 = \mu_2 = \cdots = \mu_k$ if and only if

$$H_0: \tau_1 = \tau_2 = \cdots = \tau_k = 0$$

is true and that $H_a: \mu_i \neq \mu_l$ for some i and l is true if and only if $H_a: \tau_i \neq 0$ for some i, $i = 1, 2, \ldots, k$, is true. Thus, the F test for equality of means that we presented in Section 13.3 is the test of the null hypothesis

$$H_0: \tau_1 = \tau_2 = \cdots = \tau_k = 0$$

versus the alternative that at least one $\tau_i \neq 0$.

Exercises

13.13   Let $\bar{T}_i$ denote the average of all of the responses to treatment i. Utilize the model for the completely randomized design to derive $E(\bar{T}_i)$ and $V(\bar{T}_i)$.

13.14   Refer to Exercise 13.13 and consider $\bar{T}_i - \bar{T}_l$ for $i \neq l$.

(a)   Show that $E(\bar{T}_i - \bar{T}_l) = \mu_i - \mu_l = \tau_i - \tau_l$. Note that this result implies that $\bar{T}_i - \bar{T}_l$ is an unbiased estimator of the difference in the effects of treatment i and l.

(b)   Derive $V(\bar{T}_i - \bar{T}_l)$.

13.15   Refer to the model for the completely randomized design.

(a)   Show that $H_0: \tau_1 = \tau_2 = \cdots = \tau_k = 0$ is true if and only if $H_0: \mu_1 = \mu_2 = \cdots = \mu_k$ is true.

(b)   Show that $H_a: \tau_i \neq 0$ for at least one i is true if and only if $H_a: \mu_i \neq \mu_l$ for some $i \neq l$ is true.

13.6 Proof of Additivity of the Sums of Squares and E(MST) for a Completely Randomized Design (Optional)

The proof that

$$\text{Total SS} = \text{SST} + \text{SSE}$$

for the completely randomized design is presented in this section for the benefit of those who are interested. It may be omitted without loss of continuity.

The proof utilizes elementary results on summations that appear in the exercise for Chapter 1 and the device of adding and subtracting $\bar{T}_i$ within the expression for the Total SS. Thus

$$\text{Total SS} = \sum_{i=1}^{k} \sum_{j=1}^{n_i} (y_{ij} - \bar{y})^2 = \sum_{i=1}^{k} \sum_{j=1}^{n_i} (y_{ij} - \bar{T}_i + \bar{T}_i - \bar{y})^2$$

$$= \sum_{i=1}^{k} \sum_{j=1}^{n_i} [(y_{ij} - \bar{T}_i) + (\bar{T}_i - \bar{y})]^2$$

$$= \sum_{i=1}^{k} \sum_{j=1}^{n_i} [(y_{ij} - \bar{T}_i)^2 + 2(y_{ij} - \bar{T}_i)(\bar{T}_i - \bar{y}) + (\bar{T}_i - \bar{y})^2]$$

Summing first over j, we obtain

$$\text{Total SS} = \sum_{i=1}^{k} \left[\sum_{j=1}^{n_i} (y_{ij} - \bar{T}_i)^2 + 2(\bar{T}_i - \bar{y}) \sum_{j=1}^{n_i} (y_{ij} - \bar{T}_i) + n_i (\bar{T}_i - \bar{y})^2 \right]$$

where

$$\sum_{j=1}^{n_i} (y_{ij} - \bar{T}_i) = T_i - n_i \bar{T}_i = T_i - T_i = 0$$

Consequently, the middle term in the expression for the Total SS is equal to zero.

Then summing over i, we obtain

$$\text{Total SS} = \sum_{i=1}^{k} \sum_{j=1}^{n_i} (y_{ij} - \bar{T}_i)^2 + \sum_{i=1}^{k} n_i (\bar{T}_i - \bar{y})^2 = \text{SSE} + \text{SST}$$

The first expression is SSE, the pooled sum of squares of deviations of the sample measurements about their respective means. The second is the formula for SST.

Proof of the additivity of the analysis of variance sums of squares for other experimental designs can be obtained in a similar manner. The procedure is tedious.

We now proceed with the derivation of the expected value of MST for a completely randomized design. Using the statistical model for the completely randomized design presented in Section 13.5, it follows that

$$\bar{T}_i = \frac{1}{n_i} \sum_{j=1}^{n_i} Y_{ij} = \frac{1}{n_i} \sum_{j=1}^{n_i} (\mu + \tau_i + \varepsilon_{ij}) = \mu + \tau_i + \bar{\varepsilon}_i$$

where

$$\bar{\varepsilon}_i = \frac{1}{n_i} \sum_{j=1}^{n_i} \varepsilon_{ij}$$

Note that because the ε_{ij} are independent random variables with $E(\varepsilon_{ij}) = 0$ and $V(\varepsilon_{ij}) = \sigma^2$, Theorem 5.12 implies (see Example 5.27) that $E(\bar{\varepsilon}_i) = 0$ and $V(\bar{\varepsilon}_i) = \sigma^2 / n_i$.

In a completely analogous manner, $\bar{Y}$ is given by

$$\bar{Y} = \frac{1}{n} \sum_{i=1}^{k} \sum_{j=1}^{n_i} Y_{ij} = \frac{1}{n} \sum_{i=1}^{k} \sum_{j=1}^{n_i} (\mu + \tau_i + \varepsilon_{ij}) = \mu + \bar{\tau} + \bar{\varepsilon}$$

where

$$\bar{\tau} = \frac{1}{n}\sum_{i=1}^{k}n_i\tau_i \quad \text{and} \quad \bar{\varepsilon} = \frac{1}{n}\sum_{i=1}^{k}\sum_{j=1}^{n_i}\varepsilon_{ij}$$

Note that $\bar{\tau}$ is simply a constant and again using Theorem 5.12, $E(\bar{\varepsilon}) = 0$ and $V(\bar{\varepsilon}) = \sigma^2/n$.

Therefore, with respect to the terms in the model for the completely randomized design,

$$\text{MST} = \left(\frac{1}{k-1}\right)\sum_{i=1}^{k}n_i(\bar{T}_i - \bar{Y})^2 = \left(\frac{1}{k-1}\right)\sum_{i=1}^{k}n_i(\tau_i + \bar{\varepsilon}_i - \bar{\tau} - \bar{\varepsilon})^2$$

$$= \left(\frac{1}{k-1}\right)\sum_{i=1}^{k}n_i(\tau_i - \bar{\tau})^2 + \left(\frac{1}{k-1}\right)\sum_{i=1}^{k}2n_i(\tau_i - \bar{\tau})(\bar{\varepsilon}_i - \bar{\varepsilon})$$

$$+ \left(\frac{1}{k-1}\right)\sum_{i=1}^{k}n_i(\bar{\varepsilon}_i - \bar{\varepsilon})^2$$

Because $\bar{\tau}$ and τ_i, $i = 1, 2, \ldots, k$ are constants and $E(\varepsilon_i) = E(\bar{\varepsilon}) = 0$, it follows that

$$E(\text{MST}) = \left(\frac{1}{k-1}\right)\sum_{i=1}^{k}n_i(\tau_i - \bar{\tau})^2 + \left(\frac{1}{k-1}\right)E\left[\sum_{i=1}^{k}n_i(\bar{\varepsilon}_i - \bar{\varepsilon})^2\right]$$

Note that

$$\sum_{i=1}^{k}n_i(\bar{\varepsilon}_i - \bar{\varepsilon})^2 = \sum_{i=1}^{k}(n_i\bar{\varepsilon}_i^2 - 2n_i\bar{\varepsilon}_i\bar{\varepsilon} + n_i\bar{\varepsilon}^2)$$

$$= \sum_{i=1}^{k}n_i\bar{\varepsilon}_i^2 - 2n\bar{\varepsilon}^2 + n\bar{\varepsilon}^2 = \sum_{i=1}^{k}n_i\bar{\varepsilon}_i^2 - n\bar{\varepsilon}^2$$

Because $E(\bar{\varepsilon}_i) = 0$ and $V(\bar{\varepsilon}_i) = \sigma^2/n_i$, it follows that $E(\bar{\varepsilon}_i^2) = \sigma^2/n_i$, $i = 1, 2, \ldots, k$. Similarly, $E(\bar{\varepsilon}^2) = \sigma^2/n$ and hence

$$E\left[\sum_{i=1}^{k}n_i(\bar{\varepsilon}_i - \bar{\varepsilon})^2\right] = \sum_{i=1}^{k}n_i E(\bar{\varepsilon}_i^2) - nE(\bar{\varepsilon}^2) = k\sigma^2 - \sigma^2 = (k-1)\sigma^2$$

Summarizing, we obtain

$$E(\text{MST}) = \sigma^2 + \left(\frac{1}{k-1}\right)\sum_{i=1}^{k}n_i(\tau_i - \bar{\tau})^2$$

where

$$\bar{\tau} = \frac{1}{n} \sum_{i=1}^{k} n_i \tau_i$$

Under $H_0: \tau_1 = \tau_2 = \cdots = \tau_k = 0$, $\bar{\tau} = 0$, and hence $E(\text{MST}) = \sigma^2$. Thus, when H_0 is true, MST/MSE is the ratio of two unbiased estimators for σ^2.

13.7 Estimation in the Completely Randomized Design

Confidence intervals for a single treatment mean and for the difference between a pair of treatment means based upon data obtained by using a completely randomized design (Section 13.3) are identical to those given in Chapter 8. The confidence interval for the mean of treatment i or the difference between treatments i and j are, respectively, as follows:

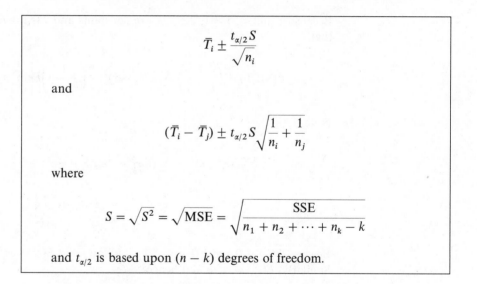

$$\bar{T}_i \pm \frac{t_{\alpha/2} S}{\sqrt{n_i}}$$

and

$$(\bar{T}_i - \bar{T}_j) \pm t_{\alpha/2} S \sqrt{\frac{1}{n_i} + \frac{1}{n_j}}$$

where

$$S = \sqrt{S^2} = \sqrt{\text{MSE}} = \sqrt{\frac{\text{SSE}}{n_1 + n_2 + \cdots + n_k - k}}$$

and $t_{\alpha/2}$ is based upon $(n - k)$ degrees of freedom.

Note that the confidence intervals just stated are appropriate for single treatment means or a comparison of a pair of means selected prior to observation of the data. The stated confidence coefficients are based on random sampling. If we were to look at the data and always compare the largest and smallest sample means, certainly the difference between the largest and smallest sample means would be expected to be larger than for a pair of means specified to be of interest before observing the data.

EXAMPLE 13.3 Find a 95% confidence interval for the mean score for teaching technique 1, Example 13.2.

Solution The 95% confidence interval for the mean score is

$$\bar{T}_1 \pm \frac{t_{.025}s}{\sqrt{6}}$$

or

$$75.67 \pm \frac{(2.093)(7.94)}{\sqrt{6}}$$

or

$$75.67 \pm 6.78$$

EXAMPLE 13.4 Find a 95% confidence interval for the difference in mean score for teaching techniques 1 and 4, Example 13.2.

Solution The 95% confidence interval is

$$(\bar{T}_1 - \bar{T}_4) \pm (2.093)(7.94)\sqrt{1/6 + 1/4}$$

or

$$-12.08 \pm 10.73$$

Hence the 95% confidence interval for $(\mu_1 - \mu_4)$ is -22.81 to -1.35. This suggests that $\mu_4 > \mu_1$.

Exercises

13.16 Refer to Exercise 13.7. As noted in the description of the experiment, the oopherecto-mized and the perimenopausal groups of women were of approximately the same age, but those in the oophorectomized group suffered from an estrogen deficiency. Form a 95% confidence interval for the difference in mean bone densities for these two groups of women. Would you conclude that the mean bone densities for the oopherectomized and perimenopausal women are significantly different? Why?

13.17 Refer to Exercise 13.5. Let μ_A and μ_B denote the mean strengths of concrete specimens prepared for mix A and mix B, respectively.

(a) Find a 90% confidence interval for μ_A.

(b) Find a 95% confidence interval for $(\mu_A - \mu_B)$.

13.18   Refer to Exercise 13.6. Let μ_A and μ_B, respectively, denote the mean scores at the end of the semester for the populations of extremely hostile students who were treated throughout that semester by method A and method B.

(a)   Find a 95% confidence interval for μ_A.

(b)   Find a 95% confidence interval for μ_B.

(c)   Find a 95% confidence interval for $(\mu_A - \mu_B)$.

13.19   Refer to Exercise 13.8.

(a)   Construct a 95% confidence interval for the mean percentage of Carbon 14 that remains in the vegetable pulp when the low level of acetonitrile is used.

(b)   Give a 90% confidence interval for the difference in mean percentages of Carbon 14 that remains in the vegetable pulp for low and medium levels of acetonitrile.

13.20   Refer to Exercise 13.9.

(a)   Give a 95% confidence interval for the mean left-turn maneuver time for buses and trucks.

(b)   Estimate the difference in mean maneuver times for small and large cars with a 95% confidence interval.

(c)   The study report on by Lu involved vehicles that passed through the intersection of Guadalupe Avenue and 38th Street in Austin, Texas. Do you think that the results in parts (a) and (b) would be valid for a "nonprotected" intersection in your hometown? Why or why not?

13.21   It has been hypothesized that treatment (after casting) of a plastic used in optic lenses will improve wear. Four different treatments are to be tested. To determine whether any differences in mean wear exist among treatments, twenty-eight castings from a single formulation of the plastic were made and seven castings were randomly assigned to each of the treatments. Wear was determined by measuring the increase in "haze" after 200 cycles of abrasion (better wear being indicated by small increases).

Treatment			
A	B	C	D
9.16	11.95	11.47	11.35
13.29	15.15	9.54	8.73
12.07	14.75	11.26	10.00
11.97	14.79	13.66	9.75
13.31	15.48	11.18	11.71
12.32	13.47	15.03	12.45
11.78	13.06	14.86	12.38

(a)   Is there evidence of a difference in mean wear among the four treatments? Use $\alpha = 0.05$.

(b)   Estimate the mean difference in haze increase between treatments B and C using a 99% confidence interval.

(c)   Find a 90% confidence interval for the mean wear for lenses receiving treatment A.

13.22 With the ongoing energy crisis, researchers for the major oil companies are attempting to find alternative sources of oil. It is known that some types of shale contain small amounts of oil that feasibly (if not economically) could be extracted. Four methods have been developed for extracting oil from shale, and the government has decided that some experimentation should be done to determine if the methods differ significantly in the average amount of oil that each is able to extract from the shale. Method 4 is known to be the most expensive method to implement, and method 1 is the least expensive, so inferences about the differences in performance of these two methods are of particular interest, also. Sixteen bits of shale (of the same size) were randomly subjected to the four methods, with the accompanying results (the units are in liters per cubic meter). All inferences are to be made with $\alpha = .05$.

Method 1	Method 2	Method 3	Method 4
3	2	5	5
2	2	2	2
1	4	5	4
2	4	1	5

(a) Assuming that the sixteen experimental units were as alike as possible, do the appropriate analysis of variance to determine if there is any significant difference among the mean amounts extracted by the four methods. Use $\alpha = .05$.

(b) Set up a 95% confidence interval for the difference in the mean amounts extracted by the two methods of particular interest. Interpret the result.

13.23 Refer to Exercise 13.10. Construct a 95% confidence interval for the mean amount of residue from DDT.

13.24 Refer to Exercise 13.11. Compare the mean dissolved oxygen content in midstream above the plant with the mean content adjacent to the plant (location 2 versus location 3). Use a 95% confidence interval.

13.25 Refer to Exercise 13.11. Compare the mean dissolved oxygen content for the two locations above the plant with the mean content slightly downriver from the plant by finding a 95% confidence interval for $(1/2)(\mu_1 + \mu_2) - \mu_4$.

13.26 Refer to Exercise 13.12.

(a) Find a 90% confidence interval for the difference in mean increase in heart rate between the 10–19 age group and the 60–69 age group.

(b) Find a 90% confidence interval for the mean increase in heart rate for the 20–39 age group.

13.8 A Statistical Model for the Randomized Block Design

The method for constructing a randomized block design was presented in Section 12.5. As previously indicated in Definition 12.7, the randomized block design is a design for comparing k treatments utilizing b blocks. The treatments are randomly assigned to the experimental units in each block in such a way that each treatment appears exactly once in each of the b blocks. Thus, the total number of observations obtained in a randomized block design is $n = bk$. Implicit in the consideration of a randomized block design is the presence of two qualitative independent variables, "blocks" and "treatments." In this section we present a formal statistical model for the randomized block design.

Statistical Model for a Randomized Block Design

For $i = 1, 2, \ldots, k$ and $j = 1, 2, \ldots, b$,

$$Y_{ij} = \mu + \tau_i + \beta_j + \varepsilon_{ij}$$

where

$Y_{ij} =$ the observation on treatment i in block j

$\mu =$ the overall mean

$\tau_i =$ the nonrandom effect of treatment i, $\left(\sum_{i=1}^{k} \tau_i = 0 \right)$

$\beta_j =$ the nonrandom effect of block j, $\left(\sum_{j=1}^{b} \beta_j = 0 \right)$

$\varepsilon_{ij} =$ random error terms such that ε_{ij} are independent normally distributed random variables with $E(\varepsilon_{ij}) = 0$ and $V(\varepsilon_{ij}) = \sigma^2$

Note that $\mu, \tau_1, \tau_2, \ldots, \tau_k$ and $\beta_1, \beta_2, \ldots, \beta_b$ are all assumed to be unknown constants. This model differs from that for the completely randomized design only in containing parameters associated with the different blocks. Because the block effects are assumed to be fixed but unknown, this model usually is referred to as the *fixed block effects* model. Another model for the randomized block design in which the β's are assumed to be random variables (called the *random block effects* model) is considered in the supplementary exercises. Our formal development in the body of this text is restricted to the fixed block effects model.

The statistical model just presented very clearly summarizes all of the assumptions made in the analysis of data in a fixed blocks effect randomized block design. Let us consider the observation made on treatment i in block j, Y_{ij}. Note that the assumptions in the model imply that $E(Y_{ij}) = \mu + \tau_i + \beta_j$ and $V(Y_{ij}) = \sigma^2$ for $i = 1, 2, \ldots, k$ and $j = 1, 2, \ldots, b$. Let us consider the observations made on treatment i, and observe that two observations receiving treatment i have means that differ only by the difference of the block effects. For example,

$$E(Y_{i1}) - E(Y_{i2}) = \mu + \tau_i + \beta_1 - (\mu + \tau_i + \beta_2) = \beta_1 - \beta_2$$

Similarly, two observations that are taken from the same block have means that differ only by the difference of the treatment effects. That is, if $i \neq i'$,

$$E(Y_{ij}) - E(Y_{i'j}) = \mu + \tau_i + \beta_j - (\mu + \tau_{i'} + \beta_j) = \tau_i - \tau_{i'}$$

Observations that are taken on different treatments and in different blocks have means that differ by the difference in the treatment effects plus the difference in the block effects because if $i \neq i'$ and $j \neq j'$,

$$E(Y_{ij}) - E(Y_{i'j'}) = \mu + \tau_i + \beta_j - (\mu + \tau_{i'} + \beta_{j'}) = (\tau_i - \tau_{i'}) - (\beta_j - \beta_{j'})$$

In the next section, we will proceed with an analysis of the data obtained from a randomized block design.

Exercises

13.27   State the assumptions underlying the analysis of variance for a randomized block design.

13.28   Let $\bar{T}_i$ denote the average of all of the responses to treatment i. Utilize the model for the randomized block design to derive $E(\bar{T}_i)$ and $V(\bar{T}_i)$. Is $\bar{T}_i$ an unbiased estimator for the mean response to treatment i? Explain why or why not.

13.29   Refer to Exercise 13.28 and consider $\bar{T}_i - \bar{T}_l$ for $i \neq l$.

   (a)   Show that $E(\bar{T}_i - \bar{T}_l) = \mu_i - \mu_l = \tau_i - \tau_l$. Note that this result implies that $\bar{T}_i - \bar{T}_l$ is an unbiased estimator of the difference in the effects of treatment i and l.

   (b)   Derive $V(\bar{T}_i - \bar{T}_l)$.

13.30   Refer to the model for the randomized block design, and let $\bar{B}_j$ denote the average of all of the responses in block j.

   (a)   Derive $E(\bar{B}_j)$ and $V(\bar{B}_j)$.

   (b)   Show that $\bar{B}_j - \bar{B}_l$ is an unbiased estimator for $\beta_j - \beta_l$ the difference in the effects of blocks j and l.

   (c)   Derive $V(\bar{B}_j - \bar{B}_l)$.

13.9 The Analysis of Variance for a Randomized Block Design

The analysis of variance for a randomized block design proceeds much like that for a completely randomized design. In this case, the total sum of squares of deviations of the response measurements from their mean may be partitioned into three parts: the sum of squares for blocks, treatments, and error.

Denote the total and average of all observations in block j as B_j and $\bar{B}_j$, respectively. Similarly, let T_i and $\bar{T}_i$ represent the total and the average for all observations receiving treatment i. Then for a randomized block design involving b blocks and k treatments, we have the following sums of squares:

$$\text{Total SS} = \text{SSB} + \text{SST} + \text{SSE}$$

$$= \sum_{j=1}^{b} \sum_{i=1}^{k} (y_{ij} - \bar{y})^2 = \sum_{j=1}^{b} \sum_{i=1}^{k} y_{ij}^2 - \text{CM}$$

$$\text{SSB} = k \sum_{j=1}^{b} (\bar{B}_j - \bar{y})^2 = \frac{\sum_{j=1}^{b} B_j^2}{k} - \text{CM}$$

$$\text{SST} = b \sum_{i=1}^{k} (\bar{T}_i - \bar{y})^2 = \frac{\sum_{i=1}^{k} T_i^2}{b} - \text{CM}$$

$$\text{SSE} = \text{Total SS} - \text{SSB} - \text{SST}$$

In the formulas

$$\bar{y} = (\text{average of all } n = bk \text{ observations}) = \frac{\sum_{j=1}^{b} \sum_{i=1}^{k} y_{ij}}{n}$$

and

$$\text{CM} = \frac{(\text{total of all observations})^2}{n} = \frac{\left(\sum_{j=1}^{b} \sum_{i=1}^{k} y_{ij} \right)^2}{n}$$

The analysis of variance for the randomized block design is presented in Table 13.5. The degrees of freedom associated with each sum of squares are

Table 13.5
ANOVA table for a randomized block design

Source	d.f.	SS	MS
Blocks	$b - 1$	SSB	$\dfrac{\text{SSB}}{b - 1}$
Treatments	$k - 1$	SST	$\dfrac{\text{SST}}{k - 1}$
Error	$n - b - k + 1$	SSE	MSE
Total	$n - 1$	Total SS	

shown in the second column. Mean squares are calculated by dividing the sum of squares by their respective degrees of freedom.

To test the null hypothesis that there is no difference in treatment means, we use the F statistic,

$$F = \frac{\text{MST}}{\text{MSE}}$$

and reject if $F > F_\alpha$ based on $v_1 = (k - 1)$ and $v_2 = (n - b - k + 1)$ degrees of freedom.

Blocking not only reduces the experimental error, it also provides an opportunity to see whether evidence exists to indicate a difference in the mean response for blocks. Under the null hypothesis that there is no difference in mean response for blocks, MSB provides an unbiased estimator for σ^2 based on $(b - 1)$ degrees of freedom. Where real differences exist among block means, MSB will tend to be inflated in comparison with MSE and

$$F = \frac{\text{MSB}}{\text{MSE}}$$

provides a test statistic. As in the test for treatments, the rejection region for the test will be

$$F > F_\alpha$$

based on $v_1 = (b - 1)$ and $v_2 = (n - b - k + 1)$ degrees of freedom.

EXAMPLE 13.5 A stimulus–response experiment involving three treatments was laid out in a randomized block design using four subjects. The response was the length of time to reaction measured in seconds. The data, arranged in blocks, are shown in Figure 13.2. The treatment number is circled and shown above each observation. Do the data present sufficient evidence to indicate a difference in

Figure 13.2
Randomized block
design for
Example 13.5

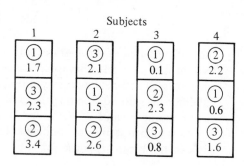

the mean response for stimuli (treatments)? Subjects? Use $\alpha = .05$ for each test and give the associated p-values.

Solution

The sums of squares for the analysis of variance are shown individually below and jointly in Table 13.6. Thus

$$CM = \frac{(\text{total})^2}{n} = \frac{(21.2)^2}{12} = 37.45$$

$$\text{Total SS} = \sum_{j=1}^{4} \sum_{i=1}^{3} (y_{ij} - \bar{y})^2 = \sum_{j=1}^{4} \sum_{i=1}^{3} y_{ij}^2 - CM = 46.86 - 37.45 = 9.41$$

$$SSB = \frac{\sum_{j=1}^{4} B_j^2}{3} - CM = 40.93 - 37.45 = 3.48$$

$$SST = \frac{\sum_{i=1}^{3} T_i^2}{4} - CM = 42.93 - 37.45 = 5.48$$

$$SSE = \text{Total SS} - SSB - SST = 9.41 - 3.48 - 5.48 = .45$$

Table 13.6
ANOVA table for
Example 13.5

Source	d.f.	SS	MS	F
Blocks	3	3.48	1.160	15.47
Treatments	2	5.48	2.740	36.53
Error	6	.45	.075	
Total	11	9.41		

We use the ratio of mean square for treatments to mean square for error to test a hypothesis of no difference in the expected response for treatments. Thus

$$F = \frac{MST}{MSE} = \frac{2.74}{.075} = 36.53$$

The critical value of the F statistic ($\alpha = .05$) for $v_1 = 2$ and $v_2 = 6$ degrees of freedom is $F_{.05} = 5.14$. Because the computed value of F exceeds the critical value, there is sufficient evidence to reject the null hypothesis and conclude that real differences do exist among the expected responses for the three stimuli. The corresponding p-value $= P(F > 36.53)$, which is such that p-value $< .005$.

A similar test may be conducted for the null hypothesis that no difference exists in the mean response for subjects. Rejection of this hypothesis would imply that subject-to-subject variability does exist and that blocking is desirable. The computed value of F based on $v_1 = 3$ and $v_2 = 6$ degrees of freedom is

$$F = \frac{MSB}{MSE} = \frac{1.16}{.075} = 15.47$$

Because this value of F exceeds the corresponding tabulated critical value, $F_{.05} = 4.76$, we reject the null hypothesis and conclude that a real difference exists in the expected response among the group of subjects. The associated p-value $= P(F > 15.47) < .005$.

Exercises

13.31   The accompanying data are yields on resistance to stain for three materials (M_1, M_2, and M_3) treated with four chemicals in a randomized block design. (A low value indicates good stain resistance.) Is there evidence of differences in mean resistance among the four chemicals? Give bounds for the p-value. What would you conclude at the $\alpha = .05$ level of significance?

	Material			
Chemical	M_1	M_2	M_3	Total
A	5	9	7	21
B	3	8	4	15
C	8	13	9	30
D	4	6	8	18
Total	20	36	28	84

$$\sum_i \sum_j y_{ij}^2 = 674, \qquad \frac{1}{12}\left(\sum_i \sum_j y_{ij}\right)^2 = 588$$

13.32   An experiment was conducted to determine the effect of three methods of soil preparation on the first-year growth of slash pine seedlings. Four locations (state forest lands) were selected and each location was divided into three plots. Because it was felt that soil fertility within a location was more homogeneous than that between locations, a randomized block design was employed, using locations as blocks. The methods of soil preparation were A (no preparation), B (light fertilization), and C (burning). Each soil preparation was randomly applied to a plot within each location. On each plot the same number of seedlings was planted, and the observation recorded was the average first-year growth (in centimeters) of the seedlings on each plot.

(a)   Conduct an analysis of variance. Do the data provide sufficient evidence to indicate a difference in the mean growth for the three soil preparations?

(b)   Is there evidence to indicate a difference in mean growth for the four locations?

	Location			
Soil Preparation	1	2	3	4
A	11	13	16	10
B	15	17	20	12
C	10	15	13	10

13.33   Dudeck and Peacock [1] reported on an experiment conducted to evaluate the performance of several coolseason grasses for winter overseeding of golf greens in northern Florida. One of the variables of interest was the distance that a golf ball would roll on a green after being rolled down a ramp (used to induce a constant initial velocity to the ball). Because the distance the ball would roll was influenced by the slope of the green and the direction in which the grass was mowed, the experiment was set up in a randomized blocks design. The blocks were determined so that the slopes of the individual plots were constant within blocks (a transit was used to ensure accuracy), and all plots were mowed in the same direction and at the same height to eliminate mowing effects. The base grass was "Tiftgreen" bermuda grass in a semidormant state. The same method of seeding and rates of application were used for all of the ryegrasses that appear in the portion of the following data. Measurements are average distances (in meters) from the base of the ramp to the stopping points for five balls rolled down the ramp and directly up the slope on each plot. Cultivars used in the study included A (Pennfine ryegrass), B (Dasher ryegrass), C (Regal ryegrass), D (Marvelgreen supreme), and E (Barry ryegrass). The grasses were planted within blocks, and resulted in the following measurements:

	Variety					
Block	A	B	C	D	E	Total
1	2.764	2.568	2.506	2.612	2.238	12.688
2	3.043	2.977	2.533	2.675	2.616	13.844
3	2.600	2.183	2.334	2.164	2.127	11.408
4	3.049	3.028	2.895	2.724	2.697	14.393
Total	11.456	10.756	10.268	10.175	9.678	52.333

(a)   Perform the appropriate analysis of variance to test whether there is sufficient evidence to indicate that the mean distance of ball roll differs for the five cultivars. Give bounds for the attained significance level. What would you conclude at the $\alpha = .01$ level of significance?

(b)   Is there evidence of a significant difference between the "blocks" used in the experiment? Test using $\alpha = .05$.

13.34   Refer to Exercise 13.22. Suppose we now find out that the sixteen experimental units were obtained in the following manner. One sample was taken from each of four locations, each individual sample was split into four parts, and then each of the methods was applied to exactly one part from each location (with the proper randomization). The data are now presented more correctly in the form shown in the accompanying table. Does this new information suggest a more appropriate method of analysis than that used in Exercise 13.22? If so, perform the new analysis and answer the question in part (a) of Exercise 13.22. Is this new information worthwhile?

Location	Method 1	Method 2	Method 3	Method 4
I	3	2	5	5
II	2	2	2	2
III	1	4	5	4
IV	2	4	1	5

13.35   Suppose that a randomized block design with b blocks and k treatments has each treatment measured *twice* in each block. Indicate how you would do the computations for an analysis of variance.

13.36   An evaluation of diffusion bonding of zircaloy components is performed. The main objective is to determine which of three elements—nickel, iron, or copper—is the best bonding agent. A series of zircaloy components are bonded using each of the possible bonding agents. Because there is a great deal of variation in components machined from different ingots, a randomized block design is used, blocking on the ingots. A pair of components from each ingot are bonded together using each of the three agents, and the pressure (in units of 1000 pounds per square inch) required to separate the bonded components is measured. The following data are obtained.

	Bonding Agent		
Ingot	Nickel	Iron	Copper
1	67.0	71.9	72.2
2	67.5	68.8	66.4
3	76.0	82.6	74.5
4	72.7	78.1	67.3
5	73.1	74.2	73.2
6	65.8	70.8	68.7
7	75.6	84.9	69.0

Is there evidence of a difference in pressure required to separate the components among the three bonding agents? Use $\alpha = 0.05$.

13.37   From time to time, one branch office of a company must make shipments to a certain branch office in another state. There are three package delivery services between the two cities where the branch offices are located. Because the price structures for the three delivery services are quite similar, the company wants to compare the delivery times. The company plans to make several different types of shipments to its branch office. To compare the carriers, each shipment will be sent in triplicate, one with each carrier. The results listed in the table are the delivery times in hours.

	Carrier		
Shipment	I	II	III
1	15.2	16.9	17.1
2	14.3	16.4	16.1
3	14.7	15.9	15.7
4	15.1	16.7	17.0
5	14.0	15.6	15.5

Is there evidence of a difference in mean delivery times among the three carriers? Give bounds for the attained significance level.

*13.38   Refer to the model for the randomized block design presented in Section 13.8.

(a)   Derive $E(MST)$.

(b)   Derive $E(MSB)$.

(c)   Derive $E(MSE)$.

Note that these quantities appear in the F statistics used to test for differences in the mean response among the blocks and among the treatments.

13.10 Estimation in the Randomized Block Design

The confidence interval for the difference between a pair of means in a randomized block design is exactly the same as for the completely randomized design, Section 13.7. It is as follows:

$$(\bar{T}_i - \bar{T}_j) \pm t_{\alpha/2} S \sqrt{\frac{2}{b}}$$

where $n_i = n_j = b$, the number of observations contained in a treatment mean, and $S = \sqrt{MSE}$. The difference between the confidence intervals for the completely randomized and the randomized block designs is that S, appearing

---

* Exercises preceded by an asterisk are optional.

in the preceding expression, will tend to be smaller than it is for the completely randomized design.

One also may construct a $(1 - \alpha)100\%$ confidence interval for the difference between a pair of block means. Each block contains k observations corresponding to the k treatments. Therefore, the confidence interval is as follows:

$$(\bar{B}_i - \bar{B}_j) \pm t_{\alpha/2} S \sqrt{\frac{2}{k}}$$

EXAMPLE 13.6 Construct a 95% confidence interval for the difference between the mean responses for treatments 1 and 2, Example 13.5.

Solution The confidence interval for the difference in mean response for a pair of treatments is

$$(\bar{T}_i - \bar{T}_j) \pm t_{\alpha/2} S \sqrt{\frac{2}{b}}$$

where, for our example, $t_{.025}$ is based upon 6 degrees of freedom. For treatments 1 and 2 we have

$$(.98 - 2.63) \pm (2.447)(.27) \sqrt{\frac{2}{4}}$$

or

$$-1.65 \pm .47$$

Exercises

13.39 Refer to Exercise 13.31.

(a) Construct a 95% confidence interval for the difference between mean responses for chemicals A and B.

(b) Construct a 95% confidence interval for the difference between mean responses for materials M_2 and M_3.

13.40 Refer to Exercise 13.32. Construct a 90% confidence interval for the difference in mean growth for methods A and B.

13.41   Refer to Exercise 13.33. Construct a 95% confidence interval for the difference in the mean distance of roll when Dasher ryegrass and Marvelgreen supreme are used for overseeding.

13.42   Refer to Exercise 13.34. Construct a 95% confidence interval for the difference between the mean amounts of oil extracted by methods 1 and 4. Compare the answer to that obtained in Exercise 13.22(b).

13.43   Refer to Exercise 13.36. Estimate the difference in mean pressures to separate components that are bonded with nickel and iron, using a 99% confidence interval.

13.11 Selecting the Sample Size

Selecting the sample size for the completely randomized or the randomized block design is an extension of the procedures of Section 8.7. We confine our attention to the case of equal sample sizes, $n_1 = n_2 = \cdots = n_k$, for the treatments of the completely randomized design. The number of observations per treatment is equal to b for the randomized block design. Thus the problem is to select n_1 or b for these two designs so as to purchase a specified quantity of information.

The selection of sample size follows a similar procedure for both designs; we will outline a general method. First the experimenter must decide on the parameter (or parameters) of major interest. Usually, this involves comparing a pair of treatment means. Second, the experimenter must specify a bound on the error of estimation that can be tolerated. Once determined, the next task is to select n_i (the number of observations in a treatment mean) or, correspondingly, b (the number of observations in a treatment mean for a randomized block design) that will reduce the half width of the confidence interval for the parameter so that it is less than or equal to the specified bound on the error of estimation. It should be emphasized that the sample size solution *always* will be an approximation, because σ is unknown and s is unknown until the sample is acquired. The best available value will be used for s to produce an approximate solution. We will illustrate the procedure with an example.

EXAMPLE 13.7   A completely randomized design is to be conducted to compare five teaching techniques in classes of equal size. Estimation of the difference in mean response on an achievement test is desired correct to within 30 test-score points, with probability equal to .95. It is expected that the test scores for a given teaching technique will possess a range approximately equal to 240. Find the approximate number of observations required for each sample in order to acquire the specified information.

Solution   The confidence interval for the difference between a pair of treatment means is

$$(\bar{T}_i - \bar{T}_j) \pm t_{\alpha/2} S \sqrt{\frac{1}{n_i} + \frac{1}{n_j}}$$

Therefore, we will wish to select n_i and n_j so that

$$t_{\alpha/2}s\sqrt{\frac{1}{n_i}+\frac{1}{n_j}}\le 30$$

The value of σ is unknown, and s is a random variable. However, an approximate solution for $n_i = n_j$ can be obtained by guessing s to be roughly equal to one-fourth of the range. Thus $s \approx 240/4 = 60$. The value of $t_{\alpha/2}$ will be based upon $(n_1 + n_2 + \cdots + n_5 - 5)$ degrees of freedom, and for even moderate values of n_i, $t_{.025}$ will approximately equal 2. Then

$$t_{.025}s\sqrt{\frac{1}{n_i}+\frac{1}{n_j}}\approx (2)(60)\sqrt{\frac{2}{n_i}}=30$$

or

$$n_i = 32, \qquad i = 1, 2, \dots, 5 \qquad\qquad \text{☙}$$

EXAMPLE 13.8   An experiment is to be conducted to compare the toxic effect of three chemicals on the skin of rats. The resistance to the chemicals was expected to vary substantially from rat to rat. Therefore, all three chemicals were to be tested on each rat, thereby blocking out rat-to-rat variability.

The standard deviation of the experimental error was unknown, but prior experimentation involving several applications of a given chemical on the same type of rat suggested a range of response measurements equal to five units.

Find a value for b such that the error of estimating the difference between a pair of treatment means is less than one unit, with probability equal to .95.

Solution   A very approximate value for s would be one-fourth of the range, or $s \approx 1.25$. Then we wish to select b so that

$$t_{.025}s\sqrt{\frac{1}{b}+\frac{1}{b}}=t_{.025}s\sqrt{\frac{2}{b}}\le 1$$

Because $t_{.025}$ will depend upon the degrees of freedom associated with s^2, which will be $(n - b - k + 1)$, we will guess $t_{.025} \approx 2$. Then

$$(2)(1.25)\sqrt{\frac{2}{b}}=1$$

or

$$b \approx 13$$

Approximately thirteen rats will be required to obtain the desired information.

The degrees of freedom associated with s^2 will be 24, based on this solution. Therefore, the guessed value of t would seem to be adequate for this approximate solution.   ∽

The sample size solutions for Examples 13.7 and 13.8 are very approximate and are intended to provide only a rough estimate of sample size and consequent costs of the experiment. The experimenter will obtain information on σ as the data are being collected and can recalculate a better approximation to n as the experiment proceeds.

Exercises

13.44   Refer to Exercise 13.5. About how many specimens per mix should be prepared to allow estimation of the difference in mean strengths for a preselected pair of specimen types to within .02 tons per square inch? Assume knowledge of the data given in Exercise 13.5.

13.45   Refer to Exercises 13.6 and 13.18 (a). Approximately how many observations would be necessary to estimate μ_A to within ten units? Use a 95% confidence coefficient.

13.46   Refer to Exercises 13.6 and 13.18 (c). Assuming equal sample sizes for each treatment, approximately how many observations from method A and method B would be necessary to estimate $\mu_A - \mu_B$ to within twenty units? Use a 95% confidence coefficient.

13.47   Refer to Exercise 13.32. How many locations would have to be used in order to estimate the difference between the mean growth of any two specified soil preparations to within one unit, with confidence coefficient .95?

13.48   Refer to Exercise 13.34 and 13.42. How many locations should be used if it is desired to estimate $\mu_1 - \mu_4$ to within .5 unit, with confidence coefficient .95?

13.12 Simultaneous Confidence Intervals for More than One Parameter

The methods of Section 13.7 can be used to construct $(1 - \alpha)100\%$ confidence intervals for a single treatment mean or for the difference between a pair of treatment means for a completely randomized design. Suppose that in the course of an analysis we wish to construct several of these confidence intervals. Although it is true that each interval will enclose the estimated parameter with probability $(1 - \alpha)$, what is the probability that *all* the intervals will enclose their respective parameters? The objective of this section is to present a procedure for forming sets of confidence intervals where the simultaneous confidence coefficient is no smaller than $(1 - \alpha)$ for an arbitrary α.

Suppose that we want to find confidence intervals $I_1, I_2, \ldots, I_m$ for parameters $\theta_1, \theta_2, \ldots, \theta_m$ so that

$$P(\theta_j \text{ in } I_j \text{ for } all \ j = 1, \ldots, m) \geq 1 - \alpha$$

This goal can be achieved by using a simple probability inequality, known as the *Bonferroni inequality*. For any events $A_1, A_2, \ldots, A_m$ we have

$$P(A_1 \cap A_2 \cap \cdots \cap A_m) = 1 - P(\bar{A}_1 \cup \bar{A}_2 \cup \cdots \cup \bar{A}_m)$$

Also, from the additive law of probability we know that

$$P(\bar{A}_1 \cup \bar{A}_2 \cup \cdots \cup \bar{A}_m) \leq \sum_{i=1}^{m} P(\bar{A}_i)$$

Therefore, it follows that

$$P(A_1 \cap A_2 \cap \cdots \cap A_m) \geq 1 - \sum_{i=1}^{m} P(\bar{A}_i)$$

Suppose that $P(\theta_j \text{ in } I_j) = 1 - \alpha_j$, and let A_j denote the event that θ_j is in I_j. Then

$$P(\theta_1 \text{ in } I_1, \ldots, \theta_m \text{ in } I_m) \geq 1 - \sum_{j=1}^{m} P(\theta_j \text{ not in } I_j) = 1 - \sum_{j=1}^{m} \alpha_j$$

If all $\alpha_j, j = 1, \ldots, m$, are chosen equal to α, we can see that the simultaneous confidence coefficient of the intervals $I_j, j = 1, \ldots, m$, could be as small as $(1 - m\alpha)$, which is smaller than $(1 - \alpha)$ if $m > 1$. A simultaneous confidence coefficient of at least $(1 - \alpha)$ can be assured by choosing each $I_j, j = 1, \ldots, m$, to have confidence coefficient $1 - (\alpha/m)$. We illustrate the use of this technique in the following example.

EXAMPLE 13.9   For the four treatments given in Example 13.2, construct confidence intervals for all comparisons of the form $\mu_i - \mu_j$, with simultaneous confidence coefficient no smaller than .95.

Solution   The appropriate $(1 - \alpha)100\%$ confidence interval for a single comparison, say $\mu_1 - \mu_2$, would be of the form

$$(\bar{T}_1 - \bar{T}_2) \pm t_{\alpha/2} S \sqrt{\frac{1}{n_1} + \frac{1}{n_2}}$$

Because there are six such differences to consider, each interval should have confidence coefficient $1 - (\alpha/6)$. Thus the corresponding t value will be $t_{\alpha/2(6)} = t_{\alpha/12}$. Because we want $1 - \alpha = .95$, or $\alpha = .05$, the t value will be

$$t_{.05/12} = t_{.004}$$

The closest table value is $t_{.005}$, so we will use this to approximate the desired result. The MSE for the data in Example 13.2 is based upon 19 degrees of freedom, so the table $t_{.005}$ value is 2.861.

Because $s = \sqrt{\text{MSE}} = \sqrt{63} = 7.937$, the intervals with simultaneous confidence coefficient approximately .95 have realizations as follows:

$$\mu_1 - \mu_2: \quad (75.67 - 78.43) \pm 2.861(7.937)\sqrt{\frac{1}{6} + \frac{1}{7}}$$

or

	-2.76 ± 12.63
$\mu_1 - \mu_3:$	4.84 ± 13.11
$\mu_1 - \mu_4:$	-12.08 ± 14.66
$\mu_2 - \mu_3:$	7.60 ± 12.63
$\mu_2 - \mu_4:$	-9.32 ± 14.23
$\mu_3 - \mu_4:$	-16.92 ± 14.66

We emphasize that the technique presented in this section guarantees simultaneous coverage probabilities of *at least* $1 - \alpha$. The actual simultaneous coverage probability can be much larger than the nominal value $1 - \alpha$. Other methods for constructing simultaneous confidence intervals can be found in the books listed in the references at the end of the chapter.

Exercises

13.49   Refer to Exercise 13.9. Construct confidence intervals for all possible differences between mean maneuver times for the three vehicle classes so that the simultaneous confidence coefficient is at least .95. Interpret the results.

13.50   Refer to Exercise 13.8. After looking at the data, a reader of the report of Wheeler *et al.*, elected to compare the mean proportions of Carbon 14 remaining in the vegetable pulp for high and low concentrations of acetonitrile, because that comparison provides the largest sample difference. How would you suggest this comparison be made?

13.51   Refer to Exercise 13.32. Construct confidence intervals on all possible differences among treatment (soil preparation) means so that the simultaneous confidence coefficient is at least .90.

13.52   Refer to Exercise 13.34. Because method 4 is the most expensive, it is desired to compare it to the other three. Construct confidence intervals for the differences $\mu_1 - \mu_4$, $\mu_2 - \mu_4$, and $\mu_3 - \mu_4$ so that the simultaneous confidence coefficient is at least .95.

13.13 Analysis of Variance Using Linear Models

The methods for analyzing linear models presented in Chapter 11 can be adapted for use in the analysis of variance. We will illustrate the method by formulating a linear model for data obtained through a completely randomized design involving $k = 2$ treatments.

Let Y_{ij} denote the random variable to be observed on the jth observation from treatment i, $i = 1, 2$. Let us define a *dummy*, or *indicator*, *variable* x as follows:

$$x = \begin{cases} 1, & \text{if } i = 2 \\ 0, & \text{otherwise} \end{cases}$$

Note that x is 0 if the observation is taken from population 1 and x is 1 if the observation is taken from population 2. If we use x as an independent variable in a linear model, we can model Y_{ij} as

$$Y_{ij} = \beta_0 + \beta_1 x + \varepsilon_{ij}$$

where ε_{ij} is a normally distributed random error with $E(\varepsilon_{ij}) = 0$ and $V(\varepsilon_{ij}) = \sigma^2$. In this model,

$$\mu_1 = E(Y_{1j}) = \beta_0$$

and

$$\mu_2 = E(Y_{2j}) = \beta_0 + \beta_1$$

Thus it follows that $\beta_1 = \mu_2 - \mu_1$, and a test of the hypothesis $\mu_2 - \mu_1 = 0$ is equivalent to the test that $\beta_1 = 0$. Our intuition would suggest that $\hat{\beta}_0 = \bar{T}_1$ and $\hat{\beta}_1 = \bar{T}_2 - \bar{T}_1$ are good estimators of β_0 and β_1; indeed, it can be shown (proof omitted) that these are the least squares estimators obtained by fitting the linear model formulated above. We illustrate the use of this technique through re-analyzing the data presented in Example 13.1.

EXAMPLE 13.10 Fit an appropriate linear model to the data of Example 13.1 and test to see if there is a significant difference between μ_1 and μ_2.

Solution

The model, as indicated earlier, is given by

$$Y_{ij} = \beta_0 + \beta_1 x + \varepsilon_{ij}$$

where

$$x = \begin{cases} 1, & \text{if } i = 2 \\ 0, & \text{otherwise} \end{cases}$$

The matrices used for the least squares estimators are then

$$Y = \begin{bmatrix} 6.1 \\ 7.1 \\ 7.8 \\ 6.9 \\ 7.6 \\ 8.2 \\ 9.1 \\ 8.2 \\ 8.6 \\ 6.9 \\ 7.5 \\ 7.9 \end{bmatrix} \quad \text{and} \quad X = \begin{bmatrix} 1 & 0 \\ 1 & 0 \\ 1 & 0 \\ 1 & 0 \\ 1 & 0 \\ 1 & 0 \\ 1 & 1 \\ 1 & 1 \\ 1 & 1 \\ 1 & 1 \\ 1 & 1 \\ 1 & 1 \end{bmatrix}$$

Hence $\quad X'X = \begin{bmatrix} 12 & 6 \\ 6 & 6 \end{bmatrix} \quad$ and $\quad (X'X)^{-1} = \begin{bmatrix} 1/6 & -1/6 \\ -1/6 & 1/3 \end{bmatrix}$

The least squares estimates are given by

$$\hat{\beta} = (X'X)^{-1}X'Y = \begin{bmatrix} 1/6 & -1/6 \\ -1/6 & 1/3 \end{bmatrix} \begin{bmatrix} 91.9 \\ 48.2 \end{bmatrix} = \begin{bmatrix} 7.28 \\ .75 \end{bmatrix}$$

Note that $\hat{\beta}_0 = 7.28 = \bar{T}_1$ and $\hat{\beta}_1 = .75 = \bar{T}_2 - \bar{T}_1$.
Further,

$$SSE = Y'Y - \hat{\beta}'X'Y = 5.8617$$

which is the same as the SSE calculated in Example 13.1. Therefore, $s^2 = SSE/(n - 2) = .58617$, and $s = \sqrt{.58617} = .7656$.
To test $H_0: \beta_1 = 0$, we construct the t statistic (see Section 11.7).

$$t = \frac{\hat{\beta}_1}{s\sqrt{c_{11}}} = \frac{.75}{.7656\sqrt{1/3}} = 1.697$$

Because we are interested in a two-tailed test, the associated p-value is $2P(t > 1.697)$ where t is based on 10 degrees of freedom. Thus, using Table 5, Appendix III, $.05 < P(t > 1.697) < .10$ and $.10 < p$-value $< .20$. Therefore, for any α value less than .1, we cannot reject H_0. That is, there is insufficient evidence to indicate that μ_1 and μ_2 differ.

This t test is equivalent to the F test of Example 13.1. In fact, the square of the observed t value is the observed F value of Example 13.1.

We will illustrate the linear model approach to a more complicated analysis of variance problem by considering a randomized block design.

EXAMPLE 13.11   An experiment was conducted to compare the effects of four chemicals, A, B, C, and D, on water resistance in textiles. Three types of material, I, II, and III, were used, with each chemical treatment being applied to one piece of each type of material. The data are given in Table 13.7. Write a linear model for this experiment and test the hypothesis that there are no differences among the treatment means. Use $\alpha = .05$.

Table 13.7
Data for
Example 13.11

		Treatments		
Material	A	B	C	D
----------	--------	------	------	------
I	10.1	11.4	9.9	12.1
II	12.2	12.9	12.3	13.4
III	11.9	12.7	11.4	12.9

Solution

In formulating the model, we will define β_0 to be the mean response for treatment A on material I, and then we will introduce a distinct indicator variable for each treatment change and each material (block) change. The model is

$$Y = \beta_0 + \beta_1 x_1 + \beta_2 x_2 + \beta_3 x_3 + \beta_4 x_4 + \beta_5 x_5 + \varepsilon$$

where

$$x_1 = \begin{cases} 1, & \text{if material II is used} \\ 0, & \text{otherwise} \end{cases}$$

$$x_2 = \begin{cases} 1, & \text{if material III is used} \\ 0, & \text{otherwise} \end{cases}$$

$$x_3 = \begin{cases} 1, & \text{if treatment } B \text{ is used} \\ 0, & \text{otherwise} \end{cases}$$

$$x_4 = \begin{cases} 1, & \text{if treatment } C \text{ is used} \\ 0, & \text{otherwise} \end{cases}$$

and

$$x_5 = \begin{cases} 1, & \text{if treatment } D \text{ is used} \\ 0, & \text{otherwise} \end{cases}$$

We want to test the hypothesis that there are no differences among treatment means, which is equivalent to $H_0: \beta_3 = \beta_4 = \beta_5 = 0$. Thus we must fit a complete and a reduced model. (See Section 11.10.)

For the complete model we have

$$Y = \begin{bmatrix} 10.1 \\ 12.2 \\ 11.9 \\ 11.4 \\ 12.9 \\ 12.7 \\ 9.9 \\ 12.3 \\ 11.4 \\ 12.1 \\ 13.4 \\ 12.9 \end{bmatrix} \quad \text{and} \quad X = \begin{bmatrix} 1 & 0 & 0 & 0 & 0 & 0 \\ 1 & 1 & 0 & 0 & 0 & 0 \\ 1 & 0 & 1 & 0 & 0 & 0 \\ 1 & 0 & 0 & 1 & 0 & 0 \\ 1 & 1 & 0 & 1 & 0 & 0 \\ 1 & 0 & 1 & 1 & 0 & 0 \\ 1 & 0 & 0 & 0 & 1 & 0 \\ 1 & 1 & 0 & 0 & 1 & 0 \\ 1 & 0 & 1 & 0 & 1 & 0 \\ 1 & 0 & 0 & 0 & 0 & 1 \\ 1 & 1 & 0 & 0 & 0 & 1 \\ 1 & 0 & 1 & 0 & 0 & 1 \end{bmatrix}$$

A little matrix algebra yields

$$SSE_2 = Y'Y - \hat{\beta}'X'Y = 1721.760 - 1721.225 = .535$$

For the reduced model we have

$$Y = \beta_0 + \beta_1 x_1 + \beta_2 x_2 + \varepsilon$$

and the corresponding X matrix has only the first three columns of the one given for the complete model. We then obtain

$$\hat{\beta} = (X'X)^{-1}X'Y = \begin{bmatrix} 10.875 \\ 1.825 \\ 1.350 \end{bmatrix}$$

and

$$SSE_1 = Y'Y - \hat{\beta}'X'Y = 1721.760 - 1716.025 = 5.735$$

It follows that

$$s_2^2 = \frac{\text{SSE}_2}{n - (k+1)} = \frac{.535}{12 - 6} = .0892$$

and

$$s_3^2 = \frac{\text{SSE}_1 - \text{SSE}_2}{k - g} = \frac{5.735 - .535}{3} = 1.733$$

The F ratio is then

$$F = \frac{s_3^2}{s_2^2} = \frac{1.733}{.0892} = 19.4$$

The tabulated F for $\alpha = .05$, $v_1 = 3$, and $v_2 = 6$ is 4.76. Hence, if we choose $\alpha = .05$, we reject the null hypothesis and conclude that the data present sufficient evidence to indicate that differences exist among the treatment means. The associated p-value is given by $P(F > 19.4) < .005$. The F test used in this example is equivalent to the one that would have been produced by the methods discussed in Section 13.9.   ∽

Although it provides a very useful technique, the linear model approach to analysis of variance calculation generally is used only when the computations are being done on a computer. The calculation formulas given earlier in the chapter are more convenient for hand calculation.

Exercises

13.53 Refer to Exercise 13.6.

(a) Answer the question posed in Exercise 13.6 by fitting complete and reduced linear models. Test using $\alpha = .05$.

(b) Use the calculations for the complete model from part (a) to test the hypothesis that there is no difference between the means for methods A and B. Test using $\alpha = .05$.

(c) Give the attained significance levels for the tests implemented in parts (a) and (b).

13.54 Refer to Exercise 13.31. Answer the question regarding possible differences among the treatment means by fitting complete and reduced models.

13.55 Refer to Exercise 13.32. Answer part (b) by constructing an F test, using complete and reduced linear models.

13.14 Summary

The completely randomized and the randomized block designs are illustrations of experiments involving one and two qualitative independent variables, respectively. The analysis of variance partitions the total sum of squares of deviations of the response measurements about their mean into portions associated with each independent variable and the experimental error. The former may be compared with the sum of squares for error, using mean squares and the F statistics, to see whether the mean squares for the independent variables are unusually large and thereby indicative of an effect on the response.

In this chapter we have presented a very brief introduction to the analysis of variance and its associated subject, the design of experiments. Experiments can be designed to investigate the effect of many quantitative and qualitative variables on a response. These may be variables of primary interest to the experimenter as well as nuisance variables, such as blocks, which we attempt to separate from the experimental error. These experiments are subject to an analysis of variance when properly designed. A more extensive coverage of the basic concepts of experimental design and the analysis of experiments will be found in the texts listed in the references.

References and Further Readings

1. Dudeck, A. E., and Peacock, C. H. "Effects of Several Overseeded Ryegrasses on Turf Quality, Traffic Tolerance and Ball Roll," *Proceedings of the Fourth International Turfgrass Research Conference*, R. W. Sheard, ed., pp. 75-81. Ontario Agricultural College, University of Guelph, Guelph, Ontario, and the International Turfgrass Society, 1981.

2. Graybill, F. *Theory and Application of the Linear Model.* N. Scituate, Mass.: Duxbury Press, 1976.

3. Guenther, W. C. *Analysis of Variance.* Englewood Cliffs, N.J.: Prentice-Hall, 1964.

4. Hicks, C. R. *Fundamental Concepts in the Design of Experiments.* 3d ed. New York: Holt, Rinehart and Winston, 1982.

5. Li, J. C. R. *Introduction to Statistical Inference.* Ann Arbor, Mich.: Edwards, 1961.

6. Mendenhall, W. *An Introduction to Linear Models and the Design and Analysis of Experiments.* Belmont, Calif.: Wadsworth, 1968.

7. Richelson, L. S.; Wahner, H. W.; Melton, L. J. 3rd; and Riggs, B. L. Excerpted from material originally appearing in "Relative Contributions of Aging and Estrogen Deficiency to Postmenopausal Bone Loss," *New England Journal of Medicine*, Vol. 311, no. 20 (1984), pp. 1273-5.

Supplementary Exercises

13.56 Three cleansing agents for the skin were used on three persons. For each person three patches of skin were exposed to a contaminant and afterward cleansed by using one of the three cleansing agents. After 8 hours the residual contaminant was measured, with the following results:

$$SST = 1.18 \qquad SSB = .78 \qquad SSE = 2.24$$

(a)   What are the experimental units and what are the blocks in this experiment?

(b)   Test the hypothesis that there are no differences among the treatment means, using $\alpha = .05$.

13.57   Refer to Exercise 13.5. Suppose that the sand used in the mixes for samples 1, 2, 3, and 4 came from pit A, that the sand used for samples 5, 6, 7, and 8 came from pit B, and that the sand for the other samples came from pit C. Analyze the data, assuming that the requirements for a randomized block are met with three blocks consisting, respectively, of samples 1, 2, 3, and 4; samples 5, 6, 7, and 8; and samples 9, 10, 11, and 12.

(a)   At the 5% significance level, is there evidence of differences in concrete strength due to the sand used?

(b)   Is there evidence, at the 5% significance level, of differences in average strength among the four types of concrete used?

(c)   Does the conclusion of part (b) contradict the conclusion that was obtained in Exercise 13.5?

13.58   Refer to Exercise 13.57. Let μ_A and μ_B, respectively, denote the mean strengths of concrete specimens prepared from mix A and mix B.

(a)   Find a 95% confidence interval for $(\mu_A - \mu_B)$.

(b)   Is the interval found in part (a) the same interval found in Exercise 13.17 (b)? Explain why or why not?

13.59   A study was initiated to investigate the effect of two drugs, administered simultaneously, in reducing human blood pressure. It was decided to utilize three levels of each drug and to include all nine combinations in the experiment. Nine high-blood-pressure patients were selected for the experiment, and one was randomly assigned to each of the nine drug combinations. The response observed was a drop in blood pressure over a fixed interval of time.

(a)   Is this a randomized block design?

(b)   Suppose that two patients were assigned to each of the nine drug combinations. What type of experimental design is this?

13.60   Refer to Exercise 13.59. Suppose that prior experimentation suggests that $\sigma = 20$.

(a)   How many replications would be required to estimate any treatment (drug combination) mean correct to within ± 10 with probability .95?

(b)   How many degrees of freedom will be available for estimating σ^2 when using the number of replications determined in part (a)?

(c)   Give the approximate half width of a confidence interval for the difference in mean response for two treatments when using the number of replications determined in part (a).

13.61   A dealer has in stock three cars (car A, car B, and car C) of the same make and model. Wishing to compare these cars in gas consumption, a customer arranged to test each car with each of three brands of gasoline (brand A, brand B, and brand C). In each trial a gallon of gasoline was added to an empty tank and the car was driven without stopping until it ran out of gasoline. The accompanying table shows the number of miles covered in each of the nine trials.

	Distance (miles)		
Brand of Gasoline	Car A	Car B	Car C
A	22.4	17.0	19.2
B	20.8	19.4	20.2
C	21.5	18.7	21.2

(a) Should the customer conclude that the three cars differ in gas mileage? Test at the $\alpha = .05$ level.

(b) Do the data indicate that the brand of gasoline affects gas mileage?

13.62 Refer to Exercise 13.61. Suppose that the gas mileage is unrelated to brand of gasoline. Carry out an analysis of the data appropriate for a completely randomized design with three treatments.

(a) Should the customer conclude that the three cars differ in gas mileage? Test at the $\alpha = .05$ level.

(b) Comparing your answer for part (a) in Exercise 13.61 with your answer for part (a), can you suggest a reason why blocking may be unwise in certain cases?

13.63 In the hope of attracting more riders, a city transit company plans to have express bus service from a suburban terminal to the downtown business district. These buses should save travel time. The city decides to perform a study of the effect of four different plans (such as a special bus lane and traffic signal progression) on the travel time for the buses. Travel times (in minutes) are measured for several weekdays during a morning rush-hour trip while each plan is in effect. The results are recorded in the table.

(a) What type of experimental design was employed?

(b) Is there evidence of a difference in the mean travel times for the four plans? Use $\alpha = 0.01$.

Plan			
1	2	3	4
27	25	34	30
25	28	29	33
29	30	32	31
26	27	31	
	24	36	

(c) Form a 95% confidence interval for the difference between plan 1 (express lane) and plan 3 (a control: no special travel arrangements).

13.64   A study was conducted to compare the effect of three levels of digitalis on the level of calcium in the heart muscle of dogs. A description of the actual experimental procedure is omitted, but it is sufficient to note that the general level of calcium uptake varies from one animal to another so that comparison of digitalis levels (treatments) had to be blocked on heart muscles. That is, the tissue for a heart muscle was regarded as a block and comparisons of the three treatments were made within a given muscle. The calcium uptakes for the three levels of digitalis, A, B, and C, were compared based on the heart muscle of four dogs. The results are as follows:

	Dogs		
1	*2*	*3*	*4*
A	C	B	A
1342	1698	1296	1150
B	B	A	C
1608	1387	1029	1579
C	A	C	B
1881	1140	1549	1319

(a)   Calculate the sums of squares for this experiment and construct an analysis-of-variance table.

(b)   How many degrees of freedom are associated with SSE?

(c)   Do the data present sufficient evidence to indicate a difference in the mean uptake of calcium for the three levels of digitalis?

(d)   Do the data indicate a difference in the mean uptake in calcium for the four heart muscles?

(e)   Give the standard deviation of the difference between the mean calcium uptake for two levels of digitalis.

(f)   Find a 95% confidence interval for the difference in mean response between treatments A and B.

13.65   Refer to Exercise 13.64. Approximately how many replications are required for each level of digitalis (how many obervations per treatment), so that the error of estimating the difference in mean response for a pair of digitalis levels is less than 20, with probability .95? Assume that additional observations would be made within a randomized block design.

13.66   A completely randomized design was conducted to compare the effect of five stimuli on reaction time. Twenty-seven people were employed in the experiment, which was conducted using a completely randomized design. Regardless of the results of the analysis of variance, it is desired to compare stimuli A and D. The reaction times (in seconds) were as shown in the accompanying table.

	Stimulus				
	A	*B*	*C*	*D*	*E*
	.8	.7	1.2	1.0	.6
	.6	.8	1.0	.9	.4
	.6	.5	.9	.9	.4
	.5	.5	1.2	1.1	.7
		.6	1.3	.7	.3
		.9	.8		
		.7			
Total	2.5	4.7	6.4	4.6	2.4
Mean	.625	.671	1.067	.920	.48

(a) Conduct an analysis of variance and test for a difference in mean rection time due to the five stimuli. Give bounds for the *p*-value.

(b) Compare stimuli *A* and *D* to see if there is a difference in mean reaction time. What can be said about the attained significance level?

13.67 The experiment in Exercise 13.66 might have been conducted more effectively by using a randomized block design with people as blocks, because we would expect mean reaction time to vary from one person to another. Hence, four people were used in a new experiment, and each person was subjected to each of the five stimuli in a random order. The reaction times (in seconds) were as shown in the accompanying table. Conduct an analysis of variance and test for differences in treatments (stimuli).

		Stimulus				
Subject		*A*	*B*	*C*	*D*	*E*
1		.7	.8	1.0	1.0	.5
2		.6	.6	1.1	1.0	.6
3		.9	1.0	1.2	1.1	.6
4		.6	.8	.9	1.0	.4

13.68 Refer to Exercise 13.33. Construct confidence intervals to compare each of the ryegrass cultivars with Marvelgreen supreme in such a way that the simultaneous confidence coefficient is at least .95. Interpret the results.

13.69 Show that

$$\text{Total SS} = \text{SST} + \text{SSB} + \text{SSE}$$

for a randomized block design, where

$$\text{SSE} = \sum_{j=1}^{b} \sum_{i=1}^{k} (Y_{ij} - \bar{B}_j - \bar{T}_i + \bar{Y})^2$$

*13.70   Consider the following model for the responses measured in a randomized block design containing b blocks and k treatments:

$$Y_{ij} = \mu + \tau_i + \beta_j + \varepsilon_{ij}$$

where

Y_{ij} = the observation on treatment i in block j

μ = the overall mean

τ_i = the nonrandom effect of treatment i, $\left(\sum\limits_{i=1}^{k} \tau_i = 0 \right)$

β_j = the random effect of block j, such that β_j are independent normally distributed random variables with $E(\beta_j) = 0$ and $V(\beta_j) = \sigma_\beta^2, j = 1, 2, \ldots, b$.

ε_{ij} = random error terms such that ε_{ij} are independent normally distributed random variables with $E(\varepsilon_{ij}) = 0$ and $V(\varepsilon_{ij}) = \sigma_\varepsilon^2$, $i = 1, 2, \ldots, k$ and $j = 1, 2, \ldots, b$.

Further, assume that the β_j and ε_{ij} also are independent. This model differs from that presented in Section 13.8 in that the block effects are assumed to be random variables, instead of fixed but unknown constants.

(a)   If the model just described is appropriate, show that observations taken from different blocks are independent of one another. That is, show that Y_{ij} and Y_{il} are independent if $j \neq l$, as are Y_{ij} and $Y_{i'l}$ if $i \neq i'$ and $j \neq l$.

(b)   Under the model just described, derive the covariance of two observations from the same block. That is, find $\text{cov}(Y_{ij}, Y_{i'j})$ if $i \neq i'$.

(c)   Two random variables that have a joint normal distribution are independent if and only if their covariance is 0. Use the result from part (b) to determine conditions under which two observations from the same block are independent of one another.

*13.71   Refer to the model for the randomized block design with random block effect given in Exercise 13.70.

(a)   Give the expected value and variance of Y_{ij}.

(b)   Let $\bar{T}_i$ denote the average of all of the responses to treatment i. Utilize the model for the randomized block design to derive $E(\bar{T}_i)$ and $V(\bar{T}_i)$. Is $\bar{T}_i$ an unbiased estimator for the mean response to treatment i? Explain why or why not. Note that $V(\bar{T}_i)$ depends on b and both σ_β^2 and σ_ε^2.

(c)   Consider $\bar{T}_i - \bar{T}_l$ for $i \neq l$. Show that $E(\bar{T}_i - \bar{T}_l) = \tau_i - \tau_l$. Note that this result implies that $\bar{T}_i - \bar{T}_l$ is an unbiased estimator of the difference in the effects of treatments i and l.

(d)   Derive $V(\bar{T}_i - \bar{T}_l)$. Note that $V(\bar{T}_i - \bar{T}_l)$ depends only on b and σ_ε^2.

*13.72   Refer to the model for the randomized block design with random block effect given in Exercise 13.70, and let $\bar{B}_j$ denote the average of all of the responses in block j.

(a)   Derive $E(\bar{B}_j)$ and $V(\bar{B}_j)$.

(b)   Derive $E(\text{MST})$.

(c)   Derive $E(\text{MSB})$.

(d)   Derive $E(\text{MSE})$.

*13.73   Refer to the model for the randomized block design with random block effect given in Exercise 13.70 and the results obtained in Exercise 13.72 (c) and (d).

(a)   Give an unbiased estimator for σ_ε^2.

(b)   Give an unbiased estimator for σ_β^2.

*13.74 Suppose that $Y_1, \ldots, Y_n$ is a random sample from a normal distrbution with mean μ and variance σ^2. The independence of $\sum_{i=1}^{n} (Y_i - \bar{Y})^2$ and $\bar{Y}$ can be shown as follows. Define an $n \times n$ matrix $\mathbf{A}$ by

$$
\mathbf{A} = \begin{bmatrix}
\dfrac{1}{\sqrt{n}} & \dfrac{1}{\sqrt{n}} & & \cdots & & \dfrac{1}{\sqrt{n}} \\[2ex]
\dfrac{1}{\sqrt{2}} & \dfrac{-1}{\sqrt{2}} & 0 & \cdots & 0 & 0 \\[2ex]
\dfrac{1}{\sqrt{2 \cdot 3}} & \dfrac{1}{\sqrt{2 \cdot 3}} & \dfrac{-2}{\sqrt{2 \cdot 3}} & 0 & \cdots & 0 & 0 \\[2ex]
\vdots & \vdots & & & & \vdots \\[2ex]
\dfrac{1}{\sqrt{(n-1)n}} & \dfrac{1}{\sqrt{(n-1)n}} & & \cdots & & \dfrac{-(n-1)}{\sqrt{(n-1)n}}
\end{bmatrix}
$$

and note that $\mathbf{A}'\mathbf{A} = \mathbf{I}$, the identity matrix. Then

$$
\sum_{i=1}^{n} Y_i^2 = \mathbf{Y}'\mathbf{Y} = \mathbf{Y}'\mathbf{A}'\mathbf{A}\mathbf{Y}
$$

where $\mathbf{Y}$ is the vector of Y_i's.

(a) Show that

$$
\mathbf{A}\mathbf{Y} = \begin{bmatrix}
\sqrt{n}\,\bar{Y} \\
u_1 \\
u_2 \\
\vdots \\
u_{n-1}
\end{bmatrix}
$$

where $u_1, u_2, \ldots, u_{n-1}$ are linear functions of $Y_1, \ldots, Y_n$. Thus

$$
\sum_{i=1}^{n} Y_i^2 = n\bar{Y}^2 + \sum_{i=1}^{n-1} u_i^2
$$

(b) Show that the linear functions $\sqrt{n}\,\bar{Y}, u_1, \ldots, u_{n-1}$, are pairwise orthogonal and hence independent under the normality assumption. See Exercise 5.81.

(c) Show that

$$
\sum_{i=1}^{n} (Y_i - \bar{Y})^2 = \sum_{i=1}^{n-1} u_i^2
$$

and conclude that this quantity is independent of $\bar{Y}$.

(d) Using the results of (c), show that

$$\frac{\sum\limits_{i=1}^{n} (Y_i - \bar{Y})^2}{\sigma^2} = \frac{(n-1)S^2}{\sigma^2}$$

has a χ^2 distribution with $(n-1)$ degrees of freedom.

*13.75 Consider a completely randomized design with k treatments. Assume that Y_{ij} is the jth response receiving treatment i and that Y_{ij} is normal with mean μ_i and variance σ^2, $j = 1, \ldots, n_i, i = 1, \ldots, k$.

(a)   Use Exercise 13.74 to justify that $\bar{Y}_1, \bar{Y}_2, \ldots, \bar{Y}_k$ are independent of SSE.

(b)   Show that MST/MSE has an F distribution with $v_1 = k - 1$ and $v_2 = n_1 + n_2 + \cdots + n_k - k$ degrees of freedom under $H_0: \mu_1 = \mu_2 = \cdots = \mu_k$. (You may assume, for simplicity, that $n_1 = n_2 = \cdots = n_k$).

chapter fourteen

ANALYSIS OF ENUMERATIVE DATA

14.1 A Description of the Experiment

Many experiments, particularly in the social sciences, result in *enumerative* (or count) *data.* For instance, the classification of people into five income brackets would result in an enumeration or count corresponding to each of the five income classes. Or we might be interested in studying the reaction of a mouse to a particular stimulus in a psychological experiment. If a mouse will react in one of three ways when the stimulus is applied and if a large number of mice were subjected to the stimulus, the experiment would yield three counts, indicating the number of mice falling in each of the reaction classes. Similarly, a traffic study might require a count and classification of the type of motor vehicles using a section of highway. An industrial process manufactures items that fall into one of three quality classes: acceptable, seconds, and rejects. A student of the arts might classify paintings in one of k categories according to style and period in order to study trends in style over time. We might wish to classify ideas in a philosophical study or style in the field of literature. The results of an advertising campaign would yield count data indicating a classification of consumer reaction. Indeed, many observations in the physical sciences are not amenable to measurement on a continuous scale and hence result in enumerative or classificatory data.

The illustrations in the preceding paragraph exhibit, to a reasonable degree of approximation, the following characteristics, which define a *multinomial experiment*:

1. The experiment consists of n identical trials.

2. The outcome of each trial falls into one of k classes or cells.

3. The probability that the outcome of a single trial will fall in a particular cell, say cell i, is $p_i(i = 1, 2, \ldots, k)$ and remains the same from trial to trial. Note that

$$p_1 + p_2 + p_3 + \cdots + p_k = 1$$

4. The trials are independent.

5. We are interested in $n_1, n_2, n_3, \ldots, n_k$, where n_i $(i = 1, 2, \ldots, k)$ is equal to the number of trials in which the outcome falls in cell i. Note that $n_1 + n_2 + n_3 + \cdots + n_k = n$.

This experiment is analogous to tossing n balls at k boxes, where each ball must fall in one of the boxes. The boxes are arranged such that the probability that a ball will fall in a box varies from box to box but remains the same for a particular box in repeated tosses. Finally, the balls are tossed in such a way that the trials are independent. At the conclusion of the experiment, we observe n_1 balls in the first box, n_2 in the second, ..., and n_k in the kth. The total number of balls is equal to

$$\sum_{i=1}^{k} n_i = n$$

Note the similarity between the binomial and the multinomial experiments and, in particular, that the binomial experiment represents the special case for the multinomial experiment when $k = 2$. The two cell probabilities, p and q, of the binomial experiment are replaced by the k cell probabilities, $p_1, p_2, \ldots, p_k$, of the multinomial experiment. The objective of this chapter is to make inferences about the cell probabilities $p_1, p_2, \ldots, p_k$. The inferences will be expressed in terms of a statistical test of a hypothesis concerning their specific numerical values or their relationship one to another.

Because the calculation of multinomial probabilities is somewhat cumbersome, it would be difficult to calculate the exact probabilities associated with type I errors for hypotheses on $p_1, p_2, \ldots, p_k$. Fortunately, we have been relieved of this chore by the British statistician Karl Pearson, who proposed a very useful test statistic for testing hypotheses concerning $p_1, p_2, \ldots, p_k$ and gave its approximate probability distribution in repeated sampling. We will outline the construction of Pearson's test statistic in the following section.

14.2 The Chi-Square Test

Suppose that $n = 100$ balls were tossed at the cells and that we knew that p_1 was equal to .1. How many balls would be expected to fall in the first cell? Referring to Chapter 5 and utilizing knowledge of the multinomial experiment, we would calculate

$$E(n_1) = np_1 = (100)(.1) = 10$$

In like manner, the expected number falling in the remaining cells may be calculated by using the formula

$$E(n_i) = np_i, \qquad i = 1, 2, \ldots, k$$

Now suppose that we hypothesize values for $p_1, p_2, \ldots, p_k$ and calculate the expected value for each cell. Certainly if our hypothesis is true, the cell counts n_i should not deviate greatly from their expected values $np_i (i = 1, 2, \ldots, k)$. Hence it would seem intuitively reasonable to use a test statistic involving the k deviations,

$$n_i - np_i, \qquad i = 1, 2, \ldots, k$$

In 1900 Karl Pearson proposed the following test statistic, which is a function of the squares of the deviations of the observed counts from their expected values, weighted by the reciprocals of their expected values:

$$X^2 = \sum_{i=1}^{k} \frac{[n_i - E(n_i)]^2}{E(n_i)} = \sum_{i=1}^{k} \frac{[n_i - np_i]^2}{np_i}$$

Although the mathematical proof is beyond the scope of this text, it can be shown that when n is large, X^2 will possess, approximately, a chi-square probability distribution in repeated sampling. We can easily show this result for the case $k = 2$, as follows. If $k = 2$, then $n_2 = n - n_1$ and $p_1 + p_2 = 1$. Thus

$$X^2 = \sum_{i=1}^{2} \frac{[n_i - E(n_i)]^2}{E(n_i)} = \frac{(n_1 - np_1)^2}{np_1} + \frac{(n_2 - np_2)^2}{np_2}$$

$$= \frac{(n_1 - np_1)^2}{np_1} + \frac{[(n - n_1) - n(1 - p_1)]^2}{np_2}$$

$$= \frac{(n_1 - np_1)^2}{np_1} + \frac{(-n_1 + np_1)^2}{np_2}$$

$$= \frac{p_2(n_1 - np_1)^2 + p_1(-n_1 + np_1)^2}{np_1(1 - p_1)} = \frac{(n_1 - np_1)^2}{np_1(1 - p_1)}$$

We have seen (Section 7.5) that

$$\frac{n_1 - np_1}{\sqrt{np_1(1 - p_1)}}$$

is approximately standard normal in distribution for large n. Thus for large n, X^2 as just given is approximately a χ^2 random variable with 1 degree of freedom. Recall that the square of a standard normal random variable has a χ^2 distribution (see Example 6.11).

Experience has shown that the cell counts n_i should not be too small in order that the chi-square distribution provide an adequate approximation to the distribution of X^2. As a rule of thumb we will require that all expected cell counts equal or exceed 5, although Cochran [2] has noted that this value can be as low as 1 for some situations.

You will recall the use of the chi-square probability distribution for testing a hypothesis concerning a population variance, σ^2, in Section 10.8. In particular, we stated that the shape of the chi-square distribution would vary depending upon the number of degrees of freedom associated with s^2, and we discussed the use of Table 6, Apprendix III, which presents the critical values of χ^2 corresponding to various right-hand tail areas of the distribution. Therefore, we must know which χ^2 distribution to use—that is, the number of degrees of freedom—in approximating the distribution of X^2, and we must know whether to use a one-tailed or two-tailed test in locating the rejection region for the test. The latter problem may be solved directly. Because large deviations of the observed cell counts from those expected would tend to contradict the null hypothesis concerning the cell probabilities $p_1, p_2, \ldots, p_k$, we would reject the null hypothesis when X^2 is large and employ a one-tailed statistical test, using the upper-tail values of χ^2 to locate the rejection region.

The determination of the appropriate number of degrees of freedom to be employed for the test can be rather difficult and therefore will be specified for the physical applications described in the following sections. In addition, we will state the principle involved (which is fundamental to the mathematical proof of the approximation) so that you will understand why the number of degrees of freedom changes with various applications. This principle states that the appropriate number of degrees of freedom will equal the number of cells, k, less 1 degree of freedom for each independent linear restriction placed upon the observed cell counts. For example, one linear restriction is present because the sum of the cell counts must equal n; that is,

$$n_1 + n_2 + n_3 + \cdots + n_k = n$$

Other restrictions will be introduced for some applications because of the necessity for estimating unknown parameters required in the calculation of the expected cell frequencies or because of the method in which the sample is collected. When unknown parameters must be estimated in order to compute X^2, a maximum-likelihood estimator should be employed. The degrees of freedom for the approximating chi-square distribution will be reduced by 1 for each parameter estimated. These cases will become apparent as we consider various practical examples.

14.3 A Test of a Hypothesis Concerning Specified Cell Probabilities: A Goodness-of-Fit Test

The simplest hypothesis concerning the cell probabilities would be one that specifies numerical values for each. In this case we are testing $H_0: p_1 = p_{1,0}$, $p_2 = p_{2,0}, \ldots, p_k = p_{k,0}$, where $p_{i,0}$ denotes a specified value for p_i. The alternative is the general one that states that at least one of the equalities does not hold.

Because the only restriction on the observations is that $\sum_{i=1}^{k} n_i = n$, the X^2 test statistic will have, approximately, a χ^2 distribution with $k - 1$ degrees of freedom.

EXAMPLE 14.1   A group of rats, one by one, proceed down a ramp to one of three doors. We wish to test the hypothesis that the rats have no preference concerning the choice of a door and therefore that

$$H_0: p_1 = p_2 = p_3 = 1/3$$

where p_i is the probability that a rat will choose door i, $i = 1, 2,$ or 3.

Suppose that the rats were sent down the ramp $n = 90$ times and that the three observed cell frequencies were $n_1 = 23, n_2 = 36,$ and $n_3 = 31$. The expected cell frequency would be the same for each cell: $E(n_i) = np_i = (90)(1/3) = 30$. The observed and expected cell frequencies are presented in Table 14.1. Note the discrepancy between the observed and expected cell frequencies. Do the data present sufficient evidence to warrant rejection of the hypothesis of no preference?

Table 14.1
Observed and
expected cell counts

	Door		
	1	*2*	*3*
Observed cell frequency	$n_1 = 23$	$n_2 = 36$	$n_3 = 31$
Expected cell frequency	(30)	(30)	(30)

Solution

The chi-square test statistic for our example will possess $(k - 1) = 2$ degrees of freedom since the only linear restriction on the cell frequencies is that

$$n_1 + n_2 + n_3 = 90$$

Therefore, if we choose $\alpha = .05$ we would reject the null hypothesis when $X^2 > 5.991$ (see Table 6, Appendix III).

Substituting into the formula for X^2, we obtain

$$X^2 = \sum_{i=1}^{k} \frac{[n_i - E(n_i)]^2}{E(n_i)} = \sum_{i=1}^{k} \frac{(n_i - np_i)^2}{np_i}$$

$$= \frac{(23 - 30)^2}{30} + \frac{(36 - 30)^2}{30} + \frac{(31 - 30)^2}{30} = 2.87$$

Because X^2 is less than the tabulated critical value of χ^2, the null hypothesis is not rejected, and we conclude that the data do not present sufficient evidence to indicate that the rats have a preference for a particular door. In this case, the p-value is given by p-value $= P(\chi^2 > 2.87)$ where χ^2 possesses a chi-square distribution with $k - 1 = 2$ degrees of freedom. Using Table 6, Appendix III, it follows that p-value > 0.10.

The χ^2 statistic also can be used to test whether sample data indicate that a particular model for a population distribution does not fit the data. An example of such a test, called the *goodness-of-fit* test, is given in the following example.

EXAMPLE 14.2   The number of accidents per week, Y, at a certain intersection was checked for $n = 50$ weeks, with the results as shown in Table 14.2. Test the hypothesis that the random variable Y has a Poisson distribution, assuming the observations to be independent. Use $\alpha = .05$.

Table 14.2
Data for
Example 14.2

y	Frequency
0	32
1	12
2	6
3 or more	0

Solution

The null hypothesis H_0 states that Y has the Poisson distribution, given by

$$p(y) = \frac{\lambda^y e^{-\lambda}}{y!} \qquad y = 0, 1, 2, \ldots$$

Because λ is unknown we must find its maximum-likelihood estimator, which turns out to be $\hat{\lambda} = \bar{Y}$ (see Exercise 9.62). For the given data $\hat{\lambda}$ has the value $\bar{y} = 24/50 = .48$.

We have, for the given data, three cells with five or more observations—the cells defined by $Y = 0$, $Y = 1$, and $Y \geq 2$. Under H_0, the probabilities for these cells are

$$p_1 = P(Y = 0) = e^{-\lambda}$$

$$p_2 = P(Y = 1) = \lambda e^{-\lambda}$$

and

$$p_3 = P(Y \geq 2) = 1 - e^{-\lambda} - \lambda e^{-\lambda}$$

These probabilities are estimated by replacing λ with $\hat{\lambda}$, which gives

$$\hat{p}_1 = e^{-.48} = .619$$

$$\hat{p}_2 = .48e^{-.48} = .297$$

and

$$\hat{p}_3 = 1 - \hat{p}_1 - \hat{p}_2 = .084$$

If the observations are independent, the cell frequencies n_1, n_2, and n_3 have a multinomial distribution with parameters p_1, p_2, and p_3. Thus $E(n_i) = np_i$, and the estimated expected cell frequencies are given by

$$\hat{E}(n_1) = n\hat{p}_1 = 30.95$$

$$\hat{E}(n_2) = n\hat{p}_2 = 14.85$$

and

$$\hat{E}(n_3) = n\hat{p}_3 = 4.20$$

Thus the test statistic is given by

$$X^2 = \sum_{i=1}^{3} \frac{[n_i - \hat{E}(n_i)]^2}{\hat{E}(n_i)}$$

which has approximately a χ^2 distribution with $(k - 2) = 1$ degree of freedom. (One degree of freedom is lost because λ had to be estimated, the other because $\sum_{i=1}^{3} n_i = n$.)

On computing X^2 we find

$$X^2 = \frac{(32 - 30.95)^2}{30.95} + \frac{(12 - 14.85)^2}{14.85} + \frac{(6 - 4.20)^2}{4.20} = 1.354$$

Because $\chi^2_{.05} = 3.841$, with 1 degree of freedom, we do not reject H_0. The data do not present sufficient evidence to contradict our hypothesis that Y possesses a Poisson distribution.

Exercises

14.1   A city expressway utilizing four lanes in each direction was studied to see whether drivers preferred to drive on the inside lanes. A total of 1000 automobiles was observed during the heavy early morning traffic and their respective lanes recorded. The results were as shown in the accompanying table. Do the data present sufficient evidence to indicate that some lanes are preferred over others? (Test the hypothesis that $p_1 = p_2 = p_3 = p_4 = 1/4$, using $\alpha = .05$.) Give bounds for the associated p-value.

Lane	Observed Count
1	294
2	276
3	238
4	192

14.2   A die was rolled 600 times, with the accompanying results. Do these data present sufficient evidence to indicate that the die is unbalanced? Test with $\alpha = .05$.

Observed Number	Frequency
1	89
2	113
3	98
4	104
5	117
6	79

14.3 After inspecting the data in Exercise 14.2, you might wish to test the hypothesis that the probability of a 6 is 1/6 against the alternative that this probability is less than 1/6.

(a) Carry out the test above, using $\alpha = .05$.

(b) What tenet of good statistical practice is violated in the test of part (a)?

14.4 The Mendelian theory states that the number of a certain type of peas falling into the classifications round and yellow, wrinkled and yellow, round and green, and wrinkled and green should be in the ratio 9:3:3:1. Suppose that 100 such peas revealed 56, 19, 17, and 8 in the respective classes. Are these data consistent with the model? Use $\alpha = .05$. (The expression 9:3:3:1 means that 9/16 of the peas should be round and yellow, 3/16 should be wrinkled and yellow, etc.)

14.5 Officials in a particular community are seeking a federal program that they hope will boost local income levels. As justification, the city claims that its local income distribution differs substantially from the national distribution and that incomes tend to be lower than expected. A random sample of 2000 family incomes was classified and compared with the corresponding national percentages. The data are shown in the table. Do the data provide sufficient evidence to indicate that the distribution of family incomes within the city differs from the national distribution? Give bounds for the associated p-value. What would you conclude with $\alpha = .05$.

Income	National Percentages	City Salary Class Frequency
More than $50,000	2	27
$25,000 to $50,000	16	193
$20,000 to $25,000	13	234
$15,000 to $20,000	19	322
$10,000 to $15,000	20	568
$5,000 to $10,000	19	482
Less than $5,000	11	174
Total	100	2000

14.6 Two types of defects, A and B, frequently are seen in the output of a certain manufacturing process. Each item can be classified into one of the four classes AB, $A\bar{B}$, $\bar{A}B$, $\bar{A}\bar{B}$, where $\bar{A}$ denotes the absence of the type A defect. For 100 inspected items the following frequencies were observed:

$$AB: 48 \qquad A\bar{B}: 18 \qquad \bar{A}B: 21 \qquad \bar{A}\bar{B}: 13$$

Test the hypothesis that the four categories, in the order listed, occur in the ratio 5:2:2:1. (Use $\alpha = .05$.)

14.7 The accompanying data show the frequency counts for 400 observations on the number of bacterial colonies within the field of a microscope, using samples of milk film. Test the hypothesis that the data fit the Poisson distribution. (Use $\alpha = .05$.)

Number of Colonies per Field	Frequency of Observation
0	56
1	104
2	80
3	62
4	42
5	27
6	9
7	9
8	5
9	3
10	2
11	0
19	1
	400

Source   C. I. Bliss and R. A. Fisher, "Fitting the Negative Binomial Distribution to Biological Data," *Biometrics*, vol. 9 (1953), pp. 176–200.

14.8   The number of accidents experienced by machinists in a certain industry was observed for a certain period of time, with the results as shown in the accompanying table. Test, at the 5% level of significance, the hypothesis that the data come from a Poisson distribution.

Accidents per Machinist	Frequency of Observation (Number of Machinists)
0	296
1	74
2	26
3	8
4	4
5	4
6	1
7	0
8	1

Source   C. I. Bliss and R. A. Fisher "Fitting the Negative Binomial Distribution to Biological Data," *Biometrics*, vol. 9 (1953), pp. 176–200.

14.4 Contingency Tables

A problem frequently encountered in the analysis of count data concerns the independence of two methods of classification of observed events. For example, we might classify a sample of people by sex and by opinion on a political issue in order to test the hypothesis that opinions on this issue are independent of sex, or we might classify patients suffering from a certain disease according to the type of medication and the rate of recovery in order to see if recovery rate depends upon the type of medication. In each of these examples we wish to investigate a *dependency* (or *contingency*) between two classification criteria.

Suppose we wish to classify defects found on furniture produced in a certain manufacturing plant according to (1) type of defect and (2) the production shift. A total of $n = 309$ furniture defects was recorded and the defects were classified as one of four types, A, B, C, or D. At the same time each piece of furniture was identified according to the production shift in which it was manufactured. These counts are presented in Table 14.3, which is known as a

Table 14.3
A contingency table

Shift	*Type of Defect*				
	A	*B*	*C*	*D*	*Total*
1	15 (22.51)	21 (20.99)	45 (38.94)	13 (11.56)	94
2	26 (22.99)	31 (21.44)	34 (39.77)	5 (11.81)	96
3	33 (28.50)	17 (26.57)	49 (49.29)	20 (14.63)	119
Total	74	69	128	38	309

contingency table. (Note: Numbers in parentheses are the estimated expected call frequencies.) Our objective is to test the null hypothesis that type of defect is independent of shift against the alternative that the two categories are dependent. That is, we wish to test H_0: column classification is independent of row classification.

Let p_A equal the unconditional probability that a defect will be of type A. Similarly, define p_B, p_C, and p_D as the probabilities of observing the three other types of defects. Then these probabilities, which we will call the *column probabilities* of Table 14.3, will satisfy the requirement

$$p_A + p_B + p_C + p_D = 1$$

In like manner, let $p_i (i = 1, 2, \text{or } 3)$ equal the row probability that a defect will have occurred on shift i, where

$$p_1 + p_2 + p_3 = 1$$

If the two classifications are independent of each other, a cell probability will equal the product of its respective row and column probabilities in accordance with the multiplicative law of probability. For example, the probability that a particular defect will occur on shift 1 and be of type A is $(p_1)(p_A)$. We observe that the numerical values of the cell probabilities are unspecified in the problem under consideration. The null hypothesis specifies only that each cell probability will equal the product of its respective row and column probabilities and therefore imply independence of the two classifications.

The analysis of the data obtained from a contingency table differs from the analysis in Example 14.1 because we must *estimate* the row and column probabilities in order to estimate the expected cell frequencies.

As we have noted, the estimated expected cell frequencies may be substituted for the $E(n_i)$ in X^2, and X^2 will continue to possess a distribution in repeated sampling that is approximated by the chi-square probability distribution.

The maximum-likelihood estimator for any row or column probability is found as follows. Let n_{ij} denote the observed frequency in row i and column j of the contingency table, and let p_{ij} denote the probability of an observation falling into this cell. If observations are independently selected, then the cell frequencies have a multinomial distribution, and the maximum-likelihood estimator of p_{ij} is simply the observed relative frequency for that cell. That is,

$$\hat{p}_{ij} = \frac{n_{ij}}{n}, \qquad i = 1, \ldots, r; j = 1, \ldots, c$$

(see Exercise 9.68).

Likewise, viewing row i as a single cell, the probability for row i is given by p_i, and hence

$$\hat{p}_i = \frac{r_i}{n}$$

(where r_i denotes the number of observations in row i) is the maximum-likelihood estimator of p_i.

By analogous arguments, the maximum-likelihood estimator of the jth-column probability is c_j/n, where c_j denotes the number of observations in column j.

Under the null hypothesis, the maximum-likelihood estimate of the expected value of n_{11} is

$$\hat{E}(n_{11}) = n(\hat{p}_1 \cdot \hat{p}_A) = n\left(\frac{r_1}{n}\right)\left(\frac{c_1}{n}\right) = \frac{r_1 \cdot c_1}{n}$$

In other words, we observe that the estimated expected value of the observed cell frequency n_{ij} for a contingency table is equal to the product of its respective row and column totals divided by the total frequency. That is,

$$\hat{E}(n_{ij}) = \frac{r_i c_j}{n}$$

The estimated expected cell frequencies for our example are shown in parentheses in Table 14.3.

We may now use the expected and observed cell frequencies shown in Table 14.3 to calculate the value of the test statistic:

$$X^2 = \sum_{j=1}^{4} \sum_{i=1}^{3} \frac{[n_{ij} - \hat{E}(n_{ij})]^2}{\hat{E}(n_{ij})}$$

$$= \frac{(15 - 22.51)^2}{22.51} + \frac{(26 - 22.99)^2}{22.99} + \cdots + \frac{(20 - 14.63)^2}{14.63}$$

$$= 19.17$$

The only remaining obstacle involves the determination of the appropriate number of degrees of freedom associated with the test statistic. We will give this as a rule, which we will attempt to justify. *The degrees of freedom associated with a contingency table possessing r rows and c columns will always equal* $(r - 1)(c - 1)$. For our example we will compare X^2 with the critical value of χ^2 with $(r - 1)(c - 1) = (3 - 1)(4 - 1) = 6$ degrees of freedom.

You will recall that the number of degrees of freedom associated with the χ^2 statistic will equal the number of cells (in this case, $k = rc$) less 1 degree of freedom for each independent linear restriction placed upon the observed cell frequencies. The total number of cells for the data of Table 14.3 is $k = 12$. From this we subtract 1 degree of freedom because the sum of the observed cell frequencies must equal n; that is,

$$n_{11} + n_{12} + \cdots + n_{34} = 309$$

In addition, we used the cell frequencies to estimate three of the four column probabilities. Note that the estimate of the fourth column probability will be determined once we have estimated p_A, p_B, and p_C, because

$$p_A + p_B + p_C + p_D = 1$$

Thus we lose $c - 1 = 3$ degrees of freedom for estimating the column probabilities.

Finally, we used the cell frequencies to estimate $(r - 1) = 2$ row probabilities, and therefore we lose $(r - 1) = 2$ additional degrees of freedom. The total number of degrees of freedom remaining will be

$$\text{d.f.} = 12 - 1 - 3 - 2 = 6$$

And, in general, we see that the total number of degrees of freedom associated with an $r \times c$ contingency table will be

$$\text{d.f.} = rc - 1 - (c - 1) - (r - 1) = (r - 1)(c - 1)$$

Therefore, if we use $\alpha = .05$, we will reject the null hypothesis that the two classifications are independent if $X^2 > 12.592$. Because the value of the test statistic, $X^2 = 19.17$, exceeds the critical value of χ^2, we will reject the null hypothesis at the $\alpha = .05$ level of significance. The associated p-value is given by p-value $= P(\chi^2 > 19.17)$. Bounds on this probability can be obtained using Table 6, Appendix III, from which it follows that p-value $< .005$. Thus, for any value of α greater than or equal to .005, the data present sufficient evidence to indicate that the proportions of the various types of defects vary from shift to shift. A study of the production operations for the three shifts would probably reveal the cause.

EXAMPLE 14.3   A survey was conducted to evaluate the effectiveness of a new flu vaccine that had been administered in a small community. The vaccine was provided free of charge in a two-shot sequence over a period of two weeks to those wishing to avail themselves of it. Some people received the two-shot sequence, some appeared only for the first shot, and the others received neither.

A survey of 1000 local inhabitants in the following spring provided the information shown in Table 14.4. Do the data present sufficient evidence to indicate a dependence between vaccine classification and the occurrence or nonoccurrence of flu?

Table 14.4
Data tabulation for
Example 14.3

	No Vaccine	One Shot	Two Shots	Total
Flu	24 (14.4)	9 (5.0)	13 (26.6)	46
No flu	289 (298.6)	100 (104.0)	565 (551.4)	954
Total	313	109	578	1000

Solution   The question asks whether the data provide sufficient evidence to indicate a dependence between the vaccine classification and the occurrence or nonoccurrence of flu. We therefore analyze the data as a contingency table.

The estimated expected cell frequencies may be calculated by using the appropriate row and column totals,

$$\hat{E}(n_{ij}) = \frac{r_i c_j}{n}$$

Thus

$$\hat{E}(n_{11}) = \frac{r_1 c_1}{n} = \frac{(46)(313)}{1000} = 14.4$$

$$\hat{E}(n_{12}) = \frac{r_1 c_2}{n} = \frac{(46)(109)}{1000} = 5.0$$

These values are shown in parentheses in Table 14.4.

The value of the test statistic X^2 will now be computed and compared with the critical value of χ^2 possessing $(r-1)(c-1) = (1)(2) = 2$ degrees of freedom. Then for $\alpha = .05$, we will reject the null hypothesis when $X^2 > 5.991$. Substituting into the formula for X^2, we obtain

$$X^2 = \frac{(24-14.4)^2}{14.4} + \frac{(289-298.6)^2}{298.6} + \cdots + \frac{(565-551.4)^2}{551.4}$$

$$= 17.35$$

Observing that X^2 falls in the rejection region, we reject the null hypothesis of independence of the two classifications. If we choose to use the attained significance level approach to making our inference, it would follow that p-value $< .005$. As is always the case, we note the agreement between our fixed α level approach to testing, and the proper interpretation of the p-value.

A comparison of the percentage incidence of flu for each of the three categories would suggest that those receiving the two-shot sequence were less susceptible to the disease. Further analysis of the data could be obtained by deleting one of three categories, the second column, for example, to compare the effect of the vaccine with that of no vaccine. This could be done either by using a 2×2 contingency table or by treating the two categories as two binomial populations and using the methods of Section 10.3. Or we might wish to analyze the data by comparing the results of the two-shot vaccine sequence with those of the combined no-vaccine–one-shot group. That is, we would combine the first two columns of the 2×3 table into one.

We have considered only the simplest hypothesis connected with a contingency table, that of independence between rows and columns. Many other hypotheses are possible, and numerous techniques have been devised to test these hypotheses. For further information on this topic, consult Conover [3] and Daniel [4].

Exercises

14.9   The following data represent a small portion of a telephone survey conducted by Jerry R. Lynn to determine the characteristics of the newspaper readership in small towns, rural areas, and farms in Tennessee, as well as the impact of advertising on that market. A sample of 1486 persons was selected from among the small towns, rural areas, and farms of Tennessee. One survey question asked whether the respondent did or did not read a newspaper. The numbers of respondents in the six community-readership categories follow:

Community	Readers	Nonreaders	Total
Urban	529	121	650
Rural	373	137	510
Farm	237	89	326
Total	1139	347	1486

Source Lynn, Jerry R., "Newspaper Ad Impact in Nonmetropolitan Markets," *Journal of Advertising Research*, 21, no. 4 (1981). Reprinted by permission.

(a) Do the data provide sufficient evidence to indicate that whether one reads a newspaper depends upon community type? Test using $\alpha = .05$.

(b) Give bounds for the associated *p*-value and interpret the result.

14.10 Reporting on a Gallup Poll, the *San Francisco Chronicle* (August 31, 1981)[†] notes that cigarette smoking is on the decline in the United States, "apparently because growing numbers of Americans have become convinced there is a causal relationship between smoking and such diseases as lung cancer, throat cancer, heart disease, and birth defects." Do smokers see less harm in smoking than nonsmokers? In response to the question, "Is smoking harmful to your health?" the percentage of smokers responding "no" (17 percent) appears to be much higher than the corresponding percentage of nonsmokers (2 percent). The percentages follow.

	Yes	No	No Opinion
Smokers	80	17	3
Nonsmokers	96	2	2

[†] *Source* Reprinted by permission. *The San Francisco Chronicle* (August 31, 1981). © 1981 Chronicle Publishing Company. All rights reserved.

The survey involved 1535 adults, approximately 35 percent of whom were smokers. Therefore, the approximate numbers in the response categories were:

	Yes	No	No Opinion	Total
Smokers	430	91	16	537
Nonsmokers	958	20	20	998
Total	1388	111	36	1535

(a) Do the cell counts shown above indicate that the response to the question, "Is smoking harmful to your health?" depends upon whether the respondent is a smoker or a nonsmoker? Test using $\alpha = .05$.

(b) Give bounds for the attained significance level.

14.11 A survey of 477 hospitals by the American College of Surgeons' Commission on Cancer produced the data in the table, which classifies women with liver tumors into six classes. The women are classified according to whether they used oral contraceptives and according to type of liver tumor. Do the data provide sufficient evidence to indicate a dependence between type of tumor and whether they used oral contraceptives? Give bounds for the p-value. What would you conclude at the $\alpha = .05$ level of significance?

| | Type of Tumor | | |
	Benign	Malignant	Total
Contraceptive users	138	49	187
Nonusers	39	41	80
Use not known	35	76	111
Total	212	166	378

Source Reprinted by permission of *Journal of American Medical Association.* J. Vana, G. Murphy, B. Aronoff, H. Baker, "Primary Liver Tumor and Oral Contraceptives," *JAMA* 238, no. 20. © 1977. All rights reserved.

14.12 The *Sociological Quarterly* for spring 1978 reports on a study of the relationship between athletic involvement and academic achievement for college students. The 852 students sampled were categorized according to amount of athletic involvement and grade point average (GPA) at graduation, with results as shown in the accompanying table.

| | Athletic Involvement | | | |
GPA	None	1–3 Semesters	4 or More Semesters	Total
Below mean	290	94	42	426
Above mean	238	125	63	426
Total	528	219	105	852

(a) Do the data present evidence that final grade point averages are dependent upon athletic involvement? (Use $\alpha = .05$.)

(b) For students with 4 or more semesters of athletic involvement, is the proportion with GPAs above the mean significantly different, at the .05 level, from the proportion with GPAs below the mean?

14.13   A study of the amount of violence viewed on television as it relates to the age of the viewer showed the accompanying result for eighty-one people. (Each person in the study could be classified, according to the person's TV viewing habits, as a low-violence or high-violence viewer.) Do the data indicate that viewing of violence is not independent of age of viewer, at the 5% significance level?

	Age		
Viewing	16–34	35–54	55 and Over
Low violence	8	12	21
High violence	18	15	7

14.14   In a study of changing evaluations of flood plain hazards, Payne and Pigram present the results of a survey of the attitudes of people at risk of flood hazards in the Hunter River Valley, Australia. One aspect of the study involves a recording of the individual action in response to the threat of a flood, characterized as low, moderate, or high cost, and the expectation of the level of flood damage, major or minor. The total number of respondents interviewed was 110. The numbers of responses in the six cost-damage categories are shown below:

	Damage		
Cost	Major	Minor	Total
Low	43	16	59
Moderate	10	28	38
High	4	9	13
Total	57	53	110

Source   Payne, Robert J., and John J. Pilgram, "Changing Evaluations of Flood Plain Hazard, The Hunter Valley, Australia," *Environment and Behavior*, 13, no. 4 (1981), pp. 461–480. © 1981 Sage Publications, Inc. with permission.

(a)   Is there sufficient evidence to indicate that the level of preparation depends upon the perceived threat of flood? Test using $\alpha = .05$.

(b)   Give bounds for the observed significance level.

14.15   Refer to Exercise 14.6. Test the hypothesis, at the 5% significance level, that the type A defects occur independently of the type B defects.

14.16   An interesting and practical use of the χ^2 test comes about in testing for segregation of species of plants or animals. Suppose that two species of plants, say A and B, are growing on a test plot. To assess whether the species tend to segregate, a researcher randomly samples n plants from the plot; the species of each sampled plant *and* the species of its *nearest* neighbor are recorded. The data are then arranged in a table, as shown here.

	Nearest Neighbor	
Sampled Plant	A	B
A	a	b
B	c	d
		n

If a and d are large relative to b and c, we would be inclined to say that the species tend to segregate. (Most of A's neighbors are of type A, and most of B's neighbors are of type B.) If b and c are large compared to a and d, we would say that the species tend to be overly mixed. In either of these cases (segregation or overmixing), a χ^2 test should yield a large value and the hypothesis of random mixing would be rejected. For each of the following cases, test the hypothesis of random mixing (or, equivalently, the hypothesis that the species of a sample plant is independent of the species of its nearest neighbor). Use $\alpha = .05$ in each case.

(a) $a = 20, b = 4, c = 8, d = 18$

(b) $a = 4, b = 20, c = 18, d = 8$

(c) $a = 20, b = 4, c = 18, d = 8$

14.5 r × c Tables with Fixed Row or Column Totals

In the previous section we have described the analysis of an $r \times c$ contingency table by using examples that, for all practical purposes, fit the multinomial experiment described in Section 14.1. Although the methods of collecting data in many surveys obviously may adhere to the requirements of a multinomial experiment, other methods do not. For example, we might not wish to randomly sample the population described in Example 14.3 because we might find that, due to chance, one category is completely missing. People who have received no flu shots might fail to appear in the sample. We might decide beforehand to interview a specified number of people in each column category, thereby fixing the column totals in advance. We would then have three separate and independent binomial experiments, corresponding to "no vaccine," "one shot," and "two shots," with respective probabilities p_1, p_2, and p_3 of a person contracting flu. A test of independence between vaccine classification and occurrence of flu is equivalent to a test of

$$H_0: p_1 = p_2 = p_3$$

(We actually are testing the equivalence of three binomial distributions.) Under this hypothesis the maximum-likelihood estimators of the expected cell frequencies are the same as in Section 14.4, namely,

$$\hat{E}(n_{ij}) = \frac{r_i c_j}{n}$$

It can be shown that the resulting X^2 will possess a probability distribution in repeated sampling that is approximated by a chi-square distribution with $(r-1)(c-1)$ degrees of freedom.

To illustrate, suppose that we wish to test a hypothesis concerning the equivalence of four binomial populations, as indicated in the following example.

EXAMPLE 14.4   A survey of voter sentiment was conducted in four midcity political wards to compare the fraction of voters favoring candidate A. Random samples of 200 voters were polled in each of the four wards, with results as shown in Table 14.5. Do the data present sufficient evidence to indicate that the fractions of voters favoring candidate A differ in the four wards?

Table 14.5
Data tabulation for
Example 14.4

		Ward			
Opinion	1	2	3	4	Total
Favor A	76 (59)	53 (59)	59 (59)	48 (59)	236
Do not favor A	124 (141)	147 (141)	141 (141)	152 (141)	564
Total	200	200	200	200	800

Solution   You will observe that the test of a hypothesis concerning the equivalence of the parameters of the four binomial populations corresponding to the four wards is identical to a hypothesis implying independence of the row and column classifications. If we denote the fraction of voters favoring A as p and hypothesize that p is the same for all four wards, we imply that the first- and second-row probabilities are equal to p and $(1-p)$, respectively. The probability that a member of the sample of $n = 800$ voters falls in a particular ward will equal one-fourth, because this was fixed in advance. Then the cell probabilities for the table would be obtained by multiplication of the appropriate row and column probabilities under the null hypothesis and be equivalent to a test of independence of the two classifications.

The estimated expected cell frequencies, calculated by using the row and column totals, appear in parentheses in Table 14.5. We see that

$$X^2 = \sum_{j=1}^{4} \sum_{i=1}^{2} \frac{[n_{ij} - \hat{E}(n_{ij})]^2}{\hat{E}(n_{ij})}$$

$$= \frac{(76 - 59)^2}{59} + \frac{(124 - 141)^2}{141} + \cdots + \frac{(152 - 141)^2}{141} = 10.72$$

The critical value of χ^2 for $\alpha = .05$ and $(r - 1)(c - 1) = (1)(3) = 3$ degrees of freedom is 7.815. Because X^2 exceeds this critical value, we reject the null hypothesis and conclude that the fraction of voters favoring candidate A is not the same for all four wards. The associated p-value is given by $P(\chi^2 > 10.72)$ when χ^2 has 3 degrees of freedom. Thus, $.01 \le p\text{-value} \le .025$.   ∽

This example was worked out in Exercise 10.75 by the likelihood ratio method. Note that the conclusions are the same.

Exercises

14.17   A study to determine the effectiveness of a drug (serum) for arthritis resulted in the comparison of two groups each consisting of 200 arthritic patients. One group was inoculated with the serum whereas the other received a placebo (an inoculation that appears to contain serum but actually is not active). After a period of time each person in the study was asked whether his or her arthritic condition improved. The accompanying results were observed. Do these data present sufficient evidence to indicate that the serum affected the condition of arthritic patients?

Condition	Treated	Untreated
Improved	117	74
Not improved	83	126

(a)   Test by using the X^2 statistic. Use $\alpha = .05$.

(b)   Test by using the z test of Section 10.3 and $\alpha = .05$. Compare your result with that in part (a).

(c)   Give bounds for the attained significance level associated with the test in part (a).

14.18   The chi-square test used in Exercise 14.17 is equivalent to the two-tailed z test of Section 10.3 provided α is the same for the two tests. Show algebraically that the chi-square test statistic X^2 is the square of the test statistic z for the equivalent test.

14.19   A manufacturer of buttons wishes to determine whether the fraction of defective buttons produced by three machines varied from machine to machine. Samples of 400 buttons were selected from each of the three machines and the number of defectives counted for each sample. The results were as shown in the accompanying table. Do these data present sufficient evidence to indicate that the fraction of defective buttons varies from machine to machine?

Machine Number	Number of Defectives
1	16
2	24
3	9

(a)   Test, using $\alpha = .05$, with a χ^2 test.

(b)   Test, using $\alpha = .05$, with a likelihood ratio test.

14.20   In 1976 H. W. Menard wrote about manganese nodules, a mineral-rich concoction found abundantly on the deep-sea floor. In one portion of his report, Menard provided data relating the magnetic age of the earth's crust to "the probability of finding manganese nodules." The data shown in the table give the number of samples of the earth's core and the percentage of those that contain manganese nodules for each of a set of magnetic-crust ages. Do the data provide sufficient evidence to indicate that the probability of finding manganese nodules in the deep-sea earth's crust is dependent upon the magnetic-age classification? Test with $\alpha = .05$.

Age	Number of Samples	Percentage with Nodules
Miocene—recent	389	5.9
Oligocene	140	17.9
Eocene	214	16.4
Paleocene	84	21.4
Late Cretaceous	247	21.1
Early and Middle Cretaceous	1,120	14.2
Jurassic	99	11.0

Source Menard, H. W., "Time, Chance and the Origin of Manganese Nodules," *Amer. Scientist*, September–October 1976. © 1976 Scientific Research Company of North America. All rights reserved.

14.21   According to the *Wall Street Journal* (October 23, 1981)*, "most bosses shun symbols of status, help take care of household tasks." The *Wall Street Journal*-Gallup survey finds

* *Source*   Reprinted by permission of *The Wall Street Journal*, © Dow Jones & Company, Inc. 1981. All rights reserved worldwide.

that 49 % of the 307 heads of the nation's largest firms included in the survey sometimes do grocery shopping. Compare this 49 % to 47 % of the executives of 309 medium-sized firms and 45 % of the executives of 208 small companies. Do the percentages suggest a difference in the proportions of the chief executives of small, medium, and large firms who sometimes do grocery shopping?

(a) Do the data provide sufficient evidence to indicate that the proportions of chief executives who sometimes do grocery shopping differ for the three categories of company size? Test using $\alpha = .05$. [Hint: Construct a 2×3 contingency table showing the numbers of executives in each of the six cell categories.] Give bounds for the attained significance level.

(b) Suppose that prior to observing the data, you had planned to compare the proportions of executives of large corporations and small corporations who sometimes do grocery shopping. Do the data provide sufficient evidence to indicate a difference in proportions for these two groups? Test using $\alpha = .05$. What is the associated p-value?

14.22 A survey was conducted to study the relationship between lung disease and air pollution. Four areas were chosen for the survey, two cities frequently plagued with smog and two nonurban areas in states that possessed low air pollution counts. Only adult permanent residents of the area were included in the study. Random samples of 400 adult permanent residents from each area gave the accompanying results.

Area	Number with Lung Disease
City A	34
City B	42
Nonurban area 1	21
Nonurban area 2	18

(a) Do the data provide sufficient evidence to indicate a difference in the proportions with lung disease for the four locations?

(b) Should cigarette smokers have been excluded from the samples? How would this affect inferences drawn from the data?

14.23 Refer to Exercise 14.22. Estimate the difference in the fractions of adult permanent residents with lung disease between cities A and B. Use a 95 % confidence interval.

14.24 A survey was conducted to investigate interest of middle-aged adults in physical fitness programs in Rhode Island, Colorado, California, and Florida. The objective of the investigation was to determine whether adult participation in physical fitness programs varies from one region of the United States to another. Random samples of people were interviewed in each state and the accompanying data were recorded. Do the data indicate differences among the rates of adult participation in physical fitness programs from one state to another? What would you conclude with $\alpha = .01$?

Participation	Rhode Island	Colorado	California	Florida
Yes	46	63	108	121
No	149	178	192	179

14.6 Other Applications

The applications of the chi-square test in analyzing enumerative data described in Sections 14.3, 14.4, and 14.5 represent only a few of the interesting classificatory problems that may be approximated by the multinomial experiment and for which our method of analysis is appropriate. By and large these applications are complicated to a greater or lesser degree because the numerical values of the cell probabilities are unspecified and hence require the estimation of one or more population parameters. Then, as in Sections 14.4 and 14.5, we can estimate the cell probabilities. Although we omit the mechanics of the statistical tests, several additional applications of the chi-square test are worth mention as a matter of interest.

For example, suppose that we wish to test a hypothesis stating that a population possesses a normal probability distribution. The cells of a sample frequency histogram would correspond to the k cells of the multinomial experiment, and the observed cell frequencies would be the number of measurements falling in each cell of the histogram. Given the hypothesized normal probability distribution for the population, we could use the areas under the normal curve to calculate the theoretical cell probabilities and hence the expected cell frequencies. Maximum-likelihood estimators must be employed when μ and σ are unspecified for the normal population, and these parameters must be estimated to obtain the estimated cell probabilities.

The construction of a two-way table to investigate dependency between two classifications can be extended to three or more classifications. For example, if we wish to test the mutual independence of three classifications, we would employ a three-dimensional "table" or rectangular parallelepiped. The reasoning and methodology associated with the analysis of both the two- and three-way tables are identical, although the analysis of the three-way table is a bit more complex.

A third and interesting application of our methodology would be its use in the investigation of the rate of change of a multinomial (or binomial) population as a function of time. For example, we might study the decision-making ability of a human (or any animal) subjected to an educational program and tested over time. If, for instance, the human is tested at prescribed intervals of time and the test is of the yes or no type, yielding a number of correct answers, y, that would follow a binomial probability distribution, we would be interested in the behavior of the probability of a correct response, p, as a function of time. If the number of correct responses was recorded for c time periods, the data would fall in a $2 \times c$ table similar to that in Example 14.4 (Section 14.5). We would then be interested in testing the hypothesis that p is equal to a constant—that is, that no learning has occurred—and we would then proceed to more interesting hypotheses to determine whether the data present sufficient evidence to indicate a gradual (say, linear) change over time as opposed to an abrupt change at some point in time. The procedures we have described could be extended to decisions involving more than two alternatives.

You will observe that our learning example is common to business, to industry, and to many other fields, including the social sciences. For example, we might wish to study the rate of consumer acceptance of a new product for various types of advertising campaigns as a function of the length of time that the campaign has been in effect. Or we might wish to study the trend in the lot fraction defective in a manufacturing process as a function of time. Both of these examples, as well as many others, require a study of the behavior of a binomial (or multinomial) process as a function of time.

The examples just described are intended to suggest the relatively broad application of the chi-square analysis of enumerative data, a fact that should be borne in mind by the experimenter concerned with this type of data. The statistical test employing X^2 as a test statistic is often called a *goodness-of-fit* test. Its application for some of these examples requires care in the determination of the appropriate estimates and the number of degrees of freedom for X^2, which, for some of these problems, may be rather complex.

14.7 Summary

The material in this chapter has been concerned with a test of a hypothesis regarding the cell probabilities associated with a multinomial experiment. When the number of observations n is large, the test statistic X^2 can be shown to possess, approximately, a chi-square probability distribution in repeated sampling, the number of degrees of freedom being dependent upon the particular application. In general we assume that n is large and that the minimum expected cell frequency is equal to or greater than 5.

Several words of caution concerning the use of the X^2 statistic as a method of analyzing enumerative-type data are appropriate. The determination of the correct number of degrees of freedom associated with the X^2 statistic is very important in locating the rejection region. If the number is specified incorrectly, erroneous conclusions might result. Also, note that nonrejection of the null hypothesis does not imply that it should be accepted. We would have difficulty in stating a meaningful alternative hypothesis for many practical applications, and therefore we would lack knowledge of the probability of making a type II error. For example, we hypothesize that the two classifications of a contingency table are independent. A specific alternative would have to specify a measure of dependence that may or may not possess practical significance to the experimenter. Finally, if parameters are missing and the expected cell frequencies must be estimated, missing parameters should be estimated by the method of maximum likelihood in order that the test be valid. In other words, the application of the chi-square test for other than the simple applications outlined in Sections 14.3, 14.4, and 14.5 will require experience beyond the scope of this introductory presentation of the subject.

References and Further Readings

1. Anderson, R. L., and Bancroft, T. A. *Statistical Theory in Research*. New York: McGraw-Hill, 1952, Chap. 12.

2. Cochran, W. G. "The χ^2 Test of Goodness of Fit." *Annals of Mathematical Statistics*, vol. 23 (1952), pp. 315–345.

3. Conover, W. J. *Practical Nonparametric Statistics*, 2d ed. New York: Wiley, 1980.

4. Daniel, W. W. *Applied Nonparametric Statistics*. Boston: Houghton Mifflin, 1978.

5. Dixon, W. J., and Massey, F. J., Jr. *Introduction to Statistical Analysis*. 3rd ed. New York: McGraw-Hill, 1968.

6. Kendall, M. G., and Stuart, A. *The Advanced Theory of Statistics*, 4th ed., vol. 2. New York: Hafner Press, 1979, Chap. 30.

Supplementary Exercises

14.25   List the characteristics of a multinomial experiment.

14.26   A survey was conducted to determine student, faculty, and administration attitudes on a new university parking policy. The distribution of those favoring or opposed to the policy was as shown in the accompanying table. Do the data provide sufficient evidence to indicate that attitudes regarding the parking policy are independent of student, faculty, or administration status?

Opinion	Student	Faculty	Administration
Favor	252	107	43
Oppose	139	81	40

14.27   Wilson and Mandelbrote (1978) reported on factors related to conviction for a criminal offense after persons have received treatment for drug abuse. One part of their study of sixty graduates of a drug rehabilitation program categorizes them according to whether they were convicted after treatment and according to years of education. The data follow.

Education	Convicted	Not Convicted	Total
16 years or more	6	18	24
15 years or less	16	20	36
Total	22	38	66

Source Reprinted by permission. *British Jounral of Criminality*, 18, no. 4. © 1978 Associated Book Publishers. All rights reserved.

The data appear to suggest that persons with less education are more prone to· conviction. Do the data indicate that whether or not one is convicted after treatment is dependent upon a person's educational level? Use the chi-square test with $\alpha = .05$. Give bounds for the p-value.

14.28 The computer, which at its advent was expected to play its primary role in scientific computation, has become essential in banking, in the recording of personal data, in merchandising, and in many areas of our daily lives. Along with this growth in computer applications have come numerous cases of computer abuse, financial fraud, theft of information, and so on. The data in the accompanying table give four different types of computer abuse that were reported and verified for the years 1970 to 1973. The frequency of computer abuses would be expected to increase as the years go by unless safeguards are found to prevent their occurrence. However, are the proportions of the four types of abuses changing over time? Test by using $\alpha = .10$. Give bounds for the p-value.

Year	Financial Fraud	Theft of Information or Property	Unauthorized Use	Vandalism	Total
1970	7	5	9	8	29
1971	22	18	6	6	52
1972	12	15	6	12	45
1973	21	15	16	9	61
Total	62	53	37	35	187

Source Parker, D. B., S. Nycum, and S. S. Oura, *Computer Abuse*. Menlo Park, Calif. Stanford Research Institute, 1973.

14.29 As part of a study of small-town police, Galliher and colleagues questioned police in towns of varying sizes to determine what types of people they watched most carefully because of their potential for criminal activity. Particularly, they were interested in determining whether the "most watched" type of person depended upon the size of the town. They interviewed police in 224 small towns and they obtained the data shown in the table. Do the data provide sufficient information to indicate a dependence between the type of persons most watched and town size? Test by using $\alpha = .05$. Give bounds for the p-value.

	Town Size		
Type of Persons Most Watched	Under 5,000	5,000–10,000	10,000–25,000
Known criminals	21	26	25
Young people	39	15	20
Strangers, and other suspicious persons	13	5	11
Bar crowds	11	2	2
Others	10	11	13
Total	94	59	71

Source   Reprinted by permission, *Journal of Police Science & Administration*, vol. 3, no. 1. © 1975, International Association of Chiefs of Police, Inc. All rights reserved.

14.30   It is often not clear whether all properties of a binomial experiment are actually met in a given application. A goodness-of-fit test is desirable for such cases. Suppose that an experiment consisting of four trials was repeated 100 times. The number of repetitions on which a given number of successes was obtained is recorded in the accompanying table. Estimate p (assuming that the experiment was binomial), obtain estimates of the expected cell frequencies, and test for goodness of fit. To determine the appropriate number of degrees of freedom for X^2, note that p was estimated by a linear combination of the observed frequencies.

Possible Results (no. of successes)	No. of Times Obtained
0	11
1	17
2	42
3	21
4	9

14.31   Counts on the number of items per cluster (or colony or group) must necessarily be greater than or equal to 1. Thus, the Poisson distribution generally does not fit these kinds of counts. For modeling counts on phenomena such as number of bacteria per colony, number of people per houshold, and number of animals per litter, the *logarithmic series* distribution often proves useful. This discrete distribution has probability function given by

$$p(y) = -\frac{1}{\ln(1-\alpha)}\frac{\alpha^y}{y}, \qquad y = 1, 2, 3, \ldots, 0 < \alpha < 1$$

where α is an unknown parameter.

(a) Show that the maximum-likelihood estimator, $\hat{\alpha}$, of α satisfies the equation

$$\bar{y} = \frac{\hat{\alpha}}{-(1 - \hat{\alpha}) \ln (1 - \hat{\alpha})},$$

where $\bar{y}$ is the mean of the sampled observations $y_1, \ldots, y_n$.

(b) The following data give frequencies of observation for counts on the number of bacteria per colony, for a certain type of soil bacteria.[†]

Bacteria per Colony	1	2	3	4	5	6	7 or more
Number of Colonies Observed	359	146	57	41	26	17	29

Test the hypothesis that these data fit a logarithmic series distribution. Use $\alpha = 0.05$. (Note that $\bar{y}$ must be approximated because we do not have exact information on counts greater than six.)

14.32 Refer to the $r \times c$ contingency table of Section 14.4. Show that the maximum-likelihood estimator of the probability for row i, p_i, is $\hat{p}_i = r_i/n$, $i = 1, 2, \ldots, r$.

*14.33 A genetic model states that the proportions of offspring in three classes should be p^2, $2p(1 - p)$, and $(1 - p)^2$ for a parameter p, $0 \leq p \leq 1$. An experiment yielded frequencies of 30, 40, and 30 for the respective classes.

(a) Does the model fit the data? (Use maximum likelihood to estimate p.)

(b) Suppose the hypothesis states that the model holds with $p = .5$. Do the data contradict this hypothesis?

*14.34 According to the genetic model for the relationship between sex and color blindness, the four categories, male and normal, female and normal, male and color blind, female and color blind, should have probabilities given by $p/2$, $(p^2/2) + pq$, $q/2$, and $q^2/2$, respectively, where $q = 1 - p$. A sample of 2000 people revealed 880, 1032, 80, and 8 in the respective categories. Do these data agree with the model? Use $\alpha = .05$. (Use maximum likelihood to estimate p.)

*14.35 Suppose that $(Y_1, \ldots, Y_k)$ has a multinomial distribution with parameters $n, p_1, p_2, \ldots, p_k$, and $(X_1, \ldots, X_k)$ has a multinomial distribution with parameters $m, p_1^*, p_2^*, \ldots, p_k^*$. Construct a test of the null hypothesis that the two multinomial distributions are identical; that is, test $H_0: p_1 = p_1^*, \ldots, p_k = p_k^*$.

*14.36 In an insecticide experiment the probability of insect survival was expected to be linearly related to the dosage D over the region of experimentation. That is, $p = 1 + \beta D$. An experiment was conducted using four levels of dosage, 1, 2, 3, and 4, and 1000 insects in each group. The resulting data were as shown in the accompanying table. Do these data

[†] *Source* C. A. Bliss and R. A. Fisher. Reprinted by permission. "Fitting the Negative Binomial Distribution to Biological Data," *Biometrics,* 9, 1953, pp. 176–200, Biometrics Society. All rights reserved.

* Exercises preceded by an asterisk are optional.

contradict the hypothesis that $p = 1 + \beta D$? (Hint: Write the cell probabilities in terms of β, and find the maximum-likelihood estimator of β.)

Dosage	No. of Survivors
1	820
2	650
3	310
4	50

chapter fifteen

NONPARAMETRIC STATISTICS

15.1 Introduction

Some experiments yield response measurements that defy quantification. That is, they generate response measurements that can be ordered (ranked), but the location of the response on a scale of measurement is arbitrary. Although experiments of this type occur in almost all fields of study, they are particularly evident in social science research and in studies of consumer preference. For example, suppose that a judge is employed to evaluate and rank the instructional abilities of four teachers or the edibility and taste characteristics of five brands of cornflakes. Because it clearly is impossible to give an exact measure of teacher competence or tastability of food, the response measurements are of a completely different character from those presented in preceding chapters. Nonparametric statistical methods are useful for analyzing this type of data.

Nonparametric statistical procedures apply not only to observations that are difficult to quantify, they are particularly useful in making inferences in situations where serious doubt exists about the assumptions that underlie standard methodology. For example, the t test for comparing a pair of means, Section 10.7, is based on the assumption that both populations are normally distributed with equal variances. The experimenter will never know whether these assumptions hold in a practical situation but often will be reasonably certain that departures from the assumptions will be small enough that the properties of the statistical procedure will be undisturbed. That is, α and β will be approximately what the experimenter thinks they are. On the other hand, it is not uncommon for the experimenter seriously to question the assumptions and wonder whether he or she is using a valid statistical procedure. This difficulty may be circumvented by using a nonparametric statistical test and thereby avoiding reliance on a very uncertain set of assumptions.

The term *nonparametric statistics* has no standard definition that is agreed upon by all statisticians. However, most would agree that nonparametric statistical methods work well under fairly general assumptions about the nature of any probability distributions or parameters that are involved in an inferential problem. As a working definition we will define parametric methods as those that apply to problems where the distribution(s) from which the sample(s) is (are) taken is (are) specified except for the values of a finite number of parameters. Nonparametric methods apply in all other instances. For example, the one-sample t test developed in Chapter 10 applies to problems in which the population is normally distributed with unknown mean and variance. Because the distribution from which the sample is taken is specified except for the values of two parameters, μ and σ^2, this t test is a parametric procedure. Alternatively, suppose that independent samples are taken from two populations and we wish to test the hypothesis that the two population distributions are identical but of unspecified form. In this case the distribution is unspecified and the hypothesis must be tested by using nonparametric methods.

Research has shown that nonparametric statistical tests are almost as capable of detecting differences among populations as the parametric methods of preceding chapters when normality and other assumptions are satisfied. They may be, and often are, more powerful in detecting population differences when the assumptions are not statisfied. For this reason many statisticians advocate the use of nonparametric statistical procedures in preference to their parametric counterparts.

15.2 A General Two-Sample Shift Model

Many times an experimenter takes observations from two populations with the objective of testing whether the populations have the same distribution. For example, if independent random samples $X_1, \ldots, X_{n_1}$ and $Y_1, \ldots, Y_{n_2}$ are taken from normal populations with equal variances and respective means μ_X and μ_Y, the experimenter may wish to test $H_0: \mu_X - \mu_Y = 0$ versus $H_a: \mu_X - \mu_Y < 0$. In this case, if H_0 is true, both populations are normally distributed with the same mean and the same variance; that is, the population distributions are identical. If H_a is true, then $\mu_Y > \mu_X$ and the distributions of X_1 and Y_1 are the same, except that the location parameter (μ_Y) for Y_1 is larger than the location parameter (μ_X)

for X_1. That is, the distribution of Y_1 is shifted to the right of the distribution of X_1 (see Figure 15.1).

This is an example of a two-sample parametric *shift* (location) *model*. The model is parametric because the distributions are specified (normal) except for the values of the parameters μ_X, μ_Y, and σ^2. The amount that the distribution of Y_1 is shifted to the right of the distribution of X_1 is $\mu_Y - \mu_X$ (see Figure 15.1). In the remainder of this section we define a shift model that applies for any distribution, normal or otherwise.

Figure 15.1
Two normal distributions with equal variances but unequal means

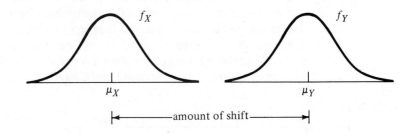

amount of shift

Let $X_1, \ldots, X_{n_1}$ be a random sample from a population with distribution function $F(x)$, and let $Y_1, \ldots, Y_{n_2}$ be a random sample from a population with distribution function $G(y)$. If we wish to test whether the two populations have the same distribution—that is, $H_0: F(z) = G(z)$ versus $H_a: F(z) \neq G(z)$, with the actual form of $F(z)$ and $G(z)$ unspecified—a nonparametric method is required. Note that H_a is a very broad hypothesis. Many times an experimenter may wish to consider the more specific alternative hypothesis that Y_1 has the same distribution as X_1 shifted by an (unknown) amount θ (see Figure 15.2); that is, that the distributions *differ in location*. Then $G(y) = P(Y_1 \leq y)$ $= P(X_1 \leq y - \theta) = F(y - \theta)$ for some unknown parameter value θ. Note that the particular form of $F(x)$ remains unspecified.

Figure 15.2
Two density functions, with the density for Y shifted θ units to the right of that for X

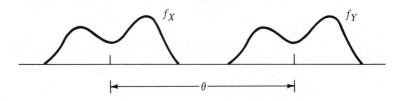

Throughout this chapter if we refer to the two-sample shift (location) model, we assume that $X_1, \ldots, X_{n_1}$ constitutes a random sample from distribution function $F(x)$ and that $Y_1, \ldots, Y_{n_2}$ constitutes a random sample from distribution function $G(y) = F(y - \theta)$ for some unknown value θ. For the two-sample shift model, $H_0: F(z) = G(z)$ is equivalent to $H_0: \theta = 0$. If θ is greater (less) than 0, then the distribution of the Y values is located to the right (left) of the distribution of the X values.

15.3 The Sign Test for a Paired Experiment

Suppose that we have n pairs of observations of the form (X_i, Y_i) and that we wish to test the hypothesis that the distribution of the X's is the same as that of the Y's versus the alternative that the distributions differ in location (see Section 15.2). Similar to what we did in Section 12.5, we let $D_i = X_i - Y_i$. One of the simplest nonparametric tests is based on the signs of these differences and, reasonably enough, is called the *sign test*. Under the null hypothesis that X_i and Y_i come from the same continuous probability distributions, the probability that D_i is positive is equal to $1/2$ (as is the probability that D_i is negative). Let M denote the total number of positive (or negative) differences. Then if the X_i's and Y_i's have the same distribution, M will have a binomial distribution with $p = 1/2$, and the rejection region for a test based on M can be obtained by using the binomial probability distribution of Chapter 3. The sign test is summarized as follows:

The Sign Test for Comparing Two Populations

Let $p = P(X > Y)$.

H_0: $p = 1/2$

H_a: $p > 1/2$ or $(p < 1/2$ or $p \neq 1/2)$

Test Statistic: M = number of positive differences where $D_i = X_i - Y_i$

Rejection Region: For H_a: $p > 1/2$, reject H_0 for the largest values of M; for H_a: $p < 1/2$, reject H_0 for the smallest values of M; for H_a: $p \neq 1/2$, reject H_0 for very large or very small values of M.

Assumptions: The pairs (X_i, Y_i) are randomly and independently selected.

The following example illustrates the use of the sign test.

EXAMPLE 15.1   The number of defective electrical fuses proceeding from each of two production lines, A and B, was recorded daily for a period of 10 days with the results shown in Table 15.1. Assume that both production lines produced the same daily output. Compare the number of defectives produced by A and B each day and let M equal the number of days when A exceeded B. Do the data present sufficient evidence to indicate that either production line produces more defectives than the other? State the null hypothesis to be tested and use M as a test statistic.

Table 15.1
Data for
Example 15.1

Day	A	B
1	172	201
2	165	179
3	206	159
4	184	192
5	174	177
6	142	170
7	190	182
8	169	179
9	161	169
10	200	210

Solution

Pair the observations as they appear in the data tabulation and let M be the number of times that the observation for production line A exceeds that for B in a given day. Under the null hypothesis that the two distributions of defectives are identical, the probability p that A exceeds B for a given pair is $p = .5$, given that there are no ties. Consequently, the null hypothesis is equivalent to a hypothesis that the binomial parameter $p = .5$.

Very large or very small values of M are most contradictory to the null hypothesis. Therefore, the rejection region for the test will be located by including the most extreme values of M that at the same time provide an α feasible for the test.

Suppose that we would like α to be somewhere on the order of .05 or .10. We would commence the selection of the rejection region by including $M = 0$ and $M = 10$ and calculate the α associated with this region, using $p(y)$ (the probability distribution for the binomial random variable, Chapter 3). With $n = 10$, $p = .05$, we have

$$\alpha = p(0) + p(10) = \binom{10}{0}(.5)^{10} + \binom{10}{10}(.5)^{10} = .002$$

Because this value of α is too small, the region will be expanded by including the next pair of M values most contradictory to the null hypothesis, $M = 1$ and $M = 9$. The value of α for this region ($M = 0, 1, 9, 10$) can be obtained from Table 1, Appendix III.

$$\alpha = p(0) + p(1) + p(9) + p(10) = .022$$

This also is too small, so we will again expand the region to include $M = 0, 1, 2, 8, 9, 10$. You can verify that the corresponding value of α is .11. We will suppose that this value of α is acceptable to the experimenter and will employ $M = 0, 1, 2, 8, 9, 10$ as the rejection region for the test.

From the data we observe that $m = 2$, and therefore we reject the null hypothesis. We conclude that sufficient evidence exists to indicate that the population distributions of number of defects are not identical. The probability of rejecting the null hypothesis when true is only $\alpha = .11$, and therefore we are reasonably confident of our conclusion.

The experimenter in this example is using the test procedure as a rough tool for detecting faulty production lines. The rather large value of α is not likely to disturb him because he can easily collect additional data if he is concerned about making a type I error in reaching his conclusion.   ∽

Attained significance levels (p-values) for the sign test are calculated as outlined in Section 10.5. Specifically, if $n = 15$, and we wish to test $H_0: p = 1/2$ versus $H_a: p < 1/2$ based on an observed value of $M = 3$, Table 1 of Appendix III can be used to determine that (because $n = 15$, $p = 1/2$)

$$p\text{-value} = P(M \le 3) = .018$$

For the two-tailed test ($H_a: p \neq 1/2$), p-value $= 2(.018) = .036$.

One problem that may occur when conducting a sign test is that the observations associated with one or more pairs may be equal and therefore result in ties. When this situation occurs, delete the tied pairs and reduce n, the total number of pairs.

You will also encounter situations where n, the number of pairs, is large. Then the values of α associated with the sign test can be obtained by using the normal approximation to the binomial probability distribution discussed in Section 7.5. You can verify (by comparison of exact probabilities with their approximations) that these approximations will be quite adequate for n as small as 25. This result is due to the symmetry of the binomial probability distribution for $p = .5$. For $n \ge 25$, the Z test of Chapter 10 will be quite adequate, where

$$Z = \frac{M - np}{\sqrt{npq}} = \frac{M - n/2}{(1/2)\sqrt{n}}$$

This statistic would be used for testing the null hypothesis $p = .5$ against the alternative $p \neq .5$ for a two-tailed test or against the alternative $p > .5$ (or $p < .5$) for a one-tailed test. The tests would utilize the familiar rejection regions of Chapter 10.

You will note that the data of Example 15.1 are the result of a paired-difference experiment. Suppose that the paired differences are normally distributed with a common variance σ^2. Will the sign test detect a shift in location of the two populations as effectively as the Student's t test? Intuitively, we would suspect that the answer is "no," and this is correct, because the Student's t test utilizes comparatively more information. In addition to giving the sign of the difference, the t test uses the magnitudes of the observations to obtain a more accurate value of the sample means and variances. Thus we might say that the sign test is not as "efficient" as the Student's t test, but this statement is meaningful only if the populations conform to the assumption just stated; that is, the differences in paired observations are normally distributed with a common variance σ_D^2. The sign test might be more efficient when these assumptions are not satisfied.

Sign Test for Large Samples $n > 25$

1. Null hypothesis: $H_0: p = .5$ (neither treatment is preferred to the other)

2. Alternative Hypothesis: $H_a: p \neq .5$ for a two-tailed test (Note: We use the two-tailed test for an example. Many analyses will require a one-tailed test.)

3. Test Statistic: $Z = \dfrac{M - n/2}{(1/2)\sqrt{n}}$

4. Rejection Region: Reject H_0 if $z \geq z_{\alpha/2}$ or if $z \leq -z_{\alpha/2}$ where $z_{\alpha/2}$ is obtained from Table 3, Appendix III.

The sign test actually tests the null hypothesis that the *median* of the D_i's is zero versus the alternative that it is different from zero. [The median of the D_i's being zero does imply that $P(D_i < 0) = P(D_i > 0)$.] If the X_i's and Y_i's have the same distribution, then the median of the D_i's will be zero, as previously discussed. However, for models other than the shift model, there are other situations in which the median of the D_i's is zero. In these instances the null hypothesis for the sign test is slightly more general than the statement that X_i and Y_i have the same distribution.

Summarizing, the sign test is a very easily applied nonparametric procedure for comparing two populations. No assumptions are made concerning the underlying population distributions. The value of the test statistic can be obtained quickly by a visual count, and the rejection region can be located easily by using a table of binomial probabilities. Furthermore, we need not know the exact values of pairs of responses, just whether $X_i > Y_i$ for each pair (X_i, Y_i). Exercise 15.3 provides an example of the use of the sign test for data of this sort.

Exercises

15.1 What significance levels between $\alpha = .01$ and $\alpha = .15$ are available for a two-tailed sign test utilizing twenty-five paired observations? (Make use of tabulated values in Table 1, Appendix III, $n = 25$.) What are the corresponding rejection regions?

15.2 For a comparison of two junior high schools, A and B, in academic effectiveness, an experiment was designed requiring the use of ten sets of identical twins, each twin having just completed the sixth grade. In each case the twins in the same set had obtained their schooling in the same classrooms at each grade level. One child was selected at random from each set and assigned to school A. The other was sent to school B. Near the end of the ninth grade, an achievement test was given to each child in the experiment. The results are shown in the accompanying table.

Twin Pair	A	B	Twin Pair	A	B
1	67	39	6	50	52
2	80	75	7	63	56
3	65	69	8	81	72
4	70	55	9	86	89
5	86	74	10	60	47

(a) Test (using the sign test) the hypothesis that the two schools are the same in academic effectiveness, as measured by scores on the achievement test, against the alternative that the schools are not equally effective. Give the attained significance level. What would you conclude with $\alpha = .05$?

(b) Suppose it were suspected that junior high school A had a superior faculty and better learning facilities. Test the hypothesis of equal academic effectiveness against the alternative that school A is superior. What is the p-value associated with this test?

15.3 New food products are frequently subjected to taste tests by a panel of judges. The judges are usually asked merely to state a preference for one food over another, so that no quantitative scale need be employed. Suppose that two new mixtures, say A and B, of an orange-flavored drink are presented to ten judges. The preferences of the judges are given in the accompanying table. Does this evidence indicate a significant difference between the tastes of A and B, at the 5% significance level?

Judge	Preference	Judge	Preference
1	A	6	A
2	A	7	B
3	A	8	A
4	A	9	B
5	A	10	A

15.4 On clear, cold nights in the central Florida citrus region, the precise location of the below-freezing temperatures is important, since the methods of protecting trees from freezing conditions are very expensive. One method of locating likely cold spots is by relating temperature to elevation. It is conjectured that, on calm nights, the cold spots will be at low elevations. The highest and lowest spots in a particular grove showed the accompanying minimum temperatures for ten cold nights in a recent winter.

Night	High Elevation	Low Elevation
1	32.9	31.8
2	33.2	31.9
3	32.0	29.2
4	33.1	33.2
5	33.5	33.0
6	34.6	33.9
7	32.1	31.0
8	33.1	32.5
9	30.2	28.9
10	29.1	28.0

(a) Is there sufficient evidence to support the conjecture that low elevations tend to be colder? (Use the sign test. Give the associated p-value.)

(b) Would it be reasonable to use a t test on the data? Explain why or why not?

15.5 A psychological experiment was conducted to compare the lengths of response time (in seconds) for two different stimuli. To remove natural person-to-person variability in the responses, both stimuli were applied to each of nine subjects, thus permitting an analysis of the difference between stimuli *within* each person.

Subject	Stimulus 1	Stimulus 2
1	9.4	10.3
2	7.8	8.9
3	5.6	4.1
4	12.1	14.7
5	6.9	8.7
6	4.2	7.1
7	8.8	11.3
8	7.7	5.2
9	6.4	7.8

(a) Use the sign test to determine whether sufficient evidence exists to indicate a difference in mean response for the two stimuli. Use a rejection region for which $\alpha \le .05$.

(b) Test the hypothesis of no difference in mean response using Student's t test.

15.6 Refer to Exercise 12.16. Using the sign test, is there sufficient evidence to conclude that completion times differ for the two populations? Use $\alpha = .10$.

15.7 The accompanying data set lists the number of industrial accidents in twelve manufacturing plants for one-week periods before and after an intensive promotion on safety.

(a) Do the data support the claim that the campaign was successful? What is the attained significance level? What would you conclude with $\alpha = .01$?

(b)   Discuss the problems associated with a parametric analysis designed to answer the question in (a).

Plant	Before	After	Plant	Before	After
1	3	2	7	5	3
2	4	1	8	3	3
3	6	3	9	2	0
4	3	5	10	4	3
5	4	4	11	4	1
6	5	2	12	5	2

15.4 The Wilcoxon Signed-Rank Test for a Paired Experiment

As in Section 15.3, assume that we have n paired observations of the form (X_i, Y_i) and that $D_i = X_i - Y_i$. Again we assume that we are interested in testing the hypothesis that the X's and the Y's have the same distribution versus the alternative that the distributions differ in location. Under the null hypothesis of no difference in the distributions of the X's and Y's, you would expect (on the average) half of the differences in pairs to be negative and half to be positive. That is, the expected number of negative differences between pairs would be $n/2$ (where n is the number of pairs). Further, it would follow that positive and negative differences of equal absolute magnitude should occur with equal probability. If we were to order the differences according to their absolute values and rank them from smallest to largest, the expected rank sums for the negative and positive differences would be equal. Sizable differences in the sums of the ranks assigned to the positive and negative differences would provide evidence to indicate a shift in location between the distributions.

To carry out the Wilcoxon test, we calculate the differences (D_i) for each of the n pairs. Differences equal to zero are eliminated, and the number of pairs, n, is reduced accordingly. Then we rank the *absolute values* of the differences, assigning a 1 to the smallest, a 2 to the second smallest, and so on. If two or more absolute differences are tied for the same rank, then the average of the ranks that would have been assigned to these differences is assigned to each member of the tied group. For example, if two absolute differences are tied for ranks 3 and 4, then each would receive rank 3.5 and the next highest absolute difference would be assigned rank 5. Then we calculate the rank sum for the negative differences and also calculate the rank sum for the positive differences. For a two-tailed test we use the smaller of these two quantities, T, as a test statistic to test the null hypothesis that the two population relative frequency histograms are identical. The smaller the value of T, the greater will be the weight of evidence favoring rejection of the null hypothesis. Hence we will reject the null hypothesis if T is less than or equal to some value, say T_0.

To detect the one-sided alternative, that the distribution of the X's is shifted to the right of that of the Y's, we use the rank sum T^- of the negative differences, and we reject the null hypothesis for small values of T^-, say $T^- \leq T_0$. If we wish to detect a shift of the distribution of the Y's to the right of the X's, we use the rank sum T^+ of the positive differences as a test statistic, and we reject small values of T^+, say $T^+ \leq T_0$.

The probability that T is less than or equal to some value, T_0, has been calculated for a combination of sample sizes and values of T_0. These probabilities, given in Table 9, Appendix III, can be used to find the rejection region for the test based on T.

For example, suppose that you have $n = 7$ pairs and wish to conduct a two-tailed test of the null hypothesis that the two population relative frequency distributions are identical. Then, with $\alpha = .05$, you would reject the null hypothesis for all values of T less than or equal to 2. The rejection region for the Wilcoxon rank-sum test for a paired experiment is always of this form: reject the null hypothesis if $T \leq T_0$ where T_0 is the critical value for T. Bounds for the attained significance level (p-level) are determined as follows. For a two-tailed test, if $T = 3$ is observed when $n = 7$, Table 9, Appendix III, indicates that H_0 would be rejected if $\alpha = .1$, but not if $\alpha = .05$. Thus, $.05 < p$-value $< .1$. For the one-sided alternative that the X's are shifted to the right of the Y's with $n = 7$ and $\alpha = .05$, H_0 is rejected if $T = T^- \leq 4$. In this case if $T = T^- = 1$, then $.01 < p$-value $< .025$. The test based on T, called the *Wilcoxon signed-rank test*, is summarized as follows.

Wilcoxon Signed-Rank Test for a Paired Experiment

H_0: The population distributions for the X's and Y's are identical.

H_a: The two population distributions differ in location (two-tailed) or H_a: the population relative frequency distribution for the X's is shifted to the right[†] of that for the Y's (one-tailed).

Test Statistic:

1. For a two-tailed test, use $T = \min(T^+, T^-)$ where T^+ = sum of the ranks of the positive differences and T^- = sum of the ranks of the negative differences.

2. For a one-tailed test (to detect the one-tailed alternative just given) use the rank sum T^- of the negative differences.

Rejection Region:

1. For a two-tailed test, reject H_0 if $T \leq T_0$, where T_0 is the critical value given in Table 9, Appendix III.

2. For a one-tailed test (as described earlier) reject H_0 if $T^- \leq T_0$.

[†] To detect a shift of the distribution of the Y's to the right of the distribution of the X's, use the rank sum T^+, the sum of the ranks of the positive differences and reject H_0 if $T^+ \leq T_0$.

EXAMPLE 15.2   Test the hypothesis of no difference in population distributions of cake density for a paired difference experiment involving six baked cakes, one using mix A and one using mix B in each pair. What can be said about the attained significance level?

Solution   The original data and differences in density (in ounces per cubic inch) for the six pairs of cakes are shown in Table 15.2.

Table 15.2
Data and their differences

A	B	Difference, $A - B$	Rank
.135	.129	.006	3
.102	.120	−.018	5
.108	.112	−.004	1.5
.141	.152	−.011	4
.131	.135	−.004	1.5
.144	.163	−.019	6

As with our other nonparametric tests, the null hypothesis to be tested is that the two population frequency distributions of cake densities are identical. The alternative hypothesis, which implies a two-tailed test, is that the distributions differ in location.

Because the amount of data is small, we will conduct our test by using $\alpha = .10$. From Table 9, Appendix III, the critical value of T for a two-tailed test, $\alpha = .10$, is $T_0 = 2$. Hence we will reject H_0 if $T \leq 2$.

Because there is only one positive difference that has rank 3, $T^+ = 3$ and $T^- = 18$, and therefore $T = 3$. The observed value of T exceeds the critical value of T, and therefore there is not sufficient evidence to indicate a difference in the two population frequency distributions of cake densities. Because we cannot reject H_0 for $\alpha = .10$, we can only say that p-value $> .10$.   ∽

Although Table 9, Appendix III, is applicable for values of n(number of data pairs) as large as $n = 50$, it is worth noting that T^+ (or T^-) will be approximately normally distributed when the null hypothesis is true and n is large (say 25 or more). This enables us to construct a large-sample Z test, where if $T = T^+$

$$E(T^+) = \frac{n(n + 1)}{4} \quad \text{and} \quad V(T^+) = \frac{n(n + 1)(2n + 1)}{24}$$

Then the Z statistic,

$$Z = \frac{T^+ - E(T^+)}{\sqrt{V(T^+)}} = \frac{T^+ - [n(n + 1)/4]}{\sqrt{n(n + 1)(2n + 1)/24}}$$

can be used as a test statistic. Thus for a two-tailed test and $\alpha = .05$, we would reject the hypothesis of identical population distributions when $|z| \geq 1.96$. For a one-tailed test that the distribution of the X's is shifted to the right (left) of the distribution of the Y's, reject H_0 when $z > z_\alpha$ ($z < -z_\alpha$).

A Large-Sample Wilcoxon Signed-Rank Test for a Paired Experiment: $n > 25$

1. **Null Hypothesis:** H_0: The population relative frequency distributions for the X's and Y's are identical.

2. **Alternative Hypothesis:** H_a: The two population relative frequency distributions differ in location (a two-tailed test) or H_a: The population relative frequency distribution for the X's is shifted to the right (or left) of the relative frequency distribution of the Y's (one-tailed tests).

3. **Test Statistic:** $Z = \dfrac{T^+ - [n(n + 1)/4]}{\sqrt{n(n + 1)(2n + 1)/24}}$

4. **Rejection Region:** Reject H_0 if $z \geq z_{\alpha/2}$ or $z \leq -z_{\alpha/2}$ for a two-tailed test. To detect a shift in the distributions of the X's to the right of the Y's, reject H_0 when $z \geq z_\alpha$. To detect a shift in the opposite direction, reject H_0 if $z \leq -z_\alpha$.

Exercises

15.8 The accompanying table gives the scores of a group of fifteen students in mathematics and art.

Student	Math	Art	Student	Math	Art
1	22	53	9	62	55
2	37	68	10	65	74
3	36	42	11	66	68
4	38	49	12	56	64
5	42	51	13	66	67
6	58	65	14	67	73
7	58	51	15	62	65
8	60	71			

(a)   Use Wilcoxon's signed-rank test to determine if the locations of the distributions of scores for these students differ significantly for the two subjects. Give bounds for the *p*-value and indicate the appropriate conclusion with $\alpha = .05$.

(b)   State specific null and alternative hypotheses for the test you conducted in (a).

15.9   Refer to Exercise 15.2. What answers are obtained if Wilcoxon's signed-rank test is used in analyzing the data? Compare these answers with the answers obtained in Exercise 15.2.

15.10   Refer to Exercise 15.4. Answer the question in part (a) by using the Wilcoxon signed-rank test.

15.11   Eight subjects were asked to perform a simple puzzle assembly task under normal conditions and under conditions of stress. During the stress condition the subjects were told that a mild shock would be delivered 3 minutes after the start of the experiment and every 30 seconds thereafter until the task was completed. Blood pressure readings were taken under both conditions. The accompanying data represent the highest reading during the experiment. Do the data present sufficient evidence to indicate higher blood pressure readings during conditions of stress? Analyze the data by using the Wilcoxon rank-sum test for a paired experiment. Give the appropriate *p*-value.

Subject	Normal	Stress
1	126	130
2	117	118
3	115	125
4	118	120
5	118	121
6	128	125
7	125	130
8	120	120

15.12   Economic indices provide measures of economic change. The June 1977 issue of *U.S. News and World Report* listed a sample of economic indices for May 1977 and May 1976 in order to provide a means for measuring economic change over the year:

	May 1977	*May 1976*
Steel products	99.3	106.4
Automobile products	123.9	118.7
Crude petroleum products	91.0	93.2
Lumber products	117.1	112.6
Freight products	84.8	82.2
Electric power products	169.5	167.0

Source   Reprinted by permission. *U.S. News and World Report* © 1977. All rights reserved.

(a) Conduct a paired difference t test to compare the mean values of these indices for the month of May in 1976 and 1977. Use $\alpha = 0.05$. What assumptions are necessary for the validity of this procedure? Why might these assumptions be doubtful?

(b) Use the Wilcoxon signed-rank test to determine whether these data provide evidence that the locations of the probability distributions of all economic indices have changed. Use $\alpha = 0.05$.

15.13 Dental researchers have developed a new material for preventing cavities, a plastic sealant, which is applied to the chewing surfaces of teeth. To determine whether the sealant is effective, it was applied in half of the teeth of each of twelve school-age children. After 5 years, the number of cavities in the sealant-coated teeth and untreated teeth were counted. The results are given in the accompanying table. Is there sufficient evidence to indicate that sealant-coated teeth are less prone to cavities than are untreated teeth? Test using $\alpha = 0.05$.

Child	Sealant-coated	Untreated
1	3	3
2	1	3
3	0	2
4	4	5
5	1	0
6	0	1
7	1	5
8	2	0
9	1	6
10	0	0
11	0	3
12	4	3

15.14 Refer to Exercise 12.17. With $\alpha = .01$, use the Wilcoxon signed-rank test to see if there was a significant loss in muck depth between 1972 and 1978.

15.15 Suppose $Y_1, \ldots, Y_n$ is a random sample from a continuous distribution function $F(y)$. It is desired to test a hypothesis concerning the median ξ of $F(y)$. Construct a test of $H_0: \xi = \xi_0$ against $H_a: \xi \neq \xi_0$, where ξ_0 is a specified constant, by making use of these tests:

(a) the sign test

(b) the Wilcoxon signed-rank test

15.16 The spokesman for an organization supporting property tax reductions in a certain section of a city stated that the median annual income for household heads in that section was $15,000. A random sample of ten household heads from that section revealed the following annual incomes:

14,800, 16,900, 18,000, 19,100, 13,200, 18,500, 20,000, 19,200, 15,100, 16,500

With $\alpha = .10$, test the hypothesis that the median income for the population from that section is $15,000 against the alternative that it is greater than $15,000.

(a)   Use the sign test.

(b)   Use the Wilcoxon signed-rank test.

15.5 The Use of Ranks for Comparing Two Population Distributions: Independent Random Samples

A statistical test for comparing two populations based on independent random samples, the *rank-sum* test, was proposed by F. Wilcoxon in 1945. Again, we assume that we are interested in testing that the two populations have the same distribution versus the shift or location alternative (see Section 15.2). Suppose that you were to select independent random samples of n_1 and n_2 observations, each from two populations, call them I and II. Wilcoxon's idea was to combine the $n_1 + n_2 = n$ observations and rank them, in order of magnitude, from 1 (the smallest) to n (the largest). Ties are treated as in Section 15.4. That is, if two or more observations are tied for the same rank, then the average of the ranks that would have been assigned to these observations is assigned to each member of the tied group. Then if the observations were selected from identical populations, the rank sums for the samples should be more or less proportional to the sample sizes n_1 and n_2. For example, if n_1 and n_2 were equal, you would expect the rank sums to be nearly equal. In contrast, if the observations in one population, say population I, tended to be larger than those in population II, then the observations in sample I would tend to receive the highest ranks and would have a larger than expected rank sum. Thus (sample sizes being equal) if one rank sum is very large (and, correspondingly, the other is very small), you may have evidence to indicate a difference between the two populations.

Mann and Whitney proposed a statistical test in 1947 that also used the rank sums of the two samples, and their test can be shown to be equivalent to the Wilcoxon test. Because the Mann-Whitney U test, and tables of critical values of U, occurs so often in the literature, we will explain its use in Section 15.6 and will give several examples of its applications. In this section we illustrate the logic of the rank-sum test and show you how the rejection region for the test and how α are determined.

EXAMPLE 15.3   The bacteria counts per unit volume are shown in Table 15.3 for two types of cultures, A and B. Four observations were made for each culture. Let n_1 and n_2 represent the number of observations in samples A and B, respectively.

For the data given in Table 15.3, the corresponding ranks are as shown in Table 15.4. Do these data present sufficient evidence to indicate a difference in the population distributions for A and B?

Table 15.3
Data for
Example 15.3

A	B
27	32
31	29
26	35
25	28

Table 15.4
Ranks

	A	B
	3	7
	6	5
	2	8
	1	4
Rank Sum	12	24

Solution

Let W equal the rank sum for sample A (for this sample, $W = 12$). Certainly very small or very large values of W will provide evidence to indicate a difference between the two population distributions, and hence W, the *rank sum*, will be employed as a test statistic.

The rejection region for a given test will be obtained in the same manner as for the sign test. We will commence by selecting the most contradictory values of W as the rejection region and will add to these until α is of sufficient size.

The minimum rank sum would include the ranks 1, 2, 3, 4, or $W = 10$. Similarly, the maximum would include 5, 6, 7, 8, with $W = 26$. We will commence by including these two values of W in the rejection region. What is the corresponding value of α?

Finding the value of α is a probability problem that can be solved by using the methods of Chapter 2. If the populations are identical, every permutation of the eight ranks will represent a sample point and will be equally likely. Then α will represent the sum of the probabilities of the sample points (arrangements) that imply $W = 10$ or $W = 26$ in sample A. The total number of permutations of the eight ranks is 8! The number of different arrangements of the ranks 1, 2, 3, 4 in sample A with the 5, 6, 7, 8 of sample B is (4!4!). Similarly, the number of arrangements that place the maximum value of W in sample A (ranks 5, 6, 7, 8) is (4!4!). Then the probability that $W = 10$ or $W = 26$ is

$$p(10) + p(26) = \frac{(2)(4!)(4!)}{8!} = \frac{2}{\binom{8}{4}} = \frac{1}{35} = .029$$

If this value of α is too small, the rejection region can be enlarged to include the next smallest and largest rank sums, $W = 11$ and $W = 25$. The rank sum $W = 11$ will conclude the ranks 1, 2, 3, 5, and

$$p(11). = \frac{4!4!}{8!} = \frac{1}{70}$$

Similarly,

$$p(25) = 1/70$$

Then

$$\alpha = p(10) + p(11) + p(25) + p(26) = 2/35 = .057$$

Expansion of the rejection region to include 12 and 24 will substantially increase the value of α. The set of sample points giving a rank of 12 will be all sample points associated with rankings of (1, 2, 3, 6) and (1, 2, 4, 5). Thus

$$p(12) = \frac{(2)(4!)(4!)}{8!} = \frac{1}{35}$$

and

$$\alpha = p(10) + p(11) + p(12) + p(24) + p(25) + p(26)$$
$$= \frac{1}{70} + \frac{1}{70} + \frac{1}{35} + \frac{1}{35} + \frac{1}{70} + \frac{1}{70} = \frac{4}{35} = .114$$

This value of α might be considered too large for practical purposes. In this case we would be better satisfied with the rejection region $W = 10, 11, 25,$ and 26.

The rank sum for the sample, $W = 12$, falls in the nonrejection region, and hence we do not have sufficient evidence to reject the hypothesis that the population distributions of bacteria counts for the two cultures are identical.

∽

15.6 The Mann-Whitney U Test: Independent Random Samples

The Mann-Whitney statistic U is obtained by ordering all $(n_1 + n_2)$ observations according to their magnitude and counting the number of observations in sample A that precede each observation in sample B. The U statistic is the sum of these counts.

For example, the eight observations of Example 15.3 are

25	26	27	28	29	31	32	35
A	A	A	B	B	A	B	B

The smallest B observation is 28, and $u_1 = 3$ observations from sample A precede it. Similarly, $u_2 = 3$ A observations precede the second B observation and $u_3 = 4$ and $u_4 = 4$ A observations precede the third and fourth B observations, respectively (32 and 35). Then

$$U = u_1 + u_2 + u_3 + u_4 = 14$$

Or you could count the number of B observations that precede each A observation and use the sum of the counts, U_B, as the U statistic. In either case, very large or small values of U will imply a separation of the ordered A and B observations and will provide evidence to indicate a difference (a shift in location) between the population distributions for A and B.

As noted in Section 15.5, it can be shown that the Mann-Whitney U statistic is related to Wilcoxon's rank sum. In fact, it can be shown (proof omitted) that

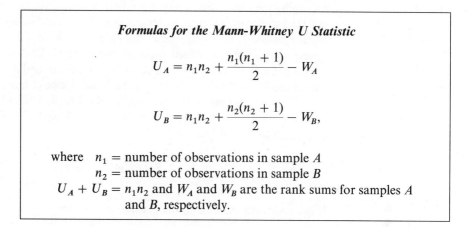

Formulas for the Mann-Whitney U Statistic

$$U_A = n_1 n_2 + \frac{n_1(n_1 + 1)}{2} - W_A$$

$$U_B = n_1 n_2 + \frac{n_2(n_2 + 1)}{2} - W_B,$$

where n_1 = number of observations in sample A
n_2 = number of observations in sample B
$U_A + U_B = n_1 n_2$ and W_A and W_B are the rank sums for samples A and B, respectively.

As you can see from the formulas for U_A and U_B, U_A will be small when W_A is large, a situation likely to occur when the population distribution of the A measurements is shifted to the right of the population distribution for the B measurements. Consequently, to conduct a one-tailed test to detect a shift in the A distribution to the right of the B distribution, you will reject the null hypothesis of no difference in the population distributions if U_A is less than some specified value, U_0. That is, you will reject H_0 for small values of U_A. Similarly, to conduct a one-tailed test to detect a shift of the B distribution to the right of the A distribution, you would reject H_0 if U_B is less than some specified value, say U_0.

Table 8, Appendix III, gives the probability that an observed value of U will be less than some specified value, say U_0. This is the value of α for a one-tailed test. To conduct a two-tailed test—that is, to detect a shift in the population distributions for the A and B measurements in either direction—we will agree to always use U, the smaller of U_A or U_B, as the test statistic and reject H_0 for $U < U_0$. The value of α for the two-tailed test will be double the tabulated value given in Table 8, Appendix III.

To see how to locate the rejection region for the Mann-Whitney U test, suppose that $n_1 = 4$ and $n_2 = 5$. Then you would consult the third table in Table 8, Appendix III, the one corresponding to $n_2 = 5$. Note that the table is constructed on the assumption that $n_1 \leq n_2$. That is, you will always identify the smaller sample as sample 1. From the table we see that, for example, $P(U \leq 2) = .0317$ and $P(U \leq 3) = .0556$. So if you want to conduct a one-tailed Mann-Whitney U test with $n_1 = 4$ and $n_2 = 5$ and would like α to be near .05, you would reject the null hypothesis of equality of population relative frequency distributions when $U \leq 3$. The probability of a type I error for the test would be $\alpha = .0556$. If you use this same rejection for a two-tailed test, that is, $U \leq 3$, α would be double the tabulated value, or $\alpha = 2(.0556) = .1112$.

When applying the test to a set of data, you may find that some of the observations are of equal value. Ties in the observations can be handled by averaging the ranks that would have been assigned to the tied observations and assigned this average to each. Thus if three observations are tied and are due to receive ranks 3, 4, 5, we would assign the rank of 4 to all three. The next observation in the sequence would receive the rank of 6, and ranks 3 and 5 would not appear. Similarly, if two observations are tied for ranks 3 and 4, each would receive a rank of 3.5, and ranks 3 and 4 would not appear.

Table 8, Appendix III, can also be used to find the observed significance level for a test. For example, if $n_1 = 5$, $n_2 = 5$, and $U = 4$, the p-value for a one-tailed test would be

$$P\{U \leq 4\} = .0476.$$

If the test were two-tailed, the p-value would be

$$2(.0476) \quad \text{or} \quad .0952.$$

The Mann-Whitney U Test

1. Null Hypothesis: H_0: The population relative frequency distributions for A and B are identical.

2. Alternative Hypothesis: H_a: The two populations relative frequency distributions are shifted in respect to their relative locations (a two-tailed test). Or H_a: The population relative frequency distribution for A is shifted to the right of the relative frequency distribution for population B (a one-tailed test).[†]

3. Test Statistic: For a two-tailed test, use U, the smaller of

$$U_A = n_1 n_2 + \frac{n_1(n_1 + 1)}{2} - W_A$$

and

$$U_B = n_1 n_2 + \frac{n_2(n_2 + 1)}{2} - W_B$$

Where W_A and W_B are the rank sums for samples A and B, respectively. For a one-tailed test, use U_A.

4. Rejection Region:

(i) For the two-tailed test and a given value of α, reject H_0 if $U \leq U_0$, where $P(U \leq U_0) = \alpha/2$. [Note: Observe that U_0 is the value such that $P(U \leq U_0)$ is equal to half of α.]

(ii) For a one-tailed test and a given value of α, reject H_0 if $U_A \leq U_0$ where $P(U_A \leq U_0) = \alpha$.

Assumptions: Samples have been randomly and independently selected from their respective populations. Ties in the observations can be handled by averaging the ranks that would have been assigned to the tied observations and assigning this average to each. Thus, if three observations are tied and are due to receive ranks 3, 4, 5, we would assign the rank of 4 to all three.

---

† For the sake of convenience, we will describe the one-tailed test as one designed to detect a shift in the distribution of the A measurements to the right of the distribution of the B measurements. To detect a shift in the B distribution to the right of the A distribution, just interchange the letters A and B in the discussion.

EXAMPLE 15.4 Test the hypothesis that there is no difference in the population distributions for the bacteria count data of Example 15.3.

Solution We have already noted that the Mann-Whitney U test and the Wilcoxon rank-sum test are equivalent, so we should reach the same conclusions as for Example 15.3. Recall that the alternative hypothesis was that there was a difference in the distributions of bacteria counts for cultures A and B and that this implied a two-tailed test. Thus because Table 8, Appendix III, gives values of $P(U \leq U_0)$ for specified sample sizes and values of U_0, we must double the tabulated value to find α. Suppose, as in Example 15.3, that we desire a value of α near .05. Checking Table 8 for $n_1 = n_2 = 4$, we find $P(U \le 1) = .0286$. When $U \le 1$ is the rejection region, α will equal $2(.0286) = .0572$ or, rounding to three decimal places, $\alpha = .057$ (the same value of α obtained for Example 15.3).

For the bacteria data the rank sums are $W_A = 12$ and $W_B = 24$. Then

$$U_A = n_1 n_2 + \frac{n_1(n_1 + 1)}{2} - W_A = (4)(4) + \frac{4(4 + 1)}{2} - 12 = 14$$

and

$$U_B = n_1 n_2 + \frac{n_2(n_2 + 1)}{2} - W_B = (4)(4) + \frac{4(4 + 1)}{2} - 24 = 2$$

Because the smaller observed value of U is 2 (just calculated), U does not fall in the rejection region. Hence, there is not sufficient evidence to show a difference in the population distributions of bacteria counts for cultures A and B. The p-value is given by $2P(U \le 2) = 2(.0571) = .1142$. $\backsim$

EXAMPLE 15.5   An experiment was conducted to compare the strength of two types of kraft papers, one a standard kraft paper of a specified weight and the other the same standard kraft paper treated with a chemical substance. Ten pieces of each type of paper, randomly selected from production, produced the strength measurements shown in Table 15.5. Test the hypothesis of no difference in the distributions of strengths for the two types of paper against the alternative hypothesis that the treated paper tends to be of greater strength.

Table 15.5
Data for
Example 15.5

	Standard, A	Treated, B
	1.21 (2)	1.49 (15)
	1.43 (12)	1.37 (7.5)
	1.35 (6)	1.67 (20)
	1.51 (17)	1.50 (16)
	1.39 (9)	1.31 (5)
	1.17 (1)	1.29 (3.5)
	1.48 (14)	1.52 (18)
	1.42 (11)	1.37 (7.5)
	1.29 (3.5)	1.44 (13)
	1.40 (10)	1.53 (19)
Rank Sums	$W_A = 85.5$	$W_B = 124.5$

Solution   In Table 15.5 the ranks are shown in parentheses alongside the $n_1 + n_2 = 10 + 10 = 20$ strength measurements and the rank sums, W_A and W_B, are shown below the columns. Because we wish to detect a shift in the distribution of the B

measurements to the right of the distribution for the A measurements, we would reject the null hypothesis of no difference in population strength distributions when W_B is excessively large. Because this situation will occur when U_B is small, we will conduct a one-tailed statistical test and reject the null hypothesis when $U_B \leq U_0$.

Suppose that we choose a value of α near .05. Then we can find U_0 by consulting the portion of Table 8, Appendix III, corresponding to $n_2 = 10$. The probability $P(U \leq U_0)$ nearest .05 is .0526 and corresponds to $U_0 = 28$. Hence, we will reject if $U_B \leq 28$.

Calculating U_B, we have

$$U_B = n_1 n_2 + \frac{n_2(n_2 + 1)}{2} - W_B = (10)(10) + \frac{(10)(11)}{2} - 124.5 = 30.5$$

As you can see, U_B is not less than $U_0 = 28$. Therefore, we cannot reject the null hypothesis. At the $\alpha = .05$ level of significance, there is not sufficient evidence to indicate that the treated kraft paper is stronger than the standard. The p-value is given by $P(U_B \leq 30.5) = .0716$.   ∽

A simplified large-sample test ($n_1 > 10$ and $n_2 > 10$) can be obtained by using the familiar Z statistic of Chapter 10. When the population distributions are identical, it can be shown that the U statistic has the following expected value and variance when $U = U_A$ (or $U = U_B$):

$$E(U_A) = \frac{n_1 n_2}{2} \quad \text{and} \quad V(U_A) = \frac{n_1 n_2 (n_1 + n_2 + 1)}{12}$$

Also, the distribution of

$$Z = \frac{U_A - E(U_A)}{\sigma_{U_A}}$$

tends to normality with mean zero and variance equal to 1 as n_1 and n_2 become large. This approximation will be adequate when n_1 and n_2 both are greater than or equal to 10. Thus for a two-tailed test with $\alpha = .05$, we would reject the null hypothesis if $|z| \geq 1.96$.

Observe that the Z statistic will reach the same conclusion as the exact U test for Example 15.5. Thus

$$z = \frac{30.5 - [(10)(10)/2]}{\sqrt{[(10)(10)(10 + 10 + 1)]/12}} = \frac{30.5 - 50}{\sqrt{2100/12}} = -\frac{19.5}{\sqrt{175}}$$

$$= -\frac{19.5}{13.23} = -1.47$$

For a one-tailed test with $\alpha = .05$ located in the lower tail of the z distribution, we will reject the null hypothesis if $z < -1.645$. You can see that $z = -1.47$ does not fall in the rejection region and that this test reaches the same conclusion as the exact U test of Example 15.5.

The Mann-Whitney U Test for Large Samples, $n_1 > 10$ and $n_2 > 10$

1. **Null Hypothesis:** H_0: The population relative frequency distributions for A and B are identical.

2. **Alternative Hypothesis, H_A:** The two population relative frequency distributions are not identical (a two-tailed test). Or H_a: The population relative frequency distribution for A is shifted to the right (or left) of the relative frequency distribution for population B (a one-tailed test).

3. **Test Statistic:** $Z = \dfrac{U_A - (n_1 n_2/2)}{\sqrt{n_1 n_2 (n_1 + n_2 + 1)/12}}$.

4. **Rejection Region:** Reject H_0 if $z > z_{\alpha/2}$ or $z < -z_{\alpha/2}$ for a two-tailed test. For a one-tailed test, place all of α in one tail of the z distribution. To detect a shift in the distribution of the A observations to the right of the distribution of the B observations, let $U = U_A$ and reject H_0 when $z < -z_{\alpha}$. To detect a shift in the opposite direction, let $U = U_A$ and reject H_0 when $z > z_{\alpha}$. Tabulated values of z are given in Table 4, Appendix III.

It may seem to you that the Mann-Whitney U test, or the equivalent Wilcoxon rank-sum test, is not very efficient because it does not appear to use all the information in the sample. Actually, theoretical studies have shown that this is not the case. Suppose, for example, that all the assumptions for a two-sample t test are met in testing $H_0: \mu_1 - \mu_2 = 0$ versus $H_a: \mu_1 - \mu_2 = d_0$, for some specific d_0. Because the two-sample t test simply tests for a difference in location (see Section 15.2), we can use the Mann-Whitney U statistic to test these same hypotheses. For a given α and β the total sample size required for the t test is approximately .95 times the total sample size required for the Mann-Whitney U. Thus, the nonparametric procedure is almost as good as the t test for the situation in which the t test is optimal. For many nonnormal distributions the nonparametric procedure requires fewer observations than a corresponding parametric procedure would require to produce the same values of α and β.

Exercises

15.17   Two plastics, each produced by a different process, were tested for ultimate strength. The measurements in the accompanying table represent breaking load in units of 1000 pounds per square inch. Do the data present evidence of a difference between the locations of the distributions of ultimate strengths for the two plastics? Solve by using the Mann-Whitney U test with a level of significance as near as possible to $\alpha = .10$.

Plastic 1	Plastic 2
15.3	21.2
18.7	22.4
22.3	18.3
17.6	19.3
19.1	17.1
14.8	27.7

15.18 The coded values for a measure of brightness in paper (light reflectivity), prepared by two different processes, are as shown in the accompanying table for samples of size 9 drawn randomly from each of the two processes. Do the data present sufficient evidence to

A	B
6.1	9.1
9.2	8.2
8.7	8.6
8.9	6.9
7.6	7.5
7.1	7.9
9.5	8.3
8.3	7.8
9.0	8.9

indicate a difference in locations of brightness measurements for the two processes? Give bounds for the attained significance level.

(a) Use the Mann-Whitney U test.

(b) Use Student's t test.

(c) Give specific null and alternative hypotheses, along with any assumptions, for the tests used in parts (a) and (b).

15.19 Fifteen experimental batteries were selected at random from a lot at pilot plant A, and fifteen standard batteries were selected at random from production at plant B. All thirty batteries were simultaneously placed under an electrical load of the same magnitude. The first battery to fail was an A, the second a B, the third a B, and so on. The following sequence shows the order of failure for the thirty batteries:

$$A\ B\ B\ B\ A\ B\ A\ A\ B\ B\ B\ B\ A\ B\ A$$
$$B\ B\ B\ B\ A\ A\ B\ A\ A\ A\ B\ A\ A\ A\ A$$

Using the large-sample theory for the U test, determine if there is sufficient evidence to conclude that the lengths of life for the experimental batteries tend to be greater than the lengths of life for the standard batteries. Use $\alpha = .05$.

15.20   Refer to Exercises 8.72 and 8.73. Is there sufficient evidence to indicate a difference in the populations of LC50 measurements for DDT and Diazinon? Give bounds for the attained significance level associated with the U statistic. What do you conclude when $\alpha = .10$?

15.21   Data collected on air pollution at eight Ohio locations seem to suggest that air pollution, caused by pollution drifting from urban areas, is greater in rural areas than in urban areas. The data, reproduced from *Environmental Midwest* (December 1976), are shown in the table.

City	Designation	*Maximum Concentration (ppm)*	*Days Exceeding Standard (%)*	*Number of Violations (Total)*
Canton	urban	.14	44	148
Cincinnati	urban	.18	44	54
Cleveland	urban	.14	26	51
Columbus	urban	.15	27	113
Dayton	urban	.13	35	114
McConnelsville	rural	.16	56	239
Wilmington	rural	.18	58	259
Wooster	rural	.17	55	262

(a)   Use the U statistic to test the null hypothesis that there is no difference in the locations of the distributions of violations between urban and rural Ohio cities.

(b)   Use the U statistic to test a similar hypothesis concerning the maximum concentration of pollutants recorded over the time period.

15.22   Cancer treatment by means of chemicals, chemotherapy, utilizes chemicals that kill both cancer cells as well as normal cells. In some instances the toxicity of the cancer drug, that is, its effect on normal cells, can be reduced by the simultaneous injection of a second drug. A study was conducted to determine whether a particular drug injection was beneficial in reducing the harmful effects of a chemotherapy treatment on the survival time for rats. Two randomly selected groups of rats, twelve rats in each group, were used for the experiment. Both groups, call them A and B, received the toxic drug in a dosage large enough to cause death, but, in addition, group B received the antitoxin that was to reduce the toxic effect of the chemotherapy on normal cells. The test was terminated at the end of 20 days, or 480 hours. The lengths of survival time for the two groups of rats, to the nearest 4 hours, are shown in the table. Do the data provide sufficient evidence to indicate that rats receiving the antitoxin tend to survive longer after chemotherapy than those not receiving the toxin? Use the Mann-Whitney U test with a value of α near .05.

Only Chemotherapy, A	Chemotherapy plus Drug, B
84	140
128	184
168	368
92	96
184	480
92	188
76	480
104	244
72	440
180	380
144	480
120	196

15.7 The Kruskal-Wallis H Test for the Completely Randomized Design

In Section 13.3 we presented an analysis of variance procedure to compare the means of k populations. The resultant F test was based on the assumption that we obtained independent samples from normal populations with equal variances. That is, as discussed in Section 15.2, we were interested in testing whether all of the populations had the same distribution versus the alternative that the populations differed in location. A key element in the development of the procedure was the quantity identified as the sum of squares for treatments, SST. As we pointed out in the discussion in Section 13.3, the larger the value of SST, the greater the weight of evidence favoring rejection of the null hypothesis that the means were all equal. In this section we present a nonparametric technique to test whether the populations differ in location. As with other nonparametric techniques discussed in this chapter, the advantage of the Kruskal-Wallis procedure is that no assumptions about the actual form of the probability distributions are required.

As in Section 13.3, we assume that independent random samples have been drawn from k populations that differ only in location. However, we need not assume that these populations possess normal distributions. To be completely general, we permit the sample sizes to be unequal and let n_i, $i = 1, \ldots, k$, represent the size of the sample drawn from the ith population. By analogy with the procedure of Section 15.5, combine all of the $n_1 + n_2 + \cdots + n_k = n$ observations and rank them from 1 (the smallest) to n (the largest). Ties are treated as in previous sections. That is, if two or more observations are tied for the same rank, then the average of the ranks that would have been assigned to these observations is assigned to each member of the tied group. Let R_i denote the sum of the ranks of the observations from population i, and $\bar{R}_i = R_i/n_i$ denote the corresponding average of the ranks. If $\bar{R}$ equals the overall average of all of the ranks, consider the rank analogue of SST, which is computed by using the ranks rather than the actual values of the measurements, that is,

$$V = \sum_{i=1}^{k} n_i(\bar{R}_i - \bar{R})^2$$

If the null hypothesis is true and the populations do not differ in locations, we would expect the $\bar{R}_i$'s to be approximately equal and the resulting value of V to be relatively small. If the alternative hypothesis is true, we would expect this to be exhibited in differences among the values of the $\bar{R}_i$'s, leading to a large value for V. Note that $\bar{R} = (\text{sum of the first } n \text{ integers})/n = [n(n+1)/2]/n = (n+1)/2$ and thus that

$$V = \sum_{i=1}^{k} n_i\left(\bar{R}_i - \frac{n+1}{2}\right)^2$$

Instead of V, Kruskal and Wallis [7] considered the statistic $H = 12V/[n(n+1)]$, which may be rewritten (see the exercises) as

$$H = \frac{12}{n(n+1)} \sum_{i=1}^{k} \frac{R_i^2}{n_i} - 3(n+1)$$

As previously noted, the null hypothesis of equal locations would be rejected in favor of the alternative that the populations differ in location if the value of H is large. Thus, the corresponding α-level test would call for rejection of the null hypothesis in favor of the alternative if $H > h(\alpha)$ where $h(\alpha)$ is such that when H_0 is true, $P[H > h(\alpha)] = \alpha$.

If the underlying distributions are continuous and there are no ties among the n observations, the null distribution of H can be (tediously) found by using the methods of Chapter 2. We can find the distribution of H for any values of k and $n_1, n_2, \ldots, n_k$ by calculating the value of H for each of the $n!$ equally likely permutations of the ranks of the n observations. These calculations have been performed and tables developed for some relatively small values of k and n_1, $n_2, \ldots, n_k$ (e.g., see Table A.7 of Hollander and Wolfe [5]).

Kruskal and Wallis showed that if the n_i's are "large," the null distribution of H can be approximated by a chi-square distribution with $k-1$ degrees of freedom. This approximation is generally accepted to be adequate if each of the n_i's is greater than or equal to 5. Our examples and exercises are all such that this "large sample" approximation is adequate. If you wish to use the Kruskal-Wallis analysis for smaller data sets where this large sample approximation is not adequate, refer to Hollander and Wolfe [5] to obtain the appropriate critical values.

We summarize the large sample Kruskal-Wallis procedure as follows.

Kruskal-Wallis H *Test for Comparing* k *Population Distributions*

H_0:   The k population distributions are identical.

H_a: At least two of the population distributions differ in location.

Test Statistic: $H = \dfrac{12}{n(n+1)} \displaystyle\sum_{i=1}^{k} \dfrac{R_i^2}{n_i} - 3(n+1)$

where
> n_i = Number of measurements in sample from population i.
> R_i = Rank sum for sample i, where the rank of each measurement is computed according to its relative size in the overall set of $n = n_1 + n_2 + \cdots + n_k$ observations formed by combining the data from all k samples.

Rejection Region: Reject H_0 if $H > \chi_\alpha^2$ with $(k-1)$ degrees of freedom.

Assumptions:
> The k samples are randomly and independently drawn. There are five or more measurements in each sample.

EXAMPLE 15.6   A quality control engineer has selected independent samples from the output of three assembly lines in an electronics plant. For each line, the output of ten randomly selected hours of production was examined for defects. Do the data in Table 15.6 provide evidence that the probability distributions of the number of defects per hour of output differ in location for at least two of the lines? Use $\alpha = .05$. Also give the p-value associated with the test.

Table 15.6
Data for
Example 15.6

	Line 1		Line 2		Line 3
Defects	*Rank*	*Defects*	*Rank*	*Defects*	*Rank*
6	5	34	25	13	9.5
38	27	28	19	35	26
3	2	42	30	19	15
17	13	13	9.5	4	3
11	8	40	29	29	20
30	21	31	22	0	1
15	11	9	7	7	6
16	12	32	23	33	24
25	17	39	28	18	14
5	4	27	18	24	16
	$R_1 = 120$		$R_2 = 210.5$		$R_3 = 134.5$

Solution

In this case $n_1 = 10 = n_2 = n_3$ and $n = 30$. Thus

$$H = \frac{12}{30(31)} \left[\frac{(120)^2}{10} + \frac{(210.5)^2}{10} + \frac{(134.5)^2}{10} \right] - 3(31) = 6.097$$

Because all of the n_i's are greater than or equal to 5, we may use the approximation for the null distribution of H and reject the null hypothesis of equal locations if $H > \chi_\alpha^2$ based on $k - 1 = 2$ degrees of freedom. We consult Table 15.6 to determine that $\chi_{.05}^2 = 5.99147$. Thus, we reject the null hypothesis at the $\alpha = .05$ level and concluded that at least one of the three lines tends to produce a greater number of defects than the others.

According to Table 6, Appendix III, the value of $H = 6.097$ would lead to a rejection of the null hypothesis if $\alpha = .05$ but not if $\alpha = .025$. Thus, $.025 < p\text{-value} < .05$.

It can be shown that (proof omitted) if we wish to compare only $k = 2$ populations, the Kruskal-Wallis test is equivalent to the two-sample Wilcoxon two-sided test presented in Section 15.5. If data is obtained from a completely randomized design with $k > 2$ treatments, but we wish to compare a particular pair of treatments, the Wilcoxon two-sample test (or the equivalent Mann-Whitney U test of Section 15.6) can be used for this purpose. Note that the analysis based on the Kruskal-Wallis H statistic does not require knowledge of the actual values of the observations. We need only know the ranks of the observations to complete the analysis. Exercise 15.24 illustrates the use of the Kruskal-Wallis analysis for such a case.

Exercises

15.23   Three different brands of magnetron tubes (the key components in microwave ovens) were subjected to stressful testing, and the number of hours each operated without repair was recorded. Although these times do not represent typical life lengths, they do indicate how well the tubes can withstand extreme stress:

	Brand	
A	B	C
36	49	71
48	33	31
5	60	140
67	2	59
53	55	42

(a)   Use the F test for a completely randomized design (Chapter 13) to test the hypothesis that the mean length of life under stress is the same for the three brands. Use $\alpha = 0.05$. What assumptions are necessary for the validity of this procedure? Is there any reason to doubt these assumptions?

(b)   Use the Kruskal-Wallis H test to determine whether evidence exists to conclude that the brands of magnetron tubes tend to differ in length of life under stress. Test using $\alpha = 0.05$.

15.24   An experiment was conducted to compare the length of time it takes a human to recover from each of the three types of influenza—Victoria A, Texas, and Russian. Twenty-one humans subjects were selected at random from a group of volunteers and divided into three groups of seven each. Each group was randomly assigned a strain of the virus and the influenza was induced in the subjects. All the subjects were then cared for under identical conditions, and the recovery time (in days) was recorded. The ranks of the results are as follows:

Victoria A	Texas	Russian
20	14.5	9
6.5	16.5	1
21	4.5	9
16.5	2.5	4.5
12	14.5	6.5
18.5	12	2.5
9	18.5	12

(a)   Do the data provide sufficient evidence to indicate that the recovery times for one (or more) type(s) of influenza tend(s) to be longer than for the other types? Give the associated p-value.

(b)   Do the data provide sufficient evidence to indicate a difference in locations of the distributions of recovery times for the Victoria A and Russian types? Give the associated p-value.

15.25   The EPA wants to determine whether temperature changes in the ocean's water caused by a nuclear power plant will have a significant effect on the animal life in the region. Recently hatched specimens of a certain species of fish are randomly divided into four groups. The groups are placed in separate simulated ocean environments that are identical in every way except for water temperature. Six months later, the specimens are weighed. The results (in ounces) are given in the table. Do the data provide sufficient evidence to indicate that one (or more) of the temperatures tend(s) to produce larger weight increases than the other temperatures? Test using $\alpha = 0.10$.

Weights of Specimens			
38°F	42°F	46°F	50°F
22	15	14	17
24	21	28	18
16	26	21	13
18	16	19	20
19	25	24	21
	17	23	

15.26   Weevils cause millions of dollars worth of damage each year to cotton crops. Three chemicals designed to control weevil populations are applied, one to each of three cotton fields. After 3 months, ten plots of equal size are randomly selected within each field and the percentage of cotton plants with weevil damage is recorded for each. Do the data in the table provide sufficient evidence to indicate a difference in location among the distributions of damage rates corresponding to the three treatments? Give bounds for the associated p-value.

A	B	C
10.8	22.3	9.8
15.6	19.5	12.3
19.2	18.6	16.2
17.9	24.3	14.1
18.3	19.9	15.3
9.8	20.4	10.8
16.7	23.6	12.2
19.0	21.2	17.3
20.3	19.8	15.1
19.4	22.6	11.3

15.27   The Kruskal-Wallis statistic $H = \dfrac{12}{n(n + 1)} \sum_{i=1}^{k} n_i \left(\bar{R}_i - \dfrac{n + 1}{2} \right)^2$. Perform the indicated squaring of each term in the sum and add the resulting values to show that $H = \dfrac{12}{n(n + 1)} \sum_{i=1}^{k} \dfrac{R_i^2}{n} - 3(n + 1)$. [Hint: recall that $\bar{R}_i = R_i/n_i$ and that $\sum_{i=1}^{k} R_i = $ sum of the first n integers $= n(n + 1)/2$.]

15.28   Assuming no ties, obtain the exact null distribution of the Kruskal-Wallis H statistic for the case $k = 3$, $n_1 = n_2 = n_3 = 2$. (Note that since the sample sizes are all equal, if ranks 1 and 2 are assigned to treatment 1; ranks 3 and 4 assigned to treatment 2; and ranks 5 and 6 assigned to treatment 3; the value of H is exactly the same if ranks 3 and 4 are assigned to treatment 1, 5 and 6 to treatment 2, and ranks 1 and 2 assigned to treatment 3. That is, for any particular set of ranks, we may interchange the roles of the k populations and obtain the same values of the H statistic. Thus, the number of cases that we must consider can be reduced by a factor of $1/k!$ Therefore, H need only be evaluated for $(6!/[2!2!2!])/3! = 15$ distinct arrangements of ranks.)

15.8 The Friedman Test for Randomized Block Designs

In Section 12.5 we discussed the merits of utilizing a randomized block design in an experiment to compare the performance of several "treatments." We assume that b blocks are used in the experiment, which is designed to compare the locations of the distributions of the responses corresponding to each of k "treatments." The analysis of variance, discussed in Section 13.9, was based on the assumptions that the observations in each block-treatment combination were normally distributed with equal variances. As in the case of the completely randomized design, SST was the key quantity in the analysis.

The Friedman test, developed by Nobel prize–winning economist Milton Friedman [3], is designed to test the null hypothesis that the probability distributions of the k treatments are identical versus the alternative that at least two of the distributions differ in location. The test is based on a statistic that is a rank analogue of SST for the randomized blocks design (see Section 13.9), and is computed in the following manner. After obtaining the data from a randomized blocks design, *for each block*, the observed values of the responses to each of the k "treatments" are ranked from 1 (the smallest in the block) to k (the largest in the block). If two or more observations in the same block are tied for the same rank, then the average of the ranks that would have been assigned to these observations is assigned to each member of the tied group. Note that "ties" need be dealt with in this manner only if they occur within the same block.

Let R_i denote the sum of the ranks of the observations corresponding to "treatment" i and $\bar{R}_i = R_i/b$ denote the corresponding average of the ranks (recall that in a randomized blocks design, each treatment is applied exactly once in each block, resulting in a total of b observations per treatment and hence bk total observations). Because ranks of 1 to k are assigned within each block, the sum of the ranks assigned in each block is $1 + 2 + \cdots + k = k(k + 1)/2$. Thus the sum of all of the ranks assigned in the analysis is $bk(k + 1)/2$. If $\bar{R}$ denotes the overall average of the ranks of all the bk observations, it follows that $\bar{R} = (k + 1)/2$. Consider the rank analogue of SST for a randomized block design given by

$$W = b \sum_{i=1}^{k} (\bar{R}_i - \bar{R})^2$$

If the null hypothesis is true and the probability distributions of the treatment responses do not differ in location, we expect the $\bar{R}_i$'s to be approximately equal and the resulting value for W to be small. If the alternative hypothesis were true, we would expect this to lead to differences among the values of the $\bar{R}_i$'s and corresponding large values of W. Instead of W, Friedman considered the statistic $F_r = 12W/[k(k + 1)]$, which may be rewritten (see the exercises) as

$$F_r = \frac{12}{bk(k + 1)} \sum_{i=1}^{k} R_i^2 - 3b(k + 1)$$

As previously noted, the null hypothesis of equal locations would be rejected in favor of the alternative that the treatment distributions differ in location for larger values of F_r. That is, the corresponding α-level test would reject the null hypothesis in favor of the alternative if $F_r > f(\alpha)$ where $f(\alpha)$ is such that when H_0 is true $P[F_r > f(\alpha)] = \alpha$.

If there are no ties among the observations within the blocks, the null distribution of F_r can be (tediously) found by using the methods of Chapter 2. For any values of b and k, the distribution of F_r is found as follows. If the null hypothesis is true, then within each block each of the $k!$ permutations of the ranks $1, 2, \ldots, k$ are equally likely. Further, because we assume that the observations in different blocks are mutually independent, it follows that each of the $(k!)^b$ possible combinations of the b sets of permutations for the within block ranks are equally likely when H_0 is true. This being the case, we can evaluate the value of F_r for each possible case and thereby give the null distribution of F_r. Selected values for $f(\alpha)$ for various choices of k and b are given in Table A.15 of Hollander and Wolfe [5]. Like the other nonparametric procedures discussed in this chapter, the real advantage of this procedure is that it can be used regardless of the form of the actual distributions of the populations corresponding to the "treatments."

As with the Kruskal-Wallis statistic, the null distribution of the Friedman F_r statistic can be approximated by a chi-square distribution with $k - 1$ degrees of freedom as long as b is "large." Empirical evidence indicates that the approximation is adequate if either b (the number of blocks) or k (the number of treatments) exceeds 5. Again, our examples and exercises deal with situations where this large sample approximation is adequate. If you need to implement a Friedman analysis for small samples, refer to Hollander and Wolfe [5] to obtain appropriate critical values.

Friedman F_r Test for a Randomized Blocks Design

H_0:   The probability distributions for the k treatments are identical.
H_a: At least two of the distributions differ in location.

Test Statistic: $F_r = \dfrac{12}{bk(k+1)} \displaystyle\sum_{i=1}^{k} R_i^2 - 3b(k+1)$

where

     b = number of blocks.
     k = number of treatments.
     R_i = sum of the ranks for the ith treatment, where the rank of each measurement is computed relative to its size within its own block.

Rejection Region: $F_r > \chi_\alpha^2$ with $(k - 1)$ degrees of freedom.

Assumptions:
     The treatments are randomly assigned to experimental units within blocks. Either the number of blocks (b) or the number of treatments (k) exceeds 5.

EXAMPLE 15.7   An experiment to compare completion times for three technical tasks was performed in the following manner. Because completion times may vary considerably from person to person, each of the six technicians was asked to perform all three tasks. The tasks were presented to each technician in a random order with suitable time lags between the tasks. Do the data in Table 15.7 present sufficient evidence to indicate that the distributions of completion times for the three tasks differ in location? Use $\alpha = .05$. Give bounds for the associated p-value.

Table 15.7
Completion times for three tasks

Technician	Task A	Rank	Task B	Rank	Task C	Rank
1	1.21	1	1.56	3	1.48	2
2	1.63	1.5	2.01	3	1.63	1.5
3	1.42	1	1.70	2	2.06	3
4	1.16	1	1.27	2.5	1.27	2.5
5	2.43	2	2.64	3	1.98	1
6	1.94	1	2.81	3	2.44	2
		$R_1 = 7.5$		$R_2 = 16.5$		$R_3 = 12$

Solution

The experiment was run according to a randomized blocks design with technicians playing the role of blocks. In this case $k = 3$ treatments are compared using $b = 6$ blocks. Because the number of blocks exceeds 5, we may use the Friedman analysis and compare the value of F_r to χ_α^2 based on $k - 1 = 2$ degrees of freedom. Consulting Table 6, Appendix III, we find $\chi_{.05}^2 = 5.99147$. For the data given in Table 15.7,

$$F_r = \frac{12}{6(3)(4)} [(7.5)^2 + (16.5)^2 + (12)^2] - 3(6)(4) = 6.75$$

Because $F_r = 6.75$, which exceeds 5.99147, we conclude at the $\alpha = .05$ level that the completion times of at least two of the three tasks possess probability distributions which differ in location.

Because $F_r = 6.75$ is the observed value of a statistic that has an approximately χ^2 distribution with 2 degrees of freedom, it follows that (approximately) $.025 < p\text{-value} < .05$.

It can be seen (see the exercises) that if we wish to compare only $k = 2$ treatments using a randomized blocks design (note that the blocks would be of size 2), the Friedman statistic is the square of the standardized sign statistic (i.e., the square of the Z statistic given in Section 15.3). Thus, for $k = 2$, the Friedman analysis is equivalent to a two-tailed sign test.

Exercises

15.29 Corrosion of different metals is a problem in many mechanical devices. Three sealers used to help retard the corrosion of metals were tested to see whether there were any differences among them. Samples of ten different metal compositions were treated with each of the three sealers and the amount of corrosion was measured after exposure to the same environmental conditions for 1 month. The data are given in the table. Is there any evidence of a difference in the abilities of the sealers to prevent corrosion? Test using $\alpha = 0.05$.

	Sealer		
Metal	I	II	III
1	4.6	4.2	4.9
2	7.2	6.4	7.0
3	3.4	3.5	3.4
4	6.2	5.3	5.9
5	8.4	6.8	7.8
6	5.6	4.8	5.7
7	3.7	3.7	4.1
8	6.1	6.2	6.4
9	4.9	4.1	4.2
10	5.2	5.0	5.1

15.30   A serious drought-related problem for farmers is the spread of aflatoxin, a highly toxic substance caused by mold, which contaminates field corn. In higher levels of contamination, aflatoxin is potentially hazardous to animal and possibly human health. (Officials of the FDA have set a maximum limit of 20 parts per billion aflatoxin as safe for interstate marketing.) Three sprays, A, B, and C, have been developed to control aflatoxin in field corn. To determine whether differences exist among the sprays, ten ears of corn are randomly chosen from a contaminated corn field and each is divided into three pieces of equal size. The sprays are then randomly assigned to the pieces for each ear of corn, thus setting up a randomized block design. The table gives the amount (in parts per billion) of aflatoxin present in the corn samples after spraying. Use the Friedman F_r test to determine whether there are differences among the sprays for control of aflatoxin. Give approximate bounds for the p-value.

	Spray		
Ear	A	B	C
1	21	23	15
2	29	30	21
3	16	19	18
4	20	19	18
5	13	10	14
6	5	12	6
7	18	18	12
8	26	32	21
9	17	20	9
10	4	10	2

15.31   A study was performed to compare the preferences of eight "expert listeners" regarding fifteen models (with approximately equal list prices) of a particular component in a stereo system. Every effort was made to ensure that differences perceived by the listeners were due to the particular component of interest and no other cause (all of the other components in the system were identical, the same type of music was used, the music was played in the same room, etc.). It is important to note that the results of the listening tests reflected the audio preferences of the judges and not judgments regarding quality, reliability, etc. Also, the results pertain only to the models of the components used in the study and not to any other models that may be offered by the various manufacturers. The data below gives the results of the listening tests. The models are depicted simply as models $A, B, \ldots, O$. Although the actual rankings of each judge are not given, under each column heading are the numbers of judges who ranked each brand of component from 1 (highest rank) to 15 (lowest rank).

	Rank														
Model	1	2	3	4	5	6	7	8	9	10	11	12	13	14	15
A	0	0	0	0	0	0	0	0	0	0	0	0	0	0	8
B	0	0	0	1	0	2	1	1	1	0	0	0	0	2	0
C	0	1	1	1	4	0	0	1	0	0	0	0	0	0	0
D	1	0	1	1	0	1	0	0	1	0	1	0	1	1	0
E	0	2	1	3	0	2	0	0	0	0	0	0	0	0	0
F	0	0	0	0	0	0	0	0	1	2	2	3	0	0	0
G	0	0	0	0	0	0	0	0	0	1	0	2	4	1	0
H	1	2	1	1	0	0	2	1	0	0	0	0	0	0	0
I	3	2	1	0	0	0	0	0	0	1	0	1	0	0	0
J	0	0	1	0	2	0	2	0	0	0	2	0	1	0	0
K	0	0	0	0	0	0	1	1	0	2	1	1	1	1	0
L	0	0	0	0	0	0	1	1	4	0	1	0	1	0	0
M	1	1	2	1	1	2	0	0	0	0	0	0	0	0	0
N	2	0	0	0	0	0	1	1	0	0	0	1	0	3	0
O	0	0	0	0	1	1	0	2	1	2	1	0	0	0	0

(a)   Use the Friedman procedure to test whether the distributions of the preference scores differ in location for the fifteen component models. Give bounds for the attained significance level. What would you conclude at the $\alpha = .01$ level of significance? (Hint: The sum of the ranks associated with the component of "model O" is $5 + 6 + 8 + 8 + 9 + 10 + 10 + 11 = 67$, other rank sums can be computed in an analogous manner).

(b)   If, prior to running the experiment, we desired to compare components of models H and G, this comparison could be made using the sign test presented in Section 15.3. Using the information just given, we can determine that model H was preferred to model G by all eight judges. Explain why or why not. Give the attained significance level if the sign test is used to compare these two brands of speakers.

(c)   Explain why there is not enough information given to use the sign test in a comparison of only the components of models H and M.

15.32 An experiment is conducted to investigate the toxic effect of three chemicals, A, B, and C, on the skin of rats. Three adjacent 1-inch squares are marked on the backs of eight rats and each of the three chemicals is applied to each rat. The squares of skin are then scored from 0 to 10, depending on the degree of irritation. The data are given in the table. Is there sufficient evidence to support the research hypothesis that the probability distributions of skin irritation scores corresponding to the three chemicals differ in location? Use $\alpha = 0.01$.

	Chemical		
Rat	*A*	*B*	*C*
1	6	5	3
2	9	8	4
3	6	9	3
4	5	8	6
5	7	8	9
6	5	7	6
7	6	7	5
8	6	5	7

15.33 Consider the Friedman statistic F_r, when $k = 2$ and $b = $ (number of blocks) $= n$. Then $F_r = \frac{2}{n}(R_1^2 + R_2^2) - 9n$. Let $M = $ the number of blocks (pairs) in which treatment one has rank 1. If there are no ties, then treatment one has rank 2 in the remaining $n - M$ pairs. Thus, $R_1 = M + 2(n - M) = 2n - M$. Analogously, $R_2 = n + M$. Substitute these values into the preceding expression for F_r and show that the resulting value is $4(M - .5n)^2/n$. Compare this result with the square of the Z statistic in Section 15.3. Note that we have demonstrated that $F_r = Z^2$.

15.34 Consider the Friedman statistic $F_r = \frac{12b}{k(k + 1)} \sum_{i=1}^{k} (\bar{R}_i - \bar{R})^2$. Square each term in the sum and show that an alternative form of F_r is $F_r = \frac{12}{bk(k + 1)} \sum_{i=1}^{k} R_i^2 - 3b(k + 1)$. [Hint: Recall that $\bar{R}_i = R_i/b$, $\bar{R} = (k + 1)/2$ and note that $\sum_{i=1}^{k} R_i = $ sum of all of the ranks $= bk(k + 1)/2$].

15.35 If there are no ties and $b = 2$, $k = 3$, derive the exact null distribution of F_r.

15.9 The Runs Test: A Test for Randomness

Consider a production process in which manufactured items emerge in sequence and each is classified as either defective (D) or nondefective (N). We have studied how we might compare the fraction defective over two equal time intervals by using the normal deviate test (Chapter 10) and extended this to a test of a hypothesis of constant p over two or more time intervals, using the chi-square test of Chapter 14. The purpose of these tests was to detect a change or trend in the fraction defective p. Evidence to indicate an increasing fraction defective might indicate the need for a process study to locate the source of difficulty. A decreasing value might suggest that a process quality control program was having a beneficial effect in reducing the fraction defective.

Trends in the fraction of defective items (or other quality measures) are not the only indication of lack of process control. A process might be causing periodic runs of defective items even though the average fraction of defective items remains constant, for all practical purposes, over long periods of time. For example, photoflash bulbs are manufactured on a rotating machine with a fixed number of positions for bulbs. A bulb is placed on the machine at a given position, the air is removed, oxygen is pumped into the bulb and the glass base is flame-sealed. If a machine contains twenty positions, and several adjacent positions are faulty (perhaps due to too much heat used in the sealing process), surges of defective lamps will emerge from the process in a periodic manner. Tests that compare the process fraction of defective items produced during equal intervals of time will not detect this periodic difficulty in the process. This periodicity, indicated by runs of defectives, is indicative of nonrandomness in the occurrence of defective items over time and can be detected by a *test for randomness*. The statistical test we present, known as the *runs test*, is discussed in detail by Wald and Wolfowitz [10]. Other practical applications of the runs test will follow.

As the name implies, the runs test is used to study a sequence of events where each element in the sequence can assume one of two outcomes, success (S) or failure (F). If we think of the sequence of items emerging from a manufacturing process as defective (F) or nondefective (S), the observation of twenty items might yield

$$S\ S\ S\ S\ S\ F\ F\ S\ S\ S$$
$$F\ F\ F\ S\ S\ S\ S\ S\ S\ S$$

We notice the groupings of defectives and nondefectives and wonder whether this grouping implies nonrandomness and, consequently, lack of process control.

Definition 15.1

> A *run* is defined to be a maximal subsequence of like elements.

For example, the first five successes is a subsequence of five like elements and it is maximal in the sense that it includes the maximum number of like elements before encountering an F. (The first four elements form a subsequence of like elements, but it is not maximal because the fifth element also could be included.) Consequently, the twenty elements are arranged in five runs, the first containing five S's, the second containing two F's, and so on.

A very small or very large number of runs in a sequence would indicate nonrandomness. Therefore, let R (the number of runs in a sequence) be the test statistic, and let the rejection region be $R \le k_1$ and $R \ge k_2$, as indicated in Figure 15.3. We must then find the probability distribution for R, $P(R = r)$, in order to calculate α and to locate a suitable rejection region for the test.

Figure 15.3
The rejection region for the runs test

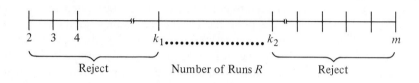

Suppose that the complete sequence contains n_1 S elements, and n_2 F elements, resulting in Y_1 S runs and Y_2 F runs, where $(Y_1 + Y_2) = R$. Then for a given Y_1, Y_2 can equal Y_1, $(Y_1 - 1)$, or $(Y_1 + 1)$. Let m denote the maximum possible number of runs. Note that $m = 2n_1$ if $n_1 = n_2$ and $m = (2n_1 + 1)$ if $n_1 < n_2$. We will suppose that every distinguishable arrangement of the $(n_1 + n_2)$ elements in the sequence constitutes a simple event for the experiment and that the sample points are equiprobable. It then remains for us to count the number of sample points that imply R runs.

The total number of distinguishable arrangements of n_1 S elements and n_2 F elements is $\binom{n_1 + n_2}{n_1}$, and therefore the probability per sample point is $1/\binom{n_1 + n_2}{n_1}$. The number of ways of achieving y_1 S runs is equal to the number of distinguishable arrangements of n_1 indistinguishable elements in y_1 cells, none of which is empty. The situation is represented in Figure 15.4. You will note that

Figure 15.4
The distribution of n_1 S elements in y_1 cells (none empty)

$$|S|SSSS|SS \ldots |SS|SSS|S|$$

this is equal to the number of ways of distributing the $(y_1 - 1)$ identical inner bars in the $(n_1 - 1)$ spaces between the S elements (the outer two bars remain fixed). Consequently, it is equal to the number of ways of selecting $(y_1 - 1)$ spaces (for the bars) out of the $(n_1 - 1)$ spaces available. This will equal

$$\binom{n_1 - 1}{y_1 - 1}$$

The number of ways of observing y_1 S runs and y_2 F runs, obtained by applying the mn rule, is

$$\binom{n_1 - 1}{y_1 - 1}\binom{n_2 - 1}{y_2 - 1}$$

This gives the number of sample points in the event "y_1 S runs and y_2 F runs." Then multiplying this number by the probability per sample point, we obtain the probability of exactly y_1 S runs and y_2 F runs:

$$p(y_1, y_2) = \frac{\binom{n_1 - 1}{y_1 - 1}\binom{n_2 - 1}{y_2 - 1}}{\binom{n_1 + n_2}{n_1}}$$

Then $P(R = r)$ equals the sum of $p(y_1, y_2)$ over all values of y_1 and y_2 such that $(y_1 + y_2) = r$.

To illustrate the use of the formula, the event $R = 4$ could occur when $y_1 = 2$ and $y_2 = 2$ with either the S or F elements commencing the sequences. Consequently,

$$P(R = 4) = 2P(Y_1 = 2, Y_2 = 2)$$

On the other hand, $R = 5$ could occur when $y_1 = 2$ and $y_2 = 3$, or $y_1 = 3$ and $y_2 = 2$, and these points are mutually exclusive. Then

$$P(R = 5) = P(Y_1 = 3, Y_2 = 2) + P(Y_1 = 2, Y_2 = 3)$$

EXAMPLE 15.8   Suppose that a sequence consists of $n_1 = 5$ S elements and $n_2 = 3$ F elements. Calculate the probability of observing $R = 3$ runs. Also, calculate $P(R \leq 3)$.

Solution   Three runs could occur when $y_1 = 2$ and $y_2 = 1$, or $y_1 = 1$ and $y_2 = 2$. Then

$$P(R = 3) = P(Y_1 = 2, Y_2 = 1) + P(Y_1 = 1, Y_2 = 2)$$

$$= \frac{\binom{4}{1}\binom{2}{0}}{\binom{8}{5}} + \frac{\binom{4}{0}\binom{2}{1}}{\binom{8}{5}} = \frac{4}{56} + \frac{2}{56} = .107$$

Next we require that $P(R \le 3) = P(R = 2) + P(R = 3)$. Accordingly, note that

$$P(R = 2) = 2P(Y_1 = 1, Y_2 = 1) = (2) \frac{\binom{4}{0}\binom{2}{0}}{\binom{8}{5}} = \frac{2}{56} = .036$$

Thus the probability of 3 or fewer runs is .143.   ∽

The values of $P(R \le a)$ are given in Table 10, Appendix III, for all combinations of n_1 and n_2 where n_1 and n_2 are less than or equal to 10. These can be used to locate the rejection regions of one- or two-tailed tests. We will illustrate with an example.

EXAMPLE 15.9   A true–false examination was constructed with the answers running in the following sequence:

$$T \; F \; F \; T \; F \; T \; F \; T \; T \; F \; T \; F \; F \; T \; F \; T \; F \; T \; T \; F$$

Does this sequence indicate a departure from randomness in the arrangement of T and F answers?

Solution   The sequence contains $n_1 = 10 \; T$ and $n_2 = 10 \; F$ answers with $y = 16$ runs. Nonrandomness can be indicated by either an unusually small or an unusually large number of runs, and consequently we will be concerned with a two-tailed test.

Suppose that we wish to use α approximately equal to .05 with .025 or less in each tail of the rejection region. Then from Table 10, Appendix III, with $n_1 = n_2 = 10$, we note that $P(R \le 6) = .019$ and $P(R \le 15) = .981$. Then $P(R \ge 16) = 1 - P(R \le 15) = .019$, and we would reject the hypothesis of randomness if $R \le 6$ or $R \ge 16$. Because $R = 16$ for the observed data, we conclude that evidence exists to indicate nonrandomness in the professor's arrangement of answers. His attempt to mix the answers was overdone.   ∽

A second application of the runs test is in detecting nonrandomness of a sequence of quantitative measurements over time. These sequences, known as *time series*, occur in many fields. For example, the measurement of a quality characteristic of an industrial product, blood pressure of a human, and the price of a stock on the stock market all vary over time. Departures in randomness in a series, caused either by trends or periodicities, can be detected by examining the deviations of the time series measurements from their average. Negative and positive deviations could be denoted by S and F, respectively, and we could then test this time sequence of deviations for nonrandomness. We will illustrate with an example.

EXAMPLE 15.10   Paper is produced in a continuous process. Suppose that a brightness measurement, Y, is made on the paper once every hour and that the results appear as shown in Figure 15.5.

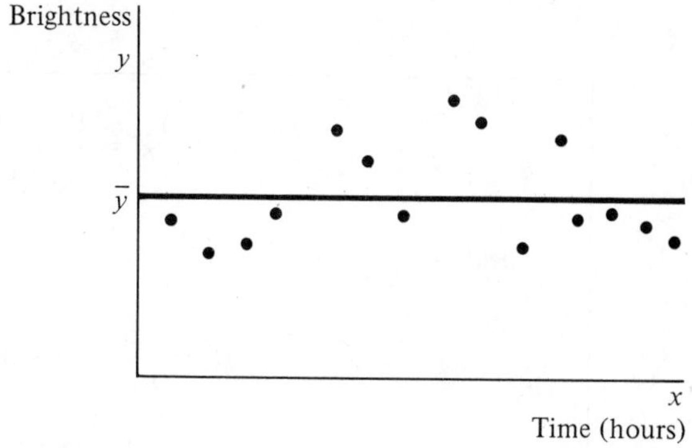

Figure 15.5
Paper brightness
versus time

The average for the fifteen sample measurements, $\bar{y}$, appears as shown. Note the deviations about $\bar{y}$. Do these data indicate a lack of randomness and thereby suggest periodicity and lack of control in the process?

Solution   The sequence of negative (S) and positive (F) deviations as indicated in Figure 15.5 is

$$S\ S\ S\ S\ F\ F\ S\ F\ F\ S\ F\ S\ S\ S\ S$$

Then $n_1 = 10$, $n_2 = 5$, and $R = 7$. Consulting Table 10, in Appendix III, $P(R \leq 7) = .455$. This value of R is not improbable, assuming the hypothesis of randomness to be true. Consequently, there is not sufficient evidence to indicate nonrandomness in the sequence of brightness measurements.

The runs test also can be used to compare two population frequency distributions for a two-sample unpaired experiment. Thus it provides an alternative to the Mann-Whitney U test (Section 15.6). If the measurements for the two samples are arranged in order of magnitude, they will form a sequence. The measurements for samples 1 and 2 can be denoted as S and F, respectively, and once again we are concerned with a test for randomness. If all measurements for sample 1 are smaller than those for sample 2, the sequence will result in $SSSS \cdots SFFF \cdots F$, or $R = 2$ runs. A small value of R will provide evidence of a difference in population frequency distributions, and the rejection region chosen would be $R \leq a$. This rejection region would imply a one-tailed statistical test. An illustration of the application of the runs test to compare two population frequency distributions will be left as an exercise.

As in the case of the other nonparametric test statistics studied in earlier sections of this chapter, the probability distribution for R tends to normality as n_1 and n_2 become large. The approximation is good when n_1 and n_2 are both greater than 10. Consequently, we may use the Z statistic as a large-sample test statistic, where

$$Z = \frac{R - E(R)}{\sqrt{V(R)}}$$

and

$$E(R) = \frac{2n_1 n_2}{n_1 + n_2} + 1$$

$$V(R) = \frac{2n_1 n_2 (2n_1 n_2 - n_1 - n_2)}{(n_1 + n_2)^2 (n_1 + n_2 - 1)}$$

are the expected value and variance of R, respectively. The rejection region for a two-tailed test, with $\alpha = .05$, is $|z| \geq 1.96$. If α is the desired probability of a type I error, for an upper-tail test, reject if $z > z_\alpha$ (for a lower-tail test reject if $z < -z_\alpha$).

Exercises

15.36   Consider a runs test based on $n_1 = n_2 = 5$ elements. Assuming H_0 to be true, use Table 10, Appendix III, to find the following:

(a) $P(R = 2)$

(b) $P(R \leq 3)$

(c) $P(R \leq 4)$

15.37   A union supervisor claims that applicants for jobs are selected without regard to race. An examination of the hiring records of the local, one that contains all male members, gave the following sequence of white and black hirings:

$$W\ W\ W\ W\ B\ W\ W\ W\ B\ B\ W\ B\ B$$

Do these data suggest a nonrandom racial selection in the hiring of the union's members?

15.38   The conditions (D for diseased, S for sound) of the individual trees in a row of ten poplars were found to be, from left to right: $S, S, D, D, S, D, D, D, S, S$. Is there sufficient evidence to indicate nonrandomness in the sequence and therefore the possibility of contagion?

15.39 Items emerging from a continuous production process were classified as defective or nondefective. A sequence of items, observed over time, was as follows:

$$D\ N\ N\ N\ N\ N\ N\ D\ D\ N\ N\ N\ N\ N\ N\ D\ D$$
$$D\ N\ N\ N\ N\ N\ D\ N\ N\ N\ D\ D\ N\ N\ N\ D\ D$$

(a) Compute the probability that $R \leq 11$, where $n_1 = 11$ and $n_2 = 23$.

(b) Do these data suggest lack of randomness in the occurrence of defectives (D) and nondefectives (N)? Use the large-sample approximation for the runs test.

15.40 A quality control chart has been maintained for a certain measurable characteristic of items taken from a conveyor belt at a certain point in a production line. The measurements obtained today in order of time are as follows:

68.2	71.6	69.3	71.6	70.4	65.0	63.6	64.7
65.3	64.2	67.6	68.6	66.8	68.9	66.8	70.1

(a) Classify the measurements in this time series as above or below the sample mean and determine (use the runs test) whether consecutive observations suggest lack of stability in the production process.

(b) Divide the time period into two equal parts and compare the means, using Student's t test. Do the data provide evidence of a shift in the mean level of the quality characteristics?

15.41 Refer to Exercise 15.18. Use the runs test to analyze the data. Compare your answer with the answer to Exercise 15.18.

15.42 Refer to Exercise 15.19. If, indeed, the experimental batteries have a greater mean life, what would be the effect on the expected number of runs? Using the large-sample theory for the runs test, test (using $\alpha = .05$) whether there is a difference in the distributions of battery life for the two populations. Give the approximate p-value.

15.10 Rank Correlation Coefficient

In the preceding sections we have used ranks to indicate the relative magnitude of observations in nonparametric tests for comparison of treatments. We will now employ the same technique in testing for a correlation between two ranked variables. Two common rank correlation coefficients are the Spearman r_s and the Kendall τ. We will present the Spearman r_s because its computation is identical to that for the sample correlation coefficient r of Chapter 11. Kendall's rank correlation coefficient is discussed in detail in Kendall and Stuart [6].

Suppose that eight elementary science teachers have been ranked by a judge according to their teaching ability, and all have taken a national teachers' examination. The data are given in Table 15.8. Do the data suggest an agreement between the judge's ranking and the examination score? Or we might express this question by asking whether a correlation exists between ranks and scores.

Table 15.8
Data for science
teachers

Teacher	Judge's Rank	Examination Score
1	7	44
2	4	72
3	2	69
4	6	70
5	1	93
6	3	82
7	8	67
8	5	80

The two variables of interest are rank and test score. The former is already in rank form, and the test scores may be ranked similarly, as shown in Table 15.9. The ranks for tied observations are obtained by averaging the ranks that the tied observations would occupy, as for the Mann-Whitney U statistic.

Table 15.9
Data of Table 15.7
with scores ranked

Teacher	Judge's Rank (x_i)	Test Rank (y_i)
1	7	1
2	4	5
3	2	3
4	6	4
5	1	8
6	3	7
7	8	2
8	5	6

The *Spearman rank correlation coefficient* r_s is calculated by using the ranks as the paired measurements on the two variables, X and Y, in the formula for r, Chapter 11. Thus

$$r_s = \frac{n \sum\limits_{i=1}^{n} x_i y_i - \left(\sum\limits_{i=1}^{n} x_i \right) \left(\sum\limits_{i=1}^{n} y_i \right)}{\sqrt{\left[n \sum\limits_{i=1}^{n} x_i^2 - \left(\sum\limits_{i=1}^{n} x_i \right)^2 \right] \left[n \sum\limits_{i=1}^{n} y_i^2 - \left(\sum\limits_{i=1}^{n} y_i \right)^2 \right]}}$$

When there are no ties in either the x observations or the y observations, this expression for r_s algebraically reduces to a simpler expression:

$$r_s = 1 - \frac{6 \sum\limits_{i=1}^{n} d_i^2}{n(n^2 - 1)} \quad \text{where} \quad d_i = x_i - y_i$$

If the number of ties is small in comparison with the number of data pairs, little error will result in using this shortcut formula. We will leave proof of this simplification as an exercise and will illustrate the use of the formula by an example.

EXAMPLE 15.11   Calculate r_s for the teacher–judge test score data.

Solution   The differences and squares of differences between the two rankings are shown in Table 15.10.

Table 15.10
Data and calculations,
Example 15.11

Teacher	x_i	y_i	d_i	d_i^2
1	7	1	6	36
2	4	5	−1	1
3	2	3	−1	1
4	6	4	2	4
5	1	8	−7	49
6	3	7	−4	16
7	8	2	6	36
8	5	6	−1	1
Total				144

Substituting into the formula for r_s, we obtain

$$r_s = 1 - \frac{6 \sum_{i=1}^{n} d_i^2}{n(n^2 - 1)} = 1 - \frac{6(144)}{8(64 - 1)} = -.714$$

The Spearman rank correlation coefficient may be employed as a test statistic to test a hypothesis of no association between two populations. We assume that the n pairs of observations, (x_i, y_i), have been randomly selected and therefore no association between the populations would imply a random assignment of the n ranks within each sample. Each random assignment (for the two samples) would represent a sample point associated with the experiment, and a value of r_s could be calculated for each. It is possible to calculate the probability that r_s assumes a large absolute value due solely to chance and thereby suggests an association between populations when none exists.

The rejection region for a two-tailed test would include values of r_s near $+1$ and near -1. If the alternative is that the correlation between X and Y is negative, we would reject H_0 for values of r_s near -1. Similarly, if the alternative is that the correlation between X and Y is positive, we would reject H_0 for large positive values of r_s.

The critical values of r_s are given in Table 11, Appendix III. Across the top of the table are recorded values of α that you might wish to use for a one-tailed test of the null hypothesis of no association between X and Y. The number of rank pairs, n, appears at the left side of the table. The table entries give the critical value r_0 for a one-tailed test. Thus $P(r_s \geq r_0) = \alpha$. For example, suppose you have $n = 8$ rank pairs and the research hypothesis is that the correlation between the ranks is positive. Then you would want to reject the null hypothesis of no association only for large positive values of r_s and would use a one-tailed test. Referring to Table 11, and using the row corresponding to $n = 8$ and the column for $\alpha = .05$, you read $r_0 = .643$. Therefore, you would reject H_0 for all values of r_s greater than or equal to .643.

If you wish to give the p-value associated with an observed value of $r = .82$, Table 11 gives that H_0 would be rejected with $\alpha = .025$ but not with $\alpha = .01$. Thus, $.01 < p\text{-value} < .025$.

The test is conducted in exactly the same manner if you wish to test the alternative hypothesis that the ranks are negatively correlated. The only difference is that you would reject the null hypothesis if $r_s \leq -.643$. That is, you just place a minus sign in front of the tabulated value of r_0 to get the lower-tail critical value. Similarly, if $r = -.82$, $.01 < p\text{-value} < .025$.

To conduct a two-tailed test, reject the null hypothesis if $r_s \geq r_0$ or $r_s \leq -r_0$. The value of α for the test will be double the value shown at the top of the table. For example, if $n = 8$ and you choose the .025 column, you will reject H_0 if $r_s \geq .738$ or $r_s \leq -.738$. The α value for the test will be $2(.025) = .05$.

The p-value associated with a two-tailed test based on an observed value of $r = .82$ is twice (because of the two tails) the one-tailed p-value; that is, $.02 < p\text{-value} < .05$.

EXAMPLE 15.12 Test the hypothesis of no association between populations for Example 15.11. Give bounds for the associated p-value.

Solution The critical value of r_s for a one-tailed test with $\alpha = .05$ and $n = 8$ is .643. Let us assume that a correlation between judge's rank and the teachers' test scores could not possibly be positive. (Low rank means good teaching and should be associated with a high test score if the judge and the test measure teaching ability.) The alternative hypothesis would be that the population rank correlation coefficient, ρ_s, is less than zero, and we would be concerned with a one-tailed statistical test. Thus α for the test would be the tabulated value .05, and we would reject the null hypothesis if $r_s \leq -.643$.

The calculated value of the test statistic, $r_s = -.714$, is less than the critical value for $\alpha = .05$. Because H_0 would be rejected for $\alpha = .05$ but not for $\alpha = .025$, the p-value associated with the test will lie in the interval $.025 < p\text{-value} < .05$. Hence, the null hypothesis would be rejected at the $\alpha = .05$ level of significance. It appears that some agreement does exist between the judge's rankings and the test scores. However, it should be noted that this agreement could exist when *neither* provides an adequate yardstick for measuring teaching ability. For example, the association could exist if both the judge and those who constructed the teacher's examination possessed a completely erroneous, but identical, concept of the characteristics of good teaching. ∽

Spearman's Rank Correlation Test

1. Null Hypothesis: H_0: There is no association between the rank pairs.

2. Alternative Hypothesis: H_a: There is an association between the rank pairs (a two-tailed test). Or H_a: The correlation between the rank pairs is positive (or negative) (a one-tailed test).

3. Test Statistic:

$$r_s = \frac{n \sum_{i=1}^{n} x_i y_i - \left(\sum_{i=1}^{n} x_i \right) \left(\sum_{i=1}^{n} y_i \right)}{\sqrt{\left[n \sum_{i=1}^{n} x_i^2 - \left(\sum_{i=1}^{n} x_i \right)^2 \right] \left[n \sum_{i=1}^{n} y_i^2 - \left(\sum_{i=1}^{n} y_i \right)^2 \right]}}$$

where x_i and y_i represent the ranks of the ith pair of observations.

4. Rejection Region: For a two-tailed test, reject H_0 if $r_s \geq r_0$ or $r_s \leq -r_0$, where r_0 is given in Table 11, Appendix III. Double the tabulated probability to obtain the value of α for the two-tailed test. For a one-tailed test, reject H_0 if $r_s \geq r_0$ (for an upper-tailed test) or $r_s \leq -r_0$ (for a lower-tailed test). The α value for a one-tailed test is the value shown in Table 11, Appendix III.

Exercises

15.43 *Tennis Industry Magazine* of Florida, a leading tennis business magazine, reported the subjective evaluation of tennis racket vibration, stiffness, and torque (twist) for twelve rackets by twenty-nine male and female tennis professionals. A summary of the subjective ratings of the professionals for each racket is shown in the table. Calculate r_s for the following pairs of variables:

(a) flex and torque.

(b) vibration from sweet-spot hits and vibration from off-center hits.

(c) torque and vibration from off-center hits.

Racket	Vibration Center (Sweet Spot) Hits	Vibration Off-Center Hits	Flex	Torque
Wilson T-4000	55	71	46	71
Davis Classic	59	71	43	65
Yamaha YFG-30	55	61	59	65
P.D.P. Fiberstaff	46	55	74	54
Aldila Cannon	50	61	66	53
Dunlop-Fort	56	65	60	60
Garcia "Pro" Royal	59	68	51	66
Head Comp. II	54	66	63	59
Yonex T-7500	51	53	69	58
F. Willys-Devastator	76	75	73	73
Wilson Kramer	58	66	55	65
Garcia 240	60	65	70	63

Source Reprinted by permission. *Tennis USA*, Feb 1976. © 1976 Chilton Company, Inc. All rights reserved.

Test to determine whether the data provide sufficient evidence to indicate a correlation between each of the pairs of variables, (a), (b), and (c). Test using a value of α near .05.

15.44 Manufacturers of perishable foods often use preservatives to retard spoilage. One concern is that too much preservative will change the flavor of the food. An experiment is conducted using portions of food products with varying amounts of preservative added. The length of time until the food begins to spoil and a taste rating are recorded for each portion of food. The taste rating is the average rating for three tasters, each of whom rated each food portion on a scale from 1 (bad) to 5 (good). Twelve measurements are shown in the following table. Use a nonparametric test to determine whether spoilage times and taste ratings are correlated. Give the associated *p*-value and indicate the appropriate conclusion for an $\alpha = .05$ level test.

Food Portion	Days until Spoilage	Taste Rating
1	30	4.3
2	47	3.6
3	26	4.5
4	94	2.8
5	67	3.3
6	83	2.7
7	36	4.2
8	77	3.9
9	43	3.6
10	109	2.2
11	56	3.1
12	70	2.9

15.45 A large corporation selects graduates for employment by using both interviews and a psychological achievement test. Interviews conducted at the home office of the company were far more expensive than the test, which could be conducted on campus. Consequently, the personnel office was interested in determining whether the test scores were correlated with interview ratings and whether the tests could be substituted for interviews. The idea was not to eliminate interviews but to reduce their number. Ten prospects were ranked during interviews and tested. The paired scores were as shown in the accompanying table.

Subject	Interview Rank	Test Score
1	8	74
2	5	81
3	10	66
4	3	83
5	6	66
6	1	94
7	4	96
8	7	70
9	9	61
10	2	86

(a) Calculate the Spearman rank correlation coefficient r_s. Rank 1 is assigned to the candidate judged to be the best.

(b) Do the data present sufficient evidence to indicate that the correlation between interview rankings and test scores is less than zero? If this evidence does exist, can we say that tests could be used to reduce the number of interviews?

15.46 A political scientist wished to examine the relationship of the voter image of a conservative political candidate and the distance in miles between the residences of the voter and the candidate. Each of twelve voters rated the candidate on a scale of 1 to 20. The data are shown in the table.

Voter	Rating	Distance
1	12	75
2	7	165
3	5	300
4	19	15
5	17	180
6	12	240
7	9	120
8	18	60
9	3	230
10	8	200
11	15	130
12	4	130

(a) Calculate the Spearman rank correlation coefficient r_s.

(b) Do these data provide sufficient evidence to indicate a negative correlation between rating and distance?

15.47 Refer to Exercise 15.8. Compute Spearman's rank correlation coefficient for these data and test $H_0: \rho_s = 0$ at the 10% level of significance.

15.48 The annual median sales prices for new single-family houses over an 8-year period are given in the accompanying table. Also given is the average effective interest rate for home mortgages over the same period. Is there evidence of a significant positive correlation between sales prices and interest rates? Use $\alpha = .05$.

Year	Median Sales Price ($\times \$1000$)	Average Interest Rate (%)
1972	27.6	7.6
1973	32.5	8.0
1974	35.9	8.9
1975	39.3	9.0
1976	44.2	9.0
1977	48.8	9.0
1978	55.7	9.5
1979	62.9	10.7

Source Data adapted from *Time*, July 23, 1979, p. 67.

15.49 Refer to Exercise 11.2. Regard both book and audit values as random variables and test for positive correlation between the two by using Spearman's rank correlation coefficient. Give bounds for the *p*-value associated with the test.

15.50 Refer to Exercise 11.4. Treating both flow-through and statistic values as random variables, test for the presence of a correlation between the two by using Spearman's rank correlation coefficient, with $\alpha = .10$.

15.11 Some General Comments on Nonparametric Statistical Tests

The nonparametric statistical tests presented in the preceding pages represent only a few of the many nonparametric statistical methods of inference available. A much larger collection of nonparametric procedures, along with worked examples, is given in the texts listed in the references (e.g., see [1], [2], [5], and [9]). For example, nonparametric estimation procedures are available for estimating location parameters and for handling the inferential problems associated with linear model.

We have indicated that nonparametric statistical procedures are particularly useful when the experimental observations are susceptible to ordering but cannot be measured on a quantitative scale. Parametric statistical procedures usually cannot be applied to this type of data. Hence all inferential procedures must be based on nonparametric methods.

A second application of nonparametric statistical methods is in testing hypotheses associated with populations of quantitative data when uncertainty exists concerning the satisfaction of assumptions about the form of the population distributions. Just how useful are nonparametric methods for this situation? Nonparametric statistical methods are rapid and often lead to an immediate decision in testing hypotheses. When experimental conditions depart substantially from the basic assumptions underlying parametric tests, the response measurements often can be transformed to alleviate the condition, but an unfortunate consequence frequently develops. That is, the transformed response no longer is meaningful from a practical point of view, and analysis of the transformed data no longer answers the objectives of the experimenter. The use of nonparametric methods often will circumvent this difficulty. Finally, note that many nonparametric methods are nearly as efficient as their parametric counterparts when the assumptions underlying the parametric procedures are true, and, as noted earlier, they could be more efficient when the assumptions are unsatisfied. These reasons suggest that nonparametric techniques play a very useful role in statistical methodology.

References and Further Readings

1.   Conover, W. J. *Practical Nonparametric Statistics*, 2d ed. New York: Wiley, 1980.

2.   Daniel, W. W. *Applied Nonparametric Statistics*. Boston: Houghton Mifflin, 1978.

3.   Friedman, M. "The Use of Ranks to Avoid the Assumption of Normality Implicit in the Analysis of Variance." *Journal of the American Statistical Association*, vol. 32 (1937), pp. 675–701.

4.   Hajek, J. *Nonparametric Statistics*. San Francisco: Holden-Day, 1969.

5.   Hollander, W., and Wolfe, D. A. *Nonparametric Statistical Methods*. New York: Wiley, 1973.

6.   Kendall, M. G., and Stuart, A. *The Advanced Theory of Statistics*, 4th ed., vol. 2. New York: Hafner Press, 1979.

7.   Kruskal, W. H., and Wallis, W. A. "Use of Ranks in One-Criterion Variance Analysis." *Journal of the American Statistical Association*, vol. 47 (1952), pp. 583–621.

8.   Savage, I. R. "Bibliography of Nonparametric Statistics and Related Topics." *Journal of the American Statistical Association*, vol. 48 (1953), pp. 844–906.

9.   Siegel, S. *Nonparametric Statistics for the Behavioral Sciences*. New York: McGraw-Hill, 1956.

10.   Wald, A., and Wolfowitz, J. "On a Test Whether Two Samples Are from the Same Population." *Annals of Mathematical Statistics*, vol. 2 (1940), pp. 147–162.

Supplementary Exercises

15.51   Clinical data concerning the effectiveness of two drugs in treating a particular disease were collected from ten hospitals. The number of patients treated with the drugs varied from one hospital to another as well as between drugs within a given hospital. The data, in percentage recovery, are shown in the table. Do the data present sufficient evidence to indicate a higher recovery rate for one of the two drugs?

	Drug A			Drug B		
Hospital	*Number in Group*	*Number Recovered*	*Percentage Recovered*	*Number in Group*	*Number Recovered*	*Percentage Recovered*
1	84	63	75.0	96	82	85.4
2	63	44	69.8	83	69	83.1
3	56	48	85.7	91	73	80.2
4	77	57	74.0	47	35	74.5
5	29	20	69.0	60	42	70.0
6	48	40	83.3	27	22	81.5
7	61	42	68.9	69	52	75.4
8	45	35	77.8	72	57	79.2
9	79	57	72.2	89	76	85.4
10	62	48	77.4	46	37	80.4

(a)   Test by using the sign test. Choose your rejection region so that α is near .10.

(b)   Why would it be inappropriate to use a Student's t test in analyzing the data?

15.52   Two gourmets rated twenty meals on a scale of 1 to 10. The data are shown in the table. Do the data provide sufficient evidence to indicate that one of the gourmets tends to give higher ratings than the other? Test by using the sign test with a value of α near .05.

Meal	*A*	*B*	*Meal*	*A*	*B*
1	6	8	11	6	9
2	4	5	12	8	5
3	7	4	13	4	2
4	8	7	14	3	3
5	2	3	15	6	8
6	7	4	16	9	10
7	9	9	17	9	8
8	7	8	18	4	6
9	2	5	19	4	3
10	4	3	20	5	5

15.53   Refer to the comparison of gourmet mean ratings, Exercise 15.52, and use the Wilcoxon signed-rank test to determine whether the data provide sufficient evidence to indicate a difference in the ratings of the two gourmets. Test by using a value of α near .05. Compare the results of this test with the results of the sign test, Exercise 15.52. Are the test conclusions consistent?

15.54   In an investigation of visual scanning behavior of deaf children, measurements of eye movement rate were taken on nine deaf and nine hearing children. From the data given, does it appear that the distributions of eye movement rates for deaf children (A) and hearing children (B) differ?

	Deaf Children (A)	Hearing Children (B)
	(15) 2.75	.89 (1)
	(11) 2.14	1.43 (7)
	(18) 3.23	1.06 (4)
	(10) 2.07	1.01 (3)
	(14) 2.49	.94 (2)
	(12) 2.18	1.79 (8)
	(17) 3.16	1.12 (5.5)
	(16) 2.93	2.01 (9)
	(13) 2.20	1.12 (5.5)
Rank sum	126	45

15.55   A comparison of reaction times (in seconds) for two different stimuli in a psychological word association experiment produced the accompanying results when applied to a random sample of sixteen people. Do the data present sufficient evidence to indicate a difference in location for the distributions of reaction times for the two stimuli? Use the Mann-Whitney U statistic and test with $\alpha = .05$. (Note: This test was conducted by using Student's t in Exercise 13.1. Compare your results.)

Stimulus 1	Stimulus 2
1	4
3	2
2	3
1	3
2	1
1	2
3	3
2	3

15.56   If (as in the case of measurements produced by two well-calibrated measuring instruments) the means of two populations are equal, it is possible to use the Mann-Whitney U statistic for testing hypotheses concerning the population variances (or more general measures of variability) as follows. Rank the combined sample. Number the ranked observations from the outside in; that is, number the smallest observation 1; the largest 2; the next to smallest, 3; the next to largest, 4; and so on. This final sequence of numbers induces an ordering on the symbols A (population A items) and B (population B items). If $\sigma_A^2 > \sigma_B^2$, one would expect to find a preponderance of A's near the first of the sequences and thus a relatively small sum of ranks for the A observations.

(a)   Given the accompanying measurements produced by well-calibrated precision instruments A and B, test at near the $\alpha = .05$ level to determine whether the more expensive instrument, B, is more precise than A. (Note that this would imply a one-tailed test.) Use the Mann-Whitney U test.

A	B
1060.21	1060.24
1060.34	1060.28
1060.27	1060.32
1060.36	1060.30
1060.40	

(b)   Test by using the F statistic of Section 10.8.

15.57   Calculate the probability that $U \leq 2$ for $n_1 = n_2 = 5$. Assume that no ties will be present and that H_0 is true.

15.58   Calculate the probability that the Wilcoxon T (Section 15.4) is less than or equal to 2 for $n = 3$ pairs. Assume that no ties will be present and that H_0 is true.

15.59   To investigate possible differences among production rates for three production lines turning out similar items, independent random samples of total production figures were obtained for 7 days for each line. The data are as follows:

Line 1	Line 2	Line 3
48	41	18
43	36	42
39	29	28
57	40	38
21	35	15
47	45	33
58	32	31

Do the data provide sufficient evidence to indicate any differences in location for the three sets of production figures, at the 5% significance level?

15.60   (a)   Suppose a company wants to study how personality relates to leadership. Four supervisors with different types of personalities are selected. Several employees are then selected from the group supervised by each, and these employees are asked to rate the leader of their group on a scale from 1 to 20 (20 signifies highly favorable). The table shows the resulting data. Is there sufficient evidence to indicate that one or more of the supervisors tend to receive higher ratings than the others? Use $\alpha = 0.05$.

| | Supervisor | | |
I	II	III	IV
20	17	16	8
19	11	15	12
20	13	13	10
18	15	18	14
17	14	11	9
	16		10

(b)   Suppose the company is particularly interested in comparing the ratings of the personality types represented by supervisors I and III. Make this comparison using $\alpha = 0.05$.

15.61   A union wants to determine its members' preferences before negotiating with management. Ten union members are randomly selected, and each member completes an extensive questionnaire. The responses to the various aspects of the questionnaire will enable the union to rank in order of importance the items to be negotiated. The rankings are shown in the table. Is there sufficient evidence to indicate that one or more of the items are preferred to the others? Test using $\alpha = 0.05$.

Person	More Pay	Job Stability	Fringe Benefits	Shorter Hours
1	2	1	3	4
2	1	2	3	4
3	4	3	2	1
4	1	4	2	3
5	1	2	3	4
6	1	3	4	2
7	2.5	1	2.5	4
8	3	1	4	2
9	1.5	1.5	3	4
10	2	3	1	4

15.62   Six groups of three children matched for IQ and age were formed. Each child was taught the concept of time by using one of three methods: lecture, demonstration, or teaching machine. The following scores indicate the students' performance when they were tested to see how well they had grasped the concept.

Group	Lecture	Demonstration	Teaching Machine
1	20	22	24
2	25	25	27
3	30	40	39
4	37	26	41
5	24	20	21
6	16	18	25

Is there sufficient evidence to indicate that the teaching methods differ in effectiveness? Give bounds for the p-value.

15.63   Calculate $P(R \leq 6)$ for the runs test, where $n_1 = n_2 = 8$ and H_0 is true.

15.64   Consider a Wilcoxon rank-sum test for the comparison of two probability distributions based on independent random samples of $n_1 = n_2 = 5$. Find $P(W \leq 17)$ assuming that H_0 is true.

*15.65   Let U denote the Mann-Whitney statistic and W_A the Wilcoxon rank-sum statistic for the sample from population A. Show that

$$U = n_1 n_2 + (1/2)n_1(n_1 + 1) - W_A$$

*15.66   Refer to Exercise 15.65.

(a)   Show that $E(U) = (1/2)n_1 n_2$ when H_0 is true.

(b)   Show that $V(U) = (1/12)[n_1 n_2(n_1 + n_2 + 1)]$ when H_0 is true, where H_0 states that the two populations are identical.

*15.67   Let T denote the Wilcoxon signed-rank test for n pairs of observations. Show that $E(T) = (1/4)n(n + 1)$ and $V(T) = (1/24)[n(n + 1)(2n + 1)]$ when the two populations are identical. Observe that these properties do not depend on whether T is constructed from negative or positive differences.

*15.68   Refer to the Spearman rank correlation coefficient of Section 15.10. Show that when there are no ties in either the x observations or the y observations, then

$$r_s = \frac{n \sum_{i=1}^{n} x_i y_i - \left(\sum_{i=1}^{n} x_i\right)\left(\sum_{i=1}^{n} y_i\right)}{\sqrt{\left[n \sum_{i=1}^{n} x_i^2 - \left(\sum_{i=1}^{n} x_i\right)^2\right]\left[n \sum_{i=1}^{n} y_i^2 - \left(\sum_{i=1}^{n} y_i\right)^2\right]}}$$

$$= 1 - \frac{6 \sum_{i=1}^{n} d_i^2}{n(n^2 - 1)}$$

where $d_i = x_i - y_i$.

* Exercises preceded by an asterisk are optional.

MATRICES AND OTHER USEFUL MATHEMATICAL RESULTS

A1.1 Matrices and Matrix Algebra

The following presentation represents a very elementary and condensed discussion of matrices and matrix operations. If you seek a more comprehensive introduction to the subject, consult the books listed in the references indicated at the end of Chapter 11.

We will define a *matrix* as a rectangular array (arrangement) of real numbers and will indicate specific matrices symbolically with bold capital letters. The numbers in the matrix, *elements*, appear in specific row-column positions, all of which are filled. The number of rows and columns may vary from one matrix to another, so we conveniently describe the size of a matrix by giving its *dimensions*—that is, the number of its rows and columns. Thus matrix $\mathbf{A}$

$$\mathbf{A}_{2 \times 3} = \begin{bmatrix} 6 & 0 & -1 \\ 4 & 2 & 7 \end{bmatrix}$$

possesses dimensions 2×3 because it contains two rows and three columns. Similarly, for

$$\mathbf{B}_{4 \times 1} = \begin{bmatrix} 1 \\ -3 \\ 0 \\ 7 \end{bmatrix} \quad \text{and} \quad \mathbf{C}_{2 \times 2} = \begin{bmatrix} 2 & 0 \\ -1 & 4 \end{bmatrix}$$

the dimensions of $\mathbf{B}$ and $\mathbf{C}$ are 4×1 and 2×2, respectively. Note that the row dimension always appears first and that the dimensions may be written below the identifying symbol of the matrix as indicated for matrices $\mathbf{A}$, $\mathbf{B}$, and $\mathbf{C}$.

As in ordinary algebra, an element of a matrix may be indicated by a symbol, $a, b, \ldots$, and its row-column position identified by means of a double subscript. Thus a_{21} would be the element in the second row, first column. Rows are numbered in order from top to bottom and columns from left to right. In matrix $\mathbf{A}$, $a_{21} = 4$, $a_{13} = -1$, and so on.

Elements in a particular row are identified by their column subscript and hence are numbered from left to right. The first element in a row is on the left. Likewise, elements in a particular column are identified by their row subscript and therefore are identified from the top element in the column to the bottom. For example, the first element in column 2 of matrix $\mathbf{A}$ is 0, the second is 2. The first, second, and third elements of row 1 are 6, 0, and -1, respectively.

The term *matrix algebra* involves, as the name implies, an algebra dealing with matrices, much as the ordinary algebra deals with real numbers or symbols representing real numbers. Hence, we will wish to state rules for the addition and multiplication of matrices as well as to define other elements of an algebra. In so doing we will point out the similarities as well as the dissimilarities between matrix and ordinary algebra. Finally, we will use our matrix operations to state and solve a very simple *matrix equation*. This, as you may suspect, will be the solution that we desire for the least squares equations.

A1.2 Addition of Matrices

Two matrices, say $\mathbf{A}$ and $\mathbf{B}$, can be added *only* if they are of the same dimensions. The sum of the two matrices will be a matrix obtained by adding *corresponding* elements of matrices $\mathbf{A}$ and $\mathbf{B}$—that is, elements in corresponding positions. This being the case, the resulting sum will be a matrix of the same dimensions as $\mathbf{A}$ and $\mathbf{B}$.

EXAMPLE A1.1 Find the indicated sum of matrices $\mathbf{A}$ and $\mathbf{B}$:

$$\underset{2 \times 3}{\mathbf{A}} = \begin{bmatrix} 2 & 1 & 4 \\ -1 & 6 & 0 \end{bmatrix} \qquad \underset{2 \times 3}{\mathbf{B}} = \begin{bmatrix} 0 & -1 & 1 \\ 6 & -3 & 2 \end{bmatrix}$$

Solution

$$\mathbf{A} + \mathbf{B} = \begin{bmatrix} 2 & 1 & 4 \\ -1 & 6 & 0 \end{bmatrix} + \begin{bmatrix} 0 & -1 & 1 \\ 6 & -3 & 2 \end{bmatrix}$$

$$= \begin{bmatrix} (2+0) & (1-1) & (4+1) \\ (-1+6) & (6-3) & (0+2) \end{bmatrix} = \begin{bmatrix} 2 & 0 & 5 \\ 5 & 3 & 2 \end{bmatrix}$$

EXAMPLE A1.2   Find the sum of the matrices

$$\mathbf{A}_{3 \times 3} = \begin{bmatrix} 1 & 0 & 3 \\ 1 & -1 & 4 \\ 2 & -1 & 0 \end{bmatrix} \quad \text{and} \quad \mathbf{B}_{3 \times 3} = \begin{bmatrix} 4 & 2 & -1 \\ 1 & 0 & 6 \\ 3 & 1 & 4 \end{bmatrix}$$

Solution

$$\mathbf{A} + \mathbf{B} = \begin{bmatrix} 5 & 2 & 2 \\ 2 & -1 & 10 \\ 5 & 0 & 4 \end{bmatrix} \qquad \qquad \text{\reflectbox{$\mathcal{S}$}}$$

Note that $(\mathbf{A} + \mathbf{B}) = (\mathbf{B} + \mathbf{A})$, as in ordinary algebra, and remember that we never add matrices of unlike dimensions.

A1.3 Multiplication of a Matrix by a Real Number

We desire a rule for multiplying a matrix by a real number, for example, $3\mathbf{A}$, where

$$\mathbf{A} = \begin{bmatrix} 2 & 1 \\ 4 & 6 \\ -1 & 0 \end{bmatrix}$$

Certainly we would want $3\mathbf{A}$ to equal $(\mathbf{A} + \mathbf{A} + \mathbf{A})$, to conform with the addition rule. Hence, $3\mathbf{A}$ would mean that each element in the $\mathbf{A}$ matrix must be multiplied by the multiplier 3, and

$$3\mathbf{A} = \begin{bmatrix} 3(2) & 3(1) \\ 3(4) & 3(6) \\ 3(-1) & 3(0) \end{bmatrix} = \begin{bmatrix} 6 & 3 \\ 12 & 18 \\ -3 & 0 \end{bmatrix}$$

In general, given a real number c and a matrix $\mathbf{A}$ with elements a_{ij}, the product $c\mathbf{A}$ will be a matrix whose elements are equal to ca_{ij}.

A1.4 Matrix Multiplication

The rule for matrix multiplication requires "row-column multiplication," which we will define subsequently. The procedure may seem a bit complicated to the novice but should not prove too difficult after practice. We will illustrate with an example. Let **A** and **B** be

$$\mathbf{A} = \begin{bmatrix} 2 & 0 \\ 1 & 4 \end{bmatrix} \qquad \mathbf{B} = \begin{bmatrix} 5 & 2 \\ -1 & 3 \end{bmatrix}$$

An element in the *ith row* and *jth column* of the product **AB** is obtained by multiplying the *ith row of* **A** by the *jth column of* **B**. Thus the element in the first row, first column of **AB** is obtained by multiplying the first row of **A** by the first column of **B**. Likewise, the element in the first row, second column would be the product of the first row of **A** and the second column of **B**. Notice that we always use the rows of **A** and the columns of **B**, where **A** is the matrix to the left of **B** in the product **AB**.

Row-column multiplication is relatively easy. Obtain the products, first-row element by first-column element, second-row element by second-column element, third by third, and so on, and then sum. Remember that row and column elements are numbered from left to right and top to bottom, respectively.

Applying these rules to our example, we obtain

$$\underset{2 \times 2\ 2 \times 2}{\mathbf{A}\ \mathbf{B}} = \begin{bmatrix} 2 & 0 \\ 1 & 4 \end{bmatrix}\begin{bmatrix} 5 & 2 \\ -1 & 3 \end{bmatrix} = \begin{bmatrix} \boxed{10} & 4 \\ 1 & 14 \end{bmatrix}$$

The first-row–first-column product would be $(2)(5) + (0)(-1) = 10$, which is located (and circled) in the first row, first column of **AB**. Likewise, the element in the first row, second column is equal to the product of the first row of **A** and the second column of **B**, or $(2)(2) + (0)(3) = 4$. The second-row–first-column product is $(1)(5) + (4)(-1) = 1$ and is located in the second row, first column of **AB**. Finally, the second-row–second-column product is $(1)(2) + (4)(3) = 14$.

EXAMPLE A1.3   Find the products **AB** and **BA**, where

$$\mathbf{A} = \begin{bmatrix} 2 & 1 \\ 1 & -1 \\ 0 & 4 \end{bmatrix} \qquad \text{and} \qquad \mathbf{B} = \begin{bmatrix} 4 & -1 & -1 \\ 2 & 0 & 2 \end{bmatrix}$$

Solution

$$\underset{3 \times 2\ 2 \times 3}{\mathbf{A}\ \mathbf{B}} = \begin{bmatrix} 2 & 1 \\ 1 & -1 \\ 0 & 4 \end{bmatrix}\begin{bmatrix} 4 & -1 & -1 \\ 2 & 0 & 2 \end{bmatrix} = \begin{bmatrix} 10 & -2 & 0 \\ 2 & -1 & -3 \\ 8 & 0 & 8 \end{bmatrix}$$

and

$$\underset{2 \times 3\ 3 \times 2}{\mathbf{B}\ \mathbf{A}} = \begin{bmatrix} 4 & -1 & -1 \\ 2 & 0 & 2 \end{bmatrix}\begin{bmatrix} 2 & 1 \\ 1 & -1 \\ 0 & 4 \end{bmatrix} = \begin{bmatrix} 7 & 1 \\ 4 & 10 \end{bmatrix}$$

Note that in matrix algebra, unlike ordinary algebra, **AB** does not equal **BA**. Because **A** contains three rows and **B** contains three columns, we can form $(3)(3) = 9$ row-column combinations and hence nine elements for **AB**. In contrast, **B** contains only two rows, **A** two columns, and hence the product **BA** will possess only $(2)(2) = 4$ elements, corresponding to the four different row-column combinations.

Furthermore, we observe that row-column multiplication is predicated on the assumption that the rows of the matrix on the left contain the same number of elements as the columns of the matrix on the right, so that corresponding elements will exist for the row-column multiplication. What do we do when this condition is not satisfied? We agree never to multiply two matrices, say **AB**, where the rows of **A** and the columns of **B** contain an unequal number of elements.

An examination of the dimensions of the matrices will tell whether they can be multiplied as well as give the dimensions of the product. Writing the dimensions underneath the two matrices,

$$\underset{m \times p}{\mathbf{A}} \ \underset{p \times q}{\mathbf{B}} = \underset{m \times q}{\mathbf{AB}}$$

we observe that the inner two numbers, giving the number of elements in a row of **A** and column of **B**, respectively, must be equal. The outer two numbers, indicating the number of rows of **A** and columns of **B**, give the dimensions of the product matrix. You may verify the operation of this rule for Example A1.3.

EXAMPLE A1.4   Obtain the product **AB**:

$$\underset{1 \times 3}{\mathbf{A}} \ \underset{3 \times 2}{\mathbf{B}} = [2 \quad 1 \quad 0] \begin{bmatrix} 2 & 0 \\ 0 & 3 \\ -1 & 0 \end{bmatrix} = [4 \quad 3]$$

Note that product **AB** is (1×2) and that **BA** is undefined because of the respective dimensions of **A** and **B**.   ∽

EXAMPLE A1.5   Find the product **AB**, where

$$\mathbf{A} = [1 \quad 2 \quad 3 \quad 4] \quad \text{and} \quad \mathbf{B} = \begin{bmatrix} 1 \\ 2 \\ 3 \\ 4 \end{bmatrix}$$

Solution

$$\underset{1 \times 4}{\mathbf{A}} \; \underset{4 \times 1}{\mathbf{B}} = \begin{bmatrix} 1 & 2 & 3 & 4 \end{bmatrix} \begin{bmatrix} 1 \\ 2 \\ 3 \\ 4 \end{bmatrix} = [30]$$

Note that this example produces a different method for writing a sum of squares.

A1.5 Identity Elements

The identity elements for addition and multiplication in ordinary algebra are 0 and 1, respectively. In addition, 0 plus any other element, say a, is identically equal to a; that is,

$$0 + 2 = 2 \qquad 0 + (-9) = -9$$

Similarly, the multiplication of the identity element 1 by any other element, say a, is equal to a; that is,

$$(1)(5) = 5 \qquad (1)(-4) = -4$$

In matrix algebra two matrices are said to be equal when all corresponding elements are equal. With this in mind we will define the identity matrices in a manner similar to that employed in ordinary algebra. Hence, if $\mathbf{A}$ is any matrix, a matrix $\mathbf{B}$ will be an identity matrix for addition if

$$\mathbf{A} + \mathbf{B} = \mathbf{A} \qquad \text{and} \qquad \mathbf{B} + \mathbf{A} = \mathbf{A}$$

It easily can be seen that the identity matrix for addition is one in which every element is equal to zero. This matrix is of interest but of no practical importance in our work.

Similarly, if $\mathbf{A}$ is any matrix, the identity matrix for multiplication is a matrix $\mathbf{I}$ that satisfies the relation

$$\mathbf{A}\mathbf{I} = \mathbf{A} \qquad \text{and} \qquad \mathbf{I}\mathbf{A} = \mathbf{A}$$

This matrix, called the *identity matrix*, is the *square matrix*

$$\underset{n \times n}{\mathbf{I}} = \begin{bmatrix} 1 & 0 & 0 & 0 & \cdots & 0 \\ 0 & 1 & 0 & 0 & \cdots & 0 \\ 0 & 0 & 1 & 0 & \cdots & 0 \\ 0 & 0 & 0 & 1 & \cdots & 0 \\ \vdots & \vdots & \vdots & \vdots & & \vdots \\ 0 & 0 & 0 & 0 & \cdots & 1 \end{bmatrix}$$

That is, all elements in the *main diagonal* of the matrix, running from top left to bottom right, are equal to 1; all other elements equal zero. Note that the identity matrix is always indicated by the symbol **I**.

Unlike ordinary algebra, which contains only one identity element for multiplication, matrix algebra must contain an infinitely large number of identity matrices. Thus we must have matrices with dimensions 1×1, 2×2, 3×3, 4×4, and so on, so as to provide an identity of the correct dimensions to permit multiplication. All will be of this pattern.

That the **I** matrix satisfies the relation

$$\mathbf{IA} = \mathbf{AI} = \mathbf{A}$$

can be shown by an example.

EXAMPLE A1.6   Let

$$\mathbf{A} = \begin{bmatrix} 2 & 1 & 0 \\ -1 & 6 & 3 \end{bmatrix}$$

Show that $\mathbf{IA} = \mathbf{A}$ and $\mathbf{AI} = \mathbf{A}$.

Solution

$$\underset{2 \times 2}{\mathbf{I}} \ \underset{2 \times 3}{\mathbf{A}} = \begin{bmatrix} 1 & 0 \\ 0 & 1 \end{bmatrix} \begin{bmatrix} 2 & 1 & 0 \\ -1 & 6 & 3 \end{bmatrix} = \begin{bmatrix} 2 & 1 & 0 \\ -1 & 6 & 3 \end{bmatrix} = \mathbf{A}$$

and

$$\underset{2 \times 3}{\mathbf{A}} \ \underset{3 \times 3}{\mathbf{I}} = \begin{bmatrix} 2 & 1 & 0 \\ -1 & 6 & 3 \end{bmatrix} \begin{bmatrix} 1 & 0 & 0 \\ 0 & 1 & 0 \\ 0 & 0 & 1 \end{bmatrix} = \begin{bmatrix} 2 & 1 & 0 \\ -1 & 6 & 3 \end{bmatrix} = \mathbf{A} \qquad \backsim$$

A1.6 The Inverse of a Matrix

For matrix algebra to be useful, we must be able to construct and solve matrix equations for a matrix of unknowns in a manner similar to that employed in ordinary algebra. This, in turn, requires a method of performing division.

For example, we would solve the simple equation in ordinary algebra,

$$2x = 6$$

by dividing both sides of the equation by 2 and obtaining $x = 3$. Another way to view this operation is to define the reciprocal of each element in an algebraic system and to think of division as multiplication by the reciprocal of an element. We could solve the equation $2x = 6$ by multiplying both sides of the equation by the reciprocal of 2. Because every element in the real number system possesses a reciprocal, with the exception of 0, the multiplication operation eliminates the need for division.

The reciprocal of a number c in ordinary algebra is a number b that satisfies the relation

$$cb = 1$$

that is, *the product of a number by its reciprocal must equal the identity element for multiplication.* For example, the reciprocal of 2 is $1/2$ and $(2)(1/2) = 1$.

A reciprocal in matrix algebra is called the *inverse* of a matrix and is defined as follows:

Definition A1.1

> Let $A_{n \times n}$ be a square matrix. If a matrix A^{-1} can be found such that
>
> $$AA^{-1} = I \qquad \text{and} \qquad A^{-1}A = I$$
>
> then A^{-1} is called the *inverse* of A.

Note that the requirement for an inverse in matrix algebra is the same as in ordinary algebra—that is, the product of A by its inverse must equal the identity matrix for multiplication. Furthermore, the inverse is undefined for nonsquare matrices, and hence many matrices in matrix algebra do not have inverses (recall that 0 was the only element in the real number system without an inverse). Finally, we state without proof that many square matrices do not possess inverses. Those that do will be identified in Section A1.9, and a method will be given for finding the inverse of a matrix.

A1.7 The Transpose of a Matrix

We have just discussed a relationship between a matrix and its inverse. A second useful matrix relationship defines the *transpose* of a matrix.

Definition A1.2

> Let $A_{p \times q}$ be a matrix of dimensions $p \times q$. Then A', called the *transpose* of A, is defined to be a matrix obtained by interchanging corresponding rows and columns of A; that is, first with first, second with second, and so on.

For example, let

$$\mathop{\mathbf{A}}_{3 \times 2} = \begin{bmatrix} 2 & 0 \\ 1 & 1 \\ 4 & 3 \end{bmatrix}$$

Then

$$\mathop{\mathbf{A}'}_{2 \times 3} = \begin{bmatrix} 2 & 1 & 4 \\ 0 & 1 & 3 \end{bmatrix}$$

Note that the first and second rows of $\mathbf{A}'$ are identical with the first and second columns, respectively, of $\mathbf{A}$.

As a second example, let

$$\mathbf{Y} = \begin{bmatrix} y_1 \\ y_2 \\ y_3 \end{bmatrix}$$

Then $\mathbf{Y}' = [y_1 \quad y_2 \quad y_3]$. As a point of interest, we observe that $\mathbf{Y}'\mathbf{Y} = \sum_{i=1}^{3} y_i^2$. Finally, if

$$\mathbf{A} = \begin{bmatrix} 2 & 1 & 4 \\ 0 & 2 & 3 \\ 1 & 6 & 9 \end{bmatrix}$$

then

$$\mathbf{A}' = \begin{bmatrix} 2 & 0 & 1 \\ 1 & 2 & 6 \\ 4 & 3 & 9 \end{bmatrix}$$

A1.8 A Matrix Expression for a System of Simultaneous Linear Equations

We will now introduce you to one of the very simple and important applications of matrix algebra. Let

$$2v_1 + v_2 = 5$$
$$v_1 - v_2 = 1$$

be a pair of simultaneous linear equations in the two variables, v_1 and v_2. We will then define three matrices:

$$\underset{2 \times 2}{\mathbf{A}} = \begin{bmatrix} 2 & 1 \\ 1 & -2 \end{bmatrix} \qquad \underset{2 \times 1}{\mathbf{V}} = \begin{bmatrix} v_1 \\ v_2 \end{bmatrix} \qquad \underset{2 \times 1}{\mathbf{G}} = \begin{bmatrix} 5 \\ 1 \end{bmatrix}$$

Note that $\mathbf{A}$ is the matrix of coefficients of the unknowns when the equations are each written with the variables appearing in the same order, reading left to right, and with the constants on the right-hand side of the equality sign. The $\mathbf{V}$ matrix gives the unknowns in a column and in the same order as they appear in the equations. Finally, the $\mathbf{G}$ matrix contains the constants in a column exactly as they occur in the set of equations.

The simultaneous system of two linear equations may now be written in matrix notation as

$$\mathbf{AV} = \mathbf{G}$$

a statement that can easily be verified by multiplying $\mathbf{A}$ and $\mathbf{V}$ and then comparing the answer with $\mathbf{G}$.

$$\mathbf{AV} = \begin{bmatrix} 2 & 1 \\ 1 & -1 \end{bmatrix} \begin{bmatrix} v_1 \\ v_2 \end{bmatrix} = \begin{bmatrix} 2v_1 + v_2 \\ v_1 - v_2 \end{bmatrix} = \begin{bmatrix} 5 \\ 1 \end{bmatrix} = \mathbf{G}$$

Observe that corresponding elements in $\mathbf{AV}$ and $\mathbf{G}$ are equal—that is, $2v_1 + v_2 = 5$ and $v_1 - v_2 = 1$. Therefore, $\mathbf{AV} = \mathbf{G}$.

The method for writing a pair of linear equations in two unknowns as a matrix equation can easily be extended to a system of r equations in r unknowns. For example, if the equations are

$$a_{11}v_1 + a_{12}v_2 + a_{13}v_3 + \cdots + a_{1r}v_r = g_1$$
$$a_{21}v_1 + a_{22}v_2 + a_{23}v_3 + \cdots + a_{2r}v_r = g_2$$
$$a_{31}v_1 + a_{32}v_2 + a_{33}v_3 + \cdots + a_{3r}v_r = g_3$$
$$\vdots \qquad \vdots \qquad \vdots \qquad \qquad \vdots \qquad \vdots$$
$$a_{r1}v_1 + a_{r2}v_2 + a_{r3}v_3 + \cdots + a_{rr}v_r = g_r$$

define

$$\mathbf{A} = \begin{bmatrix} a_{11} & a_{12} & a_{13} & \cdots & a_{1r} \\ a_{21} & a_{22} & a_{23} & \cdots & a_{2r} \\ a_{31} & a_{32} & a_{33} & \cdots & a_{3r} \\ \vdots & \vdots & \vdots & & \vdots \\ a_{r1} & a_{r2} & a_{r3} & \cdots & a_{rr} \end{bmatrix} \qquad \mathbf{V} = \begin{bmatrix} v_1 \\ v_2 \\ v_3 \\ \vdots \\ v_r \end{bmatrix} \qquad \mathbf{G} = \begin{bmatrix} g_1 \\ g_2 \\ g_3 \\ \vdots \\ g_r \end{bmatrix}$$

Observe that, once again, **A** is a square matrix of variable coefficients, whereas **V** and **G** are column matrices containing the variables and constants, respectively. Then **AV = G**.

Regardless of how large the system of equations, if we possess n linear equations in n unknowns, the system may be written as the simple matrix equation **AV = G**.

You will observe that the matrix **V** contains all the unknowns, whereas **A** and **G** are constant matrices.

Our objective, of course, is to solve for the matrix of unknowns, **V**, where the equation **AV = G** is similar to the equation

$$2v = 6$$

in ordinary algebra. This being true, we would not be too surprised to find that the methods of solution are the same. In ordinary algebra both sides of the equation are multiplied by the reciprocal of 2; in matrix algebra both sides of the equation are multiplied by $\mathbf{A}^{-1}$. Then

$$\mathbf{A}^{-1}(\mathbf{AV}) = \mathbf{A}^{-1}\mathbf{G}$$

or

$$\mathbf{A}^{-1}\mathbf{AV} = \mathbf{A}^{-1}\mathbf{G}$$

But $\mathbf{A}^{-1}\mathbf{A} = \mathbf{I}$ and $\mathbf{IV} = \mathbf{V}$. Therefore, $\mathbf{V} = \mathbf{A}^{-1}\mathbf{G}$. In other words, the solution to the system of simultaneous linear equations can be obtained by finding $\mathbf{A}^{-1}$ and then obtaining the product $\mathbf{A}^{-1}\mathbf{G}$. The solution values of $v_1, v_2, v_3, \ldots, v_r$ will appear in sequence in the column matrix $\mathbf{V} = \mathbf{A}^{-1}\mathbf{G}$.

A1.9 Inverting a Matrix

We have indicated in Section A1.8 that the key to the solution of a system of simultaneous linear equations by the method of matrix algebra rests on the acquisition of the inverse of the **A** matrix. Many methods exist for inverting matrices. The method that we present is not the best from a computational point of view, but it works very well for the matrices associated with most experimental designs and it is one of the easiest to present to the novice. It depends upon a theorem in matrix algebra and the use of *row operations*.

Before defining *row operations* on matrices, we must state what is meant by the addition of two rows of a matrix and the multiplication of a row by a constant. We will illustrate with the **A** matrix for the system of two simultaneous linear equations,

$$\mathbf{A} = \begin{bmatrix} 2 & 1 \\ 1 & -1 \end{bmatrix}$$

Two rows of a matrix may be added by adding corresponding elements. Thus if the two rows of the **A** matrix are added, one obtains a new row with elements $[(2+1) \quad (1-1)] = [3 \quad 0]$. Multiplication of a row by a constant means that each element in the row is multiplied by the constant. Twice the first row of the **A** matrix would generate the row $[4 \quad 2]$. With these ideas in mind, we will define three ways to operate on a row in a matrix:

1. A row may be multiplied by a constant.
2. A row may be multiplied by a constant and added to or subtracted from another row (which is identified as the one upon which the operation is performed).
3. Two rows many be interchanged.

Given matrix **A**, it is quite easy to see that we might perform a series of row operations that would yield some new matrix **B**. In this connection we state without proof a surprising and interesting theorem from matrix algebra; namely, there exists some matrix **C** such that

$$\mathbf{CA} = \mathbf{B}$$

In other words, a series of row operations on a matrix **A** is equivalent to multiplying **A** by a matrix **C**. We will use this principle to invert a matrix.

Place the matrix **A**, which is to be inverted, alongside an identity matrix of the same dimensions:

$$\mathbf{A} = \begin{bmatrix} 2 & 1 \\ 1 & -1 \end{bmatrix} \qquad \mathbf{I} = \begin{bmatrix} 1 & 0 \\ 0 & 1 \end{bmatrix}$$

Then perform the same row operations on **A** *and* **I** *in such a way that* **A** *changes to an identity matrix.* In doing so, we must have multiplied **A** by a matrix **C** so that $\mathbf{CA} = \mathbf{I}$. Therefore, **C** must be the inverse of **A**! The problem, of course, is to find the unknown matrix **C** and, fortunately, this proves to be of little difficulty. Because we performed the same row operations on **A** and **I**, the identity matrix must have changed to $\mathbf{CI} = \mathbf{C} = \mathbf{A}^{-1}$.

$$\mathbf{A} = \begin{bmatrix} 2 & 1 \\ 1 & -1 \end{bmatrix} \qquad\qquad \mathbf{I} = \begin{bmatrix} 1 & 0 \\ 0 & 1 \end{bmatrix}$$

$$\Big\} \text{ (same row operations) } \Big\{$$

$$\mathbf{CA} = \mathbf{I} \qquad\qquad \mathbf{CI} = \mathbf{C} = \mathbf{A}^{-1}$$

We will illustrate with the following example.

EXAMPLE A1.7   Invert the matrix

$$A = \begin{bmatrix} 2 & 1 \\ 1 & -1 \end{bmatrix}$$

Solution

$$A = \begin{bmatrix} 2 & 1 \\ 1 & -1 \end{bmatrix} \quad I = \begin{bmatrix} 1 & 0 \\ 0 & 1 \end{bmatrix}$$

Step 1. Operate on row 1 by multiplying row 1 by 1/2. (Note: It is helpful to the beginner to identify the row upon which he or she is operating because all other rows will remain unchanged, even though they may be used in the operation. We will star the row upon which the operation is being performed.)

$$\begin{matrix} * \\ \; \end{matrix} \begin{bmatrix} 1 & 1/2 \\ 1 & -1 \end{bmatrix} \quad \begin{bmatrix} 1/2 & 0 \\ 0 & 1 \end{bmatrix}$$

Step 2. Operate on row 2 by subtracting row 1 from row 2.

$$\begin{matrix} \; \\ * \end{matrix} \begin{bmatrix} 1 & 1/2 \\ 0 & -3/2 \end{bmatrix} \quad \begin{bmatrix} 1/2 & 0 \\ -1/2 & 1 \end{bmatrix}$$

(Note that row 2 is simply used to operate on row 1 and hence remains unchanged.)

Step 3. Multiply row 2 by $(-2/3)$.

$$\begin{matrix} \; \\ * \end{matrix} \begin{bmatrix} 1 & 1/2 \\ 0 & 1 \end{bmatrix} \quad \begin{bmatrix} 1/2 & 0 \\ 1/3 & -2/3 \end{bmatrix}$$

Step 4. Operate on row 1 by multiplying row 2 by 1/2 and subtracting from row 1.

$$\begin{matrix} * \\ \; \end{matrix} \begin{bmatrix} 1 & 0 \\ 0 & 1 \end{bmatrix} \quad \begin{bmatrix} 1/3 & 1/3 \\ 1/3 & -2/3 \end{bmatrix}$$

(Note that row 2 is simply used to operate on row 1 and hence remains unchanged.) Hence the inverse of **A** must be

$$A^{-1} = \begin{bmatrix} 1/3 & 1/3 \\ 1/3 & -2/3 \end{bmatrix}$$

A ready check on the calculations for the inversion procedure is available because $A^{-1}A$ must equal the identity matrix **I**. Thus

$$A^{-1}A = \begin{bmatrix} 1/3 & 1/3 \\ 1/3 & -2/3 \end{bmatrix}\begin{bmatrix} 2 & 1 \\ 1 & -1 \end{bmatrix} = \begin{bmatrix} 1 & 0 \\ 0 & 1 \end{bmatrix}$$

EXAMPLE A1.8   Invert the matrix

$$A = \begin{bmatrix} 2 & 0 & 1 \\ 1 & -1 & 2 \\ 1 & 0 & 0 \end{bmatrix}$$

and check the results.

Solution

$$A = \begin{bmatrix} 2 & 0 & 1 \\ 1 & -1 & 2 \\ 1 & 0 & 0 \end{bmatrix} \qquad I = \begin{bmatrix} 1 & 0 & 0 \\ 0 & 1 & 0 \\ 0 & 0 & 1 \end{bmatrix}$$

Step 1. Multiply row 1 by 1/2.

$$*\begin{bmatrix} 1 & 0 & 1/2 \\ 1 & -1 & 2 \\ 1 & 0 & 0 \end{bmatrix} \qquad \begin{bmatrix} 1/2 & 0 & 0 \\ 0 & 1 & 0 \\ 0 & 0 & 1 \end{bmatrix}$$

Step 2. Operate on row 2 by subtracting row 1 from row 2.

$$\begin{bmatrix} 1 & 0 & 1/2 \\ *0 & -1 & 3/2 \\ 1 & 0 & 0 \end{bmatrix} \qquad \begin{bmatrix} 1/2 & 0 & 0 \\ -1/2 & 1 & 0 \\ 0 & 0 & 1 \end{bmatrix}$$

Step 3. Operate on row 3 by subtracting row 1 from row 3.

$$\begin{bmatrix} 1 & 0 & 1/2 \\ 0 & -1 & 3/2 \\ *0 & 0 & -1/2 \end{bmatrix} \qquad \begin{bmatrix} 1/2 & 0 & 0 \\ -1/2 & 1 & 0 \\ -1/2 & 0 & 1 \end{bmatrix}$$

Step 4. Operate on row 2 by multiplying row 3 by 3 and adding to row 2.

$$\begin{bmatrix} 1 & 0 & 1/2 \\ *0 & -1 & 0 \\ 0 & 0 & -1/2 \end{bmatrix} \qquad \begin{bmatrix} 1/2 & 0 & 0 \\ -2 & 1 & 3 \\ -1/2 & 0 & 1 \end{bmatrix}$$

Step 5. Multiply row 2 by (-1).

$$\begin{bmatrix} 1 & 0 & 1/2 \\ *0 & 1 & 0 \\ 0 & 0 & -1/2 \end{bmatrix} \qquad \begin{bmatrix} 1/2 & 0 & 0 \\ 2 & -1 & -3 \\ -1/2 & 0 & 1 \end{bmatrix}$$

Step 6. Operate on row 1 by adding row 3 to row 1.

$$*\begin{bmatrix} 1 & 0 & 0 \\ 0 & 1 & 0 \\ 0 & 0 & -1/2 \end{bmatrix} \quad \begin{bmatrix} 0 & 0 & 1 \\ 2 & -1 & -3 \\ -1/2 & 0 & 1 \end{bmatrix}$$

Step 7. Multiply row 3 by (-2).

$$\begin{bmatrix} 1 & 0 & 0 \\ 0 & 1 & 0 \\ *0 & 0 & 1 \end{bmatrix} \quad \begin{bmatrix} 0 & 0 & 1 \\ 2 & -1 & -3 \\ 1 & 0 & -2 \end{bmatrix} = \mathbf{A}^{-1}$$

The seven row operations have changed the $\mathbf{A}$ matrix to the identity matrix and, barring errors of calculation, have changed the identity to $\mathbf{A}^{-1}$.

Checking, we have

$$\mathbf{A}^{-1}\mathbf{A} = \begin{bmatrix} 0 & 0 & 1 \\ 2 & -1 & -3 \\ 1 & 0 & -2 \end{bmatrix}\begin{bmatrix} 2 & 0 & 1 \\ 1 & -1 & 2 \\ 1 & 0 & 0 \end{bmatrix} = \begin{bmatrix} 1 & 0 & 0 \\ 0 & 1 & 0 \\ 0 & 0 & 1 \end{bmatrix}$$

We see that $\mathbf{A}^{-1}\mathbf{A} = \mathbf{I}$ and hence that the calculations are correct.

Note that the sequence of row operations required to convert $\mathbf{A}$ to $\mathbf{I}$ is not unique. One person might achieve the inverse by using five row operations whereas another might require ten, but the end result will be the same. However, in the interests of efficiency it is desirable to employ a system.

Observe that the inversion process utilizes row operations to change off-diagonal elements in the $\mathbf{A}$ matrix to 0s and the main diagonal elements to 1s. One systematic procedure is as follows. Change the top left element into a 1 and then perform row operations to change all other elements in the *first* column to 0. Then move to the diagonal element in the second row, second column, change it into a 1, and change all elements in the *second* column *below* the main diagonal to 0. This process is repeated, moving down the main diagonal from top left to bottom right, until all elements below the main diagonal have been changed to 0s. To eliminate nonzero elements above the main diagonal, operate on all elements in the last column, changing each to 0; then move to the next to last column and repeat the process. Continue this procedure until you arrive at the first element in the first column, which was the starting point. This procedure is indicated diagrammatically in Figure A1.1.

Matrix inversion is a tedious process, at best, and requires every bit as much labor as the solution of a system of simultaneous equations by elimination or substitution. You will be pleased to learn that we do not expect you to develop a facility for matrix inversion. Fortunately, most matrices associated with designed experiments follow patterns and are easily inverted.

Figure A1.1
Procedure for
matrix inversion

$A =$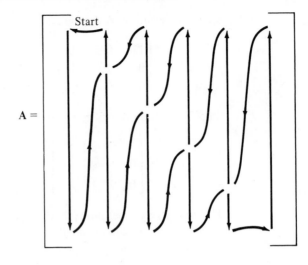

It will be beneficial to you to invert a few 2×2 and 3×3 matrices. Matrices lacking pattern, particularly large matrices, are inverted most efficiently and economically by using an electronic computer. (Programs for matrix inversion have been developed for most electronic computers.)

We emphasize that obtaining the solution for the least squares equations (Chapter 11) by matrix inversion has distinct advantages that may or may not be apparent. Not the least of these is the fact that the inversion procedure is systematic and hence is particularly suitable for electronic computation. However, the major advantage is that the inversion procedure will automatically produce the variances of the estimators of all parameters in the linear model.

Before leaving the topic of matrix inversion, we ask how one may identify a matrix that has an inverse. Reference to a discussion of linear equations in ordinary algebra should reveal the answer.

Clearly, a unique solution for a system of simultaneous linear equations cannot be obtained unless the equations are independent. Thus if one of the equations is a linear combination of the others, the equations are dependent. Coefficient matrices associated with dependent systems of linear equations do not possess an inverse.

A1.10 Solving a System of Simultaneous Linear Equations

We have finally obtained all the ingredients necessary for solving a system of simultaneous linear equations,

$$2v_1 + v_2 = 5$$
$$v_1 - v_2 = 1$$

Recalling that the matrix solution to the system of equations $\mathbf{AV} = \mathbf{G}$ is $\mathbf{V} = \mathbf{A}^{-1}\mathbf{G}$, we obtain

$$\mathbf{V} = \mathbf{A}^{-1}\mathbf{G} = \begin{bmatrix} 1/3 & 1/3 \\ 1/3 & -2/3 \end{bmatrix}\begin{bmatrix} 5 \\ 1 \end{bmatrix} = \begin{bmatrix} 2 \\ 1 \end{bmatrix}$$

Hence the solution is

$$\mathbf{V} = \begin{bmatrix} v_1 \\ v_2 \end{bmatrix} = \begin{bmatrix} 2 \\ 1 \end{bmatrix}$$

that is, $v_1 = 2$ and $v_2 = 1$, a fact that may be verified by substitution of these values in the original linear equations.

EXAMPLE A1.9 Solve the system of simultaneous linear equations

$$2v_1 + v_3 = 4$$
$$v_1 - v_2 + 2v_3 = 2$$
$$v_1 = 1$$

Solution The coefficient matrix for these equations,

$$\mathbf{A} = \begin{bmatrix} 2 & 0 & 1 \\ 1 & -1 & 2 \\ 1 & 0 & 0 \end{bmatrix}$$

appeared in Example A1.8. In that example we found that

$$\mathbf{A}^{-1} = \begin{bmatrix} 0 & 0 & 1 \\ 2 & -1 & -3 \\ 1 & 0 & -2 \end{bmatrix}$$

Solving, we obtain

$$\mathbf{V} = \mathbf{A}^{-1}\mathbf{G} = \begin{bmatrix} 0 & 0 & 1 \\ 2 & -1 & -3 \\ 1 & 0 & -2 \end{bmatrix}\begin{bmatrix} 4 \\ 2 \\ 1 \end{bmatrix} = \begin{bmatrix} 1 \\ 3 \\ 2 \end{bmatrix}$$

Thus $v_1 = 1$, $v_2 = 3$, and $v_3 = 2$ give the solution to the set of three simultaneous linear equations.

A1.11 Other Useful Mathematical Results

The purpose of this section is to provide the reader with a convenient reference to some of the key mathematical results that are used frequently in the body of the text.

The Binomial Expansion of $(x + y)^n$

Let x and y be any real numbers, then

$$(x + y)^n = \binom{n}{0} x^n y^0 + \binom{n}{1} x^{n-1} y^1 + \binom{n}{2} x^{n-2} y^2 + \cdots + \binom{n}{n} x^0 y^n$$

$$= \sum_{i=0}^{n} \binom{n}{i} x^{n-i} y^i$$

The Sum of a Geometric Series

Let r be a real number such that $|r| < 1$, and m be any integer $m \geq 1$

$$\sum_{i=0}^{\infty} r^i = \frac{1}{1-r}, \qquad \sum_{i=1}^{\infty} r^i = \frac{r}{1-r}, \qquad \sum_{i=0}^{m} r^i = \frac{1 - r^{m+1}}{1-r}$$

The (Taylor) Series Expansion of e^x

Let x be any real number, then

$$e^x = \sum_{i=0}^{\infty} \frac{x^i}{i!}$$

Some useful formulas for particular summations follow. The proofs (omitted) are most easily established by using mathematical induction.

$$\sum_{i=1}^{n} i = \frac{n(n + 1)}{2}$$

$$\sum_{i=1}^{n} i^2 = \frac{n(n + 1)(2n + 1)}{6}$$

$$\sum_{i=1}^{n} i^3 = \left(\frac{n(n + 1)}{2} \right)^2$$

Gamma Function

Let $t > 0$, then $\Gamma(t)$ is defined by the following integral:

$$\Gamma(t) = \int_{0}^{\infty} y^{t-1} e^{-y} \, dy$$

Using the technique of integration by parts, it follows that for any $t > 0$

$$\Gamma(t + 1) = t\Gamma(t)$$

and if $t = n$, where n is an integer,

$$\Gamma(n) = (n - 1)!$$

Further,

$$\Gamma(1/2) = \sqrt{\pi}$$

If $\alpha, \beta > 0$, the Beta function, $B(\alpha, \beta)$, is defined by the following integral,

$$B(\alpha, \beta) = \int_0^1 y^{\alpha - 1}(1 - y)^{\beta - 1}\, dy$$

and is related to the gamma function as follows:

$$B(\alpha, \beta) = \frac{\Gamma(\alpha)\Gamma(\beta)}{\Gamma(\alpha + \beta)}$$

COMMON PROBABILITY DISTRIBUTIONS, MEANS, VARIANCES, AND MOMENT-GENERATING FUNCTIONS

Table A2.1
Discrete distributions

Distribution	Probability Function	Mean	Variance	Moment-Generating Function
Binomial	$p(y) = \binom{n}{y} p^y (1-p)^{n-y}$; $y = 0, 1, \ldots, n$	np	$np(1-p)$	$[pe^t + (1-p)]^n$
Geometric	$p(y) = p(1-p)^{y-1}$; $y = 1, 2, \ldots$	$\dfrac{1}{p}$	$\dfrac{1-p}{p^2}$	$\dfrac{pe^t}{1-(1-p)e^t}$
Hypergeometric	$p(y) = \dfrac{\binom{r}{y}\binom{N-r}{n-y}}{\binom{N}{n}}$; $y = 0, 1, \ldots, n$ if $n \le r$, $y = 0, 1, \ldots, r$ if $n > r$	$\dfrac{nr}{N}$	$n\left(\dfrac{r}{N}\right)\left(\dfrac{N-r}{N}\right)\left(\dfrac{N-n}{N-1}\right)$	
Poisson	$p(y) = \dfrac{\lambda^y e^{-\lambda}}{y!}$; $y = 0, 1, 2, \ldots$	λ	λ	$\exp[\lambda(e^t - 1)]$
Negative binomial	$p(y) = \binom{y-1}{r-1} p^r (1-p)^{y-r}$; $y = r, r+1, \ldots$	$\dfrac{r}{p}$	$\dfrac{r(1-p)}{p^2}$	$\left[\dfrac{pe^t}{1-(1-p)e^t}\right]^r$

Table A2.2
Continuous distributions

Distribution	Probability Function	Mean	Variance	Moment-Generating Function
Uniform	$f(y) = \dfrac{1}{\theta_2 - \theta_1}; \theta_1 \le y \le \theta_2$	$\dfrac{\theta_1 + \theta_2}{2}$	$\dfrac{(\theta_2 - \theta_1)^2}{12}$	$\dfrac{e^{t\theta_2} - e^{t\theta_1}}{t(\theta_2 - \theta_1)}$
Normal	$f(y) = \dfrac{1}{\sigma\sqrt{2\pi}} \exp\left[-\left(\dfrac{1}{2\sigma^2}\right)(y-\mu)^2\right]$ $-\infty < y < +\infty$	μ	σ^2	$\exp\left(\mu t + \dfrac{t^2\sigma^2}{2}\right)$
Exponential	$f(y) = \dfrac{1}{\beta} e^{-y/\beta}, \quad \beta > 0$ $0 < y < \infty$	β	β^2	$(1 - \beta t)^{-1}$
Gamma	$f(y) = \left[\dfrac{1}{\Gamma(\alpha)\beta^\alpha}\right] y^{\alpha-1} e^{-y/\beta};$ $0 < y < \infty$	$\alpha\beta$	$\alpha\beta^2$	$(1 - \beta t)^{-\alpha}$
Chi-square	$f(\chi^2) = \dfrac{(\chi^2)^{(v/2)-1} e^{-\chi^2/2}}{2^{v/2}\Gamma(v/2)};$ $\chi^2 > 0$	v	$2v$	$(1 - 2t)^{-v/2}$
Beta	$f(y) = \left[\dfrac{\Gamma(\alpha+\beta)}{\Gamma(\alpha)\Gamma(\beta)}\right] y^{\alpha-1}(1-y)^{\beta-1};$ $0 < y < 1$	$\dfrac{\alpha}{\alpha+\beta}$	$\dfrac{\alpha\beta}{(\alpha+\beta)^2(\alpha+\beta+1)}$	does not exist in closed form

TABLES

Table 1 Binomial Probabilities

Tabulated values are $P(Y \leq a) = \sum_{y=0}^{a} p(y)$. (Computations are rounded at third decimal place.)

(a) $n = 5$

a	\multicolumn{13}{c}{P}	a
	0.01	0.05	0.10	0.20	0.30	0.40	0.50	0.60	0.70	0.80	0.90	0.95	0.99	
0	.951	.774	.590	.328	.168	.078	.031	.010	.002	.000	.000	.000	.000	0
1	.999	.977	.919	.737	.528	.337	.188	.087	.031	.007	.000	.000	.000	1
2	1.000	.999	.991	.942	.837	.683	.500	.317	.163	.058	.009	.001	.000	2
3	1.000	1.000	1.000	.993	.969	.913	.812	.663	.472	.263	.081	.023	.001	3
4	1.000	1.000	1.000	1.000	.998	.990	.969	.922	.832	.672	.410	.226	.049	4

(b) $n = 10$

a	\multicolumn{13}{c}{P}	a
	0.01	0.05	0.10	0.20	0.30	0.40	0.50	0.60	0.70	0.80	0.90	0.95	0.99	
0	.904	.599	.349	.107	.028	.006	.001	.000	.000	.000	.000	.000	.000	0
1	.996	.914	.736	.376	.149	.046	.011	.002	.000	.000	.000	.000	.000	1
2	1.000	.988	.930	.678	.383	.167	.055	.012	.002	.000	.000	.000	.000	2
3	1.000	.999	.987	.879	.650	.382	.172	.055	.011	.001	.000	.000	.000	3
4	1.000	1.000	.998	.967	.850	.633	.377	.166	.047	.006	.000	.000	.000	4
5	1.000	1.000	1.000	.994	.953	.834	.623	.367	.150	.033	.002	.000	.000	5
6	1.000	1.000	1.000	.999	.989	.945	.828	.618	.350	.121	.013	.001	.000	6
7	1.000	1.000	1.000	1.000	.998	.988	.945	.833	.617	.322	.070	.012	.000	7
8	1.000	1.000	1.000	1.000	1.000	.998	.989	.954	.851	.624	.264	.086	.004	8
9	1.000	1.000	1.000	1.000	1.000	1.000	.999	.994	.972	.893	.651	.401	.096	9

Table 1 (Continued)

(c) $n = 15$

							P							
a	0.01	0.05	0.10	0.20	0.30	0.40	0.50	0.60	0.70	0.80	0.90	0.95	0.99	a
0	.860	.463	.206	.035	.005	.000	.000	.000	.000	.000	.000	.000	.000	0
1	.990	.829	.549	.167	.035	.005	.000	.000	.000	.000	.000	.000	.000	1
2	1.000	.964	.816	.398	.127	.027	.004	.000	.000	.000	.000	.000	.000	2
3	1.000	.995	.944	.648	.297	.091	.018	.002	.000	.000	.000	.000	.000	3
4	1.000	.999	.987	.836	.515	.217	.059	.009	.001	.000	.000	.000	.000	4
5	1.000	1.000	.998	.939	.722	.403	.151	.034	.004	.000	.000	.000	.000	5
6	1.000	1.000	1.000	.982	.869	.610	.304	.095	.015	.001	.000	.000	.000	6
7	1.000	1.000	1.000	.996	.950	.787	.500	.213	.050	.004	.000	.000	.000	7
8	1.000	1.000	1.000	.999	.985	.905	.696	.390	.131	.018	.000	.000	.000	8
9	1.000	1.000	1.000	1.000	.996	.966	.849	.597	.278	.061	.002	.000	.000	9
10	1.000	1.000	1.000	1.000	.999	.991	.941	.783	.485	.164	.013	.001	.000	10
11	1.000	1.000	1.000	1.000	1.000	.998	.982	.909	.703	.352	.056	.005	.000	11
12	1.000	1.000	1.000	1.000	1.000	1.000	.996	.973	.873	.602	.184	.036	.000	12
13	1.000	1.000	1.000	1.000	1.000	1.000	1.000	.995	.965	.833	.451	.171	.010	13
14	1.000	1.000	1.000	1.000	1.000	1.000	1.000	1.000	.995	.965	.794	.537	.140	14

(d) $n = 20$

							P							
a	0.01	0.05	0.10	0.20	0.30	0.40	0.50	0.60	0.70	0.80	0.90	0.95	0.99	a
0	.818	.358	.122	.012	.001	.000	.000	.000	.000	.000	.000	.000	.000	0
1	.983	.736	.392	.069	.008	.001	.000	.000	.000	.000	.000	.000	.000	1
2	.999	.925	.677	.206	.035	.004	.000	.000	.000	.000	.000	.000	.000	2
3	1.000	.984	.867	.411	.107	.016	.001	.000	.000	.000	.000	.000	.000	3
4	1.000	.997	.957	.630	.238	.051	.006	.000	.000	.000	.000	.000	.000	4
5	1.000	1.000	.989	.804	.416	.126	.021	.002	.000	.000	.000	.000	.000	5
6	1.000	1.000	.998	.913	.608	.250	.058	.006	.000	.000	.000	.000	.000	6
7	1.000	1.000	1.000	.968	.772	.416	.132	.021	.001	.000	.000	.000	.000	7
8	1.000	1.000	1.000	.990	.887	.596	.252	.057	.005	.000	.000	.000	.000	8
9	1.000	1.000	1.000	.997	.952	.755	.412	.128	.017	.001	.000	.000	.000	9
10	1.000	1.000	1.000	.999	.983	.872	.588	.245	.048	.003	.000	.000	.000	10
11	1.000	1.000	1.000	1.000	.995	.943	.748	.404	.113	.010	.000	.000	.000	11
12	1.000	1.000	1.000	1.000	.999	.979	.868	.584	.228	.032	.000	.000	.000	12
13	1.000	1.000	1.000	1.000	1.000	.994	.942	.750	.392	.087	.002	.000	.000	13
14	1.000	1.000	1.000	1.000	1.000	.998	.979	.874	.584	.196	.011	.000	.000	14
15	1.000	1.000	1.000	1.000	1.000	1.000	.994	.949	.762	.370	.043	.003	.000	15
16	1.000	1.000	1.000	1.000	1.000	1.000	.999	.984	.893	.589	.133	.016	.000	16
17	1.000	1.000	1.000	1.000	1.000	1.000	1.000	.996	.965	.794	.323	.075	.001	17
18	1.000	1.000	1.000	1.000	1.000	1.000	1.000	.999	.992	.931	.608	.264	.017	18
19	1.000	1.000	1.000	1.000	1.000	1.000	1.000	1.000	.999	.988	.878	.642	.182	19

Table 1 (Continued)

(e) $n = 25$

| | | | | | | | P | | | | | | | |
a	0.01	0.05	0.10	0.20	0.30	0.40	0.50	0.60	0.70	0.80	0.90	0.95	0.99	a
0	.778	.277	.072	.004	.000	.000	.000	.000	.000	.000	.000	.000	.000	0
1	.974	.642	.271	.027	.002	.000	.000	.000	.000	.000	.000	.000	.000	1
2	.998	.873	.537	.098	.009	.000	.000	.000	.000	.000	.000	.000	.000	2
3	1.000	.966	.764	.234	.033	.002	.000	.000	.000	.000	.000	.000	.000	3
4	1.000	.993	.902	.421	.090	.009	.000	.000	.000	.000	.000	.000	.000	4
5	1.000	.999	.967	.617	.193	.029	.002	.000	.000	.000	.000	.000	.000	5
6	1.000	1.000	.991	.780	.341	.074	.007	.000	.000	.000	.000	.000	.000	6
7	1.000	1.000	.998	.891	.512	.154	.022	.001	.000	.000	.000	.000	.000	7
8	1.000	1.000	1.000	.953	.677	.274	.054	.004	.000	.000	.000	.000	.000	8
9	1.000	1.000	1.000	.983	.811	.425	.115	.013	.000	.000	.000	.000	.000	9
10	1.000	1.000	1.000	.994	.902	.586	.212	.034	.002	.000	.000	.000	.000	10
11	1.000	1.000	1.000	.998	.956	.732	.345	.078	.006	.000	.000	.000	.000	11
12	1.000	1.000	1.000	1.000	.983	.846	.500	.154	.017	.000	.000	.000	.000	12
13	1.000	1.000	1.000	1.000	.994	.922	.655	.268	.044	.002	.000	.000	.000	13
14	1.000	1.000	1.000	1.000	.998	.966	.788	.414	.098	.006	.000	.000	.000	14
15	1.000	1.000	1.000	1.000	1.000	.987	.885	.575	.189	.017	.000	.000	.000	15
16	1.000	1.000	1.000	1.000	1.000	.996	.946	.726	.323	.047	.000	.000	.000	16
17	1.000	1.000	1.000	1.000	1.000	.999	.978	.846	.488	.109	.002	.000	.000	17
18	1.000	1.000	1.000	1.000	1.000	1.000	.993	.926	.659	.220	.009	.000	.000	18
19	1.000	1.000	1.000	1.000	1.000	1.000	.998	.971	.807	.383	.033	.001	.000	19
20	1.000	1.000	1.000	1.000	1.000	1.000	1.000	.991	.910	.579	.098	.007	.000	20
21	1.000	1.000	1.000	1.000	1.000	1.000	1.000	.998	.967	.766	.236	.034	.000	21
22	1.000	1.000	1.000	1.000	1.000	1.000	1.000	1.000	.991	.902	.463	.127	.002	22
23	1.000	1.000	1.000	1.000	1.000	1.000	1.000	1.000	.998	.973	.729	.358	.026	23
24	1.000	1.000	1.000	1.000	1.000	1.000	1.000	1.000	1.000	.996	.928	.723	.222	24

Table 2
Table of e^{-x}

x	e^{-x}	x	e^{-x}	x	e^{-x}	x	e^{-x}
0.00	1.000000	2.60	.074274	5.10	.006097	7.60	.000501
0.10	.904837	2.70	.067206	5.20	.005517	7.70	.000453
0.20	.818731	2.80	.060810	5.30	.004992	7.80	.000410
0.30	.740818	2.90	.055023	5.40	.004517	7.90	.000371
0.40	.670320	3.00	.049787	5.50	.004087	8.00	.000336
0.50	.606531	3.10	.045049	5.60	.003698	8.10	.000304
0.60	.548812	3.20	.040762	5.70	.003346	8.20	.000275
0.70	.496585	3.30	.036883	5.80	.003028	8.30	.000249
0.80	.449329	3.40	.033373	5.90	.002739	8.40	.000225
0.90	.406570	3.50	.030197	6.00	.002479	8.50	.000204
1.00	.367879	3.60	.027324	6.10	.002243	8.60	.000184
1.10	.332871	3.70	.024724	6.20	.002029	8.70	.000167
1.20	.301194	3.80	.022371	6.30	.001836	8.80	.000151
1.30	.272532	3.90	.020242	6.40	.001661	8.90	.000136
1.40	.246597	4.00	.018316	6.50	.001503	9.00	.000123
1.50	.223130	4.10	.016573	6.60	.001360	9.10	.000112
1.60	.201897	4.20	.014996	6.70	.001231	9.20	.000101
1.70	.182684	4.30	.013569	6.80	.001114	9.30	.000091
1.80	.165299	4.40	.012277	6.90	.001008	9.40	.000083
1.90	.149569	4.50	.011109	7.00	.000912	9.50	.000075
2.00	.135335	4.60	.010052	7.10	.000825	9.60	.000068
2.10	.122456	4.70	.009095	7.20	.000747	9.70	.000061
2.20	.110803	4.80	.008230	7.30	.000676	9.80	.000056
2.30	.100259	4.90	.007447	7.40	.000611	9.90	.000050
2.40	.090718	5.00	.006738	7.50	.000553	10.00	.000045
2.50	.082085						

$$P(Y \le a) = \sum_{y=0}^{a} e^{-\lambda} \frac{\lambda^y}{y!}$$

Table 3
Poisson Probabilities

λ \ a	0	1	2	3	4	5	6	7	8	9
0.02	0.980	1.000								
0.04	0.961	0.999	1.000							
0.06	0.942	0.998	1.000							
0.08	0.923	0.997	1.000							
0.10	0.905	0.995	1.000							
0.15	0.861	0.990	0.999	1.000						
0.20	0.819	0.982	0.999	1.000						
0.25	0.779	0.974	0.998	1.000						
0.30	0.741	0.963	0.996	1.000						
0.35	0.705	0.951	0.994	1.000						
0.40	0.670	0.938	0.992	0.999	1.000					
0.45	0.638	0.925	0.989	0.999	1.000					
0.50	0.607	0.910	0.986	0.998	1.000					
0.55	0.577	0.894	0.982	0.988	1.000					
0.60	0.549	0.878	0.977	0.997	1.000					
0.65	0.522	0.861	0.972	0.996	0.999	1.000				
0.70	0.497	0.844	0.966	0.994	0.999	1.000				
0.75	0.472	0.827	0.959	0.993	0.999	1.000				
0.80	0.449	0.809	0.953	0.991	0.999	1.000				
0.85	0.427	0.791	0.945	0.989	0.998	1.000				
0.90	0.407	0.772	0.937	0.987	0.998	1.000				
0.95	0.387	0.754	0.929	0.981	0.997	1.000				
1.00	0.368	0.736	0.920	0.981	0.996	0.999	1.000			
1.1	0.333	0.699	0.900	0.974	0.995	0.999	1.000			
1.2	0.301	0.663	0.879	0.966	0.992	0.998	1.000			
1.3	0.273	0.627	0.857	0.957	0.989	0.998	1.000			
1.4	0.247	0.592	0.833	0.946	0.986	0.997	0.999	1.000		
1.5	0.223	0.558	0.809	0.934	0.981	0.996	0.999	1.000		
1.6	0.202	0.525	0.783	0.921	0.976	0.994	0.999	1.000		
1.7	0.183	0.493	0.757	0.907	0.970	0.992	0.998	1.000		
1.8	0.165	0.463	0.731	0.891	0.964	0.990	0.997	0.999	1.000	
1.9	0.150	0.434	0.704	0.875	0.956	0.987	0.997	0.999	1.000	
2.0	0.135	0.406	0.677	0.857	0.947	0.983	0.995	0.999	1.000	

Table 3
(Continued)

λ \ a	0	1	2	3	4	5	6	7	8	9
2.2	0.111	0.355	0.623	0.819	0.928	0.975	0.993	0.998	1.000	
2.4	0.091	0.308	0.570	0.779	0.904	0.964	0.988	0.997	0.999	1.000
2.6	0.074	0.267	0.518	0.736	0.877	0.951	0.983	0.995	0.999	1.000
2.8	0.061	0.231	0.469	0.692	0.848	0.935	0.976	0.992	0.998	0.999
3.0	0.050	0.199	0.423	0.647	0.815	0.916	0.966	0.988	0.996	0.999
3.2	0.041	0.171	0.380	0.603	0.781	0.895	0.955	0.983	0.994	0.998
3.4	0.033	0.147	0.340	0.558	0.744	0.871	0.942	0.977	0.992	0.997
3.6	0.027	0.126	0.303	0.515	0.706	0.844	0.927	0.969	0.988	0.996
3.8	0.022	0.107	0.269	0.473	0.668	0.816	0.909	0.960	0.984	0.994
4.0	0.018	0.092	0.238	0.433	0.629	0.785	0.889	0.949	0.979	0.992
4.2	0.015	0.078	0.210	0.395	0.590	0.753	0.867	0.936	0.972	0.989
4.4	0.012	0.066	0.185	0.359	0.551	0.720	0.844	0.921	0.964	0.985
4.6	0.010	0.056	0.163	0.326	0.513	0.686	0.818	0.905	0.955	0.980
4.8	0.008	0.048	0.143	0.294	0.476	0.651	0.791	0.887	0.944	0.975
5.0	0.007	0.040	0.125	0.265	0.440	0.616	0.762	0.867	0.932	0.968
5.2	0.006	0.034	0.109	0.238	0.406	0.581	0.732	0.845	0.918	0.960
5.4	0.005	0.029	0.095	0.213	0.373	0.546	0.702	0.822	0.903	0.951
5.6	0.004	0.024	0.082	0.191	0.342	0.512	0.670	0.797	0.886	0.941
5.8	0.003	0.021	0.072	0.170	0.313	0.478	0.638	0.771	0.867	0.929
6.0	0.002	0.017	0.062	0.151	0.285	0.446	0.606	0.744	0.847	0.916

λ	10	11	12	13	14	15	16
2.8	1.000						
3.0	1.000						
3.2	1.000						
3.4	0.999	1.000					
3.6	0.999	1.000					
3.8	0.998	0.999	1.000				
4.0	0.997	0.999	1.000				
4.2	0.996	0.999	1.000				
4.4	0.994	0.998	0.999	1.000			
4.6	0.992	0.997	0.999	1.000			
4.8	0.990	0.996	0.999	1.000			
5.0	0.986	0.995	0.998	0.999	1.000		
5.2	0.982	0.993	0.997	0.999	1.000		
5.4	0.977	0.990	0.996	0.999	1.000		
5.6	0.927	0.988	0.995	0.998	0.999	1.000	
5.8	0.965	0.984	0.993	0.997	0.999	1.000	
6.0	0.957	0.980	0.991	0.996	0.999	0.999	1.000

Table 3
(Continued)

λ \ a	0	1	2	3	4	5	6	7	8	9
6.2	0.002	0.015	0.054	0.134	0.259	0.414	0.574	0.716	0.826	0.902
6.4	0.002	0.012	0.046	0.119	0.235	0.384	0.542	0.687	0.803	0.886
6.6	0.001	0.010	0.040	0.105	0.213	0.355	0.511	0.658	0.780	0.869
6.8	0.001	0.009	0.034	0.093	0.192	0.327	0.480	0.628	0.755	0.850
7.0	0.001	0.007	0.030	0.082	0.173	0.301	0.450	0.599	0.729	0.830
7.2	0.001	0.006	0.025	0.072	0.156	0.276	0.420	0.569	0.703	0.810
7.4	0.001	0.005	0.022	0.063	0.140	0.253	0.392	0.539	0.676	0.788
7.6	0.001	0.004	0.019	0.055	0.125	0.231	0.365	0.510	0.648	0.765
7.8	0.000	0.004	0.016	0.048	0.112	0.210	0.338	0.481	0.620	0.741
8.0	0.000	0.003	0.014	0.042	0.100	0.191	0.313	0.453	0.593	0.717
8.5	0.000	0.002	0.009	0.030	0.074	0.150	0.256	0.386	0.523	0.653
9.0	0.000	0.001	0.006	0.021	0.055	0.116	0.207	0.324	0.456	0.587
9.5	0.000	0.001	0.004	0.015	0.040	0.089	0.165	0.269	0.392	0.522
10.0	0.000	0.000	0.003	0.010	0.029	0.067	0.130	0.220	0.333	0.458

	10	11	12	13	14	15	16	17	18	19
6.2	0.949	0.975	0.989	0.995	0.998	0.999	1.000			
6.4	0.939	0.969	0.986	0.994	0.997	0.999	1.000			
6.6	0.927	0.963	0.982	0.992	0.997	0.999	0.999	1.000		
6.8	0.915	0.955	0.978	0.990	0.996	0.998	0.999	1.000		
7.0	0.901	0.947	0.973	0.987	0.994	0.998	0.999	1.000		
7.2	0.887	0.937	0.967	0.984	0.993	0.997	0.999	0.999	1.000	
7.4	0.871	0.926	0.961	0.980	0.991	0.996	0.998	0.999	1.000	
7.6	0.854	0.915	0.954	0.976	0.989	0.995	0.998	0.999	1.000	
7.8	0.835	0.902	0.945	0.971	0.986	0.993	0.997	0.999	1.000	
8.0	0.816	0.888	0.936	0.966	0.983	0.992	0.996	0.998	0.999	1.000
8.5	0.763	0.849	0.909	0.949	0.973	0.986	0.993	0.997	0.999	0.999
9.0	0.706	0.803	0.876	0.926	0.959	0.978	0.989	0.995	0.998	0.999
9.5	0.645	0.752	0.836	0.898	0.940	0.967	0.982	0.991	0.996	0.998
10.0	0.583	0.697	0.792	0.864	0.917	0.951	0.973	0.986	0.993	0.997

	20	21	22
8.5	1.000		
9.0	1.000		
9.5	0.999	1.000	
10.0	0.998	0.999	1.000

Table 3
(Continued)

λ \ a	0	1	2	3	4	5	6	7	8	9
10.5	0.000	0.000	0.002	0.007	0.021	0.050	0.102	0.179	0.279	0.397
11.0	0.000	0.000	0.001	0.005	0.015	0.038	0.079	0.143	0.232	0.341
11.5	0.000	0.000	0.001	0.003	0.011	0.028	0.060	0.114	0.191	0.289
12.0	0.000	0.000	0.001	0.002	0.008	0.020	0.046	0.090	0.155	0.242
12.5	0.000	0.000	0.000	0.002	0.005	0.015	0.035	0.070	0.125	0.201
13.0	0.000	0.000	0.000	0.001	0.004	0.011	0.026	0.054	0.100	0.166
13.5	0.000	0.000	0.000	0.001	0.003	0.008	0.019	0.041	0.079	0.135
14.0	0.000	0.000	0.000	0.000	0.002	0.006	0.014	0.032	0.062	0.109
14.5	0.000	0.000	0.000	0.000	0.001	0.004	0.010	0.024	0.048	0.088
15.0	0.000	0.000	0.000	0.000	0.001	0.003	0.008	0.018	0.037	0.070

	10	11	12	13	14	15	16	17	18	19
10.5	0.521	0.639	0.742	0.825	0.888	0.932	0.960	0.978	0.988	0.994
11.0	0.460	0.579	0.689	0.781	0.854	0.907	0.944	0.968	0.982	0.991
11.5	0.402	0.520	0.633	0.733	0.815	0.878	0.924	0.954	0.974	0.986
12.0	0.347	0.462	0.576	0.682	0.772	0.844	0.899	0.937	0.963	0.979
12.5	0.297	0.406	0.519	0.628	0.725	0.806	0.869	0.916	0.948	0.969
13.0	0.252	0.353	0.463	0.573	0.675	0.764	0.835	0.890	0.930	0.957
13.5	0.211	0.304	0.409	0.518	0.623	0.718	0.798	0.861	0.908	0.942
14.0	0.176	0.260	0.358	0.464	0.570	0.669	0.756	0.827	0.883	0.923
14.5	0.145	0.220	0.311	0.413	0.518	0.619	0.711	0.790	0.853	0.901
15.0	0.118	0.185	0.268	0.363	0.466	0.568	0.664	0.749	0.819	0.875

	20	21	22	23	24	25	26	27	28	29
10.5	0.997	0.999	0.999	1.000						
11.0	0.995	0.998	0.999	1.000						
11.5	0.992	0.996	0.998	0.999	1.000					
12.0	0.988	0.994	0.997	0.999	0.999	1.000				
12.5	0.983	0.991	0.995	0.998	0.999	0.999	1.000			
13.0	0.975	0.986	0.992	0.996	0.998	0.999	1.000			
13.5	0.965	0.980	0.989	0.994	0.997	0.998	0.999	1.000		
14.0	0.952	0.971	0.983	0.991	0.995	0.997	0.999	0.999	1.000	
14.5	0.936	0.960	0.976	0.986	0.992	0.996	0.998	0.999	0.999	1.000
15.0	0.917	0.947	0.967	0.981	0.989	0.994	0.997	0.998	0.999	1.000

Table 3
(Continued)

λ \ a	4	5	6	7	8	9	10	11	12	13
16	0.000	0.001	0.004	0.010	0.022	0.043	0.077	0.127	0.193	0.275
17	0.000	0.001	0.002	0.005	0.013	0.026	0.049	0.085	0.135	0.201
18	0.000	0.000	0.001	0.003	0.007	0.015	0.030	0.055	0.092	0.143
19	0.000	0.000	0.001	0.002	0.004	0.009	0.018	0.035	0.061	0.098
20	0.000	0.000	0.000	0.001	0.002	0.005	0.011	0.021	0.039	0.066
21	0.000	0.000	0.000	0.000	0.001	0.003	0.006	0.013	0.025	0.043
22	0.000	0.000	0.000	0.000	0.001	0.002	0.004	0.008	0.015	0.028
23	0.000	0.000	0.000	0.000	0.000	0.001	0.002	0.004	0.009	0.017
24	0.000	0.000	0.000	0.000	0.000	0.000	0.001	0.003	0.005	0.011
25	0.000	0.000	0.000	0.000	0.000	0.000	0.001	0.001	0.003	0.006

	14	15	16	17	18	19	20	21	22	23
16	0.368	0.467	0.566	0.659	0.742	0.812	0.868	0.911	0.942	0.963
17	0.281	0.371	0.468	0.564	0.655	0.736	0.805	0.861	0.905	0.937
18	0.208	0.287	0.375	0.469	0.562	0.651	0.731	0.799	0.855	0.899
19	0.150	0.215	0.292	0.378	0.469	0.561	0.647	0.725	0.793	0.849
20	0.105	0.157	0.221	0.297	0.381	0.470	0.559	0.644	0.721	0.787
21	0.072	0.111	0.163	0.227	0.302	0.384	0.471	0.558	0.640	0.716
22	0.048	0.077	0.117	0.169	0.232	0.306	0.387	0.472	0.556	0.637
23	0.031	0.052	0.082	0.123	0.175	0.238	0.310	0.389	0.472	0.555
24	0.020	0.034	0.056	0.087	0.128	0.180	0.243	0.314	0.392	0.473
25	0.012	0.022	0.038	0.060	0.092	0.134	0.185	0.247	0.318	0.394

	24	25	26	27	28	29	30	31	32	33
16	0.978	0.987	0.993	0.996	0.998	0.999	0.999	1.000		
17	0.959	0.975	0.985	0.991	0.995	0.997	0.999	0.999	1.000	
18	0.932	0.955	0.972	0.983	0.990	0.994	0.997	0.998	0.999	1.000
19	0.893	0.927	0.951	0.969	0.980	0.988	0.993	0.996	0.998	0.999
20	0.843	0.888	0.922	0.948	0.966	0.978	0.987	0.992	0.995	0.997
21	0.782	0.838	0.883	0.917	0.944	0.963	0.976	0.985	0.991	0.994
22	0.712	0.777	0.832	0.877	0.913	0.940	0.959	0.973	0.983	0.989
23	0.635	0.708	0.772	0.827	0.873	0.908	0.936	0.956	0.971	0.981
24	0.554	0.632	0.704	0.768	0.823	0.868	0.904	0.932	0.953	0.969
25	0.473	0.553	0.629	0.700	0.763	0.818	0.863	0.900	0.929	0.950

	34	35	36	37	38	39	40	41	42	43
19	0.999	1.000								
20	0.999	0.999	1.000							
21	0.997	0.998	0.999	0.999	1.000					
22	0.994	0.996	0.998	0.999	0.999	1.000				
23	0.988	0.993	0.996	0.997	0.999	0.999	1.000			
24	0.979	0.987	0.992	0.995	0.997	0.998	0.999	0.999	1.000	
25	0.966	0.978	0.985	0.991	0.991	0.997	0.998	0.999	0.999	1.000

Tables

Table 4
Normal curve areas
Standard normal probability in right-hand
tail (for negative values of z areas are found
by symmetry)

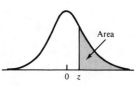

	Second decimal place of z									
z	.00	.01	.02	.03	.04	.05	.06	.07	.08	.09
0.0	.5000	.4960	.4920	.4880	.4840	.4801	.4761	.4721	.4681	.4641
0.1	.4602	.4562	.4522	.4483	.4443	.4404	.4364	.4325	.4286	.4247
0.2	.4207	.4168	.4129	.4090	.4052	.4013	.3974	.3936	.3897	.3859
0.3	.3821	.3783	.3745	.3707	.3669	.3632	.3594	.3557	.3520	.3483
0.4	.3446	.3409	.3372	.3336	.3300	.3264	.3228	.3192	.3156	.3121
0.5	.3085	.3050	.3015	.2981	.2946	.2912	.2877	.2843	.2810	.2776
0.6	.2743	.2709	.2676	.2643	.2611	.2578	.2546	.2514	.2483	.2451
0.7	.2420	.2389	.2358	.2327	.2296	.2266	.2236	.2206	.2177	.2148
0.8	.2119	.2090	.2061	.2033	.2005	.1977	.1949	.1922	.1894	.1867
0.9	.1841	.1814	.1788	.1762	.1736	.1711	.1685	.1660	.1635	.1611
1.0	.1587	.1562	.1539	.1515	.1492	.1469	.1446	.1423	.1401	.1379
1.1	.1357	.1335	.1314	.1292	.1271	.1251	.1230	.1210	.1190	.1170
1.2	.1151	.1131	.1112	.1093	.1075	.1056	.1038	.1020	.1003	.0985
1.3	.0968	.0951	.0934	.0918	.0901	.0885	.0869	.0853	.0838	.0823
1.4	.0808	.0793	.0778	.0764	.0749	.0735	.0722	.0708	.0694	.0681
1.5	.0668	.0655	.0643	.0630	.0618	.0606	.0594	.0582	.0571	.0559
1.6	.0548	.0537	.0526	.0516	.0505	.0495	.0485	.0475	.0465	.0455
1.7	.0446	.0436	.0427	.0418	.0409	.0401	.0392	.0384	.0375	.0367
1.8	.0359	.0352	.0344	.0336	.0329	.0322	.0314	.0307	.0301	.0294
1.9	.0287	.0281	.0274	.0268	.0262	.0256	.0250	.0244	.0239	.0233
2.0	.0228	.0222	.0217	.0212	.0207	.0202	.0197	.0192	.0188	.0183
2.1	.0179	.0174	.0170	.0166	.0162	.0158	.0154	.0150	.0146	.0143
2.2	.0139	.0136	.0132	.0129	.0125	.0122	.0119	.0116	.0113	.0110
2.3	.0107	.0104	.0102	.0099	.0096	.0094	.0091	.0089	.0087	.0084
2.4	.0082	.0080	.0078	.0075	.0073	.0071	.0069	.0068	.0066	.0064
2.5	.0062	.0060	.0059	.0057	.0055	.0054	.0052	.0051	.0049	.0048
2.6	.0047	.0045	.0044	.0043	.0041	.0040	.0039	.0038	.0037	.0036
2.7	.0035	.0034	.0033	.0032	.0031	.0030	.0029	.0028	.0027	.0026
2.8	.0026	.0025	.0024	.0023	.0023	.0022	.0021	.0021	.0020	.0019
2.9	.0019	.0018	.0017	.0017	.0016	.0016	.0015	.0015	.0014	.0014
3.0	.00135									
3.5	.000 233									
4.0	.000 031 7									
4.5	.000 003 40									
5.0	.000 000 287									

From R. E. Walpole, *Introduction to Statistics* (New York: Macmillan, 1968).

Table 5
Percentage points of the t distributions

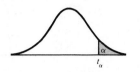

$t_{.100}$	$t_{.050}$	$t_{.025}$	$t_{.010}$	$t_{.005}$	d.f.
3.078	6.314	12.706	31.821	63.657	1
1.886	2.920	4.303	6.965	9.925	2
1.638	2.353	3.182	4.541	5.841	3
1.533	2.132	2.776	3.747	4.604	4
1.476	2.015	2.571	3.365	4.032	5
1.440	1.943	2.447	3.143	3.707	6
1.415	1.895	2.365	2.998	3.499	7
1.397	1.860	2.306	2.896	3.355	8
1.383	1.833	2.262	2.821	3.250	9
1.372	1.812	2.228	2.764	3.169	10
1.363	1.796	2.201	2.718	3.106	11
1.356	1.782	2.179	2.681	3.055	12
1.350	1.771	2.160	2.650	3.012	13
1.345	1.761	2.145	2.624	2.977	14
1.341	1.753	2.131	2.602	2.947	15
1.337	1.746	2.120	2.583	2.921	16
1.333	1.740	2.110	2.567	2.898	17
1.330	1.734	2.101	2.552	2.878	18
1.328	1.729	2.093	2.539	2.861	19
1.325	1.725	2.086	2.528	2.845	20
1.323	1.721	2.080	2.518	2.831	21
1.321	1.717	2.074	2.508	2.819	22
1.319	1.714	2.069	2.500	2.807	23
1.318	1.711	2.064	2.492	2.797	24
1.316	1.708	2.060	2.485	2.787	25
1.315	1.706	2.056	2.479	2.779	26
1.314	1.703	2.052	2.473	2.771	27
1.313	1.701	2.048	2.467	2.763	28
1.311	1.699	2.045	2.462	2.756	29
1.282	1.645	1.960	2.326	2.576	inf.

From "Table of Percentage Points of the *t*-Distribution." Computed by Maxine Merrington, *Biometrika*, Vol. 32 (1941), p. 300. Reproduced by permission of Professor E. S. Pearson.

Table 6
Percentage points of
the χ^2 distributions

d.f	$\chi^2_{0.995}$	$\chi^2_{0.990}$	$\chi^2_{0.975}$	$\chi^2_{0.950}$	$\chi^2_{0.900}$
1	0.0000393	0.0001571	0.0009821	0.0039321	0.0157908
2	0.0100251	0.0201007	0.0506356	0.102587	0.210720
3	0.0717212	0.114832	0.215795	0.351846	0.584375
4	0.206990	0.297110	0.484419	0.710721	1.063623
5	0.411740	0.554300	0.831211	1.145476	1.61031
6	0.675727	0.872085	1.237347	1.63539	2.20413
7	0.989265	1.239043	1.68987	2.16735	2.83311
8	1.344419	1.646482	2.17973	2.73264	3.48954
9	1.734926	2.087912	2.70039	3.32511	4.16816
10	2.15585	2.55821	3.24697	3.94030	4.86518
11	2.60321	3.05347	3.81575	4.57481	5.57779
12	3.07382	3.57056	4.40379	5.22603	6.30380
13	3.56503	4.10691	5.00874	5.89186	7.04150
14	4.07468	4.66043	5.62872	6.57063	7.78953
15	4.60094	5.22935	6.26214	7.26094	8.54675
16	5.14224	5.81221	6.90766	7.96164	9.31223
17	5.69724	6.40776	7.56418	8.67176	10.0852
18	6.26481	7.01491	8.23075	9.39046	10.8649
19	6.84398	7.63273	8.90655	10.1170	11.6509
20	7.43386	8.26040	9.59083	10.8508	12.4426
21	8.03366	8.89720	10.28293	11.5913	13.2396
22	8.64272	9.54249	10.9823	12.3380	14.0415
23	9.26042	10.19567	11.6885	13.0905	14.8479
24	9.88623	10.8564	12.4011	13.8484	15.6587
25	10.5197	11.5240	13.1197	14.6114	16.4734
26	11.1603	12.1981	13.8439	15.3791	17.2919
27	11.8076	12.8786	14.5733	16.1513	18.1138
28	12.4613	13.5648	15.3079	16.9279	18.9392
29	13.1211	14.2565	16.0471	17.7083	19.7677
30	13.7867	14.9535	16.7908	18.4926	20.5992
40	20.7065	22.1643	24.4331	26.5093	29.0505
50	27.9907	29.7067	32.3574	34.7642	37.6886
60	35.5346	37.4848	40.4817	43.1879	46.4589
70	43.2752	45.4418	48.7576	51.7393	55.3290
80	51.1720	53.5400	57.1532	60.3915	64.2778
90	59.1963	61.7541	65.6466	69.1260	73.2912
100	67.3276	70.0648	74.2219	77.9295	82.3581

Table 6
(Continued)

$\chi^2_{0.100}$	$\chi^2_{0.050}$	$\chi^2_{0.025}$	$\chi^2_{0.010}$	$\chi^2_{0.005}$	d.f.
2.70554	3.84146	5.02389	6.63490	7.87944	1
4.60517	5.99147	7.37776	9.21034	10.5966	2
6.25139	7.81473	9.34840	11.3449	12.8381	3
7.77944	9.48773	11.1433	13.2767	14.8602	4
9.23635	11.0705	12.8325	15.0863	16.7496	5
10.6446	12.5916	14.4494	16.8119	18.5476	6
12.0170	14.0671	16.0128	18.4753	20.2777	7
13.3616	15.5073	17.5346	20.0902	21.9550	8
14.6837	16.9190	19.0228	21.6660	23.5893	9
15.9871	18.3070	20.4831	23.2093	25.1882	10
17.2750	19.6751	21.9200	24.7250	26.7569	11
18.5494	21.0261	23.3367	26.2170	28.2995	12
19.8119	22.3621	24.7356	27.6883	29.8194	13
21.0642	23.6848	26.1190	29.1413	31.3193	14
22.3072	24.9958	27.4884	30.5779	32.8013	15
23.5418	26.2962	28.8454	31.9999	34.2672	16
24.7690	27.5871	30.1910	33.4087	35.7185	17
25.9894	28.8693	31.5264	34.8053	37.1564	18
27.2036	30.1435	32.8523	36.1908	38.5822	19
28.4120	31.4104	34.1696	37.5662	39.9968	20
29.6151	32.6705	35.4789	38.9321	41.4010	21
30.8133	33.9244	36.7807	40.2894	42.7956	22
32.0069	35.1725	38.0757	41.6384	44.1813	23
33.1963	36.4151	39.3641	42.9798	45.5585	24
34.3816	37.6525	40.6465	44.3141	46.9278	25
35.5631	38.8852	41.9232	45.6417	48.2899	26
36.7412	40.1133	43.1944	46.9630	49.6449	27
37.9159	41.3372	44.4607	48.2782	50.9933	28
39.0875	42.5569	45.7222	49.5879	52.3356	29
40.2560	43.7729	46.9792	50.8922	53.6720	30
51.8050	55.7585	59.3417	63.6907	66.7659	40
63.1671	67.5048	71.4202	76.1539	79.4900	50
74.3970	79.0819	83.2976	88.3794	91.9517	60
85.5271	90.5312	95.0231	100.425	104.215	70
96.5782	101.879	106.629	112.329	116.321	60
107.565	113.145	118.136	124.116	128.299	90
118.498	124.342	129.561	135.807	140.169	100

From "Tables of the Percentage Points of the χ^2-Distribution." *Biometrika*, Vol. 32 (1941), pp. 188–189, by Catherine M. Thompson. Reproduced by permission of Professor E. S. Pearson.

Table 7 Percentage points of the F distributions

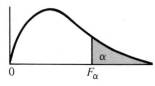

		Numerator d.f.								
Denominator d.f.	α	1	2	3	4	5	6	7	8	9
1	.100	39.86	49.50	53.59	55.83	57.24	58.20	58.91	59.44	59.86
	.050	161.4	199.5	215.7	224.6	230.2	234.0	236.8	238.9	240.5
	.025	647.8	799.5	864.2	899.6	921.8	937.1	948.2	956.7	963.3
	.010	4052	4999.5	5403	5625	5764	5859	5928	5982	6022
	.005	16211	20000	21615	22500	23056	23437	23715	23925	24091
2	.100	8.53	9.00	9.16	9.24	9.29	9.33	9.35	9.37	9.38
	.050	18.51	19.00	19.16	19.25	19.30	19.33	19.35	19.37	19.38
	.025	38.51	39.00	39.17	39.25	39.30	39.33	39.36	39.37	39.39
	.010	98.50	99.00	99.17	99.25	99.30	99.33	99.36	99.37	99.39
	.005	198.5	199.0	199.2	199.2	199.3	199.3	199.4	199.4	199.4
3	.100	5.54	5.46	5.39	5.34	5.31	5.28	5.27	5.25	5.24
	.050	10.13	9.55	9.28	9.12	9.01	8.94	8.89	8.85	8.81
	.025	17.44	16.04	15.44	15.10	14.88	14.73	14.62	14.54	14.47
	.010	34.12	30.82	29.46	28.71	28.24	27.91	27.67	27.49	27.35
	.005	55.55	49.80	47.47	46.19	45.39	44.84	44.43	44.13	43.88
4	.100	4.54	4.32	4.19	4.11	4.05	4.01	3.98	3.95	3.94
	.050	7.71	6.94	6.59	6.39	6.26	6.16	6.09	6.04	6.00
	.025	12.22	10.65	9.98	9.60	9.36	9.20	9.07	8.98	8.90
	.010	21.20	18.00	16.69	15.98	15.52	15.21	14.98	14.80	14.66
	.005	31.33	26.28	24.26	23.15	22.46	21.97	21.62	21.35	21.14
5	.100	4.06	3.78	3.62	3.52	3.45	3.40	3.37	3.34	3.32
	.050	6.61	5.79	5.41	5.19	5.05	4.95	4.88	4.82	4.77
	.025	10.01	8.43	7.76	7.39	7.15	6.98	6.85	6.76	6.68
	.010	16.26	13.27	12.06	11.39	10.97	10.67	10.46	10.29	10.16
	.005	22.78	18.31	16.53	15.56	14.94	14.51	14.20	13.96	13.77
6	.100	3.78	3.46	3.29	3.18	3.11	3.05	3.01	2.98	2.96
	.050	5.99	5.14	4.76	4.53	4.39	4.28	4.21	4.15	4.10
	.025	8.81	7.26	6.60	6.23	5.99	5.82	5.70	5.60	5.52
	.010	13.75	10.92	9.78	9.15	8.75	8.47	8.26	8.10	7.98
	.005	18.63	14.54	12.92	12.03	11.46	11.07	10.79	10.57	10.39
7	.100	3.59	3.26	3.07	2.96	2.88	2.83	2.78	2.75	2.72
	.050	5.59	4.74	4.35	4.12	3.97	3.87	3.79	3.73	3.68
	.025	8.07	6.54	5.89	5.52	5.29	5.12	4.99	4.90	4.82
	.010	12.25	9.55	8.45	7.85	7.46	7.19	6.99	6.84	6.72
	.005	16.24	12.40	10.88	10.05	9.52	9.16	8.89	8.68	8.51

Table 7   (Continued)

$$F_\alpha$$

Numerator d.f.											
10	*12*	*15*	*20*	*24*	*30*	*40*	*60*	*120*	∞	*α*	d.f.
60.19	60.71	61.22	61.74	62.00	62.26	62.53	62.79	63.06	63.33	.100	1
241.9	243.9	245.9	248.0	249.1	250.1	251.1	252.2	253.3	254.3	.050	
968.6	976.7	984.9	993.1	997.2	1001	1006	1010	1014	1018	.025	
6056	6106	6157	6209	6235	6261	6287	6313	6339	6366	.010	
24224	24426	24630	24836	24940	25044	25148	25253	25359	25465	.005	
9.39	9.41	9.42	9.44	9.45	9.46	9.47	9.47	9.48	9.49	.100	2
19.40	19.41	19.43	19.45	19.45	19.46	19.47	19.48	19.49	19.50	.050	
39.40	39.41	39.43	39.45	39.46	39.46	39.47	39.48	39.49	39.50	.025	
99.40	99.42	99.43	99.45	99.46	99.47	99.47	99.48	99.49	99.50	.010	
199.4	199.4	199.4	199.4	199.5	199.5	199.5	199.5	199.5	199.5	.005	
5.23	5.22	5.20	5.18	5.18	5.17	5.16	5.15	5.14	5.13	.100	3
8.79	8.74	8.70	8.66	8.64	8.62	8.59	8.57	8.55	8.53	.050	
14.42	14.34	14.25	14.17	14.12	14.08	14.04	13.99	13.95	13.90	.025	
27.23	27.05	26.87	26.69	26.60	26.50	26.41	26.32	26.22	26.13	.010	
43.69	43.39	43.08	42.78	42.62	42.47	42.31	42.15	41.99	41.83	.005	
3.92	3.90	3.87	3.84	3.83	3.82	3.80	3.79	3.78	3.76	.100	4
5.96	5.91	5.86	5.80	5.77	5.75	5.72	5.69	5.66	5.63	.050	
8.84	8.75	8.66	8.56	8.51	8.46	8.41	8.36	8.31	8.26	.025	
14.55	14.37	14.20	14.02	13.93	13.84	13.75	13.65	13.56	13.46	.010	
20.97	20.70	20.44	20.17	20.03	19.89	19.75	19.61	19.47	19.32	.005	
3.30	3.27	3.24	3.21	3.19	3.17	3.16	3.14	3.12	3.10	.100	5
4.74	4.68	4.62	4.56	4.53	4.50	4.46	4.43	4.40	4.36	.050	
6.62	6.52	6.43	6.33	6.28	6.23	6.18	6.12	6.07	6.02	.025	
10.05	9.89	9.72	9.55	9.47	9.38	9.29	9.20	9.11	9.02	.010	
13.62	13.38	13.15	12.90	12.78	12.66	12.53	12.40	12.27	12.14	.005	
2.94	2.90	2.87	2.84	2.82	2.80	2.78	2.76	2.74	2.72	.100	6
4.06	4.00	3.94	3.87	3.84	3.81	3.77	3.74	3.70	3.67	.050	
5.46	5.37	5.27	5.17	5.12	5.07	5.01	4.96	4.90	4.85	.025	
7.87	7.72	7.56	7.40	7.31	7.23	7.14	7.06	6.97	6.88	.010	
10.25	10.03	9.81	9.59	9.47	9.36	9.24	9.12	9.00	8.88	.005	
2.70	2.67	2.63	2.59	2.58	2.56	2.54	2.51	2.49	2.47	.100	7
3.64	3.57	3.51	3.44	3.41	3.38	3.34	3.30	3.27	3.23	.050	
4.76	4.67	4.57	4.47	4.42	4.36	4.31	4.25	4.20	4.14	.025	
6.62	6.47	6.31	6.16	6.07	5.99	5.91	5.82	5.74	5.65	.010	
8.38	8.18	7.97	7.75	7.65	7.53	7.42	7.31	7.19	7.08	.005	

Table 7   (Continued)

$$F_\alpha$$

Denominator d.f.	α	Numerator d.f.								
		1	2	3	4	5	6	7	8	9
8	.100	3.46	3.11	2.92	2.81	2.73	2.67	2.62	2.59	2.56
	.050	5.32	4.46	4.07	3.84	3.69	3.58	3.50	3.44	3.39
	.025	7.57	6.06	5.42	5.05	4.82	4.65	4.53	4.43	4.36
	.010	11.26	8.65	7.59	7.01	6.63	6.37	6.18	6.03	5.91
	.005	14.69	11.04	9.60	8.81	8.30	7.95	7.69	7.50	7.34
9	.100	3.36	3.01	2.81	2.69	2.61	2.55	2.51	2.47	2.44
	.050	5.12	4.26	3.86	3.63	3.48	3.37	3.29	3.23	3.18
	.025	7.21	5.71	5.08	4.72	4.48	4.32	4.20	4.10	4.03
	.010	10.56	8.02	6.99	6.42	6.06	5.80	5.61	5.47	5.35
	.005	13.61	10.11	8.72	7.96	7.47	7.13	6.88	6.69	6.54
10	.100	3.29	2.92	2.73	2.61	2.52	2.46	2.41	2.38	2.35
	.050	4.96	4.10	3.71	3.48	3.33	3.22	3.14	3.07	3.02
	.025	6.94	5.46	4.83	4.47	4.24	4.07	3.95	3.85	3.78
	.010	10.04	7.56	6.55	5.99	5.64	5.39	5.20	5.06	4.94
	.005	12.83	9.43	8.08	7.34	6.87	6.54	6.30	6.12	5.97
11	.100	3.23	2.86	2.66	2.54	2.45	2.39	2.34	2.30	2.27
	.050	4.84	3.98	3.59	3.36	3.20	3.09	3.01	2.95	2.90
	.025	6.72	5.26	4.63	4.28	4.04	3.88	3.76	3.66	3.59
	.010	9.65	7.21	6.22	5.67	5.32	5.07	4.89	4.74	4.63
	.005	12.23	8.91	7.60	6.88	6.42	6.10	5.86	5.68	5.54
12	.100	3.18	2.81	2.61	2.48	2.39	2.33	2.28	2.24	2.21
	.050	4.75	3.89	3.49	3.26	3.11	3.00	2.91	2.85	2.80
	.025	6.55	5.10	4.47	4.12	3.89	3.73	3.61	3.51	3.44
	.010	9.33	6.93	5.95	5.41	5.06	4.82	4.64	4.50	4.39
	.005	11.75	8.51	7.23	6.52	6.07	5.76	5.52	5.35	5.20
13	.100	3.14	2.76	2.56	2.43	2.35	2.28	2.23	2.20	2.16
	.050	4.67	3.81	3.41	3.18	3.03	2.92	2.83	2.77	2.71
	.025	6.41	4.97	4.35	4.00	3.77	3.60	3.48	3.39	3.31
	.010	9.07	6.70	5.74	5.21	4.86	4.62	4.44	4.30	4.19
	.005	11.37	8.19	6.93	6.23	5.79	5.48	5.25	5.08	4.94
14	.100	3.10	2.73	2.52	2.39	2.31	2.24	2.19	2.15	2.12
	.050	4.60	3.74	3.34	3.11	2.96	2.85	2.76	2.70	2.65
	.025	6.30	4.86	4.24	3.89	3.66	3.50	3.38	3.29	3.21
	.010	8.86	6.51	5.56	5.04	4.69	4.46	4.28	4.14	4.03
	.005	11.06	7.92	6.68	6.00	5.56	5.26	5.03	4.86	4.72

Table 7 (Continued)

$$F_\alpha$$

10	12	15	20	24	30	40	60	120	∞	α	d.f.
\multicolumn{11}{c}{*Numerator* d.f.}											

10	12	15	20	24	30	40	60	120	∞	α	d.f.
2.54	2.50	2.46	2.42	2.40	2.38	2.36	2.34	2.32	2.29	.100	8
3.35	3.28	3.22	3.15	3.12	3.08	3.04	3.01	2.97	2.93	.050	
4.30	4.20	4.10	4.00	3.95	3.89	3.84	3.78	3.73	3.67	.025	
5.81	5.67	5.52	5.36	5.28	5.20	5.12	5.03	4.95	4.86	.010	
7.21	7.01	6.81	6.61	6.50	6.40	6.29	6.18	6.06	5.95	.005	
2.42	2.38	2.34	2.30	2.28	2.25	2.23	2.21	2.18	2.16	.100	9
3.14	3.07	3.01	2.94	2.90	2.86	2.83	2.79	2.75	2.71	.050	
3.96	3.87	3.77	3.67	3.61	3.56	3.51	3.45	3.39	3.33	.025	
5.26	5.11	4.96	4.81	4.73	4.65	4.57	4.48	4.40	4.31	.010	
6.42	6.23	6.03	5.83	5.73	5.62	5.52	5.41	5.30	5.19	.005	
2.32	2.28	2.24	2.20	2.18	2.16	2.13	2.11	2.08	2.06	.100	10
2.98	2.91	2.85	2.74	2.77	2.70	2.66	2.62	2.58	2.54	.050	
3.72	3.62	3.52	3.42	3.37	3.31	3.26	3.20	3.14	3.08	.025	
4.85	4.71	4.56	4.41	4.33	4.25	4.17	4.08	4.00	3.91	.010	
5.85	5.66	5.47	5.27	5.17	5.07	4.97	4.86	4.75	4.64	.005	
2.25	2.21	2.17	2.12	2.10	2.08	2.05	2.03	2.00	1.97	.100	11
2.85	2.79	2.72	2.65	2.61	2.57	2.53	2.49	2.45	2.40	.050	
3.53	3.43	3.33	3.23	3.17	3.12	3.06	3.00	2.94	2.88	.025	
4.54	4.40	4.25	4.10	4.02	3.94	3.86	3.78	3.69	3.60	.010	
5.42	5.24	5.05	4.86	4.76	4.65	4.55	4.44	4.34	4.23	.005	
2.19	2.15	2.10	2.06	2.04	2.01	1.99	1.96	1.93	1.90	.100	12
2.75	2.69	2.62	2.54	2.51	2.47	2.43	2.38	2.34	2.30	.050	
3.37	3.28	3.18	3.07	3.02	2.96	2.91	2.85	2.79	2.72	.025	
4.30	4.16	4.01	3.86	3.78	3.70	3.62	3.54	3.45	3.36	.010	
5.09	4.91	4.72	4.53	4.43	4.33	4.23	4.12	4.01	3.90	.005	
2.14	2.10	2.05	2.01	1.98	1.96	1.93	1.90	1.88	1.85	.100	13
2.67	2.60	2.53	2.46	2.42	2.38	2.34	2.30	2.25	2.21	.050	
3.25	3.15	3.05	2.95	2.89	2.84	2.78	2.72	2.66	2.60	.025	
4.10	3.96	3.82	3.66	3.59	3.51	3.43	3.34	3.25	3.17	.010	
4.82	4.64	4.46	4.27	4.17	4.07	3.97	3.87	3.76	3.65	.005	
2.10	2.05	2.01	1.96	1.94	1.91	1.89	1.86	1.83	1.80	.100	14
2.60	2.53	2.46	2.39	2.35	2.31	2.27	2.22	2.18	2.13	.050	
3.15	3.05	2.95	2.84	2.79	2.73	2.67	2.61	2.55	2.49	.025	
3.94	3.80	3.66	3.51	3.43	3.35	3.27	3.18	3.09	3.00	.010	
4.60	4.43	4.25	4.06	3.96	3.86	3.76	3.66	3.55	3.44	.005	

Table 7 (Continued)

$$F_\alpha$$

Denominator d.f.	α	Numerator d.f. 1	2	3	4	5	6	7	8	9
15	.100	3.07	2.70	2.49	2.36	2.27	2.21	2.16	2.12	2.09
	.050	4.54	3.68	3.29	3.06	2.90	2.79	2.71	2.64	2.59
	.025	6.20	4.77	4.15	3.80	3.58	3.41	3.29	3.20	3.12
	.010	8.68	6.36	5.42	4.89	4.56	4.32	4.14	4.00	3.89
	.005	10.80	7.70	6.48	5.80	5.37	5.07	4.85	4.67	4.54
16	.100	3.05	2.67	2.46	2.33	2.24	2.18	2.13	2.09	2.06
	.050	4.49	3.63	3.24	3.01	2.85	2.74	2.66	2.59	2.54
	.025	6.12	4.69	4.08	3.73	3.50	3.34	3.22	3.12	3.05
	.010	8.53	6.23	5.29	4.77	4.44	4.20	4.03	3.89	3.78
	.005	10.58	7.51	6.30	5.64	5.21	4.91	4.69	4.52	4.38
17	.100	3.03	2.64	2.44	2.31	2.22	2.15	2.10	2.06	2.03
	.050	4.45	3.59	3.20	2.96	2.81	2.70	2.61	2.55	2.49
	.025	6.04	4.62	4.01	3.66	3.44	3.28	3.16	3.06	2.98
	.010	8.40	6.11	5.18	4.67	4.34	4.10	3.93	3.79	3.68
	.005	10.38	7.35	6.16	5.50	5.07	4.78	4.56	4.39	4.25
18	.100	3.01	2.62	2.42	2.29	2.20	2.13	2.08	2.04	2.00
	.050	4.41	3.55	3.16	2.93	2.77	2.66	2.58	2.51	2.46
	.025	5.98	4.56	3.95	3.61	3.38	3.22	3.10	3.01	2.93
	.010	8.29	6.01	5.09	4.58	4.25	4.01	3.84	3.71	3.60
	.005	10.22	7.21	6.03	5.37	4.96	4.66	4.44	4.28	4.14
19	.100	2.99	2.61	2.40	2.27	2.18	2.11	2.06	2.02	1.98
	.050	4.38	3.52	3.13	2.90	2.74	2.63	2.54	2.48	2.42
	.025	5.92	4.51	3.90	3.56	3.33	3.17	3.05	2.96	2.88
	.010	8.18	5.93	5.01	4.50	4.17	3.94	3.77	3.63	3.52
	.005	10.07	7.09	5.92	5.27	4.85	4.56	4.34	4.18	4.04
20	.100	2.97	2.59	2.38	2.25	2.16	2.09	2.04	2.00	1.96
	.050	4.35	3.49	3.10	2.87	2.71	2.60	2.51	2.45	2.39
	.025	5.87	4.46	3.86	3.51	3.29	3.13	3.01	2.91	2.84
	.010	8.10	5.85	4.94	4.43	4.10	3.87	3.70	3.56	3.46
	.005	9.94	6.99	5.82	5.17	4.76	4.47	4.26	4.09	3.96
21	.100	2.96	2.57	2.36	2.23	2.14	2.08	2.02	1.98	1.95
	.050	4.32	3.47	3.07	2.84	2.68	2.57	2.49	2.42	2.37
	.025	5.83	4.42	3.82	3.48	3.25	3.09	2.97	2.87	2.80
	.010	8.02	5.78	4.87	4.37	4.04	3.81	3.64	3.51	3.40
	.005	9.83	6.89	5.73	5.09	4.68	4.39	4.18	4.01	3.88

Table 7 (Continued)

$$F_\alpha$$

				Numerator d.f.								
10	*12*	*15*	*20*	*24*	*30*	*40*	*60*	*120*	∞	α		*d.f.*
2.06	2.02	1.97	1.92	1.90	1.87	1.85	1.82	1.79	1.76	.100		15
2.54	2.48	2.40	2.33	2.29	2.25	2.20	2.16	2.11	2.07	.050		
3.06	2.96	2.86	2.76	2.70	2.64	2.59	2.52	2.46	2.40	.025		
3.80	3.67	3.52	3.37	3.29	3.21	3.13	3.05	2.96	2.87	.010		
4.42	4.25	4.07	3.88	3.79	3.69	3.58	3.48	3.37	3.26	.005		
2.03	1.99	1.94	1.89	1.87	1.84	1.81	1.78	1.75	1.72	.100		16
2.49	2.42	2.35	2.28	2.24	2.19	2.15	2.11	2.06	2.01	.050		
2.99	2.89	2.79	2.68	2.63	2.57	2.51	2.45	2.38	2.32	.025		
3.69	3.55	3.41	3.26	3.18	3.10	3.02	2.93	2.84	2.75	.010		
4.27	4.10	3.92	3.73	3.64	3.54	3.44	3.33	3.22	3.11	.005		
2.00	1.96	1.91	1.86	1.84	1.81	1.78	1.75	1.72	1.69	.100		17
2.45	2.38	2.31	2.23	2.19	2.15	2.10	2.06	2.01	1.96	.050		
2.92	2.82	2.72	2.62	2.56	2.50	2.44	2.38	2.32	2.25	.025		
3.59	3.46	3.31	3.16	3.08	3.00	2.92	2.83	2.75	2.65	.010		
4.14	3.97	3.79	3.61	3.51	3.41	3.31	3.21	3.10	2.98	.005		
1.98	1.93	1.89	1.84	1.81	1.78	1.75	1.72	1.69	1.66	.100		18
2.41	2.34	2.27	2.19	2.15	2.11	2.06	2.02	1.97	1.92	.050		
2.87	2.77	2.67	2.56	2.50	2.44	2.38	2.32	2.26	2.19	.025		
3.51	3.37	3.23	3.08	3.00	2.92	2.84	2.75	2.66	2.57	.010		
4.03	3.86	3.68	3.50	3.40	3.30	3.20	3.10	2.99	2.87	.005		
1.96	1.91	1.86	1.81	1.79	1.76	1.73	1.70	1.67	1.63	.100		19
2.38	2.31	2.23	2.16	2.11	2.07	2.03	1.98	1.93	1.88	.050		
2.82	2.72	2.62	2.51	2.45	2.39	2.33	2.27	2.20	2.13	.025		
3.43	3.30	3.15	3.00	2.92	2.84	2.76	2.67	2.58	2.49	.010		
3.93	3.76	3.59	3.40	3.31	3.21	3.11	3.00	2.89	2.78	.005		
1.94	1.89	1.84	1.79	1.77	1.74	1.71	1.68	1.64	1.61	.100		20
2.35	2.28	2.20	2.12	2.08	2.04	1.99	1.95	1.90	1.84	.050		
2.77	2.68	2.57	2.46	2.41	2.35	2.29	2.22	2.16	2.09	.025		
3.37	3.23	3.09	2.94	2.86	2.78	2.69	2.61	2.52	2.42	.010		
3.85	3.68	3.50	3.32	3.22	3.12	3.02	2.92	2.81	2.69	.005		
1.92	1.87	1.83	1.78	1.75	1.72	1.69	1.66	1.62	1.59	.100		21
2.32	2.25	2.18	2.10	2.05	2.01	1.96	1.92	1.87	1.81	.050		
2.73	2.64	2.53	2.42	2.37	2.31	2.25	2.18	2.11	2.04	.025		
3.31	3.17	3.03	2.88	2.80	2.72	2.64	2.55	2.46	2.36	.010		
3.77	3.60	3.43	3.24	3.15	3.05	2.95	2.84	2.73	2.61	.005		

Table 7
(Continued)

F_α

Denominator d.f.	α	Numerator d.f.								
		1	*2*	*3*	*4*	*5*	*6*	*7*	*8*	*9*
22	.100	2.95	2.56	2.35	2.22	2.13	2.06	2.01	1.97	1.93
	.050	4.30	3.44	3.05	2.82	2.66	2.55	2.46	2.40	2.34
	.025	5.79	4.38	3.78	3.44	3.22	3.05	2.93	2.84	2.76
	.010	7.95	5.72	4.82	4.31	3.99	3.76	3.59	3.45	3.35
	.005	9.73	6.81	5.65	5.02	4.61	4.32	4.11	3.94	3.81
23	.100	2.94	2.55	2.34	2.21	2.11	2.05	1.99	1.95	1.92
	.050	4.28	3.42	3.03	2.80	2.64	2.53	2.44	2.37	2.32
	.025	5.75	4.35	3.75	3.41	3.18	3.02	2.90	2.81	2.73
	.010	7.88	5.66	4.76	4.26	3.94	3.71	3.54	3.41	3.30
	.005	9.63	6.73	5.58	4.95	4.54	4.26	4.05	3.88	3.75
24	.100	2.93	2.54	2.33	2.19	2.10	2.04	1.98	1.94	1.91
	.050	4.26	3.40	3.01	2.78	2.62	2.51	2.42	2.36	2.30
	.025	5.72	4.32	3.72	3.38	3.15	2.99	2.87	2.78	2.70
	.010	7.82	5.61	4.72	4.22	3.90	3.67	3.50	3.36	3.26
	.005	9.55	6.66	5.52	4.89	4.49	4.20	3.99	3.83	3.69
25	.100	2.92	2.53	2.32	2.18	2.09	2.02	1.97	1.93	1.89
	.050	4.24	3.39	2.99	2.76	2.60	2.49	2.40	2.34	2.28
	.025	5.69	4.29	3.69	3.35	3.13	2.97	2.85	2.75	2.68
	.010	7.77	5.57	4.68	4.18	3.85	3.63	3.46	3.32	3.22
	.005	9.48	6.60	5.46	4.84	4.43	4.15	3.94	3.78	3.64
26	.100	2.91	2.52	2.31	2.17	2.08	2.01	1.96	1.92	1.88
	.050	4.23	3.37	2.98	2.74	2.59	2.47	2.39	2.32	2.27
	.025	5.66	4.27	3.67	3.33	3.10	2.94	2.82	2.73	2.65
	.010	7.72	5.53	4.64	4.14	3.82	3.59	3.42	3.29	3.18
	.005	9.41	6.54	5.41	4.79	4.38	4.10	3.89	3.73	3.60
27	.100	2.90	2.51	2.30	2.17	2.07	2.00	1.95	1.91	1.87
	.050	4.21	3.35	2.96	2.73	2.57	2.46	2.37	2.31	2.25
	.025	5.63	4.24	3.65	3.31	3.08	2.92	2.80	2.71	2.63
	.010	7.68	5.49	4.60	4.11	3.78	3.56	3.39	3.26	3.15
	.005	9.34	6.49	5.36	4.74	4.34	4.06	3.85	3.69	3.56
28	.100	2.89	2.50	2.29	2.16	2.06	2.00	1.94	1.90	1.87
	.050	4.20	3.34	2.95	2.71	2.56	2.45	2.36	2.29	2.24
	.025	5.61	4.22	3.63	3.29	3.06	2.90	2.78	2.69	2.61
	.010	7.64	5.45	4.57	4.07	3.75	3.53	3.36	3.23	3.12
	.005	9.28	6.44	5.32	4.70	4.30	4.02	3.81	3.65	3.52

Table 7
(Continued)

F_α

				Numerator d.f.								
10	12	15	20	24	30	40	60	120	∞	α		d.f.
1.90	1.86	1.81	1.76	1.73	1.70	1.67	1.64	1.60	1.57	.100		22
2.30	2.23	2.15	2.07	2.03	1.98	1.94	1.89	1.84	1.78	.050		
2.70	2.60	2.50	2.39	2.33	2.27	2.21	2.14	2.08	2.00	.025		
3.26	3.12	2.98	2.83	2.75	2.67	2.58	2.50	2.40	2.31	.010		
3.70	3.54	3.36	3.18	3.08	2.98	2.88	2.77	2.66	2.55	.005		
1.89	1.84	1.80	1.74	1.72	1.69	1.66	1.62	1.59	1.55	.100		23
2.27	2.20	2.13	2.05	2.01	1.96	1.91	1.86	1.81	1.76	.050		
2.67	2.57	2.47	2.36	2.30	2.24	2.18	2.11	2.04	1.97	.025		
3.21	3.07	2.93	2.78	2.70	2.62	2.54	2.45	2.35	2.26	.010		
3.64	3.47	3.30	3.12	3.02	2.92	2.82	2.71	2.60	2.48	.005		
1.88	1.83	1.78	1.73	1.70	1.67	1.64	1.61	1.57	1.53	.100		24
2.25	2.18	2.11	2.03	1.98	1.94	1.89	1.84	1.79	1.73	.050		
2.64	2.54	2.44	2.33	2.27	2.21	2.15	2.08	2.01	1.94	.025		
3.17	3.03	2.89	2.74	2.66	2.58	2.49	2.40	2.31	2.21	.010		
3.59	3.42	3.25	3.06	2.97	2.87	2.77	2.66	2.55	2.43	.005		
1.87	1.82	1.77	1.72	1.69	1.66	1.63	1.59	1.56	1.52	.100		25
2.24	2.16	2.09	2.01	1.96	1.92	1.87	1.82	1.77	1.71	.050		
2.61	2.51	2.41	2.30	2.24	2.18	2.12	2.05	1.98	1.91	.025		
3.13	2.99	2.85	2.70	2.62	2.54	2.45	2.36	2.27	2.17	.010		
3.54	3.37	3.20	3.01	2.92	2.82	2.72	2.61	2.50	2.38	.005		
1.86	1.81	1.76	1.71	1.68	1.65	1.61	1.58	1.54	1.50	.100		26
2.22	2.15	2.07	1.99	1.95	1.90	1.85	1.80	1.75	1.69	.050		
2.59	2.49	2.39	2.28	2.22	2.16	2.09	2.03	1.95	1.88	.025		
3.09	2.96	2.81	2.66	2.58	2.50	2.42	2.33	2.23	2.13	.010		
3.49	3.33	3.15	2.97	2.87	2.77	2.67	2.56	2.45	2.33	.005		
1.85	1.80	1.75	1.70	1.67	1.64	1.60	1.57	1.53	1.49	.100		27
2.20	2.13	2.06	1.97	1.93	1.88	1.84	1.79	1.73	1.67	.050		
2.57	2.47	2.36	2.25	2.19	2.13	2.07	2.00	1.93	1.85	.025		
3.06	2.93	2.78	2.63	2.55	2.47	2.38	2.29	2.20	2.10	.010		
3.45	3.28	3.11	2.93	2.83	2.73	2.63	2.52	2.41	2.29	.005		
1.84	1.79	1.74	1.69	1.66	1.63	1.59	1.56	1.52	1.48	.100		28
2.19	2.12	2.04	1.96	1.91	1.87	1.82	1.77	1.71	1.65	.050		
2.55	2.45	2.34	2.23	2.17	2.11	2.05	1.98	1.91	1.83	.025		
3.03	2.90	2.75	2.60	2.52	2.44	2.35	2.26	2.17	2.06	.010		
3.41	3.25	3.07	2.89	2.79	2.69	2.59	2.48	2.37	2.25	.005		

Tables

Table 7
(Continued)

$$F_\alpha$$

Denominator d.f.	α	Numerator d.f.								
		1	*2*	*3*	*4*	*5*	*6*	*7*	*8*	*9*
29	.100	2.89	2.50	2.28	2.15	2.06	1.99	1.93	1.89	1.86
	.050	4.18	3.33	2.93	2.70	2.55	2.43	2.35	2.28	2.22
	.025	5.59	4.20	3.61	3.27	3.04	2.88	2.76	2.67	2.59
	.010	7.60	5.42	4.54	4.04	3.73	3.50	3.33	3.20	3.09
	.005	9.23	6.40	5.28	4.66	4.26	3.98	3.77	3.61	3.48
30	.100	2.88	2.49	2.28	2.14	2.05	1.98	1.93	1.88	1.85
	.050	4.17	3.32	2.92	2.69	2.53	2.42	2.33	2.27	2.21
	.025	5.57	4.18	3.59	3.25	3.03	2.87	2.75	2.65	2.57
	.010	7.56	5.39	4.51	4.02	3.70	3.47	3.30	3.17	3.07
	.005	9.18	6.35	5.24	4.62	4.23	3.95	3.74	3.58	3.45
40	.100	2.84	2.44	2.23	2.09	2.00	1.93	1.87	1.83	1.79
	.050	4.08	3.23	2.84	2.61	2.45	2.34	2.25	2.18	2.12
	.025	5.42	4.05	3.46	3.13	2.90	2.74	2.62	2.53	2.45
	.010	7.31	5.18	4.31	3.83	3.51	3.29	3.12	2.99	2.89
	.005	8.83	6.07	4.98	4.37	3.99	3.71	3.51	3.35	3.22
60	.100	2.79	2.39	2.18	2.04	1.95	1.87	1.82	1.77	1.74
	.050	4.00	3.15	2.76	2.53	2.37	2.25	2.17	2.10	2.04
	.025	5.29	3.93	3.34	3.01	2.79	2.63	2.51	2.41	2.33
	.010	7.08	4.98	4.13	3.65	3.34	3.12	2.95	2.82	2.72
	.005	8.49	5.79	4.73	4.14	3.76	3.49	3.29	3.13	3.01
120	.100	2.75	2.35	2.13	1.99	1.90	1.82	1.77	1.72	1.68
	.050	3.92	3.07	2.68	2.45	2.29	2.17	2.09	2.02	1.96
	.025	5.15	3.80	3.23	2.89	2.67	2.52	2.39	2.30	2.22
	.010	6.85	4.79	3.95	3.48	3.17	2.96	2.79	2.66	2.56
	.005	8.18	5.54	4.50	3.92	3.55	3.28	3.09	2.93	2.81
∞	.100	2.71	2.30	2.08	1.94	1.85	1.77	1.72	1.67	1.63
	.050	3.84	3.00	2.60	2.37	2.21	2.10	2.01	1.94	1.88
	.025	5.02	3.69	3.12	2.79	2.57	2.41	2.29	2.19	2.11
	.010	6.63	4.61	3.78	3.32	3.02	2.80	2.64	2.51	2.41
	.005	7.88	5.30	4.28	3.72	3.35	3.09	2.90	2.74	2.62

Table 7
(Continued)

$$F_\alpha$$

			Numerator d.f.								
10	12	15	20	24	30	40	60	120	∞	α	d.f.
1.83	1.78	1.73	1.68	1.65	1.62	1.58	1.55	1.51	1.47	.100	29
2.18	2.10	2.03	1.94	1.90	1.85	1.81	1.75	1.70	1.64	.050	
2.53	2.43	2.32	2.21	2.15	2.09	2.03	1.96	1.89	1.81	.025	
3.00	2.87	2.73	2.57	2.49	2.41	2.33	2.23	2.14	2.03	.010	
3.38	3.21	3.04	2.86	2.76	2.66	2.56	2.45	2.33	2.21	.005	
1.82	1.77	1.72	1.67	1.64	1.61	1.57	1.54	1.50	1.46	.100	30
2.16	2.09	2.01	1.93	1.89	1.84	1.79	1.74	1.68	1.62	.050	
2.51	2.41	2.31	2.20	2.14	2.07	2.01	1.94	1.87	1.79	.025	
2.98	2.84	2.70	2.55	2.47	2.39	2.30	2.21	2.11	2.01	.010	
3.34	3.18	3.01	2.82	2.73	2.63	2.52	2.42	2.30	2.18	.005	
1.76	1.71	1.66	1.61	1.57	1.54	1.51	1.47	1.42	1.38	.100	40
2.08	2.00	1.92	1.84	1.79	1.74	1.69	1.64	1.58	1.51	.050	
2.39	2.29	2.18	2.07	2.01	1.94	1.88	1.80	1.72	1.64	.025	
2.80	2.66	2.52	2.37	2.29	2.20	2.11	2.02	1.92	1.80	.010	
3.12	2.95	2.78	2.60	2.50	2.40	2.30	2.18	2.06	1.93	.005	
1.71	1.66	1.60	1.54	1.51	1.48	1.44	1.40	1.35	1.29	.100	60
1.99	1.92	1.84	1.75	1.70	1.65	1.59	1.53	1.47	1.39	.050	
2.27	2.17	2.06	1.94	1.88	1.82	1.74	1.67	1.58	1.48	.025	
2.63	2.50	2.35	2.20	2.12	2.03	1.94	1.84	1.73	1.60	.010	
2.90	2.74	2.57	2.39	2.29	2.19	2.08	1.96	1.83	1.69	.005	
1.65	1.60	1.55	1.48	1.45	1.41	1.37	1.32	1.26	1.19	.100	120
1.91	1.83	1.75	1.66	1.61	1.55	1.50	1.43	1.35	1.25	.050	
2.16	2.05	1.94	1.82	1.76	1.69	1.61	1.53	1.43	1.31	.025	
2.47	2.34	2.19	2.03	1.95	1.86	1.76	1.66	1.53	1.38	.010	
2.71	2.54	2.37	2.19	2.09	1.98	1.87	1.75	1.61	1.43	.005	
1.60	1.55	1.49	1.42	1.38	1.34	1.30	1.24	1.17	1.00	.100	∞
1.83	1.75	1.67	1.57	1.52	1.46	1.39	1.32	1.22	1.00	.050	
2.05	1.94	1.83	1.71	1.64	1.57	1.48	1.39	1.27	1.00	.025	
2.32	2.18	2.04	1.88	1.79	1.70	1.59	1.47	1.32	1.00	.010	
2.52	2.36	2.19	2.00	1.90	1.79	1.67	1.53	1.36	1.00	.005	

Tables

Table 8
Distribution function
of U

$P(U \leq U_0)$; U_0 is the argument; $n_1 \leq n_2$;
$3 \leq n_2 \leq 10$.

$n_2 = 3$

U_0	n_1		
	1	2	3
0	.25	.10	.05
1	.50	.20	.10
2		.40	.20
3		.60	.35
4			.50

$n_2 = 4$

U_0	n_1			
	1	2	3	4
0	.2000	.0667	.0286	.0143
1	.4000	.1333	.0571	.0286
2	.6000	.2667	.1143	.0571
3		.4000	.2000	.1000
4		.6000	.3143	.1714
5			.4286	.2429
6			.5714	.3429
7				.4429
8				.5571

Table 8
(Continued)

$n_2 = 5$

U_0	\multicolumn{5}{c}{n_1}
	1	2	3	4	5
0	.1667	.0476	.0179	.0079	.0040
1	.3333	.0952	.0357	.0159	.0079
2	.5000	.1905	.0714	.0317	.0159
3		.2857	.1250	.0556	.0278
4		.4286	.1964	.0952	.0476
5		.5714	.2857	.1429	.0754
6			.3929	.2063	.1111
7			.5000	.2778	.1548
8				.3651	.2103
9				.4524	.2738
10				.5476	.3452
11					.4206
12					.5000

$n_2 = 6$

U_0	\multicolumn{6}{c}{n_1}
	1	2	3	4	5	6
0	.1429	.0357	.0119	.0048	.0022	.0011
1	.2857	.0714	.0238	.0095	.0043	.0022
2	.4286	.1429	.0476	.0190	.0087	.0043
3	.5714	.2143	.0833	.0333	.0152	.0076
4		.3214	.1310	.0571	.0260	.0130
5		.4286	.1905	.0857	.0411	.0206
6		.5714	.2738	.1286	.0628	.0325
7			.3571	.1762	.0887	.0465
8			.4524	.2381	.1234	.0660
9			.5476	.3048	.1645	.0898
10				.3810	.2143	.1201
11				.4571	.2684	.1548
12				.5429	.3312	.1970
13					.3961	.2424
14					.4654	.2944
15					.5346	.3496
16						.4091
17						.4686
18						.5314

Table 8
(Continued)

$n_2 = 7$

U_0	n_1						
	1	*2*	*3*	*4*	*5*	*6*	*7*
0	.1250	.0278	.0083	.0030	.0013	.0006	.0003
1	.2500	.0556	.0167	.0061	.0025	.0012	.0006
2	.3750	.1111	.0333	.0121	.0051	.0023	.0012
3	.5000	.1667	.0583	.0212	.0088	.0041	.0020
4		.2500	.0917	.0364	.0152	.0070	.0035
5		.3333	.1333	.0545	.0240	.0111	.0055
6		.4444	.1917	.0818	.0366	.0175	.0087
7		.5556	.2583	.1152	.0530	.0256	.0131
8			.3333	.1576	.0745	.0367	.0189
9			.4167	.2061	.1010	.0507	.0265
10			.5000	.2636	.1338	.0688	.0364
11				.3242	.1717	.0903	.0487
12				.3939	.2159	.1171	.0641
13				.4636	.2652	.1474	.0825
14				.5364	.3194	.1830	.1043
15					.3775	.2226	.1297
16					.4381	.2669	.1588
17					.5000	.3141	.1914
18						.3654	.2279
19						.4178	.2675
20						.4726	.3100
21						.5274	.3552
22							.4024
23							.4508
24							.5000

Table 8
(Continued)

$n_2 = 8$

U_0					n_1			
	1	*2*	*3*	*4*	*5*	*6*	*7*	*8*
0	.1111	.0222	.0061	.0020	.0008	.0003	.0002	.0001
1	.2222	.0444	.0121	.0040	.0016	.0007	.0003	.0002
2	.3333	.0889	.0242	.0081	.0031	.0013	.0006	.0003
3	.4444	.1333	.0424	.0141	.0054	.0023	.0011	.0005
4	.5556	.2000	.0667	.0242	.0093	.0040	.0019	.0009
5		.2667	.0970	.0364	.0148	.0063	.0030	.0015
6		.3556	.1394	.0545	.0225	.0100	.0047	.0023
7		.4444	.1879	.0768	.0326	.0147	.0070	.0035
8		.5556	.2485	.1071	.0466	.0213	.0103	.0052
9			.3152	.1414	.0637	.0296	.0145	.0074
10			.3879	.1838	.0855	.0406	.0200	.0103
11			.4606	.2303	.1111	.0539	.0270	.0141
12			.5394	.2848	.1422	.0709	.0361	.0190
13				.3414	.1772	.0906	.0469	.0249
14				.4040	.2176	.1142	.0603	.0325
15				.4667	.2618	.1412	.0760	.0415
16				.5333	.3108	.1725	.0946	.0524
17					.3621	.2068	.1159	.0652
18					.4165	.2454	.1405	.0803
19					.4716	.2864	.1678	.0974
20					.5284	.3310	.1984	.1172
21						.3773	.2317	.1393
22						.4259	.2679	.1641
23						.4749	.3063	.1911
24						.5251	.3472	.2209
25							.3894	.2527
26							.4333	.2869
27							.4775	.3227
28							.5225	.3605
29								.3992
30								.4392
31								.4796
32								.5204

Table 8
(Continued)

$n_2 = 9$

U_0	n_1								
	1	2	3	4	5	6	7	8	9
0	.1000	.0182	.0045	.0014	.0005	.0002	.0001	.0000	.0000
1	.2000	.0364	.0091	.0028	.0010	.0004	.0002	.0001	.0000
2	.3000	.0727	.0182	.0056	.0020	.0008	.0003	.0002	.0001
3	.4000	.1091	.0318	.0098	.0035	.0014	.0006	.0003	.0001
4	.5000	.1636	.0500	.0168	.0060	.0024	.0010	.0005	.0002
5		.2182	.0727	.0252	.0095	.0038	.0017	.0008	.0004
6		.2909	.1045	.0378	.0145	.0060	.0026	.0012	.0006
7		.3636	.1409	.0531	.0210	.0088	.0039	.0019	.0009
8		.4545	.1864	.0741	.0300	.0128	.0058	.0028	.0014
9		.5455	.2409	.0993	.0415	.0180	.0082	.0039	.0020
10			.3000	.1301	.0559	.0248	.0115	.0056	.0028
11			.3636	.1650	.0734	.0332	.0156	.0076	.0039
12			.4318	.2070	.0949	.0440	.0209	.0103	.0053
13			.5000	.2517	.1199	.0567	.0274	.0137	.0071
14				.3021	.1489	.0723	.0356	.0180	.0094
15				.3552	.1818	.0905	.0454	.0232	.0122
16				.4126	.2188	.1119	.0571	.0296	.0157
17				.4699	.2592	.1361	.0708	.0372	.0200
18				.5301	.3032	.1638	.0869	.0464	.0252
19					.3497	.1942	.1052	.0570	.0313
20					.3986	.2280	.1261	.0694	.0385
21					.4491	.2643	.1496	.0836	.0470
22					.5000	.3035	.1755	.0998	.0567
23						.3445	.2039	.1179	.0680
24						.3878	.2349	.1383	.0807
25						.4320	.2680	.1606	.0951
26						.4773	.3032	.1852	.1112
27						.5227	.3403	.2117	.1290
28							.3788	.2404	.1487
29							.4185	.2707	.1701
30							.4591	.3029	.1933
31							.5000	.3365	.2181
32								.3715	.2447
33								.4074	.2729
34								.4442	.3024
35								.4813	.3332
36								.5187	.3652
37									.3981
38									.4317
39									.4657
40									.5000

Table 8
(Continued)

$n_2 = 10$

U_0	n_1									
	1	*2*	*3*	*4*	*5*	*6*	*7*	*8*	*9*	*10*
0	.0909	.0152	.0035	.0010	.0003	.0001	.0001	.0000	.0000	.0000
1	.1818	.0303	.0070	.0020	.0007	.0002	.0001	.0000	.0000	.0000
2	.2727	.0606	.0140	.0040	.0013	.0005	.0002	.0001	.0000	.0000
3	.3636	.0909	.0245	.0070	.0023	.0009	.0004	.0002	.0001	.0000
4	.4545	.1364	.0385	.0120	.0040	.0015	.0006	.0003	.0001	.0001
5	.5455	.1818	.0559	.0180	.0063	.0024	.0010	.0004	.0002	.0001
6		.2424	.0804	.0270	.0097	.0037	.0015	.0007	.0003	.0002
7		.3030	.1084	.0380	.0140	.0055	.0023	.0010	.0005	.0002
8		.3788	.1434	.0529	.0200	.0080	.0034	.0015	.0007	.0004
9		.4545	.1853	.0709	.0276	.0112	.0048	.0022	.0011	.0005
10		.5455	.2343	.0939	.0376	.0156	.0068	.0031	.0015	.0008
11			.2867	.1199	.0496	.0210	.0093	.0043	.0021	.0010
12			.3462	.1518	.0646	.0280	.0125	.0058	.0028	.0014
13			.4056	.1868	.0823	.0363	.0165	.0078	.0038	.0019
14			.4685	.2268	.1032	.0467	.0215	.0103	.0051	.0026
15			.5315	.2697	.1272	.0589	.0277	.0133	.0066	.0034
16				.3177	.1548	.0736	.0351	.0171	.0086	.0045
17				.3666	.1855	.0903	.0439	.0217	.0110	.0057
18				.4196	.2198	.1099	.0544	.0273	.0140	.0073
19				.4725	.2567	.1317	.0665	.0338	.0175	.0093
20				.5275	.2970	.1566	.0806	.0416	.0217	.0116
21					.3393	.1838	.0966	.0506	.0267	.0144
22					.3839	.2139	.1148	.0610	.0326	.0177
23					.4296	.2461	.1349	.0729	.0394	.0216
24					.4765	.2811	.1574	.0864	.0474	.0262
25					.5235	.3177	.1819	.1015	.0564	.0315
26						.3564	.2087	.1185	.0667	.0376
27						.3962	.2374	.1371	.0782	.0446
28						.4374	.2681	.1577	.0912	.0526
29						.4789	.3004	.1800	.1055	.0615
30						.5211	.3345	.2041	.1214	.0716
31							.3698	.2299	.1388	.0827
32							.4063	.2574	.1577	.0952
33							.4434	.2863	.1781	.1088
34							.4811	.3167	.2001	.1237
35							.5189	.3482	.2235	.1399
36								.3809	.2483	.1575
37								.4143	.2745	.1763
38								.4484	.3019	.1965
39								.4827	.3304	.2179

Table 8
(Continued)

$n_2 = 10$

					n_1					
U_0	1	2	3	4	5	6	7	8	9	10
40								.5173	.3598	.2406
41									.3901	.2644
42									.4211	.2894
43									.4524	.3153
44									.4841	.3421
45									.5159	.3697
46										.3980
47										.4267
48										.4559
49										.4853
50										.5147

Computed by M. Pagano, Department of Statistics, University of Florida.

Table 9
Critical values of T in the Wilcoxon matched-pairs, signed-ranks test; n = 5(1)50

One-sided	Two-sided	n = 5	n = 6	n = 7	n = 8	n = 9	n = 10
P = .05	P = .10	1	2	4	6	8	11
P = .025	P = .05		1	2	4	6	8
P = .01	P = .02			0	2	3	5
P = .005	P = .01				0	2	3

One-sided	Two-sided	n = 11	n = 12	n = 13	n = 14	n = 15	n = 16
P = .05	P = .10	14	17	21	26	30	36
P = .025	P = .05	11	14	17	21	25	30
P = .01	P = .02	7	10	13	16	20	24
P = .005	P = .01	5	7	10	13	16	19

One-sided	Two-sided	n = 17	n = 18	n = 19	n = 20	n = 21	n = 22
P = .05	P = .10	41	47	54	60	68	75
P = .025	P = .05	35	40	46	52	59	66
P = .01	P = .02	28	33	38	43	49	56
P = .005	P = .01	23	28	32	37	43	49

One-sided	Two-sided	n = 23	n = 24	n = 25	n = 26	n = 27	n = 28
P = .05	P = .10	83	92	101	110	120	130
P = .025	P = .05	73	81	90	98	107	117
P = .01	P = .02	62	69	77	85	93	102
P = .005	P = .01	55	68	68	76	84	92

Table 9
(Continued)

One-sided	Two-sided	n = 29	n = 30	n = 31	n = 32	n = 33	n = 34
P = .05	P = .10	141	152	163	175	188	201
P = .025	P = .05	127	137	148	159	171	183
P = .01	P = .02	111	120	130	141	151	162
P = .005	P = .01	100	109	118	128	138	149

One-sided	Two-sided	n = 35	n = 36	n = 37	n = 38	n = 39	
P = .05	P = .10	214	228	242	256	271	
P = .025	P = .05	195	208	222	235	250	
P = .01	P = .02	174	186	198	211	224	
P = .005	P = .01	160	171	183	195	208	

One-sided	Two-sided	n = 40	n = 41	n = 42	n = 43	n = 44	n = 45
P = .05	P = .10	287	303	319	336	353	371
P = .025	P = .05	264	279	295	311	327	344
P = .01	P = .02	238	252	267	281	297	313
P = .005	P = .01	221	234	248	262	277	292

One-sided	Two-sided	n = 46	n = 47	n = 48	n = 49	n = 50	
P = .05	P = .10	389	408	427	446	466	
P = .025	P = .05	361	379	397	415	434	
P = .01	P = .02	329	345	362	380	398	
P = .005	P = .01	307	323	339	356	373	

From "Some Rapid Approximate Statistical Procedures" (1964), 28, F. Wilcoxon and R. A. Wilcox. Reproduced with the kind permission of R. A. Wilcox and the Lederle Laboratories.

Table 10
Distribution of the total number of runs R in samples of size (n_1, n_2); $P(R \le a)$

					a				
(n_1, n_2)	2	3	4	5	6	7	8	9	10
(2, 3)	.200	.500	.900	1.000					
(2, 4)	.133	.400	.800	1.000					
(2, 5)	.095	.333	.714	1.000					
(2, 6)	.071	.286	.643	1.000					
(2, 7)	.056	.250	.583	1.000					
(2, 8)	.044	.222	.533	1.000					
(2, 9)	.036	.200	.491	1.000					
(2, 10)	.030	.182	.455	1.000					
(3, 3)	.100	.300	.700	.900	1.000				
(3, 4)	.057	.200	.543	.800	.971	1.000			
(3, 5)	.036	.143	.429	.714	.929	1.000			
(3, 6)	.024	.107	.345	.643	.881	1.000			
(3, 7)	.017	.083	.283	.583	.833	1.000			
(3, 8)	.012	.067	.236	.533	.788	1.000			
(3, 9)	.009	.055	.200	.491	.745	1.000			
(3, 10)	.007	.045	.171	.455	.706	1.000			
(4, 4)	.029	.114	.371	.629	.886	.971	1.000		
(4, 5)	.016	.071	.262	.500	.786	.929	.992	1.000	
(4, 6)	.010	.048	.190	.405	.690	.881	.976	1.000	
(4, 7)	.006	.033	.142	.333	.606	.833	.954	1.000	
(4, 8)	.004	.024	.109	.279	.533	.788	.929	1.000	
(4, 9)	.003	.018	.085	.236	.471	.745	.902	1.000	
(4, 10)	.002	.014	.068	.203	.419	.706	.874	1.000	
(5, 5)	.008	.040	.167	.357	.643	.833	.960	.992	1.000
(5, 6)	.004	.024	.110	.262	.522	.738	.911	.976	.998
(5, 7)	.003	.015	.076	.197	.424	.652	.854	.955	.992
(5, 8)	.002	.010	.054	.152	.347	.576	.793	.929	.984
(5, 9)	.001	.007	.039	.119	.287	.510	.734	.902	.972
(5, 10)	.001	.005	.029	.095	.239	.455	.678	.874	.958
(6, 6)	.002	.013	.067	.175	.392	.608	.825	.933	.987
(6, 7)	.001	.008	.043	.121	.296	.500	.733	.879	.966
(6, 8)	.001	.005	.028	.086	.226	.413	.646	.821	.937
(6, 9)	.000	.003	.019	.063	.175	.343	.566	.762	.902
(6, 10)	.000	.002	.013	.047	.137	.288	.497	.706	.864
(7, 7)	.001	.004	.025	.078	.209	.383	.617	.791	.922
(7, 8)	.000	.002	.015	.051	.149	.296	.514	.704	.867
(7, 9)	.000	.001	.010	.035	.108	.231	.427	.622	.806
(7, 10)	.000	.001	.006	.024	.080	.182	.355	.549	.743
(8, 8)	.000	.001	.009	.032	.100	.214	.405	.595	.786
(8, 9)	.000	.001	.005	.020	.069	.157	.319	.500	.702
(8, 10)	.000	.000	.003	.013	.048	.117	.251	.419	.621
(9, 9)	.000	.000	.003	.012	.044	.109	.238	.399	.601
(9, 10)	.000	.000	.002	.008	.029	.077	.179	.319	.510
(10, 10)	.000	.000	.001	.004	.019	.051	.128	.242	.414

Table 10
(Continued)

(n_1, n_2)	11	12	13	14	15	16	17	18	19	20
(2, 3)										
(2, 4)										
(2, 5)										
(2, 6)										
(2, 7)										
(2, 8)										
(2, 9)										
(2, 10)										
(3, 3)										
(3, 4)										
(3, 5)										
(3, 6)										
(3, 7)										
(3, 8)										
(3, 9)										
(3, 10)										
(4, 4)										
(4, 5)										
(4, 6)										
(4, 7)										
(4, 8)										
(4, 9)										
(4, 10)										
(5, 5)										
(5, 6)	1.000									
(5, 7)	1.000									
(5, 8)	1.000									
(5, 9)	1.000									
(5, 10)	1.000									
(6, 6)	.998	1.000								
(6, 7)	.992	.999	1.000							
(6, 8)	.984	.998	1.000							
(6, 9)	.972	.994	1.000							
(6, 10)	.958	.990	1.000							
(7, 7)	.975	.996	.999	1.000						
(7, 8)	.949	.988	.998	1.000	1.000					
(7, 9)	.916	.975	.994	.999	1.000					
(7, 10)	.879	.957	.990	.998	1.000					
(8, 8)	.900	.968	.991	.999	1.000	1.000				
(8, 9)	.843	.939	.980	.996	.999	1.000	1.000			
(8, 10)	.782	.903	.964	.990	.998	1.000	1.000			
(9, 9)	.762	.891	.956	.988	.997	1.000	1.000	1.000		
(9, 10)	.681	.834	.923	.974	.992	.999	1.000	1.000	1.000	
(10, 10)	.586	.758	.872	.949	.981	.996	.999	1.000	1.000	1.000

From "Tables for Testing Randomness of Grouping in a Sequence of Alternatives," C. Eisenhart and F. Swed, *Annals of Mathematical Statistics*. Volume 14 (1943). Reproduced with the kind permission of the Editor, *Annals of Mathematical Statistics*.

Table 11
Critical values of
Spearman's rank
correlation coefficient

n	$\alpha = .05$	$\alpha = .025$	$\alpha = .01$	$\alpha = .005$
5	0.900	—	—	—
6	0.829	0.886	0.943	—
7	0.714	0.786	0.893	—
8	0.643	0.738	0.833	0.881
9	0.600	0.683	0.783	0.833
10	0.564	0.648	0.745	0.794
11	0.523	0.623	0.736	0.818
12	0.497	0.591	0.703	0.780
13	0.475	0.566	0.673	0.745
14	0.457	0.545	0.646	0.716
15	0.441	0.525	0.623	0.689
16	0.425	0.507	0.601	0.666
17	0.412	0.490	0.582	0.645
18	0.399	0.476	0.564	0.625
19	0.388	0.462	0.549	0.608
20	0.377	0.450	0.534	0.591
21	0.368	0.438	0.521	0.576
22	0.359	0.428	0.508	0.562
23	0.351	0.418	0.496	0.549
24	0.343	0.409	0.485	0.537
25	0.336	0.400	0.475	0.526
26	0.329	0.392	0.465	0.515
27	0.323	0.385	0.456	0.505
28	0.317	0.377	0.448	0.496
29	0.311	0.370	0.440	0.487
30	0.305	0.364	0.432	0.478

From "Distribution of Sums of Squares of Rank Differences for Small Samples," E. G. Olds, *Annals of Mathematical Statistics*, Volume 9 (1938). Reproduced with the kind permission of the Editor, *Annals of Mathematical Statistics*.

Table 12
Random numbers

Line/Col.	(1)	(2)	(3)	(4)	(5)	(6)	(7)	(8)	(9)	(10)	(11)	(12)	(13)	(14)
1	10480	15011	01536	02011	81647	91646	69179	14194	62590	36207	20969	99570	91291	90700
2	22368	46573	25595	85393	30995	89198	27982	53402	93965	34095	52666	19174	39615	99505
3	24130	48360	22527	97265	76393	64809	15179	24830	49340	32081	30680	19655	63348	58629
4	42167	93003	06243	61680	07856	16376	39440	53537	71341	57004	00849	74917	97758	16379
5	37570	39975	81837	16656	06121	91782	60468	81305	49684	60672	14110	06927	01263	54613
6	77921	06907	11008	42751	27756	53498	18602	70659	90655	15053	21916	81825	44394	42880
7	99562	72095	56420	69994	98872	31016	71194	18738	44013	48840	63213	21069	10634	12952
8	96301	91977	05463	07972	18876	20922	94595	56869	69014	60045	18425	84903	42508	32307
9	89579	14342	63661	10281	17453	18103	57740	84378	25331	12566	58678	44947	05585	56941
10	85475	36857	53342	53988	53060	59533	38867	62300	08158	17983	16439	11458	18593	64952
11	28918	69578	88231	33276	70997	79936	56865	05859	90106	31595	01547	85590	91610	78188
12	63553	40961	48235	03427	49626	69445	18663	72695	52180	20847	12234	90511	33703	90322
13	09429	93969	52636	92737	88974	33488	36320	17617	30015	08272	84115	27156	30613	74952
14	10365	61129	87529	85689	48237	52267	67689	93394	01511	26358	85104	20285	29975	89868
15	07119	97336	71048	08178	77233	13916	47564	81056	97735	85977	29372	74461	28551	90707
16	51085	12765	51821	51259	77452	16308	60756	92144	49442	53900	70960	63990	75601	40719
17	02368	21382	52404	60268	89368	19885	55322	44819	01188	65255	64835	44919	05944	55157
18	01011	54092	33362	94904	31273	04146	18594	29852	71585	85030	51132	01915	92747	64951
19	52162	53916	46369	58586	23216	14513	83149	98736	23495	64350	94738	17752	35156	35749
20	07056	97628	33787	09998	42698	06691	76988	13602	51851	46104	88916	19509	25625	58104
21	48663	91245	85828	14346	09172	30168	90229	04734	59193	22178	30421	61666	99904	32812
22	54164	58492	22421	74103	47070	25306	76468	26384	58151	06646	21524	15227	96909	44592
23	32639	32363	05597	24200	13363	38005	94342	28728	35806	06912	17012	64161	18296	22851
24	29334	27001	87637	87308	58731	00256	45834	15398	46557	41135	10367	07684	36188	18510
25	02488	33062	28834	07351	19731	92420	60952	61280	50001	67658	32586	86679	50720	94953

Table 12 (Continued)

Line/Col.	(1)	(2)	(3)	(4)	(5)	(6)	(7)	(8)	(9)	(10)	(11)	(12)	(13)	(14)
26	81525	72295	04839	96423	24878	82651	66566	14778	76797	14780	13300	87074	79666	95725
27	29676	20591	68086	26432	46901	20849	89768	81536	86645	12659	92259	57102	80428	25280
28	00742	57392	39064	66432	84673	40027	32832	61362	98947	96067	64760	64584	96096	98253
29	05366	04213	25669	26422	44407	44048	37937	63904	45766	66134	75470	66520	34693	90449
30	91921	26418	64117	94305	26766	25940	39972	22209	71500	64568	91402	42416	07844	69618
31	00582	04711	87917	77341	42206	35126	74087	99547	81817	42607	43808	76655	62028	76630
32	00725	69884	62797	56170	86324	88072	76222	36086	84637	93161	76038	65855	77919	88006
33	69011	65795	95876	55293	18988	27354	26575	08625	40801	59920	29841	80150	12777	48501
34	25976	57948	29888	88604	67917	48708	18912	82271	65424	69774	33611	54262	85963	03547
35	09763	83473	73577	12908	30883	18317	28290	35797	05998	41688	34952	37888	38917	88050
36	91567	42595	27958	30134	04024	86385	29880	99730	55536	84855	29080	09250	79656	73211
37	17955	56349	90999	49127	20044	59931	06115	20542	18059	02008	73708	83517	36103	42791
38	46503	18584	18845	49618	02304	51038	20655	58727	28168	15475	56942	53389	20562	87338
39	92157	89634	94824	78171	84610	82834	09922	25417	44137	48413	25555	21246	35509	20468
40	14577	62765	35605	81263	39667	47358	56873	56307	61607	49518	89656	20103	77490	18062
41	98427	07523	33362	64270	01638	92477	66969	98420	04880	45585	46565	04102	46880	45709
42	34914	63976	88720	82765	34476	17032	87589	40836	32427	70002	70663	88863	77775	69348
43	70060	28277	39475	46473	23219	53416	94970	25832	69975	94884	19661	72828	00102	66794
44	53976	54914	06990	67245	68350	82948	11398	42878	80287	88267	47363	46634	06541	97809
45	76072	29515	40980	07391	58745	25774	22987	80059	39911	96189	41151	14222	60697	59583
46	90725	52210	83974	29992	65831	38857	50490	83765	55657	14361	31720	57375	56228	41546
47	64364	67412	33339	31926	14883	24413	59744	92351	97473	89286	35931	04110	23726	51900
48	08962	00358	31662	25388	61642	34072	81249	35648	56891	69352	48373	45578	78547	81788
49	95012	68379	93526	70765	10592	04542	76463	54328	02349	17247	28865	14777	62730	92277
50	15664	10493	20492	38391	91132	21999	59516	81652	27195	48223	46751	22923	32261	85653

Line/Col.	(1)	(2)	(3)	(4)	(5)	(6)	(7)	(8)	(9)	(10)	(11)	(12)	(13)	(14)
51	16408	81899	04153	53381	79401	21438	83035	92350	36693	31238	59649	91754	72772	02338
52	18629	81953	05520	91962	04739	13092	97662	24822	94730	06496	35090	04822	86774	98289
53	73115	35101	47498	87637	99016	71060	88824	71013	18735	20286	23153	72924	35165	43040
54	57491	16703	23167	49323	45021	33132	12544	41035	80780	45393	44812	12515	98931	91202
55	30405	83946	23792	14422	15059	45799	22716	19792	09983	74353	68668	30429	70735	25499
56	16631	35006	85900	98275	32388	52390	16815	69298	82732	38480	73817	32523	41961	44437
57	96773	20206	42559	78985	05300	22164	24369	54224	35083	19687	11052	91491	60383	19746
58	38935	64202	14349	82674	66523	44133	00697	35552	35970	19124	63318	29686	03387	59846
59	31624	76384	17403	53363	44167	64486	64758	75366	76554	31601	12614	33072	60332	92325
60	78919	19474	23632	27889	47914	02584	37680	20801	72152	39339	34806	08930	85001	87820
61	03931	33309	57047	74211	63445	17361	62825	39908	05607	91284	68833	25570	38818	46920
62	74426	33278	43972	10119	89917	15665	52872	73823	73144	88662	88970	74492	51805	99378
63	09066	00903	20795	95452	92648	45454	09552	88815	16553	51125	79375	97596	16296	66092
64	42238	12426	87025	14267	20979	04508	64535	31355	86064	29472	47689	05974	52468	16834
65	16153	08002	26504	41744	81959	65642	74240	56302	00033	67107	77510	70625	28725	34191
66	21457	40742	29820	96783	29400	21840	15035	34537	33310	06116	95240	15957	16572	06004
67	21581	57802	02050	89728	17937	37621	47075	42080	97403	48626	68995	43805	33386	21597
68	55612	78095	83197	33732	05810	24813	86902	60397	16489	03264	88525	42786	05269	92532
69	44657	66999	99324	51281	84463	60563	79312	93454	68876	25471	93911	25650	12682	73572
70	91340	84979	46949	81973	37949	61023	43997	15263	80644	43942	89203	71795	99533	50501
71	91227	21199	31935	27022	84067	05462	35216	14486	29891	68607	41867	14951	91696	85065
72	50001	38140	66321	19924	72163	09538	12151	06878	91903	18749	34405	56087	82790	70925
73	65390	05224	72958	28609	81406	39147	25549	48542	42627	45233	57202	94617	23772	07896
74	27504	96131	83944	41575	10573	08619	64482	73923	36152	05184	94142	25299	84387	34925
75	37169	94851	39117	89632	00959	16487	65536	49071	39782	17095	02330	74301	00275	48280

Table 12 (Continued)

Line/Col.	(1)	(2)	(3)	(4)	(5)	(6)	(7)	(8)	(9)	(10)	(11)	(12)	(13)	(14)
76	11508	70225	51111	38351	19444	66499	71945	05422	13442	78675	84081	66938	93654	59894
77	37449	30362	06694	54690	04052	53115	62757	95348	78662	11163	81651	50245	34971	52924
78	46515	70331	85922	38329	57015	15765	97161	17869	45349	61796	66345	81073	49106	79860
79	30986	81223	42416	58353	21532	30502	32305	86482	05174	07901	54339	58861	74818	46942
80	63798	64995	46583	09785	44160	78128	83991	42885	92520	83531	80377	35909	81250	54238
81	82486	84846	99254	67632	43218	50076	21361	64816	51202	88124	41870	52689	51275	83556
82	21885	32906	92431	09060	64297	51674	64126	62570	26123	05155	59194	52799	28225	85762
83	60336	98782	07408	53458	13564	59089	26445	29789	85205	41001	12535	12133	14645	23541
84	43937	46891	24010	25560	86355	33941	25786	54990	71899	15475	95434	98227	21824	19585
85	97656	63175	89303	16275	07100	92063	21942	18611	47348	20203	18534	03862	78095	50136
86	03299	01221	05418	38982	55758	92237	26759	86367	21216	98442	08303	56613	91511	75928
87	79626	06486	03574	17668	07785	76020	79924	25651	83325	88428	85076	72811	22717	50585
88	85636	68335	47539	03129	65651	11977	02510	26113	99447	68645	34327	15152	55230	93448
89	18039	14367	61337	06177	12143	46609	32989	74014	64708	00533	35398	58408	13261	47908
90	08362	15656	60627	36478	65648	16764	53412	09013	07832	41574	17639	82163	60859	75567
91	79556	29068	04142	16268	15387	12856	66227	38358	22478	73373	88732	09443	82558	05250
92	92608	82674	27072	32534	17075	27698	98204	63863	11951	34648	88022	56148	34925	57031
93	23982	25835	40055	67006	12293	02753	14827	23235	35071	99704	37543	11601	35503	85171
94	09915	96306	05908	97901	28395	14186	00821	80703	70426	75647	76310	88717	37890	40129
95	59037	33300	26695	62247	69927	76123	50842	43834	86654	70959	79725	93872	28117	19233
96	42488	78077	69882	61657	34136	79180	97526	43092	04098	73571	80799	76536	71255	64239
97	46764	86273	63003	93017	31204	36692	40202	35275	57306	55543	53203	18098	47625	88684
98	03237	45430	55417	63282	90816	17349	88298	90183	36600	78406	06216	95787	42579	90730
99	86591	81482	52667	61582	14972	90053	89534	76036	49199	43716	97548	04379	46370	28672
100	38534	01715	94964	87288	65680	43772	39560	12918	86537	62738	19636	51132	25739	56947

Abridged from *Handbook of Tables for Probability and Statistics*, 2nd edition, edited by William H. Beyer (Cleveland: The Chemical Rubber Company, 1968). Reproduced by permission of the publishers, The Chemical Rubber Company.

ANSWERS TO EXERCISES

chapter one

1.6 $s' = 1.11$

1.7   (a) $\bar{y} = 22.56$; $s' = 12.01$
       (b) $k = 1$: (10.55, 34.57);
            $k = 2$: (-1.46, 46.58);
            $k = 3$: (-13.47, 58.59)

1.8   (a) $\bar{y} = 3.23$; $s' = 3.10$
       (b) $k = 1$: (.13, 6.33); $k = 2$:
            (-2.97, 9.43);
            $k = 3$: (-6.07, 12.53)

1.9   (a) $\bar{y} = 4.43$; $s' = 38.72$
       (b) $k = 1$: (-34.29, 43.15); $k = 2$:
            (-73.01, 81.87); $k = 3$:
            (-111.73, 120.59)

1.10   (a) $\bar{y} = -2.49$; $s' = 10.64$; $k = 1$:
           (-13.13, 8.15); $k = 2$: (-23.77,
           18.79); $k = 3$: (-34.41, 29.43)

1.11   15.425; 3.04; 61.375

1.12   .16

1.13   .84

1.15   81.5%

1.16   (a)   2
       (c) $\bar{y} = 6.8$; $s' = 2.04$

1.17   (a)   177
       (c) $\bar{y} = 210.80$; $s' = 161.25$

1.19   231, 323

1.20   .05

1.21   (a)

k	$\mu \pm k\sigma$	approx. fraction
1	420,570	.68
2	345,645	.95
3	270,720	1.00

       (b)   .16

1.22   .025

1.24   (.5, 10.5)

1.25   1/9

1.26   at least 3/4 in interval (0, 1990),
       at least 8/9 in interval (0, 2450)

chapter two

2.5   (a)   42
       (b)   33
       (c)   18

2.6   (b)   2/3

2.7   (a)   .09
       (b)   .19

2.8   (a)   1/3
       (b)   2/5
       (c)   19/48
       (d)   49/240

2.9   (a)   .08
       (b)   .16
       (c)   .14
       (d)   .84

2.10   (d) $P(A) = 1/2$; $P(B) = 3/4$;
           $P(A \cap B) = 1/2$; $P(A \cup B) = 3/4$;
           $P(\bar{A} \cup B) = 1$

2.11   (c) $P(A) = 1/3$; $P(B) = 5/9$;
           $P(A \cup B) = 7/9$; $P(A \cap B) = 1/9$

2.12   (a)   9
       (b)   5/9
       (c)   5/9

2.13   (c)   1/2

2.14   (a)   3/5; 1/15
       (b)   14/15; 2/5

2.15   (c)   1/8

2.16   (c) $P(A) = 1/4$

2.17   (c) $P(A) = 2/9$

2.18   42

2.19   6

2.20   (a)   36
       (b)   1/6

2.21   (a)   40
       (b)   320

2.22 $9(10)^6$

2.23   720

2.24   504

2.25	1/9	2.55	.908
2.26	$2^3 = 8$	2.56	.149
2.27	18	2.57	(a) $(.02)^3$
2.28	(a) 60		(b) $(.98)^3$
	(b) .6	2.58	$(.98)^3(.02)$
2.29	3/7	2.59	(a) .095
2.30	$48\Big/\binom{52}{2}$		(b) .855
			(c) .005
2.31	1/12		(d) .955
2.32	1/56	2.60	$(.75)^4$
2.33	5/324	2.61	$1 - (.01)^3$
2.34	5/162	2.62	(a) 1/4
2.36	(a) 1/3		(b) 1/16
	(b) 1/5	2.63	.8704
	(c) 5/7	2.64	(a) 1/5
	(d) 1		(b) 2/5
	(e) 1/7		(c) 1/2
2.37	yes; yes	2.65	(a) $1/n$
2.38	(a) .000125		(b) $1/n$; $1/n$
	(b) .142625		(c) 3/7
2.39	(a) .125	2.66	7/9
	(b) .875	2.67	1/12
2.40	(a) .40	2.68	.4
	(b) .37	2.69	.00063
	(c) .10	2.70	.25
	(d) .67	2.71	.9412
	(e) .60	2.72	14/17
	(f) .33	2.73	(a) .57
	(g) .90		(b) .18
	(h) .27		(c) .3158
	(i) .25		(d) .9
2.41	no	2.74	5/7
2.45	(a) .97	2.75	(a) 2/5
	(b) .03		(b) 3/20
2.46	(a) .1	2.76	$P(Y = 0) = (.02)^3$,
	(b) .9		$P(Y = 1) = 3(.98)(.02)^2$;
	(c) .6		$P(Y = 2) = 3(.98)^2(.02)$;
	(d) 2/3		$P(Y = 3) = (.98)^3$
2.47	(a) .6	2.77	$P(Y = 0) = 1/5$;
	(b) .4		$P(Y = 1) = 3/5$;
	(c) .9		$P(Y = 2) = 1/5$
2.48	(a) .999	2.78	$P(Y = 2) = 1/15$;
	(b) .9009		$P(Y = 3) = 2/15$; $P(Y = 4) = 3/15$;
2.49	series .81; parallel .99		$P(Y = 5) = 4/15$; $P(Y = 6) = 5/15$
2.51	(a) .22	2.79	$P(Y = 2) = 3/4$; $P(Y = 3) = 1/4$
	(b) 3/22	2.81	(b) 16
2.52	(a) 1/1000	2.82	18!
	(b) 1/8000	2.83	$4\binom{13}{5}\Big/\binom{52}{5}$
2.54	$\geq .90$		

2.84 $\binom{13}{2}\binom{4}{3}\binom{4}{2}\Big/\binom{52}{5}$

2.85 270/495

2.86 (a) .4   (b) .6   (c) .25

2.87 $P(Y = 0) = .4; P(Y = 1) = .5;$
$P(Y = 2) = .1$

2.88 (a) .5
(b) .15
(c) .1
(d) .875

2.89 (a) (i) .89 (ii) .47 (iii) .68 (iv) .25
(b) no

2.90 (a) .87
(b) .31
(c) .29
(d) .94
(e) .12
(f) .21

2.91 (a) 1/66
(b) 1/33

2.92 2/11

2.93 1/8640

2.94 Design A: P(current flows) = .9801
Design B: P(current flows) = .9639

2.95 1/7

2.96 3/7

2.99 no

2.100 .5952

2.101 (a) $\binom{6}{3}(1/2)^6$
(b) $27(1/2)^{10}$

2.102 (a) $(1/4)^7$
(b) $36(1/4)^4$

chapter three

3.1 $p(0) = .2; p(1) = .7; p(2) = .1$

3.2 $p(2) = 1/6; p(3) = 1/3; p(4) = 1/2$

3.3 $p(0) = .072; p(1) = .416; p(2) = .512$

3.4 $p(0) = 1/3; p(1) = 1/2; p(3) = 1/6;$

3.5 (a) $p(2) = .1; \ p(3) = .2; \ p(4) = .3;$
$p(5) = .4$
(b) $p(y) = \begin{cases} .1, & y = 3, 4, 8, 9 \\ .2, & y = 5, 6, 7 \end{cases}$

3.6 $p(0) = .857375$
$p(1) = .135375$
$p(2) = .007125$
$p(3) = .000125$

3.7 $p(y) = (.8)^y(.2) \ y = 0, 1, 2, \ldots$

3.8 $P(X = 0) = 8/27;$
$P(X = 1) = 12/27;$
$P(X = 2) = 6/27;$
$P(X = 3) = 1/27$
$P(Y = 0) = 2744/3375;$
$P(Y = 1) = 588/3375;$
$P(Y = 2) = 42/3375;$
$P(Y = 3) = 1/3375;$
$P(X + Y = 0) = 729/3375;$
$P(X + Y = 1) = 1458/3375;$
$P(X + Y = 2) = 972/3375;$
$P(X + Y = 3) = 216/3375$

3.9 2; .642; 4; 1

3.10 2960

3.11 13,800.388

3.12 $E(Y) = 3.5; V(Y) = 35/12$

3.13 $.31

3.14 $\mu = .2; \sigma^2 = .18$

3.15 Firm I: E(profit) = \$60,000
E(total profit) = \$120,000

3.16 $E(Y) = \$8,333; \sigma = \$19,507.98$

3.17 \$120

3.18 \$14.67

3.20 13; 41

3.21 (a) $p(0) = (2/3)^4; p(1) = 2(2/3)^4;$
$p(2) = (2/3)^3; p(3) = (1/3)(2/3)^3;$
$p(4) = (1/3)^4$
(b) 1/9
(c) 4/3
(d) 8/9

3.22 (a) .1536
(b) .9728

3.23 (a) .109
(b) .999
(c) .844
(d) .589

3.24 .000

3.25 .884; .572

3.26 (a) .1681
(b) .5282

3.27 (a) $(.8)^5$
(b) $5(.6)^4(.4)$
(c) $5(.3)(.7)^4 + (.7)^5$

3.28 .992

3.31 (a) .32805; .99999
(b) 3

3.32 (c) .151
(d) .302

3.33 $E(Y) = 1; V(Y) = .9$

3.34 $185,000

3.35   3.96

3.36 $840

3.37   (a) .109
        (b) .999
        (c) .589
        (d) $\mu = 16$; $\sigma^2 = 3.2$

3.38   (a) .672
        (b) .672
        (c) 8

3.39 $1 - (.64)^3$

3.40   .072

3.41 $1/.3 = 3.33$

3.42   (a) .128
        (b) $(.8)^{10}$

3.44   (a) .009
        (b) .01

3.45 $\mu = 1.11$; $\sigma = .3514$

3.46   1/2

3.47   (a) .081
        (b) .81

3.48   .01536; .0256

3.49 $p(y) = (.06)(.94)^{y-1}$, $y = 1, 2, 3, \ldots$

3.50   125/1296

3.51   2

3.52 $E(\# \text{ on which productive well}) = 5$;
        $E(\# \text{ before productive well}) = 4$

3.54   .06

3.55   150; 4500; no

3.56   .09

3.57   (a) .04374
        (b) .99144

3.58   (a) $\mu = 1.111$; $\sigma^2 = .1235$
        (b) $\mu = 3.333$; $\sigma^2 = .3704$

3.59   .1

3.60   (a) first .4; second .24; third .144
        (b) .1728

3.61   (a) .128
        (b) .03072
        (d) $\mu = 15$; $\sigma^2 = 60$

3.62   1/12

3.63   1/42

3.64 $E(\text{repair cost}) = \$100$;
        $V(\text{repair cost}) = \$1666.67$

3.65   (a) .6
        (b) .8
        (c) .8

3.67 $\mu = 2.4$; $\sigma^2 = 1.061$

3.68 $p(0) = .2$; $p(1) = .6$; $p(2) = .2$

3.70 $N = 11$ or 12

3.71   (a) .090
        (b) .143
        (c) .857
        (d) .241

3.72   (a) .0818
        (b) $1 - 8e^{-7}$
        (c) .1277

3.73   .6288

3.74   .0521

3.75 $2.5e^{-1.5}$

3.76 $1 - (8/3)e^{-1}$

3.77 $1 - [(8/3)e^{-1}]^{10}$

3.78

	Binomial	Poisson
$p(0)$	.358	.368
$p(1)$	.378	.368
$p(2)$	.189	.184
$p(3)$	.059	.061
$p(4)$	.013	.015

3.79 $1 - (.97)^{100}$

3.80   (a) .004
        (b) yes

3.81 $61e^{-6}$

3.82 $E[Y(Y-1)] = \lambda^2$

3.83   40

3.84 $100e^{-1}$

3.85 $1300

3.91   (a) 7/3
        (b) 5/9
        (c) $p(1) = 1/6$; $p(2) = 2/6$; $p(3) = 3/6$

3.96 $[pt + (1-p)]^n$

3.97 $e^{\lambda(t-1)}$

3.98 $\lambda^3 + 3\lambda^2 + \lambda$

3.99   (a) .64
        (b) $C = 10$

3.100   (a) 805
        (b) 22.875
        (c) 759.251 to 850.749
        (d) no; 249 is more than 24 standard
            deviations below the mean

3.101   300

3.102   (85, 115)

3.103 $E(Y) = 1$; $V(Y) = .4$

3.104   (a) $p(0) = 1/8$; $p(1) = 3/8$;
            $p(2) = 3/8$; $p(3) = 1/8$
        (c) $E(Y) = 3/2$; $V(Y) = 3/4$

3.105 (a) $p(0) = .729$; $p(1) = .243$;
$p(2) = .027$; $p(3) = .001$
(c) $E(Y) = .3$; $V(Y) = .27$

3.106 (a) 24
(b) 4.898
(c) no; 40 is more than 3 standard deviations below the mean

3.107 Observed 71% is more than 16 standard deviations away from 50%

3.108 (61.03, 98.97)

3.109 $E(Y) = 100$; $V(Y) = 90$

3.110 No, using Tchebysheff's Theorem
$P(Y \geq 350) \leq .1126$

3.111 (a) 1.0
(b) .5905
(c) .1681
(d) .0312
(e) 0

3.113 (a) $n = 25$; $a = 5$
(b) $n = 25$; $a = 5$

3.114 $1 - (.99999)^{10,000}$

3.115 $E(Y) = 1$

3.116 $V(Y) = \sigma^2 = .4$

3.117 $E(Y) = 3.5$; $V(Y) = 2.917$

3.118 $3(.1)(.9)^{10}$

3.119 (a) .982
(b) $1 - (.018)^3$

3.120 (a) .9997
(b) $n = 2$

3.122 (18.35, 181.65)

3.123 (a) $e^{-\lambda p}$
(b) Poisson with mean λp

3.124 .00724

3.125 $187.20

3.126 (a) $12.5e^{-5}$; $18.5e^{-5}$

3.127 no

3.128 .0837

3.129 .9665

3.130 3

3.132 (a) .119
(b) .117

3.133 $149.09

3.134 (a) $n[1 + k(1 - .95^k)]$
(b) 5
(c) approximately $.57N$

chapter four

4.1 (a) $c = 1/2$
(b) $y^2/4$, $0 \leq y \leq 2$

(d) 3/4
(e) 3/4

4.2 (b) $2ye^{-y^2}$
(c) e^{-4}

4.3 (a) $\begin{cases} y^2/2, & 0 \leq y \leq 1 \\ y - 1/2, & 1 < y \leq 1.5 \end{cases}$
(b) .125
(c) .575

4.4 (b) $\begin{cases} y^2/2 & 0 < y < 1 \\ 2y - y^2/2 - 1 & 1 \leq y \leq 2 \end{cases}$
(c) .36
(d) 1/4

4.5 (b) $1 - b/y$, $y \geq b$
(c) $b/(b + c)$

4.6 (a) $c = 1/2$
(b) $y - y^2/4$, $0 \leq y \leq 2$
(d) 1/4

4.7 (a) $c = 3/2$
(c) $(y^3 + y^2)/2$, $0 \leq y \leq 1$
(d) 0; 0; 1
(e) 3/16
(f) 104/123

4.8 (a) $c = 1.2$
(b) $\begin{cases} .2(y + 1), & -1 < y \leq 0 \\ .6y^2 + .2y + .2, & 0 < y \leq 1 \end{cases}$
(d) 0; .2; 1
(e) .25
(f) .71

4.9 (a) $\begin{cases} 1/8, & 0 < y < 2 \\ y/8, & 2 \leq y < 4 \end{cases}$
(b) 7/16
(c) 13/16
(d) 7/9

4.10 $E(Y) = 2/3$; $V(Y) = 2/9$

4.11 $E(Y) = .708$; $V(Y) = .049$

4.12 $E(Y) = .4$; $V(Y) = .2733$

4.14 $4.65; .012

4.15 (a) $c = 105$
(b) 3/8

4.16 $E(Y) = 60$; $V(Y) = .333$

4.17 (a) $E(Y) = 2/3$; $V(Y) = 1/18$
(b) $E(X) = 220/3$; $V(X) = 20,000/9$
(c) $(-20.948, 167.614)$

4.18 4

4.19 (a) $E(Y) = 2.4$; $V(Y) = .64$
(b) $X = $ weekly cost, $E(X) = 480$;
$V(X) = 25,600$
(c) $P(X > 600) = .26$

4.20 (a) $E(Y) = 5.5$; $V(Y) = .1475$
      (b) Using Tchebysheff's Theorem
          (5, 6.268)
      (c) yes, $P(Y < 5.5) = .5775$
4.21 (a) $y, 0 < y < 1$
4.22 1/2, 1/4
4.23 3/8
4.25 (a) $k = 1/4$
      (b) $y/4 + 1/2, -2 \le y \le 2$
4.26 1/3
4.27 $E(Y) = 60$; $V(Y) = 400/12$
4.28 (a) 1/8
      (b) 1/8
      (c) 1/4
4.29 (a) .2
      (b) $\mu = 0$; $\sigma^2 = .01/12 = .00083$
4.30 $E(\text{volume}) = (6.5 \times 10^{-6})\pi$;
      $V(\text{volume}) = (7.749 \times 10^{-11})\pi^2$
4.31 (a) .3849
      (b) .3159
      (c) .3227
      (d) .1586
      (e) .3613
4.32 (a) 0
      (b) 1.1
      (c) 1.645
      (d) 2.575
4.33 .0062
4.34 $426
4.35 .073
4.36 3.000
4.37 .2266
4.38 .2660
4.39 $(.2266)^3$
4.40 (a) .9544
      (b) .83
4.41 (a) .0618
      (b) .7794
4.42 (a) .4060
      (b) 960.5
4.43 (a) .8413
      (b) 85.68
      (c) 81.48
      (d) .4364
      (e) .1886
4.44 7.301
4.47 $3(\mu^2 + \sigma^2)$
4.48 (a) .2865
      (b) .1481

4.49 .7355
4.50 (a) .1353
      (b) 460.52 cfs
4.51 1100; 2,920,000
4.52 (a) .5057
      (b) 1936
4.53 (a) .0821
      (b) .0273
4.54 1/96
4.55 e^{-1}
4.56 11.98
4.57 $2e^{-1}$
4.58 .050
4.59 $\mu = 3.2$; $\sigma^2 = 6.4$
4.60 $(y/4)e^{-y/2}, y > 0$
4.61 (0, 9.6569)
4.62 20,000; 400,000
4.63 $E(L) = 276$, $V(L) = 47,664$
4.64 $E(Y) = 3/2$; $V(Y) = 3/4$
4.65 $k = 60$
4.66 .8208
4.67 $E(Y) = 3/5$; $V(Y) = 1/25$
4.68 (a) $y^2(3 - 2y), 0 \le y \le 1$
      (c) .3960
4.70 $53.58
4.71 52/3; 29.96
4.73 (a) .75
      (b) 1/18
4.74 (a) $E(Y) = 1/2$; $V(Y) = 1/28$
      (b) $E(Y) = 1/2$; $V(Y) = 1/20$
      (c) $E(Y) = 1/2$; $V(Y) = 1/12$
      (d) Case (a) has the smallest variance
4.75 (a) $c = 105$
      (b) 3/8
4.76 (a) $1/(1 - \theta t)$
      (b) $E(Y) = \theta$; $V(Y) = \theta^2$
4.77 (a) $\exp(\mu t + \sigma^2 t^2/2)$
4.78 $\alpha\beta$; $\alpha\beta^2$
4.79 (a) $1/\sqrt{2\pi}$
      (b) $\exp(t^2/2)$
      (c) $E(Y) = 0$; $V(Y) = 1$
4.80 $m_U(t) = e^{at}m_Y(bt)$; $E(U) = a + b\mu$;
      $V(U) = b^2\sigma^2$
4.81 (a) 2/5
      (b) $1/(t + 1)$
      (c) 1
4.82 (12,351, 37,649)
4.83 .5

4.84 .962

4.85 1.0

4.86 .9502

4.87 could be!, 2000 is only .53 standard deviations above mean

4.88 276 ± 654.96

4.89 (6.39, 28.28)

4.90 $113.33

4.91 4

4.92 (a)
$$F(x) = 0, \qquad\qquad x < 0$$
$$(1/100)e^{-x/100}, 0 \le x < 200$$
$$1, \qquad\qquad x \ge 200$$
(b) 86.47

4.93 (a) $(2/\pi) \tan^{-1} y + 1/2$

(b) 0

4.94 85.36

4.95 383.65

4.96 $1 - (.927)^5$

4.97 (a) .1056

(b) .8944

(c) .1056

4.98 (a) $c = 4$

(b) 1; 1/2

(c) $(1 - t/2)^{-2}$

4.99 $\mu_1 = 0; \mu_2 = \sigma^2; \mu_3 = 0; \mu_4 = 3\sigma^4$

4.100 $\Gamma(\alpha + \beta)\Gamma(\alpha + k)/$
$[\Gamma(\alpha)\Gamma(\alpha + \beta + k)]$

4.101 $f(t) = \lambda e^{-\lambda t}, t > 0$

4.102 $e^{-2.5}$

4.103 $f(r) = (2\lambda\pi r) \exp(-\lambda\pi r^2), r > 0$

4.104 $e^{-1/2}$

4.105 $\sqrt{2}$

4.106 .0235

4.107 $k = (.4)^{1/3}$

4.108 $1 - 13e^{-4}$

4.109 $m(t) = \exp(t^2/2); 0; 1$

4.110 (a) .0045

(b) .9726

4.111 (a) $E(Y) = 598.74g;$
$V(Y) = 3.1856 \times 10^{12}g^2$

(b) $(0, 3.570 \times 10^6)$

(c) .8023

4.112 $m(t) = 1 - t^2; E(Y) = 0$

4.114 $E(Y) = \Gamma(3/2)\alpha^{1/2};$
$V(Y) = \alpha\{1 - [\Gamma(3/2)]^2\}$

4.115 (a) $e^{-2.5}$

(b) .019

4.117 (c) $1 - e^{-4}$

4.118 (a) $\dfrac{2}{\sqrt{\pi}}\left(\dfrac{m}{2KT}\right)^{-1/2}$

(b) $\dfrac{2KT}{\sqrt{\pi}} \Gamma(5/2)$

4.119 150

chapter five

5.1

Y_2	Y_1		
	0	1	2
0	1/9	2/9	1/9
1	2/9	2/9	0
2	1/9	0	0

5.2

Y_2	Y_1			
	0	1	2	3
-1	1/8	0	0	0
1	0	1/8	2/8	1/8
2	0	1/8	1/8	0
3	0	1/8	0	0

(b) $F(2, 1) = 1/2$

5.3 $\dfrac{\dbinom{4}{y_1}\dbinom{3}{y_2}\dbinom{2}{3 - y_1 - y_2}}{\dbinom{9}{3}}$

5.4 (a) $k = 4$

(b) $y_1^3 y_2^2, 0 \le y_1 \le 1, 0 \le y_2 \le 1$

(c) 9/64

5.5 (a) $k = 6$

(b) 31/64

5.6 (a) $k = 1$

(b) 2/3

5.7 (a) 29/32

(b) 1/4

5.8 (a) 7/8

(b) 1/2

5.9 (a) $e^{-1} - 2e^{-2}$

(b) 1/2

(c) e^{-1}

5.10 (a) 21/64

(b) 1/3

5.11 1/2

5.12 $(3/2)e^{-1}$

5.13 (a) $p(0) = 4/9; p(1) = 4/9; p(2) = 1/9$

(b) no

5.14 (a) $p(-1) = 1/8$; $p(1) = 4/8$;
           $p(2) = 2/8$; $p(3) = 1/8$
      (b) $1/4$

5.15 (a) $p(0) = 5/42$; $p(1) = 20/42$;
           $p(2) = 15/42$; $p(3) = 2/42$
      (b) $2/3$
      (c) $8/15$

5.16 (a) $f_1(y_1) = 2y_1, 0 \le y_1 \le 1$;
           $f_2(y_2) = 2y_2, 0 \le y_2 \le 1$
      (b) $1/4$
      (c) $2y_1, 0 \le y_1 \le 1$
      (d) $2y_2, 0 \le y_2 \le 1$
      (e) $9/16$

5.17 (a) $f_1(y_1) = 3(1 - y_1)^2, 0 \le y_1 \le 1$;
           $f_2(y_2) = 6y_2(1 - y_2), 0 \le y_2 \le 1$
      (b) $32/63$
      (c) $1/y_2, 0 \le y_1 \le y_2$
      (d) $2(1 - y_2)/(1 - y_1)^2, y_1 \le y_2 \le 1$
      (e) $1/4$

5.18 (a) $1/4$
      (b) $1/2$

5.19 (a) $f_1(y_1) =$
           $\begin{cases} 1 + y_1, & -1 \le y_1 \le 0 \\ 1 - y_1, & 0 \le y_1 \le 1 \end{cases}$
           $f_2(y_2) = 2(1 - y_2), 0 \le y_2 \le 1$
      (b) $1/3$

5.20 (a) $3/7$
      (b) $1/3$

5.21 $1/2$

5.22 (a) $f_1(y_1) = y_1 + 1/2, 0 \le y_1 \le 1$;
           $f_2(y_2) = y_2 + 1/2, 0 \le y_2 \le 1$
      (b) $3/5$
      (c) $.34375$

5.23 e^{-1}

5.24 (a) $f(y_1, y_2) = 1/y_1, 0 \le y_2 \le y_1 \le 1$
      (b) $1/2$
      (c) $\ln(2)/\ln(4)$

5.25 $1/4$

5.26 $f_1(y) = (1/2)^{y+1}, y = 0, 1, 2, \ldots$

5.27 no
5.28 no
5.29 no
5.30 yes
5.31 no
5.32 no
5.33 no
5.34 no
5.35 no
5.36 no

5.37 yes

5.38 (a) $\binom{2}{y_1}(.2)^{y_1}(.8)^{2-y_1}(.3)^{y_2}(.7)^{1-y_2}$,
           $y_1 = 0, 1, 2, y_2 = 0, 1$
      (b) $.864$

5.39 (a) $f(y_1, y_2) = (1/9)e^{-(y_1+y_2)/3}$,
           $y_1 > 0, y_2 > 0$
      (b) $1 - (4/3)e^{-1/3}$

5.40 $7/32$

5.41 (a) $1/4$
      (b) $23/144$

5.42 (a) $2/3$
      (b) $4/9$
      (c) 0

5.43 $4/3$

5.44 (a) $2/3$
      (b) $1/18$
      (c) 0

5.45 (a) $1/4$; $1/2$
      (b) $3/80$; $1/20$
      (c) $-5/4$

5.46 (a) $4/3$; $1/3$
      (b) $2/9$; $1/18$
      (c) 1
      (d) $1 \pm .81$

5.47 0
5.48 32.08
5.49 1
5.50 $1/4$
5.51 1
5.52 $-2/9$; no; Y_1 larger implies Y_2 smaller
5.53 $-1/3$
5.54 0
5.55 $1/40$; no
5.56 0; 0
5.57 $\sigma_1^2 - \sigma_2^2$; $\dfrac{\sigma_1^2 - \sigma_2^2}{\sigma_1^2 + \sigma_2^2}$
5.58 0
5.59 $\mu = 445$; $\sigma^2 = 236$
5.60 -22; 480
5.61 (a) $7/3$; $7/18$
      (b) $7/3$; $7/18$
5.62 $1/9$
5.63 $.3375$
5.64 $2/3$; $1/18$
5.65 1; 1; no
5.66 $(11.48, 52.68)$
5.67 66; 96

5.68   42; 25; no

5.69   (a) 2

      (b) 8

      (c) yes

5.70 $p_1 - p_2$;

$$\left[\frac{N-n}{n(N-1)}\right][p_1 + p_2 - (p_1 - p_2)^2]$$

5.71   140; 280; gamma density $\alpha = 70$, $\beta = 2$

5.72   (a) .0823

      (b) $n/3$; $2n/9$

      (c) $-n/9$

      (d) 0; $2n/3$

5.73 $E(C) = np_1 + 3np_2$;
$V(C) = np_1(1 - p_1) + 9np_2(1 - p_2)$
$\qquad\qquad - 6np_1p_2$

5.74   (a) .0972

      (b) .2, .072

5.75   (a) .1109

      (b) .0625

5.76   .08953

5.77   (a) 66,960

      (b) 128,209,600

5.78   (a) .046

      (b) .2262

5.79   2.5; 4.875

5.80   (a) .2759

      (b) .8031

5.82   (a) $y_2/2$

      (b) 1/4

5.83   3/2

5.84   1

5.85   3/8

5.86 $\alpha\beta$

5.87 $m_U(t) = (1 - t^2)^{-1/2}$
$E(U) = 0$; $V(U) = 1$

5.88   1/4

5.89   11/36

5.90   (a) $\dbinom{4}{y_1}\dbinom{3}{y_2}\dbinom{2}{3 - y_1 - y_2}\Big/\dbinom{9}{3}$

      (b) $\dbinom{4}{y_1}\dbinom{5}{3 - y_1}\Big/\dbinom{9}{3}$

$$\dbinom{3}{y_2}\dbinom{6}{3 - y_2}\Big/\dbinom{9}{3}$$

      (c) 9/16

5.91   (a) $f(y_1) = 3y_1^2$, $0 \le y_1 \le 1$
$f(y_2) = 3/2(1 - y_2^2)$, $0 \le y_2 \le 1$

      (b) 23/44

      (c) $f(y_1|y_2) = 2y_1/(1 - y_2^2)$,
$y_2 \le y_1 \le 1$ if $0 < y_2 \le 1$

      (d) 3/4

5.92   (a) $y_1/2$

      (b) 3/8

5.93   (a) $\dfrac{1}{\beta^2} e^{-(y_1 + y_2)/\beta}$,
$y_1, y_2 > 0$

      (b) $1 - \left(1 + \dfrac{a}{\beta}\right)e^{-a/\beta}$

5.94 $[5 - 3\ln(2)]/4$

5.95 λp

5.96   (a) $12\dbinom{n}{y}\left[\dfrac{(y + 2)!(n - y + 1)!}{(n + 4)!}\right]$

      (b) 6/5

5.97 $p(y) =$
$$\dbinom{y + \alpha - 1}{y}\left(\frac{\beta}{\beta + 1}\right)^y\left(\frac{1}{\beta + 1}\right)^\alpha,$$
$y = 0, 1, 2, \ldots$

5.98 np; $np(1 - p)$

5.101 $\mu_1 - \mu_2$; $(\sigma_1^2/n) + (\sigma_2^2/m)$

5.103   (a) $m(t_1, t_2, t_3) =$
$(p_1e^{t_1} + p_2e^{t_2} + p_3e^{t_3})^n$

      (c) $- np_1p_2$

5.104 $-\sqrt{\dfrac{p_1p_2}{(1 - p_1)(1 - p_2)}}$

chapter six

6.1   (a) $(1 - u)/2$, $-1 \le u \le 1$

      (b) $(1 + u)/2$, $-1 \le u \le 1$

      (c) $(1 - \sqrt{u})/\sqrt{u}$, $0 < u \le 1$

      (d) $E(U_1) = -1/3$; $E(U_2) = 1/3$;
$E(U_3) = 1/6$

6.2   (a) $u^2/18$, $-3 \le u \le 3$

      (b) $(3/2)(3 - u)^2$, $2 \le u \le 4$

      (c) $(3/2)\sqrt{u}$; $0 \le u \le 1$

6.3   (a) $f(u) =$
$$\begin{cases} (u + 4)/100, & -4 \le u \le 6 \\ 1/10, & 6 < u \le 11 \end{cases}$$

      (b) 5.5833

6.4   (a) $(1/12)e^{-(u-1)/12}$, $u \ge 1$

      (b) 13

6.5 $1/8\sqrt{2(u - 3)}$, $5 \le u \le 53$

6.6   (a) $f(u) = \begin{cases} u, & 0 \le u \le 1 \\ 2 - u, & 1 \le u \le 2 \end{cases}$

      (b) 1

6.7   (a) $2u$, $0 \le u \le 1$
     (b) $2/3$

6.8   (a) e^{-u}, $u \ge 0$
     (b) 1; 1

6.9   (a) $4ue^{-2u}$, $u \ge 0$
     (b) 1; $1/2$

6.10 $18u[1 - u + u(\ln u)]$, $0 \le u \le 1$

6.11 $[-\ln(1 - U)]^{1/2}$

6.12 $b/(1 - U)$

6.13 (a) $\alpha y^{a-1}/\theta^{\alpha}$ $0 \le y \le \theta$
     (b) $\theta(U)^{1/\alpha}$
     (c) $y_i = 4(u_i)^{1/2}$; 2.0785, 3.3229,
        1.5036, 1.5610, 2.4030

6.14 (a) $\alpha\beta^{\alpha}y^{-(\alpha+1)}$ $y \ge \beta$
     (b) $\beta(1 - U)^{-1/\alpha}$
     (c) $y_i = 3(1 - u_i)^{-1/2}$; 3.0087,
        3.3644, 6.2448, 3.4582, 4.7900

6.16 (a) $f(w) = (1/2)w^{-1/2}$, $0 \le w \le 1$
     (b) $f(w) = 2w$, $0 \le w \le 1$

6.20 (a) $(1/\alpha)e^{-u/\alpha}$, $u > 0$
     (b) $\Gamma[1 + (k/m)]\alpha^{k/m}$

6.21 (b) $\Gamma[1 + (k/2)]\beta^{k/2}$

6.23 (a) $\left[\dfrac{1}{\Gamma(3/2)(kT)^{3/2}}\right]w^{1/2}e^{-w/kT}$, $w > 0$
     (b) $(3/2)kT$

6.24 $1/(4\sqrt{2P})$, $162 < P < 242$

6.25 $2/(1 + u)^3$, $u > 0$

6.27 $4(80 - 31u + 3u^2)$, $4.5 \le u \le 5$

6.28 (a) $(1/\theta)e^{-u/\theta}$, $u > 0$
     (b) $E(Y) = \sqrt{\pi\theta/2}$;
        $V(Y) = \theta[1 - (\pi/4)]$

6.29 $4ue^{-2u}$, $u > 0$

6.30 gamma $(\alpha = 1, \beta = 2)$

6.31 normal; $E(U) = \mu \sum\limits_{i=1}^{n} a_i$;

     $V(U) = \sigma^2 \sum\limits_{i=1}^{n} a_i^2$

6.32 .0228

6.33 \$190.27

6.34 $ue^{-u^2/2}$, $u > 0$

6.35 binomial $(n_1 + n_2, p)$

6.36 (a) Poisson (mean $\lambda_1 + \lambda_2$)
     (b) binomial $[m, \lambda_1/(\lambda_1 + \lambda_2) = p]$

6.37 .077

6.38 .20

6.40 $m_Y(t) = \left(\dfrac{pe^t}{1 - qe^t}\right)^r$

6.43 (a) $2(1 - u)$, $0 \le u \le 1$
     (b) $2u$, $0 \le u \le 1$

6.44 (a) $F(y) = \begin{cases} 0 & y < 0 \\ (y/\theta)^n & 0 \le y \le \theta \\ 1 & y > \theta \end{cases}$
     (b) $f(y) = (ny^{n-1})/\theta^n$ $0 \le y \le \theta$

6.45 $(2/3)^5$

6.46 (a) $F(y) = \begin{cases} 0 & y < 0 \\ (3y^2 - 2y^3)^n & 0 \le y \le 1 \\ 1 & y > 1 \end{cases}$
     (b) $f(y) = 6ny(1 - y)(3y^2 - 2y^3)^{n-1}$
        $0 \le y \le 1$

6.47 (b) $1 - e^{-3.6/2.5}$

6.48 $e^{-(u-4)}$, $u \ge 4$

6.49 (a) $ne^{-n(x-\theta)}$, $x > \theta$
     (b) $\theta + (1/n)$

6.50 $m(n - 1)r^{n-2}(1 - r)$, $0 < r < 1$

6.51 (a) $1/16$
     (b) 1.6

6.53 normal; $E(U) = -\mu$; $V(U) = (5/2)\sigma^2$

6.54 $(2/3)(w^{-1/2} - w)$, $0 \le w \le 1$

6.55 gamma $(\alpha = 4, \beta = 1)$

6.56 (a) $1/(2\sqrt{u})$, $0 \le u \le 1$
     (b) $\begin{cases} 1/2, & 0 \le u \le 1 \\ 1/(2u^2), & u > 1 \end{cases}$
     (c) ue^{-u}, $u > 0$
     (d) $-\ln u$, $0 \le u \le 1$

6.57 $f(u) = 1$, $0 \le u \le 1$

6.58 $f(d) = 1$, $0 \le d \le 1$

6.59 $3/4$

6.60 $f(u) = 1/[\pi(1 + u^2)]$, $-\infty < u < \infty$

6.61 $f(u) = 1$, $0 \le u \le 1$

6.62 $f(u) = \begin{cases} 1/(4\sqrt{u}), & 0 \le u \le 1 \\ 1/(8\sqrt{u}), & 1 < u \le 9 \end{cases}$

6.63 $[1 - F(y)]^3[1 + F(y)]$

6.64 $p(u) = \begin{cases} .4156, & u = C_1 - C_3 \\ .5844, & u = C_2 - C_3 \end{cases}$

6.65 (a) $1 - e^{-6}$
     (b) gamma $\alpha = 4$, $\beta = 10$; gap
       times independent

6.66 $1 - (1/2)^4$

6.67 (a) e^{-3}
     (b) $1 - e^{-4}$

6.68 (b) $\exp[\mu + (\sigma^2/2)]$

6.69 $f(v) = (1/2\pi)(4\pi/3v)^{1/3}$, $0 \le v \le 4\pi/3$

chapter seven

7.1   .8664

7.2   44

7.3 .9876

7.4 7

7.5 (a) $\mu_1 - \mu_2$
(b) $(\sigma_1^2/m) + (\sigma_2^2/n)$
(c) $n = 18$

7.6 .0019

7.7 $P(S^2 > .065) = P(\chi^2 > 14.625) \approx .1$

7.8 (a) $v; 2v$
(b) $E(S^2) = \sigma^2; V(S^2) = 2\sigma^4/(n-1)$

7.9 $a = .656; b = 2.421$

7.10 $g_1 = -1.860(S)/3; g_2 = 1.860(S)/3$

7.12 $a = .61; b = 7.36$

7.13 (a) χ^2 with 5 d.f.
(b) χ^2 with 4 d.f.
(c) χ^2 with 5 d.f.

7.14 (a) t with 5 d.f.
(b) t with 4 d.f.
(c) F with num. d.f. = 2,
denom. d.f. = 4

7.15 .9544

7.16 $n = 157$

7.17 .0548

7.18 .9090

7.19 $n = 153$

7.20 (a) .3758
(b) no

7.21 .0217

7.22 4.47

7.23 .0062

7.25 .0062

7.26 .9876

7.27 $n = 51$

7.28 .0013

7.29 $n = 88$

7.30 .0668

7.31 .8980

7.32 (a) .9504
(b) .0708
(c) no, 30 is only 1.37 standard
deviations above the mean

7.33 .7698

7.34 (b) $n = 100$

7.35 $n = 61$

7.36 1.00

7.37 (a) .7486
(b) .729

7.38 (a) $p_1 - p_2$
(b) $[p_1(1 - p_1)/n_1] +$
$[p_2(1 - p_2)/n_2]$

7.39 .8414

7.40 .0043, reject claim,
probability too small

7.41 .0023

7.42 .0023

7.43 $\mu = 10.15$

7.44 .1251

7.45 normal [mean $= .4\mu_1 + .2\mu_2 + .4\mu_3$,
variance $= (.16\sigma_1^2/n_1) +$
$(.04\sigma_2^2/n_2) + (.16\sigma_3^2/n_3)$]

7.46 .617

7.48 $c = 183.070$

7.50 (a) F with num. d.f. = 1,
denom. d.f. = 9
(b) F with num. d.f. = 9,
denom. d.f. = 1
(c) $c = 49.04$

7.51 $v_2/(v_2 - 2)$

7.52 .9484

7.53 (b) .1587

7.57 .8413

7.58 .0668

7.59 .1587

7.61 .264

chapter eight

8.2 $a = \sigma_2^2/[\sigma_1^2 + \sigma_2^2]$

8.3 $a = [\sigma_2^2 - c]/[\sigma_1^2 + \sigma_2^2 - 2c]$

8.4 (a) $\hat{\theta}_1, \hat{\theta}_2, \hat{\theta}_3, \hat{\theta}_5$
(b) $\hat{\theta}_5$

8.5 $\hat{\theta} = \bar{Y} - 1$

8.6 (a) $\bar{Y}$
(c) $\bar{Y}^2 + \bar{Y}(4 - 1/n)$

8.7 $\hat{\theta}_3 - 9\hat{\theta}_2 + 54$

8.8 (a) $1/2$
(b) $\bar{Y} - (1/2)$
(c) $1/(12n) + 1/4$

8.9 (c) $[n^2/(n-1)](Y/n)[1 - (Y/n)]$

8.10 (b) $[(n\alpha + 1)/n\alpha] Y_{(n)}$
(c) $(2\theta^2)/[(n\alpha + 1)(n\alpha + 2)]$

8.11 (a) $\beta/[3n - 1]$
(b) $2\beta^2/[(3n-1)(3n-2)]$

8.12 (b) $\dfrac{\sqrt{(n-1)/2}\,\Gamma[(n-1)/2]}{\Gamma(n/2)} \cdot S = \hat{\sigma}$
(c) $\bar{Y} - z_\alpha \hat{\sigma}$

8.13 (a) $(1 - 2p)/(n + 2)$
(b) $[np(1 - p) + (1 - 2p)^2]/[(n+2)^2]$
(c) p near $1/2$

8.14 $E(Y_{(1)}) = \theta/(n + 1)$
$(n + 1)Y_{(1)}$ is unbiased

8.15 $MSE = \beta^2$

8.16 (a) $(4/\pi)X$ is unbiased for θ
(b) $(16/\pi^2)W$ is unbiased for θ^2

8.17 39.8 ± 4.86

8.18 $7.2\% \pm .79\%$

8.19 (a) 11.3 ± 1.54
(b) 1.3 ± 1.7
(c) $.17 \pm .08$

8.20 no

8.21 $-0.7, b = .404$

8.22 $.0327$

8.23 (a) yes
(b) no

8.24 $0.16, b = .121$

8.25 (a) $-.1, b = .0732$
(b) $-.016, b = .0828$

8.26 $98{,}550 \pm 20{,}316.3$

8.27 $.7 \pm .205$

8.28 (a) $.16 \pm .081$
(b) $.53 \pm .110$

8.29 $\hat{\lambda} = \bar{Y}; \hat{\sigma} = \sqrt{\bar{Y}/n}$

8.30 (a) 20 ± 1.265
(b) -3 ± 1.855; yes

8.31 $\hat{\theta} = \bar{Y}; \hat{\sigma} = \bar{Y}/\sqrt{n}$

8.32 1020 ± 645.10

8.33 $(1/2)(Y^2 - Y)$

8.34 (b) $\left(\dfrac{2Y}{9.48773}, \dfrac{2Y}{.710721}\right)$

8.35 (a) $(Y - 1.96, Y + 1.96)$
(b) $Y + 1.645$
(c) $Y - 1.645$

8.36 (a) $(Y^2/5.02389, Y^2/.0009821)$
(b) $Y^2/.0039321$
(c) $Y^2/3.84146$

8.37 (a) $(Y/2.24, Y/.0313)$
(b) $Y/.0627$
(c) $Y/1.96$

8.38 (b) $Y_{(n)}/[(.95)^{1/n}]$

8.39 $.629 \pm .012$

8.40 (a) $.40 \pm .016$
(b) $.55 \pm .017$
(c) $.20 \pm .018$

8.41 $5.4 \pm .277$

8.42 $(.41, .63)$

8.43 $3.7 \pm .204$; random sampling

8.44 26.2 ± 10.477

8.45 3.5 ± 4.53

8.46 (a) $.73 \pm .022$
(b) random sample

8.47 $-.18 \pm .036$

8.48 $.13 \pm .114$

8.49 $.15 \pm .064$

8.50 $(-.06, .18)$

8.51 $(-1.19, 2.21)$

8.52 (a) $7.2 \pm .751$
(b) $2.5 \pm .738$

8.53 (a) $np_1(1 - p_1) + np_2(1 - p_2) + 2np_1p_2$
(b) $(-.14, -.06)$

8.54 $(-.12, .56)$

8.55 (a) $n = 144$
(b) $n = 400$

8.56 $n = 100$

8.57 (a) $b = .0316$
(b) $n = 2537$

8.58 100

8.59 100

8.60 136

8.61 98

8.62 497

8.63 102

8.64 108

8.65 17.9 ± 2.078

8.66 60.8 ± 5.701

8.67 $(513.36, 566.64)$

8.68 $(-3.17, 11.33)$

8.69 $(9.52, 10.08)$

8.70 $(-5.72, 3.72)$

8.71 $(5.67, 12.33)$

8.72 (a) $(0, 9.76)$
(b) $(-1.52, 12.38)$

8.73 $(.03, .35)$

8.74 $(-84.39, -28.93)$

8.75 $.05 \pm .06$

8.76 $(2\bar{Y} + \bar{X}) \pm t_{\alpha/2}S\sqrt{\dfrac{4}{n} + \dfrac{3}{m}}$, where

$$S^2 = \sqrt{\dfrac{\sum\limits_{i=1}^{n}(Y_i - \bar{Y})^2 + \dfrac{1}{3}\sum\limits_{i=1}^{m}(X_i - \bar{X})^2}{n + m - 2}}$$

8.77 $(0.227, 2.194)$

8.78 $(33.784, 171.904)$

8.79 $(31.20, 90.32)$

8.80 $s^2 = .0285; (.0129, .1246)$

8.81 $(38.90, 2792.41)$

8.82 (1.41, 31.26); no

8.83 (a) $-.06 \pm .120$
      (b) $.07 \pm .091$
      (c) $.13 \pm .084$

8.84 $n_1 = n_2 = 301$

8.85 $n = 2400$

8.86 (a) $.0625 \pm .0237$
      (b) 563

8.87 $n = 256$

8.88 $n = 40,000$

8.89 $-.40 \pm .106$

8.90 $n = 768$

8.91 795 ± 7.95

8.92 (29.30, 391.15)

8.93 (9.86, 12.74)

8.94 (2.11, 5.25)

8.95 $(-.63, 6.63)$

8.96 $(-11.60, 1.60)$

8.97 $(-1.52, .02)$

8.98 39.1 ± 2.846

8.99 $.832 \pm .0153$

8.100 $.67 \pm .045$

8.101 (a) $2\sigma^4(n-1)/n^2$

8.102 $MSE(S^2) > MSE(S'^2)$ $(n > 1)$

8.103 (b) $2\sigma^4/(n_1 + n_2 - 2)$

8.104 $\left(\dfrac{S_1^2}{S_2^2}\right)\left(\dfrac{1}{F_{v_1, v_2, \alpha/2}}\right) < \dfrac{\sigma_1^2}{\sigma_2^2} <$
       $\left(\dfrac{S_1^2}{S_2^2}\right)F_{v_2, v_1, \alpha/2},$
       where $v_i = n_i - 1$, $i = 1, 2$

8.105 $2^{3/2}t_{\alpha/2}\sigma\left[\dfrac{\Gamma(n/2)}{\sqrt{n(n-1)}\Gamma\left(\dfrac{n-1}{2}\right)}\right]$

8.107 $\bar{Y} \pm t_{\alpha/2}S\sqrt{\dfrac{n+1}{n}}$

chapter nine

9.1 $1/3; 2/3; 3/5$

9.2 (b) $n^2/[8(n-2)]; n/2$

9.3 (b) $12n^2/[(n+1)^2(n+2)]$

9.5 $n - 1$

9.6 $2/n$

9.7 $1/n$

9.10 no

9.15 (b) .6826
       (c) no

9.20 $3/4$

9.21 $\alpha\beta$

9.22 no

9.23 $\bar{X}/(\bar{X} + \bar{Y})$

9.35 $\sum\limits_{i=1}^{n} \ln(Y_i)$; no

9.36 $\sum\limits_{i=1}^{n} \ln(Y_i)$

9.42 $Y_1 Y_2 \cdots Y_n$

9.43 $\dfrac{1}{n}\sum\limits_{i=1}^{n}(Y_i - \mu)^2$

9.44 yes

9.45 $\dfrac{1}{n}\sum\limits_{i=1}^{n} Y_i^2$

9.46 $3\bar{Y}[\bar{Y} + 1 - (1/n)]$

9.47 $[1 + (1/n)]Y_{(n)}$

9.48 $Y_{(1)} - (1/n)$

9.51 $\hat{\theta} = [1/(1 - \bar{Y})] - 2$; no

9.52 $\hat{\lambda} = \bar{Y}$

9.53 $\hat{\sigma}^2 = \dfrac{1}{n}\sum\limits_{i=1}^{n} Y_i^2$

9.54 $\hat{\mu} = \bar{Y}$; $\hat{\sigma}^2 = \sum\limits_{i=1}^{n}(Y_i - \bar{Y})^2/n$

9.56 $\hat{\theta} = 3\bar{Y}$; no

9.57 $\hat{\theta} = (1 - 2m_2')/(4m_2' - 1)$

9.58 $\hat{p} = 1/\bar{Y}$

9.59 $\bar{Y}$

9.60 $\bar{Y}/(3 - \bar{Y})$

9.62 (a) $\bar{Y}$
      (b) λ; λ/n

9.63 (a) $\sum\limits_{i=1}^{n} Y_i^r$
      (b) $\hat{\theta} = \dfrac{1}{n}\sum\limits_{i=1}^{n} Y_i^r$
      (c) yes

9.64 $\hat{\theta} = (1/2)(Y_{(n)} - 1)$

9.65 (a) 63
      (b) $E(\hat{\theta}) = \theta$; $V(\hat{\theta}) = \theta^2/(2n)$
      (c) 106.14

9.66 (a) $\hat{\theta} = \bar{Y}/\alpha$
      (b) $E(\hat{\theta}) = \theta$; $V(\hat{\theta}) = \theta^2/(\alpha n)$
      (d) $\sum\limits_{i=1}^{n} Y_i$
      (e) $\dfrac{2\sum Y_i}{31.4} < \theta < \dfrac{2\sum Y_i}{10.85}$

9.67 $\hat{\sigma}^2 = \dfrac{\displaystyle\sum_{i=1}^{m}(X_i - \bar{X})^2 + \sum_{i=1}^{n}(Y_i - \bar{Y})^2}{m + n}$

9.68 $\hat{p}_A = .30; \hat{p}_B = .38;$
$\hat{p}_C = .32; -.08 \pm .1641$

9.69 $\hat{\theta} = \dfrac{-n}{\sum (\ln Y_i)} - 1$

9.70 $y = 0, \hat{p} = 1/4$
$y = 1, \hat{p} = 1/4 \text{ or } 3/4$
$y = 2, \hat{p} = 3/4$

9.71 $\hat{p} = 55/200$

9.72 $Y_{(n)}/2$

9.74 $\hat{R} = \hat{p}/(1 - \hat{p})$, where $\hat{p} = (1/n)$
$\times$ (no. of defectives in sample)

9.75 s'

9.76 (a) $\hat{\theta}_1 = \bar{Y} - 1$
(b) $\hat{\theta}_2 = \min(Y_1, \ldots, Y_n)$
(c) $1/n$

9.77 $\hat{\sigma}^2 = \dfrac{1}{n} \sum (Y_i - \mu)^2$

9.78 $[1 - (1/n)]^{\Sigma Y_i}$

9.79 $e^{-t/\bar{Y}}$

9.81 (a) $\hat{N}_1 = 2\bar{Y} - 1$
(b) $E(\hat{N}_1) = N;$
$V(\hat{N}_1) = \left(\dfrac{1}{3n}\right)(N^2 - 1)$

9.82 (a) $\hat{N}_2 = \max(Y_1, \ldots, Y_n)$
(b) $\hat{N}_3 = \left(\dfrac{n+1}{n}\right)\hat{N}_2$
(c) $V(\hat{N}_3) \approx \dfrac{N^2}{n(n+2)}$

9.83 252 ± 85.193

chapter ten

10.2 (a) .032
(b) .416
(c) .021

10.3 (a) $c = 11$
(b) .596
(c) .057

10.4 (a) .9951
(b) $2p^2(1 - p) + p^3$

10.5 $z = -2.53$

10.6 $z = -4.22$

10.7 $z = -1.77$

10.8 $z = 3.65$

10.9 (a) $H_a: \mu_1 - \mu_2 > 0$
(b) one tailed
(c) $z = 4.899$, reject H_0

10.10 (a) $H_0: \mu_1 - \mu_2 = 0,$
$H_a: \mu_1 - \mu_2 \neq 0$
(b) two tailed
(c) $z = -.954$, do not reject H_0

10.11 $z = 6.390$, reject H_0

10.12 $z = -21.033$, reject H_0

10.13 (a) $z = 2.39$, reject H_0
(b) not necessarily
(c) samples random and
independent

10.14 $z = 4.51$

10.15 $z = 4.47$

10.16 .1342

10.18 $z = 1.77$

10.19 $z = 1.50$; no

10.20 $z = 2.89$; yes

10.21 $z = .49$

10.22 $z = -4.88$; yes

10.23 approx. 0

10.24 .1151

10.25 .67

10.26 $n = 309$

10.27 .025

10.28 $n = 608$

10.29 .1056

10.30 $n = 48$

10.31 $z = -1.992$; p-value $= .0233$;
reject

10.32 $z = 1.58$; p-value $= .1142$; do not
reject

10.33 (a) $z = 3.23$, reject
(b) p-value $= .001$

10.34 (a) p-value $= .1587$
(b) no
(c) $p = .0336$
(d) yes

10.35 (a) $z = 5.34$, reject
(b) p-value $< .001$

10.36 $z = 2.074$; p-value $= .0192$; reject

10.37 $z = -1.538$; p-value $= .0618$; do not
reject

10.38 $z = -1.732$; p-value $= .0836$

10.39 $z = 2.858$; p-value $= .0021$; reject

10.42 $t = -1.34$; do not reject;
p-value $> .10$

10.43 $t = 2.635; .02 < p\text{-value} < .05;$ reject

10.44 $t = 3.968;$ reject; $p\text{-value} < .005$

10.46 $t = 1.567; .10 < p\text{-value} < .20;$
do not reject

10.47 $t = 2.97;$ reject

10.48 $t = .10;$ no; $p\text{-value} > .2$

10.49 $t = .989; p\text{-value} > .1$

10.50 $t = 1.21;$ no

10.51 $t = 1.62;$ no

10.52 $t = -.647,$ do not reject

10.53 (a) $H_0: \mu_1 - \mu_2 = 0$
$H_a: \mu_1 - \mu_2 > 0$
(b) $t = 2.806,$ reject
(c) $.005 < p\text{-value} < .01$

10.54 (a) $t = -.362,$ do not reject
(b) $p\text{-value} > .2$
(c) -5.5 ± 31.878

10.55 (a) no
(b) $z = 3.262,$ reject
(c) $\chi^2 = 57.281,$ reject

10.56 $\chi^2 = 12.6;$ do not reject;
$.05 < p\text{-value} < .10$

10.60 $F = 1.922; p\text{-value} > .2;$
do not reject

10.61 $\chi^2 = 22.45; p\text{-value} < .005$

10.62 $\chi^2 = 27.36;$ no

10.63 $F = 8.03;$ yes

10.64 (a) .979
(b) .868
(c) .584
(d) .228

10.65 (a) reject H_0 if $\bar{X} > 7.82$
(b) .2611; .6406; .9131; .9909

10.66 $n = 109$

10.67 $n = 16$

10.69 (a) $U = \dfrac{2}{\beta_0} \displaystyle\sum_{i=1}^{4} Y_i$ has $\chi^2_{(24)}$

distribution under H_0: reject H_0
if $U > \chi^2_\alpha$
(b) yes

10.70 (a) reject H_0 if $\dfrac{2}{\theta_0} \displaystyle\sum_{i=1}^{n} Y_i^m > \chi^2_\alpha,$

with $2n$ degrees of freedom
(b) $n = 6$

10.71 reject if $(\sum X_i) \ln 4 +$
$(\sum Y_j) \ln(2/3) < \text{const.}$

10.72 (a) reject H_0 if $Y_{(n)} \le \theta_0 \sqrt[n]{\alpha}$
(b) yes

10.73 (a) reject H_0 if $Y_{(n)} > \theta_0;$ remainder
of critical region unspecified
(b) yes
(c) no

10.75 $-2 \ln \lambda \approx 11;$ reject
$H_0: p_1 = p_2 = p_3 = p_4$

10.76 $\chi^2 = \dfrac{(n-1)S_1^2 + (m-1)S_2^2}{\sigma_0^2}$

has $\chi^2_{(n+m-2)}$ distribution under $H_0;$
reject if $\chi^2 > \chi^2_\alpha$

10.77 (a) $\lambda = \dfrac{(\hat{\sigma}_1)^{n_1}(\hat{\sigma}_2)^{n_2}(\hat{\sigma}_3)^{n_3}}{(\hat{\sigma})^{n_1+n_2+n_3}},$ where

$\hat{\sigma}_1^2 = \dfrac{1}{n_1} \displaystyle\sum_{i=1}^{n_1}(X_i - \bar{X})^2$ and

$\hat{\sigma}^2 = \dfrac{n_1\hat{\sigma}_1^2 + n_2\hat{\sigma}_2^2 + n_3\hat{\sigma}_3^2}{n_1 + n_2 + n_3}$

(b) reject H_0 if $-2 \ln \lambda > 5.99$

10.78 (a) $\lambda = \dfrac{\bar{X}^m \bar{Y}^n}{\left(\dfrac{m\bar{X} + n\bar{Y}}{m+n}\right)^{m+n}}$

(b) $\bar{X}/\bar{Y}$ distributed as F with $2m$
and $2n$ degrees of freedom

10.81 $\lambda^{[2/(n_1+n_2+n_3)]} =$

$\dfrac{\sum(X_{i1} - \bar{X}_1)^2 + \sum(X_{i2} - \bar{X}_2)^2 + \sum(X_{i3} - \bar{X}_3)^2}{\sum\sum(X_{ij} - \bar{X})^2}$

10.82 (a) $H_0: p = .2; H_a: p > .2$
(b) $\alpha = .075$

10.83 (a) $H_0: \mu = 1100; H_a: \mu < 1100$
(b) $z < -1.645$
(c) yes; $z = -1.90$

10.84 $z = 5.24; p\text{-value approx. } 0$

10.85 $F = 2.486;$ yes; $p\text{-value} < .005$

10.86 (a) $F = 2.904;$ no
(b) (.050, .254)

10.87 $t = 9.568;$ reject

10.88 (a) $t = -2.657; .02 < p\text{-value} < .05$
(b) -4.542 ± 3.046

10.89 $T =$

$\dfrac{(\bar{X} + \bar{Y} - \bar{W}) - (\mu_1 + \mu_2 - \mu_3)}{\left\{\left(\dfrac{1+a+b}{n(3n-3)}\right)\left[\sum(X_i - \bar{X})^2 + \dfrac{1}{a}\sum(Y_i - \bar{Y})^2 + \dfrac{1}{b}\sum(W_i - \bar{W})^2\right]\right\}^{1/2}}$

with $(3n - 3)$ degrees of freedom

10.90 $t = 2.236;$ reject $H_0: \mu_1 = \mu_2$

10.91 $\lambda = \left(\dfrac{\sum (y_i - y_{(1)})}{n\theta_{1,0}}\right)^n \cdot e^{\left[-\sum (y_i - y_{(1)})/\theta_{1,0}\right] + n}$

10.92 $\lambda = \left[\dfrac{\sum (y_i - y_{(1)})}{\sum (y_i - \theta_{2,0})}\right]^n$

chapter eleven

11.1 $\hat{y} = 1.5 - .6x$

11.2 $\hat{y} = .72 + .99x$; .99; 99.72

11.3 $\hat{y} = 21.6 + 4.8x$

11.4 $\hat{y} = -.71 + .66x$; 7.15

11.5 (b) $\hat{y} = 34.0092 - .0837x$
    (d) 23.5467

11.6 $\hat{\beta}_1 = (\sum X_i Y_i)/(\sum X_i^2)$

11.7 $\hat{\beta}_1 = 2.514$

11.9 $\hat{y} = 2.1 - .6x$

11.10 $\hat{y} = -.7143 - .1429x + .1429x^2$

11.11 (a) $\hat{y} = 29.99 + 1.39x$
    (b) $\hat{y} = 36.54 + 1.39x - .20x^2$

11.12 (a) $\hat{y} = 17.609 - .229x$
    (c) $\hat{y} = 13.029$

11.13 (a) $\hat{y} = 452.11935 - 29.40184x$

11.14 $\hat{y} = 1.18 - 1.32x$

11.16 (a) $\hat{y} = 46 - .317x$
    (c) 19.033

11.17 (a) SSE = 18.29; $s^2 = 3.05$
    (b) $\hat{y} = 43.36 + 2.42x^*$;
        SSE = 18.29

11.18 (a) SSE = 56.845; $s^2 = 7.11$
    (b) $\hat{y} = 72.10 + .99x^*$;
        SSE = 56.8454

11.19 (a) SSE = 87.79; $s^2 = 10.97$
    (b) SSE = 87.64

11.20 (a) $\hat{y} = 3 + .475x$ (c) 5.025

11.22 $\operatorname{cov}(\hat{\beta}_0, \hat{\beta}_1) = \dfrac{-\bar{x}\sigma^2}{\sum (x_i - \bar{x})^2}$

11.23 SSE/n

11.24 (a) $t = 5.20$; reject
    (b) $(-.97, -.23)$

11.25 $t = -5.775$, $p < .01$; reject

11.26 (a) $t = 3.791$; $p < .01$
    (b) reject
    (c) no
    (d) $.475 \pm .289$

11.27 (a) $t = -3.08$; reject
    (b) $(-.32, -.08)$

11.28 (a) $\hat{y} = .72 + .118x$
    (b) $.118 \pm .008$

(c) $t = 4.587$; $.01 < p\text{-value} < .02$; reject

11.31 $t = \dfrac{\hat{\beta}_1 - \hat{\gamma}_1}{\sqrt{S^2\left[\dfrac{1}{\sum (x_i - \bar{x})^2} + \dfrac{1}{\sum (c_i - \bar{c})^2}\right]}}$,

where $S^2 = \dfrac{\text{SSE}_1 + \text{SSE}_2}{n + m - 4}$

11.32 (a) $t = 7.67$; reject; $t = 9.84$; reject
    (b) $t = -1.25$; do not reject

11.33 $t = -10.72$; yes

11.34 (a) $t = 17.97$; yes
    (b) (3.84, 5.84)

11.35 $t = 9.62$; yes

11.36 $t = 9.62$; $p\text{-value} < .01$

11.37 (a) $x_1 = (T_1 - 60)/10$;
        $x_2 = (P - 15)/5$;
        $x_3 = (C - 1.5)/.5$;
        $x_4 = (T_2 - 150)/50$
    (b) $y = 21.12 - 3.14x_1 - 1.21x_2 - .16x_3 - 1.27x_4$
    (c) $t = -4.19$; $p\text{-value} < .01$; reject
        $t = -1.62$; $.10 < p\text{-value} < .20$;
        do not reject
        $t = -.22$; $p\text{-value} > .20$; do not
        reject
        $t = -1.70$; $.10 < p\text{-value} < .20$;
        do not reject

11.39 $x_0 = \bar{x}$

11.40 305.11 ± 35.773

11.41 (4.67, 9.63)

11.42 (18.93, 24.95)

11.43 $x = 0$; (1.12, 1.88); $x = -2$:
    (2.03, 3.37); $x = 2$: $(-.37, .97)$

11.44 $t = 1.31$

11.45 25.395 ± 2.875

11.46 (.76, .81)

11.47 (b) (72.39, 75.77)

11.48 $x = \bar{x}$

11.49 (59.73, 70.57)

11.50 (64.19, 75.79); (68.59, 81.07)

11.51 $(-.86, 15.16)$

11.52 (.21, 55.05)

11.53 25.395 ± 10.136

11.54 (.27, .51)

11.55 (15.77, 28.10)

11.56 $F = 114.9$

11.57 (a) $F = 9.49$; reject
    (b) $F = 13.80$; reject

11.58 $F = 10.21$

11.59 $F = 12.34$; p-value $< .005$

11.60 $z = 4.69$; reject; p-value approx. 0

11.61 $z = 5.12$; reject; p-value approx. 0

11.62 $z = 1.52$; do not reject

11.63 (.875, 1.083)

11.64 (81.96, 98.80)

11.65 $t = -.58$; do not reject; $t = -3.05$; do not reject; $t = 2.53$; do not reject

11.66 $r = .9904$

11.67 (a) $\hat{\alpha}_0 = 36.70$; $\hat{\alpha}_1 = .0095$
       (b) (36.17, 37.23)

11.68 $\hat{\alpha}_0 = 3.53$; $\hat{\alpha}_1 = 2.41$

11.70 (a) $\hat{y} = -13.54 - .053x$
       (b) $t = -7.01$; reject
       (c) $.93 \pm .33$

11.71 (a) $\hat{y} = 6.52 + 10.817x$
       (b) $t = 6.32$; reject
       (c) $12.902 \pm .360$

11.72 (a) $\hat{y} = 1.4285 + .5000x_1 + .1190x_2 - .5000x_3$
       (b) $\hat{y} = 2.0715$
       (c) $t = -13.7$; reject
       (d) (1.88, 2.26)
       (e) (1.73, 2.41)

11.74 $n/2$ points at $x = -9$ and at $x = 9$

11.75 11/27

11.76 (a) $\hat{y} = 9.34 + 2.46x_1 + .60x_2 + .41x_1x_2$
       (c) 9.34; 11.80
       (d) $t = 2.63$; reject
       (e) (12.44, 13.18)
       (f) (12.03, 13.59)

11.77 $F = 1.41$; do not reject

11.79 (a) $\hat{y} = 20.33 + .033x_1 - .0225x_2 + .000125x_2^2$
       (b) $t = 2.04$; do not reject
       (c) $F = 32.33$; reject

11.81 $r = .8994$; $t = 5.04$; p-value $< .01$

11.82 $r = .89$; $t = 4.81$; p-value $< .01$; reject

chapter twelve

12.7 $n_1 = 34$; $n_2 = 56$

12.8 $n_1 = n_2 = 48$

12.9 $n = 246$; $n_1 = 92$; $n_2 = 154$

12.10 $n = 262$

12.11 three observations each at $x = 2$, $x = 5$

12.12 (a) 1.46 times as large
       (b) 2.14 times as many observations

12.14 (a) no, $t = -1.772$
       (b) $.10 < p$-value $< .20$
       (c) $-.022 \pm .031$

12.15 $t = -.35$; p-value $> .10$; do not reject

12.16 (a) $t = 2.65$; reject

12.17 (a) $\bar{d} = -3.88$; $s_d = 2.903$; $t = -5.176$; p-value $< .005$; reject
       (b) -3.88 ± 1.608
       (c) 36.93 ± 3.541
       (d) 33.046 ± 3.301

12.28 (a) μ_i
       (b) μ_i; $(1/n)(\sigma_p^2 + \sigma^2)$
       (c) $\mu_1 - \mu_2$; $2\sigma^2/n$; normal

12.31 (b) yes, $t = -1.834$
       (c) $.025 < p$-value $< .05$
       (d) $-.063 \pm .062$

12.32 112.5 ± 94.304

12.33 (b) yes, $t = 9.150$
       (c) p-value $< .01$
       (d) 106.9 ± 26.428

12.34 $k_1 = 1/4$; $k_2 = 1/2$; $k_3 = 1/4$

chapter thirteen

13.1 (a) $F = 2.93$; do not reject
      (b) $|t| = 1.71$; do not reject; $F = t^2$

13.2 (a) $F = 20.43$; $.01 < p$-value $< .025$
      (b) $F = t^2$

13.5 SSE $= .020$; $F = 2.0$; do not reject

13.6 SSE $= 498.67$; $F = 5.15$; $.025 < p$-value $< .05$ reject

13.7 SST $= .7588$; SSE $= .7462$ $F = 19.83$; $p < .005$ (approx.)

13.8 SST $= 174.1062$; SSE $= 61.1125$; $F = 47.0075$; p-value $< .005$; reject

13.9 SST $= 36.2855$; SSE $= 76.6996$; $F = 38.3164$; p-value $< .005$; reject

13.10 $F = 4.94$; yes

13.11 $F = 63.66$; yes; p-value $< .005$

13.12 $F = 0.87$; no

13.16 $-.28 \pm .102$

13.17 $2.20 \le \mu_A \le 2.30$; $-.01 \le \mu_A - \mu_B \le .18$

13.18 $67.86 \le \mu_A \le 84.14$; $55.82 \le \mu_B \le 76.48$; $-3.63 \le \mu_A - \mu_B \le 22.96$

13.19 (a) $23.965 \pm .770$
       (b) $3.502 \pm .914$

13.20 (a) $6.24 \pm .318$
       (b) $-.29 \pm .241$

13.21 (a) SST $= 36.7497$; SSE $= 60.2822$; $F = 4.88$; yes
       (b) $-.7055 \le \mu_B - \mu_C \le 4.0335$
       (c) 11.986 ± 1.0249

13.22 (a) $F = 1.32$; no
       (b) $(-.21, 4.21)$

13.23 $(.032, .050)$

13.24 $(1.39, 1.93)$

13.25 $(-.20, .49)$

13.26 (a) 2.7 ± 3.75
       (b) 27.5 ± 2.65

13.31 $F = 7$; $.01 < p$-value $< .025$

13.32 (a) $F = 10.05$; yes
       (b) $F = 10.88$; yes

13.33 (a) MST $= .1131$; MSB $= .3505$; MSE $= .0135$
       $F = 8.378$; p-value $< .005$; reject
       (b) $F = 25.96$; yes

13.34 $F = 1.40$; no

13.36 $F = 6.36$; yes

13.37 $F = 83.88$; p-value $< .005$

13.39 (a) $(-.83, 4.83)$
       (b) $(-.45, 4.45)$

13.40 $(1.61, 5.39)$

13.41 $.145 \pm .179$

13.42 $(-.23, 4.23)$

13.43 -4.8 ± 5.259

13.44 at least 50

13.45 $n = 3$

13.46 $n = 2$

13.47 $n = 16$

13.48 $n = 63$

13.49 $m = 3$; $\alpha/m = .0167$; $t_{.0084} \approx z_{.0084} = 2.39$
       $\mu_S - \mu_L$: $-.29 \pm .294$
       $\mu_S - \mu_T$: $-1.65 \pm .459$
       $\mu_L - \mu_T$: $-1.36 \pm .420$

13.51 $-6.56 \le \mu_A - \mu_B \le -.44$
       $-2.56 \le \mu_A - \mu_C \le 3.56$
       $0.94 \le \mu_B - \mu_C \le 7.06$

13.52 $-4.79 \le \mu_1 - \mu_4 \le .79$
       $-3.79 \le \mu_2 - \mu_4 \le 1.79$
       $-3.54 \le \mu_3 - \mu_4 \le 2.04$

13.53 (a) $F = 5.15$; yes

(b) $t = 1.68$; do not reject
(c) $.025 < p$-value $< .05$;
    $.1 < p$-value $< .2$

13.54 $F = 7$; reject

13.55 $F = 10.88$; reject

13.56 (b) $F = 1.05$; do not reject

13.57 SSE $= .005$; yes; yes; yes

13.58 $.02 \le \mu_A - \mu_B \le .14$; no

13.59 (a) no, completely randomized design
       (b) completely randomized design, 9 treatments, 2 observations per treatment

13.60 16; 135; 14.1

13.61 SSE $= 4.99$; no; no

13.62 $F = 7.33$; yes; blocking induces loss in degrees of freedom for estimating σ^2. Could result in slight loss of information if block to block variation is small.

13.63 (a) completely randomized design
       (b) $F = 7.79$; reject
       (c) -5.65 ± 11.72

13.64 (a)

source	d.f.	S.S	M.S.	F
Treatments	2	524,177.17	262,088.58	258.19
Blocks	3	173,415	57,805	56.95
Error	6	6,090.5	1,015.08	
Total	11	703,681.67		

       (b) 6
       (c) yes; $F = 258.19$ (p-value $< .005$)
       (d) yes; $F = 56.95$ (p-value $< .005$)
       (e) 22.529
       (f) -237.25 ± 55.13

13.65 $b = 21$

13.66 (a) SSE $= .571$; SST $= 1.212$; $F = 11.68$; p-value $< .005$
       (b) $t = -2.73$; 22 degrees of freedom; $.01 < p$-value $< .02$

13.67 $F = 27.7$; reject

13.68 $.05/4 \approx .01$;
       $\mu_A - \mu_D$: $.320 \pm .251$
       $\mu_B - \mu_D$: $.145 \pm .251$
       $\mu_C - \mu_D$: $.023 \pm .251$
       $\mu_E - \mu_D$: $-.124 \pm .251$

13.70 (b) σ_β^2
       (c) $\sigma_\beta^2 = 0$

13.71 (a) $E(Y_{ij}) = \mu + \tau_i$;
       $\text{var}(Y_{ij}) = \sigma_\beta^2 + \sigma_\varepsilon^2$

(b) $E(\bar{T}_i) = \mu + \tau_i$;
   $\text{var}(\bar{T}_i) = [\sigma_\beta^2 + \sigma_\varepsilon^2]/b$
(d) $\text{var}(\bar{T}_i - \bar{T}_l) = 2\sigma_\varepsilon^2/b$

13.73 (a) MSE
   (b) $[\text{MSE} - \text{MSB}]/k$

chapter fourteen

14.1 $X^2 = 24.48$; p-value $< .005$
14.2 $X^2 = 10.40$; do not reject
14.3   (a) $z = -2.30$; reject
   (b) formulate H_0 before collecting data
14.4 $X^2 = .658$; the data are consistent with the model
14.5 $X^2 = 173.64$; yes; p-value $< .005$
14.6 $X^2 = 1.23$; do not reject
14.7 $X^2 = 74.87$ (using 8 cells); reject
14.8 $X^2 = 55.71$ (using 4 cells); reject
14.9   (a) $X^2 = 14.496$; yes
   (b) p-value $< .005$
14.10 (a) $X^2 = 119.007$; yes
   (b) p-value $< .005$
14.11 $X^2 = 52.73$; p-value $< .005$; reject
14.12 (a) $X^2 = 13.71$; yes
   (b) $z = -2.05$; yes
14.13 $X^2 = 11.18$; yes
14.14 (a) $X^2 = 22.699$; yes
   (b) p-value $< .005$
14.15 $X^2 = 1.26$; do not reject
14.16 $X^2 = 13.99$; reject
   $X^2 = 13.99$; reject
   $X^2 = 1.36$; do not reject
14.17 (a) $X^2 = 18.53$; reject
   (c) p-value $< .005$
14.19 $X^2 = 7.19$; reject
14.20 yes; $X^2 = 38.43$
14.21 (a) no; $X^2 = .685$; p-value $> .1$
   (b) no; $X^2 = .670$; p-value $> .1$
14.22 $X^2 = 14.19$; yes
14.23 $(-.061, .021)$
14.24 $X^2 = 21.51$; reject
14.26 $X^2 = 6.18$; $.025 < p$-value $< .05$
14.27 $X^2 = 2.344$; do not reject;
   p-value $> .10$
14.28 $X^2 = 15.85$; $.05 < p$-value $< .10$
14.29 $X^2 = 18.82$; reject;
   $.01 < p$-value $< .025$
14.30 $X^2 = 8.56$; $.025 < p$-value $< .05$
14.31 $\hat{\alpha} = .7385$; $X^2 = 4.99$; do not reject

14.33 (a) $X^2 = 4$; reject H_0 that the model fits the data
   (b) $X^2 = 4$; do not reject
14.34 $X^2 = 3.26$; data agree with model
14.36 $X^2 = 74.8$; reject

chapter fifteen

15.1 $.014$, $m \le 6$ and $m \ge 19$
   $.044$, $m \le 7$ and $m \ge 18$
   $.108$, $m \le 8$ and $m \ge 17$
15.2   (a) p-value $= P(M \le 3 \text{ or } M \ge 7)$
      $= .344$; do not reject
   (b) p-value $= P(M \le 3) = .172$
15.3 $P(M \le 2 \text{ or } M \ge 8) = .11$; no
15.4   (a) $P(M \le 1) = .011$; yes
15.5   (a) $P(M \le 2 \text{ or } M \ge 7) = .18$;
      do not reject
   (b) $t = 1.65$; do not reject
15.6 $m = 1$; no
15.7   (a) p-value $= .011$; do not reject
15.8   (a) $T = 14$; p-value $< .01$; reject
15.9   (a) $T = 6$; $.02 < p$-value $< .05$
   (b) $T = 6$; $.01 < p$-value $< .025$
15.10 $T = 1$; p-value $< .005$
15.11 $T = 3.5$; $.025 < p$-value $< .05$
15.12 (a) $t = .48$; do not reject
   (b) $T = 7$; do not reject
15.13 $T^- = 11$; reject
15.14 $T = 2$; p-value $< .005$
15.16 (a) $M = 2$; p-value $= .055$; reject
   (b) $T = 6$; reject
15.17 $U = 9$; do not reject
15.18 (a) $U = 32$; p-value $= 2P(U \le 32)$
      $= .4894$
   (b) $|t| = .30$; p-value $> .2$
15.19 $z = 1.80$; yes
15.20 $z = 1.732$; p-value $= .0836$; reject
15.21 (a) $U = 0$; reject
   (b) $U = 2.5$; do not reject
15.22 $z = -3.49$; yes
15.23 (a) SST $= 2586.133$;
      SSE $= 11702.8$;
      $F = 1.33$; do not reject
   (b) $H = 1.22$; do not reject
15.24 (a) $H = 6.66$; $.025 < p$-value $< .05$
   (b) $U = 6$; p-value $= .0174$
15.25 $H = 2.03$; do not reject
15.26 $H = 19.47$; p-value $< .005$

15.28

h	$p(h)$
0	1/15
2/7	2/15
6/7	2/15
8/7	2/15
2	2/15
18/7	1/15
24/7	2/15
26/7	2/15
32/7	1/15

15.29 $F_r = 6.35$; reject

15.30 $F_r = 7.85$; $.01 < p$-value $< .025$.

15.31 (a) $F_r = 65.675$; p-value $< .005$; reject
       (b) $m = 0$; $P(M = 0) = (1/2)^8$ $= 1/256$; p-value $= 1/128$

15.32 $F_r = 1.75$; do not reject

15.35 $P(F_r = 4) = P(F_r = 0) = 1/6$
       $P(F_r = 3) = P(F_r = 1) = 1/3$

15.36 (a) .008
       (b) .040
       (c) .167

15.37 $R = 6$; no

15.38 $R = 5$; do not reject

15.39 (a) .0256
       (b) an unusually small number of runs (judged at $\alpha = .05$) would imply a clustering of defective items in time; $z = -1.95$; reject

15.40 (a) $R = 7$; do not reject with $\alpha = .10$

       (b) $t = .57$; do not reject

15.41 $R = 13$; do not reject

15.42 $z = -.37$; p-value approx. .7114

15.43 (a) $-.465$; do not reject
       (b) .708; reject
       (c) .819; reject

15.44 $r_s = -.876$; p-value $< .01$; reject

15.45 (a) $r_s = -.845$
       (b) reject

15.46 (a) $-.593$
       (b) reject

15.47 use two-tailed test; $r_s = .6768$; reject

15.48 $r_s = .9759$; yes

15.49 $r_s = 1.000$; p-value $< .005$

15.50 $r_s = .9394$; reject

15.51 (a) $m = 2$; yes
       (b) variance not constant

15.52 $m = 8$; no

15.53 $T = 73.5$; no, consistent with Exercise 15.52

15.54 $U = 0$; yes

15.55 $U = 17.5$; do not reject

15.56 (a) $U = 3$; reject
       (b) $F = 4.91$; do not reject

15.57 .0159

15.58 .375

15.59 $H = 7.154$; reject

15.60 (a) $H = 14.61$; reject
       (b) $U = 1.5$; reject

15.61 $Fr = 6.21$; do not reject

15.62 $Fr = 4.08$; p-value $> .10$

15.63 .10

15.64 1/63

Discrete Distributions

Distribution	Probability Function	Mean	Variance	Moment-generating Function
binomial	$p(y) = \binom{n}{y} p^y (1-p)^{n-y}$; $y = 0, 1, \ldots, n$	np	$np(1-p)$	$[pe^t + (1-p)]^n$
geometric	$p(y) = p(1-p)^{y-1}$; $y = 1, 2, \ldots$	$\dfrac{1}{p}$	$\dfrac{1-p}{p^2}$	$\dfrac{pe^t}{1 - (1-p)e^t}$
hypergeometric	$p(y) = \dfrac{\binom{r}{y}\binom{N-r}{n-y}}{\binom{N}{n}}$; $y = 0, 1, \ldots, n$ if $n \le r$, $y = 0, 1, \ldots, r$ if $n > r$	$\dfrac{nr}{N}$	$n\left(\dfrac{r}{N}\right)\left(\dfrac{N-r}{N}\right)\left(\dfrac{N-n}{N-1}\right)$	
Poisson	$p(y) = \dfrac{\lambda^y e^{-\lambda}}{y!}$; $y = 0, 1, 2, \ldots$	λ	λ	$\exp[\lambda(e^t - 1)]$
negative binomial	$p(y) = \binom{y-1}{r-1} p^r (1-p)^{y-r}$; $y = r, r+1, \ldots$	$\dfrac{r}{p}$	$\dfrac{r(1-p)}{p^2}$	$\left[\dfrac{pe^t}{1 - (1-p)e^t}\right]^r$